Communications in Computer and Information Science 1331

More information about this series at http://www.springer.com/series/7899

Vladimir Voevodin · Sergey Sobolev (Eds.)

Supercomputing

6th Russian Supercomputing Days, RuSCDays 2020
Moscow, Russia, September 21–22, 2020
Revised Selected Papers

Editors
Vladimir Voevodin
Research Computing Center (RCC)
Moscow State University
Moscow, Russia

Sergey Sobolev
Research Computing Center (RCC)
Moscow State University
Moscow, Russia

ISSN 1865-0929 ISSN 1865-0937 (electronic)
Communications in Computer and Information Science
ISBN 978-3-030-64615-8 ISBN 978-3-030-64616-5 (eBook)
https://doi.org/10.1007/978-3-030-64616-5

This Springer imprint is published by the registered company Springer Nature Switzerland AG
The registered company address is: Gewerbestrasse 11, 6330 Cham, Switzerland

Preface

The 6th Russian Supercomputing Days Conference (RuSCDays 2020), was held during September 21–22, 2020. The conference was dedicated to the 65th anniversary of the Research Computing Center, Moscow State University (RCC MSU) and the 100th anniversary of I.S. Berezin, the first RCC MSU director.

The conference was organized by the Supercomputing Consortium of Russian Universities and the Russian Academy of Sciences. The conference organization coordinator was RCC MSU.

Due to COVID-19 pandemic, for the first time the conference was held online. It was a new challenge for the conference organizers. However, online format provides a number of opportunities which are hard to perform offline, e.g. gathering invited speakers from all around the world. At the same time, the Organizing Committee did its best to keep the look and feel of the real-life meeting for attendees, including traditional sets of conference events – sessions, workshops, exhibitions, etc.

The conference was supported by the Russian Foundation for Basic Research and our respected partner (IBM), platinum sponsors (RSC, Merlion, Hewlett Packard Enterprise, Intel, Huawei), gold sponsors (NVIDIA, Dell Technologies in a partnership with CompTek), and silver sponsor (Xilinx). The conference was organized in a partnership with the ISC High Performance conference series.

RuSCDays was born in 2015 as a union of several supercomputing event series in Russia and quickly became one of the most notable Russian supercomputing international meetings. The conference caters to the interests of a wide range of representatives from science, industry, business, education, government, and students – anyone connected to the development or the use of supercomputing technologies. The conference topics cover all aspects of supercomputing technologies: software and hardware design, solving large tasks, application of supercomputing technologies in industry, exaflops-scale computing issues, supercomputing co-design technologies, supercomputing education, and others.

All 106 papers submitted to the conference were reviewed by three referees in the first review round. During single-blind peer reviewing, the papers were evaluated according to their relevance to the conference topics, scientific contribution, presentation, approbation, and related works description. After notification of conditional acceptance, the second review round arranged aimed at the final polishing of papers and also at the evaluation of authors' work after the referees' comments. After the conference, the 55 best papers were carefully selected to be included in this volume.

The proceedings editors would like to thank all the conference committees members, especially the Organizing and Program Committee members as well as the referees and reviewers for their contributions. We also thank Springer for producing these high-quality proceedings of RuSCDays 2020.

October 2020

Vladimir Voevodin
Sergey Sobolev

Organization

Steering Committee

V. A. Sadovnichiy (Chair)	Moscow State University, Russia
V. B. Betelin (Co-chair)	Russian Academy of Sciences, Russia
A. V. Tikhonravov (Co-chair)	Moscow State University, Russia
J. Dongarra (Co-chair)	The University of Tennessee, USA
A. I. Borovkov	Peter the Great Saint-Petersburg Polytechnic University, Russia
Vl. V. Voevodin	Moscow State University, Russia
V. P. Gergel	Lobachevsky State University of Nizhni Novgorod, Russia
G. S. Elizarov	NII Kvant, Russia
V. V. Elagin	Hewlett Packard Enterprise, Russia
A. K. Kim	MCST, Russia
E. V. Kudryashova	Northern (Arctic) Federal University, Russia
N. S. Mester	Intel, Russia
E. I. Moiseev	Moscow State University, Russia
A. A. Moskovskiy	RSC Group, Russia
V. Yu. Opanasenko	T-Platforms, Russia
G. I. Savin	Joint Supercomputer Center, Russian Academy of Sciences, Russia
A. S. Simonov	NICEVT, Russia
V. A. Soyfer	Samara University, Russia
L. B. Sokolinskiy	South Ural State University, Russia
I. A. Sokolov	Russian Academy of Sciences, Russia
R. G. Strongin	Lobachevsky State University of Nizhni Novgorod, Russia
A. N. Tomilin	Institute for System Programming, Russian Academy of Sciences, Russia
A. R. Khokhlov	Moscow State University, Russia
B. N. Chetverushkin	Keldysh Institutes of Applied Mathematics, Russian Academy of Sciences, Russia
E. V. Chuprunov	Lobachevsky State University of Nizhni Novgorod, Russia
A. L. Shestakov	South Ural State University, Russia

Program Committee

Vl. V. Voevodin (Chair)	Moscow State University, Russia
R. M. Shagaliev (Co-chair)	Russian Federal Nuclear Center, Russia

I. B. Meerov	Lobachevsky State University of Nizhni Novgorod, Russia
M. Michalewicz	University of Warsaw, Poland
L. Mirtaheri	Kharazmi University, Iran
S. G. Mosin	Kazan Federal University, Russia
A. V. Nemukhin	Moscow State University, Russia
G. V. Osipov	Lobachevsky State University of Nizhni Novgorod, Russia
A. V. Semyanov	Lobachevsky State University of Nizhni Novgorod, Russia
Ya. D. Sergeev	Lobachevsky State University of Nizhni Novgorod, Russia
H. Sithole	Centre for High Performance Computing, South Africa
A. V. Smirnov	Moscow State University, Russia
R. G. Strongin	Lobachevsky State University of Nizhni Novgorod, Russia
H. Takizawa	Tohoku University, Japan
M. Taufer	University of Delaware, USA
V. E. Turlapov	Lobachevsky State University of Nizhni Novgorod, Russia
E. E. Tyrtyshnikov	Institute of Numerical Mathematics, Russian Academy of Sciences, Russia
V. A. Fursov	Samara University, Russia
L. E. Khaymina	Northern (Arctic) Federal University, Russia
T. Hoefler	Eidgenössische Technische Hochschule Zürich, Switzerland
B. M. Shabanov	Joint Supercomputer Center, Russian Academy of Sciences, Russia
L. N. Shchur	Higher School of Economics, Russia
R. Wyrzykowski	Czestochowa University of Technology, Poland
M. Yokokawa	Kobe University, Japan

Industrial Committee

A. A. Aksenov (Co-chair)	Tesis, Russia
V. E. Velikhov (Co-chair)	National Research Center "Kurchatov Institute", Russia
A. V. Murashov (Co-chair)	T-Platforms, Russia
Yu. Ya. Boldyrev	Peter the Great Saint-Petersburg Polytechnic University, Russia
M. A. Bolshukhin	Afrikantov Experimental Design Bureau for Mechanical Engineering, Russia
R. K. Gazizov	Ufa State Aviation Technical University, Russia
M. P. Lobachev	Krylov State Research Centre, Russia
V. Ya. Modorskiy	Perm National Research Polytechnic University, Russia
A. P. Skibin	Gidropress, Russia
S. Stoyanov	T-Services, Russia

A. B. Shmelev	RSC Group, Russia
S. V. Strizhak	Hewlett-Packard, Russia

Educational Committee

V. P. Gergel (Co-chair)	Lobachevsky State University of Nizhni Novgorod, Russia
Vl. V. Voevodin (Co-chair)	Moscow State University, Russia
L. B. Sokolinskiy (Co-chair)	South Ural State University, Russia
Yu. Ya. Boldyrev	Peter the Great Saint-Petersburg Polytechnic University, Russia
A. V. Bukhanovskiy	ITMO University, Russia
R. K. Gazizov	Ufa State Aviation Technical University, Russia
S. A. Ivanov	Hewlett-Packard, Russia
I. B. Meerov	Lobachevsky State University of Nizhni Novgorod, Russia
V. Ya. Modorskiy	Perm National Research Polytechnic University, Russia
S. G. Mosin	Kazan Federal University, Russia
I. O. Odintsov	RSC Group, Russia
N. N. Popova	Moscow State University, Russia
O. A. Yufryakova	Northern (Arctic) Federal University, Russia

Organizing Committee

Vl. V. Voevodin (Chair)	Moscow State University, Russia
V. P. Gergel (Co-chair)	Lobachevsky State University of Nizhni Novgorod, Russia
B. M. Shabanov (Co-chair)	Joint Supercomputer Center, Russian Academy of Sciences, Russia
S. I. Sobolev (Scientific Secretary)	Moscow State University, Russia
A. A. Aksenov	Tesis, Russia
A. P. Antonova	Moscow State University, Russia
A. S. Antonov	Moscow State University, Russia
K. A. Barkalov	Lobachevsky State University of Nizhni Novgorod, Russia
M. R. Biktimirov	Russian Academy of Sciences, Russia
Vad. V. Voevodin	Moscow State University, Russia
T. A. Gamayunova	Moscow State University, Russia
O. A. Gorbachev	RSC Group, Russia
V. A. Grishagin	Lobachevsky State University of Nizhni Novgorod, Russia
S. A. Zhumatiy	Moscow State University, Russia
V. V. Korenkov	Joint Institute for Nuclear Research, Russia
I. B. Meerov	Lobachevsky State University of Nizhni Novgorod, Russia

D. A. Nikitenko	Moscow State University, Russia
I. M. Nikolskiy	Moscow State University, Russia
N. N. Popova	Moscow State University, Russia
N. M. Rudenko	Moscow State University, Russia
A. S. Semenov	NICEVT, Russia
I. Yu. Sidorov	Moscow State University, Russia
L. B. Sokolinskiy	South Ural State University, Russia
K. S. Stefanov	Moscow State University, Russia
V. M. Stepanenko	Moscow State University, Russia
N. T. Tarumova	Moscow State University, Russia
A. V. Tikhonravov	Moscow State University, Russia
A. Yu. Chernyavskiy	Moscow State University, Russia
P. A. Shvets	Moscow State University, Russia
M. V. Yakobovskiy	Keldysh Institutes of Applied Mathematics, Russian Academy of Sciences, Russia

Contents

Distributed and Cloud Computing

Parallel Algorithms

3D Model of Wave Impact on Shore Protection Structures and Algorithm of Its Parallel Implementation

Alexander Sukhinov, Alexander Chistyakov, and Sofya Protsenko(✉)

Don State Technical University, Rostov-on-Don, Russia
sukhinov@gmail.com, cheese_05@mail.ru, rab55555@rambler.ru

Abstract. The present research considers the three-dimensional mathematical model of wave processes that allows to study the wave influence on shore protection structures and coastal infrastructure facilities. This model is based on Navier-Stokes equations of motion in areas with dynamically changing computational domain geometry. The pressure correction method is used to approximate the hydrodynamic model. The finite-difference schemes describing the mathematical model are constructed on the basis of integro-interpolation method using a scheme with weights. The adaptive variable-triangular iterative method is used to solve the system of grid equations. The numerical algorithms and program complex for their implementation practical significance is the possibility of their application in the hydrophysical processes study in the coastal water systems, as well as to build the water environment velocity and pressure fields and the hydrodynamic effects evaluation on shore protection structures and coastal structures.

Keywords: Coastal structures · Surface gravitational waves · Mathematical model of wave impact · Three-dimensional wave processes

1 Introduction

The three-dimensional wave models in coastal marine systems construction and study are especially relevant to study now. Large scientific schools at Center for Applied Coastal Research, University of Delaware, Arizona University, University of Maryland, University of California at San-Diego, University of Texas at Austin, University of Michigan, University of North Caroline, University of Hamburg, Aachen University and others, University Le Bourge, Marseille University, Lion University, IFREMER, University of Malaga, Delft University, Norwegian University of Science and Technology are engaged in modeling wave processes. Scientific schools model hydrodynamic wave processes with the various coastal structures presence, assess the hydrodynamic effects on coastal protection structures and coastal structures in the surface waves presence.

V. Voevodin and S. Sobolev (Eds.): RuSCDays 2020, CCIS 1331, pp. 3–14, 2020.
https://doi.org/10.1007/978-3-030-64616-5_1

Existing software systems examples making it possible to simulate the hydrodynamic processes occurring in coastal marine systems are POM software package (Princeton Ocean Models), DELFT 3D-ECO, Mars3d software package, CARDINAL (Coastal Area Dynamics Investigation Algorithm. The most popular are wind-wave models third-generation WAM, SWAN (Simulation Waves Nearshore), WaveWatch. These software systems have several disadvantages. They use simplified hydrodynamic processes models, are based on the hydrostatic approximation, do not take into account spatially inhomogeneous aquatic environment motion, are not conservative, and do not take into account the complex bottom and shore topography shape.

Complexes of interconnected spatially 3D hydrodynamics models to describe these processes need to be developed. It makes possible to solve computationally time-consuming tasks with the required tens meters spatial resolution. The purpose of the work is to analyze the wave impact on the coast in the presence of structures based on the developed complex of interconnected 3D models of hydrodynamics. Computing-intensive tasks include millions grid equations for many time steps thousands. This leads to the building parallel algorithms necessity that maintain high computational efficiency for many hundreds thousands computing cores.

2 Wave Hydrodynamics Problem Statement

Spatially inhomogeneous three-dimensional wave hydrodynamics mathematical model of shallow water body includes [1–3]:

— Navier - Stokes motion equations:

$$
\begin{aligned}
u'_t + uu'_x + vu'_y + wu'_z &= -\tfrac{1}{\rho}P'_x + (\mu u'_x)'_x + \left(\mu u'_y\right)'_y + (\nu u'_z)'_z, \\
v'_t + uv'_x + vv'_y + wv'_z &= -\tfrac{1}{\rho}P'_y + (\mu v'_x)'_x + \left(\mu v'_y\right)'_y + (\nu v'_z)'_z, \\
w'_t + uw'_x + vw'_y + ww'_z &= -\tfrac{1}{\rho}P'_z + (\mu w'_x)'_x + \left(\mu w'_y\right)'_y + (\nu w'_z)'_z + g;
\end{aligned} \tag{1}
$$

— continuity equation:

$$
\rho'_t + (\rho u)'_x + (\rho v)'_y + (\rho w)'_z = 0, \tag{2}
$$

where $V = \{u, v, w\}$ is the velocity vector of the water current in a shallow water body; ρ is the aquatic environment density; P is the hydrodynamic pressure; g is the gravitational acceleration; μ, ν are coefficients of turbulent exchange in the horizontal and vertical directions; n is the normal vector to the surface describing the computational domain boundary.

— entrance (left border): $\mathbf{V} = \mathbf{V}_0,\ P'_n = 0$,

— bottom border: $\rho\mu\left(\mathbf{V}_\tau\right)'_n = -\boldsymbol{\tau},\ \mathbf{V}_n = 0,\ P'_n = 0$,

— lateral border: $(\mathbf{V}_\tau)'_n = 0,\ \mathbf{V}_n = 0,\ P'_n = 0,$
— upper border: $\rho\mu(\mathbf{V}_\tau)'_n = -\boldsymbol{\tau},\ w = -\omega - P'_t/\rho g,\ P'_n = 0,$
— surface of the structure: $\rho\mu(\mathbf{V}_\tau)'_n = -\boldsymbol{\tau},\ w = 0,\ P'_n = 0,$

where ω is the liquid evaporation intensity, $\boldsymbol{V}_n$, $\boldsymbol{V}_\tau$ are the normal and tangential components of the velocity vector, $\boldsymbol{\tau} = \{\tau_x, \tau_y, \tau_z\}$ is the tangential stress vector.

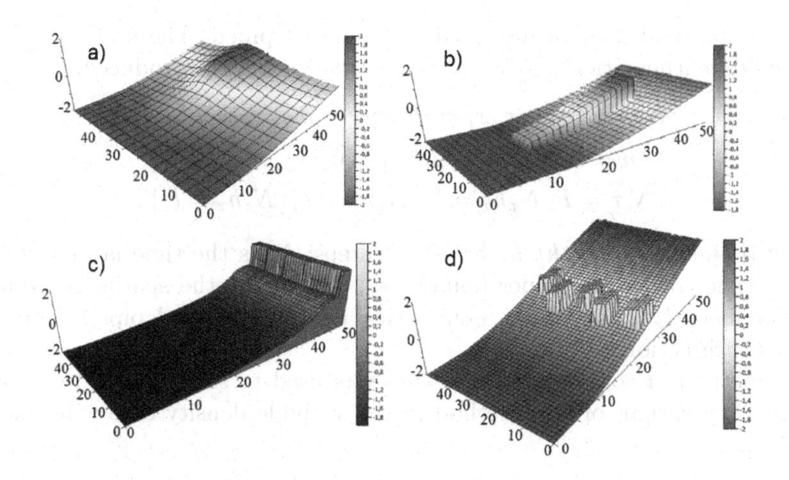

Fig. 1. The computational domain geometry: a) the bottom; b) single breakwater; c) breakwater wall; d) set of breakwaters

Figure 1 shows the computational domain depth map. Figure 2 shows the bottom surface and the coastline depth function igsolines.

Let $\boldsymbol{\tau} = \rho_a Cd_s |w| w$ be the tangential stress vector for the free surface, $Cd_s = 0,0026$, where w is the wind velocity relative to water, ρ_a is the atmosphere density, Cd_s is the dimensionless surface resistance coefficient, which depends on wind speed.

Let us set the tangential stress vector for the bottom taking into account the movement of water as follows: $\boldsymbol{\tau} = \rho Cd_b |\mathbf{V}| \mathbf{V}$, $Cd_b = gk^2/h^{1/3}$, where $k = 0,04$ is the group roughness coefficient in the Manning formula, considered in the range of 0,025 – 0,2; h = H + η is the depth of the water area, [m]; H is the undisturbed surface depth, [m]; η is the free surface elevation relative to the geoid (sea level), [m].

Let us use an approximation that making it possible to build a non-uniform in depth vertical turbulent exchange coefficient with respect to the water flow velocity measured pulsations [4]:

$$\nu = C_s^2 \Delta^2 \frac{1}{2} \sqrt{\left(\frac{\partial \overline{U}}{\partial z}\right)^2 + \left(\frac{\partial \overline{V}}{\partial z}\right)^2}, \tag{3}$$

where C_s is the dimensionless empirical constant, determined with respect to the attenuation process of homogeneous isotropic turbulence calculation; Δ is the characteristic grid scale; $\overline{U}, \overline{V}$ are the time-averaged ripple water flow velocity components in the horizontal direction.

3 Shallow Water Reservoirs Hydrodynamics Discrete Model

The computational domain inscribed in a parallelepiped. The uniform grid for the discrete mathematical model numerical realization is introduced:

$$\begin{aligned} \bar{w}_h = \{ & t^n = n\tau,\ x_i = ih_x,\ y_j = jh_y,\ z_k = kh_z; \\ & n = \overline{0..N_t},\ i = \overline{0..N_x},\ j = \overline{0..N_y},\ k = \overline{0..N_z}; \\ & N_t\tau = T,\ N_x h_x = l_x,\ N_y h_y = l_y,\ N_z h_z = l_z \}, \end{aligned}$$

where τ is time step; h_x, h_y h_z are space steps; N_t is the time layers number; T is the time coordinate upper bound; N_x, N_y N_z are the spatial coordinates nodes numbers; l_x, l_y l_z are the boundaries along the parallelepiped in the Ox, Oy and Oz direction accordingly.

The method of correction to pressure was used to solve the hydrodynamic problem. The variant of this method in the variable density case will take the form [5,6]:

$$\begin{aligned} & \tfrac{\tilde{u}-u}{\tau} + u\bar{u}'_x + v\bar{u}'_y + w\bar{u}'_z = \left(\mu\bar{u}'_x\right)'_x + \left(\mu\bar{u}'_y\right)'_y + \left(\nu\bar{u}'_z\right)'_z, \\ & \tfrac{\tilde{v}-v}{\tau} + u\bar{v}'_x + v\bar{v}'_y + w\bar{v}'_z = \left(\mu\bar{v}'_x\right)'_x + \left(\mu\bar{v}'_y\right)'_y + \left(\nu\bar{v}'_z\right)'_z, \\ & \tfrac{\tilde{w}-w}{\tau} + u\bar{w}'_x + v\bar{w}'_y + w\bar{w}'_z = \left(\mu\bar{w}'_x\right)'_x + \left(\mu\bar{w}'_y\right)'_y + \left(\nu\bar{w}'_z\right)'_z + g, \\ & P''_{xx} + P''_{yy} + P''_{zz} = \tfrac{\hat{\rho}-\rho}{\tau^2} + \tfrac{\left(\hat{\rho}\tilde{u}\right)'_x}{\tau} + \tfrac{\left(\hat{\rho}\tilde{v}\right)'_y}{\tau} + \tfrac{\left(\hat{\rho}\tilde{w}\right)'_z}{\tau}, \\ & \tfrac{\hat{u}-\tilde{u}}{\tau} = -\tfrac{1}{\hat{\rho}}\hat{p}'_x,\ \tfrac{\hat{v}-\tilde{v}}{\tau} = -\tfrac{1}{\hat{\rho}}\hat{p}'_y,\ \tfrac{\hat{w}-\tilde{w}}{\tau} = -\tfrac{1}{\hat{\rho}}\hat{p}'_z, \end{aligned} \tag{4}$$

where $V = \{u, v, w\}$ are the velocity vector components, $\{\hat{u}, \hat{v}, \hat{w}\}$, $\{\tilde{u}, \tilde{v}, \tilde{w}\}$ are the velocity vector components on the new and intermediate time layers, respectively, $\bar{u} = (\tilde{u} + u)/2$, $\hat{\rho}$ and ρ are the aqueous medium density distribution on the new and previous time layers, respectively.

The fullness of the control cells was taken into account during the discrete mathematical models of hydrodynamics construction, which makes it possible to increase the solution real accuracy in the investigated region complex geometry case by improving the boundary approximation.

Through $o_{i,j,k}$ marked the cell (i, j, k) volume of fluid (VOF) [6,7]. VOF is determined by the liquid column pressure inside this cell. If the average pressure at the nodes that belong to the cell vertices is greater than the liquid column pressure inside the cell, then the cell is considered to be full ($o_{i,j,k} = 1$). In the general case, VOF can be calculated by the following formula:

$$o_{i,j,k} = \frac{P_{i,j,k} + P_{i-1,j,k} + P_{i,j-1,k} + P_{i-1,j-1,k}}{4\rho g h_z}. \tag{5}$$

Let us introduce the coefficients q_0, q_1, q_2, q_3, q_4, q_5, q_6, describing VOF of regions located in the cell vicinity (control areas). In the case of the third kind boundary conditions $c'_n(x, y, t) = \alpha_n c + \beta_n$, the convective uc'_x and diffusion $(\mu c'_x)'_x$ transfer operators discrete analogues will take the form [7]:

$$(q_0)_i\, uc'_x \simeq (q_1)_i\, u_{i+1/2}\frac{c_{i+1}-c_i}{2h_x} + (q_2)_i\, u_{i-1/2}\frac{c_i - c_{i-1}}{2h_x},$$

$$(q_0)_i\, (\mu c'_x)'_x \simeq (q_1)_i\, \mu_{i+1/2}\frac{c_{i+1}-c_i}{h_x^2} - (q_2)_i\, \mu_{i-1/2}\frac{c_i - c_{i-1}}{h_x^2}$$

$$- \left|(q_1)_i - (q_2)_i\right| \mu_i \frac{\alpha_x c_i + \beta_x}{h_x}.$$

Similarly, approximations for the remaining coordinate directions will be recorded. The error in approximating the mathematical model is equal to $O\left(\tau + \|h\|^2\right)$, where $\|h\| = \sqrt{h_x^2 + h_y^2 + h_z^2}$. The flow conservation at the developed hydrodynamic model discrete level is proved. There are no non-conservative dissipative terms obtained as a result of discretization of the system of equations. A sufficient condition for the developed model stability [8,9] and monotony is determined with respect to the maximum principle [10], with spatial coordinates constraints on the step: $h_x < |2\mu/u|$, $h_y < |2\mu/v|$, $h_z < |2\nu/w|$ or $Re \leq 2N$, where $Re = |V| \cdot l/\mu$ is the Reynolds number [8], l is the region characteristic size $N = \max\{N_x, N_y, N_z\}$.

The system of equations (5) discrete analogs are solved by an adaptive modified alternating-triangular method of variational type [11,12].

4 Finite-Difference Schemes Based on the VOF

Consider the problem of stationary fluid flow between two coaxially infinitely long circular cylinders to verify the method efficiency, which takes into account the model domain calculated cells VOF:

$$uu'_x + vu'_y = -\rho^{-1}P'_x + \mu\Delta u, uv'_x + vv'_y = -\rho^{-1}P'_y + \mu\Delta v, \tag{6}$$
$$r_1 \leq r \leq r_2, r = \sqrt{x^2 + y^2}.$$

This problem solution looks like:

$$u(x,y) = -\frac{5y}{x^2+y^2}, v(x,y) = \frac{5x}{x^2+y^2}, P(x,y) = P(r_1) - \frac{12,5\rho}{x^2+y^2} + \rho/2. \tag{7}$$

Let us describe the test model problem numerical solution for the viscous fluid flow between two coaxial semi-cylinders ($x \geq 0$). The inner and outer cylinders radius will be taken equal: $r_1 = 5$ m, $r_2 = 10$ m. Let us enter the computational domain in a rectangle $(0 \leq x \leq 10,\ -10 \leq y \leq 10)$ with dimensions of $10 \times 20\,\text{m}^2$. Let us set the components of the velocity of the fluid: $u\,(0, y) = -5/y\,\text{m/s}$, $v\,(0, y) = 0\,\text{m/s}$ in the cylinder sectionby the plane $x = 0$. On the coarse grid (21 × 11 nodes, spatial directions steps of 1 m, time step of 0.1 s, calculation interval of 10 s), the numerical solutions errorsare most visible. Let us set the turbulent exchange coefficient $\mu = 1\,\text{m}^2/\text{s}$ and the medium density $\rho = 1000\,\text{kg/m}^3$.

Figure 2 shows the numerical solution, where fluid flows are reflected using the color palette.

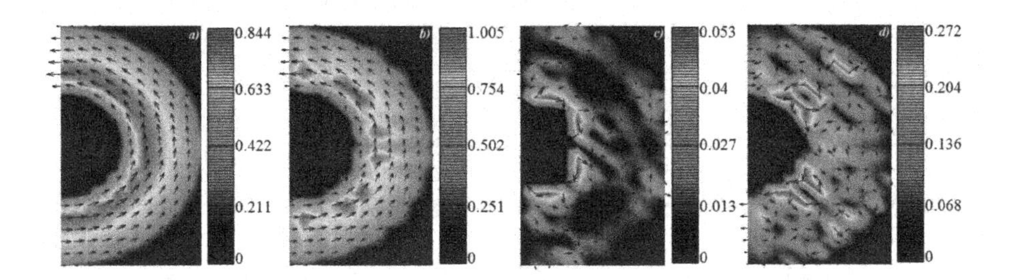

Fig. 2. The fluid flow problem numerical solution: a) the VOF case, b) the stepped interface between two media case. The difference between numerical and analytical problem solutions: c) the VOF case, d) the stepped interface between two media case.

Figure 2a shows the fluid flow between two coaxial semi-cylinders problem solution, obtained on grids that take into account the VOF, is fairly smooth. Figure 2b represents a numerical solution obtained with a interface between two media stepwise approximation, has significant errors in determining the fluid flow velocity vectors direction.

Figure 2c, 2d reflect the numerical solution error values on the grids, which take into account the cell population (the smooth boundary case), and on grids with the boundary stepped approximation. The developed schemes accuracy experimental study was carried out using the solution found on the grids of 11 × 21, 21 × 41, 41 × 81, 81 × 161 computational nodes. The table shows the fluid flow between two semi-cylinders test problem numerical solution error values for smooth and stepped boundaries.

The calculation error reaches 70% in the stepwise approximation case, using grids that take into account the VOF of cells make the error in the hydrodynamic model numerical solution caused by boundary approximation less – 6% of the problem solution. The numerical solution error calculations results analysis on a sequence of condensing grids presented in the table allows us to conclude about the difference schemes use effectiveness. It was found that the grid splitting by 8 times for each of the spatial directions does not lead to the accuracy increase the solutions obtained on grids that take into account VOF (Table 1).

Table 1. The wave hydrodynamics test problem solving error.

Grid dimensions	11 × 21	21 × 41	41 × 81	81 × 161
The error maximum value in the smooth border case, m/s	0,053	0,052	0,058	0,056
The error average value in the smooth border case, m/s	0,023	0,012	0,006	0,003
The error maximum value in the stepped boundary, m/s	0,272	0,731	0,717	0,75
The error average value in the stepped boundary case, m/s	0,165	0,132	0,069	0,056

The absence of linear error growth is achieved due to the conservativeness of difference schemes over long integration times. The implementation of the mass and momentum conservation laws studied at discrete level when developing discrete mathematical model. Otherwise, if the conservativity property is not fulfilled, non-physical sources appear, and nonlinear growth of the solution error may also occur. In practice, the absence of error accumulation was checked using calculations for longer periods. The transition process is observed after the first 3–5 waves are brought to the shore, and then the wave fluctuations tend to periodic mode.

5 The Solving Grid Equations Algorithm's Parallel Version

A software package in C++ is designed to build turbulent flows of an incompressible velocity field of the aquatic environment on high-resolution grids for predicting sediment transport and possible scenarios of changing the geometry of the bottom region of shallow water bodies. Parallel algorithms implemented in the software package for solving systems of grid equations arising during the discretization of model problems were developed using MPI technology. To solve this problem, the adaptive modified alternating-triangular method of minimum corrections was used. In parallel implementation, decomposition methods of grid domains were applied to computationally time-consuming diffusion-convection problems with respect to the architecture and parameters of a multiprocessor computing system. The calculated two-dimensional region was decomposed with two spatial variables x, y. The peak performance of the multiprocessor computing system is 18.8 teraflops. As computing nodes, 128 HP ProLiant BL685c homogeneous 16-core Blade servers of the same type were used, each being equipped with four 4-core AMD Opteron 8356 2.3 GHz processors and 32 GB RAM.

Figure 3 shows the program algorithm scheme that numerically implements the developed 2D and 3D wave hydrodynamics models.

The developed software package includes the control unit, the input unit, the construction block for the 2D and 3D water flow fields grid equations without pressure, the construction block for grid equations to pressure field and elevation function calculation, the unit for checking the presence of the structure on the surface, the unit for calculating the velocity field with regard to pressure, the block for solving grid equations on the basis of modified adaptive alternating triangular method of variational type [13], the velocity field and the pressure function output values block. The control unit contains the cycle for a time variable, and such functions as calculating the velocity field without taking into account the elevation function, calculating the elevation function, calculating the 2D velocity field, checking the presence of the surface structure and data output. The input unit contains the initial distributions for calculating the currents velocity and the level elevation function. It also contains the velocity field initial distributions, the elevation functions and the initial calculated cells VOF values.

Consider the parallel algorithm for calculating the correction vector [10]:

$$(D + \omega_m R_1) D^{-1} (D + \omega_m R_2) w^m = r^m,$$

where R_1 is the lower-triangular matrix, and R_2 is the upper-triangular matrix. To this end, the systems may be solved successively:

$$(D + \omega_m R_1) y^m = r^m, \quad (D + \omega_m R_2) w^m = D y^m.$$

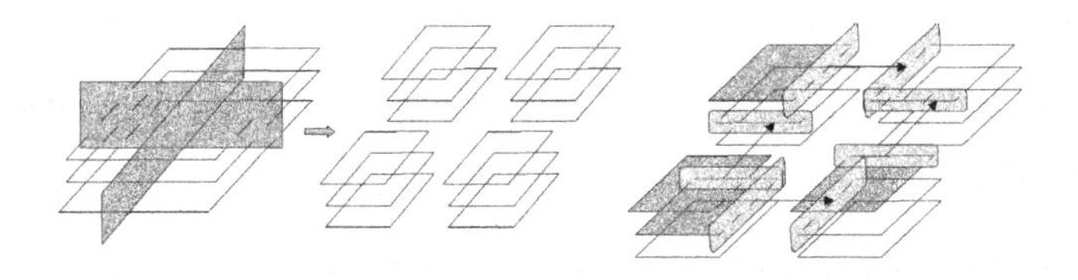

Fig. 3. The scheme for calculating the vector y^m

First, the vector y^m is calculated, and the calculation starts in the lower left corner. Then the calculation of the correction vector w^m begins from the upper right corner. Figure 3 shows the calculation of the vector y^m.

Calculating the coefficients of grid equations does not require exchanges and is done once per time step and does not significantly affect the overall efficiency of the method. The main time spent is focused on solving grid equations. In theoretical calculations, one iteration requires $50N$ arithmetic operations, where N is the grid nodes number, the residual vector calculation is completed in $14N$ operations, the calculation of the correction vector is $18N$, the scalar multiplications' calculation for calculating the iterative parameters is $16N$, and the transition to the next iteration is completed in $2N$ arithmetic operations. The parallel implementation of the correction procedure (calculation of the correction vector) is considered, since it is the most time-consuming procedure from the point of view of parallel calculations. The greatest efficiency drop is observed at this stage. The correction vector is calculated using a pipeline method, which requires a large number of exchanges and delays the start of the last calculator. One exchange is completed, which requires less time for approximately the same amount of data during calculating scalar products and the residual vector. The transition to the next iteration do not require exchanges.

The acceleration and efficiency calculating results, depending on the processors number for the adaptive alternating-triangular method parallel variant, are given in the Table 2.

Table 2. The acceleration and efficiency dependence on the processors number.

Processors number	Time, sec	Acceleration	Efficiency
1	7,490639	1	1
2	4,151767	1,804	0,902
4	2,549591	2,938	0,734
8	1,450203	5,165	0,646
16	0,882420	8,489	0,531
32	0,458085	16,351	0,511
64	0,265781	28,192	0,44
128	0,171535	43,668	0,341

The Table 2 shows that the alternating-triangular iterative method and its parallel realization algorithm on the basis of decomposition in two spatial directions can be effectively applied to solve hydrodynamic problems for a sufficiently large calculators number ($p \leq 128$).

6 Numerical Experiments Results

The developed software package allows to set the surface object complex configuration and the source of oscillations type and characteristics. Figure 5 shows the waves movement modeling results in the presence of the underwater body, taking into account the object's in the liquid complex geometry. The figure describes the level and bottom elevation functions for bottom without structures (a), for single breakwater (b), breakwater wall (c), and set of breakwaters (d).

The developed numerical algorithms for solving model problems and implementing a set of programs can be used to study hydrophysical processes in coastal water systems [16] and to find the water environment velocity and pressure field, to assess the possible impact on the coastline in the various coastal protective and shore fortifications presence.

The 3D wave hydrodynamic model development made it possible to describe the aquatic environment movement with respect to the wave propagation towards the shore. The modeling area has dimensions of 50 by 50 m and the depth of 2 m, the peak point rises above sea level by 2 m. Suppose that the liquid is at rest at the initial moment of time. When solving the posed model problem, the grid of 100 by 100 by 40 calculated nodes was used; the time step was 0.01 s.

Single breakwaters are built to protect the coastal zone from diffraction and reflection of waves. Waves can deviate from straight-line propagation and bypass

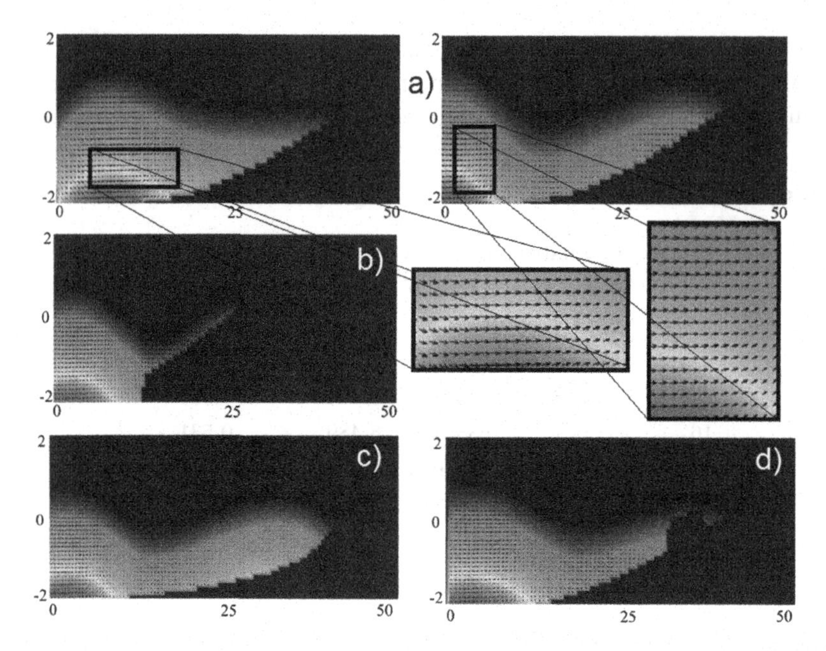

Fig. 4. Level and bottom elevation function: a) the bottom; b) single breakwater; c) breakwater wall; d) set of breakwaters

obstacles encountered in their path and penetrate the area behind them. Diffraction depends on the wavelength and the obstacle size ratio. If the wavelength is greater than the obstacle size, the waves go around it and pass on, almost without changing their structure and intensity. If the wavelength is comparable to the obstacle size, waves partially encircle it. If the wave becomes smaller, the "shadow zones" appear. If the wavelength is less than the obstacle size, waves are reflected from it, and the "shadow zones" are formed behind the obstacle (Fig. 4).

The wave field that occurs behind the obstacle can be considered as diffraction and interference combination. The diffracted wave occurs locally in the certain shadow boundary neighborhood beyond the obstacle border. The wave front breas the surface of into half-wave zones, that is, areas whose boundaries are removed from the observer at distances that differ by half the falling on the obstacle wavelength. The secondary waves coming from neighboring zones oscillate in opposite phase and therefore extinguish each other. At the same time, the waves amplitudes excited by sources, on the contrary, add up. The result is an interference wave pattern. Wave scattering occurs when part of the wave goes around an obstacle, and part of the wave is reflected from it. Diffraction and scattering processes can strongly distort the wave field structure near the shore.

Shock waves with an extremely sharp and steep front are observed in the breakwater walls presence. For a breakwater wall that is hit by a shock wave,

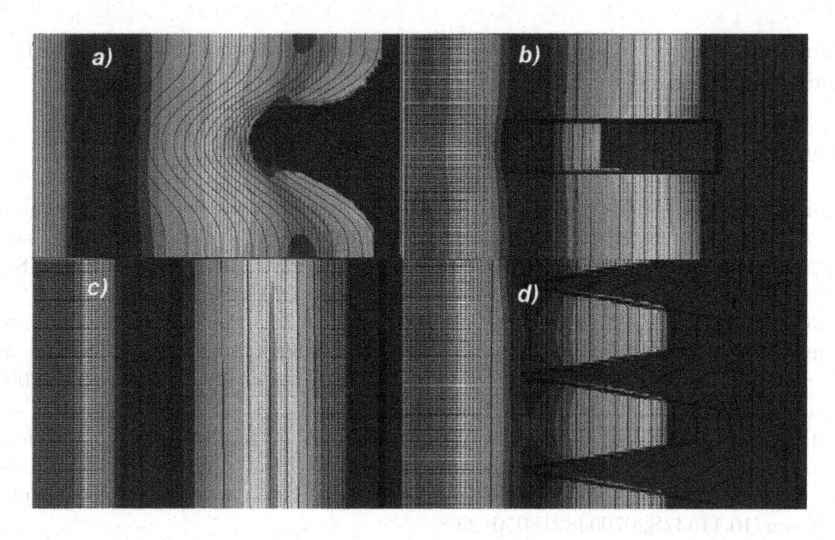

Fig. 5. The aquatic environment movement velocity vector field: a) the geometry of the bottom; b) single breakwater; c) breakwater wall; d) set of breakwaters

the pressure equal to zero before, the front arrival suddenly reaches its maximum value. Further, pressure changes are clear from the figure. It decreases and moves to the area of lowered values. The wave decreases intensity rapidly when moving away from the source. It is explained not only by the increase in the wave front area as the front diverges from the source, but also by the wave energy absorption.

7 Conclusion

The paper considers the 3D wave model in coastal marine systems the construction and study. The model takes into account the dynamically changing bottom geometry and the shore protection structures presence. 3D models are used to study the hydrophysical processes in coastal water systems, and to assess the hydrodynamic impact on shore protection structures and coastal structures in the gravitational surface waves presence. The 3D model gives the most realistic physical wave processes description near the coastline. The developed software description is given. It makes it possible to changing the location and source oscillations characteristics. It takes into account the zones shape and the coastal areas drainage and flooding intensity degree. The developed software package can be widely used for practical studies of the waves force effect on the bottom surface geometry calculation, as well as including in the surface coastal infrastructure presence. In the future, it is planned to study the bottom surface dynamics due to the sediment transport processes and wave action based on the 3D hydrodynamics model combined with the sediment transport model.

Acknowledgements. This paper was supported by the Russian Foundation for Basic Research (RFBR) grant No. 19-31-90091.

References

1. Gushchin, V.A., Sukhinov, A.I., Nikitina, A.V., Chistyakov, A.E., Semenyakina, A.A.: A model of transport and transformation of biogenic elements in the coastal system and its numerical implementation. Comput. Math. Math. Phys. **58**(8), 1316–1333 (2018). https://doi.org/10.1134/S0965542518080092
2. Gushchin V.A., Kostomarov A.V., Matyushin P.V., Pavlyukova E.R.: Direct numerical simulation of the transitional separated fluid flows around a sphere and a circular cylinder. J. Wind Eng. and Industr. Aerodyn. **90**(4:5), 341–358 (2002). https://doi.org/10.1016/S0167-6105(01)00196-9
3. Sukhinov, A.I., Chistyakov, A.E., Shishenya, A.V., Timofeeva, E.F.: Predictive modeling of coastal hydrophysical processes in multiple-processor systems based on explicit schemes. Math. Models Comput. Simul. **10**(5), 648–658 (2018). https://doi.org/10.1134/S2070048218050125
4. Protsenko, S., Sukhinova, T.: Mathematical modeling of wave processes and transport of bottom materials in coastal water areas taking into account coastal structures. MATEC Web Conf. **132**, 04002 (2017)
5. Alekseenko, E., Roux, B., et al.: Coastal hydrodynamics in a windy lagoon. Nonlinear Process. Geophy. **20**(2), 189–198 (2013). https://doi.org/10.1016/j.compfluid.2013.02.003
6. Alekseenko, E., Roux, B., et al.: Nonlinear hydrodynamics in a mediterranean lagoon. Comput. Math. Math. Phys. **57**(6), 978–994 (2017). https://doi.org/10.5194/npg-20-189-2013
7. Favorskaya, A.V., Petrov, I.B.: Numerical modeling of dynamic wave effects in rock masses. Dokl. Math. **95**(3), 287–290 (2017). https://doi.org/10.1134/S1064562417030139
8. Belotserkovskii, O.M., Gushchin, V.A., Shchennikov, V.V.: Decomposition method applied to the solution of problems of viscous incompressible fluid dynamics. Comput. Math. Math. Phys. **15**, 197–207 (1975)
9. Kvasov, I.E., Leviant, V.B., Petrov, I.B.: Numerical study of wave propagation in porous media with the use of the grid-characteristic method. Comput. Math. Math. Phys. **56**(9), 1620–1630 (2016). https://doi.org/10.1134/S0965542516090116
10. Sukhinov, A.I., Khachunts, D.S., Chistyakov, A.E.: A mathematical model of pollutant propagation in near-ground atmospheric layer of a coastal region and its software implementation. Comput. Math. Math. Phys. **55**(7), 1216–1231 (2015). https://doi.org/10.1134/S096554251507012X
11. Chetverushkin, B.N., Shilnikov, E.V.: Software package for 3D viscous gas flow simulation on multiprocessor computer systems. Comput. Math. Math. Phys. **48**(2), 295–305 (2008)
12. Davydov, A.A., Chetverushkin, B.N., Shil'nikov, E.V.: Simulating flows of incompressible and weakly compressible fluids on multicore hybrid computer systems. Comput. Math. Math. Phys. **50**(12), 2157–2165 (2010). https://doi.org/10.1134/S096554251012016X
13. Sukhinov, A.I., Nikitina, A.V., Semenyakina, A.A., Chistyakov, A.E.: Complex of models, explicit regularized schemes of high-order of accuracy and applications for predictive modeling of after-math of emergency oil spill. In: CEUR Workshop Proceedings, vol. 1576, pp. 308–319 (2016)

Different Partitioning Algorithms Study Applied to a Problem of Digital Rock Physics

Evdokia Golovchenko(✉), Mikhail Yakobovskiy, Vladislav Balashov, and Evgeny Savenkov

Keldysh Institute of Applied Mathematics RAS, Moscow, Russia
{golovchenko,lira}@imamod.ru, vladislav.balashov@gmail.com, e.savenkov@googlemail.com

Abstract. In recent years digital rock physics technology is regarded as a promising tool which can supplement traditional laboratory techniques. This technology is based on numerical experiment with direct resolution of pore space of a rock sample, which is obtained with computed microtomography. Necessity of high resolution leads to a high dimension of discrete settings (10^6–10^9 numerical cells). The work is devoted to an application of different partitioning algorithms to the problem of flow simulation within geometry of rock sample pore space. Simulation of a single-phase fluid flow within pore space of a sandstone sample with voxel representation is used to compare the partitions obtained by various methods using parallel partitioning tools ParMETIS, Zoltan, and GridSpiderPar. Average time spent on interprocess exchange during 200 time steps of the considered parallel simulation was compared when the grid was distributed over the cores in accordance with various partitions. The obtained results demonstrate advantages of some algorithms and reveal the criteria, important for the problem.

Keywords: Graph partitioning · Mesh decomposition · Voxel geometry · Pore space

1 Introduction

Currently mathematical modeling is one of the main tools which are applied in the analysis and optimization of oil and gas field development processes. For a correct description of the processes occurring in the reservoir, it is necessary to take into account processes occurring at various scales: from kilometers (the scale of an entire reservoir) to micrometers (the scale of individual pores). The practical impossibility of using detailed models for the entire reservoir leads to the need to use a hierarchy of mathematical models: separate mathematical model on each spatial scale. This paper discusses some aspects of the numerical simulation of fluid flow at pore scale. Pore space geometry is constructed on the basis of results obtained with modern microtomographic technology.

The study is conducted with support from The Ministry of Education and Science of Russian Federation, unique identifier of the Project is RFMEFI60419X0209.

V. Voevodin and S. Sobolev (Eds.): RuSCDays 2020, CCIS 1331, pp. 15–24, 2020.
https://doi.org/10.1007/978-3-030-64616-5_2

The combined use of microtomography and numerical experiment for the study of rock samples properties is the basis of the digital rock physics [1–3]. This technology can be regarded as supplement to standard laboratory experiments, which compensates some of their shortcomings.

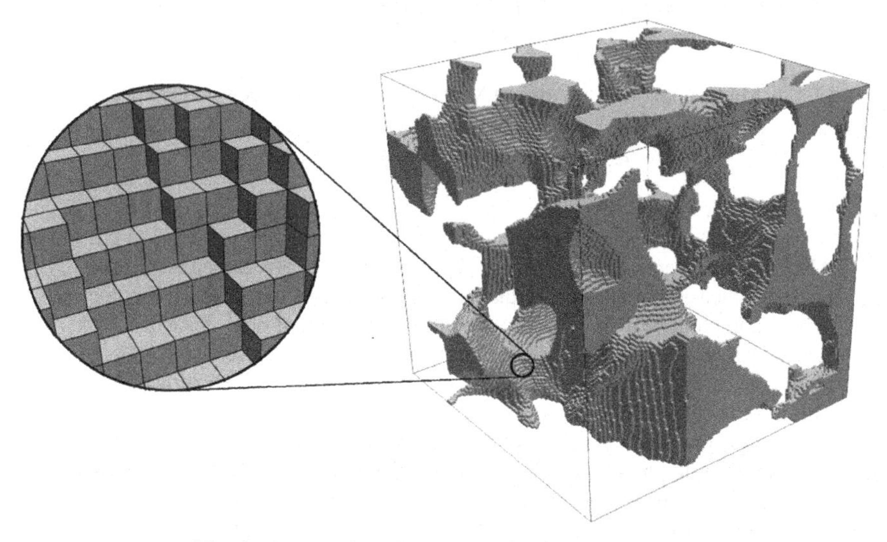

Fig. 1. A part of sandstone sample S_9 pore space [4]

Based on microtomographic images we can construct voxel (Volumetric element) representation of pore space: a three-dimensional uniform Cartesian grid is introduced, and each cell (voxel) is assigned with value "1" or "0" depending on its belonging to the rock or pore space. In Figs. 1 and 2 voxel representations of pore space geometry corresponding to different sandstones are presented [4]. Solid voxels are not shown. It is clear that the pore space geometry is very complicated.

The need for detailed resolution of pore space leads to a large voxel models: from 10^6 up to 10^9 (and sometimes even more) numerical cells. Therefore, in the problems under consideration parallel computing is used. In its turn, in preparation for parallel simulation, the problem of decomposition of the computational domain arises.

The problem of load balancing arises in parallel mesh-based numerical solution of problems of continuum mechanics, energetics, electrodynamics etc. on high-performance computing systems. Geometric parallelism is commonly used to parallelize these problems. In geometric parallelism mesh is divided among processes using its geometry. On each processor the problem is solved on its part of the mesh (subdomain). For increase of processors efficiency it is necessary to provide rational domain decomposition, taking into account the requirements of balanced mesh distribution among processors and reduction of interprocessor communications, which depend on the number of bonds between subdomains.

The problem of balanced mesh decomposition is generalized on the graph partitioning problem. In this case the graph approximates computational and communication

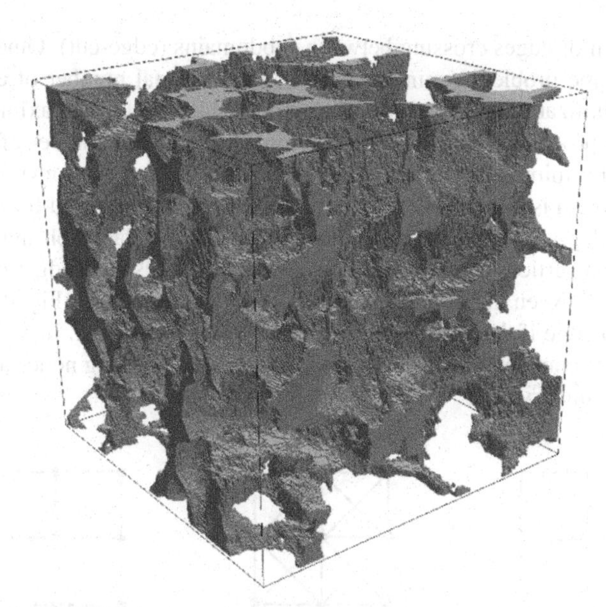

Fig. 2. A sandstone sample S_5 pore space [4].

Fig. 3. An example of a sandstone sample S_5 pore space partition.

loads. A mesh is approximated by an undirected graph G = (V, E), where V – set of vertices, E – set of edges. In this paper we consider graphs with unitary weights of vertices and edges. An optimal partition is a partition with equal sums of vertices in subdomains and minimal sum of edges crossing between subdomains (edge-cut). Other criteria are important for some problems: minimization of the maximal number of edges, cut by subdomain, minimization of the maximal communication volume (maximal number of vertices that the processor needs to send), minimization of the number of neighboring subdomains and minimization of the number of unconnected subdomains. The graph partitioning problem is NP-complete and different heuristics are used for its solving.

For each mesh graphs can be constructed in two ways: a nodal graph and a dual graph. In the nodal graph vertices correspond to the nodes of the mesh and edges correspond to the connections between the nodes. In a complete nodal graph (see Fig. 4) vertices are connected by an edge if the corresponding nodes share an element. In a reduced nodal graph vertices are connected by an edge only if the corresponding nodes are connected by an element boundary line.

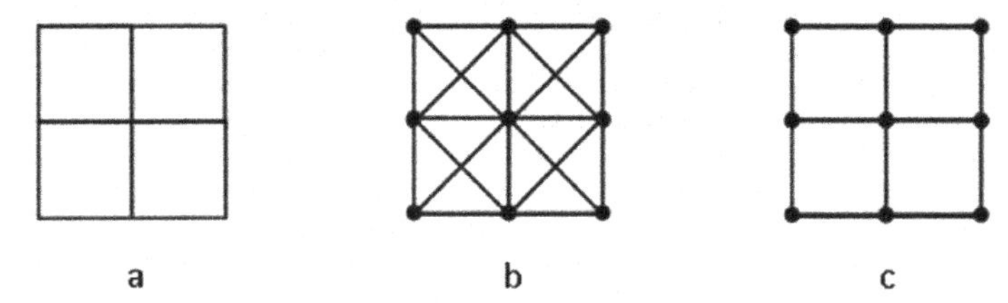

Fig. 4. Nodal graph construction: a) a mesh, b) a complete nodal graph, c) a reduced nodal graph.

In the dual graph vertices correspond to the elements of the mesh (see Fig. 5). In a reduced dual graph vertices are connected by an edge if the corresponding elements of the mesh share a common face (for 2D-mesh – an edge). In a complete dual graph vertices are connected by an edge if the corresponding elements share a common face, a common edge or a common node (for 2D-mesh – a common edge or a common node).

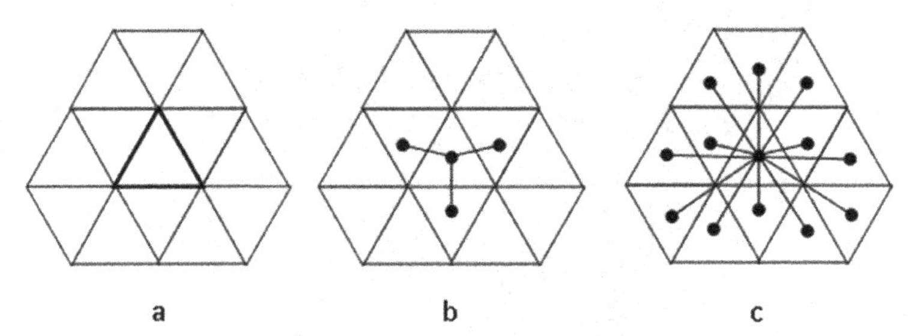

Fig. 5. Dual graph construction: a) a mesh, b) a part of a reduced dual graph for the marked element, c) a part of a complete dual graph for the marked element.

Geometric methods usually are used for a preliminary mesh partitioning among processors or for a mesh partitioning, when it is important to minimize the time of partitioning and the memory usage. Partitions made by geometric methods are more balanced than partitions formed by graph partitioning methods. Generally in geometric partitions difference in numbers of vertices in resulting subdomains is no more than one vertex.

Graph partitioning methods are used for a mesh partitioning of high quality because they take into account interprocessor communications and usually they are more complex than geometric methods.

In this paper several algorithms were compared:

- PartKway - multilevel k-way graph partitioning algorithm from the ParMETIS package [5, 6]. Figure 3 shows an example of a sandstone sample S_5 pore space partition made by this package.
- PartGeomKway - multilevel k-way graph partitioning algorithm from the ParMETIS package, making initial partitioning using a space-filling curve method [5, 6].
- GeomDecomp – the recursive coordinate bisection method from the GridSpiderPar package (Keldysh Institute of Applied Mathematics RAS) [7].
- RCB – recursive coordinate bisection method from the Zoltan package [8].
- RIB – recursive inertial bisection method from the Zoltan package [8].
- HSFC – method based on a Hilbert space-filling curve splitting from the Zoltan package [8].

Connectivity of the constructed dual graph turned out to be insufficient for the parallel incremental algorithm of graph partitioning from the GridSpiderPar package and it couldn't be compared with the other methods.

The first two methods in the list are graph partitioning methods and the last four are geometric methods.

At each stage of recursive bisection an area is divided into two parts. The resulting subareas are split the same way until there is only one subdomain left in each subarea. In recursive coordinate bisection the cut is made orthogonal to the coordinate axis along which the area is the most elongated. In recursive inertial bisection it isn't required that the cut axis was parallel to a coordinate axis. The inertial axis is calculated and usually it is the direction along which the area is the most elongated. The cut is made orthogonal to the inertial axis.

A Hilbert space-filling curve is a continuous fractal space-filling curve that maps the interval [0, 1] into a 2D or 3D area (see Fig. 6). In partitioning algorithms, using this curve, a Hilbert curve is constructed within the divided area and this curve is partitioned into required number of parts.

Multilevel algorithms consist of three phases: graph coarsening, initial partitioning and uncoarsening with refinement of the partitions. Initial partition in the considered algorithms from the ParMETIS package is made by recursive bisection, based on nested dissection and greedy partitioning refinement [9].

2 Mathematical Model of Single Phase Fluid Flow

Different partitions obtained with the described above methods are tested on the problem of single phase flow simulation within pore space of real sandstone. In this paper we use a numerical algorithm with explicit time marching and which is based on regularized

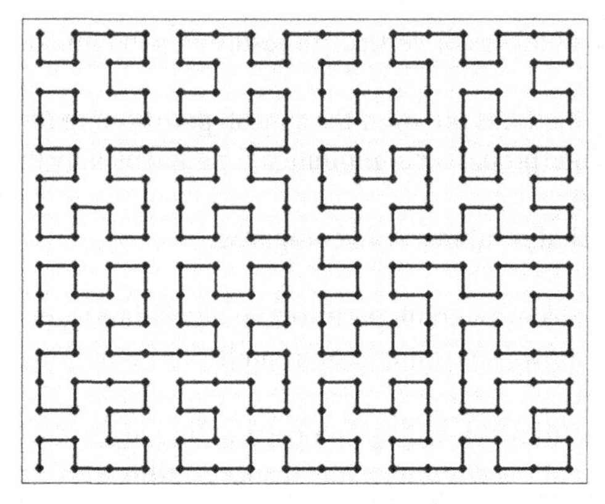

Fig. 6. A 2D Hilbert curve (4th order).

Navier–Stokes equations with quasi-hydrodynamic (QHD) method. QHD regularization is based on the assumption that the mass flux $\mathbf{j}_m$ can not be equal to average momentum of unit volume $\rho\mathbf{u}$ thus additional "regularizing velocity" appears $\mathbf{w} := \mathbf{u} - \mathbf{j}_m/\rho$. This assumption leads to arising of small dissipative terms in original equations which are proportional to small parameter τ. These terms allow one to use logically simple and easy implementable explicit conditionally stable central finite difference methods. Such methods are convenient and promising in modern parallel computations due to their good scalability properties. Here it is worth to mention that QHD-regularization terms allow one to obtain simulation results consistent with experiments for moderately rarefied gas flows.

For completeness below the QHD-regularized system for compressible viscous isothermal single phase fluid flow is presented [10–12]

$$\partial_t \rho + \nabla \cdot \mathbf{j}_m = 0, \tag{1}$$

$$\partial_t(\rho\mathbf{u}) + \nabla \cdot (\mathbf{j}_m \otimes \mathbf{u}) + \nabla p = \nabla \cdot \mathbf{\Pi}, \tag{2}$$

where ρ is mass density, $\mathbf{u}$ is fluid velocity, $\mathbf{j}_m = \rho(\mathbf{u} - \mathbf{w})$ – mass flux, $p = c_s^2\rho$ is pressure, c_s is sound speed, $\mathbf{\Pi} = \mathbf{\Pi}^{NS} + \mathbf{\Pi}^{\tau}$ is the stress tensor, $\mathbf{\Pi}^{NS} = \eta\big[(\nabla\mathbf{u} + \nabla\mathbf{u}^T) - (2/3)\mathbf{I}div\mathbf{u}\big]$ is the Navier-Stokes viscous stress tensor, $\mathbf{I}$ is the identity tensor, η is the coefficient of dynamic viscosity, $\mathbf{\Pi}^{\tau} = \rho\mathbf{u} \otimes \mathbf{w}$ is regularizing tensor, $\mathbf{w} = (\tau/\rho)(\rho\mathbf{u}\cdot\nabla\mathbf{u} + \nabla p)$, τ is the small nonnegative parameter with unit of time. If flow of dense gas or liquid is considered then on the solid boundary noslip condition $\mathbf{u} = \mathbf{0}$ is imposed. In this case QHD-terms are regarded as regularizators only. For regularizing parameter following relation is used $\tau = \alpha^* h/c_s$, where α^* is dimensionless positive number, which is selected from numerical stability considerations, h is the grid spacing.

For numerical solution of these equations explicit central finite difference scheme is used [13, 14].

3 Results

3.1 Subdomain Partitions

Dual graph with $1.9 \cdot 10^8$ vertices and $5.6 \cdot 10^8$ edges was constructed for the computational mesh and was partitioned into 140, 280 and 560 subdomains. Edge-cut is presented in Table 1 and maximal number of edges, cut by subdomain – in Table 2. The results show that graph partitioning methods make partitions with less edge-cut than geometric methods since geometric methods don't take into account interprocessor communications. The minimal numbers of cut edges are obtained by the methods PartKway and PartGeomKway from ParMETIS. Among geometric methods the best results are obtained by GeomDecomp method from GridSpiderPar and RCB method from Zoltan, the worst are made by RIB method from Zoltan package.

Table 1. Edge-cut

Methods	Subdomains		
	140	280	560
Graph partitioning			
PartKway	**$8.0 \cdot 10^6$**	**$1.3 \cdot 10^6$**	**$2.15 \cdot 10^6$**
PartGeomKway	$8.3 \cdot 10^6$	$1.3 \cdot 10^6$	$2.18 \cdot 10^6$
Geometric methods			
GeomDecomp	**$2.6 \cdot 10^6$**	$3.66 \cdot 10^6$	$4.9 \cdot 10^6$
RCB	$2.67 \cdot 10^6$	**$3.62 \cdot 10^6$**	**$4.8 \cdot 10^6$**
RIB	**$4.1 \cdot 10^6$**	**$5.5 \cdot 10^6$**	**$7.3 \cdot 10^6$**
HSFC	$3.2 \cdot 10^6$	$4.4 \cdot 10^6$	$6 \cdot 10^6$

Table 2. Maximal number of edges, cut by subdomain

Methods	Subdomains		
	140	280	560
Graph partitioning			
PartKway	$3.0 \cdot 10^4$	**$2.5 \cdot 10^4$**	$2.6 \cdot 10^4$
PartGeomKway	**$2.7 \cdot 10^4$**	$2.8 \cdot 10^4$	**$2.5 \cdot 10^4$**
Geometric methods			
GeomDecomp	**$5.2 \cdot 10^4$**	**$4.1 \cdot 10^4$**	**$2.9 \cdot 10^4$**
RCB	$5.7 \cdot 10^4$	$4.3 \cdot 10^4$	$3.1 \cdot 10^4$
RIB	**$9.6 \cdot 10^4$**	**$7.4 \cdot 10^4$**	**$4.5 \cdot 10^4$**
HSFC	$6.9 \cdot 10^4$	$5.1 \cdot 10^4$	$3.6 \cdot 10^4$

Number of connected subdomains is shown in Table 3. We can see that the greatest numbers of connected subdomains are achieved by PartKway method from ParMETIS. The least numbers of connected subdomains are obtained by RIB method from Zoltan. GeomDecomp method from the GridSpiderPar package turned out to be the best among geometric methods.

Table 3. Number of connected subdomains

Methods	Subdomains		
	140	280	560
Graph partitioning			
PartKway	99	201	**424**
PartGeomKway	100	198	408
Geometric methods			
GeomDecomp	**3**	**13**	**66**
RCB	0	12	58
RIB	**0**	**0**	**2**
HSFC	0	2	35

3.2 Exchange Times

Average time spent on interprocess exchange during 200 time steps of the considered parallel simulation on partitions into different number of subdomains is presented in Table 4. Results show that the least exchange times are achieved when running on the partitions made by the graph partitioning method PartKway from ParMETIS and the

Table 4. Average time (during 200 time steps) spent on interprocess exchange on partitions into different number of subdomains. Time is measured in 10^{-2} s.

Methods	Subdomains		
	140	280	560
Graph partitioning			
PartKway	**0.87**	**0.74**	**0.63**
PartGeomKway	0.88	0.82	0.74
Geometric methods			
GeomDecomp	1.82	**1.30**	1.05
RCB	**1.75**	1.31	**0.99**
RIB	**3.34**	**2.71**	**1.60**
HSFC	2.28	1.61	1.28

worst are on the partitions made by the inertial bisection method RIB from Zoltan. Among geometric methods the least exchange times are obtained on the partitions made by the recursive coordinate bisection methods GeomDecomp from GridSpiderPar and RCB from Zoltan. Relative difference in run times according to different methods on given number of subdomains was varied from 1.5 to 2.8 times.

The computations were carried out on supercomputer K-60 (Keldysh Institute of Applied Mathematics RAS).

4 Conclusions

Partitions, obtained by different methods, were compared using the problem of simulation of a single-phase fluid flow within pore space of a sandstone sample. Average times spent on interprocess exchange during 200 time steps of the considered parallel simulation were compared when the grid was distributed over the cores in accordance with various partitions. The partitions were made by the methods from the parallel partitioning tools ParMETIS, Zoltan, and GridSpiderPar.

Results show that the least exchange times are achieved when running on the partitions made by the graph partitioning method PartKway from the ParMETIS package and the worst are obtained on the partitions made by the inertial bisection method RIB from the Zoltan package. Among geometric methods the least exchange times are achieved on the partitions made by the recursive coordinate bisection methods GeomDecomp from the GridSpiderPar package and RCB from the Zoltan package. These results are in accordance with comparison of the partitions themselves. The criteria of minimal edge-cut, minimal maximal number of edges, cut by subdomain, and minimal number of unconnected subdomains were found to be important for exchange time reduction in the problem of simulation of a single-phase fluid flow.

References

1. Blunt, M.J.: Multiphase Flow in Permeable Media: A Pore-Scale Perspective. Cambridge University Press, Cambridge (2017)
2. Berg, C.F., Lopez, O., Berland, H.: Industrial applications of digital rock technology. J. Petrol. Sci. Eng. **157**, 131–147 (2017)
3. Dvorkin, J., Derzhi, N., Diaz, E., Fang, Q.: Relevance of computational rock physics. Geophysics **76**(5), E141–E153 (2011)
4. Imperial College London. http://www.imperial.ac.uk/earth-science/research/research-groups/perm/research/pore-scale-modelling/micro-ct-images-and-networks/
5. Karypis,G., Kumar, V.: A coarse-grain parallel formulation of multilevel k-way graph partitioning algorithm. In: Proceedings of the 8th SIAM Conference on Parallel Processing for Scientific Computing (1997)
6. Schloegel, K., Karypis, G., Kumar, V.: Parallel multilevel algorithms for multi-constraint graph partitioning. In: Bode, A., Ludwig, T., Karl, W., Wismüller, R. (eds.) Euro-Par 2000. LNCS, vol. 1900, pp. 296–310. Springer, Heidelberg (2000). https://doi.org/10.1007/3-540-44520-X_39

7. Golovchenko, E.N., Kornilina, M.A., Yakobovskiy, M.V.: Algorithms in the parallel partitioning tool GridSpiderPar for large mesh decomposition. In: Proceedings of the 3rd International Conference on Exascale Applications and Software (EASC 2015), pp. 120–125, University of Edinburgh (2015)
8. Boman, E., et al.: Zoltan: Parallel Partitioning, Load Balancing and Data-Management Services. Developer's Guide, Version 3.3 Sandia National Laboratories, 2000–2010. http://www.cs.sandia.gov/Zoltan/dev_html/dev.html
9. Karypis, G., Kumar, V.: Parallel multilevel k-way partitioning scheme for irregular graphs. SIAM Rev. **41**(2), 278–300 (1999)
10. Sheretov, Y.: Continuum Dynamics Under Spatiotemporal Averaging. RKhD, Moscow-Izhevsk (2009). [in Russian]
11. Chetverushkin, B.N.: Kinetic Schemes and Quasi-Gasdynamic System of Equations. CIMNE, Barcelona (2008)
12. Elizarova, T.G.: Quasi-Gas Dynamic Equations. Springer, Heidelberg (2009). https://doi.org/10.1007/978-3-642-00292-2
13. Balashov, V.A., Savenkov, E.B.: Direct pore-scale flow simulation using quasi-hydrodynamic equations. Dokl. Phys. **61**(4), 192–194 (2016)
14. Balashov, V.A.: Direct numerical simulation of moderately rarefied gas flow within core samples. Matem. Mod. **30**(9), 3–16 (2018)

Management of Computations with LRnLA Algorithms in Adaptive Mesh Refinement Codes

Anton Ivanov[1(✉)], Vadim Levchenko[1], Boris Korneev[1,2], and Anastasia Perepelkina[1,2]

[1] Keldysh Institute of Applied Mathematics, Moscow, Russia
aiv.racs@gmail.com, vadimlevchenko@mail.ru, korneev@kintechlab.com, mogmi@ya.ru

[2] Kintech Lab Ltd., Moscow, Russia

Abstract. The data structure for codes with adaptive mesh refinement with low overhead for data storage is developed. The computational fluid dynamics scheme is implemented on this structure with the use of the ConeFold locally recursive non-locally asynchronous algorithm. It increases the computational intensity by a recursive traversal of the parts of the dependency graph in a ConeFold. The dependencies are determined by the mesh structure, and the mesh may be coarsened or refined at the synchronization time moments. For an efficient execution in the parallel environment of the many-core and many-node systems the ConeFold task manager is used. The load balancing is performed with an account for the computation complexity of different ConeFolds.

Keywords: AMR · Grid refinement · Data structure · Z-curve · LRnLA

1 Introduction

Adaptive Mesh Refinement (AMR) [1] is used to decrease the number of calculations in different areas of computational physics. In non-uniform grids, by using larger cell size in the areas where it is applicable, the amount of data and the number of cell updates are decreased. The characteristics of studied process may change dynamically, so, in AMR, the grid adapts dynamically in the course of the simulation.

In theory, this approach may decrease the computational cost by several orders of magnitude. However, the overhead for mesh storage, neighbor search, cell traversal are significant. Thus, the issue of efficient parallel implementation of the data storage structures, mesh rebuilding, cell traversal remains a relevant task for applied software development.

For stencil based schemes, it is important to minimize the storage overhead and to improve the storage locality. This is because such problems are traditionally memory-bound, so to improve the performance efficiency of such problems,

V. Voevodin and S. Sobolev (Eds.): RuSCDays 2020, CCIS 1331, pp. 25–36, 2020.
https://doi.org/10.1007/978-3-030-64616-5_3

the arithmetic intensity should be increased, or the data should be localized in higher levels of cache hierarchy.

For optimizing the data storage, we have developed the tile-based data structure with 'light' and 'heavy' tiles [6]. Since it is tile-based and follows Z-order curve, the locality requirement is also satisfied.

To further improve the computing efficiency, we also propose the use of Locally Recursive non-Locally Asynchronous (LRnLA) algorithms [9]. The LRnLA algorithms serve the purpose of promoting memory-bound problems with local dependencies to the compute bound domain by increasing the operational intensity. The main idea is to change the traversal of the dependency graph so as to minimize the processor wait times. The scheme essentially remains the same, because the traversal rule preserves the correct dependencies.

Another issue with the AMR computation is the load balancing in the parallel environment. The grid, and the number of operations is not only non-uniform, but also may dynamically change is the calculation progress. Thus, the computational cost is not known beforehand.

LRnLA algorithms construction approach is flexible in terms of parallelization and provides good balancing [7,9]. However, here the estimation of the computation cost should be made at the mesh reformation steps.

There exist other temporal blocking approaches [11,12,18,19]. Some kind of temporal blocking had previously been implemented in other AMR frameworks [17]. The LRnLA algorithms are chosen for the temporal blocking on AMR grids, since the recursive construction of the LRnLA algorithms fit the recursive subdivision of the AMR structure. Thus, the LRnLA algorithms provide a more natural traversal of the dependency graph than even the usual step-by-step update pattern. Notably, the mesh traversal in [17] is described with a recursive space-filling curve as well.

In this paper we present our implementation of the AMR data structure, and the mesh traversal rules that correspond to recursive temporal blocking of the LRnLA algorithms. For the dynamic load balancing in the parallel environment, the computation manager is developed for the algorithm.

2 Problem Statement

2.1 System of Conservation Laws

The numerical methods of solving the partial differential equations in the divergence form are one of the natural applications of the adaptively refined grid due to the presence of discontinuities and other features of the solutions, e.g. eddies, that need more accurate local resolution [1,10].

Let the mathematical model be represented by the following system of equations

$$\frac{\partial U}{\partial t} + \frac{\partial F_i(U)}{\partial x_i} = 0 \tag{1}$$

in the domain G with given initial and boundary conditions. For example, the transport equations, the equations of fluid dynamics, the wave equations can be rewritten in the form of (1).

Let the finite volume grid be defined over the G. The coarsest grid level is Cartesian, and every cell may be refined by its 2^d–binary splitting. For most cases $d = 2$ or $d = 3$.

In this paper we suppose that the given method is explicit, has a local stencil and a CFL-type restriction on the time step. In the simplest case the first order Godunov-type finite-volume scheme [4] is expressed as follows

$$\frac{u_J(x, t + \Delta t_J) - u_J}{\Delta t_J} + \frac{1}{\Delta V_J} \sum_{i \in \mathfrak{I}_J} \mathfrak{F}_i n^i \Delta S_i = 0, \tag{2}$$

where u_J is the numerical solution in the considered cell J, Δt is the time integration step, ΔV_J is volume of the cell, $\mathfrak{F}_i$ are numerical Godunov fluxes through the faces of nearby cells with the unit normal vectors n^i. Such set of faces is denoted as $\mathfrak{I}_J$. ΔS_i is area of the i face.

3 Data Structure

3.1 Storage Space Optimization in the Tree Structure

For higher probability of local access to the neighboring cells, all cells are grouped into cubes with 2^D size, which are referred to as chunks. The chunks are further grouped into tiles, with a configurable size of 2RD cells, or $2^{(R-1)D}$ chunks (Fig. 1*a*).

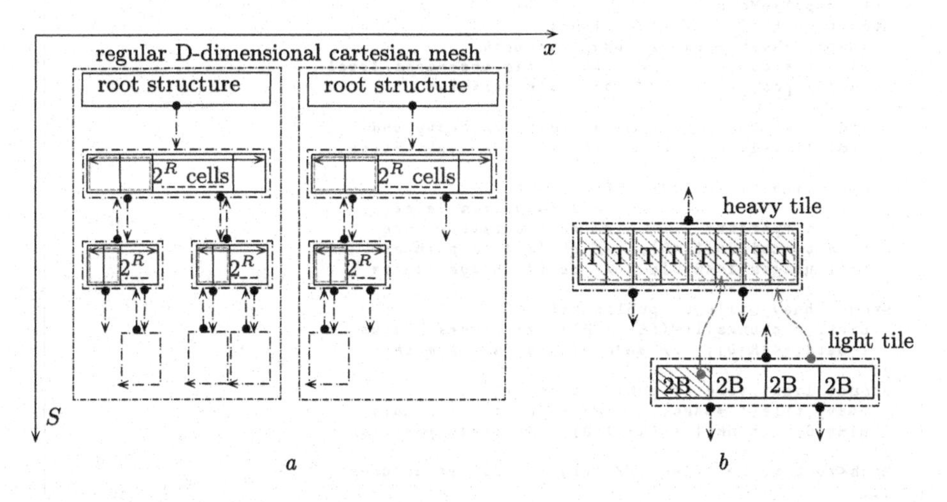

Fig. 1. Tile tree (*a*), the chunks are highlighted with green, and an interaction between the light and heavy tiles (*b*), the data of the chunk of the light tile are in the free space of the heavy tile. T is a user defined cell type (Color figure online)

Tiles are stored in an 2^D-tree structure. Tree roots are stored in a D-dimensional array on a Cartesian grid. Tiles contain data for communication with other nodes in the tree (a pointer to the parent tile and 2^D pointers to the child tiles), and also the data for the numerical scheme. In this approach, if the maximum level of subdivision of a tile is high, at one geometrical position multiple cells are defined. In most numerical schemes, only the smallest cell data are relevant at each spatial position. For example, hydrodynamical schemes with flux correction [1] require only one cell anywhere. In the schemes for mass-preserving non-uniform grid for the Lattice Boltzmann Method (LBM) [3,16], there is a layer of transition between the grids with different layers of subdivision, where two levels of subdivision of cells is required at the same place. However, to obtain a significant performance gain with AMR, the subdivision may be up to $\sim 5 \div 7$ levels. The storage of 5 levels of data at the same place is superfluous.

The concept of light tiles is introduced to deal with this problem. They do not contain the cell data, but use indirect addressing. The heavy tile is supposed to be deep in the tree structure and the cell data that it could contain would be irrelevant. Thus, instead, it contains the cell data that correspond to some light tile. Light tile contains a pointer to the heavy tile and a cube of $2^{(R-1)D}$ 2-byte numbers, which code the position of a required chunk in the heavy tile (Fig. 1*b*). The heavy tile that only contains light tile data may be allocated in the same tree structure.

The described structure is implemented as a template class `AdaptiveMesh` of the `aiwlib` library [5] with `C++-11` (Fig. 2).

```
template <typename T, // cell type
          int D,      // dimension
          int R>      // tile size
class AdaptiveMesh{
  struct root_t{   // root struct
    AdaptiveMesh *msh; // adaptive mesh
    tile_t *root;       // tile of the uppermost level
    Ind<D> pos;         // root tile position

    std::list<heavy_tile_t> htiles;  // heavy and
    std::list<light_tile_t> ltiles;  // light tiles
  };
  struct tile_t{ // base tile struct
    int S;                     // subdivision level
    tile_t *parent;            // parent tile pointer
    tile_t *childs[1<<D];      // child tile pointer
    bool usages[1<<R*D];       // cell's usage status
  };
  struct heavy_tile_t: public tile_t{
    uint2_t chunks[1<<((R-1)*D)]; // chunks status
    T data[1<<R*D]; // data stored in the mesh
  };
  struct light_tile_t: public tile_t{
    heavy_tile_t *page;       // page for cell data
    uint16_t chunks[1<<(R-1)*D];   // chunk position
  };
  Mesh<root_t, D> tiles; // Mesh of root structures
  ...
};
```

Fig. 2. The simplified code of the AMR data structure. `Ind<D>` is a D–dimensional integer vector, `Vec<D>` is a D–dimensional vector of `double` data type.

3.2 Mesh Adaptation

The `rebuild` method is used for the mesh adaptation. Its arguments are the user-defined functions—refinement criterion, the cell actions for refinement, and the cell actions for coarsening (Fig. 3). The refinement criterion argument is an object of type `Cell` and it returns 1 if the cell is to be refined, 0 if no action is to be taken, and −1 if the cell may be coarsened.

The functions for refinement and coarsening take the cell and pointer to the chunk data as arguments. It is guaranteed that in the process of rebuilding the mesh no data is deleted or moved. It may be moved after the rebuilding stage.

```
template<typename F_CHECK, typename F_SPLIT, typename F_JOIN>
void AdaptiveMesh<T, D, R>::rebuild(F_CHECK &&f_check,
                                    F_SPLIT &&f_split,
                                    F_JOIN &&f_join);
```

Fig. 3. Mesh rebuilding method.

4 Algorithms

4.1 LRnLA Algorithm ConeFold

Just like the space-time may be divided into the domains of influence, dependence, and asynchronous events, the dependency graph of a discrete computation may be subdivided into synchronous and asynchronous regions. In the schemes with finite information propagation speed, a cone of its influence may be constructed. Similarly, for a cell update N_t time steps after the start, a cone of dependence may be constructed to find areas at the initial condition that do and do not influence the result. This way, instead of iterating throughout the spacial dimensions of the domain at each time step (stepwise approach), the computation may be divided into portions that span the time-space.

In a dD1T (d-dimensional with time axis) space, one can formulate a generalized Cauchy problem: with the knowledge of the scheme update, initial and boundary conditions, find the state of the system N_T time steps later. The system state at times between $t = 0$ and $t = N_T$ is not required. Times $t = kN_T$ with integer k are called synchronization instants.

LRnLA algorithms are represented by a shape in the dependency graph space and a rule of its subdivision [7]. The shape represents a procedure of processing all the dependency graph nodes inside it. The subdivision of a shape represents the subdivision of the procedure into sub-tasks, i.e. the shapes that appear after the subdivision. By tracing the dependencies between the dependency graph nodes in the shapes, the correct order of execution, and the portions of asynchronous calculations that span the time-space may be found.

The subdivision is recursive, and it continues until some shape covers only one node. At different levels of the recursive subdivision the shapes may be

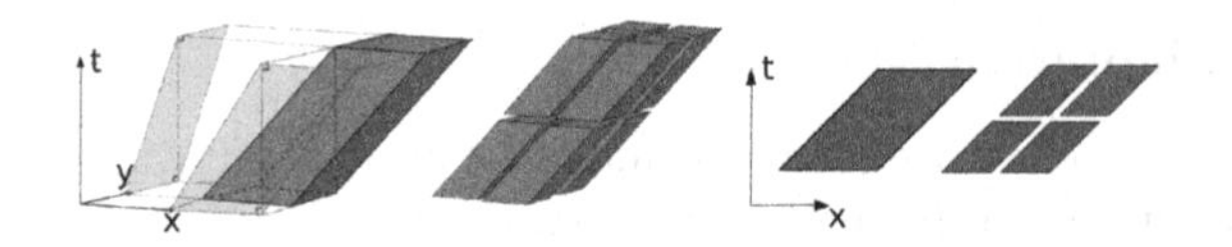

Fig. 4. ConeFold LRnLA algorithm and its subdivision

allocated to process data, localized in some level of memory, or distributed for parallel processing by different methods in the hybrid parallel systems. For this, the subdivision rules may be different at different levels of subdivision.

Here we implement the ConeFold [7,13,14] algorithm construction. It is best suited for the cube stencils [7,8]. The dD1T ConeFold of rank r is a call of 2^{d+1} ConeFolds of rank $r-1$ in a correct order (Fig. 4).

The subdivision of the ConeFold shape may be different. For example, in [8] the ConeTorre approach is implemented. The ConeTorre is a shape made of several stacked ConeFolds, and is subdivided into recursive ConeFolds or into flat planes.

4.2 LRnLA Traversal of the Non-uniform Grid

First, in a method called `C0`, the array of the tree roots is iterated over N_t times (Fig. 5a). With each root tile as a lower base, the `C1` ConeFold method is called. After N_t layers of `C1` the synchronization instant is defined: it is an instant where all data is updated up to the same time instant. The data of the whole region may be output for visualization or diagnostics. Also, the mesh rebuild operations are called only on the synchronization instants.

The `C1` method is a recursive function. It calls 2^{D+1} `C1` functions on smaller tiles. For each `C1` execution, the data that is potentially addressed in it are collected into an array denoted as *ground*. It is a cube of 2^D tiles, which includes the tiles of the lower base and the upper base of the ConeFold (Fig. 5a).

At some level of subdivision, or even at the very beginning, if at least one cell in the ground of the `C1` method is active (i.e. it is not refined further), this `C1` method becomes a `C2` ConeFold. The `C2` is subdivided further as a TorreFold LRnLA algorithm [7,15] (Fig. 5b). It is subdivided into towers of stacked ConeFold of minimal size, that is, into ConeTorres the bases of which is one cell.

4.3 ConeFold Closure

On a uniform grid, in the naive implementation, for each cell all incoming fluxes are computed depending on the u data from the two cells sharing the face. The fluxes are summed into the temporary value u_{tmp}. When all neighbors have performed the update, the u_{tmp} is copied to the current u.

The ConeFold recursion closure has access to 2^D cells. The fluxes are updated on faces as shown in Fig. 6, summed into u_{tmp} of the corresponding cells, and the $u_{\text{tmp}} \to u$ copy is executed in one cell.

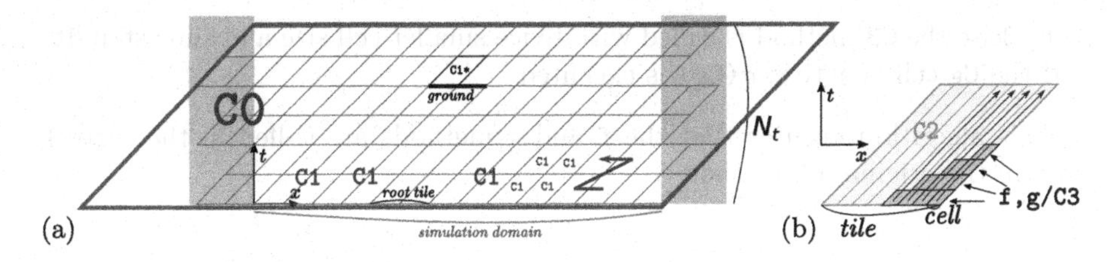

Fig. 5. ConeFold traversal. Here $N_t = 3$. The ground is pictured for one C1 method.

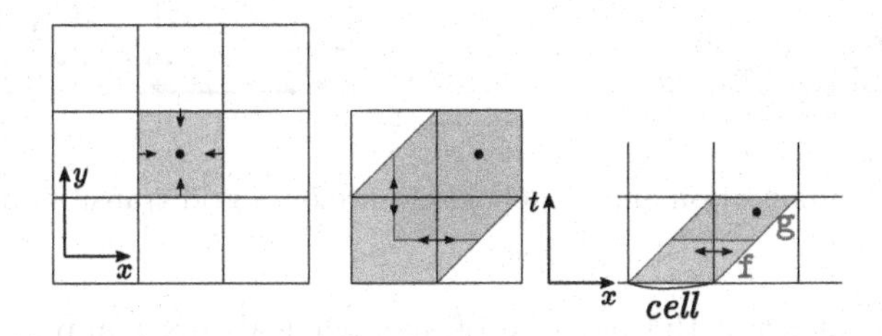

Fig. 6. A naive cell update and a ConeFold closure on a uniform grid for the chosen scheme. All required cells for the update are pictured.

This may be split into two half-steps: the flux computation operation f and the u array element update operation g. These user-defined methods may be replaced for implementation of other schemes. The f method is associated with the lower base of the ConeFold, and the g method with the upper one. This f–g closure is used on the uniform regions of the mesh.

On the grid refinement interfaces, the special recursive ConeFold is called. It is a function C3 with a following signature:

```
template<typename F, typename G, typename dT>
void C3(Cell &Cf, Cell &Cg, F &&f, G &&g, dT dt);
```

Its arguments are pointers to the two cells Cf and Cg the lower and the upper bases of the ConeFold correspondingly, as well as the two methods of the uniform closure and a time step. In case there are no grid transitions, C3 executes the two half-steps: `f(Cf, dt)}; \verbg(Cg, dt)—`. Elsewhere the following algorithm is applied:

1. If the Cf cell is active (i.e. is at the current subdivision level), f(Cf) is executed.
2. The ground1 cube of 2^D cells, which includes Cf and Cg is constructed. Some of these cells may be found to be inactive (for example, further refined).
3. Another cube ground2 is constructed inside ground1, which covers the same cube as ground1, and consists of 4^D smaller cells of the next level of refinement.
4. Execute 2^{D+1} ConeFolds, in a correct order, on the ground2 bases. Here
   (a) If the ConeFold is simple, i.e. all cells in its ground are active, only the f and g methods are called.

(b) Else the `C3` method is called with twice smaller cell size and time step `dt`.
5. If the `Cg` cell is active, `g(Cg)` is executed.

This algorithm ensures that the `f` and `g` methods are called in the correct order in the original `C3` method.

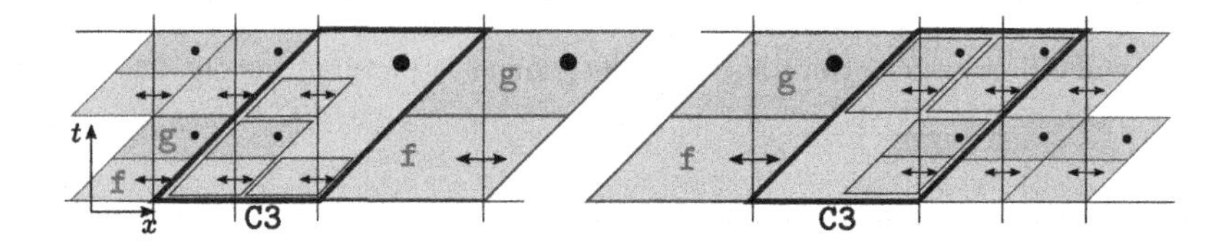

Fig. 7. Flux computation and ConeFold closure on a grid transition interface.

The rules that serve the purpose of symmetrical flux computation for the neighboring cells and, consequently, the conservation of mass, are the following:

- *Do not touch the fine neighbors.* A coarse cell does not compute fluxes on the faces, shared with the fine cells. This way, the f sub-step may omit the flux computation, as in Fig. 7, right.
- *Fine cells do all the work.* They compute fluxes, and these fluxes update the u_{tmp} value in the neighboring coarse cell as well, if it exists. This way, g functions may additionally compute fluxes, as in Fig. 7, right.

4.4 Management of Computations

The issue with implementing parallel algorithms for AMR grids is the fact that the computation load shifts dynamically in the grid. The number of operations is not known beforehand, so a smart computation manager is necessary.

The ConeFold computation manager is developed here to address the following issues:

- Load balancing in the parallel computer systems;
- Efficient use of the shared cache in systems with multicore processors;
- Allocation of data processing to the NUMA-nodes which contain these data.

Currently, the manager works for one computing node in the shared memory space and may support several NUMA-nodes, computing core domains with last level shared cache and hardware multithreading inside one core.

The manager uses a hierarchical model of the computer system, which corresponds to the the model of the HWloc [2] library. Software threads are logically merged into groups. Each group is bound to one of the HWloc processors, which are in the shared cache domain. The manager provides both static planning of the computation order and dynamic choice of the suitable task by a thread group, with a further parallel execution of it.

For static planning, the `C1` ConeFolds are chosen. After a mesh reforming, which may happen at the synchronization instants, the statistics of the estimated computational costs of such tasks are collected.

Assume the mesh size is equal to N_d root tiles in the d-axis, $d = 0..D$. Then the number of `C1` tasks in one layer is $\prod_{d=0}^{D}(N_d+1)$, and the number of `C1` ConeFolds, which require planning, is $N_t \prod_{d=0}^{D}(N_d+1)$, where N_t is the number of layers of `C1` between the synchronization instants.

Each task is uniquely identified with a set of $(D+1)$ numbers $id = (i_t, i_1 \dots i_D)$. The coordinate indexes start from the right side of the domain, since the ConeFold horizontal dependencies are directed from right to left. The process of static planning amounts to a production of one or several lists of such sets, with an account for the dependencies between the corresponding tasks. That is, each task in a list depends only on the previous tasks in the list, and influences only the following ones. According to the dependencies between `C1` ConeFolds, a list starts with a `C1` with $id = (0, N_0, \dots N_D)$ and ends with `C1` with $id = (N_t - 1, 0, \dots 0)$. The number of different possible lists is high. It is equal to the number of possible dependency graph traversals.

Depending on the hardware specifics, one of the two opposite strategies may be used. *(1) Localization.* The list traversal satisfies the requirement of maximal data reuse, for which the ConeFold stacking into towers (ConeTorre shape, following the arrows in Fig. 5b) and Z-curve is used. *(2) Asynchrony.* The list traversal satisfies the requirement of maximal asynchrony of the neighboring sub-tasks. For this, the list is sorted by the tiers of the dependency graph, $tier(id) = i_t + \sum_{d}^{D}(N_d + i_t - i_d)$. The combination of these approaches is used to move the task with maximal estimated computational cost to the start of the list, which improves the chances of the successful dynamic load balancing.

Dynamic balancing of the task execution in the parallel computing system is implemented with the atomic C++11 operations (`atomic<int>` data type), which act as semaphores. Each list of tasks is assigned with a separate atomic iterator. When a group of threads is ready for the execution of a new task, it gets an id of a unique task from the list with the help of this iterator. The list creation method guarantees that all tasks that it depends on were started beforehand. However, for the correct execution in the parallel system it is necessary to wait for the finish of these tasks. For this, the `atomic<int> A` array is used. Its size is $\prod_{d=0}^{D}(N_d + N_t)$, to account for the `C1` ConeFolds at the upper layers, which are a part of a ConeTorre that starts outside of the domain. The task denoted with id is assigned with its own atomic `A[id]` with $(i_1 + i_t, \dots i_D + i_t)$ coordinates. Before its start, the task waits for the satisfaction of the condition $A[id + S_d] \geq i_t$, $S_d \in (0, -1, -N_x \dots)$, where S_d is the shift to the atomics of the task on which the current one depends on in $D+1$ directions. After the execution is finished, `A[id]` is incremented by one.

The group of threads enters the recursive call of `C1`, and executes the `C2` method inside it according to the TorreFold algorithm [7].

For the support of the processors with several NUMA-nodes, the clustering of the root tiles takes place at the mesh reform steps. These tiles are bound to the NUMA-nodes with `hwloc`, and their data is localized in the corresponding data space. The criterion of the successful clustering is the approximate equality of the computational costs of the `C1` methods which are based on the root tiles bound to different NUMA-nodes. This way, execution lists that are bound to some specific NUMA-node, are formed. They are executed in the threads, bound to this node.

Additional special lists are constructed for balancing. If the tiles of the `C1` ground are bound to different NUMA-nodes, it goes into the balancing list for these NUMA-nodes. The balancing list can be executed by the threads assigned to any of these NUMA-nodes. `C1` ConeFolds from the top and the bottom of the overall list, i.e. from the first and the last tier of the `C0` dependency graph are also assigned into a balancing list due to the insufficient asynchrony. This list may be assigned to any thread group. The total number of lists may be up to $2^{N_{nodes}} - 1$, where N_{nodes} is the number of NUMA-nodes.

The resulting list of tasks for each NUMA-node is constructed from the current NUMA-node's list, and all the balancing lists in which this NUMA-node is involved in. Thus, to ensure the correct order of task execution, it is sufficient to sequentially choose tasks from each of the NUMA-node lists with the atomic operations. If the task is in the balancing list, then a separate atomic decides if this task was executed in another NUMA-node.

5 Simulation Results

The considered AMR algorithm was examined on a numerical simulation of the advection problem, described by the equation $\frac{\partial u}{\partial t} + a_i \frac{\partial u}{\partial x_i} = 0$, which is the particular case of the problem class described by the equation (1). As an example, we use the two-dimensional setting and the vector of transport $a = [1,\ 0]^T$. For the initial condition, the discontinuous perturbation of the domain is set having the circle shape, where $u = 1$, and zero value of $u = 0$ elsewhere. Zero values are set at the boundaries too (Fig. 8).

The numerical scheme (2) is used. For the transport problem it simplifies to the classic upwind scheme. Stability condition is chosen with $\text{CFL} = 0.9 < 1$.

The simulation is carried out on the AMR grid with three levels of refinement with the ConeFold traversal rules. In Fig. 8 we show the 2D cross-section of the simulation at a time moment when the perturbation transferred the half of the numerical domain. We obtain the solution identical to the one got by the stepwise numerical scheme with the same initial and boundary conditions.

The performance gain follows from both the memory-optimized data structure implementation, mitigating the memory-bound limit with the recursive ConeFold execution, and balanced parallel execution. The detailed performance benchmark testing is to be published elsewhere.

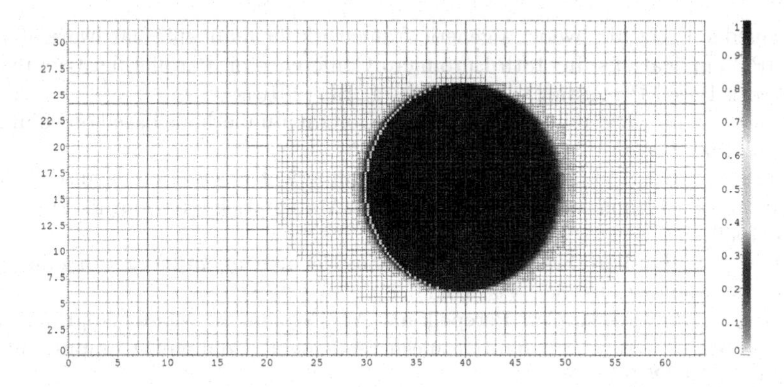

Fig. 8. Advection problem in the xy cross-section. The mesh is shown as red lines, the tiles as blue lines. Pallette for u is on the right (Color figure online)

6 Conclusion

The contribution of this work is the first successful implementation of the LRnLA temporal blocking technique for the dynamically adaptive mesh. The mesh data structure satisfies the requirement of low storage space and high locality. The LRnLA algorithm ConeFold is implemented on this data structure in the `aiwlib` library framework. The ConeFold traversal implementation is described with the standard LRnLA approach by a series of recursive calls, with different subdivision rules at some levels of recursion. The parameters of the data structure and algorithms may be adjusted so as to fit the higher levels of cache hierarchy and efficiently use the memory subsystem of the advanced computer hardware.

The computation manager is implemented with `hwloc` for parallel processing of the ConeFold sub-tasks `C1` of the root level, with an account for the fact that some of them may contain an order of magnitude more computation operations than the others. With the LRnLA approach, the subdivision into large asynchronous tasks provides the flexibility for a balanced parallel execution.

The constructed algorithm is implemented and tested to be correct for a sample scheme. The crucial components of an AMR simulation is shown to be possible to implement in the new LRnLA-AMR framework. The graph traversal is correct, and the mass conserving transition boundary is implemented. Further development of the LRnLA theory on variable meshes is expected in the following work, along with implementation of modern computational fluid dynamics schemes in the developed framework.

The work is supported by the Russian Science Foundation, grant #18-71-10004.

References

1. Berger, M.J., Colella, P.: Local adaptive mesh refinement for shock hydrodynamics. J. Comput. Phys. **82**(1), 64–84 (1989)

2. Broquedis, F., et al.: hwloc: a generic framework for managing hardware affinities in HPC applications. In: 18th Euromicro Conference on Parallel, Distributed and Network-Based Processing, pp. 180–186. IEEE (2010)
3. Chen, H., et al.: Grid refinement in lattice Boltzmann methods based on volumetric formulation. Phys. A **362**(1), 158–167 (2006)
4. Godunov, S.K.: A difference method for numerical calculation of discontinuous solutions of the equations of hydrodynamics. Matematicheskii Sbornik **89**(3), 271–306 (1959)
5. Ivanov, A.V., Khilkov, S.A.: Aiwlib library as the instrument for creating numerical modeling applications. Sci. Vis. **10**(1), 110–127 (2018)
6. Ivanov, A., Levchenko, V., Perepelkina, A., Pershin, I.: Memory-optimized tile based data structure for adaptive mesh refinement. Commun. Comput. Inf. Sci. **1129**, 64–74 (2019)
7. Levchenko, V.D., Perepelkina, A.Y.: Locally recursive non-locally asynchronous algorithms for stencil computation. Lobachevskii J. Math. **39**(4), 552–561 (2018)
8. Levchenko, V., Zakirov, A., Perepelkina, A.: GPU implementation of conetorre algorithm for fluid dynamics simulation. In: Malyshkin, V. (ed.) PaCT 2019. LNCS, vol. 11657, pp. 199–213. Springer, Cham (2019). https://doi.org/10.1007/978-3-030-25636-4_16
9. Levchenko, V.D.: Asynchronous parallel algorithms as a way to archive effectiveness of computations. J. Inf. Tech. Comp. Systems (1), 68–87 (2005). (in Russian)
10. Lutsky, A.E., Severin, A.V.: Numerical study of flow x-43 hypersonic aircraft using adaptive grids. Keldysh Institute preprints (102) (2016)
11. Malas, T., Hager, G., Ltaief, H., Stengel, H., Wellein, G., Keyes, D.: Multicore-optimized wavefront diamond blocking for optimizing stencil updates. SIAM J. Sci. Comput. **37**(4), C439–C464 (2015)
12. Nguyen, A., Satish, N., Chhugani, J., Kim, C., Dubey, P.: 3.5-D blocking optimization for stencil computations on modern CPUs and GPUs. In: Proceedings of the 2010 ACM/IEEE International Conference for High Performance Computing, Networking, Storage and Analysis, SC 2010, pp. 1–13. IEEE (2010)
13. Perepelkina, A.Y., Levchenko, V.D., Goryachev, I.A.: Implementation of the kinetic plasma code with locally recursive non-locally asynchronous algorithms. J. Physi.: Conf. Ser. **510**, 012042 (2014). IOP Publishing
14. Perepelkina, A., Levchenko, V.: LRnLA algorithm conefold with non-local vectorization for LBM implementation. In: Voevodin, V., Sobolev, S. (eds.) RuSCDays 2018. CCIS, vol. 965, pp. 101–113. Springer, Cham (2019). https://doi.org/10.1007/978-3-030-05807-4_9
15. Perepelkina, A., Levchenko, V.: Synchronous and asynchronous parallelism in the LRnLA algorithms. In: Sokolinsky, L., Zymbler, M. (eds.) PCT 2020. CCIS, vol. 1263, pp. 146–161. Springer, Cham (2020). https://doi.org/10.1007/978-3-030-55326-5_11
16. Rohde, M., Kandhai, D., Derksen, J., Van den Akker, H.E.: A generic, mass conservative local grid refinement technique for lattice-boltzmann schemes. Int. J. Numer. Meth. Fluids **51**(4), 439–468 (2006)
17. Weinzierl, T.: The peano software-parallel, automaton-based, dynamically adaptive grid traversals. ACM Trans. Math. Softw. (TOMS) **45**(2), 1–41 (2019)
18. Wolfe, M.: Loops skewing: the wavefront method revisited. Int. J. Parallel Prog. **15**(4), 279–293 (1986)
19. Wonnacott, D.G., Strout, M.M.: On the scalability of loop tiling techniques. In: IMPACT (2013). http://impact.gforge.inria.fr/impact2013/impact2013-proceedings.pdf

Multiple-Precision BLAS Library for Graphics Processing Units

Konstantin Isupov[1(✉)] and Vladimir Knyazkov[2]

[1] Vyatka State University, Kirov, Russia
ks_isupov@vyatsu.ru
[2] Penza State University, Penza, Russia
kniazkov@pnzgu.ru

Abstract. The binary32 and binary64 floating-point formats provide good performance on current hardware, but also introduce a rounding error in almost every arithmetic operation. Consequently, the accumulation of rounding errors in large computations can cause accuracy issues. One way to prevent these issues is to use multiple-precision floating-point arithmetic. This paper presents a new library of basic linear algebra operations with multiple precision for graphics processing units. The library is written in CUDA C/C++ and uses the residue number system to represent multiple-precision significands of floating-point numbers. The supported data types, memory layout, and main features of the library are considered. Experimental results are presented showing the performance of the library.

Keywords: Multiple-precision computation · Floating-point arithmetic · Residue number system · BLAS · CUDA · GPU programming

1 Introduction

It is no surprise that floating-point operations have rounding errors that occur during calculations. Such errors are natural due to the limited length of the significand in the binary32 and binary64 formats from the IEEE 754 standard [1]. For many applications, these errors do not prevent obtaining the correct results. Moreover, for some applications such as deep learning, the best option is to use lower precision formats, e.g., the 16-bit (half-precision) format [2]. However, for a rapidly growing number of scientific computing applications the natively supported IEEE 754 formats are not enough and a higher level of precision is required [3–6]. For such applications, multiple-precision libraries are used, which allow one to perform arithmetic operations on numbers represented with hundreds and thousands of digits.

This paper describes a library that provides new parallel algorithms and implementations for a number of basic linear algebra operations, like the BLAS routines [7], with multiple precision. The library, called MPRES-BLAS, is

V. Voevodin and S. Sobolev (Eds.): RuSCDays 2020, CCIS 1331, pp. 37–49, 2020.
https://doi.org/10.1007/978-3-030-64616-5_4

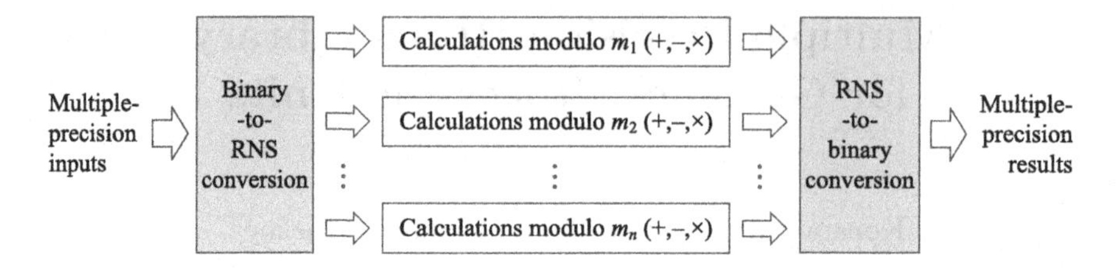

Fig. 1. Performing arithmetic operations in RNS

designed to be used on high-performance computing systems equipped with modern graphics processing units (GPUs). The library uses the residue number system (RNS) [8,9] to represent multiple-precision numbers. In the RNS, a set of moduli are given which are independent of each other. A number is represented by the residue of each modulus and the arithmetic operations are based on the residues individually as shown in Fig. 1. This introduces parallelism in arithmetic with multiple precision and makes RNS a promising number system for many-core architectures such as GPUs.

2 Related Work

One approach to get greater precision and accuracy is to use floating-point expansions, when an extended-precision number is represented as an unevaluated sum of several ordinary floating-point numbers. An example of such an expansion is the double-double format, capable of representing at least 106 bits of significand. Each double-double number is represented as an unevaluated sum of two binary64 numbers. In turn, the quad-double format is capable of representing 212 bits of significand by using four binary64 numbers. Algorithms for computing floating-point expansions are called error-free transformations [10]. Double-double arithmetic is used in the XBLAS [11] and QPBLAS [12] packages for CPUs, as well as in the QPBLAS-GPU package for GPUs [13].

The ExBLAS package [14] contains a number of optimized implementations of accurate and reproducible linear algebra operations for parallel architectures such as Intel Xeon Phi and GPUs. Reproducibility is defined as the ability to obtain a bit-wise identical result from multiple runs of the code on the same input data. To ensure reproducibility, ExBLAS uses error-free transformations and long fixed-point accumulators that can represent every bit of information of the input floating-point format (binary64). The use of long accumulators provides the replacement of non-associative floating-point operations with fixed-point operations that are associative.

The paper [15] presents highly optimized GPU implementations of the DOT, GEMV, GEMM, and SpMV operations, which are included in the BLAS-DOT2 package. In these implementations, internal floating-point operations are performed with at least twofold the precision of the input and output data precision, namely, for binary32 data, the computation is performed using the binary64 format, whereas for binary64 data, the computation is performed using the Dot2 algorithm [16], which is based on error-free transformations.

Nakata et al. proposed MPACK (aka MPLAPACK) [17], a package of multiple-precision linear algebra routines. It consists of two modules, MBLAS and MLAPACK, which are multiple-precision versions of BLAS and LAPACK for CPUs, respectively. MPACK supports several libraries like GMP, MPFR, and QD for underlying multiple-precision arithmetic. In addition, MPACK provides double-double implementations of the GEMM and SYRK routines for GPUs.

There are also a number of extended- and multiple-precision arithmetic libraries for GPUs. Support for double-double and quad-double is implemented in GQD [18], which allows one to perform basic arithmetic operations and a number of mathematical functions with extended precision. GQD mainly uses the same algorithms as the QD library for CPUs. To represent extended precision numbers, GQD uses the vector types double2 and double4 available in CUDA.

CAMPARY [19] uses n-term floating-point expansions (generalization of the double-double and quad-double formats to an arbitrary number of terms) and provides flexible CPU and GPU implementations of multiple-precision arithmetic operations. Both the binary64 and the binary32 formats can be used as basic blocks for the floating-point expansion, and the precision (expansion size) is specified as a template parameter. Generally, each addition and multiplication of n-component expansions in CAMPARY requires $3n^2 + 10n - 4$ and $2n^3 + 2n^2 + 6n - 4$ standard floating-point operations.

GARPREC [18] and CUMP [20] support arbitrary precision on GPUs using the so-called "multi-digit" format. This format stores a multiple-precision number with a sequence of digits coupled with a single exponent. The digits are themselves machine integers. The GARPREC algorithms are from David Bailey's ARPREC package for CPUs, whereas CUMP is based on the GNU MP Bignum library (GMP). In both GARPREC and CUMP, each multiple-precision operation is implemented as a single thread and an interval memory layout is used in order to exploit the coalesced access feature of the GPU.

In [21], we have proposed new multiple-precision arithmetic algorithms using the residue number system and adapted them for efficient computations with multiple-precision vectors on GPUs. Similar algorithms for dense multiple-precision matrices were then proposed and implemented. All these algorithms are used in the MPRES-BLAS library, which is discussed in this paper.

Table 1 summarizes the software packages considered in this section.

Table 1. Software for accurate and/or higher precision computations

Library	Platform	Source code	Ref
XBLAS	CPU	https://www.netlib.org/xblas	[11]
QPBLAS	CPU	https://ccse.jaea.go.jp/software/QPBLAS	[12]
QPBLAS-GPU	GPU	https://ccse.jaea.go.jp/software/QPBLAS-GPU	[13]
ExBLAS	CPU, Phi, GPU	https://github.com/riakymch/exblas	[14]
BLAS-DOT2	GPU	http://www.math.twcu.ac.jp/ogita/post-k/results.html	[15]
MPACK	CPU	https://github.com/nakatamaho/mplapack	[17]
GQD, GARPREC	GPU	https://code.google.com/archive/p/gpuprec	[18]
CAMPARY	CPU, GPU	http://homepages.laas.fr/mmjoldes/campary	[19]
CUMP	GPU	https://github.com/skystar0227/CUMP	[20]
MPRES-BLAS	GPU	https://github.com/kisupov/mpres-blas	[21]

3 Data Types and Conversions

In MPRES-BLAS, a floating-point number is an object consisting of a sign, a multiple-precision significand, an integer exponent, and some additional information about the significand. We denote such an object as follows:

$$x = \langle s, X, e, I(X/\mathcal{M})\rangle, \tag{1}$$

where s is the sign, X is the significand, e is the exponent, and $I(X/\mathcal{M})$ is the interval evaluation of the significand (additional information).

The significand is represented in the RNS by the residues $(x_1, x_2, \ldots, x_n)$ relative to the moduli set $\{m_1, m_2, \ldots, m_n\}$ and is considered as an integer in the range of 0 to $\mathcal{M} - 1$, where $\mathcal{M} = \prod_{i=1}^{n} m_i$. The residues $x_i = X \bmod m_i$ are machine integers. The binary number X corresponding to the given residues $(x_1, x_2, \ldots, x_n)$ can be derived using the Chinese remainder theorem [8] as $X = \left|\sum_{i=1}^{n} \mathcal{M}_i \left|x_i w_i\right|_{m_i}\right|_{\mathcal{M}}$, where $\mathcal{M}_i$ and w_i are the RNS constants, namely $\mathcal{M}_i = \mathcal{M}/m_i$ and w_i is the modulo m_i multiplicative inverse of $\mathcal{M}_i$. Hence, one can compute the value of a floating-point number (1) using the following formula:

$$x = (-1)^s \times \left|\sum_{i=1}^{n} \mathcal{M}_i \left|x_i w_i\right|_{m_i}\right|_{\mathcal{M}} \times 2^e.$$

Unlike the addition, subtraction, and multiplication operations that are based on the residues individually, comparison, sign determination, overflow detection, scaling, and general division are time-consuming operations in the RNS, since they require estimating the magnitude of numbers. These operations are called non-modular operations. The classic technique to perform these operations is based on the Chinese remainder theorem and consists in computing binary representations of numbers with their subsequent analysis. In large dynamic ranges this technique becomes slow. MPRES-BLAS uses an alternative method for implementing non-modular operations, which is based on computing the interval evaluation of the fractional representation of an RNS number [22]. This method is designed to be fast on modern massively parallel general-purpose computing platforms such as GPU. The interval evaluation $I(X/\mathcal{M})$ is defined by two bounds, $\underline{X/\mathcal{M}}$ and $\overline{X/\mathcal{M}}$, that localize the value of X scaled by $\mathcal{M}$. The bounds are represented in a working precision floating-point format with an extended exponent in order to avoid underflow when $\mathcal{M}$ is large. To compute $\underline{X/\mathcal{M}}$ and $\overline{X/\mathcal{M}}$, only modulo m_i operations and standard floating-point operations are required. Using interval evaluations, efficient algorithms have been developed for implementing a number of non-modular operations in the RNS [22].

In [21], basic algorithms are proposed for performing arithmetic operations with numbers of the form (1) and the following rounding error bound is given: if $\text{fl}(x \circ y)$ is the rounded result of an operation $\circ \in (+, -, \times)$, then

$$\text{fl}(x \circ y) = (x \circ y)(1 + \delta), \qquad |\delta| < \mathbf{u},$$

where $\mathbf{u} < 4/\sqrt{\mathcal{M}}$ and $\mathcal{M}$ is the RNS moduli product. Hence, the user can set arbitrary arithmetic precision by changing the used set of RNS moduli.

The C data type corresponding to a multiple-precision number of the form (1) is `mp_float_t`, defined as the following structure:

```
typedef struct {
    int digits[n];
    int sign;
    int exp;
    er_float_t eval[2];
} mp_float_t;    /* Single multiple-precision value */
```

where `er_float_t` is the C data type representing a working precision floating-point value with an extended exponent.

The `mp_float_t` type is used mainly in the host code, and MPRES-BLAS provides a set of functions for converting data between `mp_float_t` and `double`, and also between `mp_float_t` and the `mpfr_t` type from the GNU MPFR library[1].

To store an array of multiple-precision numbers in the GPU memory, MPRES-BLAS uses the data type `mp_array_t`, defined as the following structure:

```
typedef struct {
    int * digits;
    int * sign;
    int * exp;
    er_float_t * eval;
    int4 * buf;
    int * len;
} mp_array_t;    /* Array of multiple-precision values */
```

[1] https://www.mpfr.org.

The fields of this structure are as follows, where n is the precision (size of the RNS moduli set) and L is the length of the multiple-precision array:

- `digits` are the digits (residues) of significands (an integer array of size $n \times L$); all digits belonging to the same multiple-precision number are arranged consecutively in the memory;
- `sign` are the signs (an integer array of size L);
- `exp` are the exponents (an integer array of size L);
- `eval` are the interval evaluations of significands (an array of size $2L$, where the first L elements represent the lower bounds of the interval evaluations, and the second L elements represent the upper bounds);
- `buf` is the buffer (an array of size L used in arithmetic operations to transfer auxiliary variables between CUDA kernels);
- `len` is the number of items that the array holds, i.e. L; for a vector of length N, the array must contain at least $(1 + (N - 1) \times |incx|)$ items, where $incx$

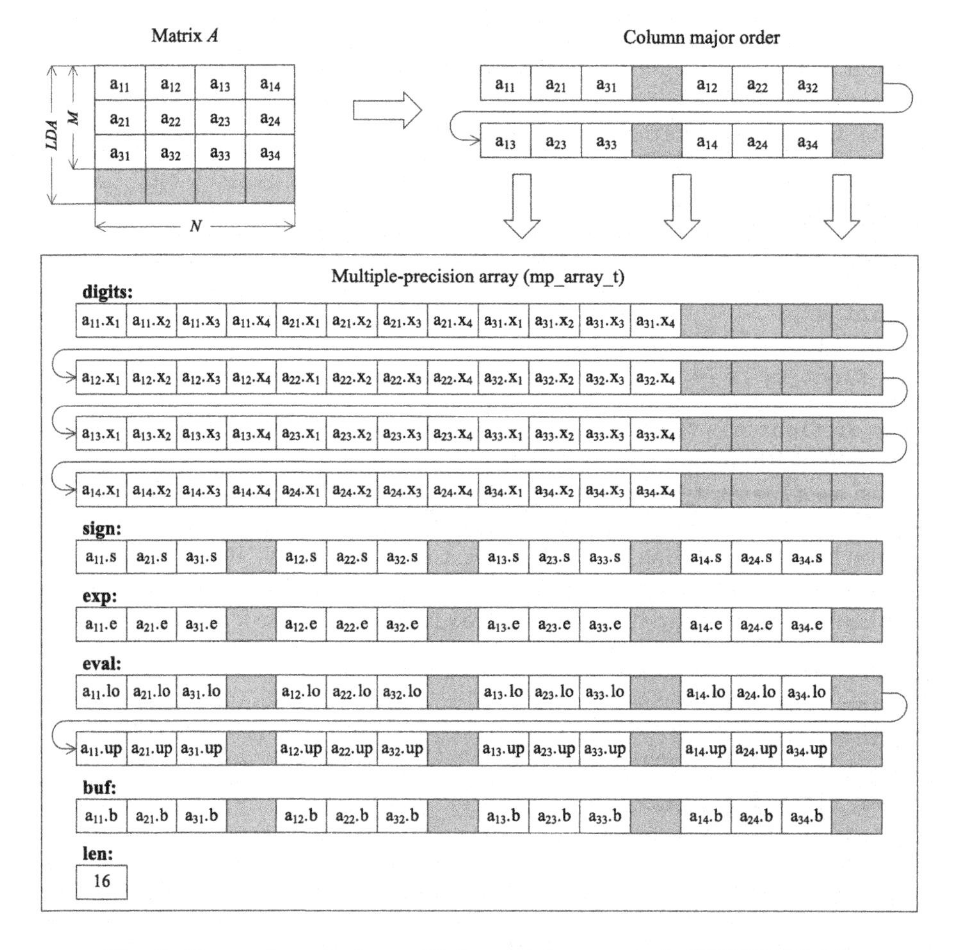

Fig. 2. Storage layout of a multiple-precision matrix in MPRES-BLAS

specifies the stride for the items; for a matrix of size $M \times N$, the array must contain at least $LDA \times N$ items, where LDA specifies the leading dimension of the matrix; the value of LDA must be at least $\max(1, M)$.

Using the `mp_array_t` structure instead of an array of `mp_float_t` structures allows ensuring coalesced memory access when dealing with multiple-precision vectors and matrices; details can be found in [21]. MPRES-BLAS provides the following helper functions for working with `mp_array_t` instances:

- `mp_array_init`: allocate a multiple-precision array in the GPU memory;
- `mp_array_clear`: release the GPU memory occupied by an array;
- `mp_array_host2device`: convert a regular multiple-precision array allocated on the host (with elements of type `mp_float_t`) to an `mp_array_t` instance allocated on the GPU;
- `mp_array_device2host`: convert an `mp_array_t` instance allocated on the GPU to a regular multiple-precision array allocated on the host.

4 Data Layout

The BLAS routines in MPRES-BLAS follow the Fortran convention of storing matrices using the column major data format. Thus, for an M by N matrix A, the address of the kth digit of the element a_{ij} in 0-based indexing is given by

$$address = k + (i + j \times LDA) \times n,$$

where n is the size of the RNS moduli set, $0 \leq k < n$, $0 \leq i < M$, and $0 \leq j < N$.

Figure 2 illustrates the column major layout of a 3×4 multiple-precision matrix using the `mp_array_t` data type. In this example, $n = 4$, i.e. the significand of each element a_{ij} consists of four digits: $X = (x_1, x_2, x_3, x_4)$. We use the dot symbol (.) to access the parts of a multiple-precision number. The symbols *lo* and *up* denote the lower and upper bounds of the interval evaluation so that $lo := \underline{X/\mathcal{M}}$ and $up := \overline{X/\mathcal{M}}$. Note that we use 1-based indexing in this example.

5 Functionality Overview

The current version of MPRES-BLAS (ver. 1.0) provides 16 GPU-based multiple-precision BLAS functions listed in Table 2. All the functions support the standard BLAS interface, except that some functions have additional arguments that are pointers to auxiliary buffers in the global GPU memory.

Implementation details, performance data, and accuracy of the ASUM, DOT, SCAL, and AXPY functions are given in [21]. Table 3 show the forward error bounds for the GEMV, GEMM, GE_ADD, and GER routines. In this table it is assumed that each matrix is a square N by N matrix.

We note that not all of the functions listed in Table 2 can be ill-conditioned and require higher numeric precision. For example, as shown in [21], the SCAL and ASUM functions have relative forward error bounds of $\mathbf{u}$ and

Table 2. Multiple-precision linear algebra operations supported by MPRES-BLAS.

ASUM — Sum of absolute values	GEMV — Matrix-vector multiplication
DOT — Dot product of two vectors	GEMM — General matrix multiplication
SCAL — Vector-scalar product	GER — Rank-1 update of a general matrix
AXPY — Constant times a vector plus a vector	GE_ADD — Matrix add and scale
AXPY_DOT — Combined AXPY and DOT	GE_ACC — Matrix accumulation and scale
WAXPBY — Scaled vector addition	GE_DIAG_SCALE — Diagonal scaling
NORM — Vector norms	GE_LRSCALE — Two-sided diagonal scaling
ROT — Apply a plane rotation to vectors	GE_NORM — Matrix norms

Table 3. Error bounds for the functions from MPRES-BLAS. According to [23], the quantity γ_N is defined as $\gamma_N := N\mathbf{u}/(1 - N\mathbf{u})$.

	Absolute error bound	Relative error bound
GEMV	$\gamma_{N+2} \cdot \lVert \lvert\alpha A\rvert \cdot \lvert x\rvert + \lvert\beta y\rvert \rVert$	$\gamma_{N+2} \cdot \lVert \lvert\alpha A\rvert \cdot \lvert x\rvert + \lvert\beta y\rvert \rVert \;/\; \lVert \alpha Ax + \beta y \rVert$
GEMM	$\gamma_{N+2} \cdot \lVert \lvert\alpha A\rvert \cdot \lvert B\rvert + \lvert\beta C\rvert \rVert$	$\gamma_{N+2} \cdot \lVert \lvert\alpha A\rvert \cdot \lvert B\rvert + \lvert\beta C\rvert \rVert \;/\; \lVert \alpha AB + \beta C \rVert$
GE_ADD	$\gamma_2 \cdot \lVert \lvert\alpha A\rvert + \lvert\beta B\rvert \rVert$	$\gamma_2 \cdot \lVert \lvert\alpha A\rvert + \lvert\beta B\rvert \rVert \;/\; \lVert \alpha A + \beta B \rVert$
GER	$\gamma_3 \cdot \lVert \lvert\alpha xy^T\rvert + \lvert A\rvert \rVert$	$\gamma_3 \cdot \lVert \lvert\alpha xy^T\rvert + \lvert A\rvert \rVert \;/\; \lVert \alpha xy^T + A \rVert$

$\mathbf{u}(N-1)/(1-\mathbf{u}(N-1))$, respectively. However, support for such functions allow one to eliminate time-consuming conversions between the IEEE 754 formats and the multiple-precision format (1) in the intermediate steps of a computation.

To achieve high performance, GPU kernels must be written according to some basic principles/techniques, stemming from the specifics of the GPU architecture [24]. One such technique is blocking. The idea is to operate on blocks of the original vector or matrix. Blocks are loaded once into shared memory and then reused. The goal is to reduce off-chip memory accesses. However, when working with multiple-precision numbers, shared memory is the limiting factor for occupancy, since the size of each number may be too large. Another problem is that the multiple-precision arithmetic algorithms contain serial and parallel sections, and while one thread computes the exponent, sign, and interval evaluation, all other threads are idle. This results in divergent execution paths within a warp and can lead to significant performance degradation. In order to resolve this problem, MPRES-BLAS follows the approach proposed in [21], according to which multiple-precision operations are split into several parts, each of which is performed by a separate CUDA kernel (*__global__* function). These parts are

- computing the signs, exponents, and interval evaluations;
- computing the significands in the RNS;
- rounding the results.

Thus, each BLAS function consists of a sequence of kernel launches. In some cases, such a decomposition may increase the total number of global memory accesses, since all intermediate results should be stored in the global memory. However, this eliminates branch divergence when performing sequential parts of

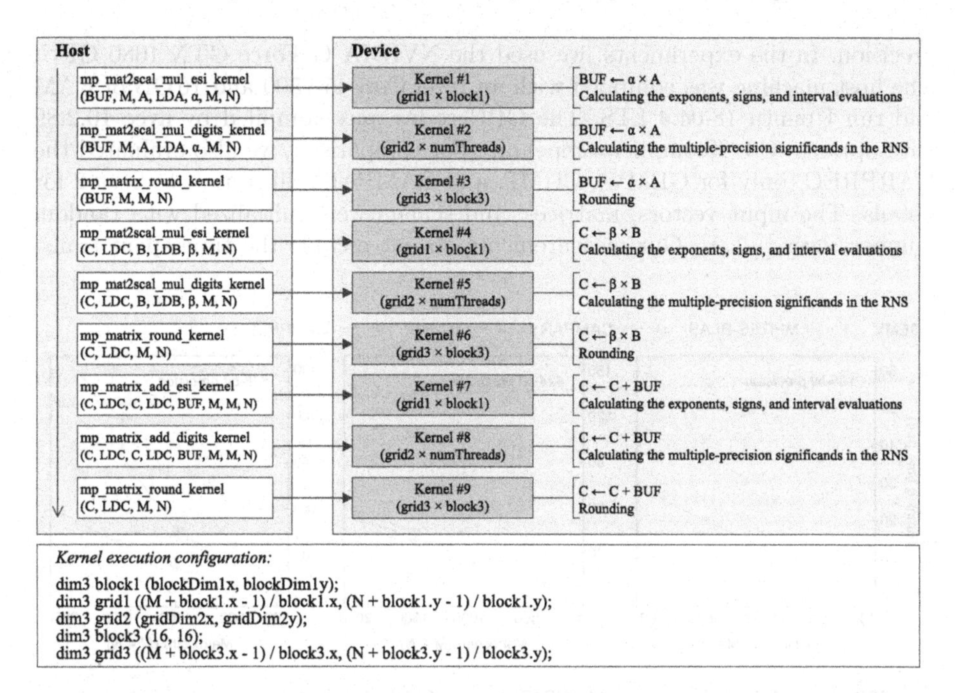

Fig. 3. Flowchart of the multiple-precision GE_ADD operation. Here *blockDim1x*, *blockDim1y*, *gridDim2x*, and *gridDim2y* are tunable parameters that are passed to the function as template arguments, while *numThreads* is automatically configured depending on the precision and hardware specification.

multiple-precision operations. Furthermore, each kernel has its own execution configuration which makes it possible to use all available GPU resources.

As an example, Fig. 3 shows a flowchart of the multiple-precision GE_ADD operation ($C \leftarrow \alpha A + \beta B$). Other BLAS operations have a similar structure. An exception is reduction operations, in which each operation with multiple precision is performed as a single thread, and communication between threads is performed using shared memory.

In MPRES-BLAS, multiple-precision vectors are processed on the GPU with 1-dimensional grids of 1-dimensional thread blocks, whereas matrices are processed with 2-dimensional grids of 1-dimensional or 2-dimensional thread blocks. Each BLAS operation has template parameters that specify the execution configuration of the kernels. For computations with multiple-precision significands, the number of threads per block is precision dependent and is configured automatically so as to ensure maximum occupancy of a streaming multiprocessor.

6 Performance Evaluation

In this section, we evaluate the performance of the GEMV, GE_ADD and GEMM functions from MPRES-BLAS for various problem sizes and levels of numeric precision. In the experiments, we used the NVIDIA GeForce GTX 1080 GPU. The host machine was equipped with an Intel Core i5-7500 and 16 GB of RAM and run Ubuntu 18.04.4 LTS. The GPU codes were compiled by nvcc 10.2.89 with options -O3 -Xcompiler=-fopenmp. For comparison, we also evaluated the GARPREC (only for GEMV), CUMP, and CAMPARY libraries; see Sect. 2 for details. The input vectors, matrices, and scalars were initialized with random numbers over $[-1, 1]$. Our measurements do not include the time spent trans-

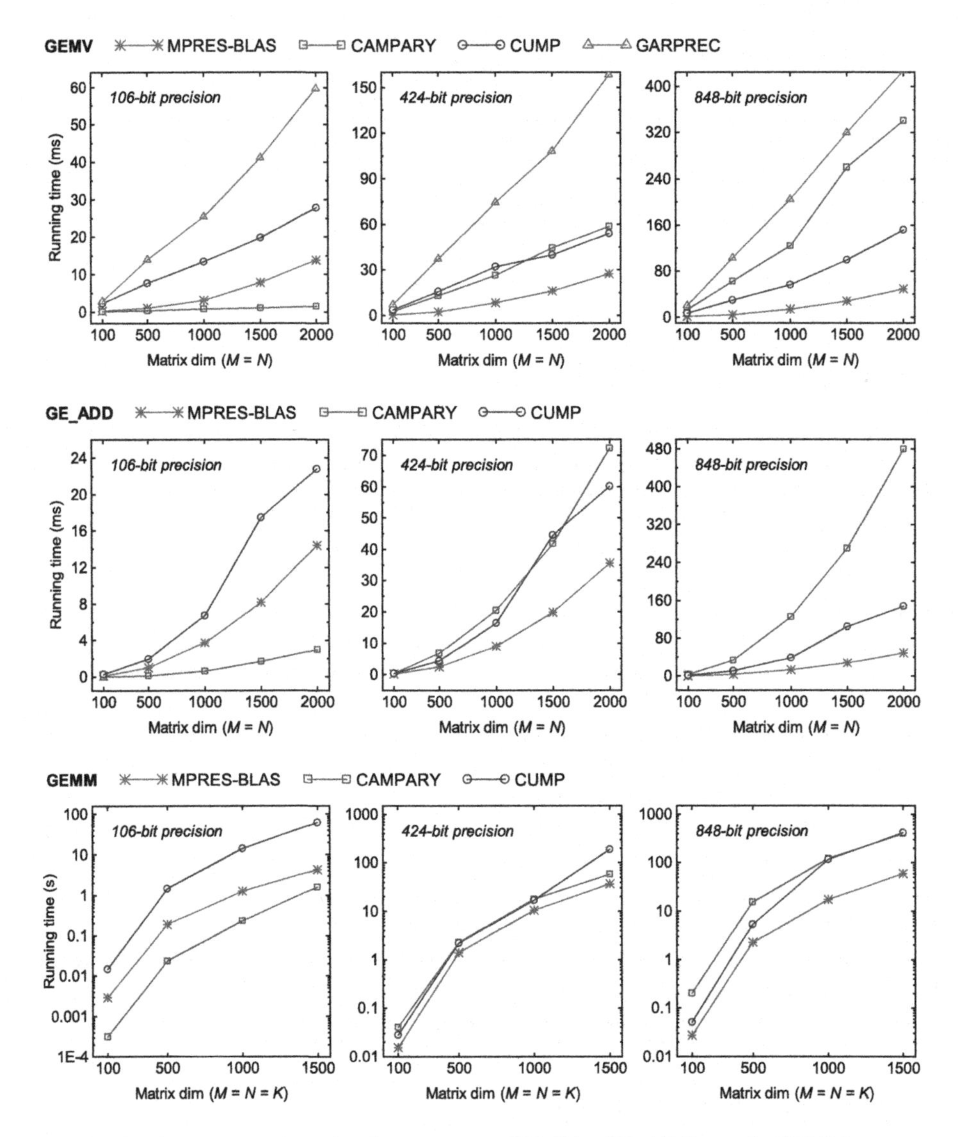

Fig. 4. Performance of multiple-precision GEMV, GE_ADD and GEMM on an NVIDIA GeForce GTX 1080 as a function of problem size.

ferring data between the CPU and the GPU, as well as time of converting data into internal multiple-precision representations.

Figure 4 shows the experimental results. In many cases, MPRES-BLAS has better performance than implementations based on existing multiple-precision libraries for GPUs. Increasing the precision of computations increases the speedup of MPRES-BLAS. At 106-bit precision, the performance of MPRES-BLAS is on average 6 times lower than that of CAMPARY. This is because the RNS-based algorithms implemented in MPRES-BLAS are designed for arbitrary (mostly large) precision and have overhead associated with calculating interval evaluations and storing intermediate results in the global memory. On the other hand, MPRES-BLAS is less dependent on precision than CAMPARY and with 848-bit precision, MPRES-BLAS performs on average 9 times better than CAMPARY.

7 Conclusions

In this paper, we have presented MPRES-BLAS, a new CUDA library of basic linear algebra operations with multiple precision based on the residue number system. The current version of MPRES-BLAS (ver. 1.0) supports 16 multiple-precision operations from levels 1, 2, and 3 of the BLAS. Some operations such as SCAL, NORM and ASUM typically do not require precision higher than the one provided by the natively supported floating-point formats. Nevertheless, support for such functions allow one to eliminate time-consuming intermediate conversions between the natively supported formats and the RNS-based multiple-precision format used.

In addition to the linear algebra kernels, MPRES-BLAS implements basic arithmetic operations with multiple precision for CPU and GPU (through the `mp_float_t` data type), so it can also be considered as a general purpose multiple-precision arithmetic library. Furthermore, the library provides a number of optimized RNS algorithms, such as magnitude comparison and power-of-two scaling, and also supports extended-range floating-point arithmetic with working precision, which prevents underflow and overflow in a computation involving extremely large or small quantities. The functionality of MPRES-BLAS will be supplemented and improved over time.

Acknowledgement. This work was supported by the Russian Science Foundation grant number 18-71-00063.

References

1. IEEE Standard for Floating-Point Arithmetic: Standard, Institute of Electrical and Electronics Engineers, New York, NY, USA (2019). https://doi.org/10.1109/IEEESTD.2019.8766229
2. Courbariaux, M., Bengio, Y., David, J.P.: Training deep neural networks with low precision multiplications (2014). https://arxiv.org/abs/1412.7024. Accessed 10 Apr 2020

3. Bailey, D.H., Borwein, J.M.: High-precision arithmetic in mathematical physics. Mathematics **3**(2), 337–367 (2015). https://doi.org/10.3390/math3020337
4. Daněk, J., Pospíšil, J.: Numerical aspects of integration in semi-closed option pricing formulas for stochastic volatility jump diffusion models. Int. J. Comput. Math. 1–25 (2019). https://doi.org/10.1080/00207160.2019.1614174
5. Feng, Y., Chen, J., Wu, W.: The PSLQ algorithm for empirical data. Math. Comp. **88**(317), 1479–1501 (2019). https://doi.org/10.1090/mcom/3356
6. Kyung, M.H., Sacks, E., Milenkovic, V.: Robust polyhedral Minkowski sums with GPU implementation. Comput. Aided Des. **67–68**, 48–57 (2015). https://doi.org/10.1016/j.cad.2015.04.012
7. Lawson, C.L., Hanson, R.J., Kincaid, D.R., Krogh, F.T.: Basic linear algebra subprograms for Fortran usage. ACM Trans. Math. Softw. **5**(3), 308–323 (1979). https://doi.org/10.1145/355841.355847
8. Omondi, A., Premkumar, B.: Residue Number Systems: Theory and Implementation. Imperial College Press, London (2007)
9. Lu, M.: Arithmetic and Logic in Computer Systems. John Wiley and Sons Inc., Hoboken (2004)
10. Shewchuk, J.R.: Adaptive precision floating-point arithmetic and fast robust geometric predicates. Discrete Comput. Geom. **18**(3), 305–363 (1997). https://doi.org/10.1007/PL00009321
11. Li, X.S., et al.: Design, implementation and testing of extended and mixed precision BLAS. ACM Trans. Math. Softw. **28**(2), 152–205 (2002). https://doi.org/10.1145/567806.567808
12. Yamada, S., Ina, T., Sasa, N., Idomura, Y., Machida, M., Imamura, T.: Quadruple-precision BLAS using Bailey's arithmetic with FMA instruction: its performance and applications. In: Proceedings of the 2017 IEEE International Parallel and Distributed Processing Symposium Workshops (IPDPSW), pp. 1418–1425 (2017). https://doi.org/10.1109/IPDPSW.2017.42
13. Quadruple precision BLAS routines for GPU: QPBLAS-GPU (2013). https://ccse.jaea.go.jp/software/QPBLAS-GPU/1.0/manual/qpblas-gpu_manual_en-1.0.pdf. Accessed 16 May 2020
14. Iakymchuk, R., Collange, S., Defour, D., Graillat, S.: ExBLAS: reproducible and accurate BLAS library. In: Proceedings of the Numerical Reproducibility at Exascale (NRE2015) Workshop at SC15 (2015)
15. Mukunoki, D., Ogita, T.: Performance and energy consumption of accurate and mixed-precision linear algebra kernels on GPUs. J. Comput. Appl. Math. **372** (2020). https://doi.org/10.1016/j.cam.2019.112701
16. Ogita, T., Rump, S.M., Oishi, S.: Accurate sum and dot product. SIAM J. Sci. Comput. **26**(6), 1955–1988 (2005). https://doi.org/10.1137/030601818
17. Nakata, M.: Poster: MPACK 0.7.0: multiple precision version of BLAS and LAPACK. In: Proceedings of the 2012 SC Companion: High Performance Computing, Networking Storage and Analysis, pp. 1353–1353 (2012). https://doi.org/10.1109/SC.Companion.2012.183
18. Lu, M., He, B., Luo, Q.: Supporting extended precision on graphics processors. In: Sixth International Workshop on Data Management on New Hardware (DaMoN 2010), pp. 19–26 (2010). https://doi.org/10.1145/1869389.1869392
19. Joldes, M., Muller, J., Popescu, V.: Implementation and performance evaluation of an extended precision floating-point arithmetic library for high-accuracy semidefinite programming. In: Proceedings of the 2017 IEEE 24th Symposium on Computer Arithmetic (ARITH), pp. 27–34 (2017). https://doi.org/10.1109/ARITH.2017.18

20. Nakayama, T., Takahashi, D.: Implementation of multiple-precision floating-point arithmetic library for GPU computing. In: Proceedings of the 23rd IASTED International Conference on Parallel and Distributed Computing and Systems (PDCS 2011), pp. 343–349 (2011). https://doi.org/10.2316/P.2011.757-041
21. Isupov, K., Knyazkov, V., Kuvaev, A.: Design and implementation of multiple-precision BLAS Level 1 functions for graphics processing units. J. Parallel Distrib. Comput. **140**, 25–36 (2020). https://doi.org/10.1016/j.jpdc.2020.02.006
22. Isupov, K.: Using floating-point intervals for non-modular computations in residue number system. IEEE Access **8**, 58603–58619 (2020). https://doi.org/10.1109/ACCESS.2020.2982365
23. Higham, N.J.: Accuracy and Stability of Numerical Algorithms, 2nd edn. SIAM, Philadelphia (2002). https://doi.org/10.1137/1.9780898718027
24. Nath, R., Tomov, S., Dong, T.T., Dongarra, J.: Optimizing symmetric dense matrix-vector multiplication on GPUs. In: Proceedings of the 2011 International Conference for High Performance Computing, Networking, Storage and Analysis, pp. 1–10 (2011). https://doi.org/10.1145/2063384.2063392. Article no. 6

New Compact Streaming in LBM with ConeFold LRnLA Algorithms

Anastasia Perepelkina[1,2](✉), Vadim Levchenko[1], and Andrey Zakirov[2]

[1] Keldysh Institute of Applied Mathematics RAS, 4 Miusskaya sq., Moscow, Russia
mogmi@narod.ru, lev@keldysh.ru

[2] Kintech Lab Ltd., 3-Ya Khoroshevskaya Ulitsa, 12, Moscow, Russia
zakirov@kintechlab.com

Abstract. The classic Lattice Boltzmann Method with a cube-shaped stencil is a memory-bound problem, and the optimization techniques aim for a goal of one load and one store per value update. We propose a data layout and a streaming pattern so that the elementary computation is an update of 8 cells in a cube, for which only the data of the same cells are required. The new streaming is symmetrical, compact, and allows split description. With it, the better localization of computation in the higher levels of the CPU memory hierarchy is possible, which leads to the use of the higher ceilings in the Roofline model. It is implemented with the ConeFold LRnLA algorithm so that the operational intensity is increased further. The compactness of the streaming allows for more asynchrony in multidimensional LRnLA decomposition. The obtained one desktop CPU performance reaches 1 billion cell updates per second.

Keywords: LRnLA Algorithms · Temporal blocking · Asynchrony · ConeFold · Lattice Boltzmann Method · Roofline model · AVX

1 Introduction

The Lattice Boltzmann Method (LBM) is a modern numerical method for computational fluid dynamics [18]. It is derived as a discretization of the Boltzmann equation with specific collision operators in the velocity space [5]. On the other hand, it is simple for programming, since it can be logically split into two basic components: an local operation of collision and operations of propagation, that amount to data accesses in the desired pattern. The optimization of the performance of this scheme is a relevant task due to its use in many fields of applied and fundamental physics, as well as because the simplicity of its logic leads to new ideas in the algorithm construction.

Among the advanced algorithms for optimization of the efficiency of LBM computation the Locally Recursive non-Locally Asynchronous (LRnLA) algorithm approach [8,9] has shown the highest performance results both on CPU [14] and GPU [6,7]. LRnLA algorithms use temporal blocking for an increase in the operational intensity, which has been applied to LBM schemes by other authors

V. Voevodin and S. Sobolev (Eds.): RuSCDays 2020, CCIS 1331, pp. 50–62, 2020.
https://doi.org/10.1007/978-3-030-64616-5_5

in [4,11,17]. The LRnLA approach is advanced in the terms that it recursively decomposes the problem so that at some levels of recursions the task may be localized in a specific level of memory, or distributed on a specific level of parallelism, so the whole hardware system is used efficiently.

One well-known step in the performance increase is the merge of the collision and propagation into one cell update [1,10,16], so that the domain is looped over not twice, but once per time step. One collision operation and Q propagation operations for a cell and its neighbors can be grouped differently.

Another direction in the optimization of the LBM simulation is the development of a data structure that allows for localized and coalesced data access. Each LBM node has a set of at least Q values that are propagated in different directions, so that grouping of data in the elements of data structure greatly affects the performance [10,19].

The definition of a data structure and an elementary cell update is a definition of a specific propagation pattern [1,3]. The development of propagation patterns is an area that is specific to the optimization of the LBM codes.

In this paper, the recent advances in the LRnLA algorithm development, as well as a new propagation pattern for the LBM scheme are presented.

Let us introduce the scheme. The scheme variant is coded with a number of dimensions in a simulation D and a number of discrete velocities Q. D3Q19 is used in the code runs in this work. A set of Q particle distribution functions (PDFs) values f_i, $i = 1..Q$, are assigned to a node in a discrete D-dimensional Cartesian grid. Each PDF corresponds to a discrete velocity $\boldsymbol{c}_i$. In D3Q19, the set is $(1,0,0)$, $(0,1,0)$, $(0,0,1)$, $(1,1,0)$, $(1,0,1)$, $(0,1,1)$, $(0,1,-1)$, $(-1,0,1)$, $(1,-1,0)$, the inverse to these, and a zero velocity. In the propagation step, the f_i values propagate into the neighboring LBM node according to the $\boldsymbol{c}_i$ direction. In the collision step, the values in one node are updated locally according to the collision operation. Here, for the performance study, the most basic BGK collision operator with a second order polynomial for the equilibrium PDF is used [18]. For this local operation, the PDF moments are collected as $m_{\alpha\beta\gamma} = \sum f_i c_{i,x}^{\alpha} c_{i,y}^{\beta} c_{i,z}^{\gamma}$. These moments represent macroscopic variables, such as particle density, fluid density, pressure tensor, etc. These moments are used in the definition of the initial conditions, since they are commonly defined from the fluid dynamics point of view. The moments are also collected for the output and analysis of the simulation results.

2 Compact LBM Streaming

One of the first optimizations of the LBM codes is the merge of the collision and streaming steps into one cell update [16]. With this, in an ideal stepwise algorithm, one load and store per value is required per one full cell update, instead of two. In the resulting streaming pattern, however, data access conflicts exist. For example, in the ‘push’ pattern, a cell, after the collision, writes data into the neighboring cells, which may overwrite the information that is needed in the collision in the neighbors. Thus, two array copies are used.

There are advanced streaming patterns with another solution for the data access conflict problem [20]. In the AA-pattern [1] and EsoTwist [3], in each elementary update, the data are read from and saved into the same location. The ConeFold-swap pattern, which was implemented with the ConeFold LRnLA algorithm in our previous work [14], ensures the correct update order so that no write conflicts occur.

Here we propose a new streaming pattern. It satisfies the "save/load to/from the same location" property, and has a further improvement: to fully update 2^D cells in a cube, the data from these same 2^D cells are required. This way, the cells are divided into fully independent groups at each update.

For comparison, the AA-pattern requires access to three cells in each direction in the even steps, and to just one cell in the odd step. EsoTwist and ConeFold-swap patterns require access to 2^d cells for each one cell update.

All these patterns can be described from the starting point of the collision operation, for which there exist Q incoming and Q outgoing PDFs. The position of the collision operation is assigned to be at the cell, where the incoming 'resting particle' PDF, f_0 with $\boldsymbol{c}_i = (0,0,0)$, is stored in. The streaming is a data access operation. In the 'push' pattern the incoming PDFs are read from the same cell, where the collision takes place, and the outgoing PDFs are written to the neighboring cells in the $\boldsymbol{c}_i$ direction. In the AA-pattern, at the even step, the incoming PDFs are read from the neighboring cells in the $-\boldsymbol{c}_i$ direction, collided, and saved into the neighboring cells into the $\boldsymbol{c}_i$ direction. At the odd steps, both incoming and outgoing PDFs are stored in the same cells.

In the proposed *compact* streaming pattern, the elementary update is defined for a group of 2^d cells in a cube. The description can be split by the coordinate axes. For each group of cells $\boldsymbol{x}_j$, $j = 1..2^d$, the *center* point is defined at $\sum \boldsymbol{x}_j/2^d$. The direction from the cell to the center point in denoted by $\boldsymbol{o}$, and the access rule for each f_i is written relative to the fact if the $\boldsymbol{c}_i$ direction coincides with the $\boldsymbol{o}$ direction (Fig. 1). The algorithm is as follows:

Read f_i from the $f_{i_{revX}}$ of the same cell if the direction of the $\boldsymbol{c}_i$ projection to the x-axis is zero or coincides with the direction of the $\boldsymbol{o}$ projection to the x-axis; else read from f_i of the cell with other x coordinate; $\boldsymbol{c}_{i_{revX}} = \boldsymbol{c}_i - \boldsymbol{e}_x(\boldsymbol{e}_x\boldsymbol{c}_i)$.

Collide all cells in the group.

Save f_i to the f_i of the same cell if the direction of the $\boldsymbol{c}_i$ projection to the x-axis does not coincide with the direction of the $\boldsymbol{o}$ projection to the x-axis; else save to the $f_{i_{revX}}$ of the cell with different x coordinate.

At the next step, the procedure is repeated in a cube shifted by 1 in all directions. The python code for the data read is in Fig. 2. In the result, in the collision procedure in cell `x`, the PDF in the `ci` direction is read from the PDF with the `cist` direction in cell `xst`.

The questions of the correct initialization procedure, data output, possibility of application of complex LBM extensions (multiphase, immersed boundary, non-uniform grids) may arise. The apparent complexity of the data storage may be resolved. During the collision phase, the data is collected in the cells in the correct order. It is a temporary state in the algorithm; but this state may be

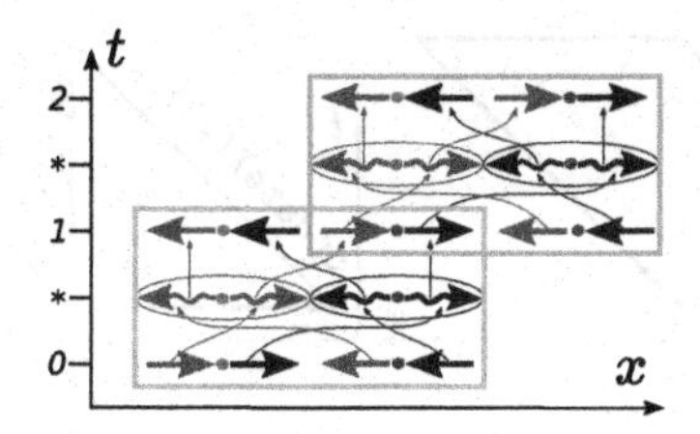

Fig. 1. Compact streaming pattern. Symbol '*' marks a temporary state where the data is collected for the collision in the right order. Left arrows denote PDF with $c_i = -1$, right arrows denote PDF with $c_i = 1$. The x-position correspond to the data position in the array. Ovals at '*' steps denote the collision operation, and thin arrows show the data flow.

```
xst = x
cist = ci
for idim in range(D):
  if ci[idim] != 0:
    if ci[idim] != o[idim]: xst[idim] = xst[idim]^1
    else: cist[idim] = - cist[idim]
```

Fig. 2. The code for the selection of the data access rule. the cell positions `x` and `xx` are relative to the first cell in the group; `ci` is the discrete velocity of the PDF f_i, required for the collision; `o` is the direction towards the center of the cell group.

a starting point for application of any LBM extension in the same way as it is formulated for the 'push' propagation pattern.

3 LRnLA Algorithm ConeFold

Consider one cell in the simulation region with size $N_x \times N_y \times N_z$ update from time t to $t+1$. To make this update, in a scheme with a local cube-shaped stencil, such as the LBM method, the data from the cells in a cube that fits the stencil are required. No other cells from the previous layer are used in this update. The dependencies may be traced to the initial conditions. The dependency region shape is a multidimensional pyramid (conoid) in the dD1T space.

This is one of the preliminary ideas for the LRnLA algorithm construction method. In this paper, the ConeFold [9,13–15] algorithm shape is used (Figs. 3, 4). In contrast to the conoid-based temporal blocking, its advantage lies in the fact that it can uniformly tile the dD1T (d-dimensions with time axis) simulation region. On one core, the performance gain is due to locality and recursive subdivision, which increases the operational intensity and allows the efficient use of all cache levels. In the parallel execution, good scaling is due to the presence of large portions of asynchronous computations.

The ConeFold is a recursive algorithm. It is a task of executing 2^{d+1} ConeFolds, according to the subdivision rule (Fig. 3). The closure is an elementary scheme update. In C++, it is implemented as a recursive function template,

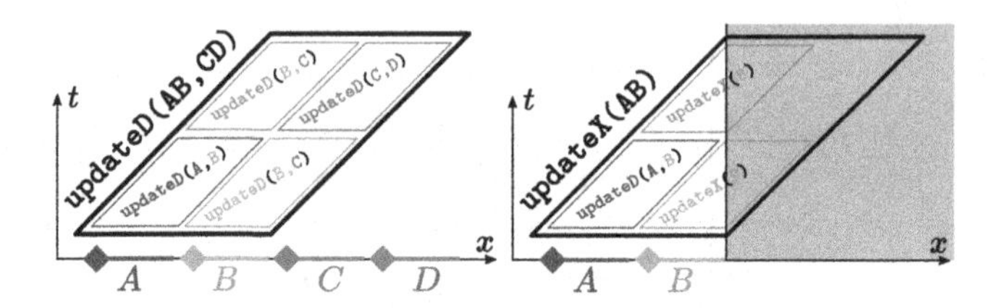

Fig. 3. ConeFold in 1D1T and its subdivision rule in the area and near the boundary. ConeFold is a task of execution the dependency graph nodes that are inside its shape in the 1D1T space. It is decomposed into 4 sub-tasks. The required data for the tasks are found by projecting the shape onto the 1D space.

parametrized by integer rank r, and the cells are stored in a Z-order curve. The data structure is also defined recursively. Here it is called a locally recursive cube: cubeLR<r,d> is a d-dimensional cube that contains 2^d cubeLR<r-1,d> cubes, and cubeLR<0,d> is a cell which contains LBM data. r is an integer number called rank. The arguments of the rank r ConeFold function are the pointers to the 2^d cubeLR<r,d> data blocks which are in the projection of the ConeFold shape. If some of the data blocks fall outside the domain, a different kind of ConeFold is called, and its closure is the application of the boundary condition.

Sample code listings are presented in Fig. 6 and Fig. 7. A d-symbol code is used to show the position of the ConeFold relative to boundary. If the d-th symbol is 'D', then the top and bottom bases of the ConeFold are inside the domain. The top base is over the right boundary in the d-th axis direction if the d-th symbol is X, the bottom base is over the left boundary in the d-th axis direction if the d-th symbol is I.

In 1D1T, the updateD at rank 0 accesses 2 cells, and performs an elementary update. At maximal rank on a domain composed of one cubeLR<MR>, two ConeFolds of maximum rank are called: updateX and updateI in this order (Fig. 5).

For the 3D1T compact streaming, since the elementary update involves a cube of 8 cells, the recursive subdivision stops at rank 1. In the updateDDD(..) at rank 1, instead of calling 16 rank 0 functions, only two are called. Rank 0 function already has access to 8 cells, so no further modification is required, and only two zero rank functions have enough data to perform 16 cell updates.

For the many-core parallelism with the ConeFold shape, the TorreFold algorithm is used (Fig. 5) [9,14,15]. It is a ConeFold shape with a different decomposition rule. At the maximal rank, the it is decomposed into high slanted prisms (ConeTorre), which are made of 2^{nLAR} ConeFolds, stacked onto each other. The size of the prism base is r = MR − nLAR. The ConeFolds in a ConeTorre are to be executed from bottom to top by one thread. The ConeFolds in adjacent ConeTorres may be processed in parallel, and the correct order may be provided by either barrier synchronization or an array of semaphores or mutexes [9,15]. The parallelism is flexible for adaptation to the modern many-core hardware. The nLAR parameter controls the degree of parallelism. At the same time, the ConeTorre base size can be adjusted to fit the local cache of the core.

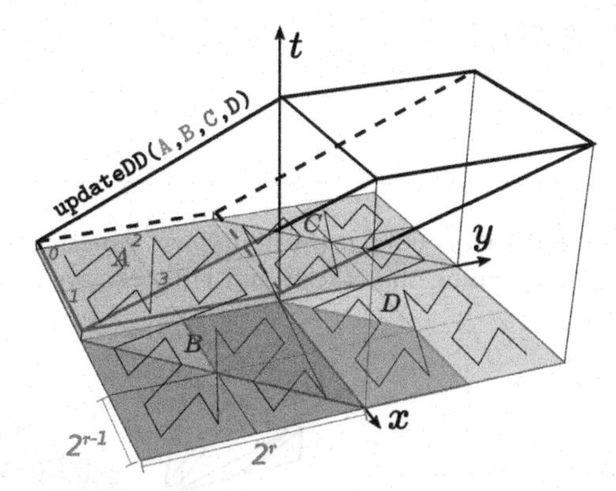

Fig. 4. ConeFold in 2D1T. ConeFold with rank r uses data in 4 data blocks, that are stored in a Z-curve (pictured in the projection below) These data blocks are Z-curve arrays with 2^r cells. The ConeFolds with rank (r-1) use data from the 4 parts of these blocks, which are Z-curve arrays with 2^{r-1} cells. The projection of one of the smaller ConeFolds that appeared after the subdivision is highlighted in pink. (Color figure online)

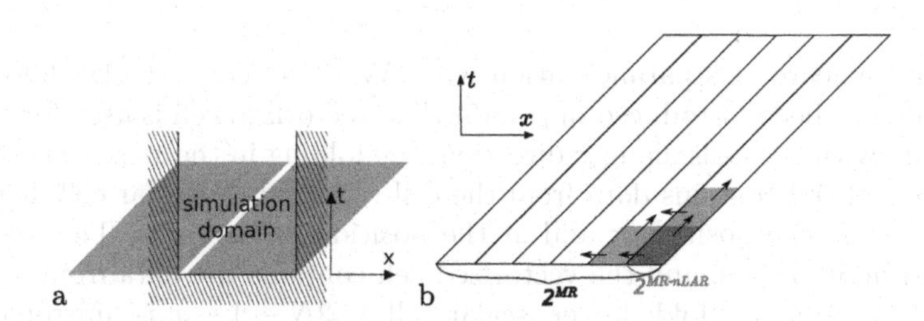

Fig. 5. (a) Two ConeFolds are called in 1D1T to progress the whole domain from $t = 0$ to $t = 2^{\text{MR}}$ (b) The TorreFold LRnLA algorithm.

```
template <int r, class TO> void updateD(cube<r>* AB, cube<r>* CD) {
  updateD(B, C);
  updateD(C, D);
  updateD(A, B);
  updateD(B, C);
}
```

Fig. 6. Sample code for 1D1T ConeFold. Function arguments reference the data blocks in Fig. 3

```
template <int r, class T0> void updateXD(cube<r>* A, cube<r>* C) {
  updateXD(A3,C1);
  updateDD(A2,A3,C0,C1);
  updateXD(C1,C3);
  updateXD(A1,A3);
  updateDD(A0,A1,A2,A3);
  updateXD(A3,C1);
}
```

Fig. 7. Sample code for 2D1T ConeFold at the boundary. Function arguments reference the data blocks in Fig. 4, assuming that the B and D blocks are outside the boundary and thus do not exist.

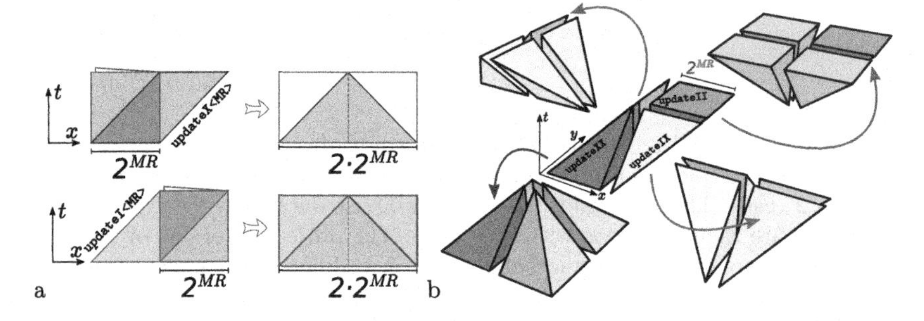

Fig. 8. By folding ConeFolds in 1D1T (a) and 2D1T (b) all required shapes for conoid-based temporal blocking may be obtained

For the synchronous parallelization with AVX256/AVX512, the 'folding' of the domain has been introduced in [14,15]. The vectorized cell is used for this. In the domain with $2N$ cells along x direction, the folding in the x-axis means that the vectorized cell contains data from the cell (referred as 'scalar cell' hereafter for contrast) at the position x and at the position $2N-1-x$. The folding is performed in all 3 axes, and the vectorized cell contains the data from 8 cells.

The data from the 'folded over' scalar cell at $2N-1-x$ is 'mirrored' in a sense that f_i with $c_i = 1$ propagates to the right in the 'normal' domain and to the left in the 'mirrored' domain. To mirror the LBM scheme, $c_{i,x}$ are multiplied by constant AVX vectors like $\{1,-1,1,-1,1,-1,1,-1\}$.

The overhead amounts to the AVX256/AVX512 intrinsic vector shuffle operations to ensure streaming at $x=0$ and $x=N$. With this, the periodic boundary conditions is implemented.

Furthermore, the folding step helps to bridge the gap from ConeFold shape to the set of pyramid-based shapes. ConeTurs represent the class of LRnLA algorithms that appear from intersection of conoids of dependency and influence in the dependency graph space [9]. In 1D1T, the `updateX` function, acting on a Z-curve array of vectorized cells, performs, in fact, the cell updates in a triangle, centered at $x=N$ (Fig. 8). The upside-down triangle is obtained with `updateI`.

Table 1. Hardware parameters

	CPU					RAM			gcc
	freqs	cores	L1	L2	L3	type	size	ch	March
Core i7-7800X	3.5 GHz	6/12	6 × 32K	6 × 1M	8.25M	DDR4-2400	128G	4	skylake
Ryzen R9 3900X	3.8 GHz	12/24	12 × 32K	12 × 512K	4 × 16M	DDR4-2666	64G	2	znver1

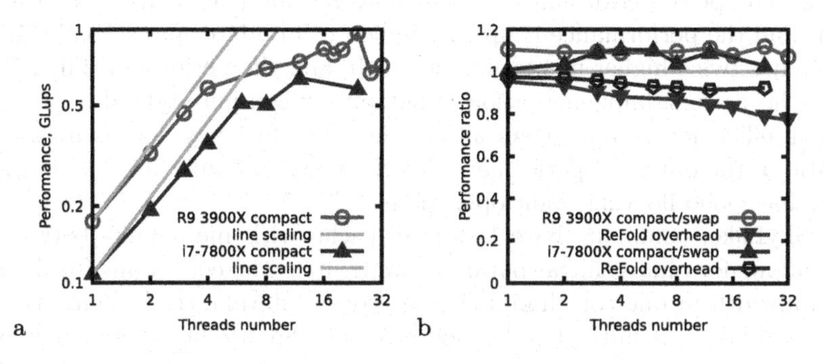

Fig. 9. Strong scaling (a), performance ratio compact/swap and ReFold overhead (b)

In 2D1T (Fig. 8), the pyramid is obtained with `updateXX`, and an upside down pyramid is obtained with `updateII`. The space between them is filled by two types of tetrahedra constructed by mirroring the `updateXI` and `updateIX`.

Now, let us consider a region with $B_x 2N$ cells in the x direction. B_x pairs may be formed. Each pair is folded (with mirroring of the second half), and `updateX` is executed on each pair. This way, in fact, B_x cones are executed. The cone base has $2N$ cells and it is centered at $x = iN$, $i = 1, 3, ..2B_x - 1$. The operations in these cones are asynchronous.

In the implementation, the scalar cells are stored in a fixed `cubeLR`, and the vectorized array is formed for each folding operation. The array is folded, processed with a ConeFold, and unfolded back onto the scalar array. This is called *ReFold* LRnLA algorithm [12,15].

4 Results and Discussion

The performance tests have been run in a cube region with 512^3 cells, periodic boundary conditions. With a mirrored vectorization ReFold algorithm, there is one `cubeLR<MR=8>` of vectorized cells to be processed. Since there is only one block, the intermediate storage in the scalar array is not required, so the vector folding operations are omitted in the performance test. The code is compiled with gcc v8.3.0 compiler. TorreFold parameter `nLArank` = 4.

The strong scaling efficiency (Fig. 9a) was tested on two modern high-end desktop computers, Intel Skylake (Core i7-7800X) and AMD Zen2 (Ryzen R9 3900X) (Table 1). Both were in the AVX-256 mode, so the cells contain vectors of 8 single-precision values. These systems are different in the number of cores

(6 and 12) and memory channels (4 and 2). The architectures and caching subsystems are different as well: Skylake has common/shared last level cache and Zen2 has distributed L3 cache.

LBM is a stencil scheme, which are generally classified as memory-bound problems [11]. With LRnLA algorithms, the limit shifts to compute-bound domain. The peak performance in the AVX-256 mode of Zen2 is almost twice higher, and the performance is up to 1 billion cell updates per second (GLUps). Skylake has twice higher memory bandwidth, and the performance in 1.5 times lower. The present implementation is not purely compute-bound as well, since the peak efficiency is lower than 33%. The peak efficiency here is measured as the ratio of the obtained performance to the peak compute limit, assuming one update has $\sim$360 floating point operations.

In SkyLake, common/share last level cache with mesh links between segments allows for almost linear parallel scaling up to the core count (6). However, low bandwidth in one core limits the one-core performance. In Zen2, the CCX distributed L3 cache mitigates this one-core performance limitation, however the linear scaling is only up to the thread count equal to the number of CCX (4). Note that to fully utilize the bandwidth of the distributed L3, in this case, the binding of processing threads to the cores of different CCX is required. Here it is implemented with the use of HWloc v2.1 library [2].

A further scaling (with the number of threads higher than the number of cores or CCX respectively) is possible with the reuse of data in the parallel execution of ConeTorre. Consider a group of ConeTorre the bases of which fall into one `cubeLR` of a certain rank, so that this `cubeLR` fits L3. If this group is started at once, they proceed, synchronizing with semaphores, up to a point that the topmost ConeFolds are executed synchronously. The exchange in these parallel ConeFolds is through the shared cache, and thus the L3 memory bandwidth becomes the actual limit of the performance.

Thus, the ConeTorres are collected into groups, and the threads that execute these ConeTorres are allocated to the shared cache domain. The threads of different groups are allocated to different shared cache domains, and are desynchronized, i.e. they process the asynchronous regions of the TorreFold.

By collecting the threads into groups of 2 and 6, in Skylake and Zen2, the additional acceleration by a factor of 1.25 and 1.65 correspondingly is obtained. The performance maximum is achieved when the number of software threads is equal to the number of the hardware threads (12 and 24 correspondingly). The performance gain with the new compact streaming pattern is 10–15% (Fig. 9b).

This concludes the tests without ReFold. Packing data into vectorized cell arrays is expected to produce some overhead, due to a large number of data access operations. This is the target mode for the code, so the next tests were performed to study this overhead. In the previous implementation [12,15], due to the complexity of this procedure, for the ConeFolds of maximum rank (an asynchronous triplet `updateXII()`, `updateIXI()` and `updateIIX()` at the second stage, and after that another triplet on the third stage), three separate repack operations were required twice. These three updates do not require all data of

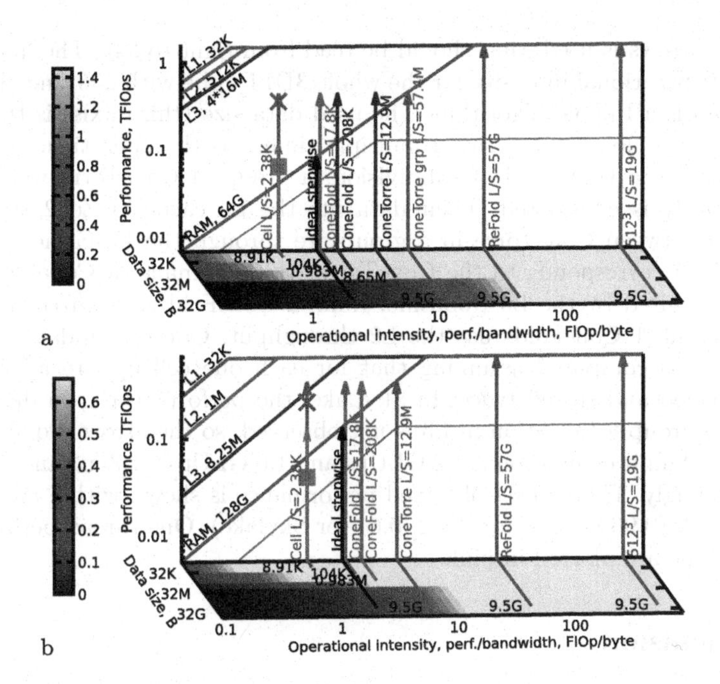

Fig. 10. RoofLine model on Ryzen R9 3900X (a) and Core i7-7800X (b)

the packed vectorized cubeLR array, since the projection of the ConeFold part that they represent involves only a part of the cube. Thus, one vectorized cubeLR may be used for these updates, and they may be processed asynchronously. The intersections exist on the boundaries of the projections, which cross the cube diagonally. The possible solution is to produce a layer of special LBM cells, in which the vectorized cells contain PDFs from different cells in a complex pattern.

With the compact streaming pattern, there are no dependencies across the diagonals. This workaround step for the diagonals is not required, and effortless parallelization between the asynchronous ConeFolds is possible. This unique feature of the compact pattern is not observed in the other streaming patterns.

The comparison of the performance with repacking operations and without them is shown in Fig. 9 with darker lines. The repacking overhead is up to ~10 and ~20% on the two systems.

The relevant algorithmic limits are studied with the RoofLine model (Fig. 10). It is plotted with an additional third axis, which shows the data storage space in which the data of the task is localized. The black arrow corresponds to the theoretical ideal stepwise implementation, for which every PDF value is loaded and stored once per time step. For the 512^3 domain, data size is 9.5 GB, so these transfers are limited by the RAM throughput. The performance limit estimation is 0.23 GLUps and 0.35 GLUps for Zen2 and SkyLake correspondingly.

In contrast, LRnLA algorithms can manifest both cache-aware and cache-oblivious properties, so the estimation of limits are more complex [9]. A sequence of colored arrows is used, that should be read from right to left. The first arrow shows the operational intensity for the whole 3D1T cube with 256 time updates. The second is a ReFold algorithm. The used data size (third axis) is the same, but the data is loaded and saved multiple times, so the operational intensity is lower. For the next level of subdivision, a group of ConeTorre which grow synchronously is considered. Their data fits the L3 cache in Zen2. Since the exchanges between ConeTorre in a group are through the L3 cache, the next arrow, which corresponds to the ConeTorre which is made of ConeFolds with rank 3, is limited by the L3 RoofLine. Rank 2 ConeFold is localized in the L2 cache, so rank 1 ConeFolds use the L2 throughput. One cell update intensity (red arrow) is computed assuming that for each one cell update all its PDF data is loaded and stored twice. In SkyLake, the performance gain due to the ConeTorre group localization in L3 is not observed, so the arrow is omitted.

The final limit estimation is 1.2 GLUps and 1.0 GLUps for Zen2 and SkyLake correspondingly. The highest obtained performance is shown with a cross mark on the graph (0.96 GLUps for Zen2, 0.64 for Skylake). One thread performance and RoofLine are plotted in pink.

5 Conclusion

The implementation of the LBM method with the compact streaming pattern and the LRnLA algorithms allows obtaining the fastest LBM code for the relevant vector CPU architectures with advanced memory hierarchy.

The compact streaming pattern currently seems to be the best approach to LBM streaming with no disadvantages, even in the stepwise codes without LRnLA. The cell updates are gathered into compact groups of 8 in 3D so that the access to the neighbors is unnecessary. With it, the computation may be localized in the faster and smaller levels of cache. It benefits the LRnLA algorithms by further reducing the dependencies between ConeFolds at maximum rank. Despite the nonintuitive data storage, it is easily coded even in 3D, and the cell data can be gathered at any time, while the data only in the local group is used for this. The recently introduced ReFold algorithm has benefited from the new streaming pattern: only one temporary vectorized array is used for processing 3 ConeFolds of maximal rank, even though their data may be in different regions in the simulation domain.

The maximum obtained performance for D3Q19 LBM BGK, single precision, is 1 GLUps on one Ryzen R9 3900X processor. This result is not only an important step in the development of the most efficient LBM codes, but it also provides an insight into how the localization of data access operations is improved by a combined consideration of the data storage and dependencies between operations.

The work is supported by the Russian Science Foundation, grant #18-71-10004.

References

1. Bailey, P., Myre, J., Walsh, S.D., Lilja, D.J., Saar, M.O.: Accelerating lattice Boltzmann fluid flow simulations using graphics processors. In: International Conference on Parallel Processing, ICPP 2009. pp. 550–557. IEEE (2009)
2. Broquedis, F., et al.: hwloc: a generic framework for managing hardware affinities in HPC applications. In: 18th Euromicro Conference on Parallel, Distributed and Network-Based Processing, pp. 180–186. IEEE (2010)
3. Geier, M., Schönherr, M.: Esoteric twist: an efficient in-place streaming algorithms for the lattice Boltzmann method on massively parallel hardware. Computation **5**(2), 19 (2017)
4. Habich, J., Zeiser, T., Hager, G., Wellein, G.: Enabling temporal blocking for a lattice Boltzmann flow solver through multicore-aware wavefront parallelization. In: 21st International Conference on Parallel Computational Fluid Dynamics, pp. 178–182 (2009)
5. Krüger, T., Kusumaatmaja, H., Kuzmin, A., Shardt, O., Silva, G., Viggen, E.M.: The Lattice Boltzmann Method: Principles and Practice. GTP. Springer, Cham (2017). https://doi.org/10.1007/978-3-319-44649-3
6. Levchenko, V., Zakirov, A., Perepelkina, A.: GPU implementation of ConeTorre algorithm for fluid dynamics simulation. In: Malyshkin, V. (ed.) PaCT 2019. LNCS, vol. 11657, pp. 199–213. Springer, Cham (2019). https://doi.org/10.1007/978-3-030-25636-4_16
7. Levchenko, V., Zakirov, A., Perepelkina, A.: LRnLA lattice Boltzmann method: a performance comparison of implementations on GPU and CPU. In: Sokolinsky, L., Zymbler, M. (eds.) PCT 2019. CCIS, vol. 1063, pp. 139–151. Springer, Cham (2019). https://doi.org/10.1007/978-3-030-28163-2_10
8. Levchenko, V.D: Asynchronous parallel algorithms as a way to archive effectiveness of computations. J. Inf. Tech. Comp. Syst. 1, 68–87 (2005). (in Russian: **АСИНХРОННЫЕ ПАРАЛЛЕЛЬНЫЕАЛГОРИТМЫКАКСПОСОБДОСТИЖЕНИЯЭФФЕКТИВНОСТИВЫЧИСЛЕНИЙ Левченко В.Д. Информационные технологии и вычислительные системы. 2005. №** 1.)
9. Levchenko, V., Perepelkina, A.: Locally recursive non-locally asynchronous algorithms for stencil computation. Lobachevskii J. Math. **39**(4), 552–561 (2018)
10. Mattila, K., Hyväluoma, J., Timonen, J., Rossi, T.: Comparison of implementations of the lattice-Boltzmann method. Comput. Math. Appl. **55**(7), 1514–1524 (2008)
11. Nguyen, A., Satish, N., Chhugani, J., Kim, C., Dubey, P.: 3.5-D blocking optimization for stencil computations on modern CPUs and GPUs. In: SC 2010: Proceedings of the 2010 ACM/IEEE International Conference for High Performance Computing, Networking, Storage and Analysis, pp. 1–13. IEEE (2010)
12. Perepelkina, A., Levchenko, V.: Enhanced asynchrony in the vectorized ConeFold algorithm for fluid dynamics modelling. Math. Model. **3**(2), 52–54 (2019)
13. Perepelkina, A.Y., Levchenko, V.D., Goryachev, I.A.: Implementation of the kinetic plasma code with locally recursive non-locally asynchronous algorithms. J. Phys: Conf. Ser. **510**, 012042 (2014). IOP Publishing
14. Perepelkina, A., Levchenko, V.: LRnLA algorithm ConeFold with non-local vectorization for LBM implementation. In: Voevodin, V., Sobolev, S. (eds.) RuSCDays 2018. CCIS, vol. 965, pp. 101–113. Springer, Cham (2019). https://doi.org/10.1007/978-3-030-05807-4_9

15. Perepelkina, A., Levchenko, V.: Synchronous and asynchronous parallelism in the LRnLA algorithms. In: Sokolinsky, L., Zymbler, M. (eds.) PCT 2020. CCIS, vol. 1263, pp. 146–161. Springer, Cham (2020). https://doi.org/10.1007/978-3-030-55326-5_11
16. Pohl, T., Kowarschik, M., Wilke, J., Iglberger, K., Rüde, U.: Optimization and profiling of the cache performance of parallel lattice Boltzmann codes. Parallel Process. Lett. **13**(04), 549–560 (2003)
17. Shimokawabe, T., Endo, T., Onodera, N., Aoki, T.: A stencil framework to realize large-scale computations beyond device memory capacity on GPU supercomputers. In: Cluster Computing (CLUSTER), pp. 525–529. IEEE (2017)
18. Succi, S.: The Lattice Boltzmann Equation: For Fluid Dynamics and Beyond. Oxford University Press, Oxford (2001)
19. Tomczak, T., Szafran, R.G.: A new GPU implementation for lattice-boltzmann simulations on sparse geometries. Comput. Phys. Commun. **235**, 258–278 (2019)
20. Wittmann, M., Zeiser, T., Hager, G., Wellein, G.: Comparison of different propagation steps for lattice Boltzmann methods. Comput. Math. Appl. **65**(6), 924–935 (2013)

Optimization of Load Balancing Algorithms in Parallel Modeling of Objects Using a Large Number of Grids

Vladislav Fofanov(✉) and Nikolay Khokhlov

Moscow Institute of Physics and Technology, Moscow Region, Dolgoprudny City, Russia
fofanov@phystech.edu, k_h@inbox.ru

Abstract. This work aims to find the optimal algorithm of computational load distribution for parallel modeling of objects using multiple grids. Reducing of calculation time will be achieved by reducing the amount of information transmitted between processes through contact surfaces.

For testing effectiveness of the algorithms, an elastic wave propagation model in a medium with many non-parallel cracks will be used.

Modeling with using of researched algorithms was carried out on the "Kurchatov Data Processing Center" programming complex, which is part of the "Kurchatov Institute" National Research Center.

Keywords: Parallel calculation · Numerical modeling · Wave propagation · Multiple grids modeling · Parallel algorithm

1 Introduction

Mathematical modeling of propagation of elastic waves is important for studying the influence of seismic phenomena on design of buildings, bridges, dams, nuclear power plants and other complex ground structures, affecting, first of all, correct choice of location for foundation laying and further construction of the structure.

In this paper, simulated object consists from propagation medium (Earth's surface) and source of elastic waves (explosion, earthquake epicenter). The problem of wave propagation in a homogeneous medium can be modeled using a single computational grid. However, more accurate models consider heterogeneity of the medium created by cracks or sections with a lower density of the medium. It is convenient to use several structural grids to model such computational domains.

Theoretically, it is possible to simulate a medium with cracks with one curvilinear grid by "rounding" its angles near with cracks. However, in this case, spatial step between the grid nodes will be rapidly decreased in these angles, which will lead to increase the Courant number at the same time step. That is, for stability of the numerical solution, it will be necessary to reduce time step. This leads to increase the number of time steps and, consequently, to increase in the calculation time of this model.

In this article, we will consider several different computational load distribution algorithms for parallel modeling of objects using multiple computational grids. We will

V. Voevodin and S. Sobolev (Eds.): RuSCDays 2020, CCIS 1331, pp. 63–73, 2020.
https://doi.org/10.1007/978-3-030-64616-5_6

compare the effectiveness of these algorithms depending on number of computing units. For this, we will use model of the propagation of elastic waves in a medium with cracks problem.

2 Overview of Existing Solutions in Field of Computational Grids Decomposition

To simulate various physical phenomena (wave propagation from an explosion, water flow modeling, deformation of solids, etc.) using modern computer systems, it is necessary to discretize used physical quantities, that is, replace the continuous functions of these quantities with a superposition of their values at some points. Discretization is necessary because the architecture of even the most modern computing systems is not able to store the value of a physical quantity on an infinite number of points for continuous modeling. One way to discretize these quantities in the numerical solution of differential and integral equations of physics is a computational grid. A computational grid is a collection of nodes (cells) that store values of a grid function. The quality of the computational grid used significantly affects the accuracy of the numerical solution of the equation.

The decomposition of any grids can initially be divided into static (decomposition is performed once before the start of calculations) and dynamic (decomposition can be performed an unlimited number of times during the calculation cycle). It makes sense to use the second in problems where the counting time of an individual node is unstable, and the load on the processes under static decomposition would change over time. Dynamic decomposition methods are implemented, for example, in the DRAMA [1] and Zoltan [4] packages. In [2], the dynamic decomposition method is considered, which focuses on the exact calculation of physical phenomena at the media boundary. In [3], a dynamic grid decomposition method is considered using a vector field constructed based on calculations on a less accurate grid. We will consider static decomposition in more detail depending on the type of computational grid.

There are two main classes of computational grids: structural and nonstructural. A grid is called non-structural if its elements are not ordered (each node have an arbitrary number of neighbors). In a software implementation, non-structural grids can be implemented as linked lists. The problem of non-structural grid decomposition usually comes down to the task of partitioning a graph that corresponds to the communication loads of the grid. In [5], a hierarchical algorithm is considered, consisting of roughening a graph, decomposing the smallest of the obtained graphs, and mapping the partition into previous graphs. The widely used Yakobovsky incremental algorithm is described in [6]. In [8], the so-called KL/FM algorithm is used, based on sequential optimization by moving the vertices of some blocks to others. An unusual, "genetic algorithm" is used to decompose grids in [14]. In the algorithm used in [15], the emphasis is more on minimizing slices between domains than on balancing the loads of these domains. An algorithm based on the recursive clustering method for decomposing grids into an arbitrary number of subdomains is described in [16]. A significant number of decomposition non-structural grids methods were implemented by software: PARTY [7], Jostle [9], METIS [11], Chaco [12], GridSpiderPar [13], and so on.

A grid is called structural if its many nodes are ordered. Each node of the structural grid has the same number of neighbors. In software implementation, the structural grid can be represented as a multidimensional array with indexing of the form (i, j, k) (in the three-dimensional case). Structural grids include rectangular, curved, nested and hierarchical. The algorithm described in this article is used to decompose structural grids, so we will consider them in more detail.

The decomposition of structural grids is much simpler than the decomposition of non-structural ones. The decomposition algorithms used for non-structural grids can also be applied to structural grids, but the result of it work will be a set of non-structural grids. From the point of view of software implementation, this is not always convenient, since it does not allow using the advantages of structural decomposition when implementing the numerical method and when exchanging data between blocks. The various subtypes of structural grids described above are identical in most properties; therefore, it is be a good idea to consider the decomposition algorithms common to them all firstly.

An important feature of using computational grids for modeling an object is the number of grids into which an object can be divided. In the presence of one grid (the simplest case), the algorithm of simple geometric decomposition is most often used: the computational domain is divided into equal (or almost equal) parts along several axes. This method, due to its simplicity, is used in many scientific works: for modeling impurity transport [17], for modeling seismic problems [18], for modeling plasma [19], and for many other problems of computational physics. The simple geometric decomposition algorithm, its strengths and weaknesses, are considered in [20–22]. Note that this method is also suitable for the case of several grids, which will be discussed later.

Often, it is convenient to use several structural grids to computational domain model. The simplest approach to parallelizing such tasks is to generate a certain number of grids corresponding to the number of processes on which calculations will be performed. In [23, 24], this approach is used to simulate the aquatic environment of the ocean. In [25], several methods for decomposing multigrid objects, including mesh generation by the number of processes, are used to model airplane wings in various states. In [26], for the most efficient grids generation by the number of processes, a four-path vector symmetry field is used to describe the local orientation of the grid on surface triangulations.

If the decomposition is carried out independently of the generation of grids, then the task becomes more complicated. The authors of [27–29] considered an algorithm consisting of dividing the largest-sized grids into blocks and their further grouping. When grouping blocks, information on the contacts between the grids is used: the block joins the smallest group in which there is already at least one block in contact with it.

In accordance with the classification described above, this paper will describe the static decomposition algorithm of multiblock structural grids, provided that the decomposition is carried out independently of the generation of grids.

3 Description of Researched Problem

For practical effectiveness verification of decomposition algorithm presented above, the problem of passing waves through a ground object was chosen.

The problem is solved by the grid-characteristic method. At first, the method of splitting by spatial coordinates is applied, as a result, we have three systems:

$$\frac{\partial \boldsymbol{u}}{\partial t} = \boldsymbol{A}_j \frac{\partial u}{\partial x_j}, j = 1, 2, 3 \tag{1}$$

Each of these systems is hyperbolic, so they can be rewritten in the form:

$$\frac{\partial \boldsymbol{u}}{\partial t} = \Omega_j^{-1} \Lambda_j \Omega_j \frac{\partial u}{\partial x_j} \tag{2}$$

where the matrix Ω_j is the matrix of eigenvectors, Λ_j is the diagonal matrix, eigenvalues.

For all coordinates, the matrix Λ has the form:

$$\Lambda = diag\{c_1, -c_1, c_2, -c_2, c_2, -c_2, 0, 0, 0\} \tag{3}$$

where c_1 = $\sqrt{}((\lambda + 2\mu)/\rho)$ is the longitudinal speed of sound in the medium, $c_2 = \sqrt{}$ (μ/ρ) is the transverse speed of sound.

After changing the variables $\nu = \Omega u$, each of the systems decomposes into nine independent scalar transport equations:

$$\frac{\partial \nu}{\partial t} + \Lambda \frac{\partial \nu}{\partial x} = 0 \tag{4}$$

One-dimensional transport equations are solved using a second-order TVD scheme. After all components of ν are transferred, the solution itself is restored:

$$u^{n+1} = \Omega^{-1} \nu^{n+1} \tag{5}$$

4 Description of Algorithms

When parallelizing tasks consisting of many structural grids, the most obvious way to decompose is dividing each grid between all processes. This method implements the most optimal distribution of computational load. However, with many numbers of grids, this algorithm may be inefficient due to the formation of a large number of blocks and, therefore, a large number of boundary nodes requiring data exchange with other processes. Next, we will call this method a simple decomposition algorithm.

Djomehri M. J. and Biswas R. describe a more advanced computational load balancing algorithm - Largest Task First with Minimum Finish Time and Available Communication Costs (LTF MFT ACC), which is based on dynamic load balancing by assigning each new block to a process with a minimum current computational load . However, this algorithm has several disadvantages, primarily the presence of a grouping of blocks obtained by dividing large grids and requires setting additional parameters for such division for users. These actions significantly complicate the implementation of the algorithm, but do not bring any big gain in the speed of calculations. A full description of this algorithm with example can be found in [29].

In this paper, we will consider a modification of the LTF MFT ACC algorithm, aimed at simplifying its implementation without significant performance losses. For theoretical assessments and further practical calculations, we will use the OpenMPI software package library used to create parallel programs on MIMD systems.

The input data for the proposed algorithm are grid sizes and the number of MPI processes used for calculation.

5 Theoretical Evaluation of Algorithms

Let our model consist of N grids, the i-th grid has M_i nodes. The calculation will occur on P processes. We will act in accordance with the following algorithm:

1. The number of nodes that send to one MPI process is calculated. We call this value "the optimal block size for one process".

$$M_{opt} = \frac{\sum_{i=1}^{N} M_i}{P} \tag{6}$$

2. All grids are divided into two groups: large and small. Grids smaller than the M_{opt} block size becomes small, and the rest becomes large.
3. For each large grid, the optimal number of processes is calculated for its processing - P_j - the quotient of the grid size and the optimal size, rounded down. The resulting number of processes is immediately allocated to the grids.

$$P_i = \left[\frac{M_i}{M_{opt}} \right] \tag{7}$$

4. Small grids need to be grouped by the number of processes left unoccupied. To do this, we will use the greedy algorithm. Groups are created, initially by the number of small grids. Step by step, the two smallest groups (by the number of nodes) are combined until the groups become evenly divided with unoccupied processes. Processes are assigned to groups. This part is used the same small grid grouping way as the LTF MFT ACC algorithm.
5. If after the previous step there were free processes (for example, there were simply no small grids initially), then they are distributed one at a time to large grids with maximum $Load_i$ value. The separation of large grids between the selected processes occurs by the method of simple geometric composition. The loading of grid processes is the grid size divided by the number of processes allocated earlier for it.

$$Load_i = \frac{M_i}{P_i} \tag{8}$$

Let us conduct a theoretical assessment of the effectiveness of this algorithm and compare it with a theoretical evaluation of the simple decomposition algorithm. Comparison with the LTF MFT ACC algorithm cannot be made due to complexity of calculating its effectiveness (it's followed from complexity of the algorithm itself).

For a theoretical assessment, the use of a computational model of a medium with cracks will be difficult. So, for these purposes we consider a simplified version. Let computational model be described by N grids and calculated on P processes. We will use following approximations:

1. The number of processes is much (an order of magnitude) greater than number of computational grids (P >> N). This condition allows us to ignore calculation time inside the process grid and interpret total model calculation time as time of data exchange between grids of various processes.

2. The grids are same size. Each of N grids consists of M nodes.
3. Grids are cubes - each side has a length $M^{\frac{1}{3}}$ nodes.
4. The conditions are such that P and $\frac{P}{N}$ are complete cubes, so, a simple decomposition will divide grids between such a number of processes equally in three axes.
5. The absence of shipments at the borders of the computational domain can be ignored.

For Simple Decomposition

Consider a separate grid. For it, each process sends and receives 6 messages per step (according to the number cube sides). In total, 6P exchanges take place in grid. The length of one such message is the area of cube side - $(\frac{M}{P})^{\frac{2}{3}}$ The size of the transmitted information in grid is $2 \cdot 6P \cdot (\frac{M}{P})^{2/3} = 12P^{1/3}M^{2/3}$. In N networks we have 6PN messages. Total size of transmitted information is $12NP^{1/3}M^{2/3}$

For Proposed Decomposition Algorithm

On one grid there are $\frac{P}{N}$ processes and, accordingly, $6\frac{P}{N}$ messaging. The size of one message is $(\frac{MN}{P})^{2/3}$. The size of all messages in grid is $2 \cdot 6\frac{P}{N} \cdot (\frac{M}{P})^{2/3} = 12\left(\frac{M^2N}{P}\right)^{1/3}$. In N grids 6P exchanges. Total size of transmitted information is $N \cdot 12\left(\frac{M^2N}{P}\right)^{1/3} = 12P^{1/3}(MN)^{2/3}$.

Thus, the number of message exchanges decreased by N times, and total amount of information sent decreased by $N^{1/3}$ times. The size of a single message is expected to grow $N^{2/3}$ times.

It should be noted that this theoretical assessment is very approximate in nature and was carried out in order to demonstrate the potential effectiveness of proposed method compared to well-known algorithm. As we will see later, practical calculations show values that are far enough from theoretical indicators. In addition, even using this algorithm to solve real problems, balancing processes load may not be optimal. It's using is advisable under the assumption that the lost time is compensated by the acceleration of information exchange.

6 Modeling Results

The testing model calculation with load balancing algorithms described above was carried out on the Kurchatov data center software package, which is part of the Kurchatov Institute national research center. The complex includes two computational clusters HPC2 and HPC4 with a total number of computing cores approximately equal to 15000.

In all cases, the computational domain is the wave propagation medium with a size of 10 km * 10 km * 1 km. In the medium at a depth of 500 m, a layer of non-parallel plane cracks is observed, uniformly distributed over the entire surface of the medium. Cracks sizes and number are variable parameters, but their ratio is always selected considering the coverage of the entire plane of the medium under consideration. A crack is a flat rectangle with equal length and width, and the height is much less than the length. Each crack is inclined along one of the coordinate axes, the inclination angle for all cracks is different. At a depth of 100 m in the center of the computational domain the source of elastic waves is placed.

The material of the propagation medium has the following parameters: density p = 4000 kg/m^3, longitudinal velocity of wave propagation Cp = 2000 m/s, transverse Ct = 1000 m/s. The crack material has parameters: p = 800 kg/m^3, Cp = 400 m/s, Ct = 200 m/s. Parameters of computational grids: spatial grid spacing = 5 m, the time step is selected based on Courant's criteria and equals to 0.03 s.

As the variable parameters, the angle of cracks (takes two values: 20 and 35°), and their number (three configurations were considered in the work: 9×9 = 81, 19×19 = 361, $49 \times 49 = 2401$) was chosen. The main variable parameter that allows us to compare efficiency of parallelization algorithms is number of processes used for modeling.

The result of each calculation is a package with files containing numerical data describing the process of wave propagation in a medium with cracks. Using special programs (for example, paraview package), you can visualize the received data. The figures show results of visualization of calculated data for the case of a medium with the number of 81 cracks (Fig. 1 and Fig. 2).

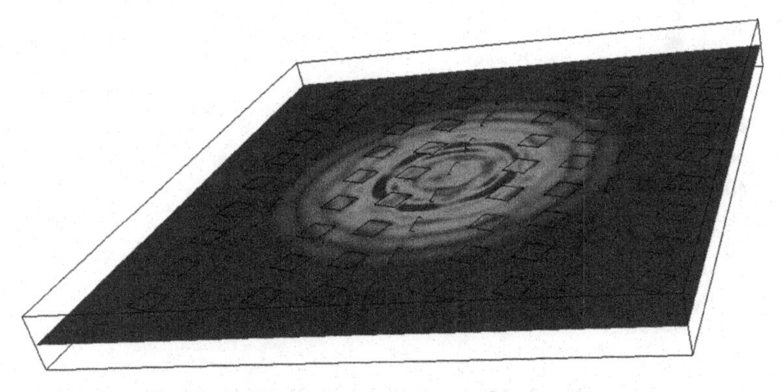

Fig. 1. Simulation with angle = 20° and time = 3c

It can be seen from images, the wave propagation occurs non-uniformly, low and high intensity regions are observed near the cracks. Due to different inclination angles of the cracks, the regions are not symmetrical with respect to the wave propagation center. The picture looks plausible, which confirms the correctness of the simulation.

7 Calculation Results

These graphs show results of using parallelization algorithms to calculate model described above. Each chart display results of calculation using all mentioned algorithms: simple decomposition, LTF MFT ACC and proposed decomposition for constant number of cracks. Graphs show calculation time dependency from number of used process. Only those values of processes number are shown that show difference in efficiency of considered algorithms. For each algorithm we have two lines: with cracks angle equal 20° and 35° (Fig. 3, Fig. 4 and Fig. 5).

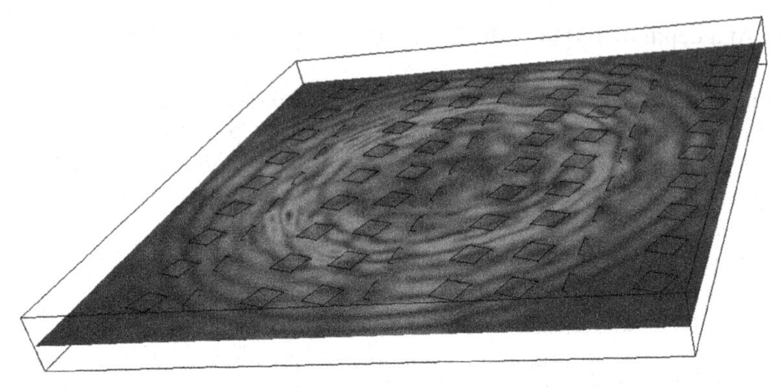

Fig. 2. Simulation with angle = 20° and time = 6c

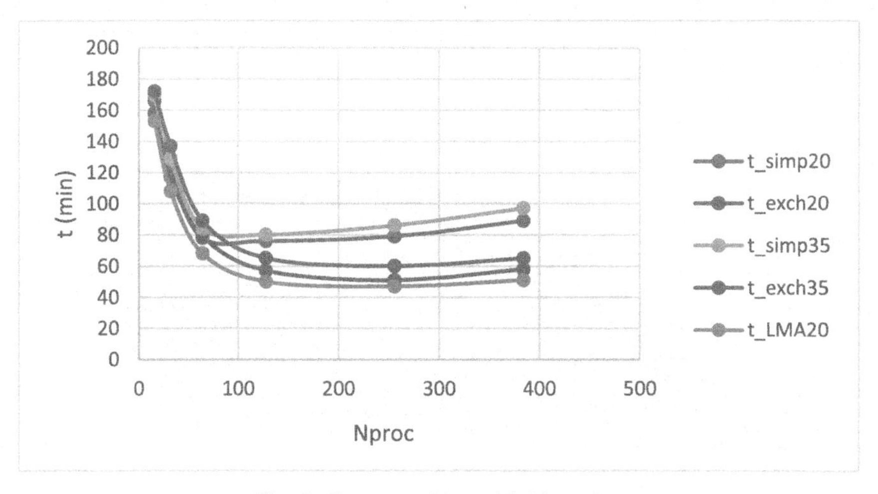

Fig. 3. Decomposition with 81 cracks.

It can be seen from graphs, the type of dependence from number of processes is similar for all algorithms and for all models. On each graph there is a minimum value of calculation time, which is explained by contribution of two main quantities - the calculation time inside the grid for each process and the data exchange time between grids of different processes. For the first term we have: $t_{int} \sim P^{-1}$, for the second: $t_{exch} \sim P^{\frac{1}{3}}$. It can be seen from graphs that the minimum for compared algorithms differs – the minimum for proposed and LTF MFT ACC algorithms is achieved with a larger number of processes and has a smaller value than the minimum for simple algorithm.

At large values of processes number, we have $t_{int} \sim 0$, respectively, the graph should correspond to our calculations made earlier, that is:

$$t_{mod} \sim V_{exch_{data}} \sim P^{\frac{1}{3}} \tag{9}$$

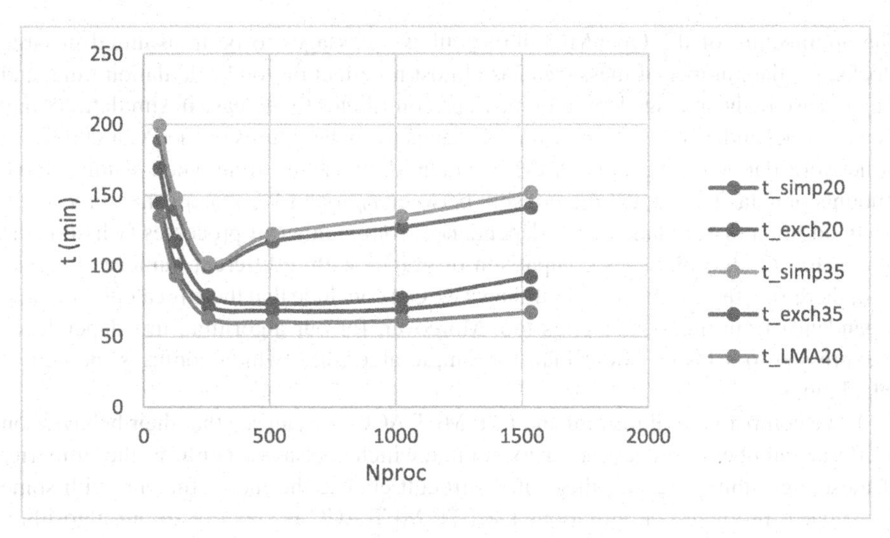

Fig. 4. Decomposition with 361 cracks.

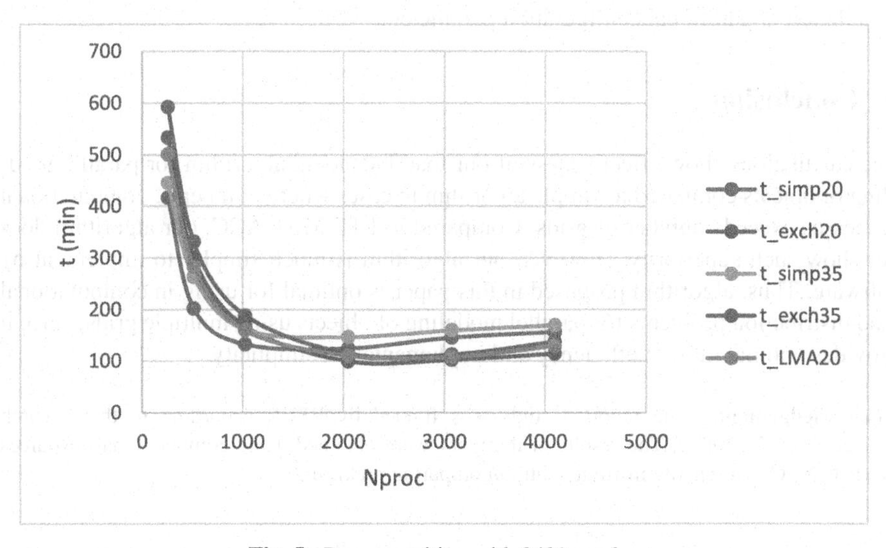

Fig. 5. Decomposition with 2401 cracks.

for all algorithms. If we look at these graphs in logarithmic coordinates, we will see that the slope coefficient for the algorithms is the same, but its average value is $k_{simp} = k_{adv} = 0.26 \sim \frac{1}{4}$, which is less than calculated coefficient.

This difference can be explained following way. In the formula above, we assume that the exchange time t_{exch} is proportional to the amount of information transmitted between all processes. This seems logically but may not match the actual OpenMPI library architecture. In fact, we have: $V_{exch\_data} = N_{msg} * \hat{l}_{msg}$, where the first factor is the number of transmitted messages, and the second is the average length of one message.

The architecture of the OpenMPI library allows messages to be transmitted in large blocks, so, the number of messages has almost no effect on total calculation time, and the increase in the average length of message contributes to increase in simulation time. However, dependence here is non-linear, thanks to such features of the PC architecture as network data caching and IOMMU, which allow saving time when sending large amounts of data. That is, the relationship between t_{mod} and V_{exch_data} is not direct.

In addition of simulation time dependence from number of processes (which is the similar for all algorithms), a comparison of graphs with a different number of cracks (and, therefore, the number of grids) shows we can conclude that there is calculation time dependence from number of grids too. Moreover, for our algorithm, this dependence is expressed to a lesser extent than for simple algorithm, which confirms theoretical calculations.

If we compare our algorithm and LTF MFT ACC, we can see that their behavior on a different number of processes is very similar, which once again confirms the similarity of these algorithms. Among them, it is difficult choose the most efficient, with some investigated problem configurations the LTF MFT ACC is superior to our algorithm, with others - conversely. However, as mentioned in chapter "Theoretical evaluation of algorithms", the LTF MFT ACC algorithm is more difficult to automate, since it requires installation of additional configuration parameters.

8 Conclusion

The calculations show effectiveness of our load balancing algorithm for parallel modeling of objects compared to simple algorithm in cases where number of computational processes exceeds number of grids. Compared to LTF MFT ACC, our algorithm does not show such superiority. However, our algorithm is much simpler to implement by software. Thus, algorithm proposed in this paper is optimal for using in computational load distribution problems for parallel modeling of objects using multiple grids, since it provides optimal ratio of efficiency and implementation simplicity.

Acknowledgements. The reported study was funded by RFBR according to the research project № 18-31-20041. The results of the work were obtained using computational resources of MCC NRC «Kurchatov Institute», http://computing.nrcki.ru/.

References

1. Devine, K., Boman, E., Heaphy, R., et al.: Zoltan data management services for parallel dynamic applications. Comput. Sci. Eng. **4**(2), 90–97 (2002)
2. Lagnese, J.E., Leugering, G.: Dynamic domain decomposition in approximate and exact boundary control in problems of transmission for wave equations. SIAM J. Control Optim. **38**(2), 503–537 (2000)
3. Cacace, S., Falcone, M.: Dynamic domain decomposition for a class of second order semilinear equations (2015)
4. Basermann, A., Clinckemaillie, J., Coupez, T., et al.: Dynamic load-balancing of finite element applications with the DRAMA library. Appl. Math. Model. **25**(2), 83–98 (2000)

5. Golovchenko, E.N.: Decomposition of computational grids for solving problems of continuum mechanics on high-performance computing systems
6. Yakobovsky, M.V.: Incremental graph decomposition algorithm
7. Preis, R., Diekmann, R.: PARTY - a software library for graph partitioning (1997)
8. Fiduccia, C.M., Mattheyses, R.M.: A linear-time heuristic for improving network partitions (1982)
9. Walshaw, C., Cross, M.: JOSTLE: parallel multilevel graph-partitioning software - an overview (2007)
10. Yakobovsky, M.V.: Distributed systems and networks
11. Karypis, G.: A software package for partitioning unstructured graphs, partitioning meshes, and computing full-reducing orderings of sparse matrices (1998)
12. Hendrickson, B., Leland, R.: The Chaco user's guide: version 2.0 (1995)
13. Golovchenko, E.N., Yakobovsky, M.V.: GridSpiderPar package of large grids parallel decomposition
14. Rama Mohan Rao, A., Appa Rao, T., Dattaguru, B.: Automatic decomposition of unstructured meshes employing genetic algorithms for parallel FEM computations. Struct. Eng. Mech. **14**(6), 625–647 (2002)
15. Aykanat, C., Cambazoglu, B.B., Findik, F., Kurc, T.: Adaptive decomposition and remapping algorithms for object-space-parallel direct volume rendering of unstructured grids. J. Parallel Distrib. Comput. **67**(1), 77–99 (2007)
16. Jones, B.W., McManus, K., Cross, M., Everett, M.G., Johnson, S.: Parallel unstructured mesh CFD codes: a role for recursive clustering techniques in mesh decomposition (1993)
17. Belikov, D.A., Starchenko A.V.: Parallelization of the impurity transfer spatial model on systems with distributed memory using one and two-dimensional domain decomposition
18. Burtsev, A.P., Popova, N.N., Kurin, E.A.: Modeling wave processes in seismic problems on the Lomonosov supercomputer using GPU
19. Bastrakov, S.I., et al.: Investigation and search for the most effective approaches to parallel plasma modeling by the method of particles in cells on cluster systems
20. Taghavi, R.: Automatic block decomposition using fuzzy logic analysis (2000)
21. Farrashkhalvat, M., Miles, J.P.: Basic Structured Grid Generation with an Introduction to Unstructured Grid Generation (2003)
22. Ali, Z., Tyacke, J., Tucker, P.G., Shahpar, S.: Block topology generation for structured multi-block meshing with hierarchical geometry handling. Procedia Eng. **163**, 212–224 (2016)
23. Luong, P., Breshears, C.P., Ly, L.N.: Coastal ocean modeling of the US west coast with multiblock grid and dual-level parallelism (2001)
24. Luong, P., Breshears, C.P., Ly, L.N.: Application of multiblock grid and dual-level parallelism in coastal ocean circulation modeling. J. Sci. Comput. **20**(2), 257–275 (2004)
25. Allen, C.B.: Automatic structured multiblock mesh generation using robust transfinite interpolation (2007)
26. Fogg, H.J., Armstrong, C.G., Robinson, T.T.: Automatic generation of multiblock decompositions of surfaces. Int. J. Numer. Methods Eng. **101**(13), 965–991 (2015)
27. Djomehri, M.J., Biswas, R.: Performance analysis of a hybrid overset multi-block application on multiple architectures. In: Pinkston, T.M., Prasanna, V.K. (eds.) HiPC 2003. LNCS, vol. 2913, pp. 383–392. Springer, Heidelberg (2003). https://doi.org/10.1007/978-3-540-24596-4_41
28. Djomehri, M.J., Biswas, R.: Performance enhancement strategies for multi-block overset grid CFD applications. Parallel Comput. **29**(11-12), 1791–1810 (2003)
29. Djomehri, M.J., Biswas, R., Lopez-Benitez, N.: Load balancing strategies for multi-block overset grid applications: NAS Technical report NAS-03-007

Parallel BIILU2-Based Iterative Solution of Linear Systems in Reservoir Simulation: Do Optimal Parameters Exist?

Igor Konshin[1,2,3,4](✉), Kirill Nikitin[1], Kirill Terekhov[1], and Yuri Vassilevski[1,3,4]

[1] Marchuk Institute of Numerical Mathematics of the Russian Academy of Sciences, Moscow, Russia
igor.konshin@gmail.com, nikitin.kira@gmail.com, kirill.terehov@gmail.com, yuri.vassilevski@gmail.com
[2] Dorodnicyn Computing Centre of FRC CSC RAS, Moscow, Russia
[3] Moscow Institute of Physics and Technology (State University), Moscow region, Dolgoprudny, Russia
[4] Sechenov University, Moscow, Russia

Abstract. Three-phase black oil model for SPE-10 test problem with highly heterogeneous permeability and porosity fields was considered. A set of linear systems by the research parallel three-phase fully implicit black oil simulator was generated. The parallel performance of the MPI-based BIILU2 linear solver with the default parameters on such systems was analyzed. The trends in the choice of optimal BIILU2-based linear solver parameters for such systems were determined.

Keywords: Numerical linear algebra · Parallel computing · Iterative linear solver · Parameters tuning

1 Introduction

The goal of the paper is to study the parallel performance of a publicly available solver on a set of matrices arising in black oil simulation. A research reservoir simulator is developed at INM RAS and is based on the open source INMOST software platform [1–3] for supercomputer simulation on general meshes. The solver BIILU2 (Block Incomplete Inverse LU-decomposition and ILU2 approximate factorization) is one of available linear solvers provided by the platform INMOST. For a review of the other solvers of the platform we refer to [3]. A comprehensive overview of parallel solvers for general sparse linear systems can be found in [4,5]. The object of the study is the MPI-version of BIILU2 solver for matrices generated by the three-phase fully implicit black oil research reservoir simulator of INM RAS. The black oil model is widely used by reservoir engineers as it describes water flooding in water–oil–gas reservoirs [6–8]. A second order monotone cell-centered finite volume disretization applicable to general

V. Voevodin and S. Sobolev (Eds.): RuSCDays 2020, CCIS 1331, pp. 74–85, 2020.
https://doi.org/10.1007/978-3-030-64616-5_7

meshes and general heterogeneous coefficients, produces nonlinear algebraic systems to be solved by the Newton method at every time step of the implicit Euler scheme [3]. The linear solver decomposes the system matrix (Jacobian) in a number of blocks which is equal to the number of MPI processes. The primary objective of the study is optimal parameters of the solver and its scalability on up to hundreds of cores of the INM parallel cluster [9] with up to thousands computational cells per core.

We address the BIILU2 solver since its combination with BiCGStab iterations provides the robust and efficient solution of linear systems generated by the black oil reservoir simulators. Although it is not the only applicable solver (parallel block ILU(k) [10] or constrained pressure residual [11] combined with a parallel solver for the elliptic pressure block are viable alternatives), BIILU2 solver is appealing by its robustness: all systems generated by the reservoir simulator can be solved with a unique choice of parameters. Therefore the search of optimal parameters is meaningful. The novelty of the paper is that in reservoir simulation it is the first use of BIILU (rather than the Additive Schwarz) framework for matrix decomposition, the second order approximate factorization ILU2 and the study of existence of opimal parameters for the BiCGStab+BIILU+ILU2 solver. The use of BIILU2 solver in other applications is discussed in [12–16].

The study is motivated by observation that for the BIILU2 solver, the iteration count grows with the increase of the number of cores. This feature stems from the domain decomposition paradigm: due to the limited number of matrix overlapping layers and absence of a coarse space correction, the elliptic block of the Jacobian matrices generates convergence dependence on the number of subdomains (processors). It is expected that the proper choice of the number of overlapping layers and the preconditioner threshold will improve the performance. To study such opportunity, one has to generate a set of systems by the research parallel three-phase fully implicit black oil simulator, to test parallel performance of the MPI-based BIILU2 linear solver with the default parameters on such systems, and to determine trends in the choice of optimal BIILU2-based linear solver parameters for such systems.

The paper is organized as follows. In Sect. 2 we introduce the three-phase black oil model and address SPE-10 test problem with highly heterogeneous permeability and porosity fields. We add 12 injector and 12 producer wells and simulate water flooding. Simulation results and formal analysis of chosen linear systems complete the chapter. In Sect. 3 the description of the set of linear systems and matrix properties is presented. In Sect. 4 we specify the parallel preconditioner based on Block Incomplete Inverse LU-decomposition and ILU2 approximate factorization. In Sect. 5 we analyze parallel performance of the solver on a set of chosen linear systems. Special attention in Sect. 6 is paid to tuning of linear solver parameters. In Sect. 7 we summarize the results of our study.

2 Black Oil Model for SPE-10 Test Problem

In the present paper we test the performance of the BIILU2 linear solver for a set of linear systems generated by the INM black oil simulator on SPE-10 data (see [17]).

The simulator uses the three-phase black oil model [6,7] in the heterogeneous porous media:

$$\begin{aligned} \frac{\partial \rho_w \varphi S_w}{\partial t} + \operatorname{div}(\rho_w \mathbf{u}_w) &= q_w, \\ \frac{\partial \rho_o \varphi S_o}{\partial t} + \operatorname{div}(\rho_o \mathbf{u}_o) &= q_o, \\ \frac{\partial (\rho_{go}\varphi S_o + \rho_g \varphi S_g)}{\partial t} + \operatorname{div}(\rho_{go}\mathbf{u}_o + \rho_g \mathbf{u}_g) &= q_g, \end{aligned} \tag{1}$$

with Darcy fluxes:

$$\mathbf{u}_\alpha = -\lambda_\alpha \mathbb{K} (\nabla p_\alpha - \rho_\alpha g \nabla z), \quad \alpha = w, o, g. \tag{2}$$

Here φ is the rock porosity, $\rho_\alpha = \rho_{\alpha,0}/B_\alpha$ is the phase density, $\rho_{go} = \rho_{g,0} Rs/B_o$ is the density of gas dissolved in the oil, $\rho_{\alpha,0}$ is the phase density at surface conditions, B_α is the phase formation volume factor, Rs is the solubility of gas in the oil at a given bubble point pressure p_b, $\mathbb{K}$ is the absolute permeability tensor ($\mathbb{K}$ is scalar in this study), $\lambda_\alpha = k_{r\alpha}/\mu_\alpha$ is the phase mobility, $k_{r\alpha}$ is the phase relative permeability, μ_α is the phase viscosity, g is the gravity term, z is the depth, and q_α is the source or sink for phase $\alpha = w, o, g$.

The phases fill all voids: $S_w + S_o + S_g = 1$. The water and gas pressures are connected to oil pressure p through the capillary pressures:

$$p_w = p_o - Pc_{ow}(S_w), \quad p_g = p_o + Pc_{go}(S_g). \tag{3}$$

There are three unknowns in the model associated with three equations (1): the oil pressure p_o, the water saturation S_w, and an unknown Y that represents either the gas saturation S_g or the bubble point pressure p_b.

We consider two computational grids for SPE-10 data set (see [17]). The size of the small computational grid is $60 \times 220 \times 25$ cells ($0.33 \cdot 10^6$ cells). The size of the large computational grid is $60 \times 220 \times 85$ cells ($1.122 \cdot 10^6$ cells). The top 35 layers of the large model is a Tarbert formation, and is a representation of a prograding near shore environment, while the bottom 50 layers represents Upper Ness which is fluvial. The coefficients of the media are very contrast. The porosity varies from $1.3 \cdot 10^{-5}$ to 0.5, see Fig. 1 (left), while the permeability varies from 10^{-3} to $3 \cdot 10^4$, see Fig. 1 (right).

The model has 24 vertical wells completed throughout formation. The Peaceman model [8] with the given bottom-hole pressure is used for the well flux calculation. Water is injected through 12 injector wells, and the mixture is produced from 12 producer wells. Wells are placed chequerwise (see Fig. 2), $p_{bh,\mathrm{inj}} = 282.68$ psi, $p_{bh,\mathrm{prod}} = 268.89$ psi.

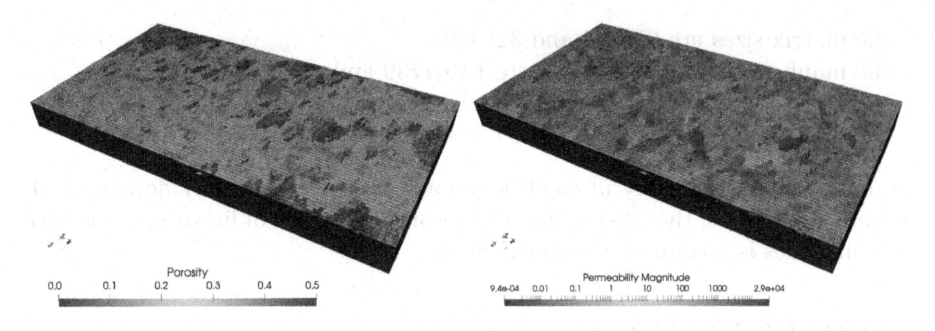

Fig. 1. Porosity and permeability distribution for SPE-10 problem.

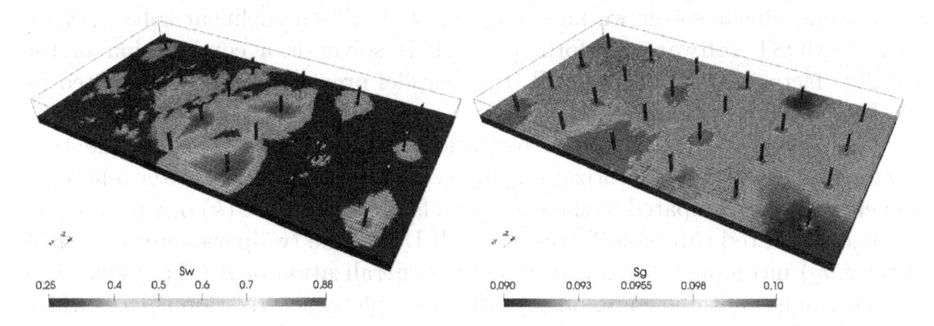

Fig. 2. Water and gas saturation for $t = 0.1$ days.

Flow coefficients were scaled by 887 times and thus the flow evolves faster. The simulation time is 0.1 days, the initial time step is 0.00001 days and maximum time step is 0.0025 days.

To demonstrate the simulated pressures and saturations, we consider the primary unknowns on the large grid at the end of simulation ($t = 0.1$ days). Accordingly, water saturation for the bottom 25 layers is presented in Fig. 2 (left) and gas saturation is given in Fig. 2 (right).

3 The Set of Linear Systems and Matrix Properties

Since the INM reservoir simulator is based on the INMOST parallel platform [2], it uses the automatic differentiation and the cell-centered finite volume method for generation of the Jacobian matrices. The names of the test Jacobian matrices "nXtYZ" encode sparse matrices arising at a Newton iteration at a timestep of the simulation. Here $X = 25, 85$ is the number of layers in the computational grid; $Y = 0, 1, 2, 3, 4, 5$ specify the simulation time 0, 10^{-5}, 10^{-4}, 10^{-3}, 10^{-2}, 10^{-1}, respectively; $Z = \mathrm{b}, \mathrm{e}$ denotes the first ("beginning") and the last ("end") Newton iteration at time Y.

The main properties of the unsymmetric matrices n25* and n85* can be formulated as follows:

- the matrix sizes are 979404 and 3283194;
- the numbers of nonzeros nz(A) are 13101739 and 43775630;
- the average numbers of non-zero entries per row are 13.4 and 13.3;
- the maximal numbers of non-zero entries in rows for both sets are 22.

All matrices are extremely ill conditioned and are not diagonally dominant M-matrices. Therefore, the efficient parallel iterative solution of linear systems with these matrices is a computational challenge.

4 BIILU2-Based Linear Solver

As a robust linear solver we have used the BIILU2-based linear solver [13,14] from INMOST software platform [1,2]. This solver is a combination of the ILU2(τ) factorization and the BIILU(q) parallel preconditioning accelerated by the BiCGStab iterations.

The second-order ILU2(τ) factorization [12], also referenced as the two-parameters ILU2(τ, τ_2) factorization [16], is used because of its greater reliability and efficiency as compared to the structural factorization ILU(k) or with the traditional truncated threshold factorization ILU(τ). The two-parameter threshold ILU2(τ, τ_2) incomplete factorization is the generalization of ILUT(τ) factorization [4] and has nothing common with the incomplete factorization ILU(k) which preserves the sparsity structure of matrix A^k [4]. For $\tau_2 = \tau^2$ the two-parameters ILU2(τ, τ_2) factorization is equivalent to the second-order triangular ILU2(τ) factorization, while for $\tau_2 = \tau$ it is equivalent to the traditional ILUT(τ) factorization.

Usually, the Additive Schwarz AS(q) method is used to construct parallel preconditioner based on triangular factorization, where the overlap parameter q is the number of layers in subdomain overlap. We have used the alternative approach based on BIILU(q) (Block Incomplete Inverse LU-decomposition).

The preconditioner based on Block Incomplete Inverse LU-decomposition (BIILU) and ILU2 approximate factorization is abbreviated as BIILU2 preconditioner. In what follows we consider the BIILU2(τ, τ_2, q) preconditioner with the extended set of parameters as well as BIILU2(τ, q) preconditioner with the basic set parameters, assuming $\tau_2 = \tau^2$.

5 Performance for the Default Set of Parameters

We have studied parallel performance of the BIILU2-based linear solver on the INM RAS cluster [9], on the set of eight nodes (cl1n005–cl1n012) of x12core segment:

- Compute Node Arbyte Alkazar+ R2Q50;
- 24 cores (two 12-core Intel® Xeon processors E5-2670v3@2.30 GHz);
- RAM: 64 GB per node;
- Operating System: SUSE Linux Enterprise Server 11 SP3 (x86_64).

We have used Intel MPI v.4.0.1 with Intel C++ v.15.0.2 compilers.

The total number of cores in x12core segment is 192. We have performed our parallel experiments on up to the maximal number of 192 cores.

The default values of main parameters of the BIILU2(τ, q) preconditioner are $q = 3$ and $\tau = 0.003$. These parameters were chosen on the base of numerous experiments for the different types of linear systems. We recall that q is the number of overlap layers for the additive preconditioner, and the factorization threshold τ control the preconditioner accuracy and sparsity. The second order ILU2(τ) factorization corresponds to the two-parameter incomplete ILU(τ_1, τ_2) factorization with $\tau_1 = \tau$ and $\tau_2 = \tau^2$. The first threshold τ_1 controls the sparsity of the preconditioner, while the second parameter τ_2 controls the preconditioner accuracy (see Sect. 4). Before the preconditioner construction, the implicit diagonal scaling of the matrix is performed. But SPE-10 matrix coefficients varies in very large scale due to the jumps in permeability (see Sect. 2). It may leads to factorization disbalance, which is expressed in too small or too large value of pivot elements.

One of the important improvements in the BIILU2 preconditioner construction is the possibility to perform preliminary iterative balancing scaling [15] of the linear system to equalize the norm of the left and the right triangles of matrix A. After this procedure the pivot values becomes more uniform.

Remark 1. We use the balancing scaling (INMOST parameter sctype=1) performing 3 balancing iterations for each linear system (INMOST parameter nitersc = 3).

In the present section we describe the results of parallel experiments on the set of p MPI cores, where $p = 1, 2, 4, 8, 16, 32, 64, 96, 128, 192$.

The following notation is used in the rest of the paper:

1. Dens denotes the relative density of the preconditioner H with respect to the original matrix A, i.e., Dens = nz(H)/nz(A);
2. #it is the number of multiplication of the coefficient matrix A by a vector;
3. T_{prec} is the time spent in the preconditioner construction;
4. T_{iter} is the time spent in the iterative solution phase;
5. T_{total} is the total solution time;
6. S is the relative speedup for the parallel run on p cores with respect to the serial run on one core.

Clearly, the value $T_{\text{total}} - T_{\text{prec}} - T_{\text{iter}}$ equals to the time spent in the construction of auxiliary data structures for parallel processing as well as for preliminary balancing of the linear system if required. The time for balancing iterations is taken into account in the total solution time T_{total}.

The numerical testing of BIILU2 preconditioned BiCGStab iterative solver assumed the zero initial guess $x_0 = 0$, and the iteration convergence criterion was set to $\|b - Ax_{\text{iter}}\|/\|b\| \le \varepsilon$, where $\varepsilon = 10^{-6}$.

The performance statistics for linear systems n25t4b and n85t4b (the beginning of the Newton iteration at the 4-th time moment $t = 0.01$) are presented in

Table 1. Performance statistics for linear systems n25t4b and n85t4b for the default set of parameters.

p	Dens	#it	T_{prec}	T_{iter}	T_{total}	S
1	1.786	25	102.272	5.298	108.707	1.00
2	1.703	28	51.368	2.879	54.784	1.98
4	1.741	34	44.778	2.196	47.273	2.30
8	1.744	41	22.343	1.590	24.104	4.51
16	1.782	50	14.687	1.176	15.970	6.81
32	1.775	55	5.420	0.777	6.264	17.35
64	1.777	56	4.826	0.532	5.517	19.70
96	1.835	62	2.350	0.491	3.058	35.55
128	1.846	71	2.030	0.424	2.672	40.68
192	1.897	64	1.071	0.408	1.781	61.04
1	1.275	27	175.732	16.617	195.937	1.00
2	1.302	105	107.871	36.144	146.093	1.34
4	1.356	62	63.315	13.942	78.572	2.49
8	1.366	74	44.424	9.140	54.146	3.62
16	1.410	101	19.185	7.852	27.401	7.15
32	1.416	97	13.414	4.642	18.296	10.71
64	1.422	113	9.626	3.797	13.697	14.31
96	1.438	111	6.729	3.948	11.056	17.72
128	1.470	109	5.106	2.968	8.452	23.18
192	1.507	87	3.215	2.165	5.841	33.55

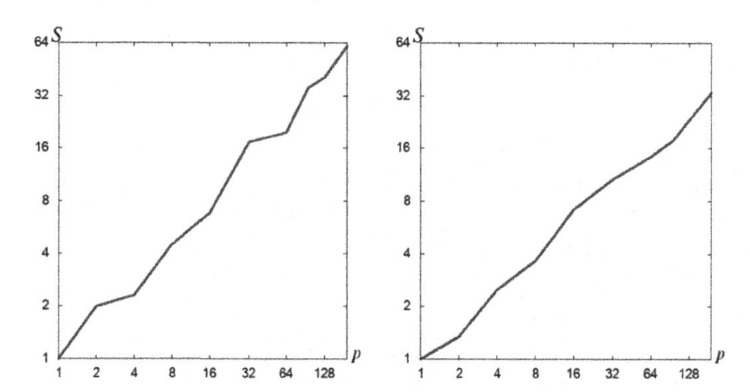

Fig. 3. Speedup for solution of linear systems n25t4b and n85t4b for the default set of parameters.

Table 1. The time spent in the preconditioner construction is much greater than the iteration time. This motivates tuning of the parameters, which will be performed in the following section. Left and right panels of Fig. 3 show the speedup for the solution of linear systems n25 and n85, respectively, for the default set of parameters at 4-th time moments for the first Newton iteration. One can observe

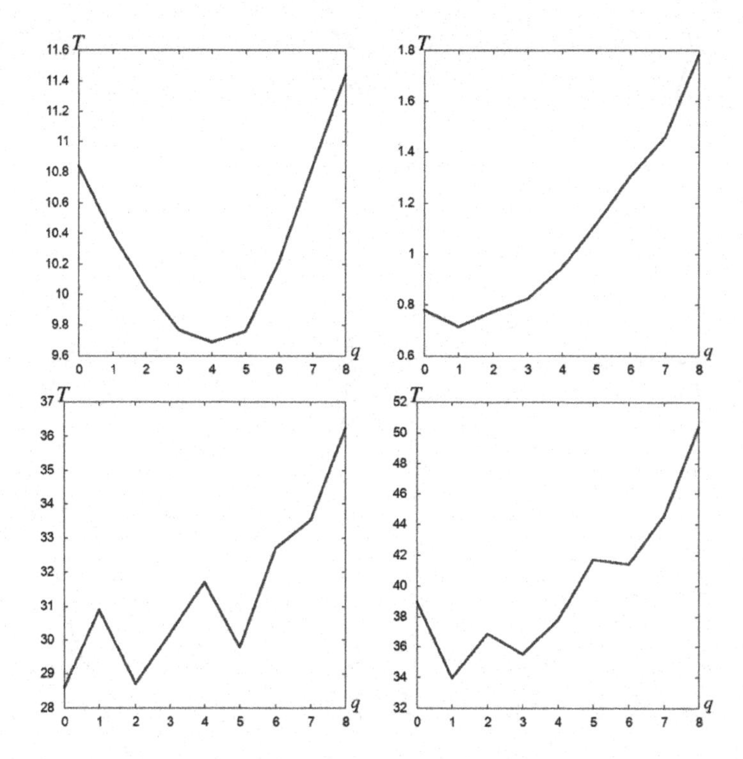

Fig. 4. Dependence of the solution time T on the overlap size q for different number of MPI cores for problems n25t0e(4) and n25t1e(192) (top); n85t4b(16) and n85t4e(16) (bottom).

quite reasonable speedup. Sometimes the speedup behaviour is not monotone, probably, due to computer cluster specifics.

In addition, it should be noted that the speedup increases with the simulation time. This is due to the flow development and the respective increase of the matrix conditioning that leads to the increase of the preconditioner density Dens. The preconditioner construction accelerates better due to greater arithmetics/communication ratio in comparison with the iterations.

6 BIILU2 Parameters Tuning

6.1 Tuning the Overlap Size q

The overlap size q influences the BIILU-type decomposition quality. The larger q, the better preconditioner quality is, the smaller number of BiCGStab iterations is, but the preconditioner construction time increases. Therefore, the above conflicting trends are to be balanced. In the present section we analyze optimal values of q minimizing the solution time.

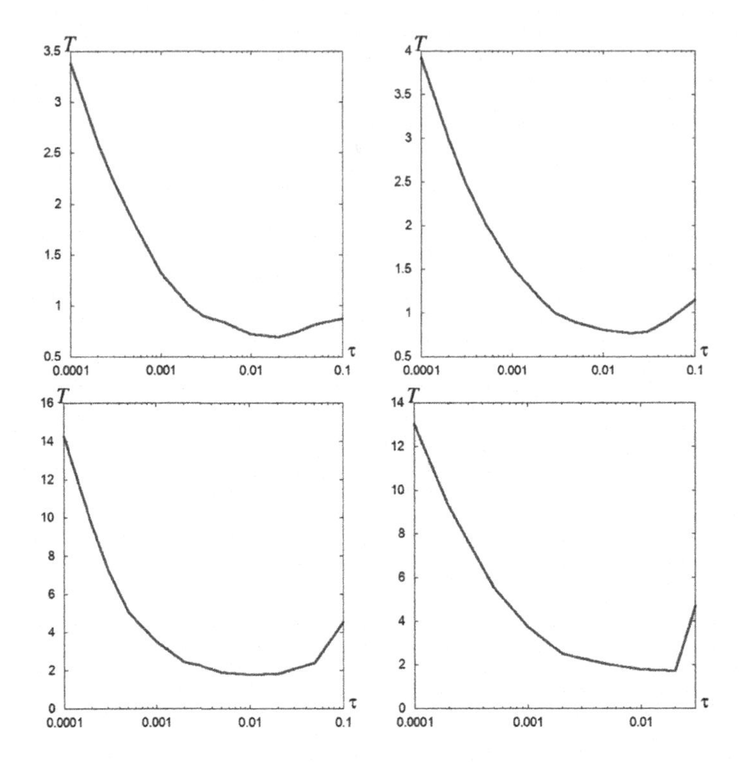

Fig. 5. Dependence of the solution time T on the threshold parameter τ for different number of MPI cores for problems n25t1b(192) and n25t2b(192) (top); n85t0b(192) and n85t0e(192) (bottom).

Figure 4 shows dependence of the solution time on parameter q. At the top and bottom rows we present the solution time of small (n25) and large (n85) problems, respectively, on different number of MPI cores. For brevity, we selected only small subset of performed experiments. The main conclusion from these experiments is that in most cases the use of the overlapping preconditioner ($q > 0$) is more efficient than the use of the nonoverlapping preconditioner ($q = 0$). Moreover, sometimes the nonoverlapping preconditioner may fail to converge leading to time step cutting. That is why we recommend to use the robust case $q = 3$.

6.2 Tuning the Threshold Parameter τ

The threshold parameter τ influences the quality of the approximate ILU2 factorization. The smaller τ, the better preconditioner quality is, the smaller number of BiCGStab iterations is, but the preconditioner construction time increases. Therefore, the above conflicting trends are to be balanced. In the present section we analyze optimal values of τ minimizing the solution time.

Figure 5 shows dependence of the solution time tuning on parameter τ. At the top and bottom rows we present the solution time of small (n25) and large

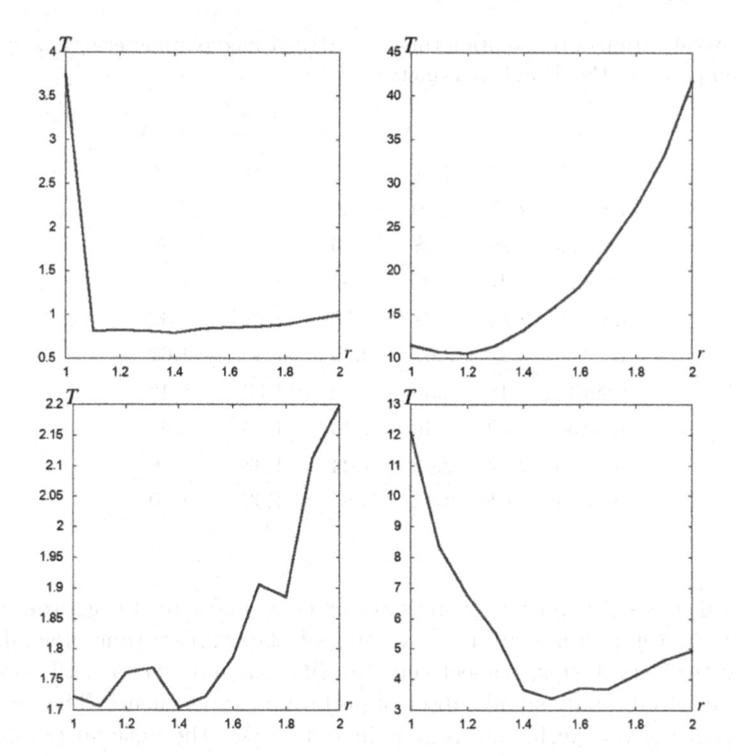

Fig. 6. Dependence of the solution time T on the factorization order parameter r for different number of MPI cores for problems n25t2b(192) and n25t2b(1) (top); n85t0b(192) and n85t4e(192) (bottom).

(n85) problems, respectively, on different number of MPI cores. For brevity, we selected only small subset of performed experiments. The main conclusion from these experiments is that the dependence on the threshold τ is smooth enough, and the exact optimum is less important. Sometimes less expensive preconditioner with large threshold τ provides better performance, but iterations may fail to converge leading to time step cutting. That is why we recommend to use sufficiently small threshold $\tau = 0.003$.

6.3 Tuning the Factorization Order r

The primary threshold τ and the secondary threshold τ_2 of the approximate two-parameter ILU2(τ, τ_2) factorization are introduced in Sect. 4. The secondary threshold $\tau_2 = \tau^2$ implies the second order ILU2 factorization, while $\tau_2 = \tau$ implies the standard first order incomplete ILUT factorization [4].

To span these types of incomplete triangular factorization, we introduce the factorization order parameter r, where $1 \le r \le 2$ and consider the secondary threshold $\tau_2 = \tau^r$. For $r = 1$ and $r = 2$ it corresponds to the first and the second order incomplete factorizations, respectively. The intermediate values $1 < r < 2$ also make sense for the analysis.

Table 2. Acceleration of the solution time due to the tuned parameters $q = 2$, $\tau = 0.01$, $r = 1.5$ compared to the default parameters.

Name	$p = 1$	$p = 4$	$p = 16$	$p = 64$	$p = 192$
n85t0b	2.00	2.27	1.97	1.41	2.16
n85t0e	2.05	2.10	1.69	1.86	4.56
n85t1b	1.95	1.68	1.75	1.48	1.80
n85t1e	2.27	2.09	1.81	1.64	1.75
n85t2b	2.10	2.31	1.73	1.74	3.32
n85t2e	2.62	2.67	2.29	1.81	1.67
n85t3b	2.48	2.30	1.39	1.53	1.42
n85t3e	2.70	2.40	2.45	1.46	2.93
n85t4b	2.52	2.02	1.24	1.33	1.26
n85t4e	3.61	3.12	3.51	1.92	1.50

Figure 6 shows dependence of the solution time on factorization order parameter r. At the top and bottom rows we present the solution time of small (n25) and large (n85) problems, respectively, on different number of MPI cores. For brevity, we selected only small subset of performed experiments. The main conclusion from these experiments is that in some cases the optimal parameter r may be chosen close to 1, with a risk of convergence failure. Our recommendation is the use of more robust value $r = 1.5$.

7 Conclusion

The open source massively parallel solver based on BIILU2 factorization is studied in terms of computational efficiency of the black oil reservoir simulation on 192 cores. The solver confirms its reliability and parallel efficiency.

The numerical tuning of the solver parameters suggests the use $q = 2$, $\tau = 0.01$, $r = 1.5$ for minimization of the solution time, although the most robust choice is the default choice $q = 3$, $\tau = 0.003$, $r = 2$. For all large systems n85* we summarize in Table 2 acceleration of the solution time T due to the use of tuned parameters $q = 2$, $\tau = 0.01$, $r = 1.5$ compared to the use of the default parameters.

Table 2 proves the importance of parameters tuning and existence of optimal parameters set in the parallel solution of linear systems arising in reservoir simulation.

Acknowledgements. The work was supported by Aramco Innovations.

References

1. Vassilevski, Y., Konshin, I., Kopytov, G., Terekhov, K.: INMOST - a software platform and graphical environment for development of parallel numerical models on general meshes. Lomonosov Moscow State University Publcation, Moscow, 144 p. (2013) (in Russian)
2. INMOST - a toolkit for distributed mathematical modeling. http://www.inmost.org/. Accessed 15 Apr 2020
3. Vassilevski, Y., Terekhov, K., Nikitin, K., Kapyrin, I.: Parallel Finite Volume Computation on General Meshes. Springer, Cham (2020). https://doi.org/10.1007/978-3-030-47232-0
4. Saad, Y.: Iterative Methods for Sparse Linear Systems, 2nd edn. Society for Industrial and Applied Mathematics (2003)
5. Benzi, M.: Preconditioning techniques for large linear systems: a survey. J. Comput. Phys. **182**(2), 418–477 (2002)
6. Aziz, K., Settari, A.: Petroleum Reservoir Simulation. Applied Sci. Publ. Ltd., London, 497 p (1979)
7. Chen Z., Huan G., Ma Y.: Computational Methods for Multiphase Flows in Porous Media. SIAM, Philadelphia, 549 p (2016)
8. Peaceman, D.W.: Fundamentals of Numerical Reservoir Simulation. Elsevier, New York, 176 p (1977)
9. INM RAS cluster. http://cluster2.inm.ras.ru/en. Accessed 15 Apr 2020
10. Bogachev, K.Y., Zhabitsky, Y.V.: The Kaporin-Konshin method of parallel implementation of block preconditioners for nonsymmetric matrices in problems of filtering a multicomponent mixture in a porous medium. Bulletin of Moscow University, Series 1: Mathematics, Mechanics, N. 1, pp. 46–52 (2010) (in Russian)
11. Wallis, J.: Incomplete Gaussian elimination as a preconditioning for generalized conjugate gradient acceleration, SPE paper 12265. In: SPE Reservoir Simulation Symposium, San Francisco, CA, USA (1983)
12. Kaporin, I.E.: High quality preconditioning of a general symmetric positive definite matrix based on its $U^TU + U^TR + R^TU$-decomposition. Numer. Linear Algebra Appl. **5**(6), 483–509 (1998)
13. Kaporin, I.E., Konshin, I.N.: Parallel solution of large sparse SPD linear systems based on overlapping domain decomposition. In: Malyshkin, V. (ed.) PaCT 1999. LNCS, vol. 1662, pp. 436–446. Springer, Heidelberg (1999). https://doi.org/10.1007/3-540-48387-X_45
14. Kaporin, I.E., Konshin, I.N.: A parallel block overlap preconditioning with inexact submatrix inversion for linear elasticity problems. Numer. Linear Algebra Appl. **9**(2), 141–162 (2002)
15. Kaporin, I.: Scaling, reordering, and diagonal pivoting in ILU preconditionings. Russ. J. Numer. Anal. Math. Modelling **22**(4), 341–375 (2007)
16. Konshin, I.N., Olshanskii, M.A., Vassilevski, Y.V.: ILU preconditioners for non-symmetric saddle-point matrices with application to the incompressible Navier-Stokes equations. SIAM J. Sci. Comput. **37**(5), A2171–A2197 (2015)
17. SPE Comparative Solution Project - dataset 2. http://www.spe.org/web/csp/datasets/set02.htm. Accessed 15 Apr 2020

Parallel Box-Counting Method for Evaluating the Fractal Dimension of Analytically Defined Curves

Ilya Pershin, Dmitrii Tumakov(✉), and Angelina Markina

Kazan Federal University, Kazan, Russia
dtumakov@kpfu.ru

Abstract. Serial and parallel implementations of the algorithm intended for calculating the number of boxes for evaluating the fractal dimension of analytical curves are considered. The algorithm contains four stages: (1) preparing the data for calculations; (2) determining the boxes into which the curve fell; (3) counting the boxes that have an intersection with a curve; (4) counting the boxes of a larger size that intersect with a curve. The acceleration of computations performed by a parallel code (on OpenMP and CUDA) with respect to calculations performed by a serial code depending on the size of the box is investigated. Numerical experiments are carried out, the results of which exhibit a significant increase in performance for GPU calculations in the case of a large number of segments of the curve. A 100-fold increase in the computational speed is obtained for a curve containing a million segments with a billion boxes (box size is 2^{-15}). The graphs depicting an increase in acceleration of parallel code performance with decreasing the box size and increasing the number of curve segments are shown. It is concluded that the efficiency of using the GPU begins with three million boxes and grows with an increase in the number of curve points.

Keywords: Box-counting method · Analytical curve · Parallelization · CUDA · OpenMP

1 Introduction

The fractal dimension is used in various fields of science and technology, for example, in geology of microstructures [1], analysis of seismic signals [2], ecology [3], as well as in engineering for modeling antennas [4]. Estimates of fractal dimensions are also used in information technology: in steganographia to allocate the features of signals [5], in the processing of large data streams [6] as well as in many other applications.

The fractal dimension itself is a metric used to characterize the filling of space with an object. To calculate the dimension, various algorithms are used [7]. The most common method is the box-counting method (BCM) that includes various modifications [8, 9]. The BCM for binary and gray scale digital images are well-explored. In addition, there exist methods for counting boxes for color two-dimensional [10] and three-dimensional images [11].

V. Voevodin and S. Sobolev (Eds.): RuSCDays 2020, CCIS 1331, pp. 86–97, 2020.
https://doi.org/10.1007/978-3-030-64616-5_8

The essence of both the BCM itself and its modifications is in counting the boxes covering the investigated image [12]. When the size of a single box reduces, the computational load increases significantly. There are not many fast BCMs. One such algorithm was proposed by Liebotich [13], which was later improved by Hou [14]. The idea of the algorithm is in a binary representation of the coordinates of each box and speeding up the counting of boxes by applying a block mask. This algorithm was optimized for the purpose of performing computations on a video card and applied to the task of calculating the dimension for three-dimensional objects [15]. A quite different algorithm is used for calculations with a uniform change in the grid step size on a logarithmic scale [16]. There are works in which an increase in computing performance is achieved through the use of the GPU [17–19].

However, the BCM calculation procedure is subject to errors arising from the arbitrary placement of the grid, and, as a consequence, incorrect counting of blocks [20–22]. This error is known as a quantization error and can vary as a function of scale, which makes the method problematic to use. This error is primarily affected by the discretization of the original image. In addition, for curved lines, errors arise due to changes in their thickness [23]. Due to these problems, an estimation of fractal dimension depends on the method used, which complicates the comparison of the results obtained by different algorithms (authors) for the same structure [24]. These remarks are, first of all, relevant for discrete curves, while, for curves defined analytically, the problems listed become less pronounced.

In the present paper, we propose a method for solving the problem of counting boxes for analytical curves. An assumption is also made that the box size may be real-valued. The initial problem is divided into four stages: (1) preparating the data for calculations; (2) determining the boxes into which the curve fell; (3) counting the boxes having an intersection with a curve; (4) counting larger boxes that intersect with the curve. All stages of solving the problem are parallelized using OpenMP and CUDA technologies.

Computational experiments are performed for Koch prefractals of the tenth iteration (for this curve, the number of segments is more than a million elements). The results of acceleration of each stage using the corresponding parallelization technology are presented. It is shown that the acceleration of individual stages on the CUDA reaches values of more than 300-fold, with a total of almost 100-fold acceleration of calculations. If one takes into account the time of copying data from the host to the device, then the acceleration of calculations on the GPU decreases to 58. It is concluded that the maximum acceleration is achieved for the minimum box sizes as well as for polylines consisting of a large number of equal-sized segments.

2 Description of Algorithm

We suppose that the input parameters of the problem are coordinates of the polyline. The box calculation algorithm consists of the following four stages:

1. Preparing the data for calculations.
2. Determining the boxes, into which the curve fell.

3. Counting the boxes having an intersection with the curve.
4. Fast algorithm for counting the number of larger boxes intersecting with the curve.

Let us consider each of the algorithms in more detail.

2.1 Stage 1. Preparing the Input Data

The input data (coordinates of vertices of a given polyline) is a sequence of $M + 1$ pairs of real numbers $\left(x^{(j)}, y^{(j)}\right)$. Coordinates of vertices are normed in such a way that their values are in the interval $[0, 1) \times [0, 1)$. The number of segments that connect the points in series is $num\_lines = M$.

For each pair of points, we construct a straight line equation $y_i = a_i x_i + b_i$, where a_i, b_i are constants. We form arrays *a* and *b* containing *num_lines* elements and fill them with the corresponding values a_i and b_i. We also form two-dimensional arrays *t* and *u* of length $2 * num\_lines$ and fill them with the corresponding pairs of values $\left\{x^{(j)}, x^{(j+1)}\right\}$ and $\left\{y^{(j)}, y^{(j+1)}\right\}$, where $x^{(j)} < x^{(j+1)}$, $y^{(j)} < y^{(j+1)}$.

In a parallel implementation, for each pair of points, the values of the arrays *a*, *b* are filled with separate threads. In a serial implementation, each pair of points is processed in a *for* loop. With OpenMP, this loop is parallelized in the usual way by adding the "#pragma omp parallel for" directive.

2.2 Stage 2. Counting the Number of Cells for Analytical Curves

The use of analytical curves actually allows considering a region in which the curve is defined as a continuous region. Thus, the classical box calculation algorithm used for discrete (pixel) regions must be applied here for continuous regions. This algorithm will be called the *continuous box counting* method or the method of *box counting in a continuous region.* Since the algorithm works with continuous regions, we assume that the box size ε is real-valued.

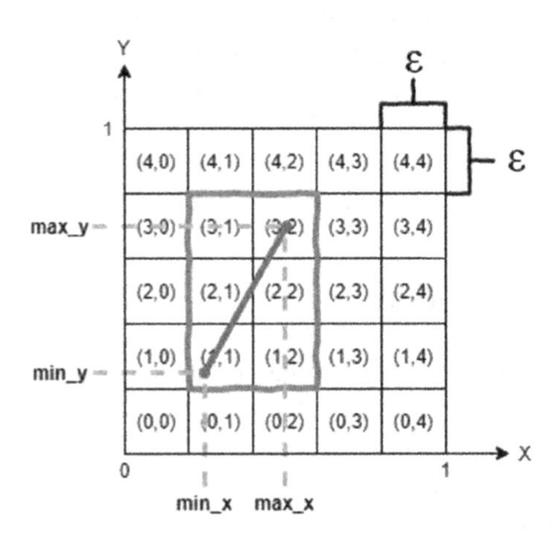

Fig. 1. Covering the lines by a grid with the step size ε.

The main idea behind applying a parallel approach to this stage is in considering each line separately and determining the boxes, with which the line intersects. Let us describe the algorithm in more detail.

First, we initialize with zeros the two-dimensional square array *boxes* of dimension *boxes_size* by *boxes_size*, where *boxes_size* $= \lceil 1/\varepsilon \rceil$. Here the notation $\lceil \ldots \rceil$ means rounding to a larger integer. The value of the element *boxes* with the index (i, j) indicates whether the box located at the intersection of the i-th row and the j-th column has an intersection with the curve. If *boxes*[i][j] = 1, then the box with index (i, j) has at least one intersection with the curve. If *boxes*[i][j] = 0, then this box has no intersections with the curve.

Next, for each line with boundary coordinates (*min_x*, *min_y*) and (*max_x*, *max_y*), we define a set of indices of boxes $D = \{(i, j)\}$, with which intersection is possible (see Fig. 1). Here, i changes from *start_i* to *end_i*, and j changes from *start_j* to *end_j*.

The extreme values of the range of indices of boxes will be calculated as

$$start\_i = \lfloor y_1/\varepsilon \rfloor, \quad end\_i = \lfloor y_2/\varepsilon \rfloor,$$
$$start\_j = \lfloor x_1/\varepsilon \rfloor, \quad end\_j = \lfloor x_2/\varepsilon \rfloor.$$

Here $\lfloor z \rfloor$ denotes the integer part of z. In the example in Fig. 1 we have that *start_i* = 1, *end_i* = 2, *start_j* = 1, *end_j* = 3.

For each box from the set D, we find the intersection with the line along the X-axis and the Y-axis. The coordinates (c_1, d_1), (c_2, d_2) for the box with index (i, j) are calculated as follows:

$$c_1 = j * \varepsilon, \quad c_2 = (j + 1) * \varepsilon,$$
$$d_1 = i * \varepsilon, \quad d_2 = (i + 1) * \varepsilon.$$

We assume that the box has an intersection with a line if and only if the following condition is satisfied

$$c_2 > x_1 \,\&\&\, c_1 \le x_2 \,\&\&\, d_2 > y_1 \,\&\&\, d_1 \le y_2$$

If the intersection condition is fulfilled for the box with index (i, j), then the corresponding element in the array *boxes* is assigned the value 1. The intersection condition is chosen so that the curve point lying on the border belongs to the box with "smaller" coordinates.

Parallelization on CUDA will be done by threads: each thread processes a separate line. If *num_lines* $\le$ 1024, then we select one block that contains the number of threads equal to *num_lines*. Otherwise, each block will contain 1024 threads, and the total number of blocks is equal to a larger integer from *num_lines*/1024. Listing 1 shows the function that implements on CUDA the determination of boxes intersecting with a polyline.

```
__global__
void CBC(float* k, float* b, float* t, float* u, unsigned
char *boxes, unsigned int boxes_size, int num_lines,
float eps) {
  unsigned int num_line = blockIdx.x * blockDim.x +
threadIdx.x;

 if (num_line < num_lines) {
  // Extreme points of the item line
  float min_x = *(t + num_line * 2);
  float max_x = *(t + num_line * 2 + 1);
  float min_y = *(u + num_line * 2);
  float max_y = *(u + num_line * 2 + 1);

  // Box number range
  int start_i = min_y / eps;
  int start_j = min_x / eps;
  int end_i = max_y / eps;
  int end_j = max_x / eps;

  float x1, x2, y1, y2;
  bool intersectX, intersectY;
  for (int i = start_i; i <= end_i; i++) {
    for (int j = start_j; j <= end_j; j++) {
      x1 = j * eps;
      x2 = (j + 1) * eps;

      if (x1 < min_x) x1 = min_x;
      if (x2 > max_x) x2 = max_x;

      y1 = *(k + num_line)*x1 + *(b + num_line);
      y2 = *(k + num_line)*x2 + *(b + num_line);

      if (y1 > y2) {
        float temp = y1;
        y1 = y2;
        y2 = temp;
      }

      intersectX = (j + 1) * eps > x1 && x2 >= j * eps;
      intersectY = (i + 1) * eps > y1 && y2 >= i * eps;

      char intersectXY = intersectX && intersectY;
      if (intersectXY > 0)*(boxes + boxes_size*i + j)=1;
    }
   }
  }
}
```

Listing 1. The kernel of CUDA for determining the boxes intersecting with the curve.

For the serial implementation of this stage on the CPU, a loop is formed over all numbers of the lines, the body of which is identical to the body of the function given in Listing 1. An implementation for OpenMP is obtained by parallelizing the outer loop.

2.3 Stage 3. Counting the Number of Boxes Intersecting with the Curve

In the previous stage, the array *boxes* was filled with either 1 (there are intersections) or 0 (no intersections). Therefore, the implementation of counting the number of boxes with intersection is carried out by summing the values of the array *boxes*.

Thus, in the CUDA, summation can be done using any reduction algorithm, and in the OpenMP, the "#pragma omp parallel for reduction" directive can be used for that purpose.

2.4 Stage 4. Fast Counting of Intersected Large Boxes

The dimension *num_boxes* of the array *boxes* is chosen equal to the degree of two (let $min\varepsilon = 2$). At each next step, we assume that the grid step $\in$ is the next power of two. Thus, each next box can be composed of four boxes of the previous step, the values of which are known to us. This method of selecting box sizes allows for more efficient implementation of calculations without having to resort to a full scan of the elements of the matrix *boxes* each time.

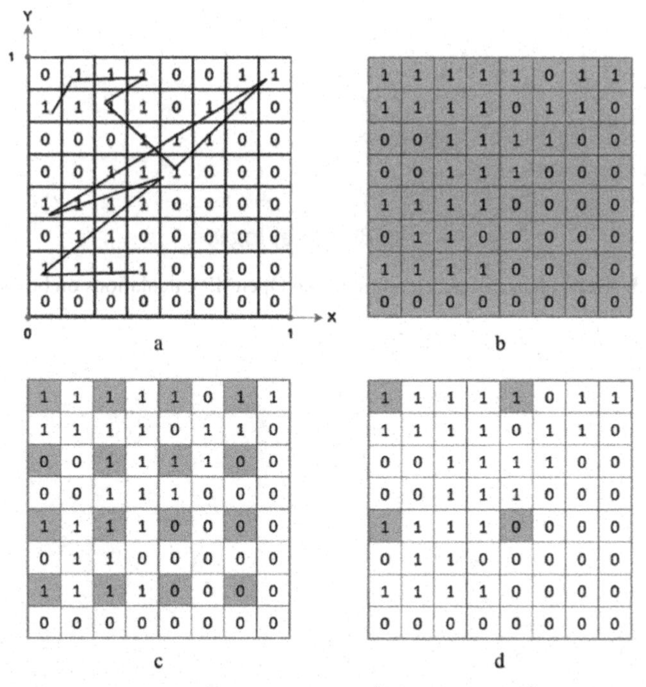

Fig. 2. (a) initial matrix *boxes*, (b) step $\in = 2^1$, (c) step $\in = 2^2$, (d) step $\in = 2^3$.

At the first iteration of the algorithm, we cover the original matrix *boxes* (see Fig. 2a), obtained as a result of the second stage of the algorithm, with a grid with a step $\in = 2^1$, that is, boxes containing 2×2 elements. If at least one element from the box $\in \times \in$ has a value not equal to 0, then we write the value 1 in the upper left element of this box. We repeat this procedure for each box. As a result, we obtain the result shown in Fig. 2b.

At the second iteration, we cover the array *boxes*, modified at the first iteration, with a grid with a step $\in = 2^2$. Boxes for this grid are made up of counted boxes of the previous iteration, so it is sufficient to consider the values of the upper left elements of the boxes from the previous iteration as elements of this box. As a result of the algorithm, we obtain an array presented in Fig. 3c. The next iteration converts the array to the form shown in Fig. 2d.

We will repeat this procedure for different values of $\in$ while the condition $\in \leq$ *num_boxes*/2 is fulfilled. Note the equality $\varepsilon = \in^{-1}$. The algorithm that implements this step on CUDA is given in Listing 2.

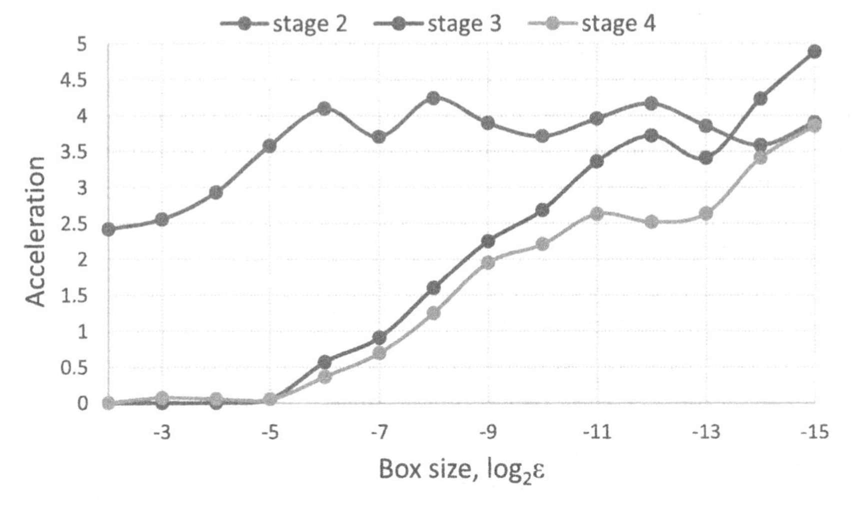

Fig. 3. Dependence of acceleration of the OpenMP calculations on box size.

```
__global__
void secondCountN(unsigned char* boxes, unsigned int box-
es_size, unsigned int* N, int eps, int iter) {
  unsigned int start_cell_i = (blockDim.y * blockIdx.y +
threadIdx.y) * eps;
  unsigned int start_cell_j = (blockDim.x * blockIdx.x +
threadIdx.x) * eps;

  unsigned int sum = 0;

  for (int y = 0; y < 2; y++) {
    for (int x = 0; x < 2; x++) {
      unsigned int i = start_cell_i + (y << iter - 1);
      unsigned int j = start_cell_j + (x << iter - 1);

      sum += *(boxes + boxes_size * i + j) > 0;
    }
  }

  if (sum > 0) {
    *(boxes + boxes_size * start_cell_i + start_cell_j)=
1;
    atomicAdd(&(*(N + iter)), 1);
  }
}
```

Listing 2. The kernel of CUDA for determining the boxes that intersect with the curve.

3 Numerical Results

Numerical experiments were conducted in Visual Studio 2019 on a computer with the characteristics shown in Table 1.

Table 1. Characteristics of computer

OS	Windows 10 Pro ×64
GPU model	Nvidia Quadro p3000
GDDR capacity	6 GB
CPU model	Intel Core i7-7700HQ 2.80 GHz
RAM	32 GB

Table 2 shows the execution time of the input data preparation stage (first stage) and the acceleration coefficients for OpenMP and CUDA relative to the serial implementation on the CPU for various values of ε. Calculations were performed for the tenth iteration of the Koch prefractal (the number of lines is 1048576). Also note that the fractal dimension values calculated by the sequential and both parallel approaches are the same. In the case of taking into account the values of $\in$ from 2^{-14} to 2^{-7}, the resulting dimension is 1.26029 (the error from the analytical value is 0.00157).

Table 2. Average execution time in ms of the first stage and acceleration relative to the serial code on the CPU for 1 million points

ε	Execution time (ms)			Acceleration	
	CPU	OpenMP	GPU	OpenMP, A_{MP}	GPU, A_G
2^{-15}	19.86	4.98	0.26	3.99	76.09
2^{-11}	17.63	5.47	0.29	3.22	61.86
2^{-7}	16.29	5.67	0.30	2.87	53.53
2^{-2}	16.76	5.72	0.31	2.93	53.42

Now let us evaluate the performance of the remaining three stages of our algorithm. At first, let us consider the acceleration of A_{MP} performance using OpenMP technology. Figure 3 presents graphs showing the acceleration coefficients of parallel OpenMP code over serial code.

It can be noted that the acceleration of the second stage (blue curve) of the algorithm monotonically increases from 2.5 to 4.1 with a decrease in the size of the box from 2^{-2} to 2^{-6} (or in other words, with an increase in the dimension of the array *boxes*), and then acceleration varies between 3.6 and 4.2. In the case of the third (red) and fourth (green curve) stages, OpenMP code acceleration emerges with a box size of 10^{-5}, and then increases to the minimum considered box sizes ($\varepsilon = 2^{-15}$), reaching 4.9 and 3.9, respectively.

Now let us consider the increase in V_G performance in GPU computing. For the second stage (blue curve), the V_G values vary from 80 to 90 for box sizes smaller than 10^{-12}, then the acceleration increases sharply to 340. As in the case of OpenMP, in the second and third stages, acceleration does not emerge immediately, but only with ε equal to 2^{-7}. After this, the V_G values increase to 53 and 196 (Fig. 4).

We will evaluate the performance of our algorithm for CUDA as a whole, taking into account all 4 stages. The A_G values for a million points are between 60 and 70 for $\varepsilon > 2^{-10}$ and slowly increase to 98 with a decrease in ε to 2^{-15}. Given the copying of an array of points from the host to the device and initializing all arrays (49.5 ms), the algorithm runs in time on the GPU more slowly for larger ε. Starting from 2^{-11}, the values of A_G sharply increase from 1 to 58.

We also consider the dependence of the growth of A_G performance on the GPU on the size of the array of points. Figure 5 shows the graphs of the dependence of the acceleration of calculations on the GPU on the number of segments of the polyline M. Starting from the values $M = 10^4$, the productivity growth will not be significant: from 53 to 58 with copying and from 90 to 98 without taking into account the time spent on copying.

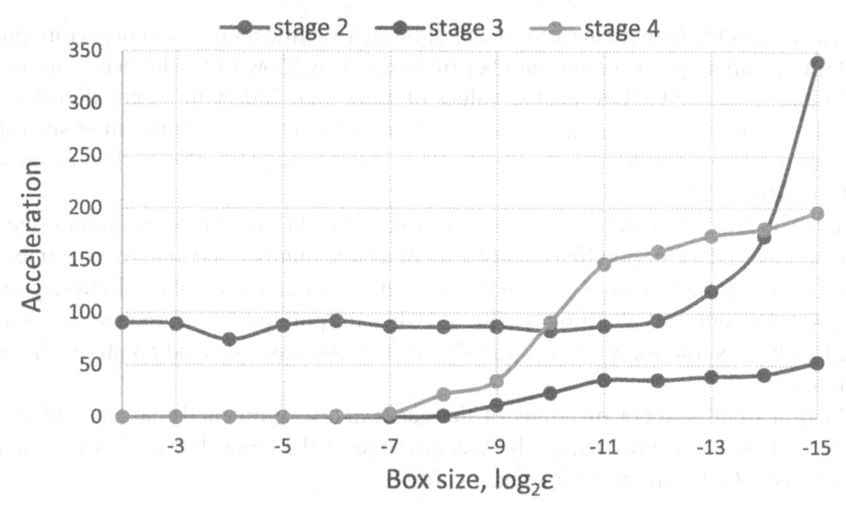

Fig. 4. Dependence of acceleration of the GPU calculations on the size of the box. (Color figure online)

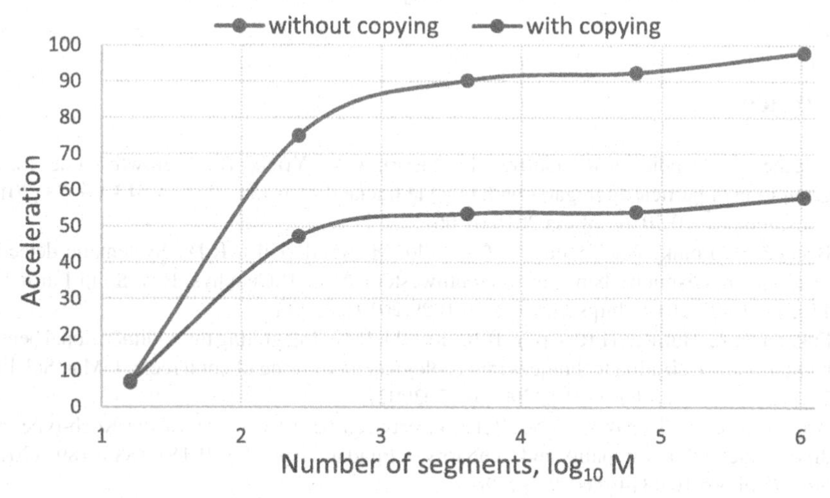

Fig. 5. Dependence of acceleration of the GPU calculations on the number of segments of the polyline. (Color figure online)

4 Conclusion

The paper proposes an algorithm for counting boxes to assess the fractal dimension of analytical curves. The algorithm contains four stages. The main advantages of the proposed approach are as follows. First, the curves are set analytically and the calculations do not contain discretization errors associated with the line thickness. Secondly, a fast algorithm for counting large boxes containing counted small boxes is applied.

The conducted numerical experiments showed a significant increase in performance for GPU computing for a large number of boxes. It is shown that the acceleration of individual stages on CUDA reaches values of more than 300, with a total almost 100-fold acceleration of the calculation speed. If we take into account the time spent on copying data from the host to the device, then the acceleration of calculations on the GPU reduces to 58.

It is concluded that the maximum acceleration is achieved for the minimum size of the boxes as well as for polylines, consisting of a large number of equal-sized segments. GPU efficiency starts with $3 \cdot 10^6$ boxes and increases as the number of curve points increases. Optimum acceleration is achieved when the polyline has segments of the same length (which is true for the Koch fractal), since in this case the load on all the threads is the same.

With a small number of points of the curve and a significantly larger number of boxes, it makes sense to change the second stage of the algorithm so that a separate thread on the GPU will process a separate box.

Acknowledgements. The work is performed according to the Russian Government.
Program of Competitive Growth of Kazan Federal University.
The reported study was funded by RFBR, project number 19-31-90129.

References

1. Guesnet, E., Dendievel, R., Jauffrès, D., Martin, C.L., Yrieix, B.: A growth model for the generation of particle aggregates with tunable fractal dimension. Phys. A **513**, 63–73 (2019). https://doi.org/10.1016/j.physa.2018.07.061
2. Beaucé, E., Frank, W.B., Paul, A., Campillo, M., vanderHilst, R.D.: Systematic detection of clustered seismicity beneath the Southwestern Alps. J. Geophys. Res. Solid Earth 124, 11531–11548 (2019). https://doi.org/10.1029/2019JB018110
3. Fukunaga, A., Burns, J.H.R., Craig, B.K., Kosaki, R.K.: Integrating three-dimensional benthic habitat characterization techniques into ecological monitoring of coral reefs. J. Mar. Sci. Eng. **7**(2) (2019). https://doi.org/10.3390/jmse7020027
4. Abgaryan, G.V., Tumakov, D.N.: Relation between base frequency of the Koch-type wire dipole, fractal dimensionality and lacunarity. J. Fundam. Appl. Sci. **9**(1S), 1885–1898 (2017). https://doi.org/10.4314/jfas.v9i1s.828
5. Mohtasham-zadeh, V., Mosleh, M.: Audio Steganalysis based on collaboration of fractal dimensions and convolutional neural networks. Multimed. Tools Appl. **78**(9), 11369–11386 (2018). https://doi.org/10.1007/s11042-018-6702-1
6. Folino, G., Guarascio, M., Papuzzo, G.: Exploiting fractal dimension and a distributed evolutionary approach to classify data streams with concept drifts. Appl. Soft Comput. **75**, 284–297 (2018). https://doi.org/10.1016/j.asoc.2018.11.009
7. Le Mehaute, A.: Fractal Geometries Theory and Applications. CRC Press, Boca Raton FL (1991)
8. Fernández-Martínez, M.: A survey on fractal dimension for fractal structures. Appl. Math. Nonlin. Sci. 1(2), 437–472 (2016). https://doi.org/10.21042/AMNS.2016.2.00037
9. Nayak, S.R., Mishra, J., Khandual, A., Palai, G.: Analysing roughness of surface through fractal dimension: a review. Image Vision Comput. **89**, 21–34 (2019)

10. Nayak, S.R., Mishra, J.: An improved method to estimate the fractal dimension of colour images. Perspect. Sci. **8**, 412–416 (2016). https://doi.org/10.1016/j.pisc.2016.04.092
11. Backes, A.R., Eler, D.M., Minghim, R., Bruno, O.M.: Characterizing 3D shapes using fractal dimension. In: Bloch, I., Cesar, R.M. (eds.) CIARP 2010. LNCS, vol. 6419, pp. 14–21. Springer, Heidelberg (2010). https://doi.org/10.1007/978-3-642-16687-7_7
12. Keller, J.M., Chen, S., Crownover, R.M.: Texture description and segmentation through fractal geometry. Comput. Vision Graph. **45**(2), 150–166 (1989). https://doi.org/10.1016/0734-189X(89)90130-8
13. Liebotich, L.S., Toth, T.: A fast algorithm to determine fractal dimensions by box counting. Phys. Lett. A **141**(8), 386–390 (1989). https://doi.org/10.1016/0375-9601(89)90854-2
14. Hou, X., Gilmore, R., Mindlin, G.B., Solari, H.G.: An efficient algorithm for fast O(N*ln(N)) box counting. Phys. Lett. A **151**, 43–46 (1990). https://doi.org/10.1016/0375-9601(90)90844-E
15. Jiménez, J., Ruiz de Miras, J.: Box-counting algorithm on GPU and multi-core CPU: an OpenCL cross-platform study. J. Supercomput. (65), 1327–1352 (2013). https://doi.org/10.03239/s11227-013-0885-z
16. Biswas, M., Ghose, T., Guha, S., Biswas, P.: Fractal dimension estimation for texture images: A parallel approach. Pattern Recogn. Let. **19**, 309–313 (1998). https://doi.org/10.1016/S0167-8655(98)00002-6
17. Ruiz de Miras, J., Jiménez Ibáñez, J.: Methodology to increase the computational speed to obtain the fractal dimension using GPU programming. In: Di Ieva, A. (ed.) The Fractal Geometry of the Brain. SSCN, pp. 533–551. Springer, New York (2016). https://doi.org/10.1007/978-1-4939-3995-4_34
18. Ruiz de Miras, J.: Fast differential box-counting algorithm on GPU. J. Supercomput. **76**(1), 204–225 (2019). https://doi.org/10.1007/s11227-019-03030-1
19. Xie, H., Wang, Q., Ni, J., et al.: A GPU-based prediction and simulation method of grinding surface topography for belt grinding process. Int. J. Adv. Manuf. Tech. 106, 5175–5186 (2020). https://doi.org/10.1007/s00170-020-04952-4
20. Foroutan-pour, K., Dutilleul, P., Smith, D.: Advances in the implementation of the box-counting method of fractal dimension estimation. Appl. Math. Comput. **105**, 195–210 (1999). https://doi.org/10.1016/S0096-3003(98)10096-6
21. Gonzato, G., Mulargia, F., Ciccotti, M.: Measuring the fractal dimensions of ideal and actual objects: implications for application in geology and geophysics. Geophys. J. Int. **142**, 108–111 (2000). https://doi.org/10.1046/j.1365-246x.2000.00133.x
22. Da Silva, D., Boudon, F., Godin, C., Puech, O., Smith, C., Sinoquet, H.: A Critical appraisal of the box counting method to assess the fractal dimension of tree crowns. In: Bebis, G., Boyle, R., Parvin, B., Koracin, D., Remagnino, P., Nefian, A., Meenakshisundaram, G., Pascucci, V., Zara, J., Molineros, J., Theisel, Holger, Malzbender, Tom (eds.) ISVC 2006. LNCS, vol. 4291, pp. 751–760. Springer, Heidelberg (2006). https://doi.org/10.1007/11919476_75
23. Pershin, I., Tumakov, D.: On optimal thickness of the curve at calculating the fractal dimension using the box-counting method. J. Comput. Theor. Nanos. **16**, 5233–5237 (2019). https://doi.org/10.1166/jctn.2019.8592
24. Miloevic, N.T., Rajkovic, N., Jelinek, H.F., Ristanovic, D.: Richardson's method of segment counting versus box-counting. In: 19th International Conference on Control Systems and Computer Science, pp. 299–305, CSCS 2013. Bucharest (2013). https://doi.org/10.1109/CSCS.2013.52

Parallel Gravitational Search Algorithm in Solving the Inverse Problem of Chemical Kinetics

Leniza Enikeeva[1,2(✉)], Mikhail Marchenko[3], Dmitrii Smirnov[3], and Irek Gubaydullin[2,4]

[1] Novosibirsk State University, Novosibirsk, Russia
leniza.enikeeva@yandex.ru
[2] Ufa State Petroleum Technological University, Ufa, Russia
irekmars@mail.ru
[3] Institute of Computational Mathematics and Mathematical Geophysics of Siberian Branch of Russian Academy of Sciences, Novosibirsk, Russia
marchenko@sscc.ru, smirnovdd@mail.ru
[4] Institute of Petrochemistry and Catalysis, RAS, Ufa, Russia

Abstract. The article describes a parallel gravitational search algorithm and its application to solving the inverse problem of chemical kinetics. The relevance of the study of metaheuristic algorithms, including the gravitational search algorithm, is given. It is shown that recently, these algorithms are becoming increasingly popular. The optimization problem is formulated on the example of solving the inverse kinetic problem. The process under study is propane pre-reforming over Ni catalyst, which is an industrially important chemical process. The description of the algorithm and its pseudocode are presented, after which the performance of the gravitational search algorithm is compared with other metaheuristic methods. The algorithm demonstrated its competitiveness, as a result of which it was applied to solve a specific industrial problem. Using this algorithm, the direct and inverse problems of chemical kinetics are solved, and the optimal values of the kinetic parameters of the reaction are found. It is proved that the model correctly describes the available experimental data.

Keywords: Global optimization · Metaheuristic algorithm · Mathematical modeling · Chemical kinetics · Gravitational search algorithm · Parallel computing technologies

1 Introduction

Optimization problems can be found in almost all engineering fields. Thus, the development of optimization techniques is very essential for engineering applications. Most of the conventional optimization techniques require the gradient information and hence they cannot be used to solve non-differentiable functions. Moreover, such techniques usually suffer from getting trapped in a local optimum in solving complex optimization problems

V. Voevodin and S. Sobolev (Eds.): RuSCDays 2020, CCIS 1331, pp. 98–109, 2020.
https://doi.org/10.1007/978-3-030-64616-5_9

with many local optima. However, many real-world engineering optimization problems are very complex, whose objective functions usually have more than one local optima. The drawbacks of conventional optimization techniques have encouraged researchers to develop better optimization methods to solve real-world engineering optimization problems [1].

Recently, many metaheuristics algorithms are successfully being applied for solving intractable problems. The term *metaheuristic* describes higher-level heuristics that are proposed for the solution a wide range of optimization problems. The appeal of using these algorithms for solving complex problems is that they obtain the best/optimal solutions even for very large problem sizes in small amounts of time. Metaheuristic algorithms provide a practical and elegant solution to many such problems and are designed to achieve approximate/optimal solutions in practical execution times for NP-Hard optimization problems [2]. A growing interest has been observed in metaheuristic methods over the last two decades. Figure 1 gives the search result of the number of related studies for the metaheuristics on google scholar website (in May 2019). The algorithms are: Genetic Algorithms (GA), Particle Swarm Optimization (PSO), Tabu Search (TS), Genetic Programming (GP), Differential Evolution (DE), Ant Colony Optimization (ACO), Simulated Annealing (SA), Artificial Bee Colony (ABC), Greedy Randomized Adaptive Search Procedure (GRASP), Variable Neighborhood Search (VNS), Firefly Algorithm (FA), Cuckoo Search (CS), Harmony Search (HS), Scatter Search (SS), Social Spider Optimization (SSO), Bacterial Foraging (BFO), Bat Algorithm (BA), Gravitational Search Algorithm (GSA) and Biogeography-based Optimization (BBO). GA and PSO have the largest numbers 1,270,000 and 263,000 related papers respectively.

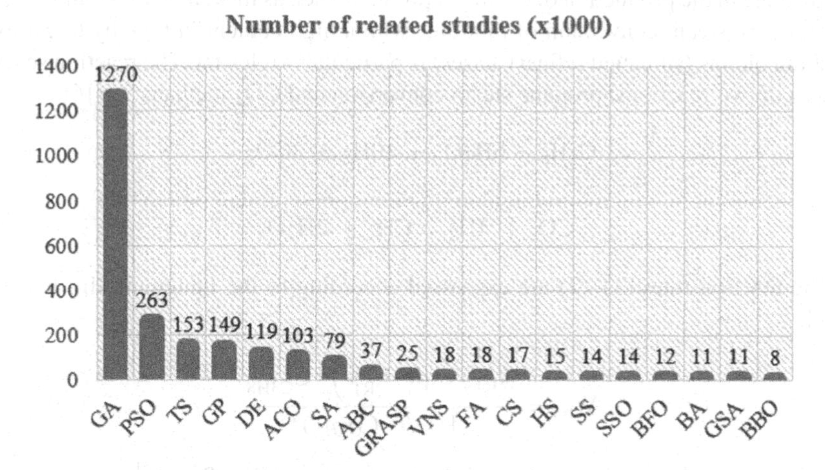

Fig. 1. The number of related papers on google scholar for the metaheuristic.

At present, metaheuristic methods have been used successfully to solve a lot of engineering optimization problems, such as multi-robot path planning, unmanned aerial vehicles navigation, the opinion leader detection in online social network, the identification of influential users in social network; the deployment of unmanned aerial vehicles, the data collection system of Internet of Things, the localization in wireless sensor network localization. In this article, we will consider the application of a metaheuristic algorithm, namely, the gravitational search algorithm, to solve the inverse problem of chemical kinetics. In recent years, some metaheuristic optimization algorithms, such as Particle Swarm Optimization [3], Genetic Algorithms [4, 5], have been applied to solve chemical kinetics problems. However, there is still no persistent conclusion to select a certain algorithm to solve inverse problems of chemical kinetics.

This research is aimed at investigating the effectiveness of metaheuristic algorithms in solving inverse kinetic problems. This work is devoted to the implementation of a parallel gravitational search algorithm for finding the kinetic parameters of an important industrial chemical process. GSA is one of the powerful metaheuristic algorithms currently available that is utilized to solve numerous applications of optimization problems. Furthermore, researchers have proposed a large diversity of methods to improve GSA, such as using enhanced operators, hybridization of GSA with other heuristic algorithms, and parameter adaptation and control schemes for GSA. The literature search showed that there is no work on applying the gravitational search algorithm to chemical kinetics problems now, which indicates the relevance of the research.

2 Mathematical Model

The process under study is propane pre-reforming into methane-rich gas over Ni catalyst, which is an industrially important chemical process. Steam conversion of hydrocarbons is widely used in the modern chemical industry to produce synthesis gas, which is an intermediate in the production of chemical products such as methanol, ammonia, etc. for oil refineries, steam conversion of mixtures with a high content of C_2+-hydrocarbons, including olefins (so-called refinery gases) is particularly relevant. The reaction scheme consists of two reactions: propane steam conversion and CO_2 methanation [6]:

$$C_3H_8 + 6H_2O \rightarrow 10H_2 + 3CO_2 \tag{1}$$

$$CO_2 + 4H_2 = CH_4 + 2H_2O \tag{2}$$

The reaction rates (1)–(2) are expressed according to the Langmuir-Hinshelwood model:

$$W_{ref} = \frac{k_{ref} \cdot \exp\left(-\frac{E_{ref}}{RT}\right) \cdot C_{C3H8}}{(1 + B \cdot C_{C3H8})^m}$$

$$W_{met} = k_{met} \cdot \exp\left(-\frac{E_{met}}{RT}\right) \cdot C_{H_2} \cdot \left[1 - \frac{P_{CH_4} P_{H_2O}^2}{K_{eq} P_{CO_2} P_{H_2}^4}\right]$$

where W_{ref} and W_{met} are the reaction rates; E_{ref} and E_{met} are the observed activation energies, J/mol; k_{ref} and k_{met} are the pre-exponential multipliers. The "ref" and "met" indexes refer to pre-forming and methanation reactions, respectively. C_{C3H8} and C_{H2} are concentrations of propane and hydrogen, mol/m^3; n is an order of reaction for propane, which varied from 0 to 2; K_{eq} is the equilibrium constant of CO_2 methanation; P_{CH4}, P_{H2O}, P_{CO2}, P_{H2} are partial pressures of the corresponding substances, bar. The mathematical model is a system of equations of material balance:

$$\begin{cases} G\frac{dy_i}{dl} = \left(v_i^{ref} W_{ref} + v_i^{met} W_{met}\right) m_i \\ 0 \le l \le L, i \in \{C_3H_8, CH_4, H_2O, H_2, CO_2\} \\ l = 0: \ y_i = y_{i0}, \end{cases} \tag{3}$$

where G is a mass flow of the mixture, kg/(m^2·sec); y_i is a mass fraction of the i-th component; ν_i is a stoichiometric coefficient of the i-th component; m_i is a molar mass of the i-th component, kg/mol; l is coordinate along the catalytic layer, m; L is a length of the catalytic layer, m. The mathematical model of chemical kinetics problems is a system of differential equations that describes the variations in substance concentrations over time according to the rates of reaction stages. The system of differential equations is a Cauchy problem containing the initial data [7]. The numerical solving of such a system of equations is a direct problem of chemical kinetics. Determining the kinetic parameters of reaction stages by comparing calculated values of substance concentrations and experimental results is an inverse problem of chemical kinetics. The mathematical problem is to minimize the functional of the deviation between calculated and experimental values. The functional of minimizations determined as the sum of absolute deviations between calculated and experimental concentrations:

$$F = \sum_{i=1}^{M} \sum_{j=1}^{N} \left| x_{ij}^{calc} - x_{ij}^{\exp} \right| \to \min, \tag{4}$$

where x_{ij}^{calc} and x_{ij}^{exp} are calculated and experimental values of component concentrations; M is the number of measuring points; N is the number of substances involved in the reaction.

3 The Framework of Proposed Algorithm

In this section, the algorithms used in this paper are briefly introduced. We consider Gravitational search algorithm for solving the inverse problems of chemical kinetics. GSA has been recently developed and found to be comparatively efficient. This algorithm is nature inspired, and population-based. A brief description of the algorithm, pseudocode, and parallel implementation will be given.

3.1 Gravitational Search Algorithm (GSA)

Rashedi et al. (2009) proposed GSA [8]. In this metaheuristic, search agents are objects and their success is proportional to their masses. The objects pull one another by the force of gravity. This force causes the movement of light agents toward heavier mass agents. The communication of agents is provided through gravitational force. The exploitation for the GSA is guaranteed by heavy masses that move slowly. Each object has a position, an inertial mass, a passive and an active gravitational mass. Each object represents a solution that is directed by setting the gravitational and inertia masses. The heaviest agent is the current best solution and other agents are attracted by this agent. GSA applies the Newtonian laws of gravitation and motion. Each object attracts every other one and the gravitational force between two objects is proportional to the product of their masses and inversely proportional to the distance between them, R. In order to be computationally effective, GSA uses the value R instead of R^2. The law of motion is that the current velocity is equal to the total sum of the fraction of its previous velocity and the change in the velocity. I.e. in an environment with N objects, the position of object i is:

$$X_i = X_i^1, \ldots, X_i^d, \ldots, X_i^n, for\, i = 1, 2, \ldots, N,$$

where X_i^d is the position of object i in the d-th dimension. The force applied to object "i" from agent "j" at time t is:

$$F_{ij}^d(t) = G(t)\frac{M_{pi}(t)M_{aj}(t)}{R_{ij}(t) + \varepsilon}\left(X_j^d(t) - X_i^d(t)\right),$$

where M_{aj} is the gravitational mass applied to agent j, M_{pi} is the passive gravitational mass applied to agent i, $G(t)$ is gravitational at time t, ε is a small constant, and $R_{ij}(t)$ is the Euclidian distance between objects i ($i = 1, 2, \ldots, N$) and j ($j = 1, 2, \ldots, N$):

$$R_{ij}(t) = \left\|X_i(t), X_j(t)\right\|_2.$$

The total force that is applied to object i in d is a random sum of d-th components of the forces from other objects:

$$F_i^d(t) = \sum_{j=1, j\neq i}^{N} rand_j F_{ij}^d(t), \tag{5}$$

where $rand_j$ is a number in [0,1]. The acceleration of the object i at time t, and in direction d-th, $a_i^d(t)$ is given as:

$$a_i^d(t) = \frac{F_{ij}^d(t)}{M_{ii}^d(t)},$$

where M_{ii} is the inertial mass of object i. The new velocity of an object is a fraction of its current velocity and its acceleration. Its position and velocity are calculates as follows:

$$v_i^d(t+1) = rand_i v_i^d(t) + a_i^d(t)$$
$$x_i^d(t+1) = x_i^d(t) + v_i^d(t+1),$$

where $rand_i$ is a uniform variable in [0,1]. The constant, G, is initialized and reduced with time to control the accuracy of the search:

$$G(t) = G(G_0, t) \tag{6}$$

A heavier mass indicates an efficient object (agent). Better solutions are represented as heavier objects that have higher attractions and move more slowly. The gravitational and inertial masses are updated by the equations given below:

$$M_{ai} = M_{pi} = M_{ii} = M_i;\ i = 1,\ 2,\ \ldots,\ N;$$

$$m_i(t) = \frac{fit_i(t) - worst(t)}{best(t) - worst(t)},\ M_i(t) = \frac{m_i(t)}{\sum_{j=1}^{N} m_j(t)},$$

where $fit_i(t)$ represents the fitness value of the agent i at time t, and, worst(t) and best(t) are defined as follows (for a minimization problem):

$$bext(t) = \min_{j \in 1,\ldots,N} fit(t),\ worst(t) = \max_{j \in 1,\ldots,N} fit_j(t)$$

The pseudocode of GSA is given in Table 1 [2].

Table 1. Pseudocode of Gravitational Search Algorithm

```
Generate initial population
while (t < iterations) do
    Calculate the fitness of all search agents
    Update G(i), best(i), worst(i) for i=1, 2, …, N.
    Calculation of acceleration and Mi(t) or each agent i
    Update velocity and position
    t = t + 1
Return the best solution
```

3.2 Parallel Implementation

Gravity search algorithms have great computational complexity, which in many applications becomes even greater due to the complexity of computing the objective function to be optimized. On the other hand, the gravitational search algorithm has a high degree of built-in parallelism, since many operations in it are performed in parallel or on individual particles. These facts make the gravitational search algorithm an almost ideal object for implementation on parallel computing systems.

There are several ways to parallelize the gravitational search algorithm: algorithms using the island model of parallelism and algorithms using the global model of parallelism.

Parallel algorithms based on the island model of parallelism are as follows: a multipopulation $S = S_1 \cup \ldots \cup S_p$ is created, consisting of the number of subpopulations (islands) S_i, equal to the number of processors P of the parallel computing system. Each processor processes its own island. We exchange data between the islands after each

t_m independent iterations in accordance with the used island communication topology. Each of the S_i islands has its own local archive sets of non-dominant functional values and corresponding points, and global archive sets of non-dominated functional values and corresponding points are stored on the host processor. The island model uses the static load balancing of the parallel computing system and provides high performance under the following conditions: the parallel computing system must have the same types of computing units; the sizes of the subpopulations are the same and quite large; A condition for completing iterations of each of the subpopulations S_i is to achieve the same number of iterations.

Parallel algorithms built on the basis of the global model of parallelism are parallel analogues of the corresponding sequential algorithms. Parallelization is carried out according to the data, and the calculations are organized according to the master-slave type. On the host processor of a parallel computing system, a gravitational search algorithm is implemented by the master process, and archive sets containing non-dominant functional values and corresponding points are also stored and processed. Each of the slave processes is assigned to be executed by one of the processors of a parallel computing system. The slave process calculates the values of the fitness function and sends the value to the master process after each iteration. Based on this data, the master process calculates the new phase coordinates of the agents and sends them to the slave processes, etc. In the global concurrency model, parallel computing can be either synchronous or asynchronous.

The global synchronous model of parallelism is based on the fact that the current iteration is completed only after the master process has received fitness function values from all slave processes. The global synchronous parallelism model corresponds to the static load balancing of the working processors of a parallel computing system. The high performance of the algorithms of the global synchronous parallelism model will be subject to the same type of computing units of a parallel computing system and the same time to calculate the values of the fitness function at any admissible point in the parameter space. Often, these conditions cannot be met.

The global asynchronous model of parallelism is based on the fact that the master process receives data from slave processes not after global synchronization, but at any time when this data is ready. Based on the information received, the master process updates the phase coordinates of the corresponding particles and immediately sends them to free slave processes to continue iterations. The global asynchronous parallelism model corresponds to the dynamic load balancing of the working processors of a parallel computing system. The global synchronous parallelism model and the global asynchronous parallelism model can be effectively implemented on MIMD systems with shared and distributed memory. Unlike the global synchronous parallelism model, the global asynchronous parallelism model cannot be effectively implemented on the GPU. The advantage of algorithms based on the global parallelism model is the simplicity of obtaining and using global information about all particles, and the disadvantage is the possible high overhead for communication.

It is recommended to use the island model of parallelism if the throughput of the communication network of the parallel computing system is low, the computational complexity of the fitness function is small and the number of particles is very large. In other cases, it is recommended to use the global concurrency model.

4 Simulation Results

4.1 Comparison of GSA with State-of-the-Art Algorithms

To evaluate the proposed GSA 5 benchmark functions were used. These functions are classical functions, which are often employed to check the optimization performance of different algorithms. All test functions have been listed in Table 2, where *D* indicates dimension of the function, *Range* is the boundary of the function's search space and *Optimal* is the global minimum. Furthermore, F1–F3 are unimodal functions, whereas F4–F5 are multimodal functions.

Table 2. Benchmark functions

No	Formulation	D	Range	Optimal
F_1	$f(x) = \sum_{i=1}^{D} x_i^2$	30	[−100,100]	0
F_2	$f(x) = \max_i\{\lvert x_i\rvert, 1 \leq i \leq D\}$	30	[−100,100]	0
F_3	$f(x) = \sum_{i=1}^{D} \lvert x_i\rvert + \prod_{i=1}^{D} \lvert x_i\rvert$	30	[−10,10]	0
F_4	$f(x) = \frac{1}{4000}\sum_{i=1}^{D} x_i^2 - \prod_{i=1}^{D} \cos\frac{x_i}{\sqrt{i}} + 1$	30	[−600,600]	0
F_5	$f(x) = -20\exp\left(-0.2\sqrt{\frac{1}{D}\sum_{i=1}^{D} x_i^2}\right) - \exp\left(\frac{1}{D}\sum_{i=1}^{D}\cos 2\pi x_i\right) + 20 + e$	30	[−100,100]	0

To compare the optimization performance among different algorithms we used the next quality indicators. Mean value and standard deviation are good indicators to measure the obtained solution quality. The smaller the mean value is, the stronger the global optimization ability of the algorithm is; the smaller the standard deviation is, the more stability the algorithm is. Table 3. The experimental results for five benchmark functions obtained by 6 metaheuristic algorithms. shows the statistical results. In this table "Mean" and "Std" indicate "mean value" and "standard deviation" respectively. Moreover, the best results are highlighted in bold.

We compared the performance between GSA and five state-of-the-art algorithms: Cuckoo Search (CS), Grey Wolf Optimizer (GWO), Whale Optimization Algorithm (WOA), Particle Swarm Optimization (PSO) and Salp Swarm Algorithm (SSA). The statistical results obtained by six algorithms have been shown in Table 3. The gravitational search algorithm shows not the best, but quite acceptable results.

Table 3. The experimental results for five benchmark functions obtained by 6 metaheuristic algorithms.

No.	Metric	CS	GWO	WOA	PSO	SSA	GSA
F_1	Mean	8.60E−25	**0.00E+00**	**0.00E+00**	8.66E−74	4.81E−09	2.02E−10
	STD	1.58E−24	**0.00E+00**	**0.00E+00**	2.21E−73	1.06E−09	1.06E−09
F_2	Mean	1.66E+01	**4.05E−75**	1.28E+01	3.24E−05	1.27E+00	−6.06E−11
	STD	4.13E+00	**1.17E−74**	1.94E+01	3.59E−05	1.53E+00	1.07E−09
F_3	Mean	1.76E−15	8.63E−179	**0.00E+00**	1.18E−31	5.24E−01	1,75E−10
	STD	5.60E−15	**0.00E+00**	**0.00E+00**	5.95E−31	7.33E−01	1.14E−09
F_4	Mean	7.45E−03	**0.00E+00**	4.14E−04	1.88E−02	6.73E−03	5.93E−09
	STD	1.68E−02	**0.00E+00**	2.27E−03	1.95E−02	7.17E−03	1.94E−08
F_5	Mean	2.00E+00	8.23E−15	**3.02E−15**	8.88E−01	2.18E+00	−2.65E−10
	STD	1.13E+00	**1.30E−15**	2.21E−15	8.87E−01	5.88E−01	5.15E−10

The following expression was used as a function of Eq. 6:

$$G = \frac{G_0}{\exp(\alpha t)}.$$

The parameters used in the GSA algorithm are as follows: population size = [100, 200], $\alpha = 20$, $G_0 = 250$.

After testing the algorithm on benchmark functions, the parallel GSA was applied to solve a real problem, namely, the inverse problem for the process of pre-reforming of liquefied petroleum gases.

4.2 Parallel GSA for Solving the Inverse Problem of Chemical Kinetics

As a result of solving the inverse problem, the values E_{ref}, E_{met}, k_{ref}, k_{met}, B and n, included in the expression of reaction rates W_{ref} and W_{met}, were optimized. When solving the inverse problem, the values of E_{ref}, E_{met}, k_{ref}, k_{met} varied, while the value of B varied from 0 to 5, and m – from 0 to 2. The values obtained are shown in Table 4.

Table 4. The values E_{ref}, E_{met}, k_{ref}, k_{met}, B and n obtained as a result of solving the inverse problem.

E_{ref}	E_{met}	k_{ref}	k_{met}	m	B
117.4	59.3	$9.9 \cdot 10^{10}$	$4.3 \cdot 10^{6}$	0.62	0.057

The obtained optimal values were used to solve the direct problem of chemical kinetics. Figure 2 shows the results of calculations for experiments on propane preforming. The model correctly describes the available experimental data.

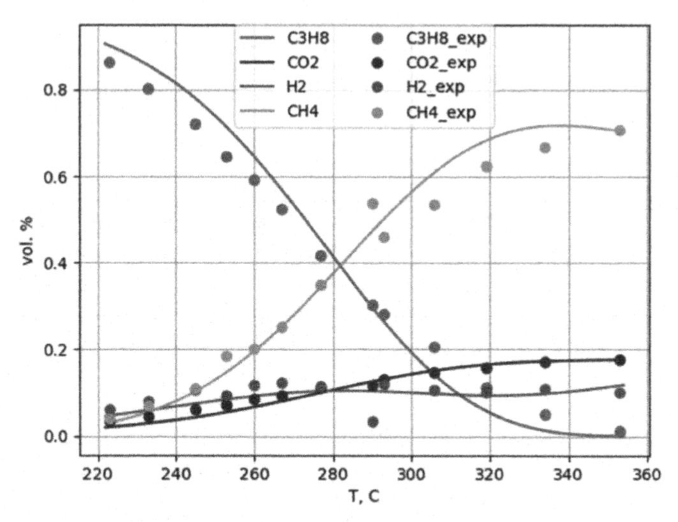

Fig. 2. Temperature dependences of the output concentrations of propane C_3H_8, methane CH_4, hydrogen H_2 and CO_2 in the process of propane preforming. Points are experimental data, lines are simulation.

To research parallel algorithms and compare them with sequential algorithms, characteristics such as speed-up and efficiency coefficients are introduced:

$$S_m = \frac{T_1}{T_m},\ E_m = \frac{S_m}{m} = \frac{T_1}{m \cdot T_m},$$

here S_m is the speed-up, E_m is the efficiency, T_m is the execution time of the parallelized program on m processors, T_1 is the execution time of the sequential program.

Figure 3 shows graphs for speed-up and efficiency of the programm.

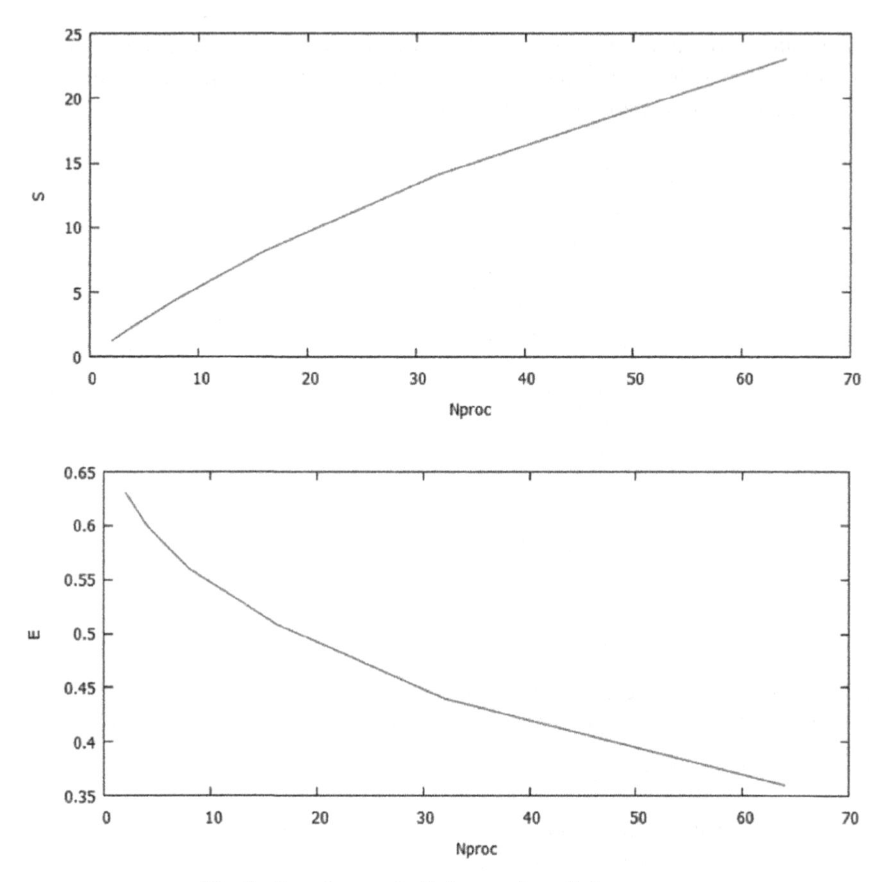

Fig. 3. Speed-up and efficiency of parallel program.

5 Conclusions

As a result of this work, a parallel algorithm of gravitational search was developed. The algorithm has been tested on five benchmark functions, and the algorithm has shown good performance. After that, the inverse problem of chemical kinetics was solved using the developed parallel GSA.

Acknowledgements. The reported study was funded by RFBR, project number 19-37-60014 (mathematical modeling) and project number 18-01-00599 (parallel implementation).

References

1. Yiying, Z., Zhigang, J.: Group teaching optimization algorithm: a novel metaheuristic method for solving global optimization problems. Expert Syst. Appl. **148**, 113246 (2020)
2. Dokeroglu, T., Sevinc, E., Kucukyilmaz, T., Cosar, A.: A survey on new generation metaheuristic algorithms. Comput. Ind. Eng. **137**(106040), 1–29 (2019)

3. Kazantsev, K.V. Bikmetova, L, Dzhikiya, O., Smolikov, M., Belyi, A.S.: Particle swarm optimization for inverse chemical kinetics problem solving as applied to n-hexane Isomerization on sulfated zirconia catalysts. Procedia Eng. **152**, 34–39 (2016)
4. Gubaydullin, I., Enikeeva, L., Naik, L.R.: Software module of Mathematical Chemistry web-laboratory for studying the kinetics of oxidation of 4-tert-butyl-phenol by aqueous solution of $H_2 0_2$ in the presence of titanosilicates. Eng. J. **20**(5), 263–270 (2016)
5. Akhmadullina, L.F., Enikeeva, L.V., Gubaydullin, I.M.: Numerical methods for reaction kinetics parameters: identification of low-temperature propane conversion in the presence of methane. Procedia Eng. **201**, 612–616 (2017)
6. Uskov, S.I., et al.: Kinetics of low-temperature steam reforming of propane in a methane excess on a Ni-based catalyst. Catal. Ind. **9**(2), 104–109 (2017)
7. Gubaydullin, I., Koledina, K., Sayfullina, L.: Mathematical modeling of induction period of the olefins hydroalumination reaction by diisobutylaluminiumchloride catalyzed with Cp2ZrCl2. Eng. J. **18**(1), 13–24 (2014)
8. Rashedi, E., Nezamabadi-Pour, H., Saryazdi, S.: Gsa: a gravitational search algorithm. Inf. Sci. **179**(13), 2232–2248 (2009)

Quantum Software Engineering: Quantum Gate-Based Computational Intelligence Supremacy

Olga Ivancova[1,2], Vladimir Korenkov[1], Nikita Ryabov[2], and Sergey Ulyanov[1,2](✉)

[1] LIT JINR, Dubna, Russia
o_ivancova@mail.ru, korenkov@jinr.ru, ulyanovsv@mail.ru
[2] Dubna State University, Dubna, Russia
ryabov_nv95@mail.ru

Abstract. A new approach to a circuit implementation design of quantum algorithmic gates for quantum massive parallel computing realization is presented. The main attention is focused on the development of design method of fast quantum algorithm operators as superposition, entanglement and interference which are in general time-consuming operations due to the number of products that have to be performed. SW&HW support toolkit of supercomputing accelerator of quantum algorithm simulation is described. The method for performing Grover's interference without product operations introduced. New quantum genetic and quantum fuzzy inference algorithm gate design considered.

Keywords: Quantum algorithm gate · Hardware architecture · Reduced quantum operations · Classical efficient simulation

1 Introduction

This report is concerned with the problem of discovering a new family of quantum algorithms (QA's). The presented method and relative hardware implement matrix operations performed in second and third step of a QA (the so-called entanglement and interference operators), providing a substantially increasing in computational speed-up with respect to the corresponding software realization of a traditional and a new quantum search algorithm (QSA). A high-level structure of a generic entanglement block that uses logic gates as analogy elements is described. Moreover, a new method for performing Grover interference without products is introduced. This model has the advantage that proving lower bounds is tractable which allows one to demonstrate provable speed-up over classical algorithms or to show that a given QA is the best possible [10, 15]. Design method of main quantum operators and hardware implementation of Grover's algorithm for fast search in large unstructured database and related topics concerning the intelligent control of an ill-defined process including search-of-minima entropy uncertainty intelligent operations is described.

V. Voevodin and S. Sobolev (Eds.): RuSCDays 2020, CCIS 1331, pp. 110–121, 2020.
https://doi.org/10.1007/978-3-030-64616-5_10

1.1 SW/HW Support of QA Accelator Computing

General Structure of QA. A QA estimates (without numerical computing) the qualitative properties of the function *f*. From a mathematical standpoint, a function *f* is the map of one logical state into another. The problem solved by a QA can be stated in the symbolic form as follows:

Input	*A function f: {0,1}ⁿ→{0,1}ᵐ*
Problem	*Find a certain property of function f*

Main goal of QA applications is the study and search of qualitative properties of functions as the solution of problem.

Figure 1 shows the general structure of QA. The main blocks in Fig. 1: i) unified operators; ii) problem-oriented operators; iii) Benchmarks of QA simulation on classical computers; and iv) quantum control algorithms based on quantum fuzzy inference (QFI) and quantum genetic algorithm (QGA) as new types of QSA.

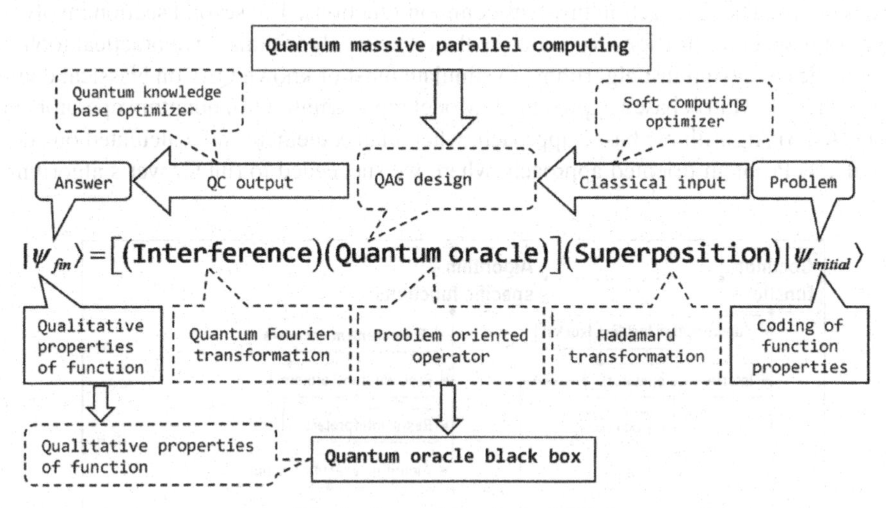

Fig. 1. General structure of QA

Superposition, entanglement (quantum oracle) and interference in quantum massive parallel computing are the main operators in QA. The superposition operator of most QAs can be expressed as following: $Sp = \left(\overset{n}{\underset{i=1}{\otimes}} H\right) \otimes \left(\overset{m}{\underset{i=1}{\otimes}} S\right)$, where n and m are the numbers of inputs and of outputs respectively. Operator S may be or Hadamard operator H or identity operator I depending on the algorithm. Numbers of outputs m as well as structures of corresponding superposition and interference operators are presented in the Table 1 for different QAs.

In general form, the structure of a QAG can be described as follows:

$$QAG = \left[\left(Int \otimes {}^{n}I\right) \cdot U_F\right]^{h+1} \cdot \left[{}^{n}H \otimes {}^{m}S\right], \tag{1}$$

Table 1. Parameters of superposition and interference operators of main quantum algorithms.

Algorithm	Superposition	m	Interference
Deutch's [1]	$H \otimes I$	1	$H \otimes H$
Deutsch-Jozsa's [2]	$^{n}H \otimes H$	1	$^{n}H \otimes I$
Grover's [4]	$^{n}H \otimes H$	1	$D_n \otimes I$
Simon's [9]	$^{n}H \otimes {}^{n}I$	n	$^{n}H \otimes {}^{n}I$
Shor's [8]	$^{n}H \otimes {}^{n}I$	n	$QFT_n \otimes {}^{n}I$

where I is the identity operator; the symbol $\otimes$ denotes the tensor product; S is equal to I or H and dependent on the problem description. One portion of the design process in Eq. (1) is the type-choice of the entanglement problem dependent operator U_F that physically describes the qualitative properties of the function f.

Structure of QA Simulation System. The software system is divided into two general sections. The first section involves common functions. The second section involves algorithm-specific functions for realizing the concrete algorithms. Five practical toolkit approaches to design fast algorithms to simulate most of known QAs on classical computers [15]: 1) Matrix based approach; 2) Model representations of quantum operators in fast QAs; 3) Algorithmic based approach, when matrix elements are calculated on "demand"; 4) Problem-oriented approach, where we succeeded to run Grover's algorithm

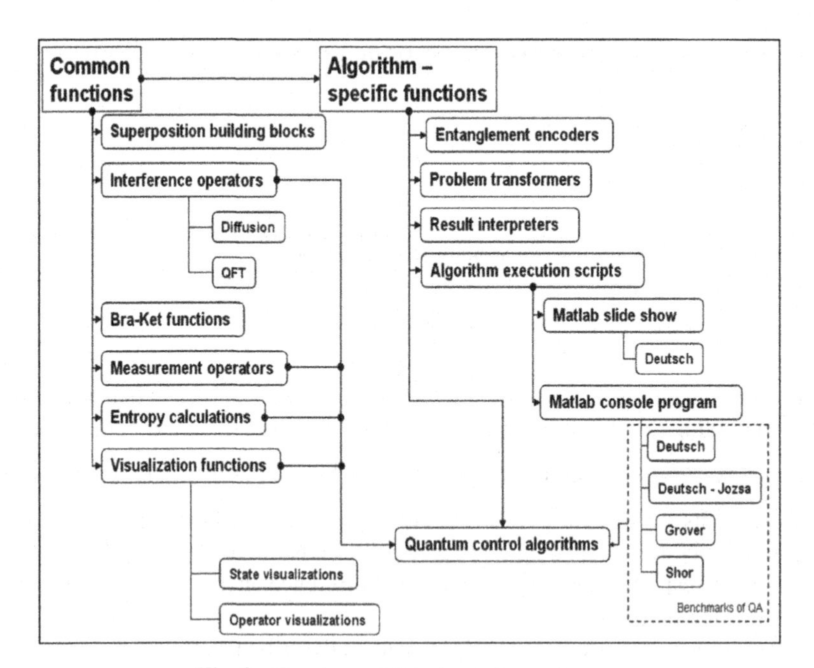

Fig. 2. Structure of QA simulation software.

with up to 64 and more qubits with Shannon entropy calculation (up to 1024 without termination condition); 5) Quantum algorithms with reduced number of operators (entanglement-free QA, and so on).

Figure 2 shows the structure platform of a software system for QA simulation.

Example: Grover's QSA. Figure 3 shows the structure of Grover's QA gate. Efficient simulation of QAs on classical computer with large number of inputs is difficult problem. For example, to operate only with 50 qubits state vector directly, it is necessary to have at

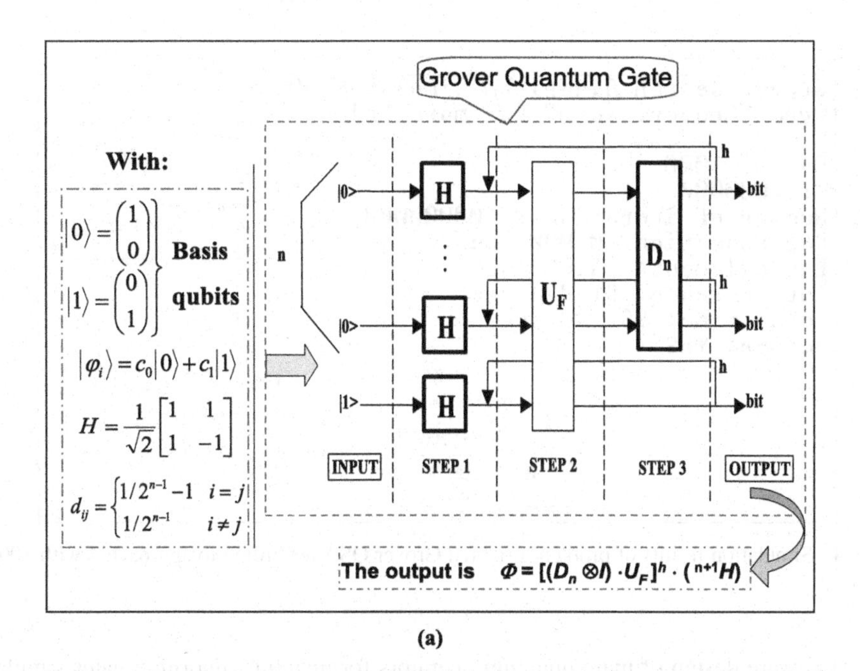

(a)

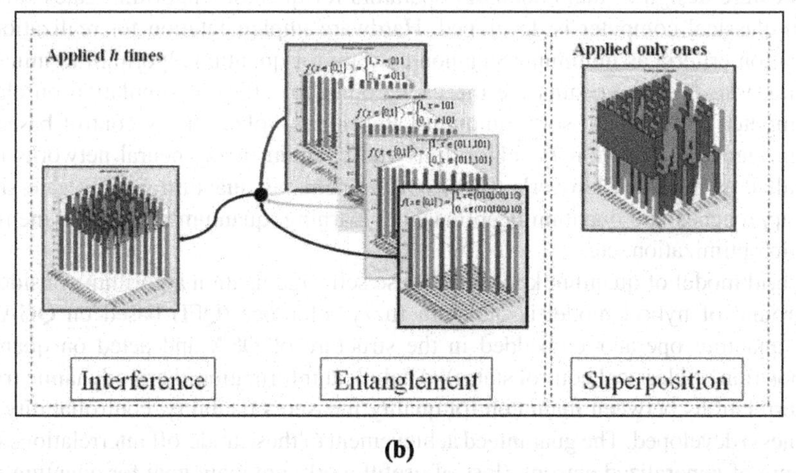

(b)

Fig. 3. (a) Quantum gate circuit; (b) QAG of Grover QSA.

least 128 TB of memory. In present report, for concrete important example as Grover's QSA, it is demonstrated the possibility to override spatiotemporal complexity, and to perform efficient simulations of QA on classical computers. Design method and hardware implementation of modular system for realization of Grover's QSA are presented in Fig. 3(a).

Evolution of Grover's QSA for three qubits (a) and quantum simulator on classical computer (b) on Fig. 3 demonstrated. Simulation results of problem-oriented Grover QSA according to approach 4 with 1000 qubits on Fig. 4 shown.

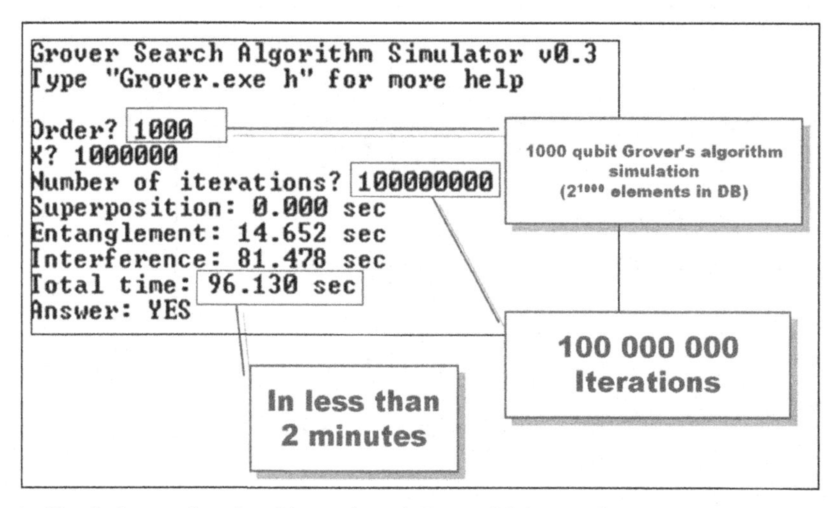

Fig. 4. Simulation results of problem-oriented Grover QSA according to approach 4 with 1000 qubits.

Hardware design of main quantum operators for quantum algorithm gates simulation on classical computer is developed. Hardware implementation for realization of information criteria as minimum Shannon entropy for quantum algorithm termination is demonstrated. These results are the background for efficient simulation on classical computer the quantum soft computing algorithms, robust fuzzy control based on quantum genetic (evolutionary) algorithms and quantum fuzzy neural networks (that can realize as modified Grover's QSA), AI-problems as quantum game's gate simulation approaches and quantum deep machine learning, quantum associative memory, quantum optimization, etc.

Hybrid model of quantum knowledge base self-organization algorithm considered. Background of hybrid model is quantum fuzzy inference (QFI) based on QGA (as a new quantum operator embedded in the structure of QSA and acted on quantum superposition of classical control states). Analytical information-thermodynamic trade-off interrelations between main control quality measures (stability, controllability and robustness) developed. The guaranteed achievement of these trade-off interrelations with minimum of generalized entropy (loss of useful work) are main goal for quantum self-organization algorithm of imperfect knowledge base (KB). The application of quantum genetic algorithm to problem solving of automatically selection an optimal type and kind of correlations in the quantum fuzzy inference discussed. The fitness function in QGA (as robustness measure of control system) is ("control object+controller") – generalized entropy. The synergetic effect as the result of the generalized search algorithm application to intelligent cognitive robotics control for example, the benchmarking system "cart – pole" (inverted pendulum) with one proportional-integral-derivative (PID) controller for Two-Degree-of-Freedom control object presented. Results of controller's behavior comparison confirm the existence of synergetic self-organization effect in the design process of robust KB on the basis of imperfect (non-robust) KB responses of fuzzy controllers. Sophisticated effect is following: from two imperfect KB with quantum approach new robust KB can be created on line using quantum correlation. In classical intelligent control based on soft computing toolkit this effect cannot be achieved.

2 Quantum Fuzzy Inference Based on Quantum Genetic Algorithm in Intelligent Robotics

Basis of control systems – PID controller, which is used in 70% of the industrial automation, but often can't cope with the task of managing and does not work well in unpredicted control situations. The use of quantum computing and QSA, as a special example, QFI, allows increasing robustness without the cost of a temporal resource – in online.

Figure 6 shows the integration of several fuzzy controllers and QFI which allows create a new quality in management – online self-organization of KB [15]. In general, the structure of a QAG based on a QGA described in (2) in the form:

$$QAG = \left[\left(Int \otimes {}^{n} I\right) \cdot U_F\right]^{h+1} \cdot [QGA]\left[{}^{n}H \otimes {}^{m} S\right]. \tag{2}$$

QGA have already been actively used in human action recognition [7] and fault diagnosis of gearbox [3]. Structure of corresponding QAG on Fig. 5 is shown.

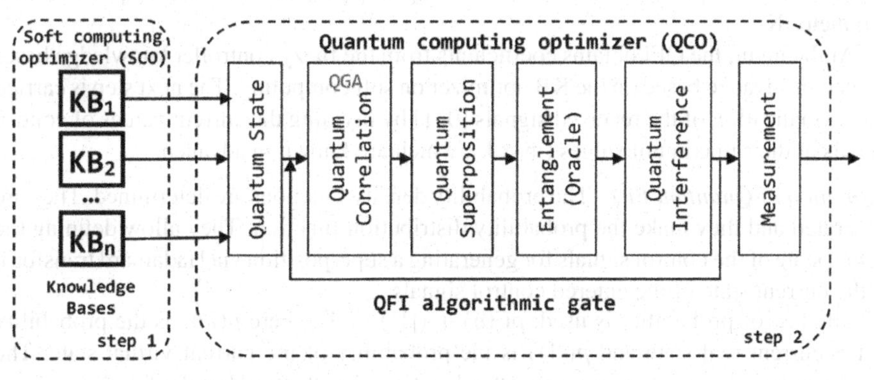

Fig. 5. QAG structure of QFI with QGA.

The first part in designing Eq. (2) is the choice of the type of the entangled state of operator U_F. The basic unit of such an intelligent control system (ICS) is the quantum genetic algorithm (QGA) (see, Fig. 6).

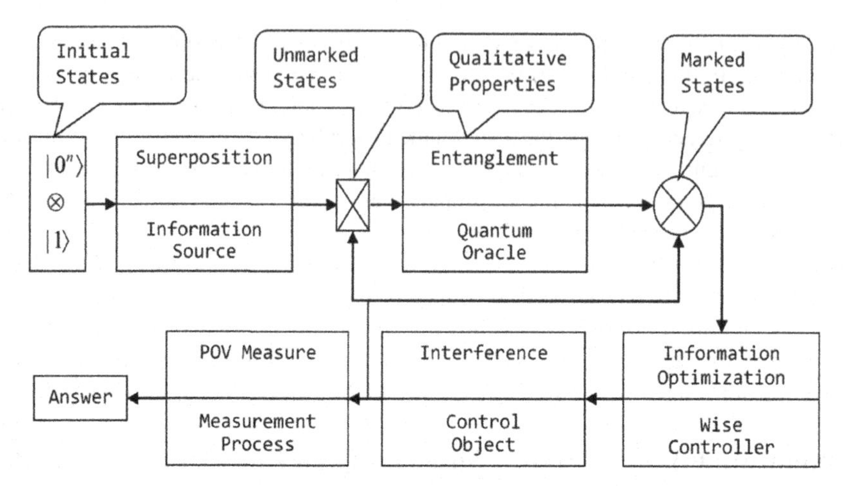

Fig. 6. Intelligent self-organizing quantum genetic algorithm for intelligent control systems.

Results of simulation show computing effectiveness of robust stability and controllability of (QFI+QGA)–controller and new information synergetic effect: from two fuzzy controllers with imperfect knowledge bases can be created robust intelligent controller (extracted hidden quantum information from classical states is the source of value work for controller [15]) in on-line. Intelligent control systems with embedding intelligent QFI-controller can be realized either on classical or on quantum processors (as an example, on D-Wave processor type).

Two classes of quantum evolution (1) are described: QGA and hybrid genetic algorithm (HGA). The QFI algorithm for determining new PID coefficient gain schedule factors K (see, below Fig. 7) consists of such steps as normalization, the formation of a quantum bit, after which the optimal structure of a QAG is selected, the state with the maximum amplitude is selected, decoding is performed and the output is a new parameter K.

At the input, the QFI obtains coefficients from the fuzzy controller knowledge bases formed in advance based on the KB optimizer on soft computing. The next step is carried out normalization of the received signals [0, 1] by dividing the current values of control signals at their maximum values (max k), which are known in advance.

Formation of Quantum Bits. The probability density functions are determined. They are integrated and they make the probability distribution function. They allow defining the virtual state of the control signals for generating a superposition via Hadamard transform of the current state of the entered control signals.

The law of probability is used: $p(|0\rangle) + (|1\rangle) = 1$, where $p(|0\rangle)$ is the probability of the current real state and $p(|1\rangle)$ is the probability of the current virtual state. The superposition of the quantum system "real state – virtual state" has the form (3).

$$|\psi\rangle = \frac{1}{2}\left(\sqrt{p(|0\rangle)}|0\rangle + \sqrt{1 - p(|0\rangle)}|1\rangle\right) \tag{3}$$

The next step is selection of the type of quantum correlation – constructing operation of entanglement. Three types of quantum correlation are considered: spatial, temporal

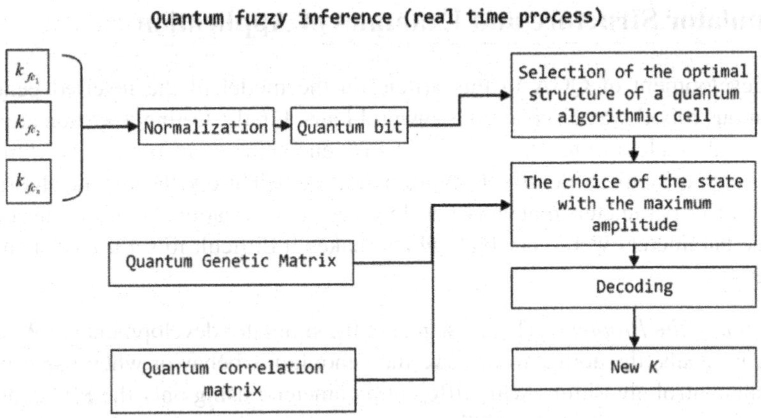

Fig. 7. Quantum fuzzy inference algorithm.

and spatial temporal. Each of them contains valuable quantum information hidden in a KB.

Quantum correlation considered as a physical computational resource, which allows increasing the successful search for solutions of algorithmically unsolvable problems. In our case, the solution of the problem of ensuring global robustness of functioning of the control object under conditions of unexpected control situations by designing the optimal structure and laws of changing the PID controller gain factors by classical control methods is an algorithmically unsolvable problem. The solution of this problem is possible based on quantum soft computing technologies [15]. The output parameters of the PID-regulators are considered as active information-interacting agents, from which the resulting controlling force of the control object is formed. In a multi-agent system, there is a new synergistic effect arising from the exchange of information and knowledge between active agents (swarm synergetic information effect) [10]. There are several different types and operators of QGAs [6].

Remark. One of the interesting ideas was proposed in 2004, taking the first steps in implementing the genetic algorithm on a quantum computer [10, 15]. The author proposed this quantum evolutionary algorithm, which can be called the reduced quantum genetic algorithm (RQGA). The algorithm consists of the following steps: 1) Initialization of the superposition of all possible chromosomes; 2) Evaluation of the fitness function by the operator F; 3) Using Grover's algorithm; 4) Quantum oracle; 5) Using of the diffusion operator Grover G; 6) Make an evaluation of the decision. The search for solutions in RQGA is performed in one operation. In this case the matrix form is the result of RQGA action.

Evolution in QGA is the result of unitary transformations that bring the state of chromosomes closer to the state of optimal chromosome with maximum fitness. For this reason, convergence to local optima is faster. QGA require fewer chromosomes. In RQGA, a population can be composed of a single chromosome in a superposition state. In such an algorithm, evolution can occur in one generation [6, 11].

3 Simulator Structure and Examples of Applications

In the development of QGA in this article on the model of the inverted pendulum (autonomous robot) was discovered a few problems. Firstly, testing a written algorithm on a robot takes a lot of time. Secondly, you may encounter an incorrectly working HW, and it is rather difficult to identify the malfunction itself. Thirdly, the genetic algorithm is the selection of parameters that work best in a particular situation, but it's quite common that these parameters were very bad, which makes it difficult to set up a dynamically unstable object.

Description of the Problem. The main goal of the simulator development is SW testing, educational goals, the ability to observe the pendulum's behavior when using various intelligent control algorithms with different parameters: using only the PID controller, adding a fuzzy controller to the ICS, using the genetic algorithm and neural network, using QGA. The simulator is interesting because it covers many areas required for its implementation. There are also many different ways of development: improvement of the 2D model or even implementation in 3D, control of the pendulum in on line (changing the parameters of the pendulum, adding various noises), making the simulator more universal for simply creating simulations of other tasks based on the prepared project.

Selection of Development Toolkit. Simulator access is as simple as possible and it is implemented as a non-typical web application. The diagram of the sequence of the user's work with the system and the interaction of the model, the presentation and the template are presented on Fig. 8.

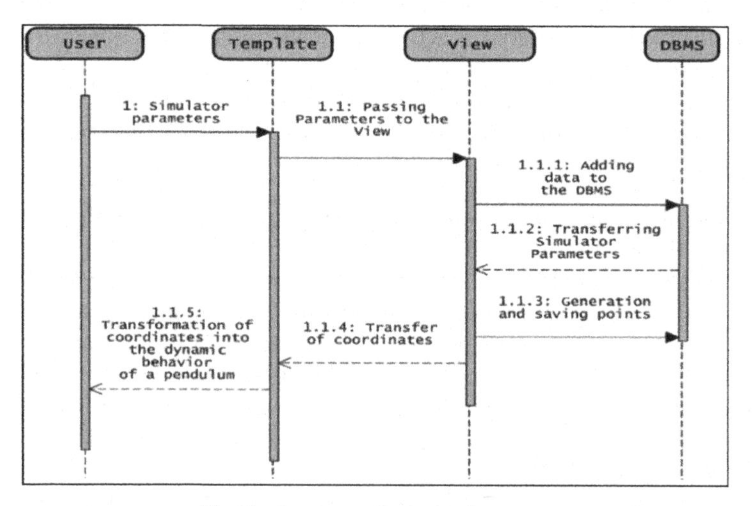

Fig. 8. Sequence diagram of system.

On Fig. 9 are demonstrated results of simulation and experimental results comparison.

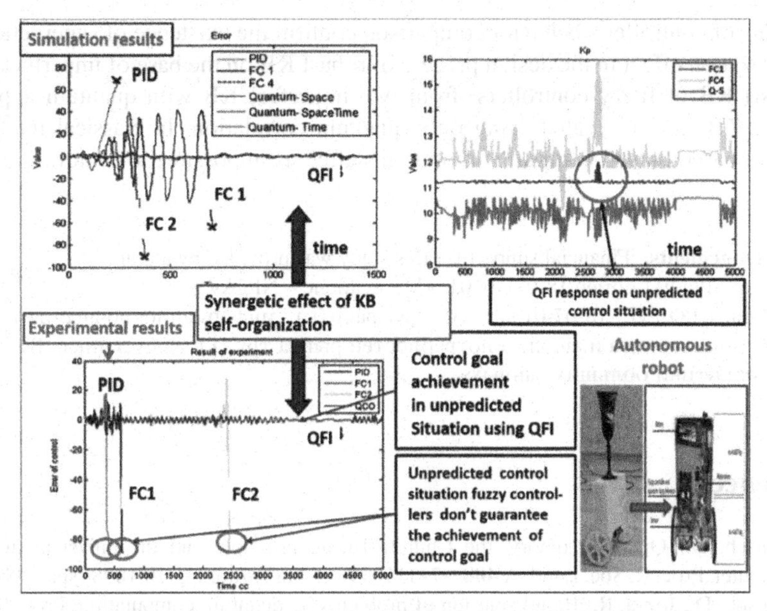

Fig. 9. Simulation & experimental results comparison for unpredicted control situation in cases of PID-controller, fuzzy controller and QFI-controller (b).

Mathematical modeling and experimental results are received for the case of unpredicted control situation and knowledge base of fuzzy controller was designing with SW of QCOptKB™ for teaching signal measured directly from control. As model of unpredicted control was the situation of feedback sensor signal delay on three times. Results of controller's behavior comparison confirm the existence of synergetic self-organization effect in the design process of robust KB on the base of imperfect (non-robust) KB of fuzzy controllers. In unpredicted control situation control error is dramatically changing and KB responses of fuzzy controllers (FC 1 and FC 2) that designed in learning situations with soft computing are imperfect and do not can achieve the control goal. Using responses of imperfect KB (as control signals for design the schedule of time dependent coefficient gain in PID-controller) in Box QFI the robust control is formed in on line. This effect is based on the existence of additional information resource that extracted by QFI as quantum information hidden in classical states of control signal (as response output of imperfect KB's on new control error) [5, 12–14].

4 Conclusions

- New circuit implementation design method of quantum gates for fast classical efficient simulation of search QAs is developed. Benchmarks of design application as Grover's QSA and QFI based on QGA demonstrated.
- Applications of QAG approach in intelligent control systems with quantum self-organization of imperfect knowledge bases are described on concrete examples.

- Results of controller's behavior comparison confirm the existence of synergetic self-organization effect in the design process of robust KB on the base of imperfect (non-robust) KB of fuzzy controllers: from two imperfect KB with quantum approach robust KB can be created using only quantum correlation. In classical intelligent control based on soft computing toolkit this effect is impossible to achieve.

Acknowledgements. Financial support for this study was provided by a grant:
• /Грант/РФФИ номер 18-02-40101: «Мегасайенс – NICA»
• /Грант/РФФИ 18-07-01359 А: «Разработка информационно-аналитической системы мониторинга и анализа потребностей рынка труда в выпускниках ВУЗов на основе аналитики больших данных».

References

1. Deutsch, D.: Quantum theory, the Church–Turing principle and the universal quantum computer. Proc. R. Soc. Lond. A **400**, 97–117 (1985). https://doi.org/10.1098/rspa.1985.0070
2. Deutsch, D., Jozsa, R.: Rapid solution of problems by quantum computation. Proc. R. Soc. Lond. A **439**, 553–558 (1992). https://doi.org/10.1098/rspa.1992.0167
3. Fen, W., Min, L., Gang, W., Xu, J., Ren, B., Wang, G.: Fault diagnosis approach of gearbox based on Support Vector Machine with improved bi-layers quantum genetic optimization. In: 13th International Conference on Ubiquitous Robots and Ambient Intelligence (URAI), Xi'an, pp. 997–1002 (2016)
4. Grover, L.K.: A fast quantum mechanical algorithm for database search. In: Proceedings, 28th Annual ACM Symposium on the Theory of Computing, p. 212 (1996)
5. Ivancova, O.V., Korenkov, V.V., Ulyanov, S.V.: Quantum software engineering. M: Kurs (2020)
6. Lahoz-Beltra, R.: Quantum genetic algorithms for computer scientists. Computers **5**(4), 31–47 (2016)
7. Liu, Y., Feng, S., Zhao, Z., Ding, E.: Highly Efficient Human Action Recognition with Quantum Genetic Algorithm Optimized Support Vector Machine. ArXiv abs/1711.09511 (2017)
8. Shor, P.W.: Algorithms for quantum computation: discrete logarithms and factoring. In: Proceedings 35th Annual Symposium on Foundations of Computer Science, Santa Fe, NM, USA, pp. 124–134 (1994) . https://doi.org/10.1109/sfcs.1994.365700
9. Simon, D.R.: On the power of quantum computation. In: Proceedings of the 35th Annual Symposium on Foundations of Computer Science, pp. 116–123 (1994)
10. Ulyanov, S.V.: Quantum soft computing in control processes design: quantum genetic algorithms and quantum neural network approaches. In: WAC (ISSCI') 2004 (5th Intern. Symp. on Soft Computing for Industry), Seville Spain, 2004, vol. 17, pp. 99–104. Springer, Heidelberg (2004)
11. Ulyanov, S.V.: Self-organizing quantum robust control methods and systems for situations with uncertainty and risk. Patent US **8788450**, B2 (2014)
12. Ulyanov, S.V: Quantum fuzzy inference based on quantum genetic algorithm: quantum simulator in intelligent robotics. In: Aliev, R.A., et al. (eds.) ICSCCW 2019, AISC 1095, pp. 78–85 (2020). https://doi.org/10.1007/978-3-030-35249-3_9

13. Ulyanov, S.V., Ryabov, N.V.: Quantum simulator for modeling intelligent fuzzy control. Nechetkie Sistemy i Myagkie Vychisleniya [Fuzzy Syst. Soft Comput.] **14**(1), 19–33 (2019). https://doi.org/10.26456/fssc49. (in Russian)
14. Ulyanov, S.V., Ryabov, N.V.: The quantum genetic algorithm in the problems of intelligent control modeling and supercomputing. Softw. Syst. **32**(2), 181–189 (2019)
15. Ulyanov, S.V.: Self-organized Robust Intelligent Control. LAP Lambert Academic Publishing (2015)

Resource-Efficient Parallel CG Algorithms for Linear Systems Solving on Heterogeneous Platforms

Nikita S. Nedozhogin[(✉)], Sergey P. Kopysov, and Alexandr K. Novikov

Udmurt State University, Izhevsk, Russia
nedozhogin07@gmail.com, s.kopysov@gmail.com,
alexander.k.novikov@gmail.com

Abstract. The article discusses the parallel implementation of solving systems of linear algebraic equations on the heterogeneous platform containing a central processing unit (CPU) and graphic accelerators (GPU). The performance of parallel algorithms for the classical conjugate gradient method schemes when using the CPU and GPU together is significantly limited by the synchronization points. The article investigates the pipeline version of the conjugate gradient method with one synchronization point, the possibility of asynchronous calculations, load balancing between the CPU and GPU when solving the large linear systems. Numerical experiments were carried out on test matrices and computational nodes of different performance of a heterogeneous platform, which allowed us to estimate the contribution of communication costs. The algorithms are implemented with the combined use of technologies: MPI, OpenMP and CUDA. The proposed algorithms, in addition to reducing the execution time, allow solving large linear systems, for which there are not enough memory resources of one GPU or a computing node. At the same time, block algorithm with the pipelining decreases the total execution time by reducing synchronization points and aggregating some messages in one.

Keywords: Parallel computing on heterogeneous platform · Hybrid parallel technologies · Conjugate gradient method · Reduction of communications

1 Introduction

Computing accelerators are contained in the computing nodes of supercomputers and are used quite successfully in solving many computing problems despite the fact that the central processor (CPU) is idle after running the core functions on the accelerator. There are several more important conditions for which the

Supported by Russian Foundation for Basic Research (RFBR) according to the research project 17-01-00402. The work was carried out with the financial support of Udmurt State University in the contest of the grants "Scientific Potential", project No. 2020-04-03.

V. Voevodin and S. Sobolev (Eds.): RuSCDays 2020, CCIS 1331, pp. 122–133, 2020.
https://doi.org/10.1007/978-3-030-64616-5_11

joint use of the CPU and accelerators (for example, GPU) in parallel computing within the same problem seems to be promising.

Each architecture of the CPU and GPU has unique features and, accordingly, is focused on solving certain tasks for which typical, for example, high performance or low latency. Heterogeneous nodes containing and sharing CPU plus GPUs can provide an effective solution to a wide range of problems or one problem for which the parallel properties of the algorithms are changed and determined to be executed on one or another computing device. We also note the high energy efficiency of the heterogeneous computing systems.

One of the computationally complex operations in numerical methods is the solution of linear systems (SLAE). Currently, many parallel algorithms have been proposed that provide high performance and scalability when solving large sparse systems of equations on modern multiprocessor with a hierarchical architecture.

The construction of hybrid solvers with a combination of direct and iterative methods for solving SLAE allows the use of several levels of parallelism [1–5]. So in [6] a hybrid method for solving systems of equations of Schur complement by preconditioned iterative methods from Krylov subspaces was built and implemented when used together the cores of central (CPU) and graphic processing units (GPU). The classical preconditioned conjugate gradient method [7] was applied for the block ordered matrix and the separation of calculations in matrix operations between the CPU and one or more GPUs, when the system of equations in Schur complement was solved in parallel.

Currently, there are several approaches for computing load scheduling of the parallel conjugate gradient method (CG) on heterogeneous Multi-CPUs/Multi-GPUs platforms. Firstly, the separate steps of the CG algorithm are executed on a CPU or GPU, such as a preconditioner [8].

Second approach is based on the data mapping (matrix block and vectors) or separate tasks are carried out of all CG steps on the CPU or GPU [6].

In [9] task scheduling on Multi-CPUs/Multi-GPUs platforms for the classical CG from the `PARALUTION` library is executed in the `StarPU`. It's unified runtime system for heterogeneous multicore architectures which is developed in the INRIA laboratory (France).

In this paper, we consider an approach that reduces the cost of data exchanging between the CPU and GPU by reducing the number of synchronization points and pipelining of computing when SLAE is solved on heterogeneous platforms.

The Krylov subspace methods are some of the most effective options for solving large-scale linear algebra problems. However, the classical Krylov subspace algorithms do not scale well on modern architectures due to the bottleneck related to synchronization of computations. Pipeline methods of the Krylov subspace [10] with hidden communications provide high parallel scalability due to the global communications overlapping with computing, performing matrix-vector and dot products. The first work on reducing communications was related to a variant of the conjugate gradient method, having one communication at each iteration [11], using the three-term recurrence relations CG [12].

The next stage of development was the emergence of s-step methods from Krylov subspaces [13], in which the iterative process in the s-block uses various bases of Krylov subspaces. As a result, it was possible to reduce the number of synchronization points to one per s iterations. However, for a large number of processors (cores), communications can still take significantly longer than computing a single matrix-vector product. In [14] it was proposed a CG algorithm using auxiliary vectors and transferring a sequential dependence between the computing of matrix-vector product and scalar products of vectors. In this approach, the latency of communications is replaced by additional calculations.

2 Pipelined Algorithm of the Conjugate Gradient Method

We consider now the pipelined version of the conjugate gradient method, which is mathematically equivalent to the classical form of the preconditioned CG method and has the same convergence rate.

Algorithm 1: Pipelined algorithm CGwO.

```
1  r = b − Ax;
2  u = M^{-1} r;
3  w = Au;
4  γ1 = (r, u); δ = (w, u);
   while ||r||2/||b||2 > ε do
5      m = M^{-1} w ;
6      n = Am;
       if (j = 0) then
7          β = 0;
       else
8          β = γ1/γ0 ;
9      α = γ1/(δ − βγ1/α) ;
10     z = n + βz; w = w − αz; s = w + βs; r = r − αs;
11     p = u + βp; x = x + αp; q = m + βq; u = u + αq;
12     γ0 = γ1;
13     γ1 = (r, u); δ = (w, u);
```

In this algorithm, the modification of the vectors r_{j+1}, x_{j+1}, s_{j+1}, p_{j+1} and matrix-vector products provides pipeline computations. The computation of dot products (line 4) can be overlapped with the computation of the product by the preconditioner (line 3) and the matrix-vector product (line 3). However, the number of triads in the algorithm increases to eight, in contrast to three for the classic version and four in [13]. In this case, a parallel computation of triads and two dot products at the beginning of the iterative process and one synchronization point is possible.

The pipelined version CG presented in this work can be used with any preconditioner. There are two ways to organize computations in the preconditioned pipelined CG, which provide a compromise between scalability and the total number of operations [15]. Thus, the CG pipeline scheme is characterized by a different order of computations, the presence of global communication, which can overlap with local computations, such as matrix-vector product and operations with a preconditioner, and the possibility of organizing asynchronous communications.

The two variants of the conjugate gradient method were compared: the classical scheme and the pipelined one. Table 1 presents the results of numerical experiments where the execution time of a sequential version of the classical CG and the CGwO pipelined scheme (Algorithm 1) executed on the CPU and GPU are shown. Note that in the variants for the GPU, joint computation of all dot products of vectors in one kernel function was implemented, independently of each other. For this, when starting the CUDA kernel, the dimension of the `Grid` hierarchy of CUDA threads was set in two-dimensional form: 3 sets of blocks, each for performing computations on its own pair of vectors. This allowed us to reduce the number of exchanges between the CPU and GPU memory, combining all the resulting scalars in one communication.

Matrices from the SuiteSparse Matrix Collection (formerly the University of Florida Sparse Matrix Collection, https://sparse.tamu.edu/) were used in the test computations. The right hand side vector was formed as a row-wise sum of matrix elements. Thus, the solution of the system $Ax = b$, dimension $N \times N$ (with the number of nonzero elements nnz) is a vector $x = (1, 1, \ldots, 1)^T$.

For systems of equations of small dimension, the solution time on the CPU according to the classical CG scheme is significantly less than the GPU execution time for the same number of iterations (see Table 1). For large systems, the costs of synchronization and forwarding between the CPU and GPU overlap with the speed of the GPU. In the pipelined version of CGwO, the computational execution costs on the GPU are reduced almost threefold for all the considered systems of equations only due to the reduction of exchanges between the GPU and the CPU in the computation of dot products.

3 CG with the Combined Use of CPU and GPU

Let us consider the application of the Algorithm 1 for the parallel solution of super-large systems of equations on computing nodes, each of which contains several CPUs and GPUs. To solve SLAEs on several GPUs, we construct a block pipelined algorithm for the conjugate gradient method. On heterogeneous platform, data exchange between different GPUs within the same computing node is carried out with OpenMP technology, and the exchange between different computing nodes is carried out by MPI technology.

For example, consider a node containing a central eight-core processor and two graphics accelerators. The number of OpenMP threads is selected by the number of available CPU cores. The first two OpenMP threads are responsible for exchanging data and running on two GPUs. Threads 2–6 provide computations on the CPU and can perform computations on a block of the SLAE matrix. The last thread provides data exchange with other computing nodes by MPI.

3.1 Matrix Partitioning

To divide the matrix A into blocks, we construct the graph $G_A(V, E)$, where $V = \{i\}$ is the set of vertices associated with the row index of the matrix (the number of vertices is equal to the number of rows of the matrix A); $E = \{(i, j)\}$ is the set of edges. Two vertices i and j are considered to be connected if the matrix A has a nonzero element with indices i and j. The resulting graph is divided into subgraphs whose number is d. For example, to split a graph, you can use the [16] layer-by-layer partitioning algorithm, which reduces communication costs due to the need to exchange only with two neighboring computing nodes. After that, each vertex of the graph is assigned its own GPU or CPU. On each computing unit, the vertices are divided into internal and boundary. The latter are connected with at least one vertex belonging to another subgraph.

After partitioning, each block A_k of the original matrix A contains the following submatrices:

- $A_k^{[i_k,i_k]}$ – matrix associated with the internal vertices;
- $A_k^{[i_k,b_k]}$, $A_k^{[b_k,i_k]}$ – matrices associated with the internal and boundary vertices;
- $A_k^{[b_k,b_l]}$ – matrix associated with the boundary vertices of the k-th and l-th blocks.

Then the matrix A can be written in the following form:

$$A = \begin{pmatrix} A_1^{[i_1,i_1]} & A_1^{[i_1,b_1]} & \cdots & 0 & 0 \\ A_1^{[b_1,i_1]} & A_1^{[b_1,b_1]} & \cdots & 0 & A_1^{[b_1,b_d]} \\ \vdots & \vdots & \ddots & \vdots & \vdots \\ 0 & 0 & \cdots & A_d^{[i_d,i_d]} & A_d^{[i_d,b_d]} \\ 0 & A_d^{[b_d,b_1]} & \cdots & A_d^{[b_d,i_d]} & A_d^{[b_d,b_d]} \end{pmatrix}.$$

We divide the matrix-vector product $n = Am$ into two components by using the obtained partition:

$$n_k^b = A_k^{[b_k,i_k]} m_k^i + \sum_{l=1}^{l \le d} A_k^{[b_k,b_l]} m_l^b, \qquad n_k^i = A_k^{[i_k,i_k]} m_k^i + A_k^{[i_k,b_k]} n_k^b. \quad (1)$$

Here k corresponds to the computing device. The block representation of the vectors involved in the algorithm is inherited from the matrix partitioning. For example, the vector m has the form $m^T = \left(m_1^i, m_1^b, \ldots, m_k^i, m_k^b, \ldots, m_d^i, m_d^b\right)$. The implementation of the matrix-vector product reduces the cost of communication between blocks at each iteration of conjugate gradient method. To perform this operation, an exchange of vectors m_k^b is required, the size of which is less than the dimension of the initial vector m.

The partitioning of the preconditioner M is carried out in a similar way.

3.2 Block Pipelined Algorithm

The matrix blocks were mapped on the available CPU and GPU with the block partitioning of the matrix and vectors. The number and size of blocks let on to map the load in accordance with the performance of the computing units, including the allocation of several blocks to one.

Algorithm 2: Block algorithm CGwO performed on k-th device

Data: Matrix partitioning into blocks $A_k^{[i_k,i_k]}$, $A_k^{[i_k,b_k]}$, $A_k^{[b_k,i_k]}$, $A_k^{[b_k,b_l]}$.

1 $r = b$;
2 $u = M^{-1}r$;
  // Parallel algorithm branches
  // $(\text{CPU} \vee \text{GPU})_k$ // CPU
3 $w_k^i = A_k^{[i_k,i_k]} \cdot u_k^i + A_k^{[i_k,b_k]} \cdot u_k^b$; Assembly of the vectors u_k^b;
4 $w_k^b = A_k^{[b_k,b_k]} \cdot u_k^b + A_k^{[b_k,i_k]} \cdot u_k^i$; $w_h^b = \sum_{l=1,l\neq k}^{l\leq d} A_k^{[b_k,b_l]} \cdot u_k^b$;
5 Copying w_h^b on the GPU_k;
6 $w_k^b = w_k^b + w_h^b$;
7 $m = M^{-1}w$; Assembly of the vectors m_k^b;
8 $\gamma_{1k} = (r_k, u_k)$; $\delta_k = (w_k, u_k)$; Assembly $\delta = \sum_k \delta_k; \gamma_1 = \sum_k \gamma_{1k}$;
  while $||r||_2/||b||_2 > \varepsilon$ **do**
9 $n_k^i = A_k^{[i_k,i_k]} \cdot m_k^i + A_k^{[i_k,b_k]} \cdot m_k^b$; $n_h^b = \sum_{l=1,l\neq k}^{l\leq d} A_k^{[b_k,b_l]} \cdot m_k^b$;
10 $n_k^b = A_k^{[b_k,b_k]} \cdot m_k^b + A_k^{[b_k,i_k]} \cdot m_k^i$; Copying n_h^b on the GPU_k;
11
12 $n_k^b = n_k^b + n_h^b$; $\beta = ((j = 0)\,?\,0 : \gamma_1/\gamma_0)$;
13 $z = n + \beta z$; $\alpha = \gamma_1/(\delta - \beta\gamma_1/\alpha)$;
14 $w = w - \alpha z$;
15 $q = m + \beta q$;
16 $s = w + \beta s$; Assembly of the vectors w_k^b;
17 $p = u + \beta p$;
18 $x = x + \alpha p$;
19 $r = r - \alpha s$; Assembly vectors m_k^b;
20 $u = u + \alpha q$;
21 $m = M^{-1}w$; Assembly $\delta = \sum_k \delta_k; \gamma_1 = \sum_k \gamma_{1k}$;
22 $\gamma_0 = \gamma_1$;
23 $\gamma_{1k} = (r_k, u_k)$; $\delta_k = (w_k, u_k)$;

Let us represent parallel block scheme of the method CGwO that is performed each k-th computing unit in the form of Algorithm 2. Two parallel branches of this algorithm are executed accordingly on the CPU and GPU/CPU. Operations performed in parallel are shown in one line of the algorithm. Vector operations on each computing unit occur in two stages, for internal and boundary nodes. The designations of the internal and boundary nodes for vectors are omitted, with the exception of the matrix-vector multiplication. Dot products are performed independently by each computing unit on its parts of vectors. The summation of intermediate scalars occurs in parallel threads responsible for communication, which is the synchronization point at each iteration of the algorithm.

In block CGwO, compared to Algorithm 1, the preconditioning step has been moved (line 5 to line 21). This is done in order to combine vector operations on the computing unit and the assembly of the vector parts of the right hand side to perform matrix-vector multiplication in preconditioning. The 13 line on the right uses the ternary operator: if $j = 0$, then $\beta = 0$, in other cases $\beta = \gamma_1/\gamma_0$. The subscript h is used for vectors that are stored only in CPU memory.

Numerical experiments on the Algorithm 2 were carried out on heterogeneous platform with various configuration of computing nodes containing several CPUs and GPUs. In the general case, the parallel computing on several heterogeneous computing nodes containing one or more CPUs and several GPUs is implemented by the combination of several technologies: MPI, OpenMP and CUDA.

Let us consider the software organization of computations using as example some cluster, which includes two computing nodes (8 CPU cores and 2 GPUs). Each computing node is associated with a parallel MPI process. In a parallel process, 9 parallel OpenMP threads are generated, which is one more than the available CPU cores. The eighth OpenMP thread is responsible for communications between different computing nodes (using MPI technology, vector assembly using the `Allgatherv` function, adding scalars `Allreduce`) and various GPUs. In the 2 Algorithm, the operations performed by this thread are presented to the right. Zero and first OpenMP threads are the host threads for one of the available GPU devices and are responsible for transfer data between the GPU-CPU (calls to asynchronous copying functions) and auxiliary computations. Each available GPU device (further considered as a computing unit) is associated with one of the parallel OpenMP threads, which is responsible for transferring data between the GPU and CPU (calls to asynchronous copy functions) and participates with the eighth treads in matrix-vector product on boundary vertices (lines 4, 9 right column). The remaining parallel threads (second to seventh) perform the calculations as a separate computing unit for their matrix block. The operations performed by computing units in the 2 Algorithm are shown on the left.

The preconditioning in lines 2, 7 and 21 implies the use of block matrix-vector multiplication of the form (1) considered above.

4 Numerical Experiments

The numerical experiments were performed on the heterogeneous partitions of cluster "Uran" on the computing nodes (CNs) of several types of Supercomputer center IMM UB RAS, Yekaterinburg, Russia. The cluster partitions with the following characteristics were used:

- partition "debug": 4 CNs tesla [31–32,46–47] with two 8-cores CPU Intel Xeon E5-2660 (2.2 GHz), cache memory is 20 MB L3 cache, RAM is 96 GB and 8 GPU Tesla M2090 (6 GB per device), network is 1 Gb/s Ethernet.

- partition "tesla[21–30]": 10 CNs with two 6-cores CPU Intel Xeon X5675 (3.07 GHz), RAM is 192 GB, cache memory is 12 MB L3 cache and 8 GPU Tesla M2090 (6 GB per device), with network is Infiniband 20 Gb/s.
- partition "tesla[33–45]": 13 CNs with two 8-cores CPU Intel Xeon E5-2660 (2.2 GHz), cache memory is 20 MB L3 cache, RAM is 96 GB, and 8 GPU Tesla M2090 (6 GB per device), network is Infiniband 20 Gb/s.
- partition "tesla[48–52]": 5 CNs with two 8-cores CPU Intel Xeon E5-2650 (2.6 GHz), cache memory is 20 MB L3 cache, RAM is 64 GB and 3 GPU Tesla K40m (12 GB per device), network is Infiniband 20 Gb/s.

The results of comparing two algorithms of the conjugate gradient method on SLAEs containing test matrices are presented in Table 1. The results are given for several types of computing nodes using a single graphics accelerator.

Table 1. Statistics of the test problems. Problem names, dimensions (N), number of nonzeros (nnz), device type (DT) and problem analysis in terms of the timing in seconds.

Matrix	N	nnz	# iter.	DT	Time, s	
					CG	CGwO
Plat362	362	5786	991	M2090	6.88E-01	**3.07E-01**
				K40m	4.13E-01	3.12E-01
1138_bus	1138	4054	717	M2090	3.81E-01	**1.84E-01**
				K40m	5.31E-01	2.01E-01
				debug	6.82E-01	1.90E-01
Muu	7102	170134	12	M2090	2.64E-01	4.68E-03
				K40m	3.31E-01	**4.55E-03**
Kuu	7102	340200	378	M2090	4.31E-01	**1.31E-01**
				K40m	4.39E-01	1.35E-01
Pres_Poisson	14822	715804	661	M2090	6.72E-01	3.13E-01
				K40m	6.346E-01	**2.73E-01**
Inline_1	503712	36816342	5642	M2090	4.74E+01	5.17E+01
				K40m	**3.06E+01**	3.37E+01
Fault_639	638802	28614564	4444	M2090	3.83E+01	4.32E+01
				K40m	**2.44E+01**	2.77E+01
				debug	2.44E+01	2.77E+01
thermal2	1228045	8580313	2493	M2090	1.35E+01	1.82E+01
				K40m	**8.33E+00**	1.18E+01
G3_circuit	1585478	7660826	592	M2090	3.43E+00	4.32E+00
				K40m	**1.94E+00**	2.92E+00
Quenn_4147	4147110	399499284	8257	M2090	5.46E+02	5.78E+02
				K40m	**3.55E+02**	3.75E+02

The matrices are ordered by increasing the order of the system of equations (N) and the number of nonzero elements (nnz). Bold indicates the best time to solve the system in each case. The pipelined algorithm CGwO showed a reduction in execution time on small SLAEs which are characterized by a small computing load, due to which a reduction in communications provides less time. Note that the classic CG algorithm was implemented based on CUBLAS, while the CGwO variant uses matrix and vector operations of its own GPU implementation.

For systems `Inline_1` and `Fault_639`, the execution time of the pipeline algorithm is 10 and 13.5% longer than the block version of CG, which is associated with additional vector operations that are not blocked by reduced communications. With a decrease in the number of iterations, for example, for solving a large system with `G3_circuit`) with an approximately equal number of equations with `thermal2`, the execution time of the CG and CgwO algorithms on one GPU increases slightly. For the system (`thermal2` and `G3_circuit`) the increase in costs becomes more significant.

Table 2 presents the results of the block variant of the algorithms for computing on several computing nodes for systems with small dimension matrices. Here are the results for 2 and 3 subdomains. Each subdomain was considered on a separate computing node. Communications were carried out using MPI technology. A significant influence of network characteristics on the performance of block methods can be seen in Table 2 for system of equations with matrix `1138_bus`. Computations for these SLAEs were performed at various computing nodes with different throughput and latency of the network. In numerical experiments on the CNs (partition "debug") connected by a Gigabit network, communication costs significantly increase the execution time of the CG algorithm.

Table 2. Time of solving by the block algorithms CG and CGwO on CPU/GPU, s.

Matrix/DT	CG/#blocks		CGwO/#blocks	
	2	3	2	3
`Plat362`/M2090	1.55E+00		**1.22E+00**	
/K40m	1.92E+00	1.56E+00	1.28E+00	1.31E+00
`1138_bus`/M2090	1.84E+00		**9.28E-01**	
/K40m	1.90E+00	1.85E+00	1.03E+00	1.04E+00
/debug	1.25E+01		**5.36E+00**	
`Muu`/M2090	6.12E-01		2.29E-01	
/K40m	6.59E-01	5.64E-01	2.89E-01	**2.88E-01**
`Kuu`/M2090	1.30E+00		**6.43E-01**	
/K40m	1.29E+00	1.36E+00	6.81E-01	7.95E-01
`Pres_Poisson`/M2090	1.55E+00		**9.57E-01**	
/K40m	1.60E+00	1.66E+00	1.02E+00	1.19E+00
`G3_circuit`/M2090	4.27E+00		3.99E+00	
/K40m	4.04E+00	3.510E+00	3.27E+00	**2.77E+00**

For example, in the variant `1138_bus` on the cluster partition "debug", the execution time of the pipeline algorithm is 3.6 times less (the line "debug" in Table 2 and any row in Table 1). Using the Infiniband 20 Gb/s communication network reduces the execution time for all presented systems of equations (lines "M2090" and "K40m").

When reducing the computational load, a decrease in the number of synchronization points and the consolidation of transfers per transaction is more pronounced. This shows a comparison of systems with matrices `Kuu` and `Muu`. Both systems have an equal number of equations and nonzero elements, but the conditionality of these matrices is significantly different and, as a consequence, the number of iterations in the conjugate gradient method is different. Table 1 shows that using the pipeline algorithm for the matrix `Muu` gives speedup by 70 times, compared with the matrix `Kuu`, where the speedup is only 2.8.

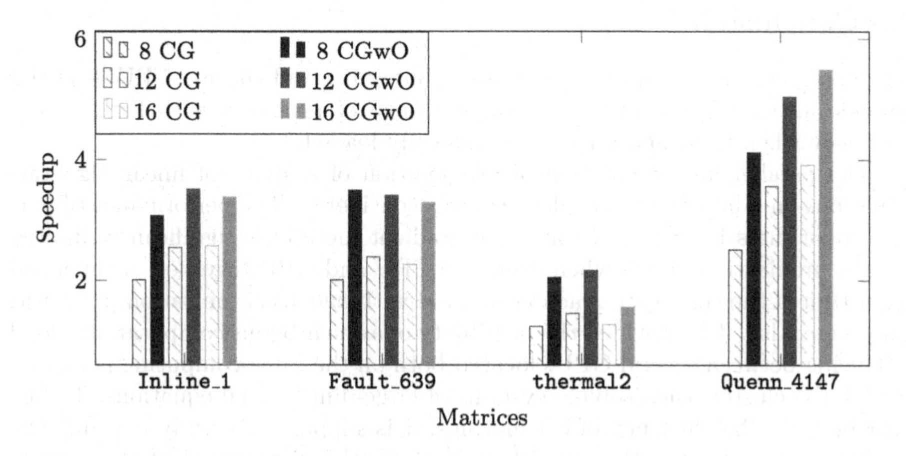

Fig. 1. Speedup of the block algorithms CG and CGwO

Figure 1 presents the results of accelerating the block algorithms of the conjugate gradient method, when divided into a larger number of blocks, accordingly 8, 12 and 16. To compute the speedup, parallel application was run repeatedly with different mapping of subdomains to several CPUs and GPUs. For example, in the case of 12 subdomains, variants were considered: 2 CPUs with 6 GPUs, 3 CPUs with 4 GPUs, 6 CPUs with 2 GPUs. The best time is shown.

The speedup was considered relative to the option on one GPU from Table 1. An application that implements this algorithm was executed in the exclusive mode of the computing node but not of the network.

As can be seen from the presented results, the pipelined CG shows the speedup greater than the classic version of conjugate gradient method. Wherein, for the largest of the considered matrices `Quenn_4147`, the speedup achieves 5.49 times, while the classical version gives 3.92 as maximum. For the strongly sparse matrix `thermal2`, block algorithms don't give high speedup (maximum is 1.56), since the computational load depends mainly on the number of the nonzero elements.

An analysis of the results showed that reducing the data size due to the matrix partitioning and reducing the synchronization points slightly decrease the impact of communication costs on the total algorithm performance. Only the use of computing nodes connected by Infiniband allowed us to get speedup when computing on several computing nodes. The matrix partitioning into blocks allowed to decrease the execution time of the pipelined block algorithm in comparison with the conjugate gradients on one node on the matrices `Inline_1`, `Fault_639` by reducing the computational load on one GPU.

Large systems `thermal2`, `G3_circuit`, solved by the block of the CGwO algorithm, as well as the reduction in communications costs and synchronization points, do not overlap the increasing costs of additional vector operations.

5 Conclusion

The heterogeneous computing platforms containing and sharing CPU + GPUs provide an effective solution to a wider range of problems with high energy efficiency when CPU and GPUs are uniformly loaded.

The parallel implementation of the solution of systems of linear algebraic equations on a heterogeneous platform was considered. The performance of parallel algorithms for classical conjugate gradient method is significantly limited by synchronization points when using the CPU and GPU together. A pipelined algorithm of the conjugate gradient method with one synchronization point was proposed. Also, it is provided the possibility of asynchronous computations, load balancing between several GPUs located both on the same computing node and for a GPU cluster when solving systems of large-dimensional equations. To further increase the efficiency of calculations, it is supposed to study not only the communication load of the algorithms but also the distributing of the computational load between the CPU and GPU. To obtain more reliable evaluation of communications costs, it is necessary to conduct a series of computational experiments on supercomputer with a completely exclusive mode of operation and a large number of heterogeneous nodes.

The following conclusions can be drawn from the analysis of data obtained during numerical experiments: the use of a pipeline algorithm reduces communication costs, but increases computational ones. For systems of small sizes or with a small number of iterations, this reduces the execution time of the algorithm when using a single GPU. For systems of large dimensions, a reduction in execution time, in comparison with CG, is possible only with a sufficiently small partition of the matrix into blocks, in which the increased computing costs overlap the communication decrease.

The proposed block algorithms, in addition to reducing the execution time, allow solving large linear systems that requires memory resources not provided by one GPU or computing node. At the same time, the pipelined block algorithm reduces the overall execution time by reducing synchronization points and combining communications into one message.

References

1. Agullo, E., Giraud, L., Guermouche, A., Roman, J.: Parallel hierarchical hybrid linear solvers for emerging computing platforms. C. R. Mec. **333**, 96–103 (2011)
2. Gaidamour, J., Henon, P.: A parallel direct/iterative solver based on a Schur complement approach. In: IEEE 11th International Conference on Computational Science and Engineering, pp. 98–105. San Paulo (2008)
3. Giraud, L., Haidar, A., Saad, Y.: Sparse approximations of the Schur complement for parallel algebraic hybrid solvers in 3D. Numer. Math. **3**, 276–294 (2010)
4. Rajamanickam, S., Boman, E.G., Heroux, M.A.: ShyLU: a hybrid-hybrid solver for multicore platforms. In: IEEE 26th International Parallel and Distributed Processing Symposium, Shanghai, pp. 631–643 (2012)
5. Yamazaki, I., Rajamanickam, S., Boman, E., Hoemmen, M., Heroux, M., Tomov, S.: Domain decomposition preconditioners for communication-avoiding Krylov methods on a hybrid CPU/GPU cluster. In: Proceedings of International Conference for High Performance Computing, Networking, Storage and Analysis (SC14), pp. 933–944 (2014)
6. Kopysov, S., Kuzmin, I., Nedozhogin, N., Novikov, A., Sagdeeva, Y.: Scalable hybrid implementation of the Schur complement method for multi-GPU systems. J. Supercomputing **69**(1), 81–88 (2014). https://doi.org/10.1007/s11227-014-1209-7
7. Hestenes, M.R., Stiefel, E.: Methods of conjugate gradients for solving linear systems. J. Res. Nat. Bur. Stan. **49**(6), 409–436 (1952)
8. Jamal, A., Baboulin, M., Khabou, A., Sosonkina, M.A.: A hybrid CPU approach GPU for the parallel algebraic recursive multilevel solver pARMS. In: 2016 18th International Symposium on Symbolic and Numeric Algorithms for Scientific Computing (SYNASC). Timisoara, pp. 411–416 (2016)
9. Kasmi, N., Zbakh, M., Mahmoudi, S.A., Manneback, P.: Performance evaluation of StarPU schedulers with preconditioned conjugate gradient solver on heterogeneous (Multi-CPUs/Multi-GPUs) architecture. IJCSNS Int. J. Comput. Sci. Netw. Secur. **17**, 206–215 (2017)
10. Cornelis, J., Cools, S., Vanroose, W.: The Communication-Hiding Conjugate Gradient Method with Deep Pipelines. https://arxiv.org/pdf/1801.04728.pdf. Accessed 14 Apr 2020
11. D'Azevedo, E.F., Romine, C.H.: Reducing communcation costs in the conjugate gradient algorithm on distributed memory multiprocessors. Technical report ORNL/TM-12192, Oak Ridge National Lab (1992)
12. Linear Algebra. Springer, Singapore (2018). https://doi.org/10.1007/978-981-13-0926-7_7
13. Chronopoulos, A.T., Gear, C.W.: s-step iterative methods for symmetric linear systems. J. Comput. Appl. Math. **25**(2), 153–168 (1989)
14. Ghysels, P., Vanroose, W.: Hiding global synchronization latency in the preconditioned Conjugate Gradient algorithm. Parallel Comput. **40**(7), 224–238 (2014)
15. Gropp, W.: Update on libraries for blue waters. http://wgropp.cs.illinois.edu/bib-/talks/tdata/2011/Stream-nbcg.pdf. Accessed 14 Apr 2020
16. Kadyrov, I.R., Kopysov, S.P., Novikov, A.K.: Partitioning of triangulated multiply connected domain into subdomains without branching of inner boundaries. Uchenye Zap. Kazanskogo Univ. Ser. Fiz. Matematicheskie Nauki **160**(3), 544–560 (2018)

Supercomputer Simulation

A Visual-Based Approach for Evaluating Global Optimization Methods

Alexander Sysoyev(✉), Maria Kocheganova, Victor Gergel, and Evgeny Kozinov

Lobachevsky State University of Nizhny Novgorod, Nizhny Novgorod, Russian Federation
{alexander.sysoyev,maria.rachinskaya,evgeny.kozinov}@itmm.unn.ru, gergel@unn.ru

Abstract. In order to evaluate efficiency of some global optimization method or compare efficiency of different methods, it is necessary to select a set of test problems, define comparison measures, and, finally, choose a way of visual presentation of the computational results. In this paper, a wide set of test optimization problems is considered including a new global constrained optimization problem generator (GCGen). Main performance measures and comparative criteria of efficiency are presented. The ways of visual presentation of computational results are suggested.

Keywords: Global optimization methods · Optimization methods comparison · Test optimization problems · Optimization problem generator · Efficiency evaluation · Operating characteristics · Visual presentation

1 Introduction

The global optimization problem, i.e. a problem of searching the extreme value of function $\varphi(y)\colon D \subset \mathbb{R}^N \to \mathbb{R}$, is discussed. Without loss of generality, the following minimization problem is considered later in the paper:

$$\varphi(y^*) = \min\{\varphi(y)\colon y \in D\}. \tag{1}$$

Here search domain D represents an N-dimensional hyperinterval:

$$D = \{y \in \mathbb{R}^N : a_i \leq y_i \leq b_i,\ i = \overline{1, N}\}.$$

Some constraints in the form of inequalities

$$\varphi_j(y) < 0,\ j = \overline{1, M}. \tag{2}$$

can be included in the optimization problem. In this case, the problem is referred to as a constrained optimization problem.

V. Voevodin and S. Sobolev (Eds.): RuSCDays 2020, CCIS 1331, pp. 137–149, 2020.
https://doi.org/10.1007/978-3-030-64616-5_12

The minimized function may have a number of peculiarities that affect the specificity and complexity of the optimization problem. For example, the function $\varphi(y)$ may be unimodal or multimodal, may have one or several global minima. The objective function may be continuous, linear, convex, separable or not have the specified characteristics. One more of the commonly used assumptions is that the minimized function satisfies the Lipschitz condition, i. e. such constant $L > 0$ exists that the following estimate takes place:

$$|\varphi(y_2) - \varphi(y_1)| \leq L\|y_2 - y_1\|,\ y_1, y_2 \in D. \tag{3}$$

Here $\|\cdot\|$ denotes the Euclidean norm in $\mathbb{R}^N$. Apart from this, the objective function may have special behavior in the vicinity of the global minimum point, for example, so-called basins or valleys. All of these characteristics have surely some impact on the process of solving the global optimization problem. More detailed description of such impact can be found, for example, in [1].

The process of developing or improving some optimization method necessarily involves series of computational experiments (tests). In the context of this work, two general goals of testing are conventionally outlined: 1) evaluating efficiency of some chosen method; 2) comparing efficiency of various methods when solving the global optimization problem. After the goal of the computational tests is determined, the researcher also faces several tasks. Firstly, a test class should be selected. Section 2 is devoted to this task. Secondly, comparison measures, or efficiency criteria, should be defined. Some important aspects of this task are reflected in the Section 3. Finally, after carrying out the computational experiments, it is important to find the most informative visual way of presenting the results. Section 4 is devoted to this task.

2 Test Classes

At present, a fairly large number of test problems are presented in the literature, both without constraints [1–5], and with constraints [6,7]. Despite this, selecting the test class when planning computational experiments is a very complex task. Such class can be composed of the similar problems or can contain problems of various types. In the first case, a class is often called a family, while in the second case it is commonly refered to as a set. In both cases, the class of problems must meet the specific needs or goals of the study. The paper [8] investigates the possible shortcomings of some test classes, as well as the details of the compatibility of the test class with the research goals. Here we dwell on some examples of test classes. These classes are successfully used when conducting computational experiments within information-statistical theory of global optimization [9].

First, consider a family of two-dimensional multiextremal problems based on the Grishagin functions [9]. An objective function as well as constraint functions (in case of the problems with constraints) are defined by the expressions

$$\varphi(y_1, y_2) = -\left\{\left(\sum_{i=1}^{7}\sum_{j=1}^{7}[A_{ij}a_{ij}(y_1, y_2) + B_{ij}b_{ij}(y_1, y_2)]\right)^2 + \left(\sum_{i=1}^{7}\sum_{j=1}^{7}[C_{ij}a_{ij}(y_1, y_2) - D_{ij}b_{ij}(y_1, y_2)]\right)^2\right\}^{\frac{1}{2}}, \quad (4)$$

where $a_{ij}(y_1, y_2) = \sin(\pi i y_1)\sin(\pi j y_2)$, $b_{ij}(y_1, y_2) = \cos(\pi i y_1)\cos(\pi j y_2)$. The search domain is a square $D = \{(y_1, y_2)\colon 0 \leq y_1, y_2 \leq 1\}$, and the parameters $A_{ij}, B_{ij}, C_{ij}, D_{ij}$ are independent random variables with uniform distribution in $[-1, 1]$. Based on the relations (5), a hundred of functions were obtained. Each of these function acts as a minimized function for one of a hundred of problems without constraints. It should be noted that such problems are multiextremal. Each problem has one and only known global minimum. The functions (5) are non-separable and satisfy the Lipschitz condition (3).

The family of test problems proposed above consists of a sufficient number of problems of equal complexity. The similar test families are often used when comparing several optimization methods or several versions of certain method (for example, a sequential version and a parallel one). In case the mentioned test goal presupposes a variety in complexity of test problems, it is proposed to use the GKLS generator. A detailed description of the generator is quite voluminous and can be found, for example, in [10]. The generator is used to derive families of multidimensional multiextremal test functions with known location of global and local minima by overriding paraboloid using polynomials. Each test class contains 100 functions and is defined by the following parameters: dimension of the problem, number of local minima, global minimum value, attraction radius of the global minimum point, distance between the global minimum point and the top of the paraboloid. By varying these parameters, one can create test families with various properties composed of the problems of different complexity [11,12]. For example, a more complex class can be created by reducing an attraction zone of the global minimum point, all else being equal.

Next, consider an example of a set of various-type unconstrained test problems. Test sets compiled in a similar way are mainly used in case it is necessary to have a large variety of problems in computational tests. For example, it happens when the scope of applicability of some optimization method is analysed: on which problems the method works at its best, on which – vice versa. In this case, it is necessary to clearly understand the characteristics and peculiarities of each problem within the set. Table 1 shows an example of a set of test problems used by the authors of the present paper. Based on the specific optimization methods researched by the authors, four main characteristics of the test problems are emphasized: the number of local and global minima, separability and problem dimension. Unless otherwise specified, the objective function of the problems in Table 1 is multiextremal, non-separable and have an only global minimum. The majority of the problems in the set have the mentioned characteristics since they are of the greatest interest in solving global optimization problems. However, the

Table 1. Example set of test problems with different peculiarities.

Problem name	Dimension	Specific characteristics
Ackley 1 Problem	$1, 2, \ldots, 32$	
Atolyan Problem	2	
Branin Problem	2	Several global minima
Camel 3,6 Hump Problem	2	
Easom Problem	2	Separable
Goldstein-Price Problem	2	
Griewank Problem	$1, 2, \ldots, 32$	
Hartman 3,6 Problem	$3, 6$	
Levy and Montavlo 1 Problem	$1, 2, \ldots, 32$	
McCormick Problem	2	
Modified Langerman Problem	$5, 10$	
Neumaier 2 Problem	4	
Neumaier 3 Problem	$2, 3, \ldots, 32$	
Paviani Problem	10	
Powell Problem	4	Unimodal
Rosenbrock Problem	$2, 3, \ldots, 32$	Unimodal
Salomon Problem	$1, 2, \ldots, 32$	
Schwefel 1.2 Problem	4	Unimodal
Schwefel 2.4 Problem	$1, 2, \ldots, 32$	Separable
Shekel 5,7,10 Problem	$5, 7, 10$	
Shubert Problem	2	Separable, several global minima

other problems are also included in the set in order to get more information when comparing method scores on different problems. Detailed description of the problems from Table 1 can be found in the works [1–3] which contain fairly complete reviews of the existing test problems.

Finally, let us propose the generator GCGen (Global Constrained optimization problem Generator). The primary description of the rules and a scheme according to which test problems are constructed can be found in [13]. The further modifications of the generator are formulated in [14]. Here let us state its most important advantages. The functions needed to formulate constrained optimization problem can be created based on some formula (thus composing some test problems family, e.g. Grishagin or GKLS functions) as well as they can be taken from some test set (e.g. two-dimensional test problems with an only global minima from Table 1). This means the problems characteristics can be varied to meet needs of computational experiments. GCGen allows formulating the constrained global optimization problems so that: 1) one could control the size of feasible domain with respect to the whole domain of the parameters variation; 2) the global minimizer of the objective function would be known a priori taking into account the constraints; 3) the global minimizer of the objective function without accounting for the constraints would be out of the feasible domain; 4) the value of the gradient at the global solution of the constrained problems is zero. Note that the location of the minimizer at the boundary of the feasible domain is an important property featuring the applied constrained optimization problems. GCGen provides the options of shifting the minimizer to the boundary of the feasible domain as well as controlling the number of the active constraints at the global solution.

Some other notes and examples of using different test classes in computational experiments when solving constrained global optimization problems can be found, for example, in [15–17].

3 Method Efficiency Criteria

When analyzing the efficiency of a certain method or conducting comparative analysis of various methods or their versions, it is recommended to have in view three following categories (see [8]): computational effort, quality of solution and method reliability.

These categories are disclosed as follows. The number of iterations and/or the running time are responsible for the *computational effort* (computational complexity) required to solve a problem. The number of iterations is often changed to similar characteristics such as number of evaluations of objective function, or functions of constraints, or derivatives of mentioned functions. While running time or memory usage demonstrates some absolute characteristic of a method, number of iterations represents relative computational effort. This becomes especially important in case an optimization problem includes some implicit functions or functions which evaluations require complex calculations (e.g. some applied problems).

The *quality of the solution* is evaluated using the achieved accuracy of an obtained solution. Note that both absolute and relative (normalized depending on the search domain size) accuracy can be considered. And besides, solution accuracy can be calculated both in terms of coordinate [18] and in terms of objective function value. The given indicators may be useful when evaluating method efficiency for some selected test problem as well as for a certain family. In the second case, the averaged values of the indicators are used (sometimes along with their variances over the family).

The *method reliability* is always evaluated according to the test results for a certain class of problems. In the paper, in order to evaluate method reliability, it is proposed to use the number (or portion) of problems solved with the method for a certain class. A problem from a selected test class is considered as solved if the method obtained a global minimum estimate which is close to the known problem solution in terms of chosen measure. When estimating method reliability based on some class of problems, it is especially important to have some visual presentation of experiments results such as operating characteristics or performance profiles (see [8]).

Note that all the efficiency criteria mentioned above represent certain final values and don't reflect the progress of achieving these values. At the same time, for example, it could be very useful to know not only achieved accuracy of some method but also an iteration number at which it was achieved. Here two ideas are proposed in order to estimate method efficiency in terms of dynamics of solving a global optimization problem. First one is a speed of getting into so-called *attraction region*. Let y^* be a global minimizer of problem (1)–(2). An attraction region is a maximum vicinity D^* of y^* in every point $y \in D^*$ of which directional derivative $\frac{\partial \varphi(y)}{\partial \mathbf{u}_y}$ of the objective function along a unit vector $\mathbf{u}_y = \frac{\overrightarrow{y-y^*}}{\|y-y^*\|}$ is positive:

$$D^* = \{y \in D \colon \forall\ 0 < \varepsilon < \|y - y^*\| \ \frac{\partial \varphi(y^* + \varepsilon \times \mathbf{u}_y)}{\partial \mathbf{u}_y} > 0\}.$$

This definition means there is no other local minima in an attraction region except global minimizer. Once some method gets into this vicinity, the methods of local optimization can be further applied. To estimate global optimization method efficiency in this context, a minimum iteration number should be marked started with which the local method never leaves the attraction region.

The second one is a *density of trial points* (see Fig. 1).

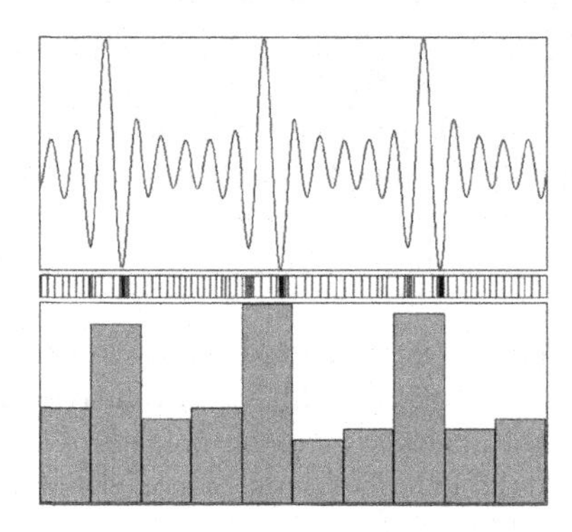

Fig. 1. Density of trial points in one-dimensional case

The density of trial points can be used to demonstrate the frequency of the trial points within the search domain for one-dimensional optimization problems. Using this graphical form, one can estimate a distribution of trial points along the search domain, observe concentration of the points in a vicinity of the global minimizer, estimate relative number of points in areas with large values of the objective function.

There are two ways to visualize the specified data for the optimization problems with a greater dimension. Firstly, a form presented in Fig. 1 can be used together with some dimension reduction scheme that reduces a multidimensional

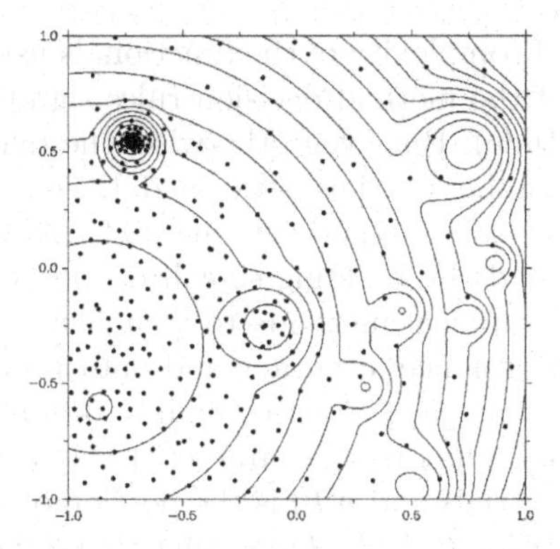

Fig. 2. Density of trial points in multidimensional case

problem to a one-dimensional one. Secondly, the trial points can be placed in a figure that contains the level lines of the objective function (see Fig. 2).

Global search algorithms for solving multidimensional applied problems are significantly computationally time-consuming. In this regard, parallelization is a typical approach to reducing computation time. When analyzing the efficiency of parallel global search algorithms, one should take into account such characteristics as acceleration and scalability, as well as the efficiency of parallel implementations.

Evaluation of these characteristics for global search algorithms is usually not an easy task. When solving problems even from certain test class (see Section 2), the values of the characteristics under consideration can vary greatly. Consequently, in order to gain reliable evaluation of the acceleration, scalability, and efficiency parameters, it is necessary to solve a series of global search problems. Note that the obtained values of the characteristics may depend on the class of problems being solved.

When solving applied problems, the time taken to select test points is usually insignificant compared to the time to calculate the values of the objective function. That's why, when comparing efficiency of different parallel implementations of global search algorithms, it is often taken into account not only reduction of computational time in a whole, but also reduction of the number of iterations needed to find solution.

4 Visual Presentation of the Computational Experiments Results

Numerical methods of global optimization are specific iterative procedures. An iteration includes the following steps: selecting a next trial point from the search domain according to some rule (trial means computing all the functionals: the objective function and constraints (if any)), checking the stop condition, performing the trial itself, updating current global minimum estimate. In applied problems, computational complexity of the functionals usually exceeds the computational complexity of the method decision rules significantly (sometimes by several orders of magnitude). But even otherwise, the main efficiency indicator for some method is the number of trials performed. Based on this, it is convenient to evaluate efficiency of certain method running with different parameters values or compare various methods by building so-called operating characteristics [9]. Note that some useful graphical methods to visualize results of the computational tests (various performance, accuracy and data profiles) are surveyed, for example, in [8]. The present paper focuses mainly on operating characteristics and their different forms. Operating characteristic is a set of pairs $(k, p(k))$, where k is the number of trials and $p(k)$ is the portion of the problems from the test set which are solved for k trials. Operating characteristics can be visually represented in the form of graphs where the number of performed trials is plotted on the abscissa-axis and the portion (or percentage, or number) of solved problems are plotted on the ordinate-axis (see Fig. 3).

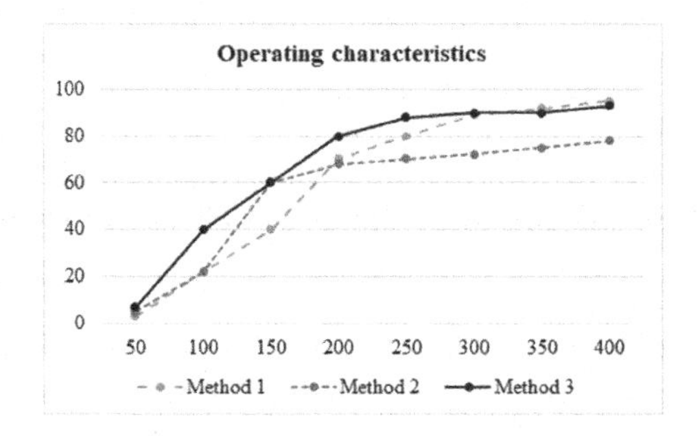

Fig. 3. Common form of operating characteristics

Operating characteristics clearly demonstrate the relative efficiency of the methods. For example, according to results in Fig. 3 method 1 solves a smaller number of problems than methods 2 and 3 for 50, 100 and 150 iterations. At 200 iterations, method 1 overtakes method 2 in the number of solved problems. After that, at 350 and 400 iterations, method 1 has the better score than method 3 as well. On the other hand, it may be possible that the maximum number of iterations that a method can do (available calculational resource) is limited, for example, to 50 iterations. In such case all of the considered methods demonstrate comparable (and low) efficiency for this test class. With a resource of 100 iterations, method 3 has a definite advantage, solving 40 problems out of 100 and twice the performance of methods 1 and 2. In general, the relative position of the graphs allows one to compare the methods efficiency: the higher the method graph is, the higher its efficiency on this test class.

Operating characteristics can also be presented in the form of bar or pie charts (see Fig. 4 and Fig. 5 respectively). In case of pie chart operating characteristics are constructed for a single method. Each sector represents a percentage

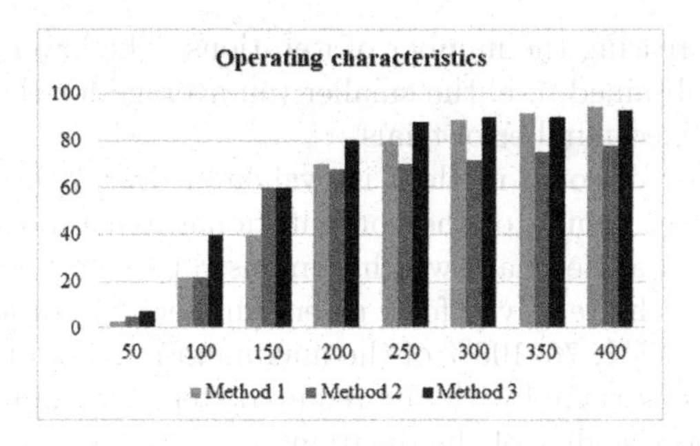

Fig. 4. Operating characteristics in the form of a bar chart

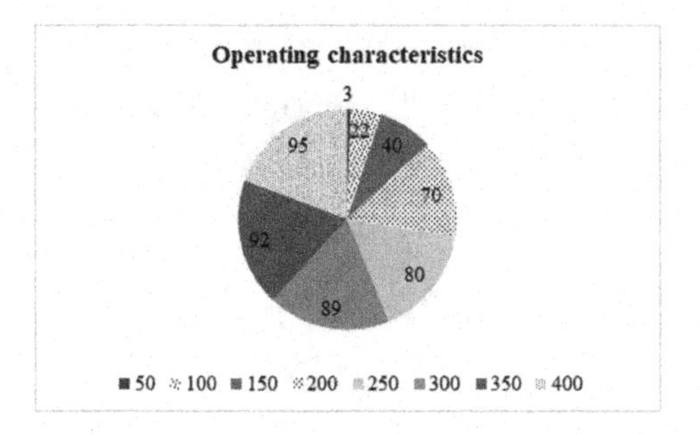

Fig. 5. Operating characteristics in the form of a pie chart

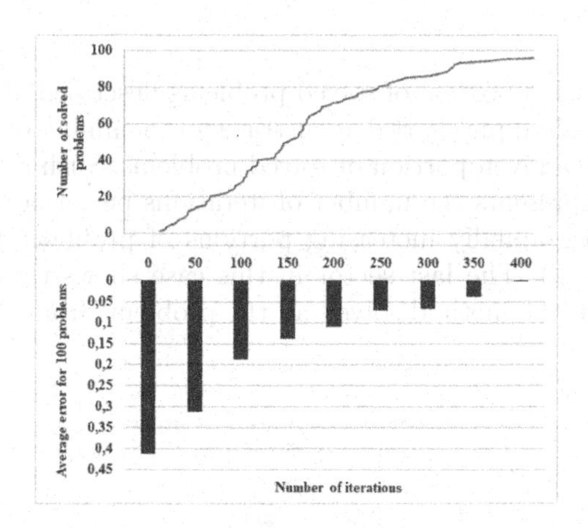

Fig. 6. Operating characteristics for average error

of the problems from the test class which are solved for corresponding number of iterations.

Operating characteristics can be supplemented with a diagram of the average error (see Fig. 6). The diagram shows how the resulting computational error decreases while increasing the number of iterations. The lower the graph is, the better results are obtained, i. e. the smaller (on average for the test class) error is reached for the fixed number of trials.

In case efficiency of some method is evaluated over a test class, one more visual tool to present results of the computational experiments may be useful (see Fig. 7). This is a pie chart which demonstrates the portion of problems from a test class which are solved for a given number of iterations (for example, 0–25%, 26–50%, 51–75%, 76–100% of the maximum number of iterations). Such a diagram can be constructed with any required step (for example, 20% or 10%) which preserves the visuality of the diagram.

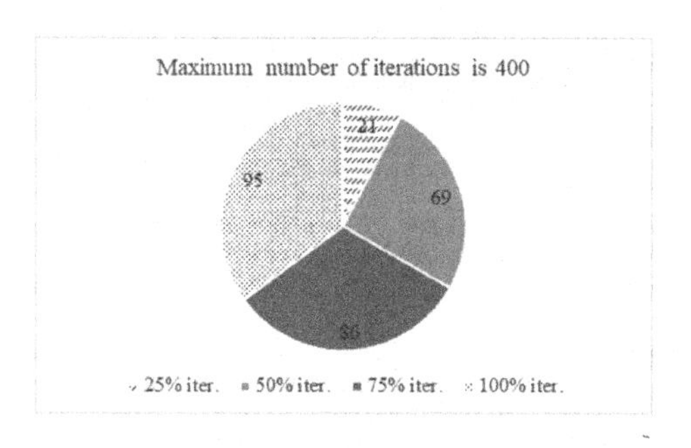

Fig. 7. Pie chart for portion of problems within test class solved for a given number of iterations

The diagram with a portion of solved problems described above may be supplemented with a paired pie chart demonstrating the number of search iterations required to achieve a given portion of solved problems. Such a diagram assumes that each sector represents the number of iterations performed by a method to solve the specified gradually increasing portions of problems from the certain test class (see Fig. 8). The last sector in this case shows the total number of iterations for which the method solved all the problems from the test class.

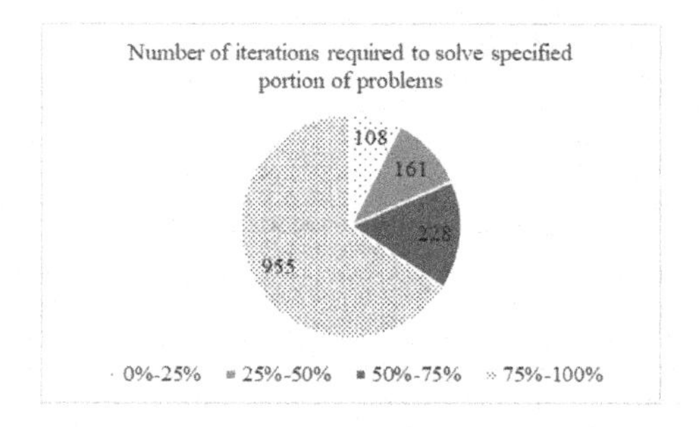

Fig. 8. Pie chart for a number of iterations performed by a method to solve given portion of problems

If two or more methods are compared according to several indicators at once it is effective to use a radar chart (see Fig. 9). In this case, the indicators values of one of the compared methods are selected as a reference point, i.e. they form a unit circle in the diagram. This selected method is considered as a base method. The indicators values of all other methods are scaled relative to the values of the base method. In order to increase visuality, it is recommended to give the same meaning to the relative position of the different indicators values toward the unit circle. For example, the less the indicator value, the better evaluation is given to the method. In this case, if the line of some analyzed method is fully located inside a unit circle, then this method is better than the base one for this test class. Conversely, if the line of the analyzed method is located outside the unit circle for all the indicators, then this method works worse than the base one for the selected class of test problems.

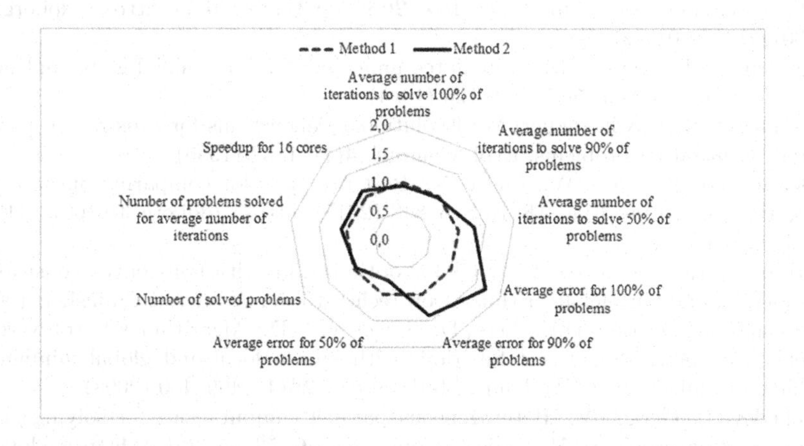

Fig. 9. Radar chart for comparing methods using several indicators

5 Conclusion

The paper highlights the most important steps of carrying computational experiments when evaluating efficiency of some global optimization method(s). An approach to perform such experiments is described. The global constrained optimization problem generator is presented. Some new efficiency criteria are proposed. The visual forms of evaluating the method efficiency are described. The proposed forms allow one to conduct comparative analysis of the different versions of one method or the different global optimization methods.

Acknowledgements. This research was supported by Russian Foundation for Basic Research (grant 19-07-00242).

References

1. Jamil, M., Yang, X.-S.: A literature survey of benchmark functions for global optimization problems. Int. J. Math. Model. Numer. Optim. **4**(2), 150–194 (2013). https://doi.org/10.1504/IJMMNO.2013.055204
2. Ali, M.M., Khompatraporn, C., Zabinsky, Z.B.: A numerical evaluation of several stochastic algorithms on selected continuous global optimization test problems. J. Glob. Optim. **31**(4), 635–672 (2005)
3. Paulavicius, R., Zilinskas, J.: Simplicial Global Optimization. Springer, New York, Briefs in Optimization, pp. 137 (2014). https://doi.org/10.1007/978-1-4614-9093-7
4. Yang, X.S.: Test Problems in Optimization. Engineering Optimization: An Introduction with Metaheuristic Applications. John Wliey & Sons, New Jersey (2010)
5. Nedělková, Z., Lindroth, P., Patriksson, M., Strömberg, A.-B.: Efficient solution of many instances of a simulation-based optimization problem utilizing a partition of the decision space. Ann. Oper. Res. **265**(1), 93–118 (2017). https://doi.org/10.1007/s10479-017-2721-y
6. Romeijn, H.E., Smith, R.L.: Simulated annealing for constrained global optimization. J. Glob. Optim. **5**(2), 101–126 (1994)
7. Michalewicz, Z., Schoenauer, M.: Evolutionary algorithms for constrained parameter optimization problems. Evol. Comput. **4**(1), 1–32 (1996)
8. Beiranvand, V., Hare, W., Lucet, Y.: Best practices for comparing optimization algorithms. Optim. Eng. **18**(4), 815–848 (2017). https://doi.org/10.1007/s11081-017-9366-1
9. Strongin, R.G., Sergeyev, Y.D.: Global optimization with non-convex constraints: sequential and parallel algorithms. Dordrecht: Kluwer Academic Publishers (2013)
10. Gaviano, M., Kvasov, D.E., Lera, D., Sergeyev, Y.D.: Algorithm 829: software for generation of classes of test functions with known local and global minima for global optimization. ACM Trans. Math. Softw. **29**(4), 469–480 (2003)
11. Lebedev, I., Gergel, V.: Heterogeneous parallel computations for solving global optimization problems. Procedia Comput. Sci. **66**, 53–62 (2015). https://doi.org/10.1016/j.procs.2015.11.008
12. Barkalov, K., Gergel, V., Lebedev, I.: Solving global optimization problems on GPU cluster. AIP Conf. Proc. **1738**, 400006 (2016). https://doi.org/10.1063/1.4952194
13. Gergel, V.: An approach for generating test problems of constrained global optimization. In: Battiti, R., Kvasov, D.E., Sergeyev, Y.D. (eds.) LION 2017. LNCS, vol. 10556, pp. 314–319. Springer, Cham (2017). https://doi.org/10.1007/978-3-319-69404-7_24
14. Gergel, V., Barkalov, K., Lebedev, I., Rachinskaya, M., Sysoyev, A.: A flexible generator of constrained global optimization test problems. AIP Conf. Proc. **2070**, 020009 (2019). https://doi.org/10.1063/1.5089976
15. Barkalov, K., Lebedev, I.: Parallel algorithm for solving constrained global optimization problems. LNCS **10421**, 396–404 (2017). https://doi.org/10.1007/978-3-319-62932-2_38
16. Gergel, V., Kozinov, E.: An approach for parallel solving the multicriterial optimization problems with non-convex constraints. LNCS **793**, 121–135 (2017). https://doi.org/10.1007/978-3-319-71255-0_10

17. Gergel, V., Barkalov, K., Lebedev, I.: A global optimization algorithm for non-convex mixed-integer problems. LNCS **11353**, 78–81 (2019). https://doi.org/10.1007/978-3-030-05348-2_7
18. Barkalov, K., Strongin, R.: Solving a set of global optimization problems by the parallel technique with uniform convergence. J. Glob. Optim. **71**(1), 21–36 (2017). https://doi.org/10.1007/s10898-017-0555-4

Adaptive Global Optimization Using Graphics Accelerators

Konstantin Barkalov[1(✉)], Ilya Lebedev[1], and Vassili Toropov[1,2]

[1] Lobachevsky State University of Nizhny Novgorod, Nizhny Novgorod, Russia
{konstantin.barkalov,ilya.lebedev}@itmm.unn.ru
[2] Queen Mary University of London, London, UK
v.v.toropov@qmul.ac.uk

Abstract. Problems of multidimensional multiextremal optimization and numerical methods for their solution are considered. The general assumption is made about the function being optimized: it satisfies the Lipschitz condition with an a priori unknown constant. Many approaches to solving problems of this class are based on reducing the dimension of the problem; i.e. addressing a multidimensional problem by solving a family of problems with lower dimension. In this work, an adaptive dimensionality reduction scheme is investigated, and its implementation using graphic accelerators is proposed. Numerical experiments on several hundred test problems were carried out, and they confirmed acceleration in the developed GPU version of the algorithm.

Keywords: Global optimization · Multiextremal functions · Reduction of dimensionality · Peano space-filling curves · Recursive optimization · Graphics accelerators

1 Introduction

A promising direction in the field of parallel global optimization (which, indeed, is true in many areas related to the software implementation of time-consuming algorithms) is the use of graphics processing units (GPUs). In the past decade, graphics accelerators have rapidly increased performance to meet the ever-growing demands of graphics application developers. Additionally, in the past few years some principles for developing graphics hardware have changed, and as a result it has become more programmable. Today, a graphics accelerator is a flexibly programmable, massive parallel processor with high performance, which is in demand for solving a range of computationally time-consuming problems [14].

However, the potential for graphics accelerators to solve global optimization problems has not yet been fully realized. Using GPUs, they basically parallelize nature-inspired optimization algorithms, which are somehow based on the idea of random search (see, for example, [5,7,17]). By virtue of their stochastic nature, algorithms of this type guarantee convergence to the global minimum only in

V. Voevodin and S. Sobolev (Eds.): RuSCDays 2020, CCIS 1331, pp. 150–161, 2020.
https://doi.org/10.1007/978-3-030-64616-5_13

the sense of probability, which differentiates them unfavorably from deterministic methods.

With regard to many deterministic algorithms of Lipschitzian global optimization with guaranteed convergence, parallel variants have been proposed [4,13,19]. However, these versions of algorithms are parallelized on CPU using shared and/or distributed memory; presently, no GPU implementations have been made. For example, [19] describes parallelization of an algorithm based on the ideas of the branch and boundary method using MPI and OpenMP.

Within the framework of this research, we consider the problems of capturing the optimum, which are characterized by a lengthy period for calculating the values of objective function in comparison with the time needed for processing them. For example, objective function can be specified using systems of linear algebraic equations, systems of ordinary differential equations, etc. Currently, graphics accelerators can be used to solve problems of this type. Moreover, an accelerator can solve several such problems at once [16]; i.e., using GPU, one can calculate multiple function values simultaneously.

Thus, calculating the optimization criterion can be implemented on GPU, and the role of the optimization algorithm (running on CPU) consists in the selection of points for conducting parallel trials. This scheme of working with the accelerator is fully consistent with the work of the parallel global search algorithm developed at the Lobachevsky State University of Nizhni Novgorod and presented in a series of papers [1–3,9–12].

2 Multidimensional Parallel Global Search Algorithm

Let's consider the problem of finding the global minimum of the N-dimensional function $\varphi(y)$ in the hyperinterval $D = \{y \in R^N : a_i \leqslant x_i \leqslant b_i, 1 \leqslant i \leqslant N\}$. We will assume that the function satisfies the Lipschitz condition with an a priori unknown constant L.

$$\varphi(y^*) = \min\{\varphi(y) : y \in D\}, \tag{1}$$

$$|\varphi(y_1) - \varphi(y_2)| \leqslant L\|y_1 - y_2\|, y_1, y_2 \in D, 0 < L < \infty. \tag{2}$$

In this work we will use an approach based on the idea of reducing dimensionality using the Peano space-filling curve $y(x)$, which continuously and unambiguously maps a segment of the real axis $[0, 1]$ onto an n-dimensional cube

$$\{y \in R^N : -2^{-1} \leqslant y_i \leqslant 2^{-1}, 1 \leqslant i \leqslant N\} = \{y(x) : 0 \leqslant x \leqslant 1\}. \tag{3}$$

The questions of numerical construction of approximations to Peano curve (*evolvents*) and the corresponding theory are discussed in detail in [21,23]. Using evolvents $y(x)$ reduces the multidimensional problem (1) to a one-dimensional problem

$$\varphi(y^*) = \varphi(y(x^*)) = \min\{\varphi(y(x)) : x \in [0, 1]\}.$$

An important property is that the relative differences of the function remain limited: if the function $\varphi(y)$ in the region D satisfies the Lipschitz condition, then the function $\varphi(y(x))$ in the interval $[0,1]$ will satisfy a uniform Hölder condition

$$|\varphi(y(x_1)) - \varphi(y(x_2))| \leqslant H|x_1 - x_2|^{\frac{1}{N}}, x_1, x_2 \in [0,1],$$

where the Hölder constant H is related to the Lipschitz constant L by the ratio $H = 2L\sqrt{N+3}$. Therefore, without limiting generality, we can consider minimizing the one-dimensional function $f(x) = \varphi(y(x)), x \in [0,1]$, which satisfies the Hölder condition.

The algorithm for solving this problem (Global Search Algorithm, GSA) involves constructing a sequence of points x_k, in which the values of objective function $z_k = f(x_k)$ are calculated. We will call the process of computing the value of a function at a single point a trial. Assume that we have $p \geqslant 1$ computational elements at our disposal and p trials which are performed simultaneously (synchronously) within a single iteration of the method. Let $k(n)$ denote the total number of trials performed after n parallel iterations.

At the first iteration of the method, the trial is carried out at an arbitrary internal point x^1 of the interval $[0,1]$. Let $n > 1$ iterations of the method be performed, during which trials were carried out at $k = k(n)$ points $x^i, 1 \leqslant i \leqslant k$. Then the trial points $x^{k+1}, \ldots, x^{k+p}$ of the next $(n+1)$th iteration are determined in accordance with the following rules.

Step 1. Renumber the points of the set $X_k = \{x^1, \ldots, x^k\} \cup \{0\} \cup \{1\}$, which includes the boundary points of the interval $[0,1]$, as well as the points of the previous trials, with the lower indices in the order of their increasing coordinate values, i.e.

$$0 = x_0 < x_1 < \ldots < x_{k+1} = 1.$$

Step 2. Assuming $z_i = f(y(x_i)), 1 \leqslant i \leqslant k$, calculate the values

$$\mu = \max_{1 \leqslant i \leqslant k} \frac{|z_i - z_{i-1}|}{\Delta_i}, \; M = \begin{cases} r\mu, \mu > 0, \\ 1, \mu = 0, \end{cases}$$

where $r > 1$ is the specified parameter of the method, and $\Delta_i = (x_i - x_{i-1})^{\frac{1}{N}}$.

Step 3. For each interval $(x_{i-1}, x_i), 1 \leqslant i \leqslant k+1$, calculate the characteristic in accordance with the formulas

$$R(1) = 2\Delta_1 - 4\frac{z_1}{M}, \; R(k+1) = 2\Delta_{k+1} - 4\frac{z_k}{M},$$

$$R(i) = \Delta_i + \frac{(z_i - z_{i-1})^2}{M^2 \Delta_i} - 2\frac{z_i + z_{i-1}}{M}, 1 < i < k+1.$$

Step 4. Arrange characteristics $R(i), 1 \leqslant i \leqslant k+1$, in descending order

$$R(t_1) \geqslant R(t_2) \geqslant \cdots \geqslant R(t_k) \geqslant R(t_{k+1})$$

and select p of the largest characteristics with interval numbers $t_j, 1 \leqslant j \leqslant p$.

Step 5. Carry out new trials at the points $x_{k+j}, 1 \leqslant j \leqslant p$, calculated using the formulas

$$x_{k+j} = \frac{x_{t_j} + x_{t_j-1}}{2}, \; t_j = 1, \; t_j = k+1,$$

$$x^{k+1} = \frac{x_{t_j} + x_{t_j-1}}{2} - \text{sign}(z_{t_j} - z_{t_j-1})\frac{1}{2r}\left[\frac{|z_{t_j} - z_{t_j-1}|}{\mu}\right]^N, 1 < t_j < k+1.$$

The algorithm stops working if the condition $\Delta_{t_j} \leqslant \varepsilon$ is satisfied for at least one number $t_j, 1 \leqslant j \leqslant p$; here $\varepsilon > 0$ is the specified accuracy. As an estimate of the globally optimal solution to the problem (1), the values are selected

$$f_k^* = \min_{1 \leq i \leq k} f(x^i), x_k^* = \arg \min_{1 \leq i \leq k} f(x^i).$$

For the rationale in using this method of organizing parallel computing see [23].

3 Dimensionality Reduction Schemes in Global Optimization Problems

3.1 Dimensionality Reduction Using Multiple Mappings

Reducing multidimensional problems to one-dimensional ones through the use of evolvents has important properties such as continuity and preserving uniform bounding of function differences with limited argument variation. However, some information about the proximity of points in multidimensional space is lost, since the point $x \in [0,1]$ has only left and right neighbors, and the corresponding point $y(x) \in R^N$ has neighbors in 2^N directions. As a result, when using evolvents, the images y', y'' that are close in N-dimensional space can correspond to rather distant preimages x', x'' on the interval $[0,1]$. This property leads to redundant calculations, because several limit points x', x'' of the sequence of the trial points generated by the method on the segment $[0,1]$, can correspond to a single limit point y in N-dimensional space.

A possible way to overcome this disadvantage is to use a set of evolvents (*multiple mappings*)

$$Y_L(x) = \{y^0(x), \; y^1(x), ..., \; y^L(x)\}$$

instead of using a single Peano curve $y(x)$ (see [22,23]). For example, each Peano curve $y^i(x)$ from $Y_L(x)$ can be obtained as a result of some shifting $y(x)$ along the main diagonal of the hyperinterval D. Another way is to rotate the evolvent $y(x)$ around the origin. The set of evolvents that have been constructed allows us to obtain for any close images y', y'' close preimages x', x'' for some mapping $y^i(x)$.

Using a set of mappings leads to the formation of a corresponding set of one-dimensional multiextremal problems

$$\min \left\{\varphi(y^l(x)) : x \in [0,1], \; \right\}, \; 0 \leqslant l \leqslant L.$$

Each problem from this set can be solved independently, and any calculated value $z = \varphi(y')$, $y' = y^i(x')$ of the function $\varphi(y)$ in the i-th problem can be interpreted as calculating the value $z = \varphi(y')$, $y' = y^s(x'')$ for any other s-th problem without repeated labor-intensive calculations of the function $\varphi(y)$. Such informational unity makes it possible to solve the entire set of problems in a parallel fashion. This approach was discussed in detail in [3].

3.2 Recursive Dimensionality Reduction Scheme

The recursive optimization scheme is based on the well-known relation

$$\min \varphi(y) : y \in D = \min_{a_1 \leqslant y_1 \leqslant b_1} \min_{a_2 \leqslant y_2 \leqslant b_2} \dots \min_{a_1 \leqslant y_N \leqslant b_N} \varphi(y), \tag{4}$$

which allows one to replace the solution of the multidimensional problem (1) with the solution of a family of one-dimensional subproblems recursively related to each other. Let's introduce a set of functions

$$\varphi_N(y_1, \dots, y_N) = \varphi(y_1, \dots, y_N), \tag{5}$$

$$\varphi_i(y_1, \dots, y_i) = \min_{a_{i+1} \leqslant y_{i+1} \leqslant b_{i+1}} \varphi_{i+1}(y_1, \dots, y_i, y_{i+1}), 1 \leqslant i \leqslant N-1. \tag{6}$$

Then, in accordance with the relation (4), the solution of the original problem (1) is reduced to the solution of a one-dimensional problem

$$\varphi_1(y_1^*) = \min\{\varphi_1(y_1), y_1 \in [a, b]\}. \tag{7}$$

However, each calculation of the value of the one-dimensional function $\varphi_1(y_1)$ at some fixed point corresponds to the solution of a one-dimensional minimization problem

$$\varphi_2(y_1, y_2^*) = \min\{\varphi(y_1, y_2) : y_2 \in [a_2, b_2]\}.$$

And so on, until the calculation of φ_N according to (5).

For the recursive scheme described above, a generalization (*block recursive scheme*) is proposed that combines the use of evolvents and a recursive scheme in order to efficiently parallelize computations.

Consider the vector y as a vector of block variables

$$y = (y_1, \dots, y_N) = (u_1, u_2, \dots, u_M),$$

where the i-th block variable u_i is a vector of sequentially taken components of the vector y, i.e. $u_1 = (y_1, y_2, \dots, y_{N_1})$,$u_2 = (y_{N_1+1}, y_{N_1+2}, \dots, y_{N_1+N_2})$,..., $u_M = (y_{N-N_M+1}, y_{N-N_M+2}, \dots, y_N)$, while $N_1 + N_2 + \dots + N_M = N$.

Using new variables, the main relation of the nested scheme (4) can be rewritten as

$$\min_{y \in D} \varphi(y) = \min_{u_1 \in D_1} \min_{u_2 \in D_2} \dots \min_{u_M \in D_M} \varphi(y), \tag{8}$$

where the subdomains $D_i, 1 \leqslant i \leqslant M$, are projections of the original search domain D onto the subspaces corresponding to the variables $u_1, 1 \leqslant i \leqslant M$.

The formulas that determine the method for solving problem (1) based on relations (8) generally coincide with the recursive scheme (5)–(7). One need only replace the original variables $y_i, 1 \leqslant i \leqslant N$, with block variables $u_1, 1 \leqslant i \leqslant M$. In this case, the fundamental difference from the original scheme is the fact that the block scheme has nested subproblems

$$\varphi_i(u_1, \ldots, u_i) = \min_{u_{i+1} \in D_{i+1}} \varphi_{i+1}(u_1, \ldots, u_i, u_{i+1}), 1 \leqslant i \leqslant M-1, \qquad (9)$$

which are multidimensional, and to solve them a method of reducing dimensionality based on Peano curves can be applied.

3.3 Adaptive Dimensionality Reduction Scheme

Solving the resulting set of subproblems (9) can be organized in various ways. The obvious method (elaborated in detail in [11] for the nested optimization scheme and in [1] for the block nested optimization scheme) is based on solving subproblems in accordance with the recursive order of their generation. However, in this case there is a significant loss of information about the target function.

Another approach is an adaptive scheme in which all subproblems are solved simultaneously, which makes it possible to more fully take into account information about a multidimensional problem and thereby to speed up the process of solving it. In the case of one-dimensional subproblems, this approach was theoretically substantiated and tested in [10,12], and in the paper [2] a generalization of the adaptive scheme for multidimensional subproblems was proposed.

The adaptive dimensionality reduction scheme changes the order in which subproblems are solved: they will be solved not one by one (in accordance with their hierarchy in the task tree), but simultaneously, i.e., there will be a number of subproblems that are in the process of being solved. Under the new scheme:

- to calculate the value of the function of the i-th level from (9), a new $(i+1)$th level problem is generated in which trials are carried out, after which a new generated problem is included in the set of existing problems to be solved;
- the iteration of the global search consists in choosing p (the most promising) problems from the set of existing problems in which trials are carried out; points for new trials are determined in accordance with the parallel global search algorithm from Section 2;
- the minimum values of functions from (9) are their current estimates based on the accumulated search information.

A brief description of the main steps of a block adaptive dimensionality reduction scheme is as follows. Let the nested subproblems in the form (9) be solved using the global search algorithm described in Section 2. Then each subproblem (9) can be assigned a numerical value called the characteristic of this problem. As such, we can take the maximum characteristic $R(t)$ of the intervals formed in this problem. In accordance with the rule for calculating characteristics, the higher the value of the characteristic, the more promising the subproblem is in the continued search for the global minimum of the original problem (1).

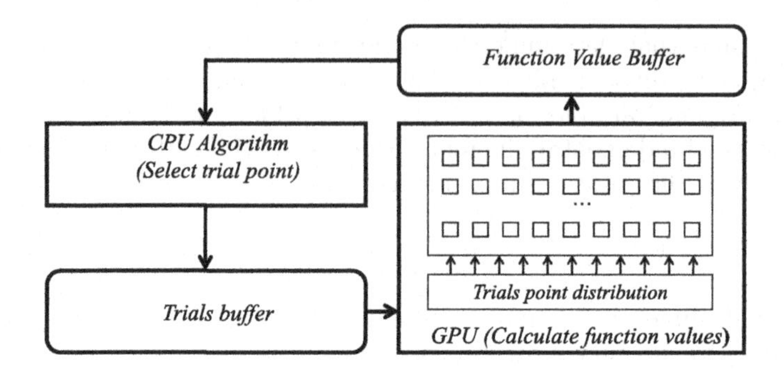

Fig. 1. Scheme of information exchanges in the GPU algorithm

Therefore, at each iteration, subproblems with the maximum characteristic are selected for conducting the next trial. The trial either leads to the calculation of the value of the objective function $\varphi_1(y)$ (if the selected subproblem belonged to the level $j = M$), or generates new subproblems according to (9) for $j \leqslant M-1$. In the latter case, the newly generated problems are added to the current set of problems, their characteristics are calculated, and the process is repeated. The optimization process is completed when the root problem satisfies the condition for stopping the algorithm that solves this problem. Some results pointing in this direction are presented in [2].

4 GPU Implementation

4.1 General Scheme

In relation to global optimization methods, an operation that can be efficiently implemented on GPU is the parallel calculation of many values of the objective function at once. Naturally, this requires implementing a procedure for calculating the value of a function on GPU. Data transfers from CPU to GPU will be minimal: one need only transfer the coordinates of the trial points to GPU, and get back the function values at these points. Functions that determine the processing of trial results in accordance with the algorithm, and require work with large amount of accumulated search information, can be efficiently implemented on CPU.

The general scheme for organizing calculations using GPU is shown in Fig. 1. In accordance with this scheme, steps 1–4 of the parallel global search algorithm are performed on CPU. The coordinates of the p trial points calculated in step 4 of the algorithm are accumulated in the intermediate buffer and then transmitted to GPU. On GPU the function values are calculated at these points, after which the trial results (again through the intermediate buffer) are transferred to CPU.

4.2 Organization of Parallel Computing

To organize parallel calculations, we will use a set of evolvents and a block adaptive dimensionality reduction scheme. We take a small number of nesting

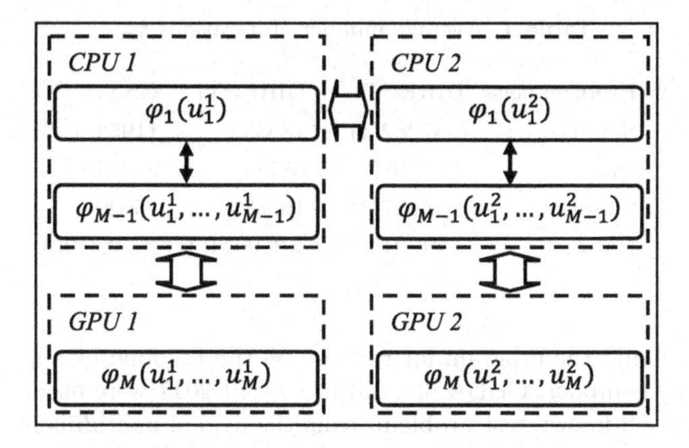

Fig. 2. Diagram of parallel computing on a cluster

levels, in which the original large dimensionality problem is divided into 2–3 nested lower dimensional subproblems. We will use multiple evolvents only at the upper level of nesting, corresponding to the variable u_1. This subproblem will be reduced to a set of one-dimensional problems that will be solved in parallel, each in a separate process. Trial results at point x obtained for the problem solved by a specific processor are interpreted as the trial results in the remaining problems (at the corresponding points $u_1^1, ..., u_1^s$). Then, applying an adaptive scheme to solve the nested subproblems (9), we get a parallel algorithm with a wide degree of variability.

Figure 2 shows the general scheme of organizing computations using several cluster nodes and several GPUs. In accordance with this scheme, nested subproblems $\varphi_i(u_1, ..., u_i) = \min_{u_{i+1} \in D_{i+1}} \varphi_{i+1}(u_1, ..., u_i, u_{i+1})$ with $i = 1, \ldots, M-2$ are solved using CPU only. The values of the function are not calculated directly in these subproblems: the calculation the function value $\varphi_i(u_1, ..., u_i)$ is a solution to the minimization problem at the next level. The subproblem of the last $(M-1)$-th level

$$\varphi_{M-1}(u_1, ..., u_{M-1}) = \min_{u_M \in D_M} \varphi_M(u_1, ..., u_M)$$

differs from all the previous subproblems; it calculates the values of the objective function, since $\varphi_M(u_1, ..., u_M) = \varphi(y_1, ..., y_N)$. This subproblem transfers data between CPU and GPU.

5 Numerical Experiments

The numerical experiments were carried out using the Lomonosov supercomputer (Lomonosov Moscow State University). Each supercomputer node included two quad-core processors Intel Xeon X5570, two NVIDIA Tesla X2070 and 12

Table 1. Average number of iterations k_{av}

N	Problem class	DIRECT	DIRECTl	GSA
4	Simple	>47282(4)	18983	11953
	Hard	>95708(7)	68754	25263
5	Simple	>16057(1)	16758	15920
	Hard	>217215(16)	>269064(4)	>148342(4)

Gb RAM. To build the program for running on the Lomonosov supercomputer, the GCC 4.3.0 compiler, CUDA 6.5 and Intel MPI 2017 were used.

Note that well-known test problems from the field of multidimensional global optimization are characterized by a short time of calculating the values of the objective function. Therefore, in order to simulate the computational complexity inherent in applied optimization problems [18], the calculation of the objective function in all experiments was complicated by additional calculations that do not change the form of the function and the location of its minima (summing the segment of the Taylor series). In the experiments carried out the average time for calculating the function values was 0.01 s, which exceeds the latency of the network or the data transfer time between CPU and GPU.

In the paper [8] a GKLS generator is described that allows one to generate multiextremal optimization problems with previously known properties: the number of local minima, the size of their regions of attraction, the global minimum point, the value of the function in it, etc.

Below are the results of a numerical comparison of three sequential algorithms: DIRECT [15], DIRECTl [6] and Global Search Algorithm (GSA) from Section 2. A numerical comparison was carried out on the Simple and Hard function classes of dimension 4 and 5 from [8]. The global minimum y^* was considered found if the algorithm generated a trial point y^k in the δ-neighborhood of the global minimum, i.e. $\left\|y^k - y^*\right\| \leqslant \delta$. The size of the neighborhood was chosen (in accordance with [20]) as $\delta = \|b - a\| \sqrt[N]{\Delta}$, here N is the dimension of the problem to be solved, a and b are the boundaries of the search domain D, the parameter $\Delta = 10^{-6}$ for $N = 4$ and $\Delta = 10^{-7}$ for $N = 5$. When using the GSA method for the Simple class, the parameter $r = 4.5$ was selected, for the Hard class $r = 5.6$; the parameter for constructing the Peano curve was $m = 10$. The maximum number of iterations allowed was $K_{max} = 10^6$.

Table 1 shows the average number of iterations, k_{av}, that the method performed when solving a series of problems from these classes. The symbol " > " reflects a situation where not all problems of the class were solved by any method whatsoever. This means that the algorithm was halted because the maximum allowed number of K_{max} iterations was reached. In this case, the value $K_{max} = 10^6$ was used to calculate the average value of the number of iterations, k_{av}, which corresponds to the lower estimate of this average value. The number of unsolved problems is indicated in parentheses.

Table 2. Speedup on CPU

	Iteration speedup				Time speedup			
	$N = 4$		$N = 5$		$N = 4$		$N = 5$	
	Simple	Hard	Simple	Hard	Simple	Hard	Simple	Hard
$p = 2$	6,6	3,2	2,1	6,6	3,2	1,5	0,7	2,1
$p = 4$	21,5	10,0	6,9	19,2	5,2	2,4	0,9	2,3

Table 3. Speedup on one GPU

	Iteration speedup				Time speedup			
	$N = 4$		$N = 5$		$N = 4$		$N = 5$	
	Simple	Hard	Simple	Hard	Simple	Hard	Simple	Hard
$p = 64$	5,1	3,9	1,2	9,3	2,2	1,9	0,5	4,0
$p = 256$	19,9	15,0	11,9	39,6	8,3	6,9	3,4	11,0
$p = 1024$	15,6	52,9	22,1	105,7	5,2	20,0	2,2	10,2

Table 4. Speedup on two GPUs

	Iteration speedup				Time speedup			
	$N = 4$		$N = 5$		$N = 4$		$N = 5$	
	Simple	Hard	Simple	Hard	Simple	Hard	Simple	Hard
$p = 64$	11,3	6,8	8,2	19,8	2,4	1,6	1,3	3,2
$p = 256$	39,0	31,5	25,4	57,1	8,0	7,0	2,3	2,4
$p = 1024$	128,4	83,5	98,4	267,9	20,4	11,0	2,8	5,2

As can be seen from Table 1, the sequential GSA surpasses DIRECT and DIRECT*l* methods in all classes of problems in terms of the average number of iterations. At the same time, in the 5-Hard class, none of the methods solved all the problems: DIRECT failed to solve 16 problems, DIRECTl and GSA – 4 problems each.

Let us now evaluate the acceleration achieved using parallel GSA using an adaptive dimensionality reduction scheme based on the number p of cores used. Table 2 shows the speedup of the algorithm that combines multiple evolvents and an adaptive scheme for solving a series of problems on CPU, compared to the sequential launch of the GSA method. Two evolvents and, accordingly, two processes were used; each process used p threads, and calculations were performed on a single cluster node.

Table 3 shows the speedup obtained when solving a series of problems on one GPU using an adaptive scheme compared to a similar launch on CPU using 4 threads. Table 4 shows the speedup of the algorithm combining multiple evolvents and an adaptive scheme when solving a series of problems on two GPUs

Table 5. Speedup on six GPUs

	Iteration speedup	Time speedup
$p = 64$	30,8	1,9
$p = 256$	92,7	1,5
$p = 1024$	597,0	2,5

compared to an adaptive scheme on CPU using 4 threads. Two evolvents and, accordingly, two processes were used; each process used p threads on each GPU; all computations were performed on a single cluster node.

The last series of experiments has been carried out on 20 six-dimensional problems from the GKLS Simple class. Table 5 shows the speedup of the algorithm combining multiple evolvents and the adaptive scheme when solving the problems on 3 cluster nodes (using 2 GPUs per node, 6144 GPU threads in all) compared to the adaptive scheme on CPU using 4 threads.

6 Conclusion

In summary, we observe that the use of graphics processors to solve global optimization problems shows noteworthy promise, because high performance in modern supercomputers is achieved (mainly) through the use of accelerators.

In this paper, we consider a parallel algorithm for solving multidimensional multiextremal optimization problems and its implementation on GPU. In order to experimentally confirm the theoretical properties of the parallel algorithm under consideration, computational experiments were carried out on a series of several hundred test problems of different dimensions. The parallel algorithm demonstrates good speedup, both in the GPU and CPU versions.

Acknowledgments. This study was supported by the Russian Science Foundation, project No. 16-11-10150.

References

1. Barkalov, K., Gergel, V.: Multilevel scheme of dimensionality reduction for parallel global search algorithms. In: OPT-i 2014–1st International Conference on Engineering and Applied Sciences Optimization, Proceedings, pp. 2111–2124 (2014)
2. Barkalov, K., Lebedev, I.: Adaptive global optimization based on nested dimensionality reduction. Adv. Intel. Syst. Comput. **991**, 48–57 (2020)
3. Barkalov, K., Lebedev, I., Sovrasov, V.: Comparison of dimensionality reduction schemes for parallel global optimization algorithms. Commun. Comput. Inform. Sci. **965**, 50–62 (2019)
4. Evtushenko, Y., Malkova, V., Stanevichyus, A.A.: Parallel global optimization of functions of several variables. Comput. Math. Math. Phys. **49**(2), 246–260 (2009)

5. Ferreiro, A., Garcia, J., Lopez-Salas, J., Vazquez, C.: An efficient implementation of parallel simulated annealing algorithm in GPUs. J. Glob. Optim. **57**(3), 863–890 (2013)
6. Gablonsky, J.M., Kelley, C.T.: A locally-biased form of the DIRECT algorithm. J. Glob. Optim. **21**(1), 27–37 (2001)
7. Garcia-Martinez, J., Garzon, E., Ortigosa, P.: A GPU implementation of a hybrid evolutionary algorithm: GPuEGO. J. Supercomput. **70**(2), 684–695 (2014)
8. Gaviano, M., Kvasov, D., Lera, D., Sergeyev, Y.: Software for generation of classes of test functions with known local and global minima for global optimization. ACM Trans. Math. Softw. **29**(4), 469–480 (2003)
9. Gergel, V., Barkalov, K., Sysoyev, A.: A novel supercomputer software system for solving time-consuming global optimization problems. Numer. Algebra Control Optim. **8**(1), 47–62 (2018)
10. Gergel, V., Grishagin, V., Gergel, A.: Adaptive nested optimization scheme for multidimensional global search. J. Glob. Optim. **66**(1), 35–51 (2016)
11. Gergel, V., Grishagin, V., Israfilov, R.: Local tuning in nested scheme of global optimization. Procedia Comput. Sci. **51**(1), 865–874 (2015)
12. Grishagin, V., Israfilov, R., Sergeyev, Y.: Convergence conditions and numerical comparison of global optimization methods based on dimensionality reduction schemes. Appl. Math. Comput. **318**, 270–280 (2018)
13. He, J., Verstak, A., Watson, L., Sosonkina, M.: Design and implementation of a massively parallel version of DIRECT. Comput. Optim. Appl. **40**(2), 217–245 (2008)
14. Hwu, W.: GPU Computing Gems, Emerald edn. Morgan Kaufmann, San Francisco (2011)
15. Jones, D.R.: The DIRECT global optimization algorithm. In: The Encyclopedia of Optimization, pp. 725–735. Springer, Heidelberg (2009)
16. Kindratenko, V. (ed.): Numerical Computations with GPUs. Springer, New York (2014)
17. Langdon, W.: Graphics processing units and genetic programming: an overview. Soft Comput. **15**(8), 1657–1669 (2011)
18. Modorskii, V., Gaynutdinova, D., Gergel, V., Barkalov, K.: Optimization in design of scientific products for purposes of cavitation problems. In: AIP Conference Proceedings **1738** (2016)
19. Paulavičius, R., Žilinskas, J., Grothey, A.: Parallel branch and bound for global optimization with combination of lipschitz bounds. Optim. Method. Softw. **26**(3), 487–498 (2011)
20. Sergeyev, Y., Kvasov, D.: Global search based on efficient diagonal partitions and a set of Lipschitz constants. SIAM J. Optim. **16**(3), 910–937 (2006)
21. Sergeyev, Y.D., Strongin, R.G., Lera, D.: Introduction to Global Optimization Exploiting Space-filling Curves. Springer Briefs in Optimization, Springer, New York (2013)
22. Strongin, R.: Algorithms for multiextremal mathematical programming problems employing the set of joint space-filling curves. J. Glob. Optim. **2**, 357–378 (1992)
23. Strongin, R.G., Sergeyev, Y.D.: Global Optimization with Non-convex Constraints Sequential and Parallel Algorithms. Kluwer Academic Publishers, Dordrecht (2000)

Application of a Novel Approach Based on Geodesic Distance and Pressure Distribution to Optimization of Automated Airframe Assembly Process

Tatiana Pogarskaia[1(✉)], Maria Churilova[1], and Elodie Bonhomme[2]

[1] Peter the Great St.Petersburg Polytechnic University, Saint Petersburg, Russia
pogarskaya.t@gmail.com, m_churilova@mail.ru
[2] Airbus SAS, Toulouse, France
elodie.bonhomme@airbus.com

Abstract. The paper is devoted to the parallel version of a new approach to temporary fastener pattern optimization. The main goal of the research is to improve the existing technological processes by reducing the number of installed temporary fasteners without compromising the assembly quality. Besides being combinatorial, the considered problem is complicated by the need to solve the contact problem on hundreds of input initial gaps for further analysis of the residual gap in the assembly after fastening. Commonly used heuristic methods become inapplicable to the regarded case due to their iterative based logic and the need for multiple calculations of the objective function value.

The proposed approach avoids iterative calculations and is based on calculation of the force distribution, needed to close an initial gap between joined parts, and geodesic distance between fasteners over the joint surfaces. The approach is applied and compared to a greedy algorithm previously used for the optimization of the robotized assembly process for an Airbus A350 S19 section.

Keywords: Aircraft assembly · Optimization · Geodesic distance · Fastening element · Supercomputing · Task parallelism

1 Introduction

Aircraft assembly process implies a wide range of manual labor operations resulting in the failure of airplane manufacturers to meet ever-increasing demand. One of the possible ways to accelerate the process is to optimize number and positions (pattern) of temporary fastening elements for each step of the assembly, as it is a key point of joining large-size airplane parts by riveting and, at the same time, one of the most time-consuming manual operations. Due to various random uncertainties while manufacturing process, riveted parts are usually not brought to full contact. For this reason, temporary fasteners are installed to join the parts and prevent the gap opening while drilling and reaming to avoid vibrations and swarf ingestion between parts.

V. Voevodin and S. Sobolev (Eds.): RuSCDays 2020, CCIS 1331, pp. 162–173, 2020.
https://doi.org/10.1007/978-3-030-64616-5_14

Therefore, it is most important to provide the best assembly quality by a minimal number of fasteners. At the same time, the same temporary fastener pattern is used for all the aircraft of one series in the assembly line. Consequently, the initial gap field between joined parts is generally unknown a priori; only the statistical information is available. To evaluate fastener pattern quality one has to calculate the residual gap after fastening. It leads to the necessity of contact problem solving. All these features in conjunction with the combinatorial nature of the problem make the problem of fastener pattern optimization very specific and complicated.

The considered problem is largely related to combinatorial geometry ones and therefore a range of similar optimization problems regarded earlier in literature was carried out by heuristic methods. The genetic algorithm is commonly used: in [1] it was used to optimize the positions of temporary fasteners with predetermined initial gap between the assembled parts; in [2] it was used to determine the optimal number of weld points and their location; in [3] its modification was proposed to optimize the welding order. Another method of the group, simulated annealing, is performed in [4] for optimization of wind turbines location with restrictions on the minimum distance between them. However, the genetic algorithm and other heuristic methods require numerous calculations of the objective function, which in our case is the most time-consuming part.

The problem of fastener pattern optimization was regarded in some works. In [5], the authors optimized the distribution of forces in the fasteners to avoid stress concentration during assembly. Similar ideas were used in [6], where the positions of fasteners were not determined in advance; instead, the authors considered the minimum allowable distance between adjacent elements. A method based on the use of matrices of influence coefficients was proposed in [7] to optimize the location of welding points during sheet panels assembly. It allowed to avoid multiple calculations of the objective function, however, the mechanical properties of the parts were not taken into account. The Method of Influence Coefficients and variation analysis is used to simulate the assembly and contact interaction and solve a various number of problems arising in the process (determining the necessary forces for high-quality assembly, tolerance analysis etc.) [8–10]. However, the disadvantage of this technique is that it is possible to take into account the contact interaction of the assembled parts only at predefined points.

2 The Assembly Process

In this paper, the optimization of a fastener pattern is considered for one of the main steps of the A350 fuselage section 19 (S19) manufacturing process. It is the continuation of study [11]. The section and the corresponding finite element model (FEM) are represented in Figs. 1 and 2. Area for optimization is highlighted in the red rectangle. As it is noted in [12], the section represents quite flexible skin made of composite materials and connected over several frames. The assembly stage scheme is presented in Fig. 2[1] (left),

[1] Since it is impossible to disclose the real A350-900 models and details of the assembly process, we use the artificial models and fastening patterns that resemble the actual ones for illustrative purpose. Similarly, all the numerical values, given in this paper, in particular related to assembly requirements and specifications, have been changed.

the relatively stiff plate called splice is joined to the skin stiffed by horizontal beam and vertical frame.

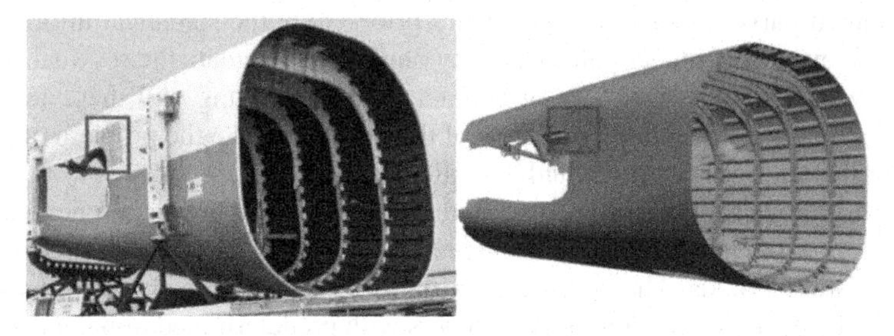

Fig. 1. A350 S19 (left) (www.airbus.com) and the corresponding FEM (right).

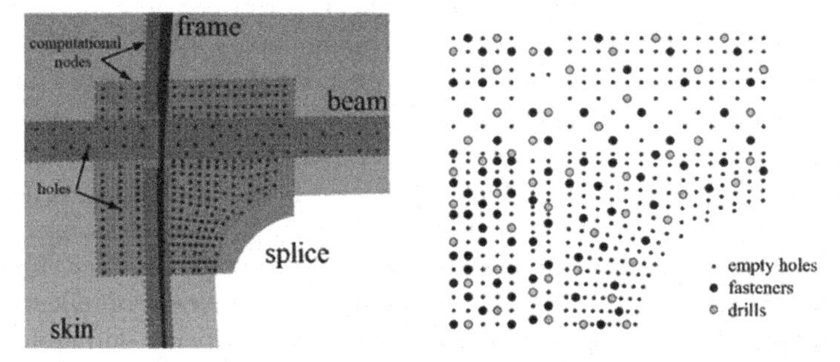

Fig. 2. The junction area (left): skin (gray), splice (orange), beam (violet) and frame (green). Holes for fasteners (black) and computational nodes (red); The area for optimization (right): the positions of drilled holes (green) and temporary fasteners (black). (Color figure online)

The splice plate is joined with the skin, the beam and the frame by some temporary fasteners. Their positions are marked in Fig. 2 (right) with large black circles. After fastening, a drilling robot sequentially reams predefined holes for the final rivets installation. The drilled holes are marked in Fig. 2 (right) with green circles. During reaming, the drill force is acting on each joined part one-by-one as the drill goes through, starting from the skin, then splice and afterwards the beam if the hole is located on it. The frame includes no holes. The maximum gap opening occurs when the drill is acting on a lower part (splice or beam) in the stack. The relatively large gap opening can cause hole eccentricity and result in poor assembly quality after riveting. Therefore, the technology requires the gap to be not larger than some predefined value (e.g. 0.3 mm) during all reaming operations. To fulfill this requirement, the temporary fasteners are used, keeping all the parts together.

It is clear that the more fasteners are installed, the less will be the gap between the joined parts during reaming operations. On the other hand, the presence of many fasteners complicates the navigation and positioning of drilling robot. Each fastener is installed and removed manually, which is also time-consuming. The problem is to minimize the number of temporary fasteners, taking into account the requirement on the maximum allowable gap. Fasteners can be installed to any empty hole in Fig. 2 (right), except the ones to be reamed.

3 Contact Problem Formulation

Optimization of a fastener pattern is impossible without a specific criterion of its quality. As far as one of the main characteristics of a joint assembly is the gap between the parts (in our case, during reaming), it is necessary to evaluate it for each modification of the regarded pattern. Determination of the deformed stress state of the assembly with applied loads from the fasteners and drill is based on solving the contact problem. It can be formulated using the standard finite element modeling technique, in a discrete variation form and reformulated to reduce the problem dimension [13, 14]. Only the main idea of the method is presented here, as all the details of numerical algorithms, as well as validation tests, are described in [15].

The considered contact problem has several features:

1. The contact may occur only in predefined region (*junction area*) and any procedures for detection of the possible contact zone are not required.
2. Loads are applied only in holes inside the junction area.
3. The installed fasteners restrict relative tangential displacements of the parts in the junction area, what makes them negligible in comparison with normal ones. This circumstance justifies implementation of node-to-node contact model instead of more complicated surface-to-surface model.
4. Friction forces between assembled parts can be excluded from consideration due to small relative tangential displacements.
5. The stationary problem is considered.
6. Stress state of a part is described by the linear elasticity theory.

We will give here briefly the main idea of the methodology used for numerical solving of the described contact problem. The approach makes it possible to reduce the initial contact problem to the quadratic programming one and perform computations only in the considered junction area taking into account mechanical properties of the whole joint. The dimension of the problem significantly reduces as number of finite element nodes in junction area is usually much less than in the full FEM.

Let $u = \{u_i\}_{i=1,n_u}$ be the vector of displacements of finite element model nodes (*computational nodes*) in the junction area. The reduced problem for determination of u is derived as described in [15]

$$\min_{Au \leq G}\left(\frac{1}{2}u^T Ku - f^T u\right), \tag{1}$$

where f is the vector of loads (from installed fasteners or drill) in junction area; G is the initial gap vector (the normal distance between nodes that may come to contact); K is the reduced stiffness matrix; A is the linear operator defining the normal and tangential directions to contact surface and pairs of contact nodes. The residual gap after applying all loads is calculated as $g_{\text{res}} = G - Au$.

Matrices A and K characterize the junction overall topology, parts mechanical properties, positions of fixations etc. Using this methodology, the initial FEM in junction area is substituted by the special assembly model based on quadratic programming problem (1) and including:

- holes located in junction area where fasteners can be installed or removed;
- positions of installed fastening elements as a set of occupied holes;
- initial gap between parts (Fig. 3). A set of such gap fields (*gap cloud*) can be generated based on statistical analysis of the available measurements and tolerances [16].

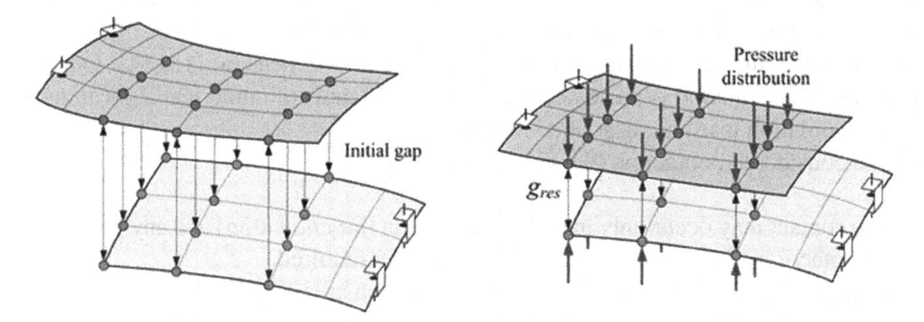

Fig. 3. Initial gap (left) and pressure distribution needed to obtain residual gap g_{res} between two parts (right).

For solving the described contact problem, the specialized software complex ASRP (Assembly Simulation of Riveting Process) is used. The ASRP is described in more detail in [20]. The optimization algorithms that are discussed further are implemented in the ASRP Simulator, which is a standalone application, written in C++. Simulator is the central part of ASRP software complex. It is designed for the riveting process simulation; it allows calculating gaps between assembled parts, absolute displacements, reaction forces caused by contact in junction area, loads in fastening elements needed to achieve contact, etc. For parallel computations the cluster version of ASRP Simulator is used.

4 Optimization Problem

The considered assembly stage involves 48 reaming operations that are done one-by-one with the same 76 pre-installed temporary fasteners. The other 283 holes remain empty and valid for the fasteners to be moved to. According to the current technology, the gap between parts while reaming operations must be no large than predefined value equal to 0.3 mm. With predefined fastener pattern the contact problem is solved for each of the reamed holes, taking into account the load from the drill. The maximum gap during reaming occurs in the reamed hole, thus, there is no need to control the gap value in other computational nodes.

For preliminary analysis, the initial gap between all parts is assumed to be zero. Computations show that the gap requirement is violated in 19 reamed holes (*violated holes*), which are shown in Fig. 4 alongside with residual gap values. The goal of the optimization is to install the minimal possible number of temporary fasteners without increasing the number of violated holes.

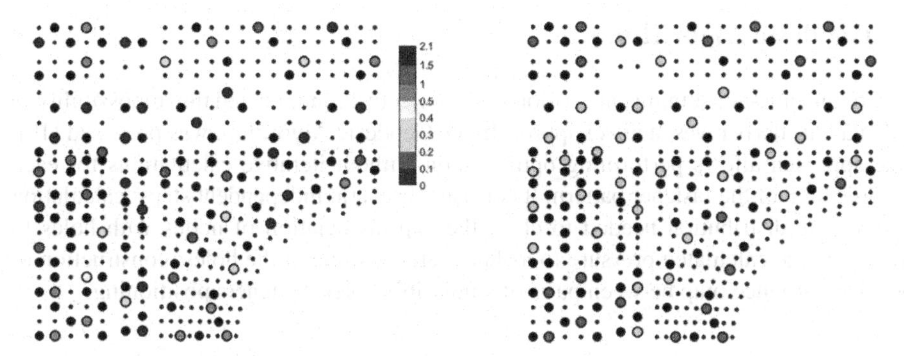

Fig. 4. Residual gaps in reamed holes, mm (left) and violated holes marked by red (right) [11]. (Color figure online)

We introduce a new nomenclature for better description. Let $R = \{r_i\}_{i=1,n_r}$ be a set of reamed holes of dimension n_r. Number of violated holes is defined as n_v. The set of all holes available for fastening (not reamed ones) is denoted as $H = \{h_i\}_{i=1,n_h}$, where n_h is the total number of available holes. Therefore, the fastener pattern of n_f fasteners is described by a set of occupied holes $H^o = \{h_i^o\}_{i=1,n_f}$, $H^o \subset H$.

The optimization procedure proposed in [11] is involves two steps: reducing the number of fasteners one-by-one, keeping the number of violated holes (reduction) and rearranging fasteners to reduce gaps in reamed holes (rearrangement). These steps are repeated one after another until no improvements in the pattern are found. The optimization criterion is chosen to minimize the sum of gaps in the reamed holes, but it can be changed to another assembly characteristic. The algorithms are described in detail in [11]. Particular attention must be given to the fasteners rearrangement procedure.

Rearrangement over a set of given positions is a complicated combinatorial problem and the global optimum can be found by exhaustive search but number of holes (order of hundreds) for fastener installation does not allow its implementation. In works [17, 20] the Local Variations Algorithm (iterative local exhaustive search) was proposed for optimization of fastener positions for upper and lower panel of wing-fuselage airframe junction. In [11] the same algorithm was used to rearrange fasteners for the discussed S19 assembly process optimization. The number of temporary fasteners was reduced by 30% without significant loss of quality. Despite its simplicity and successful applications, Local Variations Algorithm (LVA) is very time-consuming because it requires many objective function calculations. For S19 it took about four days of computations on supercomputer (Peter the Great Saint-Petersburg Polytechnic University Supercomputing Center, Tornado cluster with 612 Intel Xeon E5-2697v3 X2 nodes, 64G memory) to calculate three reduction-rearrangement steps only for zero initial gap.

To speed-up computations, a novel approach named Geodesic Algorithm (GA) was proposed to replace the rearrangement iterative procedure. It is based on calculation of force distribution, needed to close the gap between joined parts, and geodesic distance between holes for fastener positioning. In following sections, we will discuss its application to the considered optimization problem.

5 Geodesic Algorithm

Due to the time-consuming calculation of the objective function and the impossibility to calculate its derivatives, a novel approach, the Geodesic Algorithm, was proposed [19]. The approach allows performing optimization without iterative calculations for each modification of the fastener pattern. It consists of several independent steps: calculation of pressure distribution needed to close the gap; association of nodes with holes to transform the calculated pressure to a characteristic value in the holes; construction of geodesic distance map between pairs of admissible holes; fasteners positioning.

5.1 Calculation of Pressure Distribution and Association of Nodes with Holes

Calculation of pressure in computational nodes needed to close the initial gap between parts up to a predefined value δ is a preliminary step of the procedure. To accomplish that and to take into account the drill force, constraints of quadratic problem (1) with zero vector of loads is reformulated as

$$\min_{0 \leq g_{res} \leq \delta} \left(\frac{1}{2} u^T K u - f_D^T u \right), \quad (2)$$

where $g_{res} = G - Au, f_D$ is a vector of applied forces that includes non-zero value only in computational nodes corresponding to one of the reamed holes. The pressure distribution satisfying the constraint in (2) is obtained using the Lagrange multipliers.

However, the considered computational process is complicated by the need to ream 48 holes with the same fastener pattern. It results in the pressure distribution calculation for each reaming operation separately. After all different distributions are obtained, the total pressure field is calculated as l^{∞}-norm over the computational nodes. It is a general characteristic of the hole reaming process used further for fastener positioning.

For further fastener installation pressure field in nodes is substituted by pressure field in holes. Each computational node in the model is associated with the closest hole as it is shown in Fig. 5. For each hole h_i there is a set of corresponding nodes, denoted as U_i.

The pressure p_i in hole h_i is calculated as a sum of pressure values in associated nodes from the set U_i. The collected values are used to construct a pressure vector with n_h total number of holes. In this way, a total pressure in a region of computational nodes becomes a characteristic of corresponding hole located in it.

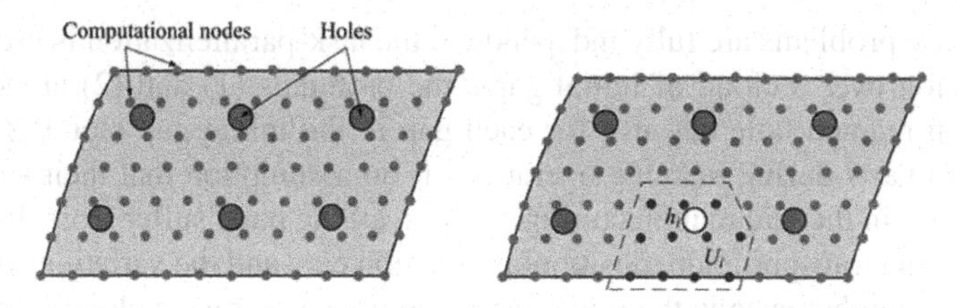

Fig. 5. Set of computational nodes associated with hole h_i.

5.2 Geodesic Distance Map Construction and Fastener Positioning

Geodesic distance is used to take into account the shape of the joined parts and to provide the best coverage of the junction area with the fastener pattern. The distance map of all holes is calculated by Dijkstra's algorithm with binary heap as a priority queue for storing a set of unvisited vertices, which was tested on different joint models [18].

Let us consider a problem of N fastener positioning. The first fastener is installed in the hole with the maximum value of calculated pressure. The geodesic distance is weighted with corresponding pressure value. Positions for the other fasteners are found one by one as the farthest (in terms of the weighted geodesic distance) empty hole from already fastened ones. Let us denote the set of empty holes as $E = \{eh_i\}_{i=1,n_{eh}}$ and the set of fastened holes as $F = \{fh_i\}_{i=1,n_{fh}}$. The position of the next fastener is found as

$$\arg\max_{eh \in E} \left(\sum_{i=1}^{n_{fh}} p(eh)\gamma(fh_i, eh) \right), \tag{3}$$

where $p(eh)$ is the calculated pressure in hole eh, $\gamma(fh_i, eh)$ is the geodesic distance between fastened hole fh_i and end empty hole eh. Such a modification of the geodesic distance map means that holes located in regions with larger pressure become farther from the fastened holes and hence more appropriate for new fastener installation.

For considered S19 section assembly optimization problem, we use the described Geodesic Algorithm on a rearrangement step to find a new pattern after fastener reduction. The obtained results are discussed in the following section. Therefore, we can optimize the pattern not only for one initial gap, but also for a set of gaps, what was impossible with the LVA. In this case the problem (2) is solved for all initial gaps in cloud $\{g_i\}_{i=1,n_g}$, where n_g is a number of gaps in the cloud. The pressure p in each computational node is then calculated as an l^{∞}-norm via all n_g calculated pressure fields.

5.3 Parallelization

There are two most time-consuming parts of the optimization procedure: solving the contact problem (1) for each reamed hole separately to calculate the number of violated holes while fastener reduction; solving the dual problem (2) for each reamed hole separately to calculate the pressure field for Geodesic Algorithm.

Thus, these problems are fully independent, the task-parallelization is used. In case of optimization over a cloud of initial gaps, the problems (1) and (2) are solved not only for each reamed hole but also for each gap in the initial gap cloud. Zero initial gaps between parts during reaming operations is an assumption that their surfaces are ideally smooth. In the real manufacturing process, all the parts suffer from dimensional variations during transportation, positioning, fixation etc., and the variations differ from one aircraft to another even in the same series. Optimization over a cloud of initial gaps based on measurements made on the assembly line can provide better results as it takes into account more detailed information.

A cloud $\{g_i\}_{i=1,n_g}$ of $n_g = 100$ initial gaps based on real measurements was used for the optimization. To calculate the objective function, a set of $n_r{\cdot}n_g$ problems is solved – the contact problem for each reamed hole on each initial gap field (n_r holes and n_g gaps in cloud). This set of problems is used as an input data for task parallelization to solve problems (1) and (2). All computations were done in Peter the Great Saint-Petersburg Polytechnic University Supercomputing Center (Tornado cluster with 612 Intel Xeon E5-2697v3 X2 nodes, 64G memory), using up to 20 nodes with 28 threads on each. Thus, a set of $n_r{\cdot}n_g$ problems was divided to load the nodes uniformly.

6 Results

For S19 assembly optimization problem three reduction-rearrangement steps are done to obtain a suboptimal solution. If using LVA in rearrangement step, the iterations continue until the algorithm has converged to some local optimum, objective function is the sum of residual gaps in all reamed holes. For GA we find a new pattern with reduced number of fasteners at each rearrangement step with no iterations.

As can be seen in Fig. 3 (left), "the worst" violated holes are located at the upper and lower borders. LVA is based on the idea of local exhaustive search and a found improvement of the fastener pattern is saved and the algorithm continues for the next iteration. It means that LVA will proceed to put fasteners around the reamed holes with maximum gap values until free positions are left and the final pattern will be non-uniform and inconvenient for a drilling robot. To avoid the situation, LVA was forbidden to move several fasteners located in the upper two and the lower two hole lines. At the same time, GA uses all the available holes and the obtained pattern is more uniform. The final optimized patterns are presented in Fig. 6 (left). The optimization by GA is done for zero initial gap between all parts, as well as over the cloud of initial gaps.

The pattern obtained by GA includes both lower numbers of fasteners and violated holes (Fig. 6). Due to the absence of iterations, GA requires much less time to optimize the fastener positions (about 50 times faster than LVA, Table 1) and makes it possible to perform optimization over a cloud of initial gaps.

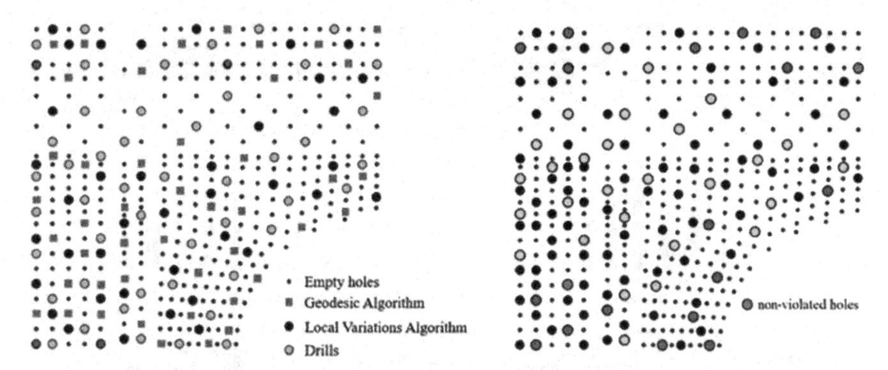

Fig. 6. Fastener patterns obtained with LVA (black) and GA (red) after 3 reduction-optimization steps for zero initial gap (left). The fastener pattern obtained with GA decreased gap in 2 holes up to the acceptable value of 3 mm (blue circles, right). (Color figure online)

Table 1. Optimization results.

	Initial pattern	LVA	GA	GA on cloud
Number of fasteners	76	53	50	55
Number of violated holes	53	19	17	17
Total gap in reamed holes	16.77 mm	17.13 mm	16.44 mm	18.16 mm
Computational time	–	~4 days	~110 min	~250 min

Verification of the obtained pattern is done on the cloud of 200 initial gaps (used in work [11]). Contact problem (1) is solved for each initial gap of the cloud, what provides a set of residual gaps and shows how the fastener pattern and the initial gap influence the gap opening during drilling. The distribution of the gap values in each reamed hole is gathered to estimate the regarded pattern quality. Figure 7 presents several examples of the obtained histograms for initial and optimized fastener patterns.

As can be seen in Fig. 7.1 and 7.2, the patterns obtained by GA and LVA are very close and provide nearly the same gap distribution in the reamed holes. At the same time (Fig. 7.3 and 7.4), the pattern optimized by GA over the initial gap cloud of 100 gaps (gap clouds used for optimization and for verification do not intersect) decreases the range of gap values in some of the reamed holes.

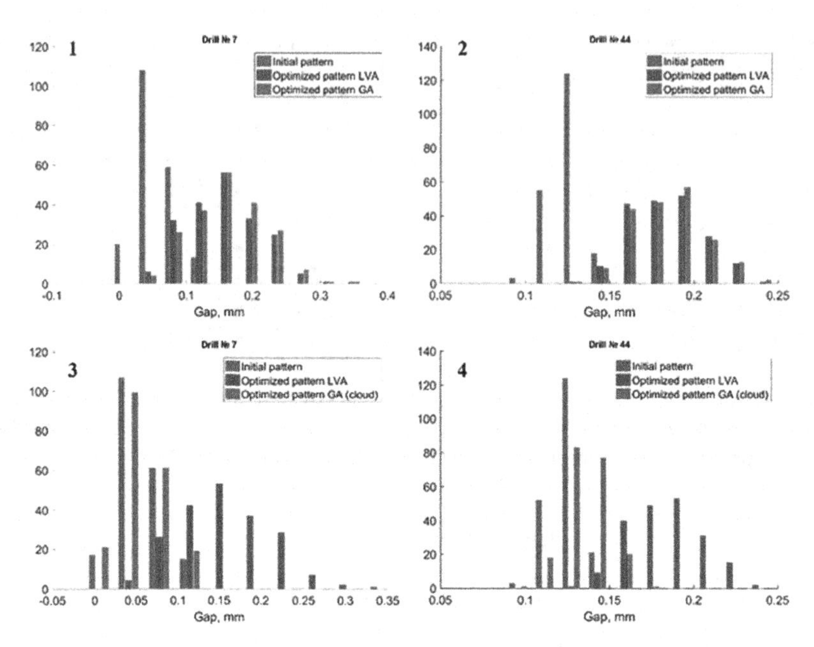

Fig. 7. Gap distributions in different drilling points for initial and optimized fastener patterns.

7 Conclusion

As a continuation of work [11], the new Geodesic Algorithm was implemented for the fastener pattern optimization in one A350 fuselage section assembly process. The proposed algorithm allows to perform optimization 50 times faster than the Local Variations used before due to its non-iterative procedure. Moreover, it does not require additional limitations on fastener positions because it does not result in grouping fasteners around the reamed holes with highest gap values. Achieved computational speed-up allows to optimize fastener pattern over a cloud of initial gaps and take into account statistics of real measurements.

The results of the work are obtained using computational resources of Peter the Great St. Petersburg Polytechnic University Supercomputing Center (www.spbstu.ru).

This research work was supported by the Academic Excellence Project 5-100 proposed by Peter the Great St. Petersburg Polytechnic University.

References

1. Yang, D., Qu, W., Ke, Y.: Evaluation of residual clearance after prejoining and pre-joining scheme optimization in aircraft panel assembly. Assembly Autom. **36**(4), 376–387 (2016)
2. Vdovin, D.S.: Optimization of the location of welding points on the body structures of the bearing systems of wheeled machines. BMSTU J. Mech. Eng. **4**, 34–39 (2007)
3. Shelley Xie, L., Hsieh, C.: Clamping and welding sequence optimization for minimizing cycle time and assembly deformation. Int. J. Mater. Prod. Technol. **17**(5–6), 389–399 (2002)

4. Lückehe, D., Kramer, O., Weisensee, M.: Simulated annealing with parameter tuning for wind turbine placement optimization. In: Proceedings of the LWA 2015 Workshops: KDML, FGWM, IR, and FGDB, Trier, Germany, pp. 108–119 (2015)
5. Chickermane, H., Gea, H.C., Yang, R.-J., Chuang, C.-H.: Optimal fastener pattern design considering bearing loads. Struct. Optim. **17**(2), 140–146 (1999)
6. Oinonen, A., Tanskanen, P., Bjork, T., Marquis, G.: Pattern optimization of eccentrically loaded multi-fastener joints. Struct. Multidiscipl. Optim. **40**(1), 597–609 (2009)
7. Cai, W.: Fixture optimization for sheet panel assembly considering welding gun variations. Proc. Inst. Mech. Eng. Part C: J. Mech. Eng. Sci. **222**(2), 235–246 (2008)
8. Wämfjord, K., Lindkvist, L., Söderberg, R.: Tolerance simulation of compliant sheet metal assemblies using automatic node-based contact detection. In: ASME IMECE 2008, vol. 14, pp. 35–44 (2008)
9. Lorin, S., Lindau, B., Lindkvist, L., Söderberg, R.: Efficient compliant variation simulation of spot-welded assemblies. J. Comput. Inf. Sci. Eng. **19**(1), 011007 (7 p.) (2019)
10. Lorin, S., Lindau, B., Tabar, R.S., Lindkvist, L., Wärmefjord, K., Söderberg, R.: Efficient variation simulation of spot-welded assemblies. In: ASME International Mechanical Engineering Congress and Exposition, Proceedings (IMECE) 2018, vol. 19, no. 1, p. 011007 (7 p.) (2019)
11. Lupuleac, S., Shinder, J., Churilova, M., Zaitseva, N., et al.: Optimization of automated airframe assembly process on example of A350 S19 splice joint. SAE Technical paper 2019-01-1882 (2019)
12. Sanchez, J., et al.: Co-cured shell & integrated RTM door-frame. In: ECCM16 - Seville, Most innovative CFRP Solutions for A350XWB RFE S19 Airframe (2014)
13. Wriggers, P.: Computational contact Mechanics, 2nd edn. Springer, Heidelberg (2006). https://doi.org/10.1007/978-3-540-32609-0
14. Petukhova, M.V., Lupuleac, S.V., Shinder, Y.K., Smirnov, A.B., Yakunin, S.A., Bretagnol, B.: Numerical approach for airframe assembly simulation. J. Math. Ind. **4**(1), 1–12 (2014)
15. Lupuleac, S., Smirnov, A., Churilova, M., Shinder, J., Bonhomme, E.: Simulation of body force impact on the assembly process of aircraft parts. In: ASME 2019 International Mechanical Engineering Congress and Exposition, vol. 2B: Advanced Manufacturing, IMECE2019-10635, 9 p. (2020)
16. Lupuleac, S., et al.: Simulation and optimization of airframe assembly process. In: ASME 2018 International Mechanical Engineering Congress and Exposition, vol. 2: Advanced Manufacturing, IMECE2018-87058, 10 p. (2019)
17. Lupuleac, S., Pogarskaia, T., Churilova, M., Kokkolaras, M., Bonhomme, E.: Optimization of fastener pattern in airframe assembly. Assembly Autom. **40**(5), 723–733 (2020)
18. Popov, N.P., Pogarskaia, T.A. Geodesic distance numerical computation on compliant mechanical parts in the aircraft industry. J. Phys. Conf. Ser. **1326**(1), 012026 (2019)
19. Pogarskaia, T.A., Lupuleac, S., Bonhomme, E.: Novel approach to optimization of fastener pattern for airframe assembly process. In: Procedia CIRP, vol. 93, 7 p. (2020)
20. Zaitseva, N., Pogarskaia, T., Minevich, O., Shinder, J.: Simulation of aircraft assembly via ASRP. SAE Technical paper 2019-01-1887 (2019)

Application of Supercomputing Technologies for Numerical Implementation of an Interaction Graph Model of Natural and Technogenic Factors in Shallow Water Productivity

Alexander Sukhinov[1], Alla Nikitina[2,3,4](✉), Alexander Chistyakov[1,4], Alena Filina[3,4], and Vladimir Litvinov[1,5]

[1] Don State Technical University, Rostov-on-Don, Russia
sukhinov@gmail.com, cheese_05@mail.ru, litvinovvn@rambler.ru
[2] Southern Federal University, Rostov-on-Don, Russia
nikitina.vm@gmail.com
[3] Supercomputers and Neurocomputers Research Center, Taganrog, Russia
j.a.s.s.y@mail.ru
[4] Science and Technology University "Sirius", Sochi, Russia
[5] Azov-Black Sea Engineering Institute of Don State Agrarian University, Zernograd, Russia

Abstract. The paper covers the research and numerical implementation of an graph model of natural and technogenic factors' interaction in shallow water productivity. Based on it, the analysis of pulse propagation in computing environment from the vertices is performed in the context of research situation of valuable fish degradation of the Azov Sea that are subject to excessive commercial fishing withdrawal. The model takes into account the convective transport, microturbulent diffusion, taxis, catch, and the influence of spatial distribution of salinity, temperature and nutrients on changes in plankton and fish concentrations. Discrete analogue of proposed model problem of water ecology, included in software complex, were developed using schemes of second order of accuracy taking into account the partial filling of computational cells. The adaptive modified alternately triangular method was used for solving the system of grid equations of large dimension, arising at model discretization. Effective parallel algorithms were developed for numerical implementation of biological kinetics problem and oriented on multiprocessor computer system and NVIDIA Tesla K80 GPU with the data storage format modification. Due to it, the production processes of biocenose populations of shallow water were analyzed in real and accelerated time.

Keywords: Graph model · Shallow water productivity · Parallel algorithm · Modified data storage format · GPU · Software

V. Voevodin and S. Sobolev (Eds.): RuSCDays 2020, CCIS 1331, pp. 174–185, 2020.
https://doi.org/10.1007/978-3-030-64616-5_15

1 Introduction

Quantitative research and mathematical modeling of plankton and fish dynamics has more than half a century of history. This research interest is due to the essential role of aquatic organisms in the functioning of water ecosystems. The balance in ecosystems, development of mathematical models in biophysics and ecology, as well as the solving problems of planning and rational use of bioresources, were researched in papers by Rosenberg G.S., Odum H.T., Riznichenko G.Yu. [1], etc. The papers by Ilyichev V.G. [2] are devoted to the research and implementation of discrete mathematical models that implement mechanisms for searching the evolutionarily stable migration routes for fish populations. The dynamic programming based on the construction of the Bellman function is used at solving the optimization problems. Simulation methods of interaction processes of biological populations taking into account the taxis, diffusion, Alle and Ferhulst effects are developed in papers by Luck R.F., Zhou S.-R., Tyutyunov Yu.V. [3]. However, despite a significant number of publications, many effects that are essential for improving the simulation accuracy and reliability of changes in ichthyological processes of artificial and natural waters were not taken into account earlier at construction mathematical models.

Universal software can be use including for simulation the biological kinetics processes. Some of them are focused on multiprocessor systems, but it versatility is to use the limited number of models, algorithms and methods to variety of different cases. In addition, only individual modules are often parallelized in them. For operational forecasts of ecological situation of shallow water, it is necessary to design software, oriented on high-performance computing systems for simulation of hydrobiological processes in water and their possible consequences in limited time. Therefore, it's necessary to develop algorithms for adapting individual modules of the Azov3D software complex [4], developed by the author's team for the NVIDIA CUDA architecture.

2 Problem Statement

Creation and application of ecosystem models is an integral part of comprehensive research of ecological systems to assess its condition, research the pollution transport and transformation, eutrophication and reproduction processes, including fish productivity of water, solution the problems of rational exploitation of its resources, preservation of water quality and model predictions further development of ecosystem under the influence natural and technogenic factors .

Using cognitive methods of system analysis, applicable with high-quality information, we can analyze information about the contradictory influence of various factors [5]. The approach to structuring expert and monitoring data in the form of a cognitive weighted graph, which can be used to develop and substantiate a mathematical model, is described in [6], where natural and technogenic factors are as vertexes of the cognitive digraph. They represent as concepts that determine the processes in ecosystem and affect the dynamics of fish stocks, where v_1 is a state of the spawning part of commercial fish; v_2 is an annual replenishment of juvenile fish; v_3 is the natural (compensatory) loss of generation; v_4 are favorable conditions for puberty; v_5 is a specific efficiency of natural reproduction; v_6 is the scale of artificial production; v_7 is a level of commercial exploitation of fish bioresources; v_8 is a biomass of dominant species of forage benthos; v_9 is the oxygen availability of eggs in spawning area; v_{10} is the sea level transgression; v_{11} is the number of the main natural enemies of juveniles; v_{12} is the available length of spawning migration routes (Fig. 1a).

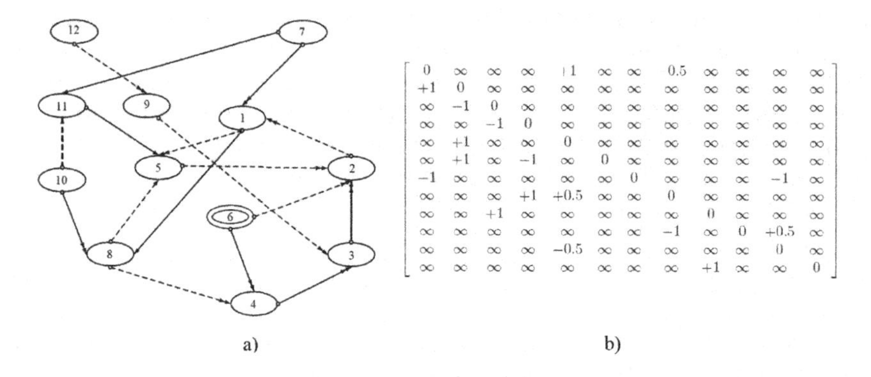

Fig. 1. Cognitive digraph of the influence of factors (a); the matrix of weights (b)

The weighted sign graph formalism is an extension of the $G(Y, E)$ digraph representation, which is supplemented by a set of parameters of vertices V, where each vertex y_i corresponds to a dimensionless parameter-concept $v_i \in V$, and the arc transformation functional $F(V \times V, E)$, which determines the sign (or weight) corresponding to the digraph arc. Weights can be matched to the ribs by defining a universal scale of interactions for the entire situation under consideration $(-u_{max}, \ldots, u_0, \ldots, +u_{max})$. The functionality of the Four kinds of influence transfer will be set in the conversion functionality $B = \{-1, -0.5, 0.5, 1\}$ in two for each arc "+" (the dotted arc) and "−" (the solid arc) (see Fig. 1a).

Negative effects of fishing and natural changes in sea level of resources are different, because the influence of $F(v_{10}, v_{11})$ is defined as "weak" and $u = -0.5$ with one arrow tip at the arc, strong – with two tips (see Fig. 1a). The weights' matrix of cognitive digraph has the form as in Fig. 1b. Vertices in such graph model are divided into dependent and free from the influence of other concepts, such as the value of fishing withdrawals and sea level fluctuations.

A multi-species model of interaction between plankton and commercial fish-detritofag of the Azov Sea (for example, *Mugil soiuy Basilewsky*, *Mugil cephalus*) was developed to research the dynamics of shallow water biohydrocenosis (Azov Sea), based on the models by Ricker W.E., Rosenzweig M.L., MacArthur R.H., Ginzburg V.L., Holling K.S., Tyutyunov Yu.V., Ilyichev V.G., Latun V.S.:

$$\frac{\partial S_i}{\partial t} + div(\mathbf{U}S_i) = \mu_i \Delta S_i + \frac{\partial}{\partial z}\left(\nu_i \frac{\partial S_i}{\partial z}\right) + \psi_i, \tag{1}$$

$$\psi_1 = g_1(S_1,S_3) - \delta_1 S_1 S_2 - \sigma_1 S_1 S_5 - \lambda_1 S_1, \psi_2 = g_2(S_1,S_2) - \delta_2 S_2 - \lambda_2 S_2,$$
$$\psi_3 = \gamma_3\lambda_4 S_4 - g_3(S_1,S_3) + B(\tilde{S}_3 - S_3) + f, \psi_4 = \lambda_1 S_1 - g_4(S_4,S_5) + \lambda_2 S_2 - \lambda_4 S_4,$$
$$\psi_5 = g_5(S_4,S_5) - \delta_5 S_5 - \lambda_5 S_5,$$

where S_i is the concentration of i-th component, $i = \overline{1,5}$; ψ_i is a chemical-biological source (runoff) or a term describing aggregation (conglutination-uncon-glutination) if the corresponding component is a suspension; i is a type of substance, $i = \overline{1,5}$: 1 is phytoplankton concentration (P), 2 is zooplankton (Z), 3 is nutrient (S), 4 is detritus (D), 5 is commersial fish concentration (F); $\mathbf{u}$ is the velocity field of water flow; $\mathbf{U} = \mathbf{u} + \mathbf{u}_i$ is the rate of convective mass transfer; $\mathbf{u}_i$ is the deposition rate of i-th component under the gravity, $i \in \overline{1,4}$; g_i are trophic functions for substances $i \in \overline{1,5}$. Let's assume that $g_1(S_1,S_3) = \gamma_1\alpha_3 S_1 S_3$, $g_2(S_1,S_2) = \gamma_2\delta_1 S_1 S_2$, $g_3(S_1,S_3) = \alpha_3 S_1 S_3$, $g_4(S_4,S_5) = \beta_4 S_4 S_5$, $g_4(S_1,S_4,S_5) = (\gamma_5\beta_4 S_4 + \xi_5\sigma_1 S_1)S_5$, where α_3 is the consumption ratio of S by P; $\gamma_1,\gamma_2,\gamma_5$ are transfer coefficients of trophic functions; γ_3 is a fraction of S in the biomass of P; λ_1 is the coefficient taking into account the mortality and metabolism of P; δ_1 is the loss of X by eating out of Z; λ_2,λ_5 are elimination (mortality) rates of Z,F respectively; δ_2 is the loss of Z by eating out of fish; δ_5 is the loss of F by eating out of fish and catching; $\tilde{S}_3$ is the maximum possible concentration of S; f is the pollution source function; B is the specific rate of the supply of S nutrients; λ_4 is the decomposition ratio of detritus; β_4 is the consumption rate of organic residues of S; σ_1 is the loss ratio of P as a result of consuming it by S; ξ_5 is the transfer coefficient of concentration growth of S due to P; μ_i,ν_i are diffusion coefficients in horizontal and vertical directions of component $i \in \overline{1,5}$.

The computational domain $\overline{G}$ (Azov Sea) is a closed area bounded by the undisturbed water surface Σ_0, the bottom $\Sigma_H = \Sigma_H(x,y)$ and the cylindrical surface σ for $0 < t \le T_0$. $\Sigma = \Sigma_0 \cup \Sigma_H \cup \sigma$ is a piecewise smooth border of G; $\mathbf{n}$ is the surface normal vector; $\mathbf{U_n}$ is a normal component of the water velocity vector to the Σ.

Let's define initial conditions

$$S_i = S_{i0}\big|_{t=0}(x,y,z), i = \overline{1,5} \tag{2}$$

and boundary conditions:

$$S_i = 0 \text{ on } \sigma \text{ if } \mathbf{U_n} < 0; \frac{\partial S_i}{\partial \mathbf{n}} = 0 \text{ on } \sigma \text{ if } \mathbf{U_n} \ge 0; \tag{3}$$

$$\frac{\partial S_i}{\partial z} = \varphi(S_i) \text{ on } \Sigma_0; \frac{\partial S_i}{\partial z} = -\varepsilon_i S_i \text{ on } \Sigma_H, i = \overline{1,5},$$

where $\varepsilon_1,\varepsilon_2,\varepsilon_3,\varepsilon_4,\varepsilon_5$ are non-negative constants; $\varepsilon_1,\varepsilon_3,\varepsilon_5$ take into account the lowering of plankton and fish to the bottom and their flooding; $\varepsilon_2,\varepsilon_4$ take into account the absorption of nutrients and detritus by bottom sediments; φ is the given function.

The mathematical model (1)–(3) takes into account the convective transport, microturbulent diffusion, deposition of substances under gravity, the influence of salinity and temperature of water environment, plankton and fish taxis, catch. Calculation results based on the Azov Sea hydrodynamics model [?], [7] we used as input data for developed model (1)–(3) taking into account the wind impact, river flows (Don, Kuban, Mius and about 40 small watercourses), water exchange with other basins, bottom relief, complex shape of the coastline, friction on the bottom, temperature, salinity, evaporation and precipitation, Coriolis force.

The mathematical model (1)–(3) can be represented as a digraph structuring multivariate analytical conclusions about the interaction of biotic and abiotic factors in the ecosystem, presented as a set of related concepts (Fig. 2).

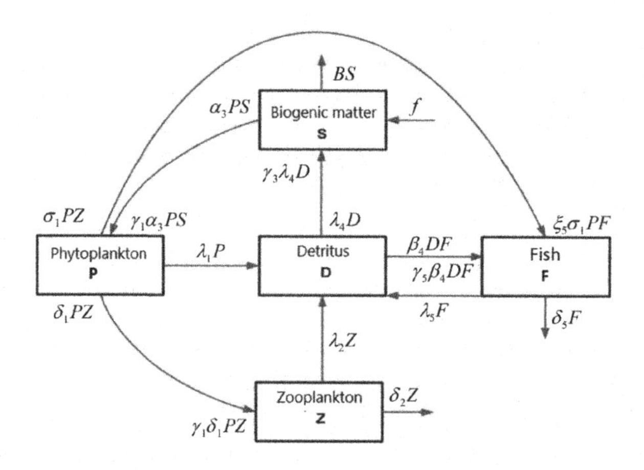

Fig. 2. Graph of $X - Z - S - D - F$ model

Compared to the model in the form (1)–(3), calculating by dispersed rules of quantities' ratio, the digraph (Fig. 2) is a visual result of objective analytical comparison and notation.

3 Solution Method and Parallel Implementation

Each equation of the system (1)–(3) after linearization was represented as the convection-diffusion equation. Schemes of high-order of accuracy taking into account the partial filling of computational cells were used for model discretization. The Couette-Taylor flow problem were used to estimate the error of the method and test the model hydrodynamic calculations. Let us use the splitting schemes into physical processes and the method of partial filling of computational cells. The adaptive modified alternately triangular method (MATM) of variational type was used to solve the obtained grid equations [8,9].

The error values of the Couette-Taylor flow problem numerical solution on grids depending on the radius are given in Fig. 3 (a – 21×41 nodes; b – 81×161 nodes) (the error in the case of a smooth boundary is denoted by red circles, in the case of a stepped boundary – by blue rings). Figure 3 represents the fact that the increasing of dimensions of computational grids does not lead to improve the accuracy and to reduce the linear dimensions of border region, where the manifest defects of the decisions related to the rough boundary approximation in the case of stepwise approximation boundary for water flow environment problem.

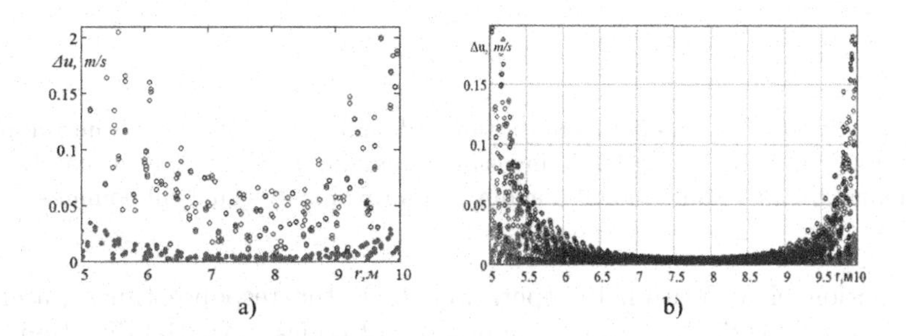

Fig. 3. Dependences of the error in solving the problem of water flow on the radius on various calculation grids

Note that the numerical solution error of model hydrodynamics problems caused by the boundary approximation does not exceed 6% of problem solution at using grids taking into account the "fullness" of cells.

For numerical implementation of the developed model of biological kinetics (1)–(3), we developed parallel algorithms, focused on a multiprocessor computer system (MCS) and NVIDIA Tesla K80 GPU.

Parallel Implementation on MCS. We describe parallel algorithm with various types of domain decomposition for solving the problem (1)–(3) on MCS with technical parameters: the peak performance is 18.8 TFlops; 8 computational racks; the computational field is based on the HP BladeSystem c-class infrastructure with integrated communication modules, power supply and cooling systems; 512 single-type 16-core HP ProLiant BL685c Blade servers are used as computational nodes, each of which is equipped by four 4-core AMD Opteron 8356 2.3 GHz processors and 32 GB RAM; the total number of computational nodes is 2048; the total amount of RAM is 4 TB.

The *k-means* method was used for geometric partition of computational domain for uniform loading of MCS calculators (processors). The result of the *k-means* method is shown in Fig. 4.

Theoretical estimates of acceleration and efficiency of the developed parallel algorithm [10]:

$$E^t = S^t/p = \chi\Big/\left(1 + \left(\sqrt{p} - 1\right)\left(\frac{\alpha}{N_z} + \frac{\beta p}{t_0}\left[t_n\left(\frac{1}{N_x} + \frac{1}{N_y}\right) + \frac{t_x\sqrt{p}}{N_x N_y}\right]\right)\right),$$

where χ is the ratio of the number of computing nodes to the total number of nodes (computing and fictitious); p is the total number of processors; t_0 is

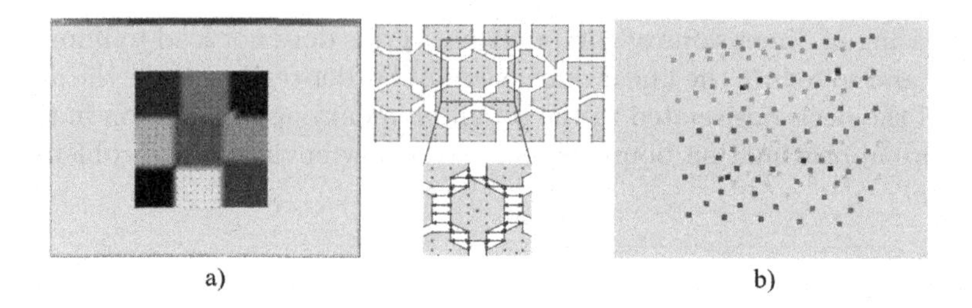

a) b)

Fig. 4. Partition of the model computational domain of regular shape into approximately equal subdomains: a) two-dimensional domain, 9 and 20 approximately equal subdomains; b) three-dimensional domain, 6 approximately equal subdomains

the execution of an arithmetic operation; t_x is the response times (latency); $\alpha = 18/25$; $\beta = 4/25$; N_x, N_y, N_z are number of nodes in spatial directions.

Results of parallel implementation the proposed algorithm for solution the problem (1)–(3) were compared and presented in Table 1, where t, S, E are operating time, acceleration and efficiency of the algorithm; S^t, E^t are theoretical estimates of acceleration and efficiency of the algorithm.

Table 1. Comparison of acceleration and efficiency of algorithms

p	t, s	S^t	S	E^t	E
1	6.073	1.0	1.0	1.0	1.0
2	3.121	1.181	1.946	0.59	0.973
4	1.811	2.326	3.354	0.582	0.839
8	0.997	4.513	6.093	0.654	0.762
16	0.620	8.520	9.805	0.533	0.613
32	0.317	15.344	19.147	0.480	0.598
64	0.184	25.682	33.018	0.401	0.516
128	0.117	39.013	51.933	0.305	0.406

According to the Table 1, the developed algorithm based on the *k-means* method can be effectively use for solving hydrodynamic problems at sufficiently large number of computing nodes.

The efficiency estimation of parallel algorithm, developed on the basis *k-means* method with using the Teil criterion, shown that its efficiency was increased on 10 – 20 % at solving problem in the form (1) – (3) compared to the standard algorithm [11].

Parallel Implementation on GPU. For numerical implementation of proposed interrelated mathematical model of biological kinetics, we developed parallel algorithms which will be adapted for hybrid computer systems using the

NVIDIA CUDA architecture. The NVIDIA Tesla K80 GPU has the high computing performance and supports all modern closed (CUDA) and open (OpenCL, DirectCompute) technologies. The NVIDIA Tesla K80 specifications: the GPU frequency is 560 MHz; the GDDR5 video memory is 24 GB; the video memory frequency is 5000 MHz; the video memory bus digit capacity is 768 bits. The NVIDIA CUDA platform characteristics: Windows 10 (x64) operating system; CUDA Toolkit v10.0.130; Intel Core i5-6600 3.3 GHz processor; DDR4 of 32 GB RAM; the NVIDIA GeForce GTX 750 Ti video card of 2GB; 640 CUDA cores.

We developed the modified storage format of the sparse matrix with a repeating sequence of elements for solution the biological kinetics problem (1) – (3). The modification of the Compressed Sparse Rows (CSR) format to improve the efficiency of data storage with a repeating sequence of CSR1S elements was performed for modeling continuous biological kinetics processes by the finite difference method on uniform grids. In this case, it's enough to change them in an array that preserves a repeating sequence to change the differential operator, instead of repeatedly finding and replacing values of non-zero elements in an array [11]. We compared the memory capacity required for CSR and CSR1S formats:

$$P_{csr} = N_{nz}B_{nz} + (N_{nz} + R + 1)B_{idx},$$

$$P_{csr1s} = B_{nz}[N_{nz}(k_i + 1) - N_{seq}(k_i k_r R + k_r R + 1) - k_i(k_r R - R - 1)],$$

where R is the number of matrix rows; R_{seq} is the number of matrix rows that contain a repeating sequence of elements; N_{nz} is the number of non-zero matrix elements; N_{seq} is the number of elements in a repeating sequence; B_{nz} is the memory capacity to store one non-zero element; B_{idx} is the memory capacity to store one index; $k_r = R_{seq}/R$, $k_i = B_{idx}/B_{nz}$.

For solution of grid equations at discretization of model (1) – (3) in CSR format on GPUs using CUDA technology, efficient function libraries were developed. The developed algorithm uses the CSR1S modified data storage format with further conversion to CSR format to solve the resulting system of linear algebraic equations (SLAE) on GPU using NVIDIA CUDA technology. For this, the algorithm of matrix conversion from CSR1S format to CSR format in minimal time was designed and tested with five times repetition and fixing the average value of calculation time. As a result of experimental research the dependence of processing time of the matrix conversion algorithm on the number of elements of the repeating sequence N_{seq} and the ratio of matrix rows, containing the sequence, to the total number of matrix rows, we determined that the algorithm using parallel computing on GPU with the TPL (Task Parallel Library) library [12] is more efficient than the serial algorithm and the algorithm using NVIDIA CUDA platform. The matrix dimension was 10^6. Dependence graphs of matrix conversion time from CSR1S to CSR format by sequential (a) and parallel (b) algorithms from k_r and N_{seq} are given in Fig. 5.

Graphs of algorithms' efficiency are given in Fig. 6: sequential implementation (Fig. 6a) and parallel implementation using the NVIDIA CUDA platform (Fig. 6b). Analysis of these graphs shown that the algorithm using NVIDIA

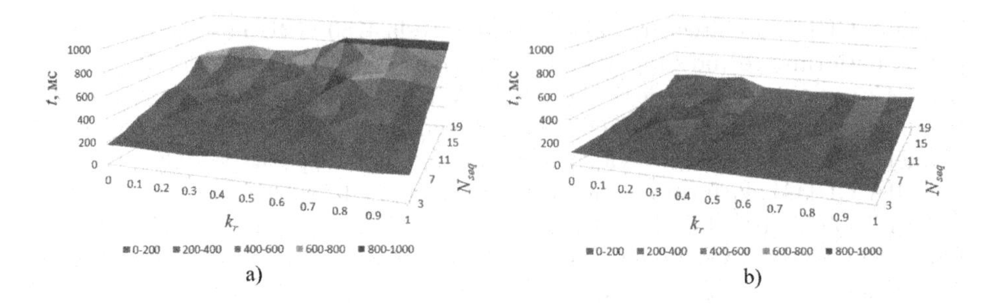

Fig. 5. Matrix conversion runtime from CSR1S to CSR format

CUDA technology is more efficient at $N_{seq} > 7$. In this case, the point of equal efficiency decreases, starting from $k_r = 0.7$.

Therefore, we can calculate the minimum k_r value, above which the second algorithm will be more efficient than the first, substituting the N_{seq} value in it. It's necessary to implement an improved iterative method that directly operates with data in the CSR1S format.

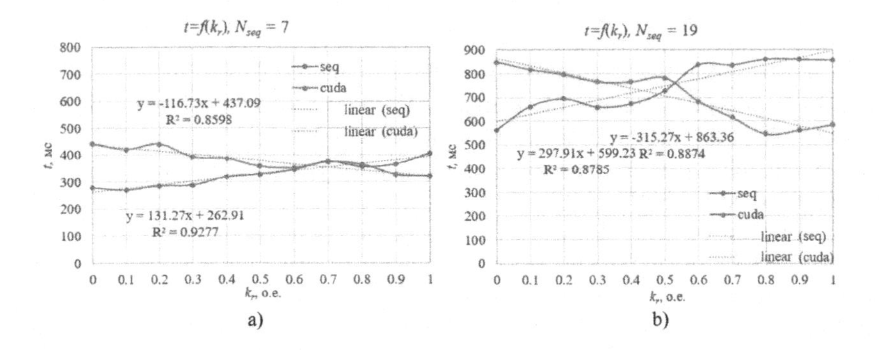

Fig. 6. Graphs of algorithms' efficiency

Analysis of the CUDA architecture characteristics showed the algorithms for numerical implementation of the developed mathematical model of hydrobiological processes could be used for creation high-performance information systems.

4 Description of Software Complex

The "Azov3D" software complex (SC) was developed for solution the biological kinetics problem with MCS and GPU implementation. It includes the control module, oceanological and meteorological databases, application library for solving hydrobiology grid problems, integration with various geoinformation systems (GIS), Global Resource Database (GRID) for geotagging and access to satellite data collection systems, NCEP/NCAR Reanalysis database (Fig. 7).

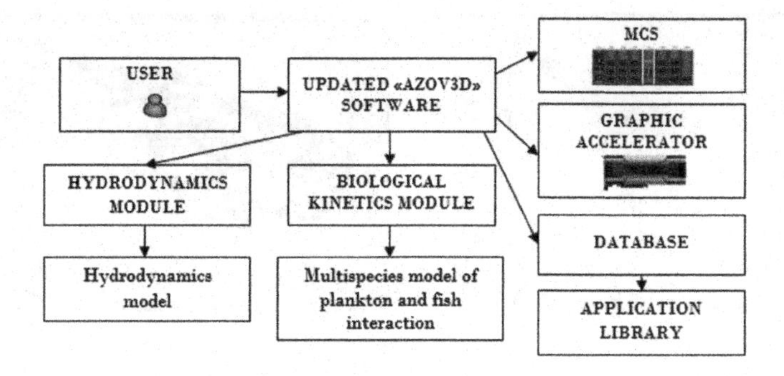

Fig. 7. Scheme of software complex

Since 2000, the authors performed the expedition researches in the Azov Sea. Based on it, the constantly updated database was created and used for calibration and verification of the model (1) – (3). During the expedition, the species composition and concentrations of plankton and fish populations, the regime of biogenic substances (determining the limiting biogenic element for simulated populations), the effect of external hormonal regulation (the effect of algae metabolites on phyto-and zooplankton concentrations), the dependence of plankton and fish growth rates in the Azov Sea on external and internal factors (water flow fields, salinity, temperature, etc.), and their interactions were researched for mathematical modeling of hydrobiological processes in shallow water. Numerical implementation of the proposed model (1) – (3) was performed on continuously thickening rectangular grids by dimensions 351x251x46, 702x502x92 nodes, etc. Physical dimensions of the computational domain: the width is 231 km; the length is 343 km; the surface area is 37605 km^2. The time interval is 150 d from May to September – a typical period of phytoplankton development, which is the main food base of zooplankton and fish.

Calculation results of pollution biogenic substance concentrations (S_3) for biogeocenosis evolution problem in shallow water (1) – (3) are given in Fig. 8a (initial distribution of water flow fields for the Northern wind; N is the number of iteration): $\mu_3 = 5 \cdot 10^{-10}$, $\nu_3 = 10^{-10}$, $B = 0.001$, $\tilde{S}_3 = 1$, $f = 3$; $N = 118$.

Using the developed SC, we researched the mechanism of suffocation zones formation in shallow waters [4]. Simulation results of possible scenarios of the Azov Sea ecosystem development (changes in phytoplankton concentrations) are given in Fig. 8b (initial distribution of water flow fields for the Eastern wind, $N = 382$). Parameter values: $\mu_1 = 5 \cdot 10^{-9}$; $\nu_1 = 10^{-9}$; $\delta_1 = 0.01$; $\sigma_1 = 0.005$; $\lambda_1 = 0.001$. Changes in detritus (S_4) and commersial fish (S_5) concentrations for time interval $T = 155$ days are given in Fig. 9, 10 (maximum concentration values are indicated by white color).

According to the presented results, the detritus concentration is decreased starting from 61 d, which means a decrease of bottom sediment concentration in the Central-Eastern part of the Azov Sea [4].

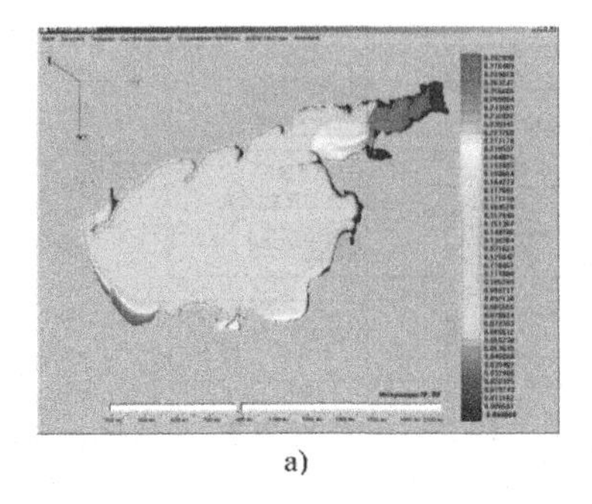

a)

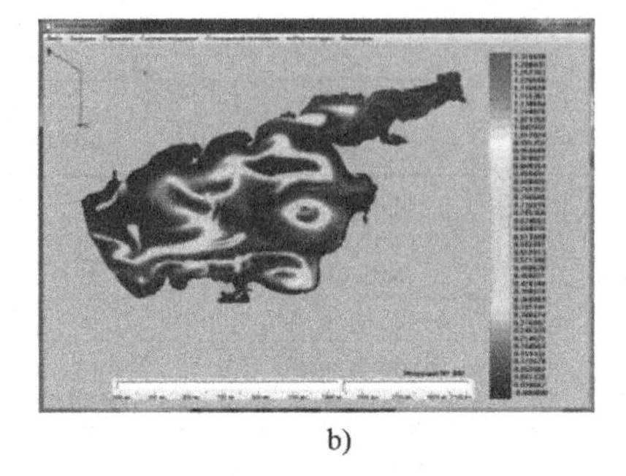

b)

Fig. 8. Pollution biogenic matter (a) and phytoplankton (b) concentrations

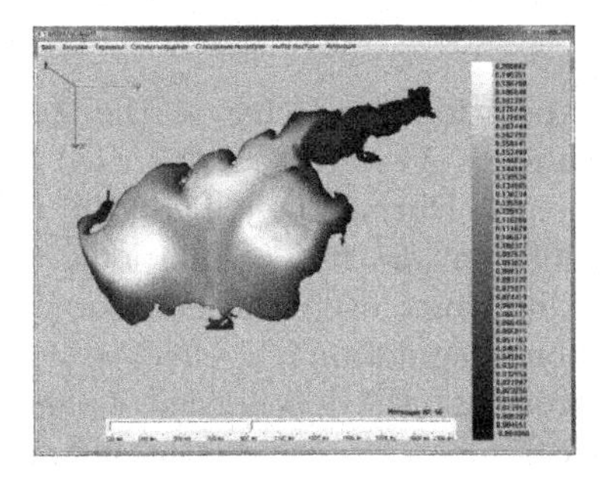

Fig. 9. Distribution of detritus concentration

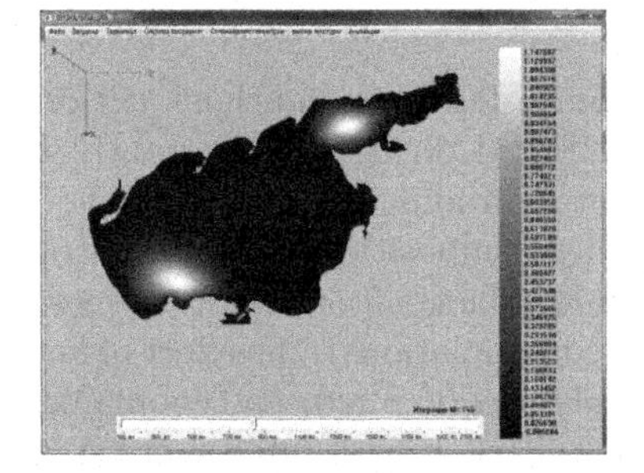

Fig. 10. Distribution of commersial fish concentrations

5 Conclusion

Analysis the dynamics of commercial fish populations at high and low numbers taking into account competition for resources, taxis, catch, spatial distribution of nutrients and detritus was performed on the basis of the multi-species model of plankton and commercial fish interaction. For biological problems, the prospects of methodology modification for applying cognitive graphs allow us to formalize the situation at qualitative level in the form of a set of concepts and directions for transmitting influence along various paths. The model discretization using the method of the partial fullness of computational cells allowed to significantly reducing the solution error at computational domain of complex shape. MATM was used as the main method for solving the system of grid equations in view of its highest convergence rate. Effective parallel algorithms were developed for numerical implementation of biological kinetics problem on MCS and NVIDIA Tesla K80 GPU with the data storage format modification. Based on it, the reproduction processes of biocenose populations have been analyzed in real and accelerated time.

Acknowledgement. The reported study was funded by RFBR, project number 19-31-51017.

References

1. Riznichenko, G.Y.: Mathematical models in biophisics and ecology. Izhevsk, p. 183 (2003)
2. Matishov, G.G., Ilyichev, V.G.: On optimal exploitation of water resources. The concept of internal prices. Rep. Acad. Sci. **406**(2), 249–251 (2006). (in Russian)
3. Tyutyunov, Y.V., Titova, L.I., Senina, I.N.: Prey-taxis destabilizes homogeneous stationary state in spatial Gause-Kolmogorov-type model for predator-prey system. Ecol. Complex. **31**, 170–180 (2017). https://doi.org/10.1016/j.ecocom.2017.07.001
4. Gushchin, V.A., Sukhinov, A.I., Nikitina, A.V., Chistyakov, A.E., Semenyakina, A.A.: A model of transport and transformation of biogenic elements in the coastal system and its numerical implementation. Comput. Math. Math. Phys. **58**(8), 1316–1333 (2018). https://doi.org/10.1134/S0965542518080092
5. Avdeeva, Z.K., Kovriga, S.V., Makarenko, D.I., Maksimov, V.I.: Cognitive approach in the control science. Probl. Manag. **3**, 2–8 (2007). (in Russian)
6. Perevarukha, A.Y.: Graph model of interaction of anthropogenic and biotic factors for the productivity of the Caspian Sea. Vestnik SamSU. Nat. Sci. Ser. **10**(132), 181–198 (2015)
7. Sukhinov, A.I., Chistyakov, A.E., Shishenya, A.V., Timofeeva, E.F.: Predictive modeling of coastal hydrophysical processes in multiple-processor systems based on explicit schemes. Math. Models Comput. Simul. **10**(5), 648–658 (2018). https://doi.org/10.1134/S2070048218050125
8. Marchuk, G.I., Sarkisyan, A.S.: Mathematical modelling of ocean circulation. Science, p. 297 (1988)
9. Konovalov, A.N.: The theory of alternating-triangular iterative method. Siberian Math. J. **43**(3), 552–572 (2002). (in Russian)
10. Gergel, V.P.: High-Performance Computing for Multiprocessor Multicore Systems, p. 544. Publishing house of Moscow University, Moskow (2010). (in Russian)
11. Voevodin, V.V., Voevodin, V.B.: Parallel computing. SPB. BHV-Petersburg, p. 608 (2002). (in Russian)
12. Parallel programming and the TPL library. https://metanit.com/sharp/tutorial/12.1.php

Developing Efficient Implementation of Label Propagation Algorithm for Modern NVIDIA GPUs

Ilya V. Afanasyev[1,2] and Dmitry I. Lichmanov[1,2](✉)

[1] Research Computing Center of Moscow State University, Moscow, Russia
afanasiev_ilya@icloud.com, dimlichmanov@gmail.com
[2] Moscow Center of Fundamental and Applied Mathematics, Moscow, Russia

Abstract. Systems equipped with modern NVIDIA GPUs nowadays are capable of solving many real-world problems, including large-scale graph processing. Efficiently implementing graph algorithms on GPUs is a challenging task, since modern real-world graphs have irregular structure. Current paper describes approaches, which can be used for developing efficient implementation of label propagation algorithm, frequently used to solve graph clustering and community detection problems. Compared to already existing GPU-based implementations, new optimization techniques have been proposed, including graph preprocessing, efficient load-balancing, using unified memory for out-of-core graph processing, and several others. The performance of the developed implementation has been evaluated on synthetic and medium-scaled real-world graphs, resulting in up to 2 times better results compared to existing approaches.

Keywords: Graph algorithms · Label propagation · Performance analysis · NVIDIA GPU · CUDA

1 Introduction

Developing efficient implementations of various graph algorithms is an extremely important problem of modern computer science, since graphs are frequently used to model many real-world objects in different application fields. For example, graphs are used for social network and web-graphs analysis, solving infrastructure and socio-economic problems, and many others. As many modern real-world graphs have extremely large size and tend to extend, parallel computing is essential for accelerating graph computations. Modern shared-memory systems equipped with NVIDIA GPUs form a promising family of architectures, which allow to significantly accelerate graph-processing, since GPUs are equipped with high-bandwidth memory and provide a large resource of parallelism.

Community structure is one of the most important graph features. If graph is divided into relatively independent communities, it can be represented much easier, since different community clusters can be managed as if they were single vertices. In addition, graph clustering is frequently used to simplify graph

V. Voevodin and S. Sobolev (Eds.): RuSCDays 2020, CCIS 1331, pp. 186–197, 2020.
https://doi.org/10.1007/978-3-030-64616-5_16

processing by mapping communities to different supercomputer nodes, thus minimizing edge cuts. In addition, detecting communities is very important problem of social networks analysis.

Label propagation is a fundamental graph algorithm, which detects communities using network structure alone as its guide, and does not require a pre-defined objective function or prior information about the communities. In this paper, we propose a novel implementation of label propagation algorithm for NVIDIA GPU architecture. We comprehensively study existing algorithms and implementation approaches, and based on the provided analysis, propose a novel modification and implementation of label propagation algorithm, which has the following advantages compared to the existing implementations.

1. Since label propagation algorithm is iterative, it usually requires to execute a large amount of iterations before termination and providing final result. In order to minimize runtime of each iteration, the proposed algorithm utilizes graph preprocessing, aimed to significantly increase the efficiency of graph traversals.
2. During each iteration of label propagation algorithm, only a part of graph vertices can be processed, which will be referred as "active" vertices further in the paper. Different approaches used for generation of active vertices subsets are described in this paper. The suggested criterias allow to significantly reduce the number of processed vertices, which results into less amount of data requested form memory, and, in addition, allows to achieve better convergence.
3. Global memory of modern GPUs is limited and usually does not allow to store large real-world graphs. The proposed implementation uses GPU Pascal (and newer) unified memory to enable easy and efficient out-of-core graph-processing.
4. In order to effectively traverse graph vertices and edges, the proposed implementation uses many existing graph algorithms optimizations, including virtual warps for efficient load balancing, prefetching of hub-vertices in GPU caches, and many others.

   As a result, the proposed in this paper implementation is significantly faster compared to the existing ones.

2 Target Architectures

In this paper NVIDIA Compute Unified Device Architecture (CUDA) programming model is used, which supports GPU programming in C and C++ languages. On the hardware side GPU consists of a set of identical Streaming Multiprocessors (SM), each of which has multiple CUDA cores, capable of performing various types of operations. CUDA programming model allows users to define kernels - special functions, which are executed on GPUs, and grids - configurations of the kernels, which define the amount of threads used to run the kernels. NVIDIA GPUs employ Single Instruction Multiple Thread (SIMT) computational model, when each 32 threads are organized in groups (warps), which execute the same instruction on any given moment of time. Each thread has its own instruction address counter and register state, and executes instructions over its own data (data-driven parallelism).

GPUs memory hierarchy consists of multiple levels: global (device) memory, L2 cache, L1 cache, texture cache, and shared memory. Data between device and host memory is transferred via NVLINK bus, which has 40 GB/s (2 sub-lanes uplink), or PCI-Express, which has up to 32 GB/s of theoretical bandwidth.

Naturally, the amount of global memory is not enough to store large objects, including many real-world graphs. In order to enable out-of-core graph processing, when required data is dynamically loaded into device memory during kernel execution, unified memory can be used. Unified memory provides a single memory address space, in which pointers are visible from both host and device code. Unified memory uses a page migration mechanism. It supports Translation Lookaside Buffers (TLBs) for each SM (for two SMs in P100), so when GPU tries to access a non-resident page, it generates a fault message and corresponding page is loaded into global memory from the host. In the latest architectures (Pascal and Volta) - each SM locks its TLB with every fault until the fault is resolved.

For the performance evaluation in this paper, NVIDIA P100 GPU has been used. P100 GPU is equipped with global memory of 16 GB capacity, one-chip configurable L1 cache and shared memory (with a total capacity of 64 KB), and L2 cache size of 4096 KB size. Each of 56 SMs has 64 CUDA cores, resulting in 3584 CUDA cores per device. In our configuration IBM Power 8 CPUs are used as host processors for 2 P100 GPUs, connected via NVLINK with 80 GB/s bidirectional bandwidth.

Since CUDA-programming is relativity sophisticated, many GPU-oriented libraries have been proposed to provide essential building-blocks for various parallel applications. In this paper, several parts of label propagation algorithms are implemented using such libraries, including thrust [5], modernGPU [4], cuRand [16].

3 State of the Art

Initially, sequential label propagation algorithm was introduced by Raghavan et al. [17]. The algorithm consists of multiple iterations and can be described in the following way. Original labels for each vertex are unique. At the beginning of each iteration, nodes of the graph are arranged in random order. After that, each vertex collects labels from adjacent vertices, chooses the most frequent label among the collected ones, and updates its own label using the most frequent label. The updating process can either be synchronous or asynchronous. In the case of synchronous updates, node v on i-th iteration updates its label based on the label values from $(i-1)$-th iteration, while asynchronous updates are based on label values from i-th iteration. Algorithm terminates when no labels have been changed on the last iteration.

In addition, semi-synchronous methods have been proposed in [8,9]. These methods can be described as a hybrid of synchronous and asynchronous schemes: graph vertices are partitioned into multiple sets, and labels from different sets are updated asynchronously, while labels from the same sets are updated synchronously. Semi-synchronous method shows a slightly better convergence, but implementing these methods on GPUs is difficult.

At the moment of this writing, multiple implementations of Label Propagation algorithm for NVIDIA GPUs architecture exists. Methods proposed in [13,18] have limited performance on power-law graphs, since they do not take into account load balancing. For example, many vertices usually have low degrees, while few others have extremely high degrees. Thus, it is important to distribute work, required for processing both these types of vertices, as equally as possible. In order to gather labels more rapidly, [13] implementation uses label counters, stored in registers, shared memory or global memory of the GPU, depending on level frequency: more frequent labels are stored in faster memory.

Approach, proposed in this paper is largely based on [12] implementation - one of fastest existing GPU algorithm at the moment of this writing. Implementation [12] uses a set of load-balanced array operations, provided in ModernGPU [4] library. This implementation uses synchronous updating process, and the convergence is fast enough due to well-organized workload balancing.

4 Input Graphs

This paper includes a comprehensive comparative performance and efficiency analysis of the developed implementations using synthetic and real-world graphs with different characteristics. Synthetic graphs allow to easily scale various input graphs parameters (such as graph size), while real-world graphs allow to more accurately evaluate the performance on real problems. Synthetic RMAT [7] and uniform-random [10] graphs are used in this work, while real-world graphs are taken from KONEKT [2] and SNAP [1] collections.

5 Implementation Details

5.1 Graph Storage Format

Adjacency list is the most common representation for sparse graphs. Because the implementation of adjacency lists relies on pointers, it is not suitable for GPUs, since coping two-dimensional pointer arrays into device memory is inefficient. Thus, Compressed Sparse Row (CSR) format is frequently used for GPU implementations. In CSR format each vertex of the graph stores destination pointers to different segments of edges array. For directed graphs, only destinations of incident edges are stored, to allow efficient gathering labels of neighbours in pull direction [6].

CSR format is a very convenient format for label propagation algorithm, since it allows to easily access neighbors of only a given subset of graph vertices without loading any excessive data about other vertices or edges, what is essential for several of the proposed optimizations.

Approach proposed in this paper utilizes graph preprocessing, which is used to increase the efficiency of graph traversal in cases when convergence rate is low, and many algorithms iterations are executed. All vertices in a graph are sorted based on the number of outgoing edges in the descending order. This sorting has $O(V * log|V|)$ complexity, which is comparable to the complexity and execution time of a single label propagation iteration. Furthermore, it is possible to speed up this sorting by implementing it on GPU using, for example, thrust sort implementation.

5.2 Algorithm Structure

As it was mentioned in previous sections, the proposed algorithm is largely based on [12] approach, since it is one of the fastest GPU implementations at the moment of this writing. However, current paper significantly extends [12] approach with multiple optimizations and algorithm modifications, described in this section. First, however, we are going to describe overall structure of the modified algorithm.

1. At the beginning of each iteration, a criteria of choosing vertices, which will be processed on current iteration is applied. This criteria allows to generate a reduced version of vertices array - active frontier of vertices, used at the current iteration.
2. For the set of active vertices, all adjacent labels are gathered using optimized Advance primitive from VGL library [3]. The result this operation is an array of edges with a significantly smaller size, compared to the number of edges in the original graph.
3. Gathered labels are sorted via segmented sort, implemented in ModernGPU library.
4. Boundaries array of edges array size, containing only 0 and 1, is created: 1 is placed in $i-th$ position if $i-th$ label is different from $i+1-th$ one. Also, 1 is set if $i-th$ and $i+1-th$ labels refer to different label segments. In other cases, 0 is placed. Boundaries array is used to exclude repeating labels and replace them with corresponding frequencies.
5. Boundaries array is scanned using exclusive scan primitive from Thrust library. In the scanned array, a lot of neighbour elements have the same values, thus the reduced scan array is created. In reduced scan, elements of the scanned array are replaced with indexes of their last entry into the scanned array.
6. Since some labels could be included in some segments several times, in reduced scan array there will be less elements, than in scanned. Futher algorithm steps are based on a reduced scan array, which should have its own segment boundaries. New boundaries for each vertex segment can be received from scanned array, since this array already has implicit information about repeated labels.

7. The reduced scan array is used to compute frequencies of each label: i-th label it can be received by subtracting i-th element of the reduced scan array from $(i+1)$-th one. All frequencies the stored in frequencies array.
8. After that, a segmented reduce operation is applied to the segment boundaries and frequencies arrays. Segmented reduce returns an index in reduced scan array with the most frequent label, and label value can be easily obtained by referring to the corresponding element of the reduced scan - the index in the gathered array.

Gathering Labels Optimization. Gathering neighboring labels for a specific subset of graph vertices is implemented via optimized graph primitive "advance" from VGL framework, which is developed and supported by authors of this paper. VGL framework provides an optimized set of graph primitives, including advance, compute and filter, similar to Gunrock [19] API. However, VGL framework has multiple differences compared to other GPU frameworks: most importantly it relies on graph preprocessing, which allows to significantly improve both locality of indirect memory accesses and parallel workload balance.

Advance primitive applies user defined operations edge_op, vertex_preprocess_op, vertex_postprocess_op to each vertex and its adjacent edge of the specified frontier – a user-defined subset of graph vertices. Thus, on each iteration a new frontier, consisting of active vertices is generated, and advance primitive is applied to this frontier in order to collect neighboring labels.

Advance primitive uses a variety of GPU optimizations, required for efficient graph-processing: virtual warps [11], CSR-segmentation and clustering [20] of graph vertices, using CUDA-streams for processing groups of vertices with different numbers of outgoing edges simultaneously, and using texture cache to improve performance on indirect memory accesses.

Frontier Implementation. A subset of graph vertices (frontier) is also implemented in VGL framework. Currently sparse frontiers are implemented as an array of vertex IDs, while dense frontiers are implemented as an array of flags. Generation of sparse frontiers is implemented via thrust::copy_if primitive.

Using Unified Memory for Out-of-Core Processing. Since modern NVIDIA GPUs have limited amount of global memory, it is not possible to avoid implementing out-of-core processing when working with many real-world large-scale graphs. Recently introduced CUDA unified memory can be used to solve this problem. P100 GPU has 16 GB of global memory, while implemented LP algorithm requires using 3 additional arrays of size $|E|$ and 4 arrays of size $|V|$, excluding the input graph itself, which has $|V| + |E|$ size. Thus, synthetic graphs, which have more than 2^{24} vertices and 32 average outgoing edges for each vertex, can not be processed with all data structures located in device memory.

Unified memory allows to store temporarily unnecessary data inside memory of host CPU, which is usually large enough to store many real-world graphs (in our case 256 GB), and load necessary pages only when streaming multiprocessors of GPU request corresponding data. Unfortunately, working with unified memory often causes some difficulties. The access speed to pages, non resident in global memory, is limited by bandwidth of the interconnect(NVLINK or PCIe). Since in such cases SM locks a mechanism of virtual page translation provided by TLB, significant overhead occurs. There are some ways to reduce such overhead. If the pattern of data usage is clearly defined and predictable, it is reasonable to prefetch the required data just before device kernel needs it for the computation. The prefetching works a little slower compared to simple memory copy operations, as it needs to traverse a list of pages and update corresponding mappings in the CPU and GPU page tables.

In addition, it is possible to allocate a small amount of frequently used data permanently in GPU memory in order to avoid unnecessary data migrations. The developed implementation uses this approach to treat several array of size V, which contain indirectly accessed labels and borders of the edge segments.

5.3 Profiling the Developed Implementation

Different parts (computational steps) of each iteration of the described algorithm have unique properties and computational features. Some computational steps may require more computing resources and consequently may have longer runtime compared to other parts. These computational steps have higher optimization priority, and investigating their performance in details allows to understand if the overall performance of the algorithm can be further improved, and, more importantly, how.

Special profiling tools are used to compare different computational steps. The proposed implementation uses CUDA Profiling Tools [15]: **nvprof** and **nvvp**. Nvprof is used to collect and view profiling data from the command-line or to form a profiling file for visual analysis. There are two different modes of profiling. The first model generates a trace of GPU program, including all data transfers and the kernel timeline of application. Another mode is based on collecting hardware GPU metrics, such as L1/L2 cache utilization, occupancy of SM, the number of transactions to all types of memory, and many others.

Performance-wise three different types of kernels exists: memory-bound, compute-bound and latency-bound. Memory bound kernels perform a lot of accesses to different levels of memory hierarchy, and thus their performance is limited with bandwidth of a certain level of memory hierarchy, since all the required accesses can not be handled simultaneously. Compute-bound kernels heavily utilize different types of computational CUDA cores. The performance of latency-bound kernels is usually significantly limited because CUDA-threads have to wait for some resources of data. Most graph algorithms kernels are usually either memory-bound or latency-bound.

Table 1 demonstrates that three most time-consuming kernels take up to 80% of total execution time of the whole application. Segmented sort kernel consists of 3 steps, and thus, 3 different kernels: load-balancing, sorting and merging sorted parts. Another time-consuming kernel is gather, which is used to collect adjacent labels and thus performs a lot of indirect memory accesses. Gather kernels is also split into different parts, since vertices with different numbers of outgoing edges can be processed by either grid, block, warp, virtual warp or thread. A relative runtime of these parts is determined by the structure of input graph – specifically, the outgoing degree distribution. The results presented in Table 1 have been obtained by power-law RMAT graphs.

The presented table indicates correlation between computational complexity and runtime of the most time-consuming kernels. However, the it doesn't provide any valuable data about performance limitations of each kernel, which requires to use more detailed profiling methods based on the hardware performance counters.

Table 1. Relative runtimes of the three most time-consuming kernels.

Kernel name	Percent of runtime	Algorithm step	Computational complexity
Segmented sort	54.1%	3	$O(\log^2(\lvert E\rvert))$
Gather	**20.0 %**	**2**	$O(\lvert E\rvert)$
Warp per vertex	10.0%		
Block per vertex	7.2%		
Virtual warp per vertex	1.7%		
Grid per vertex	0.8%		
Thread per vertex	0.3%		
Extract boundaries	6.6%	4	$O(\lvert E\rvert)$

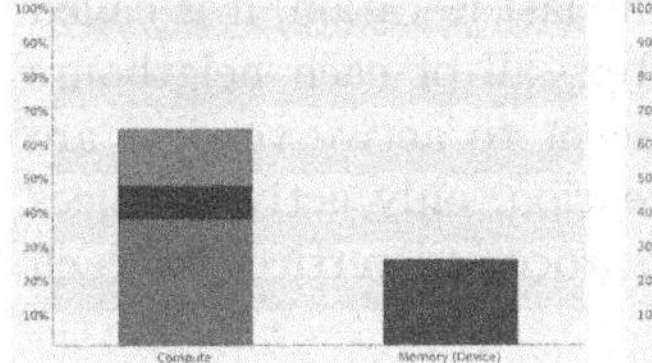

Fig. 1. Segmented sort

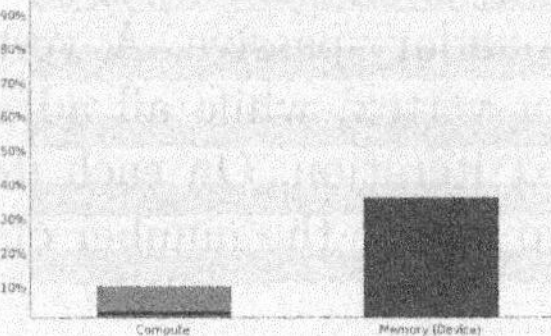

Fig. 2. Gather (warp per vertex)

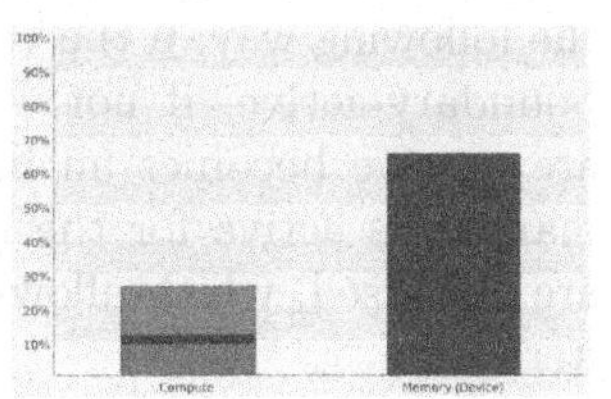

Fig. 3. Extract boundaries

Figures 1, 2, 3 present a detailed analysis of three most time-consuming kernels from Table 1. The profiling information is organized in the following way: if the left bar is high - the kernel is compute-bound, if the right - it is memory bound, while both bars being low indicates that the kernel is latency-bound. The segmented sort kernel uses shared memory for data exchanges between different threads. Since shared memory is located close to streaming multiprocessors, memory bus can hardly be a bottleneck for that kernel, while load-balancing search has significantly higher computational complexity compared to other kernels, which together makes the kernel compute-bound.

All gather kernels are memory bound, since they performs a lot of indirect memory accesses in order to collect labels from different graph vertices. Vertex clusterization in advance primitive allows this kernel to avoid being latency-bound and efficiently use texture and L2 caches for many power-law graphs.

Extract boundaries kernel is memory bound, since threads of this kernel load two elements of the input array from global memory, compare them and then store a result back into the global memory.

The provided profiling results demonstrate, that optimization techniques, implemented in VGL advance primitive and ModernGPU [4] library functions, utilize GPU hardware resources effectively. Thus, it is necessary to reduce algorithm complexity, so reducing number of active vertices seems to be one of the most important optimizations that can be implemented.

6 Performance Evaluation

6.1 Active Front Criterions

Since the most time-consuming kernels utilize hardware resources efficiently, it is crucial to reduce the number of vertices processed on each iteration of the algorithm, thus minimizing the amount of computations and data transfers being performed. In the existing GPU implementations (like [12]), the number of processed vertices does not change between iterations. So at first this naive method – all active – was recreated to be a starting point for future optimizations.

The second approach – active-passive – chooses vertices to of the frontier in the following way: if the vertex changes its label at the last iteration, it is called boundary-active, if not – boundary-passive. A vertex, all of each neigtbours are passive becomes an inner vertex, while all adjacent to active vertices are marked as active for the next iteration. On each itearion, only active vertices are traversed, which allows to reduce the number of processed vertices for each iteration.

To achieve the fast convergence and halfway cluster a graph, in idea of early exit criterion is added - active front contains only vertices which changed label on the previous iteration. This approach is further called "previous iteration".

Finally, a hybrid approach, further called "changed recently", is composed of the previous two: it removes a vertex from an active front if it hasn't changed its label during last n iterations. Parameter n shouldn't be small in order to generate better-quality clustering, however, it also should not be too big in order to significantly reducing size of active front.

6.2 Performance Results

In order to evaluate the performance of the proposed implementation, a sever with 2 IBM POWER8 CPUs has been used. Each CPU has a clock rate of up to

5.00 GHz and 256 GB of memory space with an ECC-control. In addition, 2 NVIDIA Tesla P100 GPUs with 16 GB global memory are installed into the servers, connected with CPU and each other via NVLINK.

First, the performance was evaluated for medium-scaled synthetic RMAT graphs; each graph had 32 average outgoing edges, while the number of vertices varied from 2^{20} to 2^{25}. The performance was evaluated using Traversed Edges Per Second (TEPS) metric [14]. Figure 4 shows average MTEPS value for each iteration of LP algorithm, while Fig. 5 shows performance in MTEPS for the whole algorithm.

Both figures provide information about four proposed approaches of selecting active vertices on each iteration. The frontier of active vertices is significantly smaller when it is formed only from vertices which changed their labels at previous iteration (previous iteration). This approach turned out to be very fast-converging, but the number of final community clusters can be larger, and thus the result can be less accurate.

It is challenging to compare two approaches used to form a front of active vertices. All-active approach demonstrates better result than a partial front approach. The first one works faster when it is possible to avoid using managed memory and significant time is spent to select such vertices and reduce correctly to get a temporary graph representation. As the scale of the graph increases, data transfers from CPU to GPU and back are required. So moving data about vertices which don't participate in propagation seems costly.

More sophisticated approach to remember the number of label updates for each vertex (changed recently) also demonstrates promising results. It was evaluated with parameter N equal to 5.

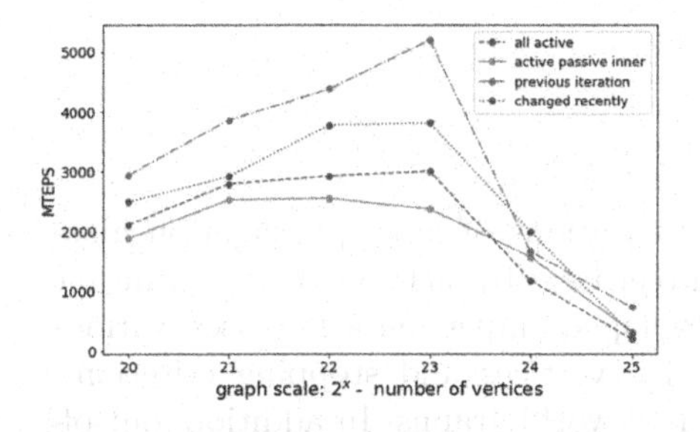

Fig. 4. Average performance (in MTEPS) for each iteration of LP algorithm for RMAT graphs.

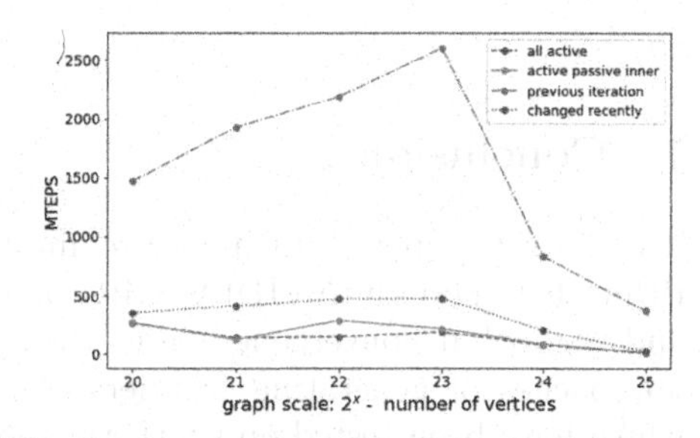

Fig. 5. Performance (in MTEPS) for a whole LP algorithm in RMAT graphs.

In order to evaluate the performance of the proposed approaches on real-world datasets, Orkut and Livejournal social network graphs have been used. Figure 6 compare MTEPS metrics of three versions of the developed algorithm with [12] results. Performance of [12] implementation was evaluated on a machine with NVIDIA Titan X GPU. Since this implementation does not have open source codes, it was impossible for us to evaluate its performance in our testing environment. However, it is possible to roughly estimate performance improvement, which is obtained by better hardware, available at the moment of this writing. Peak Memory bandwidth of Titan X GPU is 480 GB/s, while Tesla P100 has 720 GB/s memory bandwidth. Naturally, in graph processing, effective memory bandwidth values are significantly lower compared to the peak values, because of indirect accesses to 4-byte values significantly bottlenecking the performance. Usually, memory bandwidth of optimized graph algorithms is achieves approximately 50–60% of peak bandwidth values, so [12] time on our machine would be 10–15% faster due to the hardware difference. Nevertheless, our implementation has better convergence rate and clustering likelihood beats from [12] due to the implemented optimization techniques, such as optimized gather step and reducing the amount of work by using sparse frontiers. Also, on Fig. 7 the proposed implementation is compared to label propagation algorithm implemented in Gunrock, version 0.5. Gunrock [19] is a stable and powerful substrate for GPU-based graph-centric research and development.

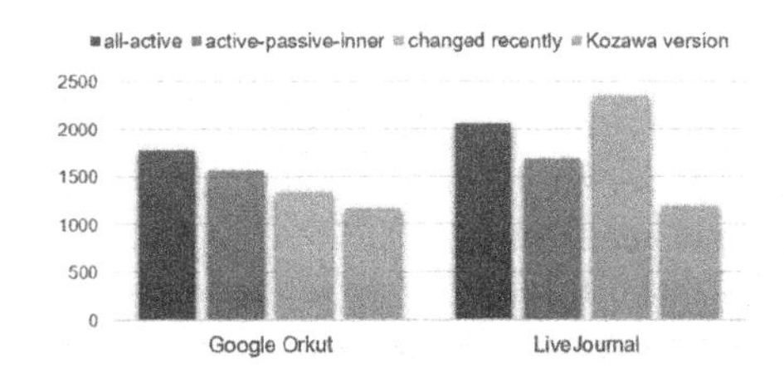

Fig. 6. Real-world graphs evaluation

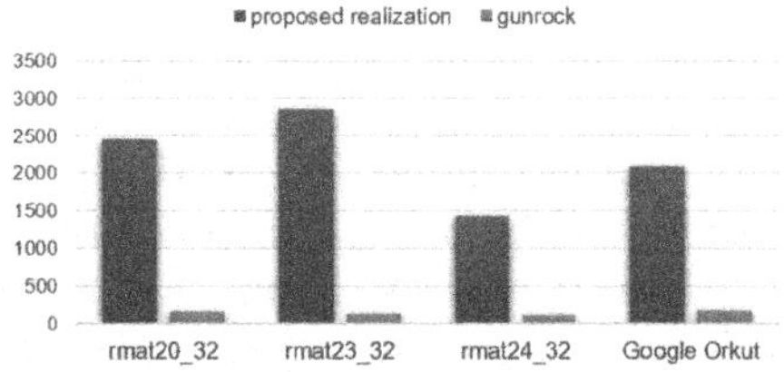

Fig. 7. Comparison with gunrock

7 Conclusion

This paper proposed an improved implementations of label propagation algorithm for modern NVIDIA GPU architecture. To achieve better runtime and algorithm convergence rate, the developed implementation uses various approaches of generating frontiers of active vertices and stopping criterions, which have been tested on synthetic and real-world graphs. In addition, out-of-core graph processing using unified memory was implemented to allow processing large-scale graphs. In addition, several optimizations to gather step have been proposed. The performance of the developed implementation has been compared to the fastest existing implementations: taking into account some differences in hardware characteristics, we managed to reduce label propagation algorithm runtime in approximately 1.2–2 times. Also we managed to beat gunrock implementation in several times.

The reported study was funded by RFBR, project number 19-37-90002. The research is carried out using the equipment of the shared research facilities of HPC computing resources at Lomonosov Moscow State University.

References

1. Stanford Large Network Dataset Collection - SNAP. https://snap.stanford.edu/data/
2. The Koblenz Network Collection - KONECT. http://konect.uni-koblenz.de
3. Vector graph library (VGL) (2020). https://github.com/afanasyev-ilya/VectorGraphLibrary
4. Baxter, S.: Moderngpu wiki (2016). https://github.com/moderngpu/moderngpu/wiki
5. Bell, N., Hoberock, J.: Thrust: A productivity-oriented library for CUDA. In: GPU Computing Gems Jade edition, pp. 359–371. Elsevier (2012)
6. Besta, M., Podstawski, M., Groner, L., Solomonik, E., Hoefler, T.: To push or to pull: on reducing communication and synchronization in graph computations. In: Proceedings of the 26th International Symposium on High-Performance Parallel and Distributed Computing, pp. 93–104 (2017)
7. Chakrabarti, D., Zhan, Y., Faloutsos, C.: R-MAT: a recursive model for graph mining. In: Proceedings of the 2004 SIAM International Conference on Data Mining, pp. 442–446. SIAM (2004)
8. Cordasco, G., Gargano, L.: Label propagation algorithm: a semi-synchronous approach. Int. J. Soc. Netw. Min. **1**(1), 3–26 (2012)
9. Duriakova, E., Hurley, N., Ajwani, D., Sala, A.: Analysis of the semi-synchronous approach to large-scale parallel community finding. In: Proceedings of the Second ACM Conference on Online Social Networks, pp. 51–62 (2014)
10. Erdds, P., Réwi, A.: On random graphs I, vol. 6, p. 18 (1959)
11. Hong, S., Kim, S.K., Oguntebi, T., Olukotun, K.: Accelerating CUDA graph algorithms at maximum warp. ACM SIGPLAN Not. **46**(8), 267–276 (2011)
12. Kozawa, Y., Amagasa, T., Kitagawa, H.: GPU-accelerated graph clustering via parallel label propagation. In: Proceedings of the 2017 ACM on Conference on Information and Knowledge Management, pp. 567–576 (2017)
13. Mišić, M., Drašković, D., Šubelj, L., Bajec, M.: Parallel implementation of the label propagation method for community detection on the GPU
14. Murphy, R.C., Wheeler, K.B., Barrett, B.W., Ang, J.A.: Introducing the graph 500. Cray Users Group (CUG) **19**, 45–74 (2010)
15. NVIDIA: Cuda 10.2 profiler reference (2019). https://docs.nvidia.com/cuda/profiler-users-guide/index.html
16. NVIDIA: Curand reference (2019). https://docs.nvidia.com/cuda/curand/index.html
17. Raghavan, U.N., Albert, R., Kumara, S.: Near linear time algorithm to detect community structures in large-scale networks. Phys. Rev. E **76**(3), 036106 (2007)
18. Soman, J., Narang, A.: Fast community detection algorithm with GPUs and multicore architectures. In: 2011 IEEE International Parallel & Distributed Processing Symposium, pp. 568–579. IEEE (2011)
19. Wang, Y., Davidson, A., Pan, Y., Wu, Y., Riffel, A., Owens, J.D.: Gunrock: a high-performance graph processing library on the GPU. In: Proceedings of the 21st ACM SIGPLAN Symposium on Principles and Practice of Parallel Programming, pp. 1–12 (2016)
20. Zhang, Y., Kiriansky, V., Mendis, C., Zaharia, M., Amarasinghe, S.P.: Optimizing cache performance for graph analytics. ArXiv abs/1608.01362 (2016)

Drop Oscillation Modeling

Lev Shchur[1,2(✉)] and Maria Guskova[1,2]

[1] Landau Institute for Theoretical Physics, Chernogolovka, Russia
lev@landau.ac.ru
[2] National Research University Higher School of Economics, Moscow, Russia
{lshchur,mguskova}@hse.ru

Abstract. The classical problem of oscillations of liquid droplets is a good test for the applicability of computer simulation. We discuss the details of our approach to a simulation scheme based on the Boltzmann lattice equation. We show the results of modeling induced vibrations in a chain of three drops in a closed tube. In the initial position, the central drop has formed as an ellipsoid, out of the spherical equilibrium form. The excitation of vibrations in the left and right droplets depends on the viscosity of the surrounding fluid and the surface tension. Droplets are moving out of the initial position as well. We discuss the limits of the applicability of our model for the study of such a problem. We will also show the dynamics of the simulated process.

Keywords: Lattice Boltzmann Method · Computer simulations · Drop oscillations · Surface tension · Viscosity · Hybrid computing architecture CPU/GPGPU

1 Introduction

The collective motion of the drops in the liquid confined in the complex geometry is a challenging problem of the supercomputer simulations. The interest is two-fold. Firstly, there are many applications in the manufacturing [1], printing [2], oil recovery [3], and cyber-physical [4] systems, among many others. Secondly, it is crucial for the simulations of the flows in the veins connected with the exact drug delivery [5] or problems of tumor cells spreading [6].

Lattice Boltzmann method (LBM) [7] is suitable for the simulation of multicomponent fluid in the complex environment [8]. The LBM is a linear method and can be easily partitioned in space and can be realized in the very efficient massively parallel simulations using supercomputer capabilities.

In the paper, we report the effect of the eigenfrequencies on the collective motion. We simplify the problem and choose the minimal setup of three drops with similar properties, and analyze the drop swing and movement of the drops influenced by the oscillation of another drop. Oscillations originated because of the initial form of one of the drops. We use the symmetric set up with the three drops of the same size and properties immersed in another fluid. At the

V. Voevodin and S. Sobolev (Eds.): RuSCDays 2020, CCIS 1331, pp. 198–206, 2020.
https://doi.org/10.1007/978-3-030-64616-5_17

initial time, the central drop excited in the eigenfrequency oscillation having the ellipsoid form while another two drops (placed at the left and right sides) are in the non-excited state, i.e., with the spherical shape. The volumes of all three drops are the same. The development of the oscillations and movement does depend on the fluid viscosity and surface tension. The simulation shows that the side drops first squeezed and then starts moving out of the center. This effect can initiate the instability of the chain of the drops in the experiment [9].

Simulations based on the Palabos development platform [10] and performed with MPI on the supercomputer cluster.

The LBM method used in simulations is described briefly in Sect. 2. The geometry and parameters of simulations are given in Sect. 3. Section 4 give some details of the program code and computations. Simulations presented in the Sect. 5 and discussion of results is presented in the Sect. 6.

2 Lattice Boltzmann Equations

We use the Shan–Chen method for multiple component fluid flows [11]. The time and three-dimensional space (D3) is discrete and measured in units of Δt and Δx, correspondingly. We use three dimensional D3Q27 representation of 27 velocities c_i pointing from the center of the cube to 8 vertices, to the middle of 12 edges, and to the middle of 6 faces, and to 1 center (zero velocity), i.e. $i = 0, 1, 2, \ldots, 26$ (see, f.e., Refs. [12] and [13]). Distribution functions are defined $f_i^j(\boldsymbol{x}, t)$ for all 27 velocities c_i and for each of two components of fluid ($j = 1, 2$) at lattice position $\boldsymbol{x}$ and evaluated in time by the equations

$$f_i^j(\boldsymbol{x} + \boldsymbol{c}_i \Delta t, t + \Delta t) - f_i^j(\boldsymbol{x}, t) = \Omega_i^j(f) + S_i^j, \tag{1}$$

with the collision operator $\Omega_i^j(\boldsymbol{x}, t)$. The collision term $S_i^j = \boldsymbol{F}^j \cdot \boldsymbol{c}_i$ controls the strength of the interaction potential between fluid components $\{j\}$, the force F^j is defined through the Shan–Chen potential [11]. Collision operator is written in the BGK form [14]

$$\Omega_i^j(f) = -\frac{f_i^j - \tilde{f}_i^j}{\tau} \Delta t, \tag{2}$$

with equilibrium distribution function [13]

$$\tilde{f}_i^j(\boldsymbol{x}, t) = \mathrm{w}_i \rho^j(\boldsymbol{x}, t) \left(1 + \frac{\boldsymbol{u} \cdot \boldsymbol{c}_i}{c_s^2} + \frac{(\boldsymbol{u} \cdot \boldsymbol{c}_i)^2}{2c_s^4} - \frac{\boldsymbol{u} \cdot \boldsymbol{u}}{2c_s^2}\right), \tag{3}$$

where sound speed $c_s = \Delta x / (\Delta t \sqrt{3})$.

3 Physical Setup

Three drops are placed in the box of linear sizes 500, 250, and 250 in the directions x, y, and z, correspondingly. The centers of the drops are in the initial positions: the left drop $x_l, y_l, z_l = 175, 125, 125$, the central drop $x_c, y_c, z_c =$

250, 125, 125, the right drop $x_r, y_r, z_r = 325, 125, 125$. All drops have the same volume of $4/3\pi R^3$ with $R = 30$. At the first moment, the right and left drops are the spheres, and the central drop is the ellipsoid with the z-axes enlarged to $2R$ while keeping the same volume.

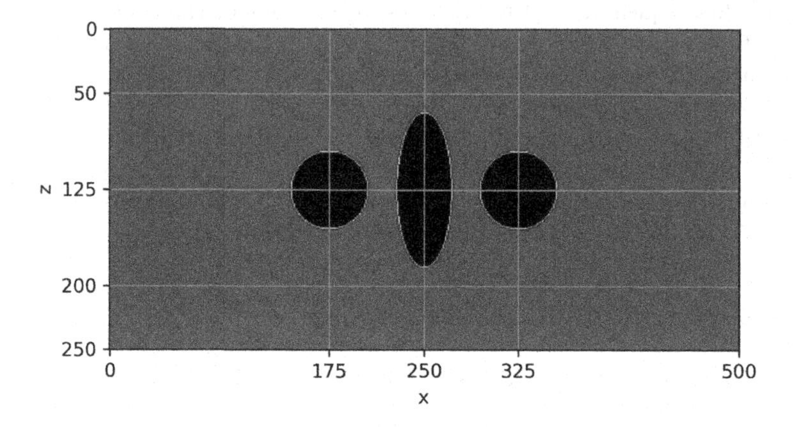

Fig. 1. Cross-section of the initial state of the drops, $y = 125$.

Leaving alone, the central drop with the initial ellipsoid state will oscillate with the component $n = 2$ of the eigenfrequency, according to the Rayleigh formula [15,16]

$$\omega_2^2 = \frac{24}{3\rho_D + 2\rho_F} \frac{\sigma}{R^3}, \tag{4}$$

where σ is the surface tension, and ρ_D and ρ_F is the density of the drop and the surrounding fluid. In simulations, the period of oscillations is about $1000\Delta t$ with the relaxation parameter $w = 4/3$, and oscillations practically damped after 2–3 periods. The details of the single drop oscillation presented in the paper [17].

In the following, we are interested in the orchestrated oscillations of all three drops caused by the swing of the central drop with the initial state exciting the frequency (4). The cross-section at $y = 125$ of the initial state of the simulations is presented in the Fig. 1.

The density ρ_D (equivalent to ρ^1 in the Expr. (1)) of the fluid component inside the drop is set as unity with the small excess $\rho_D = 1.01$, and the density of the surrounding fluid ρ_F (equivalent to ρ^2 in the Expr. (1)) is set as unity with the small excess $\rho_D = 1.01$, and the initial velocities are set up to zero. The collision constant $G = 2.7$ controls the interface region of two fluids [13]. Both fluids have the equal parameters of the viscosity ν associated with the values of the relaxation parameter $w_1 = w_2 = w = 1/\tau$

$$\nu = c_s^2 \left(\frac{1}{w} - \frac{\Delta t}{2} \right). \tag{5}$$

Simulations have been performed for the number of viscosity values, see the Table 1.

Table 1. Values of the viscosity ν and associated values of the relaxation parameter w used in simulations, see Expr. (5)

w	16/11	4/3	100/77	200/157	400/317	16/13
ν	0.625	0.083	0.090	0.095	0.0975	0.104

4 Computational Details

Program code is based on the Palabos C++ development platform [10]. Data defined through classes of type MultiBlock, which are the 3-dimensional matrices in our case. MultiBlockLattice3D data structure defines a Lattice Boltzmann cell, with double-precision floating-point numbers. Each cell contains 27 distribution functions. Calculation of the left part in Exp. (1) is named streaming, and calculation of the first term in the right part is called a collision. Computations of the streaming and the BGK collision are local, and all twenty-seven functions f_i can be computed in parallel. Technically, we solve two sets of the LBM Eqs. (1–3), one for the fluid inside the drop and another one for the fluid outside the drop. Each fluid represented by the own MultiBlockLattice data structure. Therefore the streaming process in the left part Expr. (1) and collision with the operator (2) can be calculated in parallel for each of the fluids. The data processor calculates the Shan-Chen collision dynamics between two fluids as the second term in the right part of Expr. (1). This is a non-local operation, and this part of computations avoid the parallelization. The local properties of the BGK collision and of the distribution function streaming allows the additional possibility for parallelization – the lattice partitioning in the space.

Both fluids exist in the whole domain of simulations, and each fluid represented by 250 by 250 by 500 cells. Those fluid associated with the drop has a normal density $\rho_D = 1.01$ inside the drop and negligible density 0.0001 outside the drop. Contrary, the density of surrounding fluid has density $\rho_F = 1.01$ outside the drop and negligible density 0.0001 inside the drop. The time of the life for the drop with such a density gradient is order of magnitude larger than the total simulation time.

The parallel implementation uses MPI with the library MPICH version 3.2.1 [18]. The simulations in the next section have been done using Intel Xeon Gold 6152 2.1 GHz CPU with onboard memory DDR4 2.666 GHz 768 GB RAM. The program saves the whole computation field to the hard-disk every fifty steps of time, and it takes about 125 sec for this cycle at 22 CPU cores.

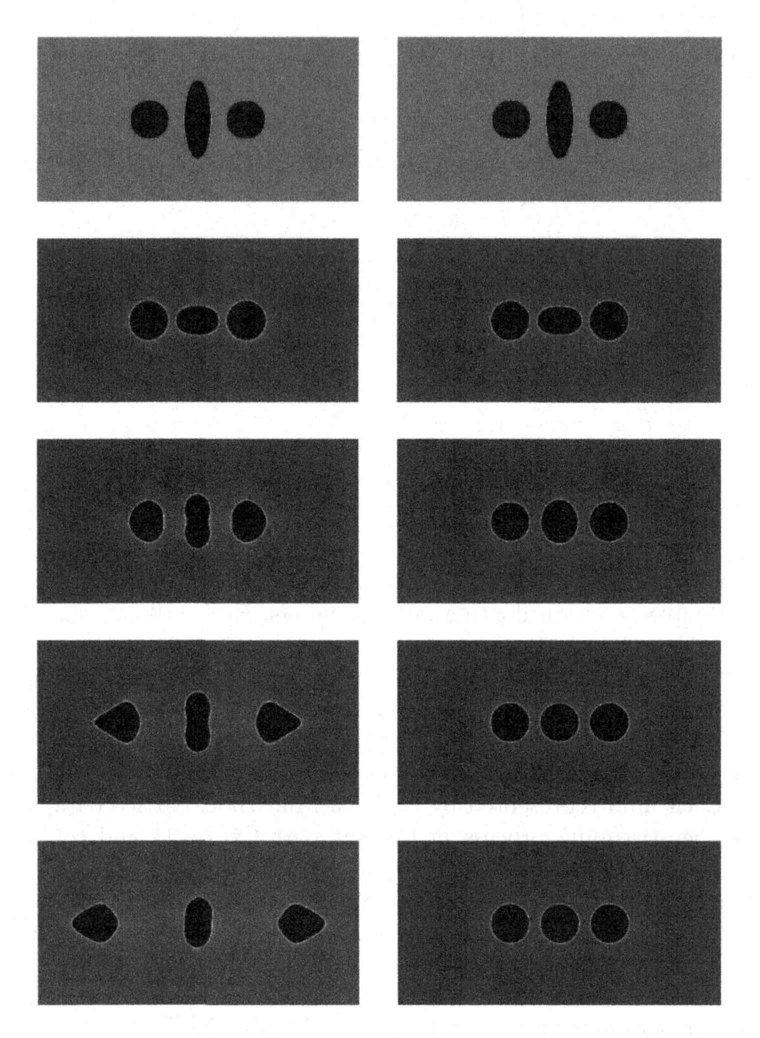

Fig. 2. Cross-section of the drops, $y = 125$ at the value of viscosity $\nu \approx 0.083$ (the left column) and $\nu \approx 0.104$ (the right column) at different times, measured in the units $\Delta t = 1$, from top to bottom $t = 50, 1000, 2000, 3000$, and 4000. The axis are the same as in Fig. 1.

5 Drop Chain Movement

There are experiments (see, f.e., the review [9]) in which the chain of equidistantly injected drops can occasionally break the order. In some cases, one of the drops goes out of the chain. Our idea is that the broke of the symmetry in the chain can be attributed to some imperfection in the drop form at the moment of injection. This imperfection will trigger the oscillations of the drop, which should follow to the spectrum of the Rayleigh frequencies [15]. The idea is to check can the oscillations in one of the drops influence the others or not. The most simple question of that kind is how the oscillations of the central drop have influenced on the left and right drops? We perform simulations of three drops to answer the question.

We simulate the three drops behavior for the number of viscosities shown in the Table 1, and for the same initial conditions formulated in the Sect. 3. Figure 2 shows the dynamics of the drops for two typical cases, $\nu = 0.083$ and $\nu = 0.104$, and which can be considered as "low" and "high" viscosities in the drop dynamics. Note that the difference in the viscosities is only 20%, and it turns quite enough for the observation of two quite different behavior.

The left column in Fig. 2 demonstrates the dynamics of the drops at the lower viscosity, $\nu = 0.083$, as the snapshot at the number of time moments. The top figure shows drops at the very beginning step, $t = 50$; the next figure is taken at the time $t = 1000$, close to the half-period of oscillation [17]. Oscillation is damped by viscosity, and drop also elongates in the perpendicular to the figure direction, keeping the volume constant. At that time, there is no visible distortion of the side drops. At the next figure corresponding to approximately the full period of ω_2-oscillations of the free drop, one can see the apparent effect of the hydrodynamic interactions. The left and right drops squeezed at the sides close to the center; in addition, drops start to sharpen the outer sides due to the internal movement of fluid inside the drops. It is visible that the central drop is no more in the ω_2 regime as the higher mode of oscillations is already present at the time $t = 2000$. The central drop forms the dumbbell-like form due to the reflection of the surrounding fluid from the side drops. It is interesting that about the same time, the new phenomena visible in the figures in the left column in Fig. 2, it is the flow of the drops out of the center. The flow consists of the complicated excitations, with the conic form closer to the box, and with the movement of the center of mass of the left and right drops toward the boundary.

The last two bottom figures in Fig. 2 show the movement of the side drops out of the center one, with the sharp conic form of the outer sides. We have to mention here that we use bounce-back boundary conditions at the box, and there are reflections of the waves in the surrounding fluid because of that. There is a reflection of the drop from the boundary at the time ≈ 3700 (not shown in the figure). We should one more time to notice that due to the symmetry, the right drop has the same behavior as the left one.

The observed effects seem to depend on the viscosity as one can see from the right column of figures in Fig. 2, where initial oscillations with the frequency ω_2 does not visibly influence the right and left drops. The oscillation of the central drop is dumped almost entirely by viscosity on the simulated time scale. One can see that in contrast to the left column of figures with the lower viscosity, there is no visible excitation of the higher modes, and central drop have the ellipsoid form at the time of the full period $t = 2000$.

Finally, at the bottom figure, drops seem to form an equilibrium picture with the drops of equal radii R and with the distance between drops close to $R/2$. Nevertheless, the visible equilibrium is not the final stage of the dynamics. One should look for longer times.

We present in the Fig. 3 the position of the left side of the left drop for the number of values of the relaxation parameter w (consult the Table 1 for the corresponding values of viscosities). Indeed, while simulate drops at the relaxation parameter $w = 16/13$, which corresponds to the right column of figures in the Fig. 2 father in time, the left side of the left drop is starting moving to the left. It is not surprising, however, taking into account the friction properties of the fluid [19].

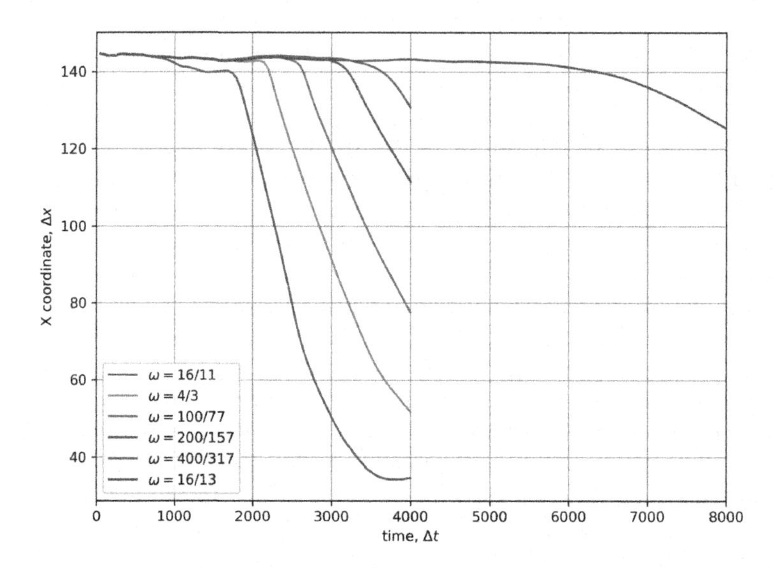

Fig. 3. The time dependence of the left side of the left drop for the different values of the fluid viscosity.

6 Conclusion

Simulations demonstrate the dynamics of the drop chain while exiting oscillations in one of the drops. The level of excitations does depend on the fluid viscosity, which can trigger the visible distortion of other drops, or not. Anyway, some drop movement can be visible at the longer observation times for the large values of viscosities.

We present computer simulation of the oscillations of three drops, with the central drop is in the first exited ellipsoid state, and the left and right drops initially rest in the spherical states. We simulate the system in the box with the bounce-back boundary conditions, and we found the influence of the boundaries is not essential for the most time of simulations. We found the visible excitations of the left and right drops, and even excitation of the higher harmonic of the central drop for the small enough fluid viscosity. For the significant values of the viscosity, there is relatively fast dumping of the central drop oscillations and no visible excitation of the neighboring drops. The developed setup can be used for the simulation of the movement of the long chain of drops in the channel in order to explain the instability of the chains observed in the experiments [9].

We also find that symmetry of the drop chain against the boundaries may influence the dynamics at the longer observation times. We use an even number of cells in all directions. Therefore the center of the central drop is not in the middle of the box, and reflections from the bounce-back boundaries lead to the attractive force in the diagonal direction. To avoid that, we check the dynamics in the box with an odd number of cells. We found there are no attractive forces from the boundaries in that case.

We also check the possible influence of the level of the velocity discretization, the D3Q19 model, with 19 velocities. We do not find any visible difference at the time of the dynamic observation.

Acknowledgment. We are thankful to S. Succi for the discussion, which triggers our interest in the problem of the drop chain stability.

Cluster Manticore of Science Center in Chernogolovka and the Supercomputing facility of the National Research University Higher School of Economics have been used for simulations. The work is carried out according to the project of the Russian Science Foundation 19-11-00286 and partially according to the RFBR project 20-07-00145.

References

1. Simonelli, M., et al.: Towards digital metal additive manufacturing via high-temperature drop-on-demand jetting. Addit. Manuf. **30**, 100930 (2019)
2. Ben-Barak, I., et al.: Drop-on-Demand 3D printing of lithium iron phosphate cathodes. J. Electrochem. Soc. **166**, A5059 (2019)
3. Yuan, D., Moghanloo, R.G.: Nanofluid pre-treatment, an effective strategy to improve the performance of low-salinity waterflooding. J. Petrol. Sci. Eng. **165**, 978 (2018)
4. Ibrahim, M., Chakrabarty, K., Zeng, J.: BioCyBig: a cyberphysical system for integrative microfluidics-driven analysis of genomic association studies. IEEE Trans. Big Data (2016). https://doi.org/10.1109/TBDATA.2016.2643683
5. Esmaeili, S.: An artificial blood vessel fabricated by 3D printing for pharmaceutical application. Nanomed J. **6**, 183 (2019)
6. Freitas, V.M., Hilfenhaus, G., Iruela-Arispe, M.L.: Metastasis of circulating tumor cells: speed matters. Dev. Cell **45**, 3 (2018)
7. Succi, S.: The Lattice Boltzmann Equation: For Complex States of Flowing Matter. Oxford University Press, Oxford (2018)
8. Shan, X., Doolen, G.: Multicomponent lattice-Boltzmann model with interparticle interaction. J. Stat. Phys. **81**, 379 (1995)
9. Beatus, T., Bar-Ziv, R.H., Tlusty, T.: The physics of 2D microfluidic droplet ensembles. Phys. Rep. **512**, 103 (2012)
10. Latt, J., et al.: Palabos: Parallel Lattice Boltzmann Solver. Comput. Math. Appl. (2020, in Press). https://doi.org/10.1016/j.camwa.2020.03.022

11. Shan, X., Chen, H.: Lattice Boltzmann model for simulating flows with multiple phases and components. Phys. Rev. E **47**, 1815 (1993)
12. Suga, K., Kuwata, Y., Takashima, K., Chikasue, R.: A D3Q27 multiple-relaxation-time lattice Boltzmann method for turbulent flows. Comput. Math. Appl. **69**, 518 (2015)
13. Krüeger, T., Kusumaatmaja, H., Kuzmin, A., Shardt, O., Silva, G., Viggen, E.M.: The Lattice Boltzmann Method, Principles and Practice. Springer, Cham (2017)
14. Bhatnagar, P.L., Gross, E.P., Krook, M.: A model for collision processes in gases. I. Small amplitude processes in charged and neutral one-component systems. Phys. Rev. **94**, 511 (1954)
15. Rayleigh, L.: On the capillary phenomenon of jets. Proc. R. Soc. London **29**, 71 (1879)
16. Lamb, H.: Hydrodynamics. Dover, New York (1932)
17. Guskova, M., Shchur, V., Shchur, L.: Simulation of drop oscillation using the lattice Boltzmann method. Lobachevskii J. Math. **41**(6), 992–995 (2020)
18. https://www.mpich.org/2017/11/11/mpich-3-2-1-released/
19. Landau, L.D., Lifshitz, E.M.: Course of Theoretical Physics VI: Fluid Mechanics. Pergamon Press, Oxford (1982)

High-Performance Simulation of High-Beta Plasmas Using PIC Method

Igor Chernykh[(✉)], Vitaly Vshivkov, Galina Dudnikova, Tatyana Liseykina, Ekaterina Genrikh, Anna Efimova, Igor Kulikov, Ivan Chernoshtanov, and Marina Boronina

Institute of Computational Mathematics and Mathematical Geophysics SB RAS, Novosibirsk, Russia
{chernykh,vsh,kulikov,boronina}@ssd.sscc.ru, gdudnikova@gmail.com, tatiana.liseykina@gmail.com, mesyats@gmail.com, anna.an.efimova@gmail.com, cherivn@ngs.ru

Abstract. In this paper, we present the new parallel PIC code for numerical simulation of plasma physics problems. We used previously developed [1] hybrid model of plasma with a new realization of numerical method which is adapted for vectorization for modern CPUs. We tested our code on a very important high-energy physics problem which is connected with thermonuclear synthesis. Numerical simulation of different types of diamagnetic traps can be used for prototyping of magnetic trap devices for compact fusion reactors. As a result of this work, we propose some parallel tests of our code and numerical simulation of the diamagnetic "bubble".

Keywords: Parallel computing · Computational plasma physics · Numerical methods

1 Introduction

Numerical simulation of modern physical problems became impossible without parallel computing. For example, full modeling of diamagnetic traps, which is very important for thermonuclear synthesis, can be done on 50K+ cores. The theory predicts, that the stationary diamagnetic "bubble" occupies a domain with sizes 10×100 cm. The thin boundary layer of the "bubble" is connected with the ion Larmor radii and is ~ 1 cm. Since a rough description of the layer requires 10 spatial grid nodes, the whole computational domain requires minimum $N_r = 100, N_z = 1000$ grid nodes. The stationary structure of the "bubble' implies long evolution times $\sim 10^4 \omega_{ci}^{-1}$. The time resolution with 10^4 steps for one unit leads to 10^8 timesteps and correspondingly long performance times. The achievement of better accuracy with the grid refinement is also limited with the conditional stability since the maximal timestep is defined by the spatial step. The permanent particle injection leads to the introducing of the new particles in the domain, so the further evolution times require a higher number of particles. The injected particles are distributed in the domain highly non-uniformly, most

V. Voevodin and S. Sobolev (Eds.): RuSCDays 2020, CCIS 1331, pp. 207–215, 2020.
https://doi.org/10.1007/978-3-030-64616-5_18

of them is concentrated in the centre of the domain. Besides, each cell must contain background ions, say, j per cell. Thus the number of the background particles is jN_rN_z and it increases in case of the grid refinement [2]. All these facts describe the problems of the high-accuracy computations and a need in a parallel algorithm to tackle them [3–11].

2 Mathematical Model

We consider a cylindrical mirror trap of the radius R and the length L. At the initial moment the domain filled with the background hydrogen plasma ($Z_i = 1$) with the density n_0. The magnetic field in the trap is created by the solenoid and two coaxial round coils with the current. The coils create magnetic mirrors. The magnetic field in the center of the region is H_0. The beam particle source is located at the center of the trap $(0, L/2)$.

The following hybrid model of plasma is considered [1]. The ions are described with kinetic Vlasov equation:

$$\frac{\partial f_i}{\partial t} + \mathbf{v}\frac{\partial f_i}{\partial \mathbf{r}} + \frac{\mathbf{F}_i}{m_i}\frac{\partial f_i}{\partial \mathbf{v}} = 0, \tag{1}$$

$$\mathbf{F}_i = e(\mathbf{E} + \frac{1}{c}[\mathbf{v}, \mathbf{H}]) + \mathbf{R}_i. \tag{2}$$

Here $(\mathbf{r}, \mathbf{v})$ is the coordinate and the velocity, $f_i(t, \mathbf{r}, \mathbf{v})$ is the ion distribution function. Index i corresponds to the ions: m_i is the ion mass, e is the ion charge (hydrogen plasma). $\mathbf{R}_i$ is the ion-electron friction force.

In the Particle-in-Cell method the Vlasov equation is replaced by the equations of its characteristics. The equations for the ion trajectories $(\mathbf{r}_\mathrm{i}, \mathbf{v}_i)$ coincide with the equations of the characteristics of kinetic Vlasov equation:

$$\frac{d\mathbf{r}_i}{dt} = \mathbf{v}_i, \quad m_i\frac{d\mathbf{v}_i}{dt} = e(\mathbf{E} + \frac{1}{c}[\mathbf{v}_\mathrm{i}, \mathbf{H}]) + \mathbf{R}_i. \tag{3}$$

For the electron component, the magneto-hydrodynamics equations are used in the limit of massless electrons. Plasma assumed quasi-neutral ($n_e = n_i$), so we can denote the density as n.

$$-e\left(\mathbf{E} + \frac{1}{c}[\mathbf{V}_e, \mathbf{H}]\right) - \frac{\nabla p_e}{n} + \mathbf{R}_e = 0, \tag{4}$$

$$n\left(\frac{\partial T_e}{\partial t} + (\mathbf{V}_e\nabla)T_e\right) = (\gamma - 1)(Q_e - \operatorname{div}\mathbf{q}_e - p_e\operatorname{div}\mathbf{V}_e). \tag{5}$$

There n, $\mathbf{V}_e$, T_e are the density, the velocity and the temperature of the electron component of plasma. The electron-ion friction force is $\mathbf{R}_e = -m_e\frac{\mathbf{V}_e - \mathbf{V}_i}{\tau_{ei}}$, $\mathbf{R}_i = -\mathbf{R}_e$ [16], τ_{ei} is the characteristic time of electron-ion collisions, $p_e = nT_e$ is the pressure of the electrons, $Q_e = j^2/\sigma$ is the electron heating caused by the ion-electron friction, $\sigma = ne^2\tau_{ei}/m_e$, $\mathbf{q}_e = -\lambda\nabla T_e$ is the heat flux, λ is the coefficient of the thermal conductivity, $\gamma = 5/3$ is the heat capacity ratio for ideal gas.

The Maxwell equations for the electromagnetic fields $\mathbf{E}$ and $\mathbf{H}$ are used. The considered processes are non-relativistic and low-frequency processes, therefore, the displacement currents $\partial\mathbf{E}/\partial t$ are neglected.

$$\mathrm{rot}\,\mathbf{H} = \frac{4\pi}{c}\mathbf{j}, \quad \mathrm{rot}\,\mathbf{E} = -\frac{1}{c}\frac{\partial\mathbf{H}}{\partial t}, \tag{6}$$

$$\mathbf{j} = en(\mathbf{V}_i - \mathbf{V}_e). \tag{7}$$

The ion average velocity $\mathbf{V}_i$ and the density n are defined by the distribution function $f_i(t,\mathbf{r},\mathbf{v})$

$$n(\mathbf{r}) = \int f_i(t,\mathbf{r},\mathbf{v})d\mathbf{v}, \quad \mathbf{V}_i(\mathbf{r}) = \frac{1}{n_i(\mathbf{r})}\int \mathbf{v}f_i(t,\mathbf{r},\mathbf{v})d\mathbf{v}. \tag{8}$$

3 Numerical Method

For the definition of the dimensionless variables we use the characteristic values H_0 and n_0. H_0 is the magnetic field in the center of the trap, n_0 is the background plasma density. The characteristic time is $t_0 = 1/\omega_{ci}$, where $\omega_{ci} = eH_0/cm_i$ is the ion cyclotron frequency. The characteristic length is $L_0 = c/\omega_{pi}$, $\omega_{pi} = \sqrt{4\pi n_0 e^2/m_i}$ is the ion plasma frequency. The characteristic velocity is Alfven velocity $V_A = H_0/\sqrt{4\pi m_i n_0}$.

The calculation cycle with the dimensionless equations is the following.

1. The Lagrangian stage: the coordinates $\mathbf{v}_i$ and the velocities $\mathbf{v}_i$ of ions $\mathbf{r}_i$ are calculated.

$$\frac{d\mathbf{r}_i}{dt} = \mathbf{v}_i, \quad \frac{d\mathbf{v}_i}{dt} = \mathbf{E} + [\mathbf{v}_i,\mathbf{H}] - \kappa(\mathbf{V}_i - \mathbf{V}_e), \tag{9}$$

$$\kappa = cm_e/eH_0\tau_{ei}.$$

2. The beginning of Eulerian stage: the density n and mean ion velocity $\mathbf{V}_i$ are calculated.
   To calculate the average functions in the PIC method, the particle shape function R is required. We use the CIC shape function R_{CIC}, which corresponds to the bilinear interpolation. Index j denotes the number of the model ion particle.

$$n(\mathbf{r}) = \sum_j R_{CIC}(\mathbf{r}-\mathbf{r}_j), \quad \mathbf{V}_i(\mathbf{r}) = \frac{1}{n}\sum_j \mathbf{v}_j R_{CIC}(\mathbf{r}-\mathbf{r}_j). \tag{10}$$

3. The electron velocity $\mathbf{V}_e$:

$$\mathbf{V}_e = \mathbf{V}_i - \frac{\mathrm{rot}\,\mathbf{H}}{n}. \tag{11}$$

4. The electric field $\mathbf{E}$:

$$\mathbf{E} = -[\mathbf{V}_e,\mathbf{H}] - \frac{\nabla p_e}{2n} + \kappa(\mathbf{V}_i - \mathbf{V}_e), \tag{12}$$

5. The magnetic field $\mathbf{H}$:

$$\frac{\partial \mathbf{H}}{\partial t} = -\operatorname{rot} \mathbf{E}, \tag{13}$$

6. The electron temperature T_e:

$$n\left(\frac{\partial T_e}{\partial t} + (\mathbf{V}_e \nabla) T_e\right) = (\gamma - 1)\left(2\kappa \frac{\mathbf{j}^2}{n} + \kappa_1 \operatorname{div} \nabla T_e - p_e \operatorname{div} \mathbf{V}_e\right), \tag{14}$$

$$\kappa_1 = 4\pi e\lambda / H_0 c.$$

At the initial moment, the electric field $\mathbf{E} = 0$ and the magnetic field $\mathbf{H_0}$ is defined as the field of the solenoid and two coils with a current located near the external r-boundary and creating magnetic mirrors. For the initial magnetic field, the simple iteration scheme is used. The initial velocities of the background ions are zero. The density of the background particles is constant $n(r,z) = n_0$. Particles are constantly injected into the center of the trap at the $(r,z) = (0, L/2)$.

At the axis $z = 0$ the boundary condition $g(r)|_{r=0} = 0$ is used for all r- and φ-components of the functions. The boundary condition of the symmetry is used for z-component of the functions. On the outer boundaries for all functions, the boundary conditions are $\frac{\partial g(t,r,z)}{\partial r}|_{r=R} = 0$, $\frac{\partial g(t,r,z)}{\partial z}|_{z=0,z=L} = 0$. For the particles the open boundary conditions are used.

As we consider an axially symmetric case, we use the cylindrical coordinate system in the two-dimensional domain and eliminate derivatives $\partial/\partial\varphi$ from the equations. We use particle-in-cell method to solve the Eqs. (9). The values of particle data $\mathbf{r}_i, \mathbf{v}_i$ are defined on the Lagrangian mesh. The staggered uniform grids are used to solve the equations for all variables (10)–(14). All numerical schemes for solving the Eqs. (9)–(14) are described in detail in [17].

Since in Eqs. (9), (12), (14) the currents appear with a factor as $\mathbf{j}/n$, we substitute $\mathbf{U} = \mathbf{j}/n$ and use the arrays for $\mathbf{U}$ during the computations. The only equation requiring values $\mathbf{j}$ is (11) and we use the multiplication $n\mathbf{U}$ there. The result of the changes is decrease of the number of the operations performed by the processors and increase of the algorithm performance. The definitions $\mathcal{E}^* = -[\mathbf{V_e}, \mathbf{H}]$, $\mathcal{E}_\nabla = -\nabla p_e/2n$, $\mathcal{E} = \mathcal{E}^* + \mathcal{E}_\nabla$ yield $\mathbf{E} = \mathcal{E} + \kappa\mathbf{U} = \mathcal{E}^* + \mathcal{E}_\nabla + \kappa\mathbf{U}$. Here $\mathcal{E}_{\nabla\varphi} = 0$ and we use only two arrays for $\mathcal{E}_\nabla$.

3.1 Parallel Implementation

The parallel implementation of the algorithm is based on the mixed decomposition. The computational domain is evenly cut along the axis z into np_0 subdomains. Each subdomain is assigned to a group of cores. The particles in the subdomain are assigned to the cores of the group and are distributed uniformly among them (see Fig. 1). From np cores the ones with ranks $0..np_0 - 1$ are named main cores and perform the computations on the grid. Each core processes the ion velocities and coordinates, then formes an array from the particles leaving the subdomain and sends the data to one of the cores of the neighbour group. The local rank of the receiving core is defined as $mod(n+1, C_g)$, where C_g is number of cores in the group number g. After receiving its particles each core computes the contribution to the density and average velocities on the grid. Then each main core gathers the data from its group using *MPI_Reduce* function and continues the computations of the currents, electron velocities, electric and magnetic field, temperature with the corresponding boundary exchanges with the neighbour main cores. In the end of the cycle, each core uses *MPI_Bcast* function to copy the new electric and magnetic fields to the cores of its group.

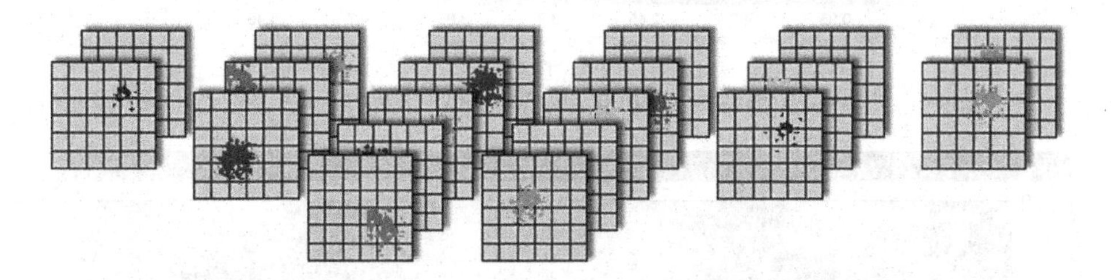

Fig. 1. The cores distribution for the mixed decomposition.

The number of the cores in group is static and defined before the computation. One the first cycle step 1 or 2 cores from the number of $np - np_0$ cores are added to the central groups, $g = (np_0 - 1)/2$ for odd np_0 and $g = np_0/2 - 1$ and $g = np_0/2$ for even np_0. On the second step, cores are added to the central 3 or 4 groups correspondingly, and so on. If the number of not-distributed cores on step k is smaller then $2k - 1$ or $2k$ correspondingly, the cycle goes again to step 1. At the end of the cycle, the more central is the subdomain the more cores it has in its group. The consequent uniform distribution of the particles in the cores within a group allows achieving a good load balancing.

3.2 Numerical Results

The tests are performed on a model problem with the computational domain $[0 : R] \times [0 : L], R = 10, L = 30$. The magnetic field in the initial time moment is defined by the two coils with currents and solenoid field, so the mirror ratio on the axis is 2. The magnetic field on the axis $H(0, L/2) = 1$. The beam density is $5 \cdot 10^2$, the injecting ion speed is $v_0 = 0.2$, the ion temperature is 0.01. These parameters correspond to the dimensional values $R = 228$ cm, $L = 684$ cm, $H(0, L/2) = 2$ kG. The dimensional background ion density is 10^{12}cm^{-3}, the beam density $5 \cdot 10^{14}\text{cm}^{-3}$, the average beam speed $4.3 \cdot 10^7\text{cm/s}$, the ion temperature $10eV$. The normalizing parameters are: $H_0 = 2$ kG, $n_0 = 10^{12}\text{cm}^{-3}$, $V_A = 4.4 \cdot 10^8\text{cm/s}$, $L_0 = 23\,\text{cm}$, $t_0 = 5.5 \cdot 10^{-8}\,\text{s}$.

Figure 2 and Fig. 3 demonstrate the absolute values of the magnetic field at the initial time moment and at t = 30. The magnetic fields is expelled from

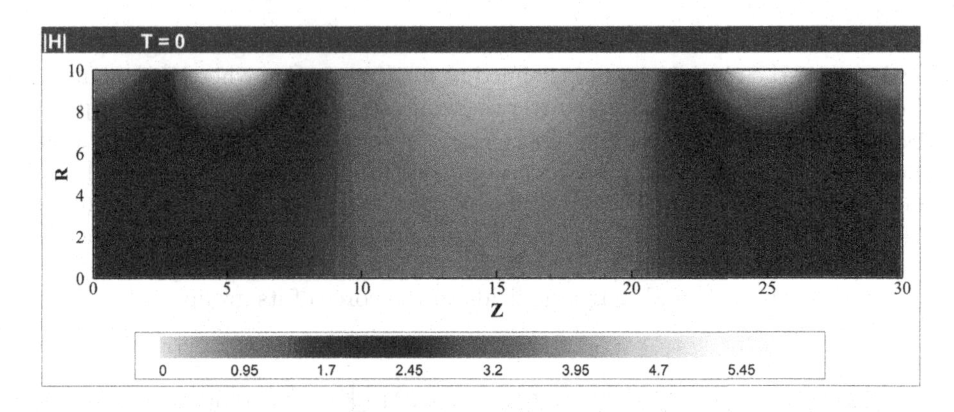

Fig. 2. Absolute value of the magnetic field at t = 0.

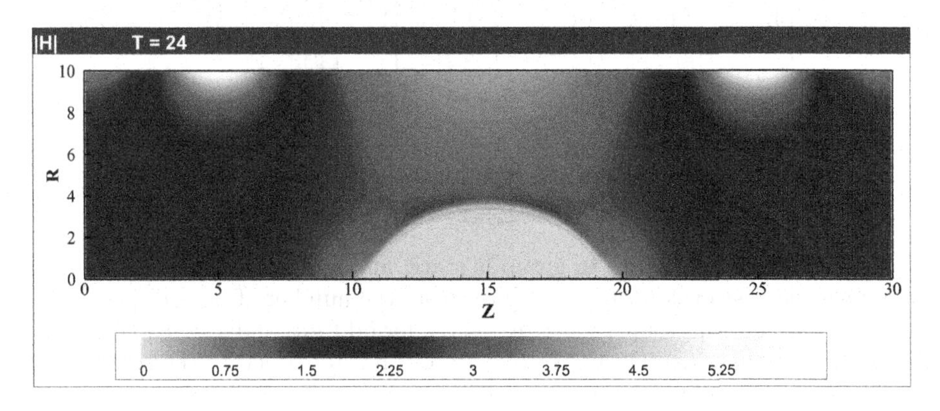

Fig. 3. Absolute value of the magnetic field at t = 30.

the central region due to the injection of the particles, the magnetic field inside the "bubble" does not exceed $5\%H_0$. The initially motionless background ions increase their speeds and move away from the region of the "bubble". Figure 4 demonstrates the coordinates (r, z) of the background ions. The injected ions move without a significant changes in velocities in the region of the expelled field, the thin boundary layer of the bubble has a high magnetic field and leads to the particle rotation in it. In Fig. 5 the coordinates (r, z) of the injected ions are shown by the dots, the color represents the value of radial velocity V_r. The results corroborate the idea of possibility to confine the plasma in the diamagnetic regime and the workability of the proposed model.

For this case the grid with $N_r = 100$ and $N_z = 300$ was used, the time step was $\tau = 10^{-5}$, the number of the time steps was $N_\tau = 4.8 \cdot 10^6$. The number of background ions was $J_b = 1.2 \cdot 10^5$, the total number of ions at the end of the computation was $J_p = 1.3 \cdot 10^6$.

We used Siberian Supercomputer Center NKS-1P facilities for the numerical experiments. Siberian Supercomputer Center supercomputer consists of three

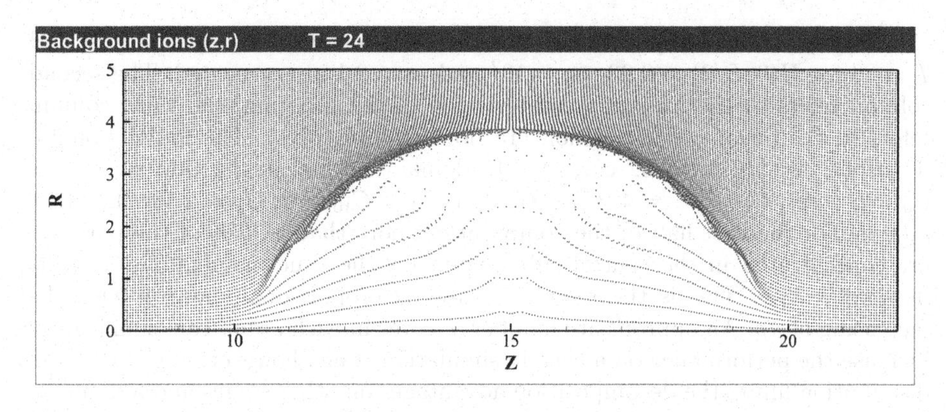

Fig. 4. The background ion coordinates (z, r).

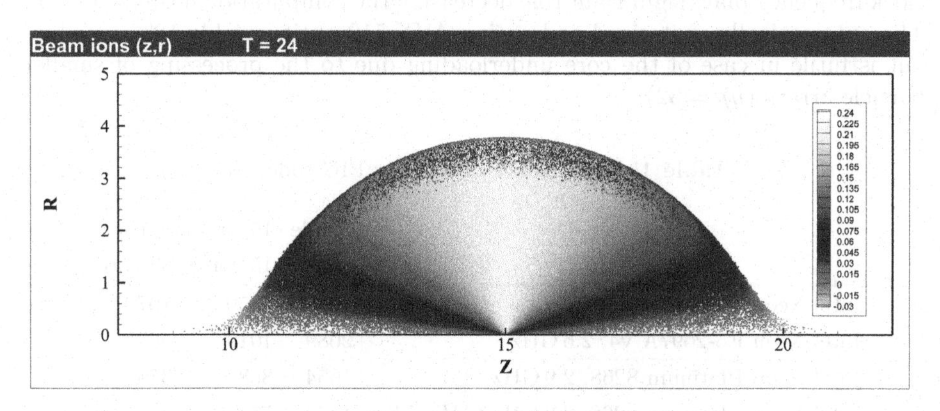

Fig. 5. The injected ion coordinates (z, r).

types of nodes which are based on Intel Xeon Phi 7290 (KNL architecture), Intel Xeon E5-2697A v4 (Broadwell architecture), Intel Xeon Platinum 8268 (CascadeLake architecture) CPUs. The numerical experiments for the parameters above took $\sim$ 22 h using 128 cores Intel Xeon E5-2697A v4 (Broadwell architecture) in 60 groups. In this case $C_{29} = C_{30} = 11$ cores, $C_{28} = C_{31} = 9$, $C_{27} = C_{32} = 7$, $C_26 = C_{33} = 5$, $C_{25} = C_{34} = 4$, $C_{24} = C_{35} = 3$, $C_{23} = C_{36} = 2$ and $C_g = 1$ for the rest of groups g. The average number of particle in the central groups $g = 29$ and $g = 30$ at $T = 24$ was $1.8 \cdot 10^4$, for g = 23..28, 31..36 it was $\sim 1.3 \cdot 10^4$. However, the cores $g = 22$ and $g = 37$ contain a higher number of particles $\sim 1.6 \cdot 10^4$: the injected particles have reached the subdomains with only one core per group. The particle processing takes $\sim$ 96% of the computation time, the grid computations take $\sim$ 3%. The computation of the contribution of the average velocities and the density of the ions takes $\sim$ 10%. Since the simulation affects only the initial stage of the injection, the particle mostly move in the radial direction and a small number of them moves away to the neighbour cores: the particles exchanges take $\sim$ 0.1% of the computational time. The grid exchanges are also of a small value.

In Table 1 the computational times in secs for 50×150 spatial grid and $T = 80$ for $H(0, L/2) = 0.03$, $n_0 = 10^2$ and $v_0 = 0.1$ are presented. The second column refers to the monoprocessor version of the program, the third column refers to the linear decomposition (15 domains, 15 cores). The fourth column describes the case with 32 cores for 15 domains, the central group $g = 7$ has $C_g = 6$ cores, $C_6 = C_8 = 4$, $C_5 = C_9 = 3$, $C_4 = C_{10} = 2$, $C_g = 1$ for the other groups. For the first half of the computation more than $\sim 47\%$ of the particles are concentrated in the central subdomain, for the time moment $T = 80$ the number is 33%. This is the reason of the non-proportional speedup with the increasing of np. An appropriate number of cores in each group can significantly increase the performance on a certain simulation stage, however, any static core distribution limits the decomposition advantages on other stages and a dynamic load-balancing is needed. Since the core communications are not intensive, the clock frequency plays significant role decreasing the computational times [12–15]. The automatically vectorized code using AVX-512 option yields 12% speedup, but is futile in case of the core underloading due to the processing of smaller particle arrays ($np = 32$).

Table 1. Performance of the new PIC code, sec

	Single core	15 subdomains	
		15 cores	32 cores
Intel Xeon Phi 7290, 1.5 GHz	10300	4650	1070
Intel Xeon E5-2697A v4, 2.6 GHz	2089	1019	582
Intel Xeon Platinum 8268, 2.9 GHz	1654	868	379
Intel Xeon Platinum 8268, 2.9 GHz (AVX 512)	1472	773	378

4 Conclusions

Vectorization is the process of converting an algorithm from operating on a single value at a time to operate on a set of values at one time. Modern CPUs provide direct support for vector operations where a single instruction is applied to multiple data (SIMD). For example, a CPU with a 512-bit register could hold 8 double-precision values and do a single calculation 8 times faster than executing a single instruction at a time. The latest CPUs have more than one 512 bit vector operations unit. Combine this with threading and multi-core CPUs leads to orders of magnitude performance gains. In this paper, we presented new PIC code for numericalsimulation of plasma physics problems. This code has high potential for vectorization. We achieved 12% speedup with autovectorization key of Intel C++ compiler. The numerical tests demonstrate the workability of the proposed model and the possibility to confine plasma in the diamagnetic regime. The mixed decomposition provides a significant speadup for a correct core distribution among subdomains.

Acknowledgments. The authors was supported by the Russian Science Foundation (project no. 19-71-20026).

References

1. Y.A. Berezin, G.I. Dudnikova, T.V. Liseikina, M.P. Fedoruk Modeling of Unsteady Plasma Processes. NSU Publishing (2017)
2. Lotov, K.V., Timofeev, I.V., Mesyats, E.A., Snytnikov, A.V., Vshivkov, V.A.: Note on quantitatively correct simulations of the kinetic beam-plasma instability. Phys. Plasmas **22**, 024502 (2015)
3. Dickman, D.O., Morse, R.L., Nielson, C.W.: Numerical simulation of axisymmetric, collisionless, finite-β plasma. Phys. Fluids **12**(8), 1708–1716 (1969)
4. Brettschneider, M., Killeen, J., Mirin, A.A.: Numerical simulation of relativistic electrons confined in an axisymmetric mirror field. J. Comput. Phys. **11**(3), 360–399 (1973)
5. Killeen, J.: Computer models of magnetically confined plasmas. Nucl. Fusion **16**(5), 841–864 (1976)
6. Turner, W.C., et al.: Field-reversal experiments in a neutral-beam-injected mirror machine. Nucl. Fusion **19**,1011 (1979). https://doi.org/10.1088/0029-5515/19/8/002
7. Byers, J.A.: Computer simulation of field reversal in mirror machines. Phys. Rev. Lett. **39**(23), 1476–1480 (1977)
8. Belova, E.V., Jardin, S.C., Ji, H., Yamada, M., Kulsrud, R.: Numerical study of tilt stability of prolate field-reversed configurations. Phys. Plasmas **7**(4996), 4996–5006 (2000)
9. Belova, E.V., Davidson, R.C., Ji, H., Yamada, M.: Advances in the numerical modeling of field-reversed configurations. Phys. Plasmas **13**(5), 056115 (2006)
10. Lifschitz, A.F., Farengo, R., Arist, N.R.: Monte Carlo simulation of neutral beam injection into a field reversed configuration. Nucl. Fusion **42**(7), 863–875 (2002)
11. Takahashi, T.: Hybrid simulation of neutral beam injection into a field-reversed configuration. J. Plasma Phys. **72**(6), 891–894 (2006)
12. Kulikov, I.: GPUPEGAS: A New GPU-accelerated Hydrodynamic Code for Numerical Simulations of Interacting Galaxies. Astrophys. J. Suppl. Ser. **214**(1), 12 (2014)
13. Kulikov, I.M., Chernykh, I.G., Snytnikov, A.V., Glinskiy, B.M., Tutukovm, A.V.: AstroPhi: a code for complex simulation of dynamics of astrophysical objects using hybrid supercomputers. Comput. Phys. Commun. **186**, 71–80 (2015)
14. Kulikov, I., Chernykh, I., Tutukov, A.: A new hydrodynamic model for numerical simulation of interacting galaxies on intel xeon phi supercomputers. J. Phys. Conf. Ser. **719**, 012006 (2016)
15. Glinsky, B., Kulikov, I., Chernykh, I.: The co-design of astrophysical code for massively parallel supercomputers. Lect. Notes Comput. Sci. **10049**, 342–353 (2017)
16. Braginskii, S.I.: Transport processes in plasma, In: Leontovich, M.A. (ed.) Reviews of Plasma Physics, vol. 1, pp. 205–311 (1965)
17. Genrikh, E.A., Boronina, M.A., Dudnikova, G.I.: Hybrid model of the open plasma trap. In: AIP Conference Proceedings, vol. 2164, pp. 110003 (2019)

Implementation of SL-AV Global Atmosphere Model with 10 km Horizontal Resolution

Mikhail Tolstykh[1,2,3](✉), Gordey Goyman[1,2,3], Rostislav Fadeev[1,2,3], and Vladimir Shashkin[1,2,3]

[1] Marchuk Institute of Numerical Mathematics Russian Academy of Sciences, Moscow, Russia
m.tolstykh@inm.ras.ru, gordeygoyman@gmail.com, rost.fadeev@gmail.com, vvshashkin@gmail.com
[2] Hydrometcentre of Russia, Moscow, Russia
[3] Moscow Institute of Physics and Technology, Dolgorpudny, Russia

Abstract. Huge computer resources needed to promptly compute the 24-h global weather forecast dictate the necessity to optimize the numerical algorithms of the model and their parallel implementation. We present some experience gained while implementing the new high-resolution version of the SL-AV global atmosphere model for numerical weather prediction at parallel systems with many thousands of processor cores. Unlike our previous scalability studies, we need to minimize the elapsed time of the forecast at given processor cores number which is currently about 4000. The results of optimizations are shown for two Roshydromet high-performance computer systems.

Keywords: Numerical weather prediction · Massively parallel computations · Combination of MPI · OpenMP technologies

1 Introduction

The growth in resolution of numerical weather prediction (NWP) models is considered as one of the main ways to increase the accuracy of weather forecast. The typical horizontal resolution of current NWP models is 7–20 km with 60–140 grid points in vertical direction [1]. The computational domain size is currently about 10^9. The number of prognostic variables in such models is about 10, with the tendency to include new variables, i.e. small atmospheric constituents. So the numerical weather prediction requires huge computer resources. World operational weather forecasting centers have computer systems with typical peak performance of 3–8 Pflops occupying the first hundreds of Top 500 list (as of December 2019, [2]). Obviously, an atmosphere model used for prediction should be able to use such a computational power efficiently. All the centers perpetually work on optimization of their models codes as this allows to use higher model resolution or larger ensemble size at a given supercomputer [3]. In Russia, the peak performance of Roshydromet supercomputer system Cray XC40 is about 1.3 Pflops (465th position in Top500 as of December 2019) so this task is more important than for other centers. The global operational numerical weather prediction model at Hydrometcentre of Russia is the SL-AV model developed at Marchuk Institute of Numerical Mathematics RAS

V. Voevodin and S. Sobolev (Eds.): RuSCDays 2020, CCIS 1331, pp. 216–225, 2020.
https://doi.org/10.1007/978-3-030-64616-5_19

and Hydrometcentre of Russia [4]. This model is employed for medium range weather forecasts and in the long-range probabilistic prediction system. Model algorithms and structure are considered in [5]. The dynamical core of this model [6] is based on the semi-implicit semi-Lagrangian method. Many subgrid-scale processes parameterizations algorithms are developed by ALADIN/LACE consortium [7, 8]. CLIRAD SW [9] and RRTMG LW [10] algorithms are applied to parameterize shortwave and longwave radiation respectively. The INM RAS-MSU multilayer soil model [11] is incorporated. The combination of one-dimensional MPI decomposition and OpenMP loop parallelization is employed in the parallel program complex, details of the parallel algorithms are given in [12]. The code is written in Fortran language using dialects from 77 to 2003 and consists of several hundred thousand lines.

A lot of work on SL-AV parallel code optimization was carried out. It was shown in [12] that this code is able to use 13608 cores with the efficiency slightly higher than 50%, for grid dimensions of $3024 \times 1513 \times 126$. The profiling analysis [12] had revealed the parts of the model code that need further improvements. The most important was the semi-Lagrangian advection accounting for 38% of total elapsed time while using 9072 processor cores. It was found that this part had worst scalability and its proportion in whole elapsed time increased with the growth of processor core number.

Recently, the new version of SL-AV model for medium range weather forecast has been implemented, SLAV10. It has higher horizontal resolution than mentioned above, about 0.1°, the grid dimensions being $3600 \times 1946 \times 104$.

We describe in this paper recent modifications of SLAV10. The goal of these modifications is to reduce the elapsed time of model forecast while using about 4000 processor cores. This is the amount of cores that can be used for operational model runs in practice. Section 2 outlines the changes while the results of numerical experiments are presented in Sect. 3.

2 Implementation Improvements

2.1 Initial Code Profiling

All the experiments were carried out at two high performance systems installed at Roshydromet's Main Computing Center. The first one is Cray XC40 system. This system consists of 936 nodes with two Intel Xeon E2697v4 18-core CPUs and 128 GB memory. All the nodes are connected with Cray ARIES interconnect. The peak performance is 1.29 PFlops. The second system is T-Platforms V6000 consisting of 99 nodes with two Intel Xeon Gold 6148 20-core processors, 192 GB memory, and 30 nodes having the same two Intel Xeon processors plus two NVIDIA Tesla V100 accelerators. All nodes are connected with HDR100 InfiniBand interconnect. Only the nodes without accelerators were used in our experiments.

Both systems share the same parallel file system.

The elapsed time necessary to compute the 24-h forecast with the new SLAV10 model version using about 4000 processor was initially 42 min at Cray XC40 system. This is too much as the operational requirements dictate the limitation of no more than 20 min. First, we have measured SL-AV model profile for the abovementioned grid resolution with dimensions $3600 \times 1946 \times 104$ using 3888 cores of Cray XC40 (108

36-cores nodes, using 6 MPI processes each having 6 OpenMP threads at every node) and T-Platforms V6000 system (97 40-cores nodes, using 8 MPI processes and 5 OpenMP threads at every node). These combinations of MPI and OpenMP were found optimal for the SLAV10 code at respective computer systems.

The profiling results are shown in Fig. 1 where "SL advection" denotes the semi-Lagrangian advection part of the code, "Parameterizations" stands for parameterizations for all subgrid-scale processes (shortwave and longwave radiation, deep and shallow convection, planetary boundary layer, gravity wave drag, microphysics et al.), "Dynamics" refers to all the computations in the dynamical core except for semi-Lagrangian advection and solvers for Helmholtz equation, wind speed reconstruction and horizontal diffusion [5], and "Elliptic solver" corresponds to the abovementioned solvers in Fourier space in longitude space.

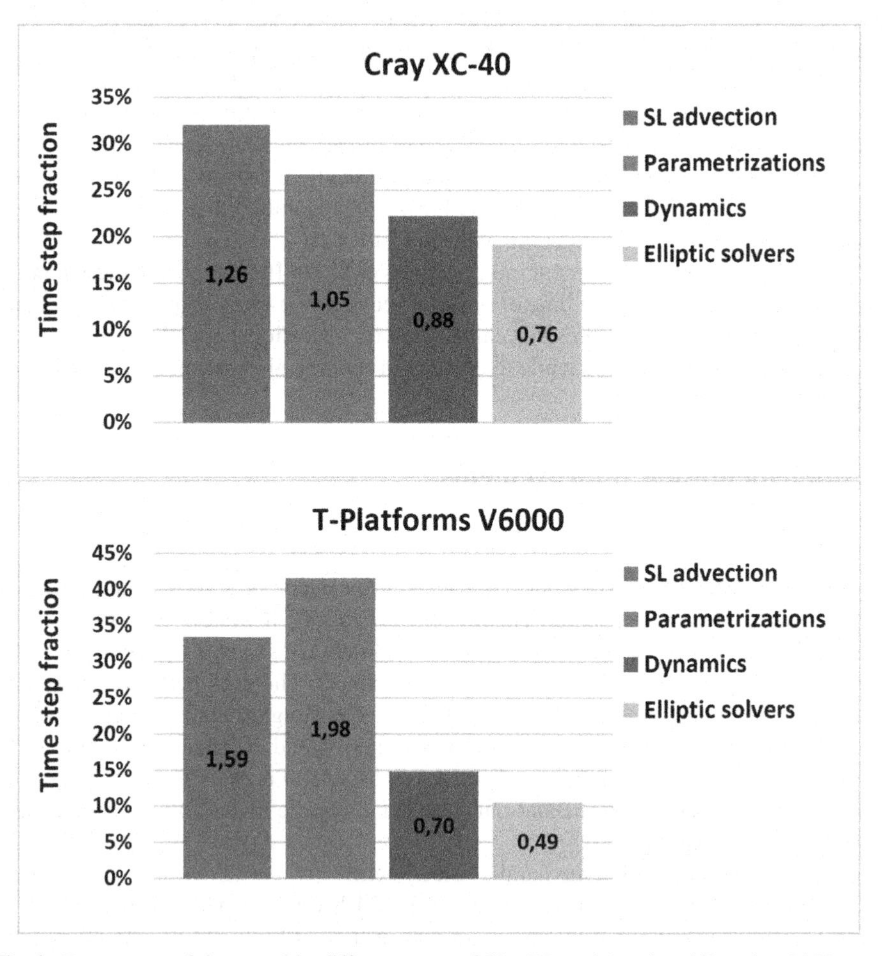

Fig. 1. Percentage of time used in different parts of SL-AV model code while using 3888 cores of Cray XC40 (top) and 3880 cores of T-Platforms V6000 (bottom) systems before optimizations. Number inside the column denotes the wall-clock time of respective code part (in seconds).

One can see that most time consuming blocks are the semi-Lagrangian advection and parameterizations of subgrid-scale processes. The share of the semi-Lagrangian advection in the whole time step for the 3600 × 1946 × 104 grid while using 3888 cores is close to this proportion for 3024 × 1513 × 126 grid when 4536 cores of Cray XC40 are used [12]. The part of the SLAV10 time step occupied by parameterizations calculations is about 26%, these calculations took 38% for 3024 × 1513 × 126 grid under the same conditions. Further we concentrate on optimizations of these parts of model code.

Note that initial elapsed time per time step at Cray XC40 system is less by 0.8 s than at T-Platforms V6000 despite the fact that V6000 uses modern Skylake-family processors while previous Broadwell-family Intel processors are used in Cray XC40 system. These results are quite different from timings of current operational SLAV20 model with grid dimensions 1600 × 866 × 51 otherwise having the similar code to SLAV10 where V6000 shows acceleration of approximately 20% with respect to Cray XC40 system (14 Cray 36-core nodes were used as compared to 11 40-core nodes of V6000 system).

2.2 Single Precision in Semi-Lagrangian Advection

Some world weather prediction centers, such as the European Center for Medium-Range Weather Forecasts, now use single precision floating point computations in their atmosphere models instead of double precision [3]. It was shown in [3] that a smart application of single precision in the model code does not deteriorate forecast quality but enables savings of about 40% of elapsed time. We have started to implement single precision in the most time-consuming part of the SL-AV model that scales worse than other parts, according to [12], the semi-Lagrangian advection block. As can be seen above (Fig. 1), it accounts for more than 30% of elapsed time at both systems considered. On average, 56% of this time is spent on parallel exchange phase and about 32% on trajectory calculations and interpolation phase.

The semi-Lagrangian advection algorithm [13] allows to use quite large time steps, typically 5–10 times larger than ones defined by the CFL condition. This algorithm first solves the kinematic equation for trajectories to determine a departure point at previous time step for every arrival grid point at current time step. Then the advection equation is solved in the Lagrangian form by interpolating the advected variables at the departure points. Currently, 1D parallel decomposition (in MPI) is still used in SL-AV model so the parallel implementation of this algorithm requires halo exchanges (exchanges of latitudinal bands adjacent to the process boundaries) with the width determined by the position of the furthest departure point and interpolation stencil. One of possible approaches to speed up this part of the model with the minimal source code changes is the use of single precision for the arrays containing values of grid functions to be exchanged and interpolated. The use of single precision allows to half the size of transferred data and reduces memory load during computations.

In order to implement the possibility to switch precision of the semi-Lagrangian buffer array, we use Fortran user-defined KIND parameter and MPI datatype parameter, which are compile-time constants defined in one module.

Profiling of the SL-AV model after the introduction of the abovementioned changes in the SL-AV model shows that there is 45% runtime decrease in the parallel communication phase and about 27% reduction of the interpolation and trajectory computation time. Importantly, the forecast scores almost have not changed.

2.3 Controlling Vector Length in Parameterizations Computations

Calculating tendencies (right-hand sides of atmosphere dynamics equations) due to parameterizations of subgrid scale processes is one of the most time-consuming parts of the atmospheric model, and SL-AV model in particular as can be seen from Fig. 1. Details on the implementation of the physics parameterizations block in the SL-AV model are given in [5]. Briefly, computations in this block are carried out in the vertical direction only, thus no parallel communications are needed. We use OpenMP loop parallelization along longitudinal direction. Since we use *(vertical levels, longitudes, latitudes)* global arrays indices arrangement in the most part of the model and the use of *(longitudes, vertical levels, latitudes)* arrangement is more efficient [12] in parameterizations block, each OpenMP thread creates local temporary arrays containing copies of variables necessary for calculations with the first two dimension being interchanged. The length of these local arrays along first dimension is $\frac{NLON}{N_{omp}}$, where NLON is the number of grid points along longitudinal direction (3600 in our case) and N_{omp} is the number of OpenMP threads.

The fact that we already use thread-local temporary arrays allows us to adjust the length of the first dimension of these arrays in order to improve data localization, thus increasing efficiency of the fast cache-memory usage (reduce the number of cache misses). Therefore, the thread's computational domain is splitted up into blocks with the length along the first dimension equals to $\frac{NLON}{N_{omp} \cdot Nb}$, where *Nb* is a number of blocks, and all the computations in the physics parametrizations part of the source code are carried out for such a smaller block at a time.

We have investigated the effect of vector length adjustment on performance of the parametrizations block with the 10 km model configuration using 108 nodes and 6 OMP-threads at Cray XC40 system. Figure 2 illustrates the change of CPU time spent in physics parametrizations part of the source code depending on the blocks at each OMP-thread compared to the use of one block. The upper part of Fig. 2 depicts the results for SLAV10 model. The measurements for current operational model SLAV20 with grid dimensions $1600 \times 866 \times 51$ using 27 nodes with 9 MPI processes each running 4 OpenMP threads at every node are shown at the bottom of this figure (hence NLON = 1600 in this case). Here we distinguish physics parameterizations without the use of solar radiation parameterization (called at every time step of the model) and with them (called every half an hour of the model time). One can see that it is optimal to use 25 blocks with the 42% and 25% decrease in CPU time for the physics parameterizations calls with and without radiation parameterization for SLAV10. The same number, 25 blocks is optimal for SLAV20 model where these reductions are 33% and 22% respectively.

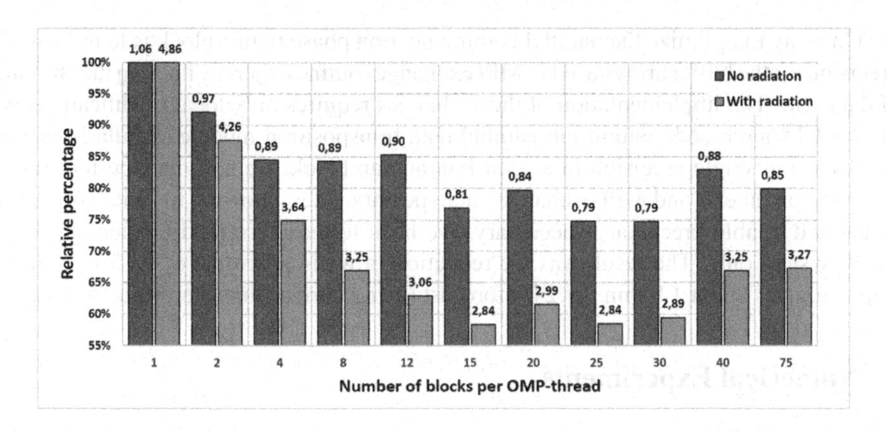

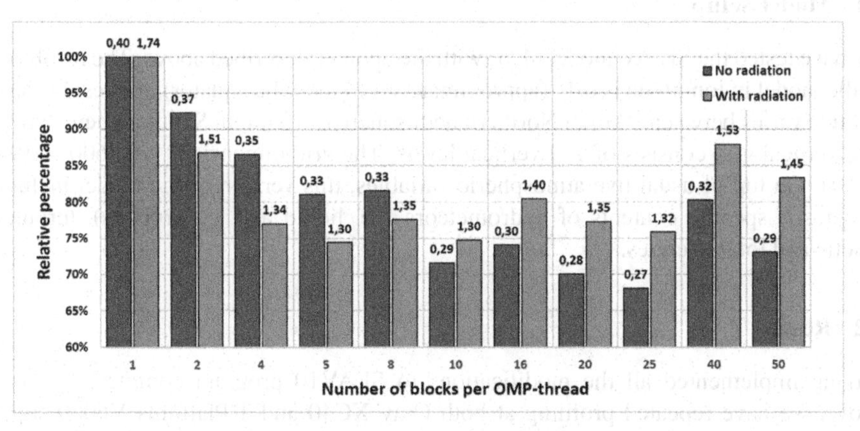

Fig. 2. Reduction of the CPU time spent in the physics parametrization part of the SL-AV10 model (top) and SLAV20 model (bottom) at Cray XC40 system. The figures above the bars show the time in seconds.

2.4 Single Precision Parallel Data Transposition

In the SL-AV model, there is a set of elliptic problems to be solved at the every time step. Namely, Helmholtz problem resulted from the semi-implicit time scheme, wind velocity reconstruction problem and implicit hyper diffusion equation. The solution algorithm of these equations is based on the use of fast Fourier transforms along longitudes and block Thomas algorithm. Practical implementation of this method includes parallel data transposition phase, which is effectively a global redistribution of data between processes and requires all-to-all type parallel communication, accounting for the substantial part of this part of source code elapsing time.

One way to optimize the parallel communication phase in this block is to use single precision for the buffer arrays used in MPI exchange routines, thereby halving the amount of data sent. The implementation of these changes requires only local modifications of the model source code within the parallel data transposition module. Similarly to the case of using single precision in a semi-Lagrangian block, we introduce buffer arrays precision parameter and MPI exchange datatype parameter, allowing to switch between single and double precision if necessary. We have tested these modifications at Cray XC40 system only. The result was the reduction in wall-clock time of 0.17 s per time step that gives about 1.8 min per 24 h forecast (using current time step value of 135 s).

3 Numerical Experiments

3.1 Model Setup

We have tested the SL-AV model [4, 5] with the updates described above. The resolution of the model in longitude is 0.1° (approximately 11 km at the equator), the resolution in latitude varies between 7 km in Northern hemisphere to 13 km in Southern hemisphere. The vertical grid consists of 104 vertical levels. The grid dimensions are $3600 \times 1946 \times 104$. On top of usual five atmospheric variables, this version of the model includes prognostic specific contents of hydrometeors (i.e. liquid and ice droplets), turbulent kinetic and total energies.

3.2 Results

Having implemented all the modifications to SLAV10 program complex described above, we have repeated profiling at both Cray XC40 and T-Platfoms V6000 supercomputer systems. The results are shown in Fig. 3. One can see that untouched blocks of elliptic solvers and dynamics part at V6000 system and dynamics part at Cray XC40 system retain their elapsed time at every time step prompting their further optimizations. At the same time, semi-Lagrangian advection part and parameterizations for subdgrid scale processes have significantly reduced their elapsed time. Now the wall-clock time to complete 24-h forecast time 3888 processor cores has been reduced from 42 min to 32.5 min at Cray XC40 system that will be a primary system for operational usage of SLAV10 model.

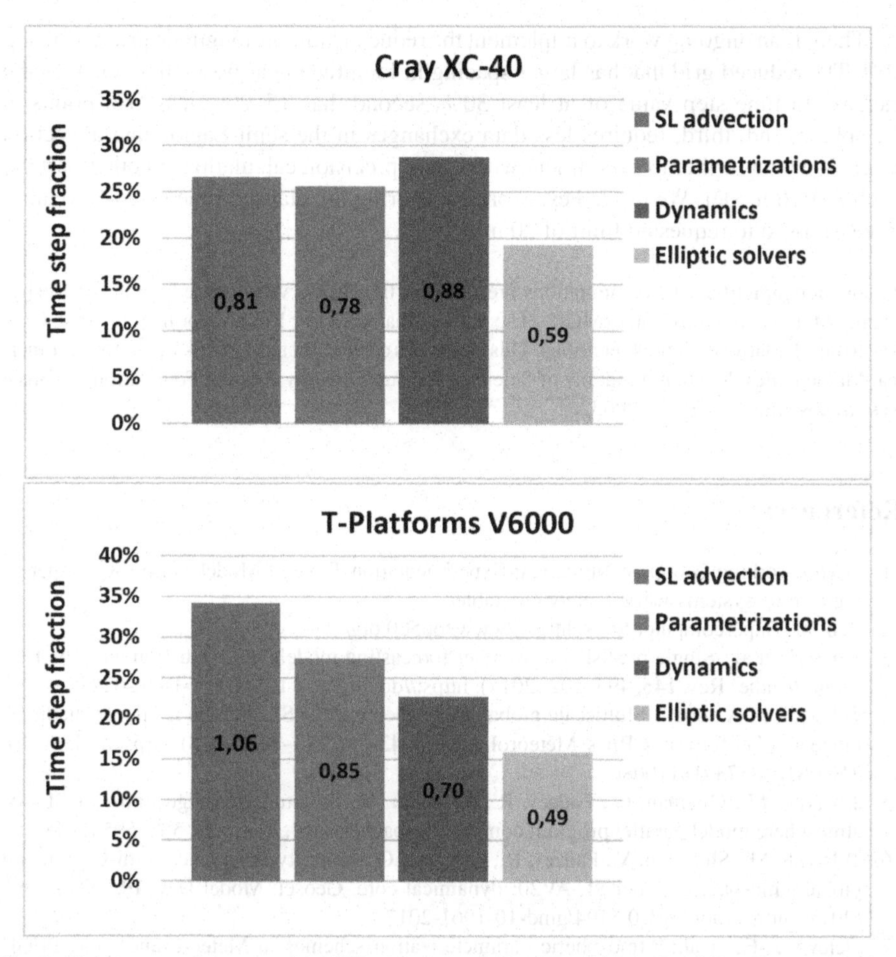

Fig. 3. Percentage of time used in different parts of SL-AV model code while using 3888 cores at Cray XC40 (top) and 3880 cores at T-Platforms V6000 (bottom) after optimizations.

4 Conclusions

We have carried out several works on reducing the elapsed time of the potential future operational Russian medium-range weather forecast model, SLAV10, having the horizontal resolution of about 10 km, and 104 vertical levels. The following works have been implemented: switching most time-consuming part to single precision floating points operations, optimizing the vector size for the block for parameterizations of subgrid scale processes, and reduction of volume in data transpositions by making them single precision instead of double precision. These modifications have reduced the elapsed time of 24-h weather forecast with SLAV10 model by 22.5% at Cray XC40 system, from 42 min to 32.5 min.

The change of precision from double to single can be considered in many codes as this is relatively easy to implement.

There is an ongoing work to implement the reduced latitude-longitude grid following [14]. The reduced grid that has larger spacing in longitude near poles, first, enables the increase in time step value of at least 30%, second, has 15–20% less grid points at the sphere, and, third, requires less data exchanges in the semi-Lagrangian advection block. We plan further to experiment with single precision calculations in other parts of the SLAV10 model. We hope these works will bring the elapsed time of 24-h weather forecast down to requested limit of 20 min.

Acknowledgements. The computations presented in this article were carried out at the Roshydromet Main Computer Center (MCC). The authors thank Sergey Loubov for help in using Cray XC40 and T-Platforms V6000 facilities. This study was carried out at Marchuk Institute of Numerical Mathematics, Russian Academy of Sciences and supported by Russian Foundation for Basic Researches grant No. 19-31-90032.

References

1. WMO Working Group on Numerical Experimentation Table of Models. http://wgne.meteoinfo.ru/nwp-systems-wgne-table/wgne-table/
2. TOP500 Supercomputer Sites. https://www.top500.org/
3. Váňa, F., et al.: Single precision in weather forecasting models: an evaluation with the IFS. Mon. Weather Rev. **145**, 495–502 (2017). https://doi.org/10.1175/MWR-D-16-0228.1
4. Tolstykh, M.A., et al.: Multiscale global atmosphere model SL-AV: the results of medium-range weather forecasts. Russ. Meteorol. Hydrol. **43**(11), 773–779 (2018). https://doi.org/10.3103/S1068373918110080
5. Tolstykh, M., Goyman, G., Fadeev, R., Shashkin, V.: Structure and algorithms of SL-AV atmosphere model parallel program complex. Lobachevskii J. Math. **39**, 587–595 (2018)
6. Tolstykh, M., Shashkin, V., Fadeev, R., Goyman, G.: Vorticity-divergence semi-Lagrangian global atmospheric model SL-AV20: dynamical core. Geosci. Model Dev. **10**, 1961–1983 (2017). https://doi.org/10.5194/gmd-10-1961-2017
7. Geleyn, J.-F., et al.: Atmospheric parameterization schemes in Meteo-France's ARPEGE N.W.P. model. In: Parameterization of Subgrid-Scale Physical Processes, ECMWF SEMINAR PROCEEDINGS, pp. 385–402. RCMWF, Reading (1994)
8. Gerard, L., Piriou, J.-M., Brožková, R., Geleyn, J.-F., Banciu, D.: Cloud and precipitation parameterization in a meso-gamma-scale operational weather prediction model. Mon. Weather Rev. **137**, 3960–3977 (2009). https://doi.org/10.1175/2009MWR2750
9. Tarasova, T., Fomin, B.: The use of new parameterizations for gaseous absorption in the CLIRAD-SW solar radiation code for models. J. Atmos. Ocean. Technol. **24**, 1157–1162 (2007). https://doi.org/10.1175/JTECH2023.1
10. Mlawer, E.J., Taubman, S.J., Brown, P.D.: RRTM, a validated correlated-k model for the longwave. J. Geophys. Res. **102**, 16663–16682 (1997). https://doi.org/10.1029/97jd00237
11. Volodin, E.M., Lykossov, V.N.: Parameterization of heat and moisture transfer in the soil-vegetation system for use in atmospheric general circulation models: 1 formulation and simulations based on local observational data. Izvestiya, Atmos. Ocean. Phys. **34**, 402–416 (1998)
12. Tolstykh, M., Goyman, G., Fadeev, R., Shashkin, V., Lubov, S.: SL-AV model: numerical weather prediction at extra-massively parallel supercomputer. In: Voevodin, V., Sobolev, S. (eds.) RuSCDays 2018. CCIS, vol. 965, pp. 379–387. Springer, Cham (2019). https://doi.org/10.1007/978-3-030-05807-4_32

13. Staniforth, A., Cote, J.: Semi-Lagrangian integration schemes for atmospheric models – a review. Mon. Weather Rev. **119**, 2206–2233 (1991)
14. Fadeev, R.: Algorithm for reduced grid generation on a sphere for a global finite-difference atmospheric model. Comput. Math. Math. Phys. **53**, 237–252 (2013). https://doi.org/10.1134/S0965542513020073

INMOST Platform for Parallel Multi-physics Applications: Multi-phase Flow in Porous Media and Blood Flow Coagulation

Kirill Terekhov[1], Kirill Nikitin[1], and Yuri Vassilevski[1,2(✉)]

[1] Marchuk Institute of Numerical Mathematics of the Russian Academy of Sciences, Moscow, Russia
kirill.terehov@gmail.com, nikitin.kira@gmail.com, yuri.vassilevski@gmail.com

[2] Sechenov University, Moscow, Russia

Abstract. INMOST (Integrated Numerical Modeling Object-oriented Supercomputing Technologies) is an open-source platform for fast development of efficient and flexible parallel multi-physics models. In this paper we review capabilities of the platform and present two INMOST-based applications for parallel simulations of multi-phase flow in porous media and clot formation in blood flow. For a more detailed description we refer to [1].

The finite volume (FV) method is the popular approach to spatial discretizations on general meshes (*i.e.* meshes composed of general polyhedral cells), especially for geophysical and biomedical applications where local mass conservation is vital. INMOST provides a complete set of tools for development of FV discretizations for linear and nonlinear problems: automatic differentiation tool for assembly of the nonlinear residual and corresponding Jacobian and Hessian matrices, iterative solvers of nonlinear systems arising from PDEs discretization, parallel solvers for sparse linear algebraic systems.

The platform also provides a technology for development of numerical models on general unstructured grids. It includes parallel mesh data structures, low-level infrastructure for reading, writing, creating, manipulating and partitioning of distributed general meshes.

The synergy of INMOST platform and efficient FV discretizations for systems of PDEs on general meshes produces a powerful tool for supercomputing simulations.

Keywords: Parallel computing · General meshes · Finite volume method · Flow and transport in porous media · Blood flow coagulation

1 Distributed General Grids

INMOST platform provides a full support of 2D or 3D consistent general grids (any two cells may share one vertex or one entire edge or one entire face), providing storage and adjacencies for the following basic element types:

V. Voevodin and S. Sobolev (Eds.): RuSCDays 2020, CCIS 1331, pp. 226–236, 2020.
https://doi.org/10.1007/978-3-030-64616-5_20

- *Vertex* with coordinates in space;
- *Edge* is defined by a set of vertices;
- *Face* is defined by a set of edges;
- *Cell* is defined by a set of faces.

For each mesh element the associated data (dense or sparse) is accessible via *Tag*. Dense data are associated with all elements, while sparse data are given on some elements. Data can have fixed or variable size, as well as various data types: bulk (single character), integer, double, a reference to an element, double with single or multiple variations, a reference to an element of another mesh. Variation is represented by a sparse vector consisting of indices and coefficients that represent partial derivatives. This type of data is used to store intermediate results of Jacobian calculation for automatic differentiation tools. Data are accessible directly in the memory through provided classes or may be copied to provided arrays.

A computational mesh can be loaded from files of supported formats (*e.g.* vtk/pvtk, gridecl or internal parallel mesh format pmf) or generated from scratch. Basic operations for mesh generation are addition of new elements and deletion of existing elements. Addition or deletion of an element is performed by modification of its adjacency connections. Another important mesh generation tool is local mesh modification. Two types of high-level mesh modification routines are provided by INMOST: uniting a set of elements into a single element and splitting an element into subelements.

Deletion, insertion and modification of a mesh element may produce topological errors (an inconsistent mesh). For this reason, mesh topological correctness tests are performed during mesh modifications, which allow the user to check the following errors: duplication of elements, element degeneracy, wrong order of edges within a face, wrong face orientation, face non-planarity, mesh inconsistency, slivers (a face containing all the cell nodes), duplication of adjacencies, disconnected adjacencies, improper dimensionality of adjacencies, etc.

Organization of the distributed mesh data within INMOST platform is based on usage of several processors, domain decomposition and overlapping grids with ghost cells. Every processor stores a part of the whole mesh. Partitioning the mesh into parts is performed by a load balancing algorithm. FV discretizations require a shell of neighboring cells around every cell. In order to provide the shell of neighbors for cells residing near the interface boundary of the local mesh part, we form additional overlapping layers of cells. Different discretization stencils require different types and widths of overlapping layers. These additional cells are exact copies of cells that reside on another processor that shares the interface with the host processor. We call these cells as *ghost* cells. In INMOST, each mesh element is assumed to have exactly one processor owner. An element that resides on its owner processor and has corresponding ghost copies on other processors, is called *shared* element. Each shared element also stores the list of processors it is copied to. The main difference between the ghost element and the normal element is that data in the ghost element should be actualized after update in the normal element. This entails inter-processor communication that moves data from shared cells to their ghost copies.

The main steps for obtaining a distributed mesh are:

- **Initialization.** For each element we define the following data:
  - *State*, which can be: "owned" (owned exclusively by current processor), "shared", or "ghost",
  - *Owner*, which is the owner processor,
  - *Processors*, which is an array of processors that have a copy of the element;
- **Synchronization.** For any two processors p_1, p_2, the number of "ghost" elements on p_1 with *Owner* equal to p_2 should be equal to the number of "shared" elements with *Processors* containing p_1 on p_2 and vice versa;
- **Formation of layers of ghost elements.** To create layers of ghost elements, we should define the set of faces that lay on the interface between local mesh parts corresponding to two processors p_1 and p_2. This set of faces is the starting set each processor from which the first layer of adjacent cells is composed. The second layer (if required) is formed by next adjacent cells etc.

Given the number of ghost layers and the mesh connectivity (neighbors through the faces, edges or vertices), INMOST functions automatically compute and distribute ghost cells and organize exchanges of elements data from processors-owners to processors possessing copies of elements [2].

INMOST provides flexible interface for mesh repartitioning and redistribution. The user may choose one of internal partitioners, external partitioners Zoltan [3] and Parmetis [4], or provide a partitioning map explicitly.

In case of local modifications of a mesh distributed among given processors, new distribution of mesh cells should correlate with the former distribution, in order to minimize the communication load and computational work for mesh reconstruction. We employ the K-means clustering algorithm [5].

Refinement or coarsening of a general polyhedral mesh distributed among several processors, require accurate handling. We assume that modifications are synchronized between processors, *i.e.* the modification on different processors should not produce topologically inconsistent and geometrically non-matching meshes. Once a distributed mesh is modified, parallel consistency of the modified mesh has to be recovered.

2 Linear Solvers

The internal linear solvers incorporated into INMOST platform are based on the combination of the following components:

- **Parallel iterative solver.** The basic iterative method of INMOST is the preconditioned biconjugate gradient stabilized method BiCGStab(ℓ) [6]. This method optionally performs ℓ BiCG steps and fits a polynomial function to accelerate convergence of preconditioned residual. The BiCGStab(ℓ) parallelization is straightforward as it only requires to accumulate sums of scalar products computed on each processor and synchronize vector elements after matrix-vector multiplication. Parallel preconditioning is based on the combination of an incomplete LU factorization and the restricted Additive Schwarz method with user-specified overlapping parameter.

- **Preprocessing.** Prior the factorization, the matrix is preprocessed by reordering and rescaling in three steps:
  - **Non-symmetric permutation for the static choice of pivoting sequence.** The result is the reordering that maximizes the product of elements on diagonal and the re-scaling that leads to I-dominant matrix [7].
  - **Symmetric permutation for fill-in reduction.** This allows to reduce the number of non-zeros in the triangular factors. Two variants of such permutations are available in INMOST. The first variant is based on reverse Cuthill-Mckee reordering [8]. The second variant relies on the node dissection algorithm from METIS library [9].
  - **Rescaling to improve the condition number and dropping strategy.** Two rescaling strategies from INMOST improve the condition number of the matrix to be factorized. The first strategy is the iterative equilibration of row and column norms to unity [10,11]. The second strategy is the transformation of the matrix into I-dominant matrix which has unit values on its diagonal and off-diagonal entries with modulii not exceeding 1. The I-dominant matrix rescaling is the by-product of the algorithm that finds the maximal transversal and can be further improved iteratively.
- **Preconditioning.** INMOST provides two variants of incomplete LU-factorization. These are row-wise incomplete LU method and Crout incomplete LU method [12]. Both factorization methods use the second-order dual-threshold dropping strategy with τ_1 and τ_2 parameters following [13]. The idea of the second-order factorization method is to have two versions of triangular factors L, $\hat{L}$ and U, $\hat{U}$. Matrices L and U contain elements whose modulii exceed the first threshold τ_1 and $\hat{L}$ and $\hat{U}$ contain elements whose modulii exceed the second threshold $\tau_2 \leq \tau_1$. During factorization, elimination is performed with products of L with U and $\hat{U}$, and products of U with L and $\hat{L}$, *i.e.* contributions of the product of $\hat{L}$ and $\hat{U}$ are ignored. This allows us to improve significantly the quality of L and U factors approaching them closer to the triangular factors of the exact factorization. Once the factorization is completed, $\hat{L}$ and $\hat{U}$ factors are abandoned and the iterations proceed with the preconditioner involving factors L^{-1} and U^{-1}, which speeds up significantly the iterative convergence.
- **Pivoting.** For dynamic pivoting in the Crout LU method we adopt the multilevel strategy proposed in [14]. If during the Crout LU factorization on level l a small diagonal pivot is encountered or the estimated condition number of the inverse factors $|L^{-1}|$ and $|U^{-1}|$ is too large, the computed factorization of the current row and column is abandoned and delayed to the next level. The subsequent elimination is performed without consideration of the delayed rows and columns. When the factorization is completed, the matrix is reordered symmetrically to place all the delayed elements to the end of the matrix.

3 Nonlinear Solvers

Nonlinear systems of algebraic equations may appear in different cases: nonlinear PDE, implicit time discretization or use of the nonlinear spacial discretization schemes. Below we consider several nonlinear solvers implemented within INMOST platform.

INMOST platform includes a module for automatic differentiation. The primary objective of the module is to facilitate development of solvers that require the matrices of the first derivatives (Jacobian matrix). Jacobian construction is required for discretizations with the implicit time integration of unsteady equations or steady-state equations with nonlinear operators.

Jacobian matrix is constructed automatically during arithmetic operations with data. The automatic construction is based on the chain rule, which declares that despite the complexity of a differentiated expression, the final expression for the derivatives is a sum of partial derivatives with certain coefficients.

The most popular nonlinear solver is the **Newton method**. Its implementation is easy within the INMOST platform. Given an assembly procedure for the residual $\mathcal{R}^k$, INMOST provides parallel assembly for the sparse Jacobian with the automatic differentiation module. The resulting system is efficiently solved with available linear solvers for systems with distributed sparse matrices:

$$\mathcal{J}_k \Delta \mathbf{x}_k = -\mathcal{R}_k, \quad \mathbf{x}_{k+1} = \mathbf{x}_k + \Delta \mathbf{x}_k. \tag{1}$$

The Newton method, however, requires a good initial guess and a smooth enough (*i.e.* with Lipschitz continuous derivative) function to converge. The Newton iterations may easily diverge or dangle around the true solution for a large class of problems involving non-differentiable functions (*i.e.* modulus operation, upstream-differencing) or functions with inflection points.

There are two possible ways for improvement of the Newton method. Introduction of parameter α into the update procedure, $\mathbf{x}_{k+1} = \mathbf{x}_k + \alpha \Delta \mathbf{x}_k$ leads to backtracking and line-search methods. Inexact computation of $\mathcal{J}_k$ leads to a family of Quasi-Newton methods including Picard iteration, Broyden update and trust-region methods.

The **line-search method** consists in finding such a parameter α, that the new residual norm $\|\mathcal{R}\left(\mathbf{x}_k + \alpha \Delta \mathbf{x}_k\right)\|$ is minimal and declares $\mathbf{x}_{k+1} = \mathbf{x}_k + \alpha \Delta \mathbf{x}_k$ as the next iterative guess. The method does not change the direction of the provided update $\Delta \mathbf{x}_k$. The line-search method is very useful if the computation of the residual is considerably less expensive than assembly of the Jacobian matrix.

Let $\mathbf{x}_{k-m}, \ldots, \mathbf{x}_k \in \mathbb{R}^N$ represent last m iterates for a nonlinear solver with residuals $\mathcal{R}_{k-m} = \mathcal{R}(\mathbf{x}_{k-m}), \ldots, \mathcal{R}_k = \mathcal{R}(\mathbf{x}_k) \in \mathbb{R}^N$. Then the estimate for the next iteration $\mathbf{x}_{k+1}^{AA}$ using the **Anderson acceleration method** is computed as a convex linear combination with coefficients α_j and mixing parameter γ [15].

$$\mathbf{x}_{k+1}^{AA} = \sum_{j=k-m}^{k} \alpha_j \mathbf{x}_j + \gamma \tilde{R}_{k+1}, \;\; \tilde{R}_{k+1} = \sum_{j=k-m}^{k} \alpha_j \mathcal{R}_j, \;\; \left\|\tilde{R}_{k+1}\right\| \to \min, \;\; \sum_{j=k-m}^{k} \alpha_j = 1. \tag{2}$$

Usually the Anderson acceleration method is not faster than the quadratically convergent Newton method [16]. However, it can improve significantly the Newton method if it is stalling or diverging. Also it is suitable for Quasi-Newton methods.

The method changes the direction of the Newton update. The line-search method may improve the Anderson acceleration. The efficient strategy is to perform line-search between solution provided by the Newton method $\mathbf{x}_{k+1}^{N}$ and the solution accelerated by the Anderson method $\mathbf{x}_{k+1}^{AA}$ as proposed in [17].

4 Finite Volume Method and Applications

Cell-centered finite volume (FV) discretizations are appealing for the approximate solution of boundary value problems since they are locally conservative and applicable to general meshes. The wide diversity of FV methods is provided by different approaches to flux approximation.

For our applications we implement such flux approximations that produce monotone schemes. Monotonicity of a numerical scheme may imply either solution positivity or the discrete maximum principle. Positivity is important for concentrations, density, energy, absolute temperatures etc. The discrete maximum principle guarantees absence of artificial extrema in the numerical solution, for numerical pressure (head) this implies absence of non-physical Darcy flows from cells with lower pressure to cells with higher pressure.

A monotone scheme can be built using a simple linear two-point flux discretization which, however, does not provide flux approximation in the general case of polyhedral meshes and tensor diffusion/permeability coefficients. Nonlinear flux discretizations generate monotone FV schemes at the cost of scheme nonlinearity, even if it is applied to a linear PDE.

First we briefly remind the FV discretization of the diffusion problem in a polyhedral domain Ω for a scalar unknown $p \in H^1(\Omega)$:

$$\begin{aligned} -\mathrm{div}\,(\mathbb{K}\nabla p) &= g, \text{ in } \Omega, \\ \alpha p + \beta \mathbf{n} \cdot (\mathbb{K}\nabla p) &= \gamma, \text{ on } \partial\Omega, \end{aligned} \tag{3}$$

where $\mathbb{K}$ is a piecewise-constant symmetric positive-definite diffusion tensor, α, β, γ are parameters of the boundary condition on the domain boundary $\partial\Omega$ with the outer normal unit vector $\mathbf{n}$. Combination $\alpha = 1, \beta = 0$ represents Dirichlet boundary conditions, combination $\alpha = 0, \beta = 1$ represents Neumann boundary conditions.

The cell-centered FV method is based on Stokes theorem applied to the divergence operator in (3) for each cell T:

$$-\int_T \mathrm{div}\,(\mathbb{K}\nabla p)\,\mathrm{d}T = -\oint_{\partial T} \mathbf{n}\cdot(\mathbb{K}\nabla p)\,\mathrm{d}S = \sum_{f\in\mathcal{F}(T)} -\int_f \mathbf{n}\cdot(\mathbb{K}\nabla p)\,\mathrm{d}S = \sum_{f\in\mathcal{F}(T)} q|f|, \tag{4}$$

where ∂T is the boundary of cell T and $\mathcal{F}(T)$ is the set of faces of T, $|f|$ is the area of face f, $q|f| = -\int_f \mathbf{n} \cdot (\mathbb{K}\nabla p)\,\mathrm{d}S$ is the total flux across f. The key element of the finite volume method is approximation of the averaged flux density q on internal and boundary faces, which determines the properties of the finite volume method.

Assuming that the discrete solution is collocated at cell centers and denoting by $\mathcal{P}_f$ the set of cells in a flux discretization stencil associated with face f, we get the following formula of the discrete flux computation for face f

$$q_h = \sum_{T_j \in \mathcal{P}_f} M_{f,j} p_j + G_f, \tag{5}$$

where G_f denotes possible contributions of the boundary condition.

Flux approximation (5) provides the FV discretization of (3):

$$\sum_{f \in \mathcal{F}(T_i)} |f| \chi_{T_i,f} q_h = \sum_{f \in \mathcal{F}(T_i)} |f| \chi_{T_i,f} \left(\sum_{T_j \in \mathcal{P}_f} M_{f,j} p_j + G_f \right) = \int_{T_i} g \mathrm{d}T, \tag{6}$$

where $\chi_{T_i,f}$ is either 1 or -1 depending on the mutual orientation of the outer normal for T_i and $\mathbf{n}_f$.

The equalities (6) define the system of N algebraic equations with N unknowns p_j, $j = 1, \ldots, N$. If coefficients $M_{f,j}$ do not depend on p_j, then the above equations are linear and (6) may be written in the matrix form

$$\mathbb{A}\mathbf{p} = \mathbf{g}, \tag{7}$$

where vector $\mathbf{p}$ is composed of p_j, and matrix $\mathbb{A}$ is assembled from coefficients in (5) multiplied by $|f|\chi_{T_i,f}$. Otherwise, (6) is the system of nonlinear equations

$$\mathbb{A}(\mathbf{p})\mathbf{p} = \mathbf{g}(\mathbf{p}), \tag{8}$$

where entries of matrix $\mathbb{A}$ and right-hand side $\mathbf{g}$ may depend on $\mathbf{p}$.

If the solution of (8) exists, it can be found by available nonlinear solvers (see Sect. 3):

- Picard iteration $\mathbb{A}(\mathbf{p}^k)\mathbf{p}^{k+1} = \mathbf{g}(\mathbf{p}^k)$,
- Anderson acceleration of Picard iterations [18],
- Newton method with analytically computed Jacobian or Jacobian due to automated differentiation.

INMOST provides a module for assembly of a complex multi-physics model. The idea of the module is to provide basic functionality that allows one to split the problem into physical processes. Each physical process is represented by a sub-model. Each sub-model is responsible for introduction of unknowns and assembly of the residual associated with a sub-model physical process. The sub-models are coupled and solved together as single model. Coupling between two physical processes introduces coupling terms into equations involving unknowns of both processes which have to be accessed.

If we consider the Jacobian matrix of such multi-physics model assembled of multiple sub-models, then each sub-model represents the diagonal blocks of the Jacobian and coupling between sub-models completes off-diagonal blocks.

An example of geophysical application is the multi-phase flow given by the black-oil model [19]. The model describes the flow in the water-oil (two-phase) or water-oil-gas (three-phase) system. The water is not mixed with two other phases, but the gas can dissolve into the oil phase. For more details on the numerical models we refer to [1].

In order to test a parallel performance, we simulate the secondary recovery problem on the Norne field [20] with one water injection well and two production wells. The problem runs for 100 modelling days with time step not exceeding 1 day. The parameters are taken from SPE9 test.

Table 1. Performance of the models on different numbers of processors.

Processors	1	8	16	32	64	128
Two-phase model						
Assembly time	1028	172	89	53	30	19
speed-up	–	6x	12x	20x	35x	55x
Solution time	1544	383	255	135	71	65
speed-up	–	4x	6x	12x	22x	24x
Total time	2581	558	345	189	101	84
speed-up	–	5x	8x	14x	26x	31x
Equations	89830	11229	5615	2808	1404	702
Three-phase model						
Assembly time	2783	449	252	152	88	59
speed-up	–	6x	11x	19x	32x	47x
Solution time	5368	3295	1656	472	325	154
speed-up	–	2x	3x	11x	17x	35x
Total time	8171	3749	1911	626	415	213
speed-up	–	2x	4x	13x	20x	38x
Equations	134745	16844	8422	4211	2106	1053

The parallel performances of the simulations are demonstrated in Table 1. The black-oil problem contains strong elliptic component and the performance of the linear solver deteriorates faster. For good performance this problem requires a specific solver [21] that can extract the elliptic part of the problem and apply the multigrid solver for it. In this problem the assembly of the matrix does not ideally scale, since we have to perform certain operations on overlap which may become significant with large number of processors. Still it remains reasonable to increase number of processors to reduce the total computational time.

An example of biomedical application is simulation of clot formation in blood flow. The purpose of multi-physics modelling of blood flow coagulation is to predict thromboembolism in blood vessels and heart chambers.

The dynamics of the blood plasma is described by the incompressible Navier-Stokes equations. The blood plasma flow is coupled with reaction equation for fibrin polymer, nonlinear advection-diffusion for platelets flow and unsteady reaction-advection-diffusion equations for seven other components [1,23].

The resulting multi-physics model combines the model of blood plasma flow and the model of the simplified coagulation cascade, assuming that:

- blood plasma is incompressible Newtonian fluid, complex rheology of blood is ignored;
- polymerized fibrin does not move with the flow, i.e. the clot cannot detach and travel with the flow;
- the blood vessel boundaries are non-porous and rigid, their poro-elasticity is ignored.

The set of kinetic reactions between components of the flow and parameters of these reactions are taken from [22,23]. Some parameters were obtained from literature, the other parameters were identified using a thrombine-generation model or were fitted to the experiment of blood plasma flow through microvessels [24].

Table 2. Performance of the models on different numbers of processors.

Processors	1	8	16	32	64	128
Assembly time	2781	461	253	148	90	46
Speed-up	–	6x	11x	19x	31x	60x
Solution time	2156	214	117	80	46	29
Speed-up	–	10x	18x	27x	47x	74x
Total time	4938	680	373	234	141	77
Speed-up	–	7x	13x	21x	35x	64x
Equations	318500	39812	19906	9954	4977	2489

The multi-physics model of blood flow coagulation is validated across an *in vitro* experiment in microfluidic capillaries [24]. In this experiment, the flow of Platelet-Rich blood Plasma (PRP) is driven by a constant pressure drop in a microfluidic capillary. Here we consider only parallel performance of the simulation, see Table 2. The model scales very well with the growth of the number of processors. The reason for the good scalability is that the main contribution to the arithmetical work is given by the reactions term which is computed on each processor independently.

5 Summary

INMOST is the parallel platform for solving multi-physics problems on general meshes. We addressed the main features and capabilities of the platform, and presented two examples of parallel application of INMOST in geophysical and biomedical problems.

Acknowledgements. This work has been supported by the RAS Research program No. 26 "Basics of algorithms and software for high performance computing" and RFBR grant 18-31-20048 "Mathematical models of coronary blood flows and thrombogenic processes in cardiac pathologies".

References

1. Vassilevski, Y., Terekhov, K., Nikitin, K., Kapyrin, I.: Parallel Finite Volume Computation on General Meshes. Springer, Heidelberg (2020). https://doi.org/10.1007/978-3-030-47232-0
2. Vassilevski, Y., Konshin, I., Kopytov, G., Terekhov, K.: INMOST - Program Platform and Graphic Environment for Development of Parallel Numerical Models on General Meshes. Moscow University Publishing, Moscow (2013). (in Russian)
3. Boman, E.G., Çatalyürek, Ü.V., Chevalier, C., Devine, K.D.: The Zoltan and Isorropia parallel toolkits for combinatorial scientific computing: partitioning, ordering and coloring. Sci. Program. **20**(2), 129–150 (2012)
4. Karypis, G., Schloegel, K., Kumar, V.: Parmetis. Parallel graph partitioning and sparse matrix ordering library. Version, 2 (2003)
5. Hartigan, J.A., Manchek, A.W.: Algorithm as 136: a k-means clustering algorithm. J. Roy. Stat. Soc. Ser. C (Appl. Stat.) **28**, 100–108 (1979)
6. Sleijpen, G.L., Fokkema, D.R.: BiCGStab (l) for linear equations involving unsymmetric matrices with complex spectrum. Electron. Trans. Numer. Anal. **1**(11), 2000 (1993)
7. Olschowka, M., Neumaier, A.: A new pivoting strategy for Gaussian elimination. Linear Algebra Appl. **240**, 131–151 (1996)
8. Cuthill, E., McKee, J.: Reducing the bandwidth of sparse symmetric matrices. In: Proceedings of the 1969 24th National Conference, pp. 157–172 (1969)
9. Karypis, G., Kumar, V.: Metis-unstructured graph partitioning and sparse matrix ordering system, version 2.0 (1995)
10. Soules, G.W.: The rate of convergence of Sinkhorn balancing. Linear Algebra Appl. **150**, 3–40 (1991)
11. Kaporin, I.E.: Scaling, reordering, and diagonal pivoting in ILU preconditionings. Russ. J. Numer. Anal. Math. Model. **22**(4), 341–375 (2007)
12. Li, N., Saad, Y., Chow, E.: Crout versions of ILU for general sparse matrices. SIAM J. Sci. Comput. **25**(2), 716–728 (2003)
13. Kaporin, I.E.: High quality preconditioning of a general symmetric positive definite matrix based on its $U^tU+U^tR+R^tU$-decomposition. Numer. Linear Algebra Appl. **5**(6), 483–509 (1998)
14. Bollhöfer, M., Saad, Y.: Multilevel preconditioners constructed from inverse-based ILUs. SIAM J. Sci. Comput. **27**(5), 1627–1650 (2006)
15. Anderson, D.G.: Iterative procedures for nonlinear integral equations. J. ACM (JACM) **12**(4), 547–560 (1965)

16. Evans, C., Pollock, S., Rebholz, L.G., Xiao, M.: A proof that Anderson acceleration improves the convergence rate in linearly converging fixed-point methods (but not in those converging quadratically). SIAM J. Numer. Anal. **58**(1), 788–810 (2020)
17. Sterck, H.D.: A nonlinear GMRES optimization algorithm for canonical tensor decomposition. SIAM J. Sci. Comput. **34**(3), A1351–A1379 (2012)
18. Lipnikov, K., Svyatskiy, D., Vassilevski, Y.: Anderson acceleration for nonlinear finite volume scheme for advection-diffusion problems. SIAM J. Sci. Comput. **35**(2), 1120–1136 (2013)
19. Chen, Z., Huan, G., Ma, Y.: Computational Methods for Multiphase Flows in Porous Media. SIAM, Philadelphia (2006)
20. Norne - the full Norne benchmark case, a real field black-oil model for an oil field in the Norwegian Sea. https://opm-project.org/?page_id=559
21. Lacroix, S., Vassilevski, Y., Wheeler, J., Wheeler, M.: Iterative solution methods for modeling multiphase flow in porous media fully implicitly. SIAM J. Sci. Comput. **25**(3), 905–926 (2003)
22. Bouchnita, A.: Mathematical modelling of blood coagulation and thrombus formation under flow in normal and pathological conditions. PhD thesis, Universite Lyon 1 - Claude Bernard; Ecole Mohammadia d'Ingenieurs - Universite Mohammed V de Rabat - Maroc. (2017)
23. Bouchnita, A., Terekhov, K., Nony, P., Vassilevski, Y., Volpert, V.: A mathematical model to quantify the effects of platelet count, shear rate, and injury size on the initiation of blood coagulation under venous flow conditions. PLOS ONE **15**(7), e0235392 (2020). https://doi.org/10.1371/journal.pone.0235392
24. Shen, F., Kastrup, C.J., Liu, Y., Ismagilov, R.F.: Threshold response of initiation of blood coagulation by tissue factor in patterned microfluidic capillaries is controlled by shear rate. Arterioscler. Thromb. Vasc. Biol. **28**(11), 2035–2041 (2008)

Kirchhoff-Type Implementation of Multi-Arrival 3-D Seismic Depth Migration with Amplitudes Preserved

Alexandr Pleshkevich[1], Anton Ivanov[2(✉)], Vadim Levchenko[2], and Sergey Khilkov[2]

[1] RosGeo, Central Geophysical Expedition, Moscow, Russia
psdm3d@yandex.ru
[2] Keldysh Institute of Applied Mathematics, Moscow, Russia
aiv.racs@gmail.com, vadimlevchenko@mail.ru, ezz666@gmail.com

Abstract. We present implemented algorithm of asymptotic 3d prestack seismic migration, that takes into account multi-arrival and caustics of ray Green's functions (GF). Seismic migration is a crucial stage in processing of the seismic prospecting data. It requires dozens of thousands and in some cases hundreds of thousands of hours of processor time. The seismic migration problem naturally decomposes into two parts. The first one is computation of the GFs for all sources and the second one is summation. In order to compute a GF with multipath ray propagation taken into account one has to use adaptive grids to approximate the wave front. It is also necessary to implement a special "capture" procedure that calculates parameters at points of the uniform grid. This procedure also deals with singularities arising on caustic surfaces. An efficient implementation of the summation part requires to move lots of GF fragments between cluster nodes. We have employed described algorithms in a program and tested it on synthetic datasets and a few real world projects.

This research was initiated and sponsored by Central Geophysical Expedition JSC of Rosgeo.

Keywords: Seismic imaging · Multi-arrival seismic migration · HPC · Aiwlib

1 Introduction

Among all the tools applied for oil and gas exploration at the moment seismic prospecting is the most accurate and widely used method to study the structure of the inhomogeneous Earth's medium. Seismic imaging implementation based on reflected and scattered seismic waves from the depth geological structures is known as a seismic migration. Special seismic surveys are performed in order to acquire seismic data. Geophones on the surface register scattered and reflected waves from the depth inhomogeneities. These waves are produced by

V. Voevodin and S. Sobolev (Eds.): RuSCDays 2020, CCIS 1331, pp. 237–248, 2020.
https://doi.org/10.1007/978-3-030-64616-5_21

special shots in shallow wells as sources. For now the usual area of the survey starts from $10^{2\div3}$ km^2 and corresponding dataset size is around $10^{2\div3}$ GB. The seismic migration process demands to solve numerous problems of wave propagation in heterogeneous medium. Thus asymptotic ray methods in the context of the acoustic approximation overrule the field. It is common practice to use a single-beam assumption. It states that there is only one ray from the source to an arbitrary point in the medium, the one which passes through the point at the shortest time. In complex media if caustic occurs the assumption breaks down and the resulting image can degrade [3]. Therefore to improve the image one should discard the single-beam assumption and take multipath propagation into account. The efficient implementation of seismic migration based on this approach is the main topics of this paper.

2 Problem Setup

There are two widely known asymptotic methods to calculate seismic wave propagation which take into account multi-arrival and thus can capture caustics. Those are Maslov's canonical operator and Gaussian Beams summation method [1,9,10,12]. The use of Maslov's canonical operator method has no efficient numerical implementation. Gaussian beams migration still is not common among industrial seismic migration codes. We present a new approach [5–7] to solve the problem. In the following time with help of this method a 3D prestack seismic migraration program for multi-fold data was developed. In our approach we solve Eikonal equation with aid of system of Hamilton's equation for bicharacteristics. Thus the wave front and ray amplitudes and travel times become multi-valued functions.

We used an asymptotic integral solution for the problem,

$$f(\boldsymbol{r}) = \frac{1}{16\pi^2} \sum_{m,n} \iint\limits_{S_m} \iint\limits_{S_n} A_s^m A_g^n \widehat{H}^{I_m+I_n} \dot{u}(\boldsymbol{s}, \boldsymbol{g}, \tau_s^m(\boldsymbol{s}) + \tau_g^n(\boldsymbol{g}))\, d\boldsymbol{s}\, d\boldsymbol{g}. \quad (1)$$

Here $f(\boldsymbol{r})$ stands for image value at the point $\boldsymbol{r}$, S_m and S_n are surface areas touched by eikonal lobe m and n coming from a point source located at point $\boldsymbol{r}$. The wave lobes have KMAH indices I_m and I_n (number of $\pi/2$ phase shifts due to caustics) at S_m ans S_n correspondingly. $A^{m,n}$ are the ray amplitudes which contain only integrable singularities. $\widehat{H}$ means Hilbert transform, τ_s and τ_g signify the times that rays take to get from depth point $\mathbf{r}$ to source $\mathbf{s}$ and receiver $\mathbf{g}$, which form dense grids on the surface.

A typical seismic migration project requires about 10^9 calculations of the integral (1). It leads to dramatically high computational cost of migration problem. Moreover for the most straightforward approach to recalculate time consuming ray tracing from the depth point for every integral the problem cost raises beyond the capabilities modern supercomputers. To decrease the number of ray tracings we have to perform we trace rays from surface points and use reciprocity relation [4] to obtain parameters required by integral (1).

Taking into account reciprocity relation for the medium with a constant acoustic impedance one can derive ray amplitudes as

$$A = \frac{1}{v_0}\sqrt{\frac{v_r \sin\theta}{|J|}}, \qquad J = \left(\frac{\partial \mathbf{r}}{\partial \varphi}, \frac{\partial \mathbf{r}}{\partial \theta}, \frac{\partial \mathbf{r}}{\partial \tau}\right). \tag{2}$$

Here by v_0 we denote wave velocity at a source/receiver point and v_r obviously stands for velocity at depth point $\boldsymbol{r}$. Angles θ, φ represent the spherical coordinates for the direction of a ray coming from the source/receiver surface point to the depth point $\mathbf{r}$. τ is the time ray takes to travel that path. Finally we use notation J for the Jacobian of coordinates transformation from (θ, φ, τ) system of the wave front into a Cartesian system $\mathbf{r}$.

The main drawback of tracing from the surface is the vanishing of Jacobian near caustic surfaces. It makes the amplitude A infinite, which produces a singularity in the image integral. Fortunately this singularity is integrable and we take measures to account for it in our numerical method.

Also for quasi-regular sources and receivers systems like onshore seismic acquisition the number of tracings may be reduced even farther to the number of unique surface coordinates for sources and receivers positions. In common practice the spread deviates from regular grid but sources and receivers positions can be shifted to the regular grid nodes by introduction of kinematic corrections. Thus the solution of seismic migration problem splits into two sequential stages. The first one is the ray tracing that produces ray Green's functions (GF). Those GFs represented in integral (1) by Jacobian, travel times, KMAH indices, and surfaces corresponding to eikonal lobes. The second stage which called summation implements the calculation of the integral (1) with help of special quadrature formulas. The whole data size transfer between stages reaches hundreds terabytes, since every ray GF may have up to gigabyte of data. An efficient way to store and transfer the data is the main question to solve by an algorithm and an implementation of a prestack seismic migration as a program.

3 Approximation of the Wave Front on Adaptive Triangular Grid During Ray Tracing

The modified Hamilton's system in an inhomogeneous isotropic medium with a velocity distribution $v = v(\mathbf{r})$ is parametrized by travel time τ. It is given by:

$$\dot{\mathbf{r}} = v^2 \mathbf{p},$$
$$\dot{\mathbf{p}} = \frac{v^2}{2}\nabla\left(\frac{1}{v^2}\right).$$

Here $\mathbf{r}$ is a ray position and $\mathbf{p}$ is directed orthogonal to wave front $\tau = \text{const}$. In an isotropic medium this direction is tangent to a ray trajectory. The absolute value of the vector $|\mathbf{p}|$ is reciprocal to the speed at the point. The system of ODEs is solved with help of Runge–Kutta 4-th order method. The absolute value of $\mathbf{p}$ is kept correct by renorming the vector to $1/v$ on every step of the method.

The ray tracing is performed in a volume bounded from above by a cone which has an apex in the rays source. In lateral direction it is bounded by a rectangle with size determined by the maximum distance between source and receiver plus aperture (which is usually $4 \div 5$ km). The depth of raytracing as general doesn't exceed 10 km. When a ray passes trough a boundary it's marked to be discarded shortly after.

In addition to isotropic velocity model we also implemented VTI and TTI anisotropic models, which are described with Tompsen parameters [11]. All models are set on uniform 3d grids and are continued outside with the values on the boundary.

Even for isotropic medium we have to calculate velocity v and its' gradient ∇v four times in arbitrary points on each time step of Runge-Kutta method. Since we have velocities in the nodes of Cartesian grid and method requires continuous values of v and ∇v it is necessary to use an interpolation of at least the third order. That implies the stencil size $4 \times 4 \times 4$ points and over a hundred flops for getting every v, ∇v pair.

In order to reduce computational cost we use independent trilinear interpolation of velocity v and its gradient ∇v. This approach lead to mismatch between values of v and ∇v inside the grid cell. Nevertheless it occurs to be accurate enough for ray tracing on a typical velocity model for an inhomogeneous medium and time steps around 10 ms (for τ). It is possible to use refined grid to reach the sufficient accuracy. To reduce computation time the interpolation was implemented with use of `AVX` vector instructions.

At the starting point rays directions distributed on triangulation of a sphere produced by recursive decomposition of a pentakis dodecahedron's faces [2]. This approach allows as to reduce by half number of initial rays in comparison with the traditional spherical coordinates grid. In addition it also does not have singularities on poles, which especially important since needle shaped triangles near the pole significantly decrease the accuracy of the calculation.

The wave front is approximated by unstructured triangular grid, with rays in each node.

We have to refine the grid to achieve required accuracy. Several different refinement algorithms were tested in the end we decided to use quite simple one, described below.

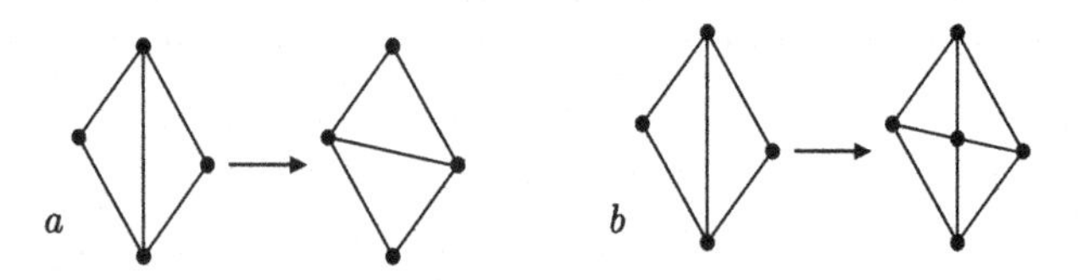

Fig. 1. Flip (*a*) and split (*b*) of an edge during the ray grid adaptation

In general we want to keep edges smaller then some threshold length. Therefore we are trying to split all edges which are longer than a critical length l_{cr}.

In order to split the edge we need to place a new ray so that its projection along the trajectory on a quadrangle formed by edge's two adjacent triangles is near the quadrangle's center (Fig. 1*(a)*). We trace the new rays all the way from the source, and use a shooting method which optimizes the mentioned distance to center by means of conjugated gradients method. Since we like to avoid overlapping of triangles on initial directions sphere we limit new point position to the interior of two triangles adjacent to the splitting edge. Thus we have to trace a few rays with different initial parameters. The number of rays essentially increases for edges which are close to caustic surfaces. For complex medium models the major part of ray tracing time may be spent on shooting.

If some part of the wave front is stretched enormously along some direction (for example due to head wave) the shooting algorithm still leads to excessive number of triangles (more than 10^6) with large aspect ratio. Thus we use an additional way of grid adaptation, the edge flip (Fig. 1*(b)*).

The edge flipping allows us to approximate complex heavily stretched parts of the wave front with aid of relatively small number of triangles. It comes at cost of notable complications in grid maintaining algorithms. Let's remember that for each ray we locally have normal to the wave front. If our approximation is good, it is also close to normal of the triangle. Hence we assume that edge flip makes triangulation better if changed triangles become close to interpolated normals at their middles.

Now we have all to describe the refining algorithm. In the beginning we put all edges that longer than threshold $l_{\rm cr}$ into a queue in the length decreasing order. Starting from the largest edge we try to improve it by flipping or splitting if flipping failed. All new large edges are also immediately inserted in the queue so that decreasing order is maintained. If the rare case that edge cannot be split it is removed from queue. The algorithm stops when the queue is empty.

Due to a high computational cost of the refining algorithm and its incompatibility with correct calculation of KMAH indices the ray tracing was divided into two stages. In the first one, which we call α–tracing, we perform the ray tracing and the grid refinement. We also save refinement history in the packed form for the next stage. The second stage (β–tracing) uses the refinement history to trace rays and calculate GFs in the points of Cartesian grid. We make α–tracing only once for each surface point, however simple β–tracing may be calculate multiple times.

The single wave front calculation is single-threaded.

4 The Green's Functions Calculation

The trajectory of a triangle with rays in each vertex forms the beam. For each time step of ray tracing the wave front consists of the beam fragments sliced by time step interval. Any fragment is formed by two triangular slices of the beam by constant time surfaces and lines connecting corresponding vertices (Fig. 2).

In order to calculate GFs it is required to find all the nodes of Cartesian grid that fall into the fragment. For every found node we have to obtain the time when the front touches the point, KMAH index, averaged ray amplitude, the direction of a ray coming to the point and the width of seismic data time averaging. The procedure of getting all those parameters we will farther call "capture". We also will refer to required parameters bundle as a track.

During β–tracing we trace all the rays saved in α–tracing history. We also keep record of all beams regardless of the time step they were created in α–tracing. Jacobians and KMAH indices are calculated for all beams, but the capture procedure is called only for the ones that were present on this time step in α–tracing.

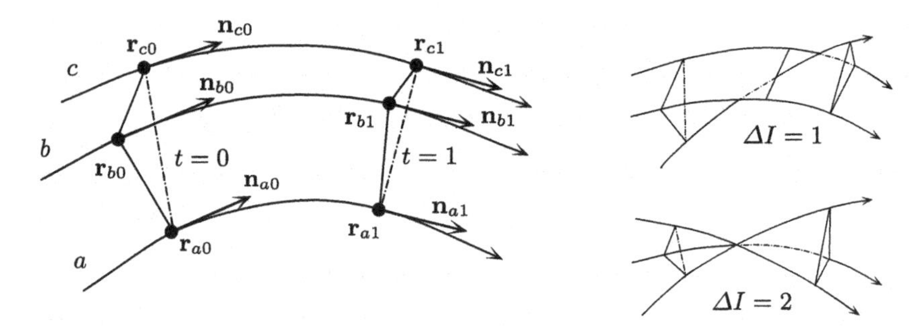

Fig. 2. A fragment of a beam for one time step and a beam behaviour near a caustic surface

For a beam we use the notation in which indices a, b and c correspond to the rays (Fig. 2). Index 0 is reserved for the values before the time step, and index 1 for values after the time step. Thus for a time step the corresponding beam fragment has initial ray coordinates $\mathbf{r}_{a0}$, $\mathbf{r}_{b0}$ and $\mathbf{r}_{c0}$. The final rays coordinates for the step are $\mathbf{r}_{a1}$, $\mathbf{r}_{b1}$, $\mathbf{r}_{c1}$. Initial and final ray propagation directions are $\mathbf{n}_{a0}$, $\mathbf{n}_{b0}$, $\mathbf{n}_{c0}$ and $\mathbf{n}_{a1}$, $\mathbf{n}_{b1}$, $\mathbf{n}_{c1}$ correspondingly. Let us introduce the parameter $t \in [0, 1)$, where $t = 0$ is linked to the beginning of the step end $t = 1$ is linked to its end. Then we can introduce linear approximation

$$\mathbf{r}(t) = \mathbf{r}_0 + (\mathbf{r}_1 - \mathbf{r}_0)\,t, \qquad \mathbf{n}(t) = \mathbf{n}_0 + (\mathbf{n}_1 - \mathbf{n}_0)\,t.$$

Than Jacobian for transformation from $\mathbf{n}, t$ to $\mathbf{r}$ for this fragment approximates as triple product

$$\mathcal{J}(t) = \frac{J}{\sin\theta} \approx \frac{1}{\Delta\Omega_0}\Big((\mathbf{r}_a - \mathbf{r}_b),\, (\mathbf{r}_b - \mathbf{r}_c),\, \langle\mathbf{n}\rangle\Big) \equiv P_0 + P_1 t + P_2 t^2 + P_3 t^3,$$

where $\Delta\Omega_0$ is the initial area of beam cross section, $\langle\mathbf{n}\rangle = \big(\mathbf{n}_a + \mathbf{n}_b + \mathbf{n}_c\big)/3$.

The index increment during the step equals to the number of roots in cubic polynomial

$$P(t) = P_0 + P_1 t + P_2 t^2 + P_3 t^3 = 0, \qquad t \in [0, 1).$$

It is worth mentioning that multiple root should be interpreted as multiple number of increases. To take the rounding errors into account we assume that $P(t)$ has at least one root if the polynomial value drops below threshold $|P(t)| < 10^{-8}$.

We have Jacobian approximation in time, but we also need to interpolate its values along the wave front. To do that we interpolate the polynomial coefficients into the ray grid. After that we can obtain the Jacobian at rays we can interpolate it linearly to any point of a beam.

In capturing procedure it is necessary to find Cartesian grid points that lays inside a beam and to calculate the track for each point found. We presume that the point belongs to beam interior if it falls into any one triangular cross section for some value of $t \in [0, 1)$.

In order to obtain all the interior points efficiently we construct two-dimensional convex hull of a beam (points $\mathbf{r}_{0a,b,c}, \mathbf{r}_{1a,b,c}$) along the lateral directions and finding the depth interval. After that in double loop we check all points falling into Cartesian product of hull and interval. For each point we check if it lies in the beam fragment using the following algorithm:

1. Find value of parameter t for which beam cross section lays in the same plane as a Cartesian grid point $\mathbf{r}_i$. For the point $\mathbf{r}^i$ it could be done by solving the equation
$$\left(\mathbf{r}_a(t) - \mathbf{r}^i,\, \mathbf{r}_b(t) - \mathbf{r}^i,\, \mathbf{r}_c(t) - \mathbf{r}^i\right) = 0,$$
which is just the cubic equation in t,
$$Q(t) = Q_0 + Q_1 t + Q_2 t^2 + Q_3 t^3 = 0. \tag{3}$$
We find all its roots numerically by means of Newton method.
2. For all roots $\{t_j\}$ of the Eq. (3) which lies in the interval $[0, 1)$ we check if point $\mathbf{r}^i$ falls in triangle $\mathbf{r}_a(t_j)$, $\mathbf{r}_b(t_j)$, $\mathbf{r}_c(t_j)$. Any positive answer means that point is inside the beam.

The Eq. (3) also helps us compute the time τ ray takes to get to point $\mathbf{r}^i$. It is found as
$$\tau = (n_t + t)h,$$
where n_t is the number of time step and h is the time step length.

The hardest part is to calculate the averaged ray amplitude. The ray amplitude formula (2) may contain integrable singularity that we integrate along the bin (the cell off sources/receivers grid). To obtain accurate result we approximate Jacobian as
$$\mathcal{J} \approx \mathcal{J}_0 + \mathcal{J}_1\xi, \qquad \mathcal{J}_0 \geq= 0, \qquad \mathcal{J}_1 \geq 0,$$
where ξ is the surface coordinate directed along the Jacobian gradient **calculated along the surface**, the value $\mathcal{J}_0$ is the Jacobian value in the depth point for the ray that connects it to the source.

We use the following formula to obtain the integral of ray amplitude over the bin:

$$\langle A\rangle_i = \Delta\frac{\cos\theta_s}{\sqrt{v_s}}\int\limits_{-\frac{\Delta}{2}}^{\frac{\Delta}{2}}\frac{\mathcal{H}\left(\left[\mathcal{J}_0+\mathcal{J}_1\xi\right]^{-\frac{1}{2}}\right)}{\left[\mathcal{J}_0+\mathcal{J}_1\xi\right]^{-\frac{1}{2}}}d\xi = \frac{\Delta\cos\theta_s}{\mathcal{J}_1\sqrt{v_s}\frac{1}{2}}\left(\mathcal{J}_0+\mathcal{J}_1\xi\right)^{\frac{1}{2}}\Bigg|_{\xi_{\min}}^{\frac{\Delta}{2}}, \quad (4)$$

where $\mathcal{H}$ is the Heaviside step function, $\Delta = \sqrt{S_i}$ stands for the linear size of the bin and $\xi_{\min}$ denotes the caustic surface position,

$$\xi_{\min} = \max\left(-\frac{\Delta}{2}, -\frac{\mathcal{J}_0}{\mathcal{J}_1}\right) = -\min\left(\frac{\Delta}{2}, \frac{\mathcal{J}_0}{\mathcal{J}_1}\right).$$

In order to calculate the $\mathcal{J}_1$ coefficient we trace four additional rays that starts from surface points distributed in cross around central ray (which defines $\mathcal{J}_0$) and directed in the same direction as the central one.

This modification slightly sophisticates the present data structure. For each ray in our front approximating grid we trace four additional rays–satellites. It results in four additional versions of the way front. Rays–satellites are considerably "light". We do not perform for them mesh refinement, indices calculations and the most important capture procedure. In the end they insignificantly increase the memory consumption during ray tracing and almost do not affect the amount of calculations.

It is necessary to make an additional averaging of ray amplitudes in group of the rays that comes to a point with same KMAH index and similar arrival times τ. The problem occurs due to large number of folds on the ray grid occurring near caustic surfaces. This averaging procedure take place after ray tracing. It substitute the group of tracks with one that has minimal arrival time in the group and geometrical mean ray amplitude. The width of averaging window (the threshold arrival time difference in the group) is the external parameter which we pick up empirically.

5 The Implementation of the Numerical Integration

When we calculate integral (1) on one bin it is important to account for a time wave front takes to move through the bin [8]. This effect may be taken into consideration by introduction of a sliding time widow filter on the seismic signal. The width of the window is obtained as

$$\Delta^{\tau} = \left|\mathbf{\Delta}^{\tau}_{s} + \mathbf{\Delta}^{\tau}_{g}\right|, \qquad \mathbf{\Delta}^{\tau}_{s,g} = \left(\frac{\Delta_x n^{s,g}_{0x}}{v_{s,g}}, \frac{\Delta_y n^{s,g}_{0y}}{v_{s,g}}\right),$$

where $\Delta_{x,y}$ is the linear bin sizes, $n^{s,g}_{0x,y}$ are directions of the ray propagation from the source/receiver, $v_{s,g}$ is velocity measured at source/receiver point. Thus components of the vector $\mathbf{\Delta}^{\tau}_{s,g}$ has different signs.

A Green's function for source s or receiver g is known on the 3d Cartesian grid in $\mathbf{r}$ space. Lateral coordinates of points in the grid correspond to lateral coordinates of the image f. However vertical step of GFs Cartesian grid Δ_z^G is several times larger than one of the image. The step of coarse grid may be scaled down for small depths to improve the image quality in the top parts.

Each node of coarse grid may get multiple rays from both source and receiver in view of multi-arrival nature of acoustic wave. Interpolation between nodes $z_{a,b}$ of the coarse grid into the node of the fine one at z is one of a special kind. It does not require to tell multi-valued GFs layers (bound by the same KMAH index value) apart. It works as follows:

$$f(z) \approx w_a \sum_i^{N_a^s} \sum_j^{N_a^g} \langle A \rangle_{a,i}^s \langle A \rangle_{a,j}^g \, \widehat{T}^{\Delta_{a\,i,j}^{\tau}} \, \widehat{H}^{I_{a,i}^s + I_{a,j}^g} \, \dot{u} \left(\tau_{a,i}^s + \tau_{a,j}^g + \delta_a \right) \tag{5}$$
$$+ w_b \sum_i^{N_b^s} \sum_j^{N_b^g} \langle A \rangle_{b,i}^s \langle A \rangle_{b,j}^g \, \widehat{T}^{\Delta_{b\,i,j}^{\tau}} \, \widehat{H}^{I_{b,i}^s + I_{b,j}^g} \, \dot{u} \left(\tau_{b,i}^s + \tau_{b,j}^g - \delta_b \right),$$

where $w_a = (z_b - z)/\Delta_z^G$, $w_b = 1 - w_a$ are the weights for linear interpolation in depths, $N_{a,b}$ is the number of tracks in top and bottom nodes of the coarse grid cell, $\delta_{a,b}$ are the signal delays.

$$\delta_a = \frac{4 w_a \Delta_z^G}{v_a + v_c}, \qquad \delta_b = \frac{4 w_b \Delta_z^G}{v_b + v_c}, \qquad v_c = w_a v_b + w_b v_a,$$

Here $v_{a,b}$ are velocities at top an bottom node of coarse grid cell.

It is possible to write Hilbert transformation $\widehat{H}^1 \dot{u}(t)$ as convolution with discrete antisymmetric kernel H_i whereby

$$\widehat{H}^2 \dot{u}(t) \approx -\dot{u}(t),$$
$$\widehat{H}^3 \dot{u}(t) \approx -\mathcal{H}^1 \dot{u}(t),$$
$$\widehat{H}^4 \dot{u}(t) \approx \dot{u}(t),$$

etc. Hence it to calculate the interpolation efficiently it is sufficient to prepare arrays of data $\widehat{H}^1 \dot{u}$ for every seismic road u.

6 The Features of the Program

As we mentioned before the image calculation consists of two stages. The first one accumulates statistics for all sources and receivers. It includes α– and β–tracings which are computed for several (one for each thread) points simultaneously on the same cluster node. We save results of α–tracing on this stage for the upcoming application. During the first stage β–tracing results are only GFs sizes required for the efficient scheduling.

Obviously the second stage produces the image. In our program we use nodes RAM to store ray GFs. The size of all GFs data often exceeds available memory size of the cluster, therefore on the calculations scheduling phase (the first phase of the second stage) the problem is broken down to smaller tasks. Those tasks are connected only through the seismic image. To be precise seismic image is the sum of separate tasks outputs.

We use stencil notation to decompose the problem into tasks. The stencil is a relatively small part of the whole seismic data amount, which is defined by a set of associated sources and receivers. Those sets are formed to be small enough that task's GFs fit into cluster RAM. On the other hand the task also should use each GF as many times as possible.

Kinematic corrections help us to shift sources and receivers coordinates to Cartesian grid points on the surface. Let's call basic cross stencil the line of receivers and the line of sources orthogonal to it. The number of sources and receivers in the stencil is limited by summation aperture which depends on the maximum distance between source an receiver in the seismic data set. The collection of basic cross stencils which has centers inside some rectangle defines the task.

Since the smaller set of seismic data corresponds to the smaller number of GFs the RAM size requirement may be significantly reduced. For the optimal scheduling one can achieve linear decrease of required RAM amount for large tasks which reduces to squared root of tasks number for the worst case of small tasks. In the latter we have to recalculate GFs which lead to increasing the runtime. Thought usually the GFs calculations still takes around $10 \div 20\%$ of the runtime.

We also divide all images and GFs into tiles. The tile is a small rectangular along the lateral direction fragment which goes all the way to the maximum depth. A task consists of two parts as well. In the first part we calculate all GFs for all sources and receivers associated with the task. Every computational unit (CU) obtains its own portion of all GFs number with help of dynamic balancing. After that is done CUs exchange the tiles so every CU gets all the tiles it needs to calculate its image tile. For the typical seismic reflection spread the numerical cost of summation is proportional to squared number of GFs.

The size of single GF may be too large for some computational systems. It results in small task size and overall efficiency drop due to frequent recalculations of GFs. In order to solve this problem GF may be split into 64 tiles. By appending 64 bits mask to each beam we can mark which fragments the beam contribute to. The marking happens while we collect β–tracing statistics. Afterwards we can discard all the beams that do not contribute to the fragment and calculate the standalone fragment of GF without visible efficiency loss. This improvement allows us to enlarge the task size for the same RAM size and thus increase the program performance.

The smallest task size which for which additional cost is less than a half of the integration time requires storage of several thousands GFs and thus several terabytes of RAM. Each task updates all the seismic image (1). Since its size is small (dozens of gigabytes) comparing to GFs size we can store it in RAM along with redundant copies. That allows us not only perform tasks sequentially but also have some tolerance for faults. Thus after the fault it is necessary to recalculate only the task that failed.

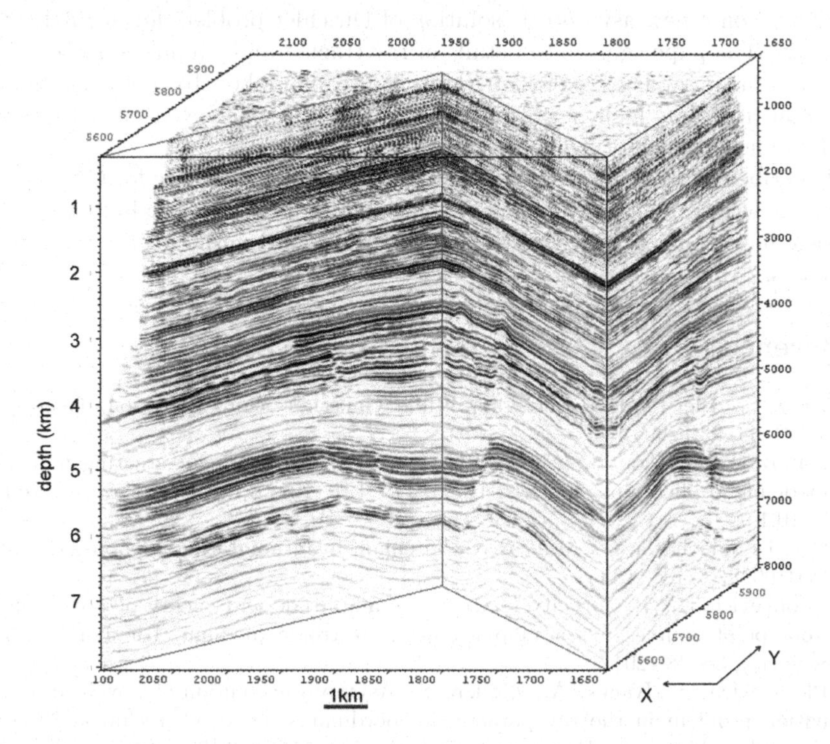

Fig. 3. The seismic image by amplitude–preserving multiarrival 3D depth migration, Timan–Pechora region

The original setup (1) has high parallel capacity. We us three-trier parallel algorithm that ensures high localisation of processed data. It also is efficient on the systems with the distributed memory. The top level of the algorithm uses MPI and is efficient for distribution work between cluster nodes. The middle level utilizes OpenMP which fits the multicore systems with shared memory. And the lowest one uses the processor vector extensions SSE and AVX. On the first stage of seismic imaging the top level accommodates tracing of beams, the middle level is used to parallelize tile processing and the lowest accelerates the velocity model access. On the second stage the top level parallelize the tiles, the middle deals with seismic data and the lowest one is again utilized for interpolation. Thus during the whole seismic image process we utilise all available parallel levels of computational system.

The Fig. 3 displays the sample of seismic image for Timan-Pechora region produced by our code.

7 Conclusion

We presented an original algorithm of multi-arrival 3-D seismic depth migration. It is based on a new asymptotic solution of Dirichlet problem for acoustic wave equation. The paper goes into details of ray tracing and summation algorithms revealing the main aspects of their work. It also shows how to create summation algorithm that has high performance and fault tolerance even on cluster with small amount of memory

Described algorithms are implemented as a software package for GNU/Linux os. It was written in c++ language making a good use of `aivlib` library [2]. The package has been successively tested on international synthetic and real seismic data sets.

References

1. Gray, S., Bleistein, N.: True-amplitude Gaussian-beam migration. Geophysics **2**(74), 11–23 (2009)
2. Ivanov, A., Khilkov, S.: Aiwlib library as the instrument for creating numerical modeling applications. Sci. Vis. **10**(1), 110–127 (2018). https://doi.org/10.26583/sv.10.1.09
3. Liu, J., Palacharla, G.: Multiarrival kirchhoff beam migration. Geophysics **76**(5), WB109–WB118 (2011)
4. Pleshkevich, A.: Reciprocity relation, the amplitude and energy of waves energy from point sources in an inhomogeneous isotropic medium. Russian Geophys. **S**(Spec.), 38–48 (2012)
5. Pleshkevich, A., Ivanov, A., Khilkov, S.: Asymptotic solution of wavefield continuation problem in the ray parametric coordinates. In: SEG Technical Program Expanded Abstracts, Houston, USA, pp. 5551–5555 (2017). https://doi.org/10.1190/segam2017-17633541.1
6. Pleshkevich, A., Ivanov, A., Levchenko, V., Khilkov, S.: Multiarrival amplitude-preserving prestack 3D depth migration. Russian Geophys., 76–84 (2017). http://geofdb.com/en/articles/view?id=1905
7. Pleshkevich, A., Ivanov, A., Levchenko, V., Khilkov, S., Moroz, B.: Efficient parallel implementation of multi-arrival 3D prestack seismic depth migration. Supercomput. Front. Innov. **6**(1), 4–8 (2019)
8. Pleshkevich, A., Turchaninov, V.: Efficient quadrature schemes with averaging window for integration of wave fields and their application to the Kirchhoff seismic migration. Russian Geophys. **S**(Spec.), 57–65 (2012)
9. Popov, M., Semtchenok, N., Popov, P., Verdel, A.: Depth migration by the Gaussian beam summation method. Geophysics **2**(75), 81–93 (2010). https://doi.org/10.1190/1.3361651
10. Protasov, M., Tcheverda, V., Pravduhin, A.: 3D true-amplitude anisotropic elastic Gaussian beam depth migration of 3D irregular data. J. Seismic Explor. **28**(2), 121–146 (2019)
11. Thomsen, L.: Weak elastic anisotropy. Geophysics **51**, 1954–1966 (1986)
12. Yang, J., Huang, J., Wang, X., Li, Z.: Amplitude-preserved Gaussian beam migration based on wave field approximation in effective vicinity under rugged topography condition. In: SEG Technical Program Expanded Abstracts, pp. 3852–3856 (2014)

Mathematical Modeling of Sustainable Coastal Systems Development Scenarios Based on Game-Theoretic Concepts of Hierarchical Management Using Supercomputer Technologies

Yulia Belova[1(✉)], Alexander Chistyakov[1], Alla Nikitina[2,3], and Vladimir Litvinov[1,4]

[1] Don State Technical University, Rostov-on-Don, Russia
yvbelova@yandex.ru, cheese_05@mail.ru
[2] Southern Federal University, Rostov-on-Don, Russia
[3] "Supercomputers and Neurocomputers Research Center" Co. Ltd., Taganrog, Russia
nikitina.vm@gmail.com
[4] Azov-Black Sea Engineering Institute of Don State Agrarian University, Zernograd, Russia
litvinovvn@rambler.ru

Abstract. The work is devoted to mathematical modeling of the scenario for managing the sustainable development of coastal systems using the example of the Azov Sea. The dynamic problem of minimizing the costs of maintaining the ecosystem of the reservoir in a given state, which is interpreted as a requirement for sustainable development, is being solved. The mathematical model of the interaction of two phytoplankton types takes into account the influence of abiotic factors, such as salinity and temperature, on the algae growth, their absorption and excretion of nutrients, as well as the transition of these nutrients from one form to another. For the numerical implementation of the proposed interconnected mathematical models of biological kinetics, parallel algorithms have been developed that are adapted to multiprocessor and hybrid computing systems using the NVIDIA CUDA architecture. An analysis of the characteristics of the CUDA architecture showed the applicability of the algorithms for the numerical implementation of the developed mathematical models of hydrobiology to create high-performance information systems. The constructed software complex allows to simulate the problem of reducing eutrophication and toxicity of the coastal system under consideration by displacing harmful blue-green algae with a cultivated strain of green alga using high performance computing systems.

Keywords: Mathematical model · Sustainable development · Phytoplankton · Biogeochemical cycle · Azov Sea · Parallel algorithm · Graphic accelerator

V. Voevodin and S. Sobolev (Eds.): RuSCDays 2020, CCIS 1331, pp. 249–260, 2020.
https://doi.org/10.1007/978-3-030-64616-5_22

1 Introduction

Nonlinear mathematical models that most adequately describe the real processes occurring in water play an important role in the mathematical modeling of hydrobiological processes [1–3]. As a rule, multispecies models of hydrobiological processes are not used due to the considerable complexity of their implementation; nonlinear interaction of hydrobionts together with hydrodynamic processes that determine the spatial distribution of nutrients, temperature, salinity and hydrobionts are not taken into account. Despite the increased number of applications, methods for researching the state of nonlinear ecosystems in shallow water bodies remain insufficiently developed, and an additional mathematical apparatus requires further development. The toxic blue-green alga *Aphanizomenon flos-aquae* was chosen as one of the modeling objects, which, according to expeditionary data, has a wide distribution and high concentration in the territory of the Azov Sea (5–6 million individuals in m^3 of water). These algae are rapidly growing in the desalinated zone of the Azov Sea; when they die, they release toxins into the water that may act on the human nervous system. Oxygen reproduced by algae during photosynthesis is not retained in water: nanoalgae consume it more than they produce, which leads to mass death of benthic, planktonic, and neustonic animals [4]. To combat blue-green algae, various modern methods of biological, physico-chemical treatment of surface water, etc., as well as the method of algolization of water bodies (tested on the Izhevsk, Voronezh reservoirs) with the strain of green alga *Chlorella Vulgaris BIN* are used to improve the quality of their waters. When implementing the concept of sustainable development of the aquatic ecosystem, the method of cognitive structuring [5,6] and game theory [7] are widely used. Environmental monitoring is one of the precautionary methods for analyzing the sustainable development of the biogeocenosis of a reservoir. For its efficient implementation, it becomes necessary to develop and numerically implement three-dimensional mathematical models of biological kinetics using supercomputer technologies [8].

2 Problem Statement

Consider an industrial enterprise (IE), located on the shore of a reservoir, which introduces green algae, for example, *Chlorella vulgaris BIN* algae in order to displace blue-green (toxic) algae *Aphanizomenon flos-aquae*. The objective of the enterprise is to minimize the costs associated with the biological rehabilitation of the reservoir on the shore of which it is located. The concentration of green (useful) algae at the initial time is the control variable. The objective function of the IE has the form:

$$\Phi_0 = m_0 \sum_{i=1}^{N} F_1 (x_i, y_i, z_i, 0)$$

$$+ m \sum_{t=1}^{T} \sum_{i=1}^{N} \max \left(0, F_2^* (x_i, y_i, z_i, t) - F_2 (x_i, y_i, z_i, t)\right) \rightarrow \min, \tag{1}$$

where (x_i, y_i, z_i), $i = \overline{1, N}$ – points of the three-dimensional area at which algal concentrations are calculated; N – the number of these points; F_1 – green algae concentration; F_2 – blue-green algae concentration; F_2^* – maximum permissible concentration (MPC) of blue-green algae; m – price of exceeding the MPC of blue-green algae; m_0 – price of introducing a unit of green algae biomass.

The first term in (1) reflects the costs of introducing green algae, the second - the total fine imposed on the IE for exceeding the MPC of blue-green algae at calculated points in the spatial domain. Condition $F_2(x_i, y_i, z_i, t) \leq F_2^*(x_i, y_i, z_i, t)$, $\forall t$, $(x_i, y_i, z_i) \in \overline{G}$ is a requirement of sustainable development, mandatory for the subject of management. The objective function (1) is considered with restrictions on controls at any point in the area $\overline{G}$:

$$0 \leq F_1(x_i, y_i, z_i, 0) \leq F_{1\,\max}, \tag{2}$$

where $F_{1\,\max}$ – maximum green algae concentration.

The development dynamics and interaction of the blue-green alga and the green alga, as well as the transformation of nutrients forms, are described by a system of non-stationary equations of convection-diffusion of the parabolic type with non-linear source functions and lower derivatives, the form of which for each model block q_i

$$\frac{\partial q_i}{\partial t} + u\frac{\partial q_i}{\partial x} + v\frac{\partial q_i}{\partial y} + w\frac{\partial q_i}{\partial z} = div\,(kgradq_i) + R_{q_i}, \tag{3}$$

where q_i – concentration of i-th component, $i = \overline{1, 8}$, [mg/l]; $\{u, v, w\}$ – components of water flow velocity vector, [m/s]; k – turbulent exchange coefficient, [m^2/s]; R_{q_i} - chemical-biological source, [mg/(l·s)].

In Eq. (3) the index i indicates the type of substance (Table 1).

Table 1. Nutrients in the model dynamics of phytoplankton

Number	Designation	Title
1	F_1	green algae *Chlorella vulgaris BIN*
2	F_2	bluegreen algae *Aphanizomenon flos-aquae*
3	PO_4	Phosphates
4	POP	Suspended organic phosphorus
5	DOP	Dissolved organic phosphorus
6	NO_3	Nitrates
7	NO_2	Nitrites
8	NH_4	Ammonium

Chemical-biological sources are described by the following equations

$$R_{F_i} = C_{F_i}(1 - K_{F_iR})q_i - K_{F_iD}q_i - K_{F_iE}q_i, i = \overline{1, 2},$$

$$R_{PO_4} = \sum_{i=1}^{2} s_P C_{F_i} (K_{F_iR} - 1) q_i + K_{PN} q_4 + K_{DN} q_5,$$

$$R_{POP} = \sum_{i=1}^{2} s_P K_{F_iD} q_i - K_{PD} q_4 - K_{PN} q_4,$$

$$R_{DOP} = \sum_{i=1}^{2} s_P K_{F_iE} q_i + K_{PD} q_4 - K_{DN} q_5,$$

$$R_{NO_3} = \sum_{i=1}^{2} s_N C_{F_i} (K_{F_iR} - 1) \frac{f_N^{(1)} (q_6, q_7, q_8)}{f_N (q_6, q_7, q_8)} \cdot \frac{q_6}{q_7 + q_6} q_i + K_{23} q_7,$$

$$R_{NO_2} = \sum_{i=1}^{2} s_N C_{F_i} (K_{F_iR} - 1) \frac{f_N^{(1)} (q_6, q_7, q_8)}{f_N (q_6, q_7, q_8)} \cdot \frac{q_7}{q_7 + q_6} q_{F_i} + K_{42} q_8 - K_{23} q_7,$$

$$R_{NH_4} = \sum_{i=1}^{2} s_N C_{F_i} (K_{F_iR} - 1) \frac{f_N^{(2)} (q_8)}{f_N (q_6, q_7, q_8)} q_i$$

$$+ \sum_{i=1}^{2} s_N (K_{F_iD} + K_{F_iE}) q_i - K_{42} q_8,$$

where K_{F_iR} – specific respiration rate of phytoplankton; K_{F_iD} – specific rate of phytoplankton dying; K_{F_iE} – specific rate of phytoplankton excretion; K_{PD} - specific speed of autolysis *POP*; K_{PN} – phosphatification coefficient *POP*; K_{DN} – phosphatification coefficient *DOP*; K_{42} – specific rate of oxidation of ammonium to nitrites in the process of nitrification; K_{23} – specific rate of oxidation of nitrites to nitrates in the process of nitrification, s_P, s_N – normalization coefficients between the content of *N*, *P* in organic matter.

The growth rate of phytoplankton is determined by the expressions:

$$C_{F_{1,2}} = K_{NF_{1,2}} f_T (T) f_S (S) min \left\{ f_P (q_3), f_N (q_6, q_7, q_8) \right\},$$

where K_{NF} – maximum specific growth rate of phytoplankton.

Dependences of temperature and salinity:

$$f_T (T) = \exp \left(-\alpha \left(\frac{T - T_{opt}}{T_{opt}} \right)^2 \right), \; f_S (S) = \exp \left(-\beta \left(\frac{S - S_{opt}}{S_{opt}} \right)^2 \right),$$

where T_{opt} – optimal temperature for this type of phytoplankton, α – coefficient that determines the width of the interval of survival of phytoplankton depending on temperature, S_{opt} – optimal salinity for this type of phytoplankton; β – coefficient that determines the width of the interval of survival of phytoplankton depending on salinity.

Functions describing biogen content

– for phosphorus $f_P (q_3) = \frac{q_3}{q_3 + K_{PO_4}}$,

where K_{PO_4} – half-saturation constant of phosphates;

– for nitrogen $f_N(q_6, q_7, q_8) = f_N^{(1)}(q_6, q_7, q_8) + f_N^{(2)}(q_8)$,

$$f_N^{(1)}(q_6, q_7, q_8) = \frac{(q_6 + q_7)\exp(-K_{psi}q_8)}{K_6 + (q_6 + q_7)}, f_N^{(2)}(q_8) = \frac{q_8}{K_{NH_4} + q_8},$$

where K_{NO_3} – half-saturation constant of nitrates, K_{NH_4} – half-saturation constant of ammonium, K_{psi} – coefficient of ammonium inhibition.

For system (3), it is necessary to specify the vector field of water flow velocities, as well as the initial values of the concentration functions q_i

$$q_i(x, y, z, 0) = q_i^0(x, y, z), \quad (x, y, z) \in \overline{G}, t = 0, i = \overline{1, 8}. \tag{4}$$

Let the boundary Σ of a cylindrical domain G is the sectionally smooth and $\Sigma = \Sigma_H \cup \Sigma_o \cup \sigma$, where Σ_H – water bottom surface, Σ_o – unperturbed surface of the water environment, σ – lateral (cylindrical) surface. Let $\mathbf{u_n}$ – the normal component of the water flow velocity vector to the Σ surface, $\mathbf{n}$ – outer normal vector to the Σ. Let assume the concentrations q_i in the form:

– at the lateral boundary:

$$q_i = 0, \; on\,\sigma, \; if\,\mathbf{u_n} < 0, \; i = \overline{1, 8}; \tag{5}$$

$$\frac{\partial q_i}{\partial n} = 0, \; on\,\sigma, \; if\,\mathbf{u_n} \geq 0, \; i = \overline{1, 8}; \tag{6}$$

– at Σ_o:

$$\frac{\partial q_i}{\partial z} = 0, \; i = \overline{1, 8}; \tag{7}$$

– at Σ_H:

$$k\frac{\partial q_i}{\partial z} = \varepsilon_{1,i}q_i, \; i = 1, 2, \; k\frac{\partial q_i}{\partial z} = \varepsilon_{2,i}q_i, \; i = \overline{3, 8}, \tag{8}$$

where $\varepsilon_{1,i}$, $\varepsilon_{2,i}$ – sedimentation rate of algae and nutrients to the bottom.

3 Algorithm for Solving the Problem of the Coastal System Algolization

1. The type and values of all input functions and model parameters are set.
2. The initial control of the IE is set (initial values $F_1(x_i, y_i, z_i, 0)$, $i = \overline{1, N}$).
3. The system of equations (3)–(8) is numerically solved under known initial conditions.
4. The value Φ_0 is calculated according to (1) for given values $F_1(x_i, y_i, z_i, 0)$, $i = \overline{1, N}$.
5. The calculated value Φ_0 is compared with the maximum. The best control in the sense of (1) at the moment is saved.
6. If the number of iterations is not ended, then the next IE control is chosen (new values of the quantities $F_1(x_i, y_i, z_i, 0)$, $i = \overline{1, N}$). Go to paragraph 3 of the algorithm.

This algorithm in a finite number of steps allows to find an approximation to the IE optimal control. When choosing a new current IE control, direct ordered sorting methods with constant or variable step are used. After finding the approximate solution, it is possible to further refine it by the bitwise approximation method. Moreover, the uncertainty interval in determining the IE strategy is $2V/(N+1)$, where $V = \max_{1 \le t \le T} \int_G dG$, N – the number of partition points of the area of admissible IE control. The error in finding the optimal control of the subject is $\delta = V/(N+1)$.

Thus, for a finite number of iterations, the proposed algorithm allows to construct an approximate solution to the model or to conclude that there is no solution. The reliability and effectiveness of the proposed algorithm follows from the properties of the method of direct ordered enumeration with a constant step during simulation calculations.

4 Parallel Implementation of the Model Problem

A discrete analogue of the proposed model problem of aquatic ecology is developed on the basis of using schemes with weights, taking into account the partial fullness of the calculated cells [9]. The concentration distributions of two phytoplankton species (green and blue-green) were obtained taking into account the influence of external and internal factors on the dynamics of their development. The modifed alternating triangular method (MATM) was used for solving the grid equations [10]. Methods for solution such problems with a better convergence rate are described in comparison with the known methods [11].

4.1 Parallel Algorithm for MCS

We described the parallel algorithm for solving the problems (3)–(8). Parallel implementation of the developed algorithm is based on MessagePassingInterface (MPI) technologies. The peak performance of a multiprocessor computer system (MCS) is 18.8 TFlops. As computing nodes, 128 HP ProLiant BL685c homogeneous 16-core Blade servers of the same type were used, each of which is equipped with four quad-core AMD Opteron 8356 2.3 GHz processors and 32 GB RAM.

When parallel implementation of the developed numerical algorithms, the decomposition methods of computational domains for various computationally labour hydrophysics problems were used taking into account the architecture and parameters of MCS (Table 2). In Table 2: A, E are acceleration and efficiency of parallel algorithm; p is the number of processors. Note that the efficiency is the ratio of acceleration to the number of quad-core processors of computing system [12].

It was established that the maximum acceleration was 63 times on 512 processors (see Table 2).

Table 2. Acceleration and efficiency of parallel MATM algorithm

p	Time, s	A (practical)	E (practical)	A_t (theoretical)
1	3.700073	1	1	1
2	1.880677	1.967	0.984	1.803
4	1.265500	2.924	0.944	3.241
8	0.489768	7.555	0.731	7.841
16	0.472151	7.837	0.490	9.837
32	0.318709	11.610	0.378	14.252
64	0.182296	20.297	0.363	26.894
128	0.076545	48.338	0.317	55.458
256	0.063180	58.563	0.229	65.563
512	0.058805	62.921	0.123	72.921

4.2 Parallel Algorithm for NVIDIA Tesla K80

For numerical implementation of proposed interrelated mathematical water hydrobiological models, we developed parallel algorithms. It will be adapted for hybrid computer systems using the NVIDIA CUDA architecture which will be use for mathematical modeling of phytoplankton production and destruction process.

The NVIDIA Tesla K80 computing accelerator has the high computing performance and supports all modern both the closed (CUDA) and open technologies (OpenCL, DirectCompute). The NVIDIA Tesla K80 specifications: the GPU frequency is 560 MHz; the GDDR5 video memory is 24 GB; the video memory frequency is 5000 MHz; the video memory bus digit capacity is 768 bits. The NVIDIA CUDA platform characteristics: Windows 10 (x64) operating system; CUDA Toolkit v10.0.130; Intel Core i5-6600 3.3 GHz processor; DDR4 of 32 GB RAM; the NVIDIA GeForce GTX 750 Ti video card of 2 GB; 640 CUDA cores.

Using GPU with CUDA technology is required to address the effective resource distribution at solving the system of linear algebraic equations (SLAE), occuring at discretization of the developed model (3)–(8). The dependence of the SLAE solution time on the matrix dimension and the number of nonzero diagonals was obtained for implementation the corresponding algorithm (see Fig. 1). Due to it, in particular, we can choose the grid size and determine the solution time of SLAE based on the amount of nonzero matrix diagonals.

When integrating process of parallel algorithms for solving slows on the GPU into the previously developed software, it became necessary to estimate the time for solving slows depending on the dimension of the tape slows for a different number of non-zero elements in the main matrix diagonal. As a result of the conducted research, a number of regression lines with corresponding determination coefficients were obtained, indicating an acceptable degree of approximation.

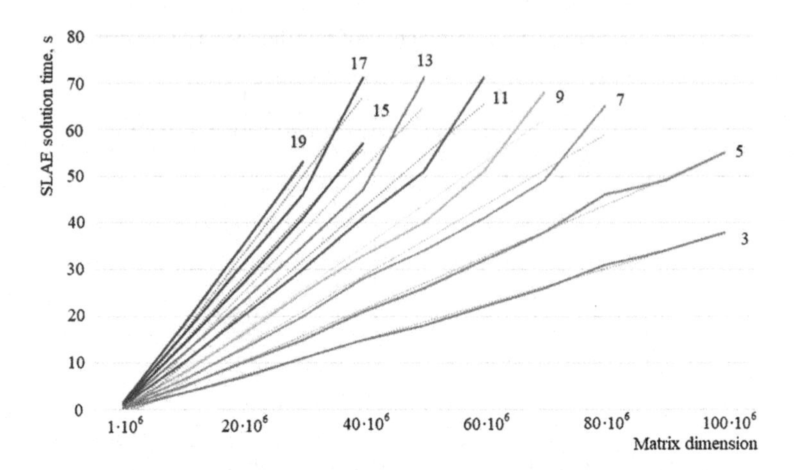

Fig. 1. Graph of SLAE solution time dependence on the order of a square matrix with a given number of nonzero diagonals (from 3 to 19 nonzero diagonals)

As a result of developed parallel algorithm for solving the biological kinetics problem, it became necessary to modify the element storage format of a sparse matrix from a system of grid equations. Elements of this matrix for internal nodes are repeating sequence. In the case of high-dimensional problems, standard storage leads to inefficient memory usage. The paper proposes a modification of the CSR format to increase the efficiency of data storage with a repeating sequence of CSR1S elements. The proposed format is effectively used for modeling continuous hydrobiological processes using the finite difference method. In additionally, to change a differential operator instead of repeatedly searching and replacing values in an array of nonzero elements, it is enough to just change them in an array that preserves a repeating sequence.

Table 3. Characteristics of data storage formats

Storage format characteristic	CSR format	CSR1S format
Number of arrays	3	5
The array size of nonzero elements	$N_{nz}B_{nz}$	$(N_{nz} - N_{seq}R_{seq})B_{nz}$
The array size of column indices where nonzero elements are located	$N_{nz}B_{idx}$	$(N_{nz} - N_{seq}R_{seq})B_{idx}$
The array size of the first nonzero row elements indices	$(R+1)B_{idx}$	$(R - R_{seq} + 1)B_{idx}$
The array size to store a repeating sequence	–	$N_{seq}B_{nz}$
The array size for storage the indices of columns in which the first elements of repeating sequence are located	–	$R_{seq}B_{idx}$

Notations in the Table 3: R is the number of rows in matrix; R_{seq} is the number of rows in matrix with the repeating sequence of elements; C is the number of columns in matrix; N_{nz} is the number of nonzero matrix elements; N_{seq} is the number of elements of the repeating sequence; B_{nz} is the memory capacity for storage one nonzero element; B_{idx} is the memory capacity for storage one index.

We estimated the required memory capacity in CSR and CSR1S formats: $P_{csr} = N_{nz}B_{nz} + N_{nz}B_{idx} + (R+1)B_{idx} = N_{nz}B_{nz} + (N_{nz} + R + 1)B_{idx}$; $P_{csr1s} = (N_{nz} - N_{seq}(R_{seq}+1))B_{nz} + (N_{nz} - R_{seq}(N_{seq}+1) + R + 1)B_{idx}$.

Denote the ratio of lines containing the repeating sequence to the total number of lines: $k_r = R_{seq}/R$.

Then, $P_{csr1s} = (N_{nz} - N_{seq}(k_r R + 1))B_{nz} + (N_{nz} - k_r R(N_{seq}+1) + R + 1)B_{idx}$.

Effective function libraries [Cuda docs] have been developed to solve SLAEs in CSR format on GPUs using the CUDA technology. The proposed approach is to use the CSR1S format at model processing with further conversion to the CSR format to solve the resulting SLAE on GPU using NVIDIA CUDA technology. In this regard, the problem of development an algorithm for matrix conversation from CSR1S to CSR format in minimum time was occurred.

Computational experiments were performed with fivefold repetition and fixing the average value of calculation time. Experimental research of the dependence of the execution time of the conversion algorithm on N_{seq} and the k_r coefficient for a sequential implementation, as well as parallel implementation using the TPL library and parallel implementation using the NVIDIA CUDA platform was performed. N_{seq} ranged from 3 to 19 in increments of 2; k_r ranged from 0 to 1 in increments of 0.1; the matrix dimension is 10^6.

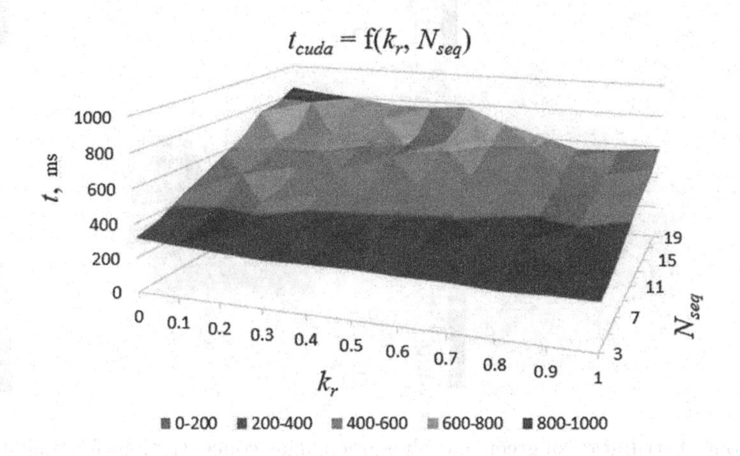

Fig. 2. Graph of time dependence of matrix conversion from CSR1S format to CSR format by an algorithm using technology NVIDIA CUDA

An analysis of the experiments shows that an algorithm using parallel computing on a CPU is more efficient than a sequential algorithm and an algorithm using NVIDIA CUDA. The character of the plane inclination in Fig. 2 shows a decrease in the computation time with increasing kr and Nseq, which allows it to be used asynchronously when working with matrices with high kr and Nseq, thereby unloading the central processor to perform other tasks.

5 Modeling the Phytoplankton Dynamics in the Azov Sea

The experimental software complex (SC) was designed for mathematical modeling of shallow water ecosystems on the example of the Azov Sea on MCS and graphic accelerator. Through the "Azov3d" SC we can construct the operational flow forecast turbulence of water environment – the velocity field on grids with high resolution. The SC was used for calculation the three-dimensional velocity vector of the Azov Sea. The SC involves the numerical implementation of model problems of hydrodynamics and biological kinetics.

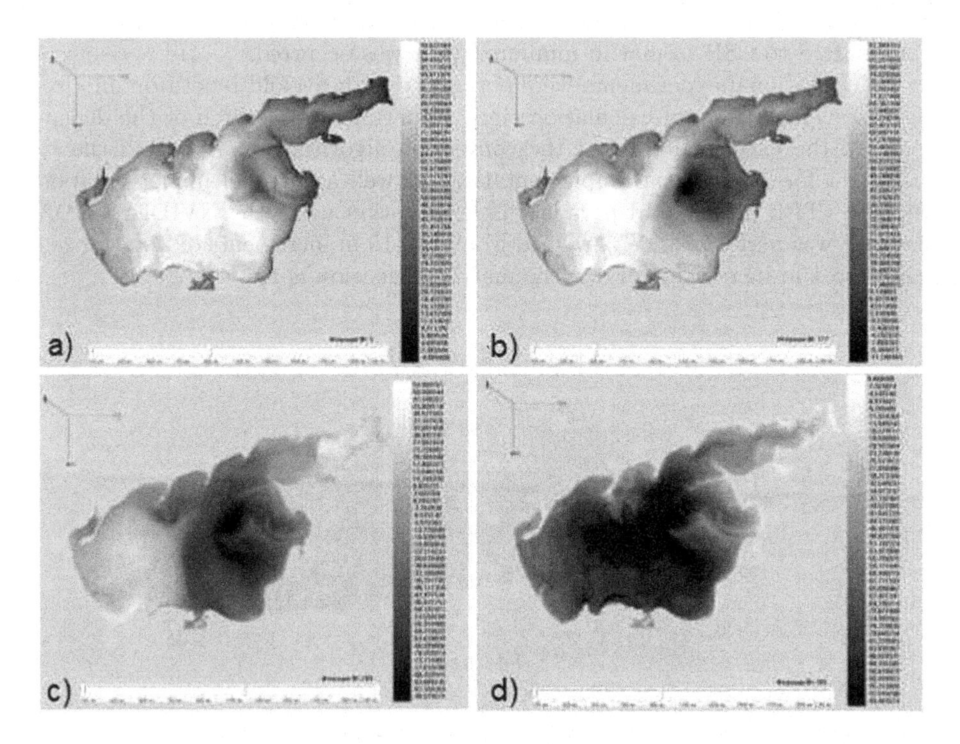

Fig. 3. Joint distribution of green and blue-green algae concentrations for time intervals a) T = 5 days, b) T = 39 days, c) T = 70 days, d) T = 122 days (Color figure online)

Describe a possible scenario for managing sustainable development in the biological rehabilitation of the Azov Sea by modeling the process of its algolization *Chlorella vulgaris BIN*. The strain of this green algae is cultivated in the wastewater of domestic, industrial and agricultural enterprises. Algolization of the reservoir must be carried out in the ice period in order to equalize the starting conditions for both native species and acclimatized strains of chlorella. For successful acclimatization of chlorella in the Azov Sea as a percentage of the total number of phytoplankton cells, it should be 10% in spring and 30% in summer. When solving the problems of biological rehabilitation, the following physical dimensions of the simulated region (the Azov Sea) were taken into account: the surface area under study was 37605 km^2, the length was 343 km, the width was 231 km, a numerical experiment was carried out on a sequence of condensing grids: computational nodes.

Figure 3 describes the displacement mechanism of the toxic blue-green algae *Aphanizomenon flos-aquae* by green algae and the results of applying the sustainable development management scenario for the biological rehabilitation of the Azov Sea by simulating the process of its algolization with the algae *Chlorella Vulgaris BIN.* White color shows the maximum concentration of toxic blue-green algae, black color shows maximum concentration of green algae. The figures also reflect the influence of abiotic factors (salinity and temperature [13]) on the development of two types of phytoplankton, the absorption of phosphates and nitrogen forms by phytoplankton, the transition of phosphorus and nitrogen forms from one to another.

6 Conclusions

In this paper, mathematical modeling of the problem of eutrophication reduction in the coastal system, such as the Azov Sea, using the concept of sustainable development management, was carried out. The considered three-dimensional mathematical model (3)–(8) of the interaction of two phytoplankton types takes into account the influence of abiotic factors, such as salinity and temperature, on the algae growth, their nutrients absorption and excretion, as well as the transformation of these nutrients from one form to another.

The numerical implementation of the model was performed on a multiprocessor computer system with distributed memory. We obtained theoretical estimates for the speedup and efficiency of parallel algorithms. Experimental software was designed for mathematical modeling of possible development scenarios of shallow waters. The example of the Azov Sea was considered in this regard. Decomposition methods of grid domains for computationally intensive diffusion-convection problems were employed for the parallel implementation, taking into account the architecture and parameters of the MCS. The parallel algorithm was developed for data distribution among processors.

Due to the use of MCS and NVIDIA Tesla K80 computing accelerator, the calculation time for the solution of the model problem decreased, while maintaining the required accuracy for modeling of hydrobiological processes in shallow waters. Note that this fact is one of primary importance in water ecology problems.

Acknowledgements. The reported study was funded by RFBR, project number 20-01-00421.

References

1. Williams, B.J.: Hydrobiological Modelling. University of Newcastle, New South Wales (2006)
2. Nikitina, A., Belova, Y., Atayan, A.: Mathematical modeling of the distribution of nutrients and the dynamics of phytoplankton populations in the Azov Sea, taking into account the influence of salinity and temperature. In: AIP Conference Proceedings, vol. 2188, no. 050027. American Institute of Physics Inc., US (2019). https://doi.org/10.1063/1.5138454
3. Alekseenko, E., Roux, B., Sukhinov, A., Kotarba, R., Fougere, D.: Nonlinear hydrodynamics in a mediterranean lagoon. J. Comput. Math. Math. Phys. **57**(6), 978–994 (2017). https://doi.org/10.5194/npg-20-189-2013
4. Sukhinov A.I., Sukhinov A.A.: Reconstruction of 2001 ecological disaster in the Azov Sea on the basis of precise hydrophysics models. In: Parallel Computational Fluid Dynamics, Multidisciplinary Applications, Proceedings of Parallel CFD 2004 Conference, pp. 231–238. Elsevier, Amsterdam (2005). https://doi.org/10.1016/B978-044452024-1/50030-0
5. Riznichenko, G.Y.: Mathematical Models in Biophisics and Ecology. Institute of Computer Research, Moscow-Izhevsk (2003)
6. Avdeeva, Z.K., Kovriga, S.V., Makarenko, D.I., Maksimov, V.I.: Cognitive approach in the control science. J. Prob. Manage. **3**, 2–8 (2007). (in Russian)
7. Sukhinov, A.I., et al.: Game-theoretic regulations for control mechanisms of sustainable development for shallow water ecosystems. Autom. Remote Control **78**(6), 1059–1071 (2017). https://doi.org/10.1134/S0005117917060078
8. Sukhinov, A.I., Chistyakov, A.E., Nikitina, A.V., Filina, A.A., Belova, Y.V.: Application of high-performance computing for modeling the hydrobiological processes in shallow water. In: Voevodin, V., Sobolev, S. (eds.) RuSCDays 2019. CCIS, vol. 1129, pp. 166–181. Springer, Cham (2019). https://doi.org/10.1007/978-3-030-36592-9_14
9. Samarskiy, A.A.: Theory of Difference Schemes. Nauka, Moscow (1989)
10. Konovalov, A.N.: The theory of alternating-triangular iterative method. J. Siberian Math. J. **43**(3), 552 (2002)
11. Sukhinov, A.I., Chistyakov, A.E.: Adaptive modified alternating triangular iterative method for solving grid equations with a non-self-adjoint operator. J. Math. Mod. Comput. Simul. **4**(4), 398–409 (2012). https://doi.org/10.1134/S2070048212040084
12. Chetverushkin, B., et al.: Unstructured mesh processing in parallel CFD project GIMM. In: Parallel Computational Fluid Dynamics, pp. 501–508. Elsevier, Amsterdam (2005)
13. Sukhinov, A.I., Belova, Y.V., Filina, A.A.: Parallel implementation of substance transport problems for restoration the salinity field based on schemes of high order of accuracy. In: CEUR Workshop Proceedings, vol. 2500. CEUR-WS (2019)

Nonlinear Bending Instabilities Accompanying Clump and Filament Formation in Collisions of Nebulae

Boris Rybakin[1](✉) and Valery Goryachev[2]

[1] Department of Gas and Wave Dynamics, Moscow State University, Moscow, Russia
rybakin1@mail.ru
[2] Department of Mathematics, Tver State Technical University, Tver, Russia
gdv.vdg@yandex.ru

Abstract. It is known that the nucleation of stars in the universe occurs, as a rule, in areas of the interstellar medium, where nebulae - gas and dust clouds consisting of molecular hydrogen, can collide with each other, changing their state during dynamic interaction with gravitational and magnetic fields. The gravitational-turbulent description of these processes is quite common in explaining the reasons for the creation of pre-stellar regions as a consequence of the collision of molecular clouds. We adhere to this approach with some simplification, assuming that the main factor in the dynamic transformations of gas formations is the influence on the collision process of mainly the kinetic energy of the clouds, separating these effects from the effects of gravitational collapse and from the influence of magnetic fields. To simulate gas-dynamic processes of different scales, a parallel numerical code was developed using grids with improved resolution, which was used in a numerical experiment on high-performance computers. The simulation showed that sharp changes in the distribution of matter in the shock core of new formations can be triggered by the Kelvin-Helmholtz instability and nonlinear thin-shell instability, which lead to sharp perturbations of the gas density in the resulting clumps, outer shells and gas filaments, with density fluctuations in the external interstellar medium. A predictive analysis of the appearance of possible pre-stellar zones during the evolution of new cloud formations is given.

Keywords: Parallel computing · Instability · Molecular Cloud-Cloud collision

1 Introduction

The formation of pre-stellar zones as a result of collisions of molecular clouds, burst expansion of the matter of stellar remnants after supernova explosions, consequences of stellar wind and radiation from star clusters in planetary nebulae, is theoretically intractable and algorithmically difficult with a possible simulation of astrophysics. Modeling of such processes plays an important role in understanding the evolution of galaxies and the diversity of nebulae in the Universe. In the case with molecular Cloud-Cloud Collision (CCC), pre-stellar areas begin to forms at transition stage of impact and mutual

V. Voevodin and S. Sobolev (Eds.): RuSCDays 2020, CCIS 1331, pp. 261–272, 2020.
https://doi.org/10.1007/978-3-030-64616-5_23

penetration of nebulae accompanied by multi-scale dissipative interplay between turbulent, radiative and magnetic fields of molecular gas in the interstellar medium (ISM). This process is accompanied by the condensation of interstellar matter in the resulting gas clusters under conditions of gravitational collapse.

Cloud collisions within the interstellar medium are widespread events. Among publications on this theme, one can note references in [1] and a review of the recent astrophysical discoveries and the numerical modeling of some cases of collisions [2]. One of the most recent articles describes nebulae collisions and shock-compressed clumps and filaments origination that serve a trigger of pre-stellar areas formation in the cluster NGC 3603 [3]. The data treatment in this study revealed more two cases of massive star formation by CCC scenario in the star cluster Westurlund 2 [4]. Among different star formation models, the magnetic and gravitational-turbulent models have recently been distinguished. The first better explains the appearance of individual stars, and the second - their ensembles. On the whole, science is moving towards a unification of these two paradigms, because, most likely, turbulence, gravity, a magnetic field, and the effect of all this in a complex are of importance. But all this together is quite may difficult to model, and to get a convincing result, one need very powerful supercomputer systems for numerical simulation.

MC's collisions have been frequently studying by computational fluid dynamics approach. Gas dynamics modeling continues to be improved using modern smooth particle hydrodynamics (SPH) approach and the adaptive grid refinement (AMR) Euler methods. Most simulations have started with the pioneering work on two-dimensional two clouds collision modeling performed in [5]. Studies [6–8] showed that a change in the shape of the resulting nebulae is the result of regular perturbations of the layers between the clouds of gas and ISM and sporadic rearrangement of the compressed core inside the collision zone. More detailed simulation is often performed taking into account the influence of self-gravity and magnetic fields on these processes [9, 10] and [11]. New improvements were done by the heating and cooling functions addition in the energy equation. They gave a possibility to study the details of cloud collisions and peculiarities formation of filaments and clumps compressed heterogeneously. In many works [12, 13], researchers tested the effect of instability on the formation of shock layers in collisions between opposing hypersonic gas flows. The conditions for the occurrence of Kelvin-Helmholtz instability and Nonlinear Thin Shell instability (NTSI) over active collision core layers under the action of strong magnetic and radiation fields were studied. In [13, 15], the ideas of Vishnyak [14] about the nonlinear instability of thin shells were actively tested for the possibility of describing oscillating gas perturbations of shock layers on velocity discontinues surfaces of contrary streams. These mechanisms may be additional drivers of star formation activity in pre-stellar areas in colliding streams.

Among different models used in astrophysics, magnetic and gravitational-turbulent models have recently been distinguished. The gravitational-turbulent pre-star formation model combines the assumptions about the reason for the creation of stars as a result of a collision of nebulae. In presented paper we selected a hydrodynamics turbulent model for mutual cloud collisions, with an emphasis on taking into account only the kinetic energy aspect of molecular cloud collisions, in order to analyze only this effect, separating it from self-gravitation.

2 Numerical Modeling

The numerical simulation of the MCS collision presented in the paper is a further simulation of CCC processes occurring in the interstellar medium, started in [16, 17].

We modeled head-on and glancing MC_1/MC_2 collisions characterized by mutual penetration of dissimilar in size and radial density distribution molecular clouds of initially spherical forms, with the relative velocities of collisions being varied. To simulate flows of nebulae interplay classical approach of gas dynamics is used.

Momentum and energy transfer are described by a set of Euler equations:

$$\frac{\partial \rho}{\partial t} + \nabla \cdot (\rho \boldsymbol{u}) = 0, \tag{1}$$

$$\frac{\partial \rho \boldsymbol{u}}{\partial t} + \nabla \cdot (\rho \boldsymbol{u} \otimes \boldsymbol{u}) + \nabla P = 0, \tag{2}$$

$$\frac{\partial \rho E}{\partial t} + \nabla \cdot [(E + P)\boldsymbol{u}] = 0. \tag{3}$$

In these equations $E = \rho\left(e + \boldsymbol{u}^2/2\right)$ is the total gas energy, e – the internal energy, ρ is the gas density, $\mathbf{u} = (u, v, w)$ – velocity vector. The total energy E and gas pressure P are related through the ideal gas closure $P = (\gamma - 1)\rho e$, where adiabatic index γ is equal to 5/3. The lateral and output boundaries of the computational domain are defined as a periodical, and the boundary conditions are set as open for primitive variables.

2.1 Parallel Realization Code

To carry out the calculations on sufficiently detailed regular grids, parallel algorithms were developed for solving problems of non-stationary gas dynamics. In parallel computations to speed up the calculations, OpenMP technologies were used for the CPUs and CUDA for the GPUs. Used OpenMP was configured using Intel VTune Amplifier XE. This toolkit allows you to profile applications directly on the nodes of the cluster computing system. To determine the quality of parallelization of CPU calculation, Light Weight Hotspots test was used. Some routines used in numerical code were calculated on graphics processing units. For parallelization on GPUs, Client Utilities & Framework (CUF) core technology was used. Data management in the code with CUF kernels was carried out explicitly. Arrays on the device side were declared with attributes: *device*, *pinned*. Data was transferred from the host to the device and vice versa manually, rather than using assignment operators.

The numerical algorithm was based on the application of the method of splitting calculations by physical coordinates. From a four-dimensional array of conservative variables, a one-dimensional section of the original array was cut out for each spatial variable and distributed between the processors. At each time step of integration, the values determined at the centers of the grid cells were calculated, and then the fluxes through the boundaries for pressure waves moving to the right and waves moving to the left were calculated. The numerical practice revealed that for not very large mesh nodes sizes (up to 1024 x 512 x 512 nodes) parallelization on the GPU gave greater acceleration of operations than parallelization using OpenMP. As the grid size increases, data does not fit in the GPU memory. Therefore, performance is slowed down due to data transfer on the relatively slow PSI-E bus.

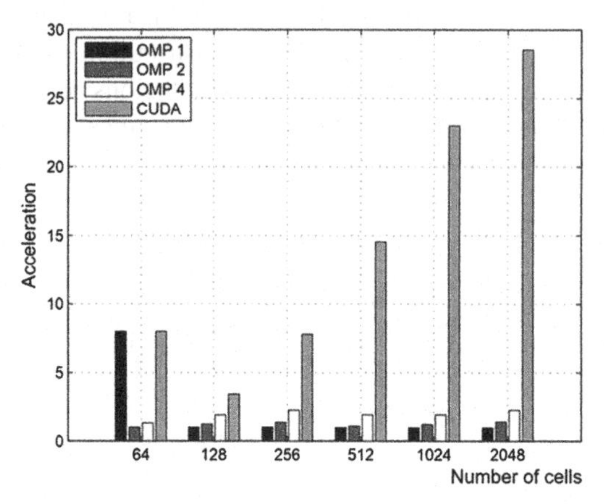

Fig. 1. Test acceleration in calculations using CUDA and OpenMP.

The quality of computing performance improvement was analyzed when solving test problems with the goal of debugging the calculation technique. The acceleration in calculations for two-dimensional test problem performed on GPUs compared with calculations performed using OpenMP technology on one, two, and four processor cores. Acceleration was checked in the variants with the largest calculation time. It can be seen from the results that, for a small computational grid, the use of OpenMP technology does not give any acceleration, but rather slows down the computation. However, the small size of the grid does not have a significant negative effect inherent in graphic cards when working with small amounts of data. With an increase in the number of computing nodes on the GPU, there is a significant acceleration of calculation - more than 25 times, compared with a work single core of the central processor. The diagram in Fig. 1 shows such acceleration in calculations.

In serial numerical experiment Eulerian Eqs. (1)–(3) were solved on regular high-resolution mesh with an adaptive Roe solver using the schemes of total variation diminishing (TVD) type. The nonlinear, second-order accurate TVD approach provides a high resolution capturing of shocks and prevents from unphysical oscillations. An application programming interface OpenMP for parallelization is employed. To take into account the instabilities influence on new nebular morphing and strong form distortion, calculations with four and more billion nodes meshes were used. Some computations are done with graphics accelerators NVIDIA K40 and CUDA for PGI Fortran. In-house code IGMC and features of its parallel realization are discussed in [16, 17]. An extensive set of solver utilities and in-house postprocessing system HDVIS with new parallelization options were used to analyze large output of numerical experiment results.

2.2 Initial Data and Parameters

Main physical parameters of MCs/ISM matter have been used in the same way as in [18, 19] и [20]. The interstellar medium consists of relatively warm matter with $T_{ism} = 10^4$ K, the temperature T_{cl} of cold MCs is equal 10^2 K. The ambient gas density of the outer cloud medium $\rho_{ism} = 2.15 \times 10^{-25}$ g·cm^{-3}, the initial gas density in the cloud centers ρ_{cl} is set in the range from 5.375×10^{-24} g·cm^{-3} and to 1.075×10^{-22} g·cm^{-3}. The computing area used for the problem under consideration is a box of $1.6 \times 0.8 \times 0.8$ pc dimensions.

The initial distribution of matter density inside the clouds physically non-limited by outer boundary plays an important role in the mass impulse action. Initial density for overlapping gas layers in the interstellar medium and cloud boundary layers are regulated by parameters of its radial profile distribution taken from [18, 19].

The modeling has been performed according to different CCC scenarios. In numerical experiments, oppositely moving clouds of different mass, size and radial density distribution collided with each other at relative velocities of 5–25 km·s^{-1}. Relative colliding velocities U_{cl} of each MC were assigned as 5.88, 11.77, 23.54 km·s^{-1}. Initial density contrast χ, that is equal to ratio of clouds density and interstellar medium density - ρ_{cl}/ρ_{ism} designated for values between MCs centers and ISM, was assigned to each cloud as 25, 100, and 500 respectively. Colliding velocity and density contrast were combined in a numerical experiment. MC's masses range from 0.3 to 1.2 $M_{\odot}$ respectively.

3 Modeling Analysis

The main goal of the simulation was to study the nebulae morphology and compression dynamics of matter in the core of colliding MCs as a determinant factor of a pre-stellar area forming that originates in a transitional zone between divergent in time clouds.

The evolution of MCs collision process simulated in head-on scenarios share common stages as it is shown in Fig. 2, where one can see clumps and filaments formation at initial and final stages of nebulae evolution. For the case with contrast ratio 100/25 the cloud MC_1 penetrates through more light and friable MC_2, and move to the right. Mutual penetration of nebulae is accompanied by a rapid change of pressure in a contact shock layer, and energy - density spatial reallocation inside with the creation of slightly denser stratum between cloud and rarefied interstellar medium border. The accretion and conical formation of the cavity in the newly formed object are accompanied by a slight decrease in the initial outer diameter of the clouds. This is due to the redistribution of gas in the radial direction, from the center to the conditional shell with some compression density value in outer clouds layers. An increasing part of the gas substance goes into a lens-like compression zone. During evolution the shape of the clouds is deforming; their bubble-like shell is compressing and reducing in size. In a reciprocal collision, this change can be repeated in time.

Fig. 2. Evolution of isodensities layers in cloud-cloud collision. Isosurfaces: $\chi = 2, 5000, 10000$ are shown at time stages 0.05–0.48 Myr. Density contrast ratio is 100/25, relative colliding velocity - 11.77 km·s^{-1}.

The formation of an arched, highly compressed gas shock layer with a strongly deformed wrinkled surface begins almost immediately after initialization of collision. The process is accompanied by the clumps growth and breakup of fragments, new clumps generation and spread-out of filaments.

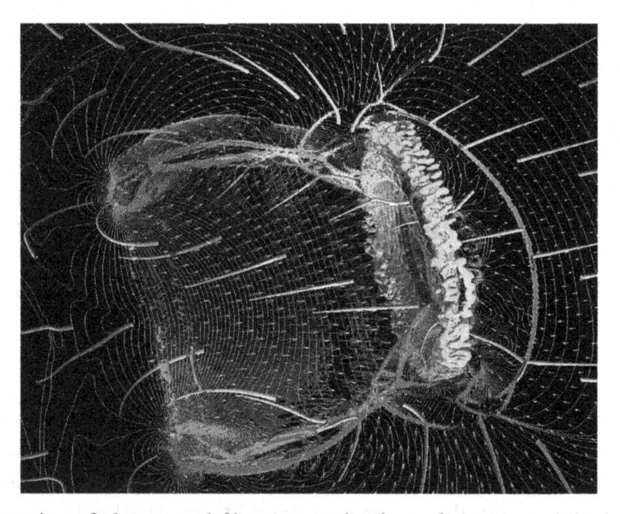

Fig. 3. The formation of clumps and filaments at the time of rupture and the beginning of the divergence of cloud fragments. Isosurfaces: $\chi = 100$ and 10000; isotachs contours U_x and velocity field with motion paths projected in the meridional plane at $t = 0.3$ Myr are shown. Density contrast ratio is 100/25, colliding velocity – 23.54 km·s^{-1}.

The end of the process of mutual penetration depends on the initial mass of the clouds and spatial redistribution of gas density during the process. The direction of movement of more compacted new-formation is determined by the mass ratio of the clouds. At the stage of the final penetration of the left cloud into the right one, the shell of the latter is completely torn. Stretchable compacted clumps begin to take the form of filaments.

The process of mutual penetration of cloud masses is accompanied by the appearance of pulsating density changes on the outer surfaces of the clouds and on the surface of the velocity rupture between the oncoming gas flows (stagnation points surface), where instability arises due to sharp velocity changes by direction in the gas mixing layers. First oscillations are triggered by Kelvin-Helmholtz instability (KHI), and then nonlinear bending instability in shock layer is added to this process. Arise of pulsation is accompanied by a stretch of filaments in a post-collision area and gas turbulization in outer shells, a continuation of clump shaping due to strengthening of KHI, spatial growth of finger structures, acceleration of clumps ablation into outer ISM and subsequent of their disappearance. Characteristic waves of instability arise at the boundaries of ISM and cloud outer layers. Typical vortices with kinks mirrored this process can be seen in respective illustrations in Fig. 4 and others shown below.

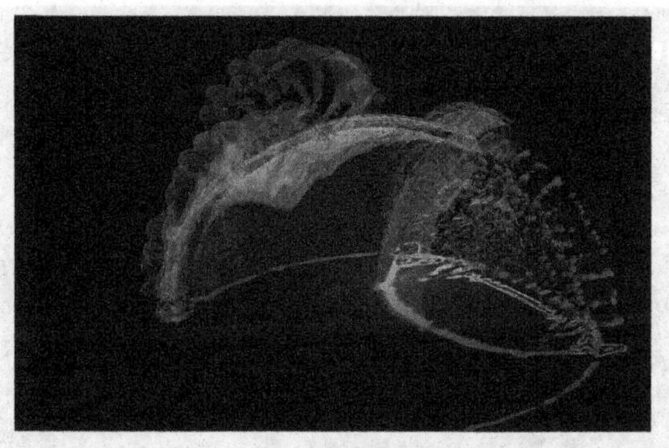

Fig. 4. Isodensity contours in space and in a middle plane for density contrast level: $\chi = 5$ (green), 500 (yellow), 10000 (red) at t = 0.43 Myr. Density contrast ratio is 100/25, relative colliding velocity - 11.77 km·s^{-1}. (Color figure online)

The shaping of outward MCs surfaces accelerates the generation of vortices inside the cloud formation, which is reflected in appearance of rolls and ripples on the outer surfaces and in the space of the growing cavity inside the penetration channel.

As for bow-shock formation, the normal stress accumulated between local stream jets located on the contact spots of ram surface is provoked by KHI instability too. Density and velocity fields in a shock-compressed core of colliding clouds are quite intermittent. Perturbations initiated by extremely strong contractions of clouds' cores become clearly observable under the analysis of temporal pulsations of gas density field that accompany mutual collision. The main gas normal tensions are accumulated over the stagnation surface of contrary gas flows in narrow channels between clumps and blobs originated above stagnation points. Redistribution of these changeable clumps is shown in Figs. 3 and 4, via rendered contours of density contrast in space and in the meridional section of the condensed core. They sporadically change their localization on the arcs and radii of a lens-like core. This is observed in all animations of dynamic visualization scenes with varying degrees of detail.

Isosurfaces of the density field inside the shock layer look like a sieve with a continuously changing net of filaments. As the relative velocity of the clouds increases, the velocity field at the collision surface becomes more uniform at the center and more discontinuous on periphery, with complex interference of faint shock waves in bending thin layers, and discontinues layers with clumps included.

Perhaps the decisive role in starting this process can be played by the NTSI (nonlinear instability of thin shell) - bending-mode instability, the mechanism of which is reasonably described in [14, 15]. This type of instability can manifest itself in a sharp increase in the amplitude of gas oscillations in the compressed layers in the outer layers of the clouds by an amount comparable in magnitude to the thickness of the shock layer in the nebula core. NTSI and KHI are factors favorable for the formation of finger and needle-like structures here. The kinetic energy of gas is dissipated in a shock layer by pulsing extrusions inside this layer. Eventually, these formations turn into a net of filaments and clumps in the stretched transitional area after the impact.

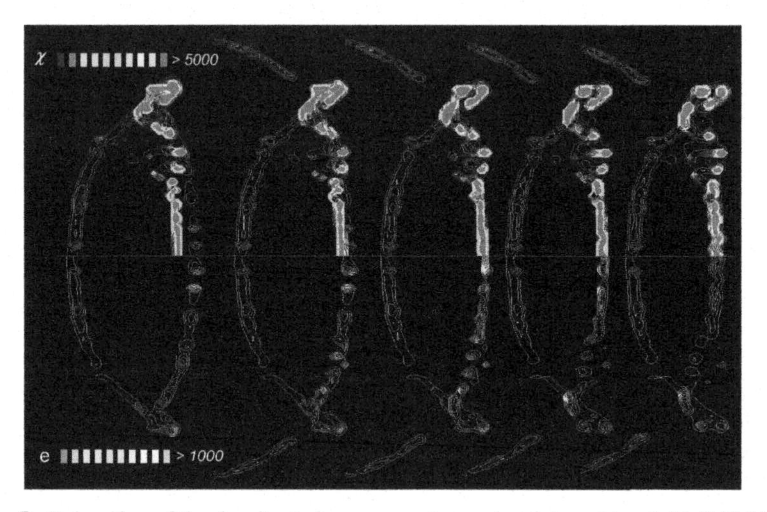

Fig. 5. Relocation of the density and energy contours at an interval t = 0.43–0.47 Myr.

An analysis of the distribution of the clumps density and internal energy along the lines of the gradient temperature change, which repeats (in the meridional plane) the bends of the compressed disk, shows that the mechanism of bending pulsations here is the same as proposed in [14]. Under the conditions of CCC simulated in [15], the temperature and density fields have opposite signs for derivatives of temperature and density, that varies in the width of the densified and cooling gas in critical areas of collisions slabs. In our studies, similar patterns of such changes are observed too, but which differ in the rate of change of these quantities, but not in sign. In other areas, the picture corresponds to a change in the corresponding characteristics of radiation heating or cooling in [13]. This is illustrated by the graphs in Figs. 5 and 6.

Our simulation was performed without taking into account self-gravity. However, the results of the calculation of gas-dynamic fields can be used to estimate the places of occurrence of possible pre-stellar area. After the passage of one cloud through another,

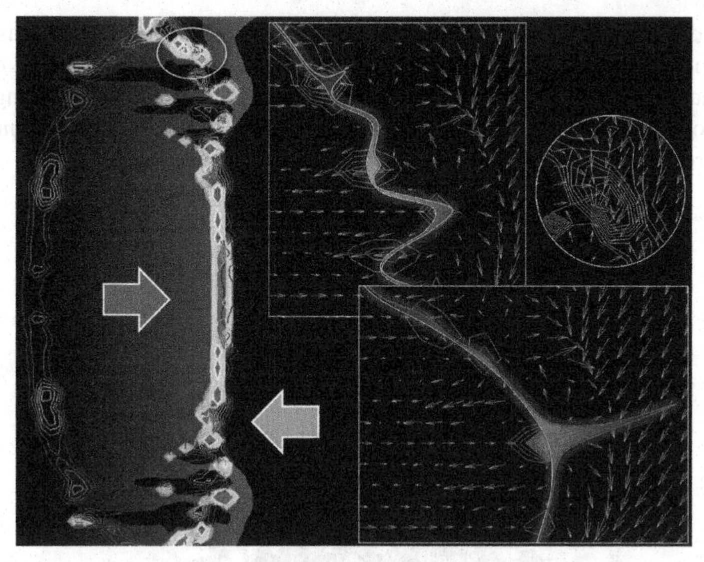

Fig. 6. Distribution of energy and density along the gradient internal energy line, between clumps that determine the periodic bending of the corrugated lens-like core.

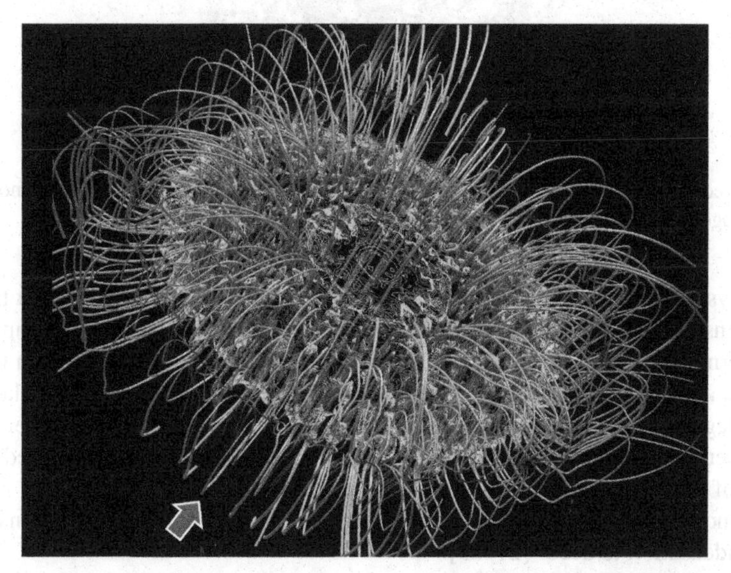

Fig. 7. Gas pathlines in formed nebular and energy distribution contours, associated with isotachs $U_X = 0.01$ at the stage t = 0.44 Myr. Density contrast ratio is 100/25, colliding velocity - 11.77 km·s^{-1}.

the gas density in the clumps that appeared outside the rupture region can reach the highest density values compared with the values observed throughout the entire evolutionary period. The beginning of the pre-stellar medium condensation is possible in those places where the divergence of the velocity field has the largest negative values. It was revealed that the nucleation of clumps is associated with zones of a strong gradient of energy located near the stagnation point set surface of the central stream in colliding nebulae (Fig. 7). Joint consideration of these factors can be used to estimate the boundaries of possible pre-stellar zone. Sample of such prediction is shown in Fig. 8.

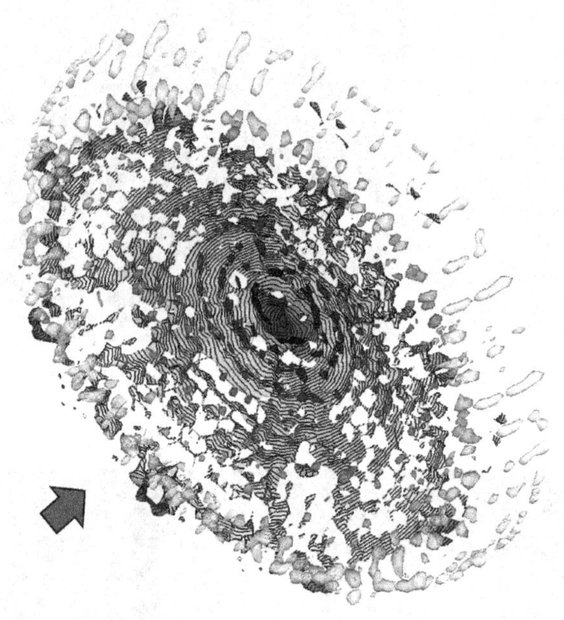

Fig. 8. Predictive zone of gas matter condensation, clumps and filaments with density more 10^{-20} g·cm^{-3} together with field of negative divergence for velocity are shown.

Analysis of the time sequence of changes in the density fields shows that the maximum density level in gas clumps and filaments is achieved in the zone of rupture and it is determined by the kinetic energy of the collision. It linearly depends on the mass of clouds and non-linearly on the relative collision velocity according to the law of the second degree. Density in the transformation zone can be thousand fold higher than the initial average value in colliding clouds. The numerical experiments revealed that the density of matter in theirs clumps varies in the range of 10^{-21}–10^{-19} $g{\cdot}cm^{-3}$, which corresponds to the generally accepted level for the pre-stellar conglomerations and for stars condensation to occur [21–23].

4 Conclusion

1. In-house parallel modeling code and tools for visualizing big data have been developed.
2. Effects of matter compression, turbulization and ablation in new nebula formation after of the molecular clouds collision have been compared for in-head collision using different initial and parametric data.

3. It was revealed how gas medium transformation consequences of MCs collision are provoked and magnified via Kelvin-Helmholtz and Nonlinear Thin Shell Instability in oscillated bent lens-like core during the collision.
4. Sporadical impulse factors can play a crucial role in the generation of the bow-shock layer.
5. Shock deformations and waved cloud core concussions induce the occurrence of filamentous structures and post-shock gas clumps in the transition zone where pre-stellar conditions may arise.

Acknowledgements. This work has been supported by RFBR Grant 19-29-09070.

References

1. Dobbs C.L., et al.: Formation of molecular clouds and global conditions for star formation. In: Beuther, H., Klessen, R., Dullemont, C., Henning, Th. (eds.) Protostars & Planets VI, pp. 3–26. Univ. of Arizona Press (2014)
2. Special Issue: Star formation triggering by cloud-cloud collision. Publ. Astron. Soc. Jpn – PASJ **70**(SP2) (2018)
3. Fukui, Y., Ohama, A., Hanaoka, N., et al.: Molecular clouds toward the super star cluster NGC 3603. Possible evidence for a cloud-cloud collision in triggering the cluster formation. Astrophys. J. **780**(36), 13 (2014)
4. Ohama, A., Dawson, J.R., Furukawa, N., et al.: Temperature and density distribution in the molecular gas toward Westurlund 2: further evidence for physical associations. Astrophys. J. **709**, 975–982 (2010)
5. Habe, A., Ohta, K.: Gravitational instability induced by a cloud-cloud collision: the case of head-on collision between clouds with different sizes and densities. Publ. Astron. Soc. Jpn. **44**, 203–226 (1992)
6. Anathpindika, S.V.: Collision between dissimilar clouds: stability of the bow-shock and the formation of pre-stellar cores. Mon. Not. R. Astron. Soc. **405**, 1431–1443 (2010)
7. Takahira, K., Shima, K., Tasker, E.J., Habe, A.: Formation of massive, dense cores by cloud-cloud collisions. PASJ **70**(SP2), 1–24 (2018)
8. Haworth, T.J., et al.: Isolating signatures of major cloud–cloud collisions – II. The lifetimes of broad bridge features. MNRAS **454**, 1634–1643 (2015)
9. Torii, K., et al.: Triggered O star formation in M20 via cloud-cloud collision: comparisons between high-resolution CO observations and simulations. Astrophys. J. **835**(142), 1–12 (2017)
10. Wu, B., Tan, J.C., Nakamura, F., Van Loo, S, Christie, D., Collins, D.: GMC collisions as triggers of star formation. II. 3D turbulent, magnetized simulations. Astrophys. J. **835**(137), 1–23 (2017)
11. Shima, K., Tasker, E.J., Federrath, C., Habe, A.: Does feedback help or hinder star formation? The effect of photoionization on star formation in giant molecular clouds. PASJ **70**(SP2), 1–19 (2018)
12. Folini, D., Walder, R.: Supersonic turbulence in shock-bound interaction zones I. Symmetric settings. Astron. Astrophys. **459**, 1–19 (2006)
13. Klein, R.I., Woods, D.T.: Bending mode instabilities and fragmentation in interstellar cloud collisions: a mechanism for complex structure. Astrophys. J. **497**, 777–799 (1998)

14. Vishniac, E.T.: Nonlinear instabilities in shock-bounded slabs. Astrophys. J. **428**, 186–208 (1994)
15. McLeod, A.D., Whitworth, A.P.: Simulations of the non-linear thin shell instability. MNRAS **431**, 710–721 (2013)
16. Rybakin, B.P., Shider, N.I.: Development of parallel algorithms for solving the problems of gravitational gas dynamics. J. Vychisl. Metody Programm. **11**(4), 388–394 (2010)
17. Rybakin, B., Goryachev, V.: Modeling of density stratification and filamentous structure formation in molecular clouds after shock wave collision. Comput. Fluids **173**, 189–194 (2018)
18. Nakamura, F., McKee, Ch.F., Klein, R.I., Fisher, R.T.: On the hydrodynamic interaction of shock waves with interstellar clouds. II. The effect of smooth cloud boundaries on cloud destruction and cloud turbulence. Astrophys. J. **164**, 477–505 (2006)
19. Pittard, J.M., Falle, S.A.E.G., Hartquist, T.W., Dyson, J.E.: The turbulent destruction of clouds. MNRAS **394**, 1351–1378 (2009)
20. Melioli, C., De Gouveia, D.P.E., Raga, A.: Multidimensional hydro dynamical simulations of radiative cooling SNRs-clouds interactions: an application to starburst environments. Astron. Astrophys. **443**, 495–508 (2005)
21. Kainulainen, J., Federrath, C., Henning, T.: Unfolding the laws of star formation: the density distribution of molecular clouds. Science **344**(6180), 183–185 (2014)
22. Lada, C.J., Lombardi, M., Alves, J.F.: On the Star formation rates in molecular clouds. Astrophys. J. **724**, 687–693 (2010)
23. Federrath, C., Banerjee, R., Clark, P.C., Klessen, R.S.: Modeling collapse and accretion in turbulent gas clouds: implementation and comparison of sink particles in AMR and SPH. APJ **713**, 269–290 (2010)

Numerical Forecast of Local Meteorological Conditions on a Supercomputer

Alexander Starchenko(✉), Sergey Prokhanov, Evgeniy Danilkin, and Dmitry Lechinsky

Tomsk State University, Tomsk, Russia
{starch,viking,ugin}@math.tsu.ru, 360fip182@gmail.com

Abstract. A high-resolution mesoscale meteorological model for forecasting and studying weather events and surface air quality in an urbanized area or a large industrial or transportation hub is presented. An effective semi-implicit second-order finite volume method with parallel implementation on multiprocessor computing system was developed to solve the equations of the model. The results of testing the parallel program on the supercomputer Cyberia of Tomsk State University demonstrated its high efficiency. The approach developed was successfully applied to predicting heavy precipitation events and an urban heat island effect.

Keywords: Numerical weather forecast · Mesoscale model · Semi-implicit difference schemes · Parallel computations · Two-dimensional decomposition · MPI

1 Introduction

Currently, mathematical modeling is widely used in environmental studies along with experimental methods. Mathematical modeling allows studying and predicting the development of atmospheric, hydrological, and geological processes on the Earth without affecting them. Many mathematical models of continuum mechanics, hydrometeorology, ecology, etc. Are based on inhomogeneous unsteady three-dimensional convection-diffusion equations. These models describe changes in such parameters of phenomena researched as density, velocity components, temperature, concentration, and turbulence. Modern mathematical models are usually a system of several convective-diffusion equations, algebraic relations used as a closure, and initial and boundary conditions.

Finite difference, finite volume, and finite element methods are used to solve the systems of differential equations. Very fine meshes of several million of nodes are used in these computations. Moreover, in prognostic modeling of atmospheric and hydrological processes the solution should be provided in a very short time.

Multiprocessor and multi-core computers speed up the process of obtaining the numerical solution [1,2]. The effectiveness of the pure MPI parallelization

V. Voevodin and S. Sobolev (Eds.): RuSCDays 2020, CCIS 1331, pp. 273–284, 2020.
https://doi.org/10.1007/978-3-030-64616-5_24

in comparison with hybrid MPI+OpenMP parallelization by means of the UM (Unified Model of UK Met Office) atmosphere model as a test case is considered in [3]. The efficiency of running a parallel program on a fixed number of processes, but with a varying number of cores on each processor was also investigated. This investigation is of particular importance because in computational problems, special attention is paid to the work with memory and an increase in the number of active cores on the processor leads to fierce competition between them for the processor resources. It is also shown that pure MPI parallelization is more efficient over the entire range of the number of processes considered (up to 1536). At the same time, using only 12 of 16 cores on each processor increases computation efficiency by 10–15%. This is due to the following factors: increase of memory bandwidth on active core, reduction of communication costs with a decrease in the number of MPI processes and active cores, and core frequency increase in turbo boost mode [3].

Both pure MPI and Hybrid MPI+OpenMP parallelization of a pollution transport unit in atmospheric and ocean models is considered in [4]. The paper shows that with a small number of MPI processes involved, further increasing their number is appropriate. In this case, additional speedup is achieved by placing local array in a faster processor cache. If there is a fairly large number of MPI processes involved, increasing the number of OpenMP threads, on the contrary, becomes a more effective method to speed up calculations. This paper focuses on the fact that the hybrid MPI+OpenMP approach is the most universal in terms of using the resources of modern supercomputers, allowing to optimize the number of MPI processes and the number of OpenMP threads taking into account the task and architecture of computing elements.

The purpose of this work is to develop an effective parallel method for solving the equations of the prognostic meteorological mesoscale model [5] using the Message Passing Interface (MPI) technology. The parallel implementation of the model was successfully tested by means of predicting such local meteorological phenomena as heavy rain and snow and urban heat island.

2 Mathematical Formulation and a Numerical Method

2.1 Mesoscale Meteorological Model

The basic equations of the atmospheric boundary layer are obtained from Reynolds-averaged differential equations of hydrothermodynamics [6] in a terrain-following coordinate system under the following assumptions:

1. Mesoscale density variations are quasi-stationary. The Boussinesq approximation is used to represent the buoyancy force in the equation for the vertical velocity component.
2. Molecular diffusion is assumed to be negligible in relation to the turbulent transfer.
3. Moisture microphysics in the atmospheric boundary layer, shortwave and longwave radiation are taken into account.

The mathematical model includes the following Eqs. [5]:

Continuity equation

$$\frac{\partial(\rho u)}{\partial x} + \frac{\partial(\rho v)}{\partial y} + \frac{\partial(\rho w)}{\partial z} = 0. \tag{1}$$

Momentum equations

$$\begin{aligned}\rho\left(\frac{\partial u}{\partial t} + u\frac{\partial u}{\partial x} + v\frac{\partial u}{\partial y} + w\frac{\partial u}{\partial z}\right) = -\frac{\partial p}{\partial x} + \rho f v \\ + \frac{\partial}{\partial x}\left(K_H\frac{\partial u}{\partial x}\right) + \frac{\partial}{\partial y}\left(K_H\frac{\partial u}{\partial y}\right) + \frac{\partial}{\partial z}\left(K_Z^m\frac{\partial u}{\partial z}\right);\end{aligned} \tag{2}$$

$$\begin{aligned}\rho\left(\frac{\partial v}{\partial t} + u\frac{\partial v}{\partial x} + v\frac{\partial v}{\partial y} + w\frac{\partial v}{\partial z}\right) = -\frac{\partial p}{\partial y} - \rho f u \\ + \frac{\partial}{\partial x}\left(K_H\frac{\partial v}{\partial x}\right) + \frac{\partial}{\partial y}\left(K_H\frac{\partial v}{\partial y}\right) + \frac{\partial}{\partial z}\left(K_Z^m\frac{\partial v}{\partial z}\right);\end{aligned} \tag{3}$$

$$\begin{aligned}\rho\left(\frac{\partial w}{\partial t} + u\frac{\partial w}{\partial x} + v\frac{\partial w}{\partial y} + w\frac{\partial w}{\partial z}\right) = -\frac{\partial p}{\partial z} - \rho g \\ + \frac{\partial}{\partial x}\left(K_H\frac{\partial w}{\partial x}\right) + \frac{\partial}{\partial y}\left(K_H\frac{\partial w}{\partial y}\right) + \frac{\partial}{\partial z}\left(K_Z^m\frac{\partial w}{\partial z}\right).\end{aligned} \tag{4}$$

Here t is time, u, v, w are the longitudinal, latitudinal, and vertical components of the averaged wind speed vector in the directions of the Cartesian axis Ox, Oy, Oz respectively (Ox is the eastward axis, Oy is the northward axis, Oz is the upwards axis), ρ is the density, f is the Coriolis parameter, K_H is the horizontal diffusion coefficient, K_Z^m is the coefficient of vertical diffusion of momentum, g is the gravitational acceleration, p is pressure.

Energy balance equation

$$\begin{aligned}\rho\left(\frac{\partial \theta}{\partial t} + u\frac{\partial \theta}{\partial x} + v\frac{\partial \theta}{\partial y} + w\frac{\partial \theta}{\partial z}\right) = \frac{\partial}{\partial x}\left(K_H\frac{\partial \theta}{\partial x}\right) \\ + \frac{\partial}{\partial y}\left(K_H\frac{\partial \theta}{\partial y}\right) + \frac{\partial}{\partial z}\left(K_Z^h\frac{\partial \theta}{\partial z}\right) + \frac{\theta}{c_p T}\left(Q_{rad} - \rho L_w \Phi\right).\end{aligned} \tag{5}$$

Here T is absolute temperature, θ is potential temperature, $\theta = T(p_0/p)^{R/c_p}(1 + 0.61q_V)$, c_p is heat capacity of air at a constant pressure, R is the gas constant, Q_{rad} is heating (cooling) of the atmosphere due to radiation propagating in the atmosphere, $\rho L_w \Phi$ is temperature change due to phase transitions of water in the atmosphere, K_Z^h is the coefficient of vertical diffusion of heat and moisture, L_w is vaporization heat, q_V is the specific humidity.

Equation of state

$$p = \rho RT; R = R_0\left[\frac{1 - q_V}{M_{air}} + \frac{q_V}{M_{H_2O}}\right]. \tag{6}$$

Moisture microphysics scheme

To model the moisture microphysics processes, the 6-class moisture microphysics scheme WSM6 [7], developed by Hong and Lim for the well-known mesoscale meteorological model WRF [2], is used in this approach. It considers six states of the moisture in the atmosphere (water vapor, cloud moisture, rain moisture, ice particles, snow, graupel (hail)). For every parameter of the moisture state, a transfer equation is used. Along with the advective transfer, various parameterizations of physical processes leading to phase transitions of the moisture states considered are accounted for.

The basic equations of WSM6 scheme are:

$$\rho\left(\frac{\partial q_j}{\partial t}+u\frac{\partial q_j}{\partial x}+v\frac{\partial q_j}{\partial y}+w\frac{\partial q_j}{\partial z}\right)+\frac{\partial\left(\rho V_j q_j\right)}{\partial z}$$
$$=\frac{\partial}{\partial x}\left(K_H\frac{\partial q_j}{\partial x}\right)+\frac{\partial}{\partial y}\left(K_H\frac{\partial q_j}{\partial y}\right)+\frac{\partial}{\partial z}\left(K_Z^h\frac{\partial q_j}{\partial z}\right)+\rho\Phi_j; \tag{7}$$
$$j=V,C,R,S,I,G.$$

Here q_V, q_C, q_R, q_S, q_I, q_G are the mass concentrations of water vapor (Vapor), cloud moisture (Cloud), rain moisture (Rain), snow (Snow), ice crystals (Ice), and graupel (Graupel) in the atmosphere. V_j is the deposition rate of j-th component ($V_V = V_C = 0$) [7]. Diffusion is taken into account only for gaseous components.

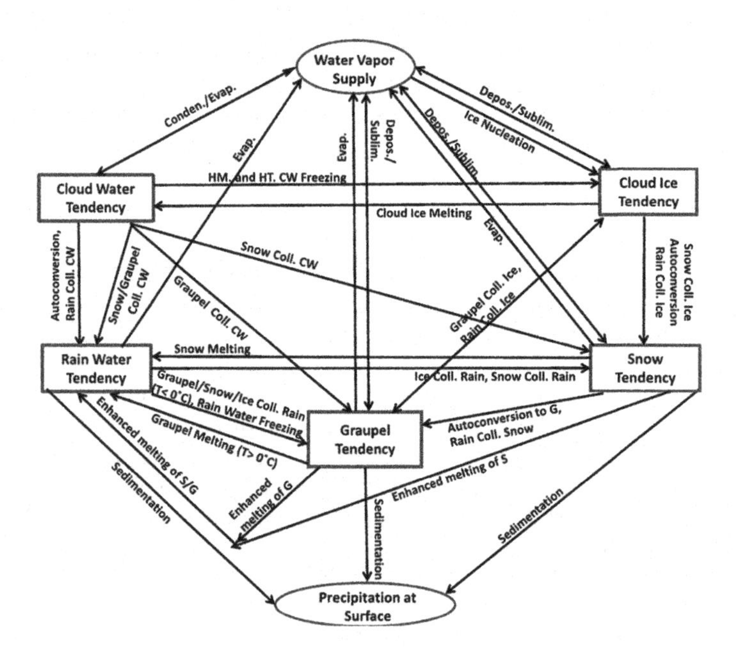

Fig. 1. Flowchart of the moisture microphysics processes in the WSM6 scheme [8]

The source terms Φ_j in the Eq. (7) for j-th class represent the parameterization of atmospheric moisture microphysics in accordance with Fig. 1. Some transitions take place when the air temperature is above zero, other when the temperature is below zero. WSM6 scheme (7) considers such processes as the capture of some components by others (accretion), melting of ice crystals, snow, graupel, evaporation/condensation of rain drops or cloud moisture, deposition/sublimation of graupel or snow, cloud-to-rain (ice crystals-to-snow or snow-to-graupel) autoconversion, evaporation/melting of snow, rain drop freezing with the formation of graupel etc.

A two-parameter turbulence model is used to close the Eqs. (1)–(7). It consists of equations for kinetic energy and turbulence scale [5], as well as algebraic relations to determine the eddy diffusion coefficient. The horizontal diffusion coefficients are estimated with Smagorinsky formulation.

Initial conditions for (1)–(7) are set based on the results of interpolation of meteorological parameters calculated with SL-AV global scale model [9], which is used by the Hydrometeorological Research Center of the Russian Federation for operational weather forecasting.

Boundary conditions on the upper boundary (~10 km) of the computational domain are:

$$\frac{\partial u}{\partial z} = \frac{\partial v}{\partial z} = 0; \; \frac{\partial q}{\partial z} = \frac{\partial q_j}{\partial z} = 0; \; w = 0; \; \frac{\partial \theta}{\partial z} = \gamma.$$

Radiation conditions [10] are used at the lateral boundaries. They take into account the trends in the mesoscale parameters based on the trends of the parameters of the global scale weather prediction model SL-AV of the Hydrometcenter of Russia [9];

Conditions corresponding to the main relations of the Monin–Obukhov similarity theory are set on the bottom boundary (Earth's surface) [11]. According to this theory, the flows of dynamic, thermodynamic, and turbulent parameters in the surface layer of the atmosphere are defined by the Obukhov length scale formulation: $L = -\frac{\theta V_*^3}{g\kappa\left(\theta' w'\right)_w}$. To find Earth surface temperature θ_w , one-dimensional unsteady heat transfer equation is solved taking into account the heat flow through the surface associated with the radiant heating (cooling) and turbulent fluxes of heat and moisture on the surface. The temperature of the lower soil layer at a depth of 2 m was also taken into account.

2.2 Numerical Method

The coordinate transformation (8) was applied to (1)–(7) before they were numerically solved by the finite volume method.

$$\begin{cases} \xi = x; \\ \eta = y; \\ \sigma = H\frac{z-h(x,y)}{H-h(x,y)}. \end{cases} \tag{8}$$

where H is the upper boundary of the research area, $h(x,y)$ is the terrain's surface elevation above sea level. Transformation (8) maps a three-dimensional domain with a curvilinear bottom boundary to a parallelepiped.

The differential problem obtained after the transformation (8) was approximated with the finite volume method and semi-implicit finite difference schemes. The main idea of this approach is to divide the spatial domain into disjoint, adjacent finite volumes so that each of them contains one grid point (i,j,k). A uniform in horizontal directions grid refining with a constant rate k towards the Earth's surface was used in computations.

After sampling research area, each differential equation of the mathematical model (1)–(7) is integrated by each finite volume and by the elementary time interval $(t^{m+1}-t^m)$, where m is the number of time step. Velocity components were defined at the finite volume faces (u at the faces perpendicular to the Ox axis, v at those perpendicular to the Oy axis, w to $O\sigma$), and scalar characteristics were defined in the cell centers. Piecewise polynomial approximations for dependent quantities were used to calculate integrals. Convective terms in the transfer equations were approximated with monotonized linear upwind (MLU) scheme of Van Leer [12].

The explicit Adams-Bashforth method and the implicit Crank-Nicolson method were used in this work to ensure the second-order approximation in time.

$$\begin{gathered}\Phi_h^{m+1}=\Phi_h^m+\Delta t_m\left(\frac{3}{2}L_h\left(\Phi_h^m\right)-\frac{1}{2}L_h\left(\Phi_h^{m-1}\right)\right)\\+\Delta t_m\left(\frac{1}{2}\Lambda_h\left(\Phi_h^{m+1}\right)+\frac{1}{2}\Lambda_h\left(\Phi_h^m\right)\right)+\Delta t_m\left(\frac{3}{2}S_h\left(\Phi_h^m\right)-\frac{1}{2}S_h\left(\Phi_h^{m-1}\right)\right).\end{gathered}\tag{9}$$

Here $\Phi_h^m=\left\{\Phi_{i,j,k}^m\right\}$ is the grid approximation of the scalar Φ defined by the (2–5), (7) ($\Phi=u,v,w,\theta,q_j$); L_h is the finite-difference approximation of the convective-diffusive operator in (2–5), (7) except for the vertical diffusion along the axis $O\sigma$; Λ_h is the approximation of the diffusion $\frac{\partial}{\partial z}\left(K_z\frac{\partial\Phi}{\partial z}\right)$, in vertical; $S(\Phi)$ are the source terms of (2–5), (7). The choice of method of time approximation of the differential problem is related to the fact that the Adams-Bashforth approximation scheme for the vertical diffusion leads to a more strict convergence condition. In addition, this approximation will not significantly complicate the computational algorithm of solving the resulting systems of linear algebraic equations. As a result, the following system of linear algebraic equations, which are conditionally numerically stable and are solved by a Tridiagonal Matrix Algorithm (TDMA) along vertical grid lines, is obtained.

To link the velocity and pressure vector fields, a predictor-corrector method was used at each step. In this scheme, the components of the velocity vector are first predicted with difference schemes of the form (9) with the value of the pressure gradient calculated from the values p_h^m from the m-th time step. Then, after solving the elliptic difference equation for pressure correction $p_h'=p_h^{m+1}-p_h^m$ the intermediate velocity and pressure fields are corrected so that the corrected values of the velocity components exactly satisfy the difference analogue of the continuity Eq. (1).

2.3 Parallel Algorithm

The decomposition of the domain chosen as the main parallelization strategy. All the values of grid function $\left\{\Phi_{i,j,k}^{m+1}\right\}$, $\left\{\Phi_{i,j,k}^{m}\right\}$, $\left\{\Phi_{i,j,k}^{m-1}\right\}$ belonging to different subdomains were sent to different processor elements (PE) along with the selected grid subdomain [13] (Fig. 2).

After the decomposition stage, when the processed data is distributed among PEs to build a parallel algorithm, we proceed to the stage of establishing connections between the units where calculations will be performed in parallel – communication planning. Due to the semi-implicit difference scheme pattern used, the values of the grid function from the neighboring adjacent processor element are required to calculate the next time step values at the near-boundary points of each subdomain. Therefore, ghost cells are created on each processor element to store data from a neighboring computing node (Fig. 2). These boundary values are sent to ensure uniformity of calculations in the internal points of the grid subdomain [13]. Grid pattern of the difference scheme (9) will contain two points on each side of the point with a number (i, j, k) because the MLU scheme was used to approximate convective terms. The MPI standard was used to send the values of grid functions calculated on each processor element to other processor elements and get data from them required for further computations.

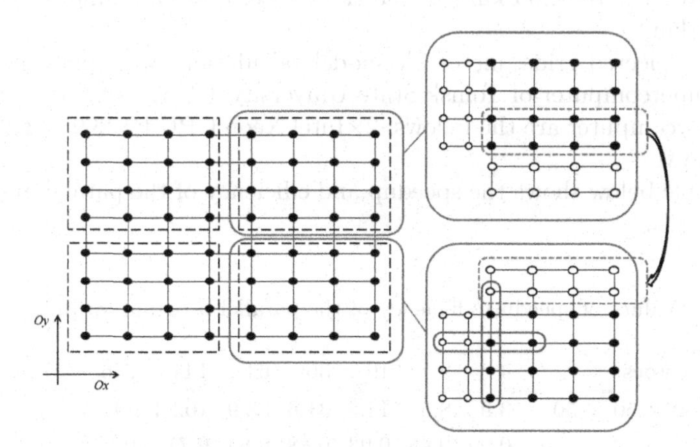

Fig. 2. Two-dimensional decomposition of the grid domain along the Ox and Oy directions and arrangement of data transfer between the subdomains

In this work, the TDMA method was used simultaneously along each vertical grid line with coordinates (i, j) to solve the difference Eq. (9). The line-by-line iterative Seidel method [13,14] was used to solve the difference equation for pressure correction p'_h. It relies on the red-black ordering of the computational grid points for each horizontal level k and implicit representation of the values of the grid function p'_h in the points $(i, j, k+1)$, (i, j, k), $(i, j, k-1)$:

$$\begin{gathered} -ab_{i,j,k}\left(p'\right)_{i,j,k-1}^{l+1} + ap_{i,j,k}\left(p'\right)_{i,j,k}^{l+1} - at_{i,j,k}\left(p'\right)_{i,j,k+1}^{l+1} = \\ = ae_{i,j,k}\left(p'\right)_{i+1,j,k}^{l} + an_{i,j,k}\left(p'\right)_{i,j+1,k}^{l} + aw_{i,j,k}\left(p'\right)_{i-1,j,k}^{l} \\ + as_{i,j,k}\left(p'\right)_{i,j-1,k}^{l} + b_{i,j,k}; i = \overline{1, Nx}; j = \overline{1, Ny}; k = \overline{1, Nz}. \end{gathered} \tag{10}$$

l is the number of iteration.

Parallel implementation of the solution of the SLAE (10) during calculations showed the independence of the iterative process convergence rate from the number of processor elements used. It is important that this implementation of the algorithm on a multiprocessor computing system completely preserves the property of the sequential algorithm and scales it to the available number of computing nodes very well.

Figure 3 and Table 1 show the speedup and efficiency of the parallel program of the mesoscale meteorological model for a local research area on various grids during two days of modelling: 50 × 50 × 50 points; 98 × 98 × 50 points; 194 × 194 × 50 points. The choice of grids for computations is determined by the size of the research area. In the first case, it is an area where a city with population of 1 million residents is located (the step size of the horizontal grid is 1–2 km, the area is 50–100 km), in the second case, it is an industrial district with a large number of enterprises with a large settlement in the center (the size of the research area is ~100–200 km), in the third case, it is an administrative region (up to 400 km).

For the selected grids, mesoscale model calculations were performed on the Cyberia supercomputer of Tomsk State University. Characteristics of the nodes of the supercomputer are the follows: 2×Intel Xeon CPU E5-2695v 3@2.30 Ghz, 256 Gb RAM.

The table below shows the speedup and efficiency of the parallel program for different grids.

Table 1. Values of speedup/efficiency of the parallel program for different grids

Nprocs	4	9	16	36	64	144	256	576
50 × 50 × 50	3.9	8.5	14.8	31.6	53.9	102.4	145.5	–
	0.99	0.95	0.93	0.88	0.84	0.71	0.57	–
98 × 98 × 50	3.9	8.7	15.3	33.0	57.9	121.5	199.6	319.7
	0.99	0.97	0.96	0.92	0.90	0.84	0.78	0.55
194 × 194 × 50	4.0	8.9	15.6	33.7	60.2	130.6	226.4	447.8
	1.00	0.99	0.98	0.94	0.94	0.91	0.88	0.78

Figure 3 shows that parallel algorithm demonstrates good scalability and high efficiency for a fairly large number of processes (up to 256) on large grids. The speedup of the parallel program ensure fulfillment of the requirement of prognostic calculations for research and early weather prediction: to get a forecast for the next day for 1 h of program operation time on a supercomputer.

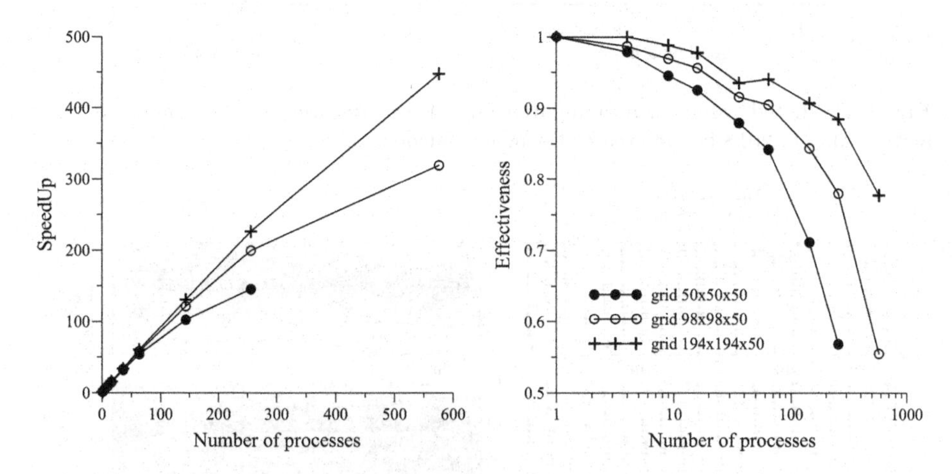

Fig. 3. Speedup and efficiency of the parallel program for calculating mesoscale meteorological processes over the airport

3 Results

Meteorological conditions over the city of Tomsk (85.0°E, 56.5°N, city center) and Bogashevo airport (85.2°E, 56.3°N) were studied with the approach presented.

The results of computations were compared with the measurements made by the instruments of the Shared Research Facilities 'Atmosphere' of V.E. Zuev Institute of Atmospheric Optics of the Siberian Branch of the RAS (https://www.iao.ru/ru/structure/juc) for different seasonal conditions [15].

The ability of the mesoscale model to predict precipitation in the conditions of the Siberian region was also evaluated by computations made for selected historical dates in 2016–2018, when heavy precipitation was observed on the territory of Bogashevo airport and registered by standard weather observations.

Figure 4 demonstrates that the first period of time is characterized by precipitation in the form of snow. Observations in the airport (http://rp5.ru) indicate continuous snow precipitation during the day with an enhancement in the time intervals from 7:00 to 9:30, and from 17:00 to 18:30. For the second date, snowfall was recorded by meteorologists until 10:30, then, until 17:00, only liquid precipitation was observed. After 17:00, there were both light snow and rain. In general, the model satisfactorily predicted the pattern of precipitation changes observed during the day for both dates.

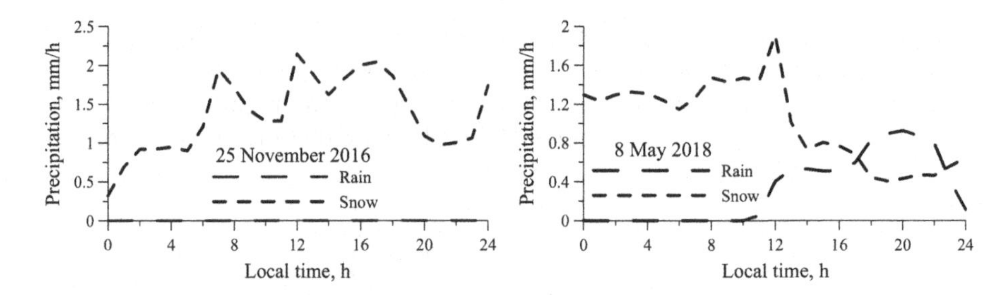

Fig. 4. Predicted values of accumulated for 1 h precipitation on November 25, 2016 and on May 8, 2018 in the area of Bogashevo airport

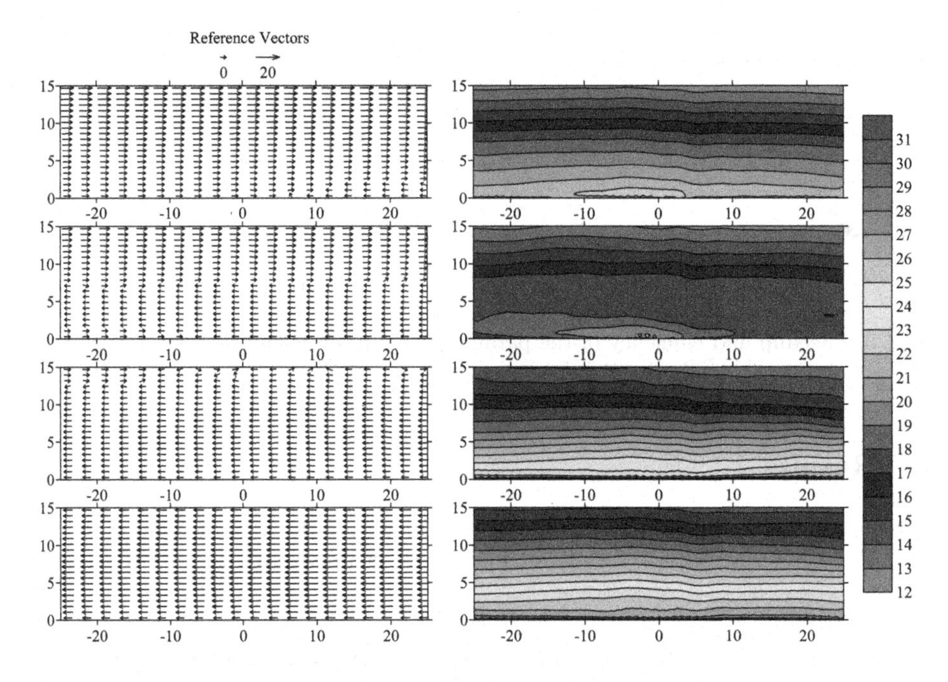

Fig. 5. Isotherms and velocity vector field in the vertical plane passing in the direction "West-East" through the center of Tomsk (56.5° N and 85° E) for August 18, 2015 at 0, 6, 12, 18 (from top to bottom) o'clock of local time calculated with the mesoscale model. Horizontal dimensions of the area are in kilometers, while vertical dimensions are in hundreds of meters

The capability of the mesoscale model to predict urban heat island were tested for different weather conditions in Tomsk [15]. The results of computations made for August 18, 2015 are given below. During this day, warm, dry, mostly cloudless or clear weather was observed, with a weak North-East-East wind of 1 m/s. Against the background of a high atmospheric pressure, the diurnal variation of air temperature (maximum 27 °C, minimum 13 °C) and relative humidity (50–96%) is well-defined. The results of computations are shown in

Fig. 5. It is clear that the urban heat island appeared in the night and early morning hours. It is evident in the morning (intensity about 2 °C, horizontal dimensions - 35 km, vertical ones - 400 m), and less evident – at 00 o'clock (increment of temperature – 1 °C, horizontal dimensions – 15 km, vertical ones – about 200 m). Due to the Eastern transfer, its position is shifted to the West relative to the city center.

4 Conclusion

The basic equations of the mesoscale meteorological model that is being developed for studying and predicting local weather conditions over a limited area are presented. A numerical method to solve the differential problem is proposed. It is based on the finite volume method, which ensures implementation of integral conservation laws, and semi-implicit difference schemes that are second-order accurate in time and space. The predictor–corrector procedure was used to calculate the velocity vector field satisfying the finite difference continuity equation. The TDMA and the line-by-line Seidel methods with red-black ordering of the grid nodes in the x-y plane were used to solve the grid equations obtained as a result of approximation. Two-dimensional latitude-longitude decomposition of the domain is the basis for an effective parallel algorithm of solving the problem with a good scaling degree. Computational experiments performed on the supercomputer Cyberia of Tomsk State University showed that the developed parallel algorithm ensures the requirement for prognostic calculations of local weather conditions. The parallel implementation of the mesoscale meteorological model was successfully applied to predicting heavy precipitation and studying the urban heat island phenomenon. The further development of the mesoscale meteorological model is focused on introducing modern parameterizations for physical processes into it and applying hybrid parallel computing technologies to speedup computations.

Acknowledgements. The computations presented in this article were carried out at the Interregional Supercomputer Center of the Tomsk State University. This study was carried out at Tomsk State University and supported by the Russian Science Foundation (project No. 19-71-20042).

References

1. Powers, J.G., et al.: The weather research and forecasting model: overview, system efforts and future directions. Bull. Am. Meteorol. Soc. **98**(8), 1717–1737 (2017). https://doi.org/10.1175/BAMS-D-15-00308.1
2. Ridwan, R., Kistijantoro, A.I., Kudsy M., Gunawan, D.: Performance evaluation of hybrid parallel computing for WRF model with CUDA and OpenMP. In: 2015 3rd International Conference on Information and Communication Technology (ICoICT), Nusa Dua, pp. 425–430. IEEE Press (2015). https://doi.org/10.1109/ICoICT.2015.7231463

3. Bermous, I., Steinle, P.: Efficient performance of the Met Office Unified Model v8.2 on Intel Xeon partially used nodes. Geosci. Model Dev. **8**, 769–779 (2015). https://doi.org/10.5194/gmd-8-769-2015
4. Mortikov, E.V.: Software implementation of impurity transfer unit in climate models based on hybrid programming MPI-OpenMP. In: Russian Supercomputing Days: Proceedings of the international conference, pp. 521–529. Publishing House of Moscow State University, Moscow (2016). https://doi.org/10.29003/m680.RussianSCDays
5. Starchenko, A.V., Bart, A.A., Bogoslovskiy, N.N., Danilkin, E.A., Terenteva, M.A.: Mathematical modelling of atmospheric processes above an industrial centre. In: Proceedings of SPIE, 20th International Symposium on Atmospheric and Ocean Optics: Atmospheric Physics, Novosibirsk, vol. 9292, pp. 929249-1–929249-30. SPIE (2014). https://doi.org/10.1117/12.2075164
6. Pielke, R.: Mesoscale Meteorological Modeling. Academic Press, San Diego (2002)
7. Hong, S.-Y., Lim, J.-O.J.: The WRF single-moment 6-class microphysics scheme (WSM6). J. Korean Meteorol. Soc. **42**(2), 129–151 (2006)
8. Bao, J.-W., Michelson, S.A., Grell, E.D.: Pathways to the production of precipitating hydrometeors and tropical cyclone development. Mon. Weather Rev. **144**(6), 2395–2420 (2016). https://doi.org/10.1175/MWR-D-15-0363.1
9. Tolstykh, M.A.: Semi-Lagrangian high-resolution atmospheric model for numerical weather prediction. Russ. Meteorol. Hydrol. **4**, 1–9 (2001)
10. Carpenter, K.: Note on the paper "radiation condition for the lateral boundaries of limited-area numerical models" by M. Miller, A. Thorpe (107, 615–628). J. R. Meteorol. Soc. **108**, 717–719 (1982)
11. Monin, A.S., Obukhov, A.M.: Basic laws of turbulent mixing in the surface layer of the atmosphere. Tr. Akad. Nauk SSSR Geofiz. **24**, 163–187 (1954)
12. Van Leer, B.: Towards the ultimate conervative difference scheme. II. Monotonicity and conservation combined in a second order scheme. J. Comput. Phys. **14**, 361–370 (1974)
13. Starchenko, A.V., Bertsun, V.N.: Parallel Computing Methods. Publishing house, Tomsk (2013)
14. Ortega, J.O.: Introduction to Parallel and Vector Solution of Linear System. Plenum Press, New York (1988)
15. Starchenko, A.V., Bart, A.A., Kizhner, L.I., Odintsov, S.L., Semyonov, E.V.: Numerical simulation of local atmospheric processes above a city. In: Proceedings of SPIE, 25th International Symposium on Atmospheric and Ocean Optics: Atmospheric Physics, Novosibirsk, vol. 11208, pp. 112088H-1–112088H-9. SPIE (2019). https://doi.org/10.1117/12.2541630

Parallel Efficiency of Time-Integration Strategies for the Next Generation Global Weather Prediction Model

Vladimir Shashkin[1,2,3](✉) and Gordey Goyman[1,2,3]

[1] G.I. Marchuk Institute of Numerical Mathematics, RAS, 8, Gubkina st., Moscow, Russia
v.shashkin@inm.ras.ru, gordeygoyman@gmail.com
[2] Hydrometeorological center of Russia, 11-13, Bol.Predtechenskiy per., Moscow, Russia
[3] Moscow Institute of Physics and Technology, 9, Institutskiy per., Dolgoprudny, Moscow Region, Russia

Abstract. Next generation global numerical weather prediction models will have horizontal resolution of 3–5 km that lead to the problem of about 10^{10} degrees of freedom. To meet operational requirements for medium-range weather forecast, $O(10^4)$-$O(10^5)$ processor cores have to be used efficiently. The non-hydrostatic equation set will be used so models have to treat efficiently a number of fast-propagating wave families. Therefore, time-integration scheme is crucial for the scalability and computational efficiency.

The next generation global atmospheric model is currently under development at INM RAS and Hydrometcentre of Russia. To choose the time integration scheme for the new model, we evaluate scalability and efficiency of several options (including exponential propagation integrator) using linearized equation set. We also present a parallel framework developed for the solution of atmospheric dynamic equations on the spherical cube grid.

Keywords: Numerical weather prediction · Time-stepping scheme · Exponential propagation integrator · Parallel efficiency · Cubed-sphere

1 Introduction

Current global atmospheric models for medium-range weather forecast (3–10 days) have horizontal grid spacing greater or equal to 10 km. The next generation models that are currently developed by most of forecast centers will have grid spacing of about 3–5 km. This will allow to reproduce mesoscale meteorological phenomena that are responsible for the significant part of extreme weather events. Previously, this was available only for regional atmospheric models. Removing errors due to boundary conditions will increase useful leadtime and coverage of extreme events forecast.

V. Voevodin and S. Sobolev (Eds.): RuSCDays 2020, CCIS 1331, pp. 285–296, 2020.
https://doi.org/10.1007/978-3-030-64616-5_25

A number of problems should be solved to enable practical global weather forecast with horizontal resolution finer than 10 km. The first is scalability challenge. The models have to use efficiently $O\left(10^4\right)$-$O\left(10^5\right)$ processor cores to meet operative requirements for the forecast calculation speed [4]. The second is using non-hydrostatic Euler equation set to deal with mesoscale phenomena. This greatly increases stiffness and non-linearity of resulting discrete equations as compared to the previously used equations with the assumption of vertical hydrostatic balance. The third problem is the convergence of the meridians to the pole points in the latitude-longitude geographic grid. Latitude-longitude grid with horizontal spacing of several kilometers at the equator will have typical longitudinal spacing of tens of meters near the poles. Thus, quasi-uniform spaced spherical meshes that are usually unstructured meshes or non-orthogonal curvilinear semi-structured or overset grids are to be used.

The computational performance and accuracy of the whole atmospheric model crucially depends on the properties of the time-integration scheme used. The scheme should optimize the cost of one step to step-size ratio. Practically, this implies that scheme must be implicit at least for the vertically-propagating sound waves, which are responsible for the biggest absolute eigenvalues of discrete equation system with typical CFL of order 10^3. From the accuracy point of view, time integration scheme should not considerably retard the inertia-gravity waves that are of the primary importance for mesoscale and middle atmosphere meteorology [11].

Considerable research on the time-integration schemes for the atmospheric models was done in the last years [16]. Mostly, implicit-explicit (IMEX) Runge-Kutta type schemes with the implicitly treated vertical sound operator were studied (HEVI schemes) [8]. Classical semi-implicit schemes that treat implicitly all horizontal and vertical fast wave terms are not abandoned [5,15]. Recently, the progress in using exponential propagation integrators [10] was achieved in the context of atmospheric modeling [14]. Exponential propagation integrators are interesting for their high accuracy. Using exponentiation of model operator matrix allows to achieve stability and wave propagation properties determined only by the spatial discretization scheme at any time-step size.

In this article, we investigate the parallel properties of time-schemes for the next generation numerical weather prediction global atmospheric model that is currently developed in INM RAS and Hydrometcentre of Russia. HEVI and exponential schemes are considered and compared with baseline fourth order 4 stage explicit Runge-Kutta scheme. The test problem is linearized Euler equations set on the sphere which describes horizontally and vertically propagating sound and gravity waves. The equations are discretized on the quasi-uniform cubed-sphere grid. The parallel framework for computations on the cubed-sphere grid is also presented.

The article is organized as follows: we present the linearized Euler equations in Sect. 2. The exponential time-integration method and IMEX HEVI scheme are presented in Sect. 3 and Sect. 4 respectively. The cubed-sphere grid and equations discretization are shortly described in Sect. 5, Sect. 6 presents the parallel framework for computations including the algorithm for the domain decomposition. Test results are given in Sect. 7, conclusions are drawn in Sect. 8.

2 Linearized Euler Equations

The Euler equations system in the shallow atmosphere approximation [11] and absence of Coriolis force write as

$$\frac{\mathrm{d}\boldsymbol{v}}{\mathrm{d}t} = -C_p\theta\nabla P, \quad \frac{\mathrm{d}w}{\mathrm{d}t} = -C_p\theta\frac{\partial P}{\partial z} - g, \tag{1}$$

$$\frac{\mathrm{d}P}{\mathrm{d}t} = -\frac{R}{C_v}P\left(\nabla\cdot\boldsymbol{v} + \frac{\partial w}{\partial z}\right), \quad \frac{\mathrm{d}\theta}{\mathrm{d}t} = 0,$$

where $\boldsymbol{v}$ is horizontal velocity vector, $\theta = T(p_0/p)^\kappa$ is potential temperature, T is temperature, p is pressure, p_0 is reference constant, $\kappa = R/C_p$, R is dry air gas constant, C_p is heat capacity at constant pressure, $P = (p/p_0)^\kappa$ is Exner pressure, w is vertical velocity, g is gravity acceleration, C_v is heat capacity at constant volume, ∇ and $\nabla\cdot$ are horizontal gradient and divergence operators.

We linearize equations set (1) with respect to the background state $\boldsymbol{v}_b = 0$, $w_b = 0$, $P_b = P_b(z)$, $\theta_b = \theta_b(z)$ in the hydrostatic balance $\theta_b\partial P_b/\partial z = -g$. The resulting linear system is as follows ($\boldsymbol{v}$, w, P and θ are now deviations from the background state):

$$\frac{\partial\boldsymbol{v}}{\partial t} = -C_p\theta_b\nabla P, \tag{2}$$

$$\frac{\partial w}{\partial t} = -C_p\theta\frac{\partial P_b}{\partial z} - C_p\theta_b\frac{\partial P}{\partial z}, \tag{3}$$

$$\frac{\partial P}{\partial t} = -\frac{R}{C_v}P_b\left(\nabla\cdot\boldsymbol{v} + \frac{\partial w}{\partial z}\right) - w\frac{\partial P_b}{\partial z}, \tag{4}$$

$$\frac{\partial\theta}{\partial t} = -w\frac{\partial\theta_b}{\partial z}. \tag{5}$$

3 Exponential Propagation Time-Integration Scheme

Consider non-linear autonomous equation system in the form

$$\frac{\partial y}{\partial t} = A(y) = Ly + N(y), \tag{6}$$

where y is an arbitrary state vector, A is non-linear operator with linear part L and non-linear residual N. One can write the following analytical expression

$$y(t+\Delta t) = \exp(\Delta tL)\left(y(t) + \int_0^{\Delta t}\exp(-\tau L)y(t+\tau)\mathrm{d}\tau\right). \tag{7}$$

The idea of exponential propagation integrators is to adopt Adams-Bashfort or Runge-Kutta explicit approximation for $y(t+\tau)$ in the integral term [6].

In the case when the norm of L is much bigger than the norm of N (which is typical for the atmospheric modeling) this will result in very accurate time scheme with time-step not limited by CFL condition. We don't need any approximations of integral term in our case, because we consider the linear system.

The main problem of exponential propagation integrator implementation is the calculation of action of large sparse matrix exponent-like function on the state vector. Based on the recent progress in computational linear algebra [13,17], the works [9,14] developed the exponential integration scheme for the shallow water equations on the sphere which is claimed to be as efficient as the semi-implicit scheme. We use Krylov subspace method with incomplete orthogonalization to compute the action of matrix exponential following these works. The algorithm to compute $q = \exp(L\Delta t)y$ can be summarized as follows:

$\beta = \|\boldsymbol{y}\|$
$\boldsymbol{q}_1 = \boldsymbol{y}/\beta$
for $i = 2, M+1$ **do**
  $\boldsymbol{q}_i = L\boldsymbol{q}_{i-1}$
  for $j = \max(1, i-2), i-1$ **do**
    $\mathsf{H}_{j,i-1} = (q_i \cdot q_j)$
    $\boldsymbol{q}_i = \boldsymbol{q}_i - \boldsymbol{q}_j(\boldsymbol{q}_i \cdot \boldsymbol{q}_j)$
  end for
  if $i < M+1$ **then**
    $\mathsf{H}_{i,i-1} = \|\boldsymbol{q}_i\|$
    $\boldsymbol{q}_i = \boldsymbol{q}_i/\|\boldsymbol{q}_i\|$
  end if
end for
$\mathsf{E} = \exp(\Delta t \mathsf{H})$ { exponential of small $M \times M$ matrix}
$\boldsymbol{q} = 0$
for $i = 1, M$ **do**
  $\boldsymbol{q} = \boldsymbol{q} + \beta \mathsf{E}_{i,1}\boldsymbol{q}_i$
end for

The required size of Krylov subspace M is about $3CFL_{max}$.

This algorithm require two scalar products and one vector norm computation per Krylov vector. We compute them locally at each MPI-process and then calculate global sums simultaneously to avoid latency of global communications initialization. The orthogonalization process formulae are modified correspondingly.

4 Horizontally Explicit Vertically Implicit Time Integration Scheme

The idea of horizontally explicit vertically implicit time-integration methods is to split the RHS of system of Eqs. (1) into two operators: V containing terms responsible for sound waves propagation in vertical (mainly, the terms with vertical derivatives) and H containing all other terms. Then, one of implicit-explicit (IMEX) Runge-Kutta (RK) methods with double Butcher table is applied (eg. [3,12]).

$(I - \Delta tV)$-type operator needs to be inversed for each vertical column of model cells at each stage of the method. Solution of only vertically coupled equation systems do not spoil the parallel properties of the scheme in the typical case, when 1D or 2D domain decomposition in horizontal is used. At the same time, the most severe CFL restriction for vertically propagating sound waves is circumvented.

We use 3-rd order ARS343 scheme [3] with 4 explicit and 3 implicit stages. Equations. (2)–(5) are split so that H includes only RHS of horizontal wind Eq. (2) and V contains all other terms. Although V contains horizontal divergence term, the implicitly treated $\boldsymbol{v}$ is found from horizontal wind equation and excluded from Exner pressure equation. The resulting vertically coupled linear system is reduced to one equation for w with tridiagonal matrix, which is solved using Gauss-elimination.

5 Spatial Discretizations

We use the cubed-sphere grid [19], obtained by the gnomonic (central) projection of the grid on the cube faces to the inscribed sphere [18] (see Fig. 1(a)). The grid on the cube's faces is equally spaced rectangular grid in angular coordinates, such that $x = \tan\alpha$, $y = \tan\beta$, $\alpha, \beta \in [-\pi/4, \pi/4]$.

The covariant components of horizontal gradient ∇P are given by $\partial P/\partial\alpha$ and $\partial P/\partial\beta$. Horizontal divergence, using contravariant components $\tilde{u}$, $\tilde{v}$ of $\boldsymbol{v}$ is written as

$$\nabla \cdot \boldsymbol{v} = \frac{1}{G}\left(\frac{\partial G\tilde{u}}{\partial\alpha} + \frac{\partial G\tilde{v}}{\partial\beta}\right), \tag{8}$$

where G is square root of metric tensor determinant (more details available in [18]).

Equations (2), (4) are discretized in horizontal using C-staggered grid [2], second order formulae are used for ∇ and $\nabla\cdot$ operators. Transformation between the covariant and contravariant horizontal vector components are carried out using bilinear interpolation. In vertical, we discretize Eqs. (2)–(5) on staggered Charney-Phillips grid where P and $\boldsymbol{v}$ are defined at cell centers, w and θ are defined at cell interfaces (including lower and upper boundaries). Charney-Phillips grid discretization gives the best numerical dispersion properties for all atmospheric wave families [20].

6 Parallel Framework on the Cubed-Sphere Grid

Development of a new global atmospheric model is a long process, which usually takes about 5–10 years before the model is put into operational usage. Therefore, it is reasonable to develop an own software infrastructure that takes into account the features of a grid, computational methods and a specific model scope. Such a software infrastructure provides model engineers and users with key building blocks for the model development and maintenance (data structures, parallel communication routines, model I/O and diagnostics, etc.). For instance, such modeling frameworks are developed at the European Centre for Medium-Range Weather Forecasts [7] and at the United Kingdom's Met Office [1].

One of the fundamental components of such infrastructure is a parallel framework that defines the data storage structure and parallel processes interaction scheme. In this paper, we present the parallel framework called ParCS (Parallel Cubed Sphere) that is planned to be used for the further development of the new atmospheric model. ParCS is an object-oriented parallel infrastructure library written in Fortran programming language that provides algorithms and structures for the domain decomposition, parallel exchanges and data storage at the logically-rectangular computational grids.

The key abstract concepts of the ParCS library are:

- Tile: the elementary unit of a grid partition which is a rectangular section of the grid non-overlapping with other tiles;
- Partition: data structure containing the list of tiles and their distribution among parallel processes;
- Grid_function: a container of given quantity discrete values;
- Exchange: an object responsible for parallel communication between MPI-processes with a given dependency pattern.

The computational domain within this framework is decomposed into the set of tiles, which are defined via local vertex coordinates i, j and the cubed sphere panel index. The number of tiles may exceed the number of computational units. These tiles are distributed among MPI-processes thereby forming a domain partition. Each MPI-process can handle different number of tiles, which allows, if necessary, to perform static computational work balancing. Halo zones (overlap regions) between tiles can be used to enable operations requiring data outside the tile.

The Grid_function object implements distributed storage of a given scalar or vector field values with a prescribed width of a halo region. Each process initializes arrays of Grid_function instances, each instance responsible for the storage at a given tile and contains 3D array as well as some meta data.

Finally, the Exchange class provides interfaces for parallel communication operations between MPI-processes. To perform a halo-exchange of a given width or a gather/scatter exchange, an instance of this class is initialized and it can invoke parallel communication operation repeatedly as needed.

We use simple partitioning strategy for this work with the number of tiles and MPI-processes equal to $6pq$, where p and q – integers (ideally $p = q$). Each panel of the cubed sphere grid is partitioned into pq tiles using standard 2D geometric domain decomposition. Figure 1(b) illustrates example of such partition with $6 \cdot 8 \cdot 4$ tiles.

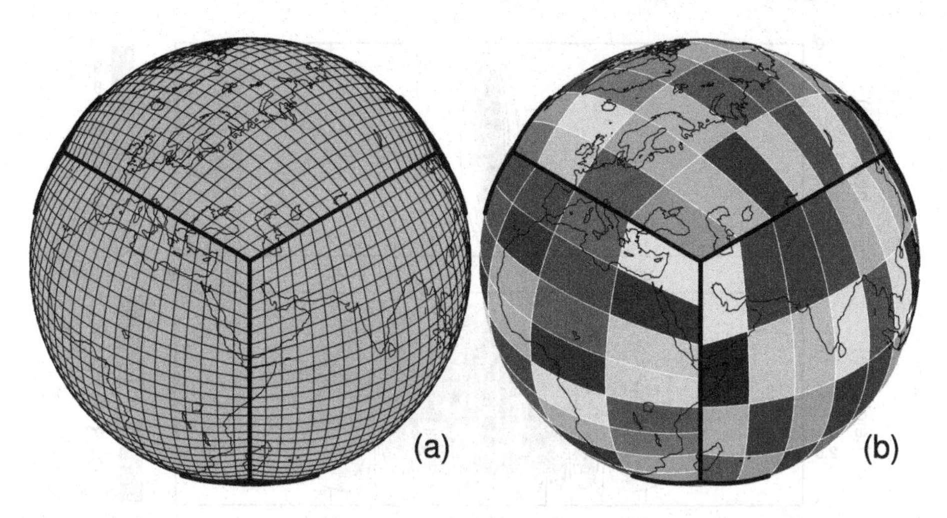

Fig. 1. (a) Cubed sphere grid with 24 cells along the cubes edge (C24). (b) Partition of cubed-sphere grid between 192 MPI-processes (the colors can repeat). (Color figure online)

7 Test Results

We run gravity wave propagation test from Dynamical Core Model Intercomparison Project (DCMIP) [21] to investigate parallel properties of Krylov-exponential and HEVI time-integration schemes. Four stage fourth order explicit Runge-Kutta scheme is used as baseline for comparison. All schemes with all parameters sets produce very similar results (see Fig. 2) that are close to the reference simulations. The tests were carried out at the Roshydromet Main Computer Center (MCC) Cray XC40 system.

The test is carried out at three grids: C128L10, C512L10, C512L200, where numbers after C and L denote the number of grid points along cube edge and the number of vertical levels correspondingly The number of degrees of freedom is about 3.9×10^6 for C128L10, 6.3×10^7 for C512L10 and 1.2×10^9 for C512L200. The Earth radius is reduced by the factor of 125 in the C128L10 experiment according to the test recommendations [21]. We reduce the radius by factor of 31.25 in C512 experiments to keep characteristic horizontal grid-spacing of about 600 m. The vertical grid spacing is 1000 m in the L10 experiments and 50 m in L200 experiment. C512L200 configuration is the closest to the real atmospheric model.

The stability of the methods in L10 experiments is limited by the horizontally propagating gravity and sound waves (that is rather untypical for the atmospheric models). The time-step used is 1 s for the RK4 and HEVI methods and 10 s for the exponential method (CFL of about 1 and 10 at the smallest grid cell respectively). The time-step sizes choice is nearly optimal in terms of computation speed for each scheme. Krylov subspace dimension of 30 is used in exponential method. The stability in L200 experiment is limited by the verti-

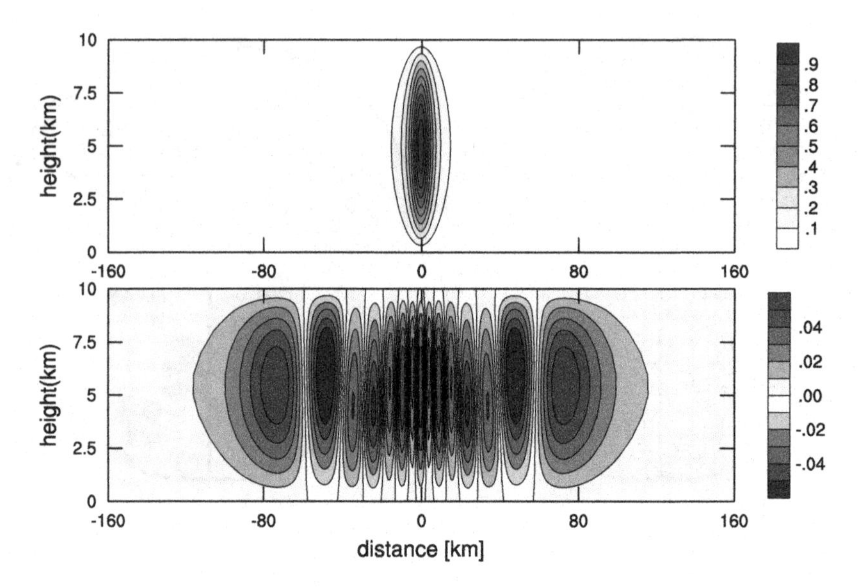

Fig. 2. Equatorial vertical cross-section of potential temperature deviation (K) from background state in gravity wave test case. Upper panel – initial conditions, lower panel – numerical solution at $t = 2400\,\mathrm{s}$.

cally propagating sound waves. The time-step size is 0.2 s for the RK4 scheme, 1 s for the HEVI method and 1.5 s for exponential method (vertical CFL 1.3, 6.9, 10.3 respectively).

The speed of computations for all three grid configurations is plotted in Fig. 3. All three methods accelerates superlinearly in the low dimensional configuration of C128L10 until the tile size of 16×16 points in horizontal (384 cores). Super-linear acceleration is caused by the better cache use efficiency with the smaller number of degrees of freedom per processor. With the greater number of cores computation speed still increase, but less efficiently. RK4 method stops accelerating with horizontal tile size of 12×16 points (1152 cores) that seems reasonable for finite-difference discretization. Exponential and HEVI methods slowly accelerate up to 1536 cores (8×8 tiles). Better scalability in the HEVI case as compared to RK4 is most likely the result of greater ratio of computations to data exchanges. Acceleration of EXP method between 1152 and 1536 cores is the result of better CPU cash use in the final sum of Krylov algorithm as tile sizes decrease. However, the acceleration of EXP method at 1536 cores relative to 24 cores is smaller than the RK4 one.

Figure 3 proves the efficient acceleration of HEVI and RK4 methods in C512 configurations up to 4992 processor cores that corresponds to the horizontal tile size of about 16×19 points. We believe the methods can use efficiently the greater number of cores, but only 4992 cores are currently available for tests. The exponential method in C512L10 configuration stops accelerating with number of cores greater than 4224 as the time spent for collective communication

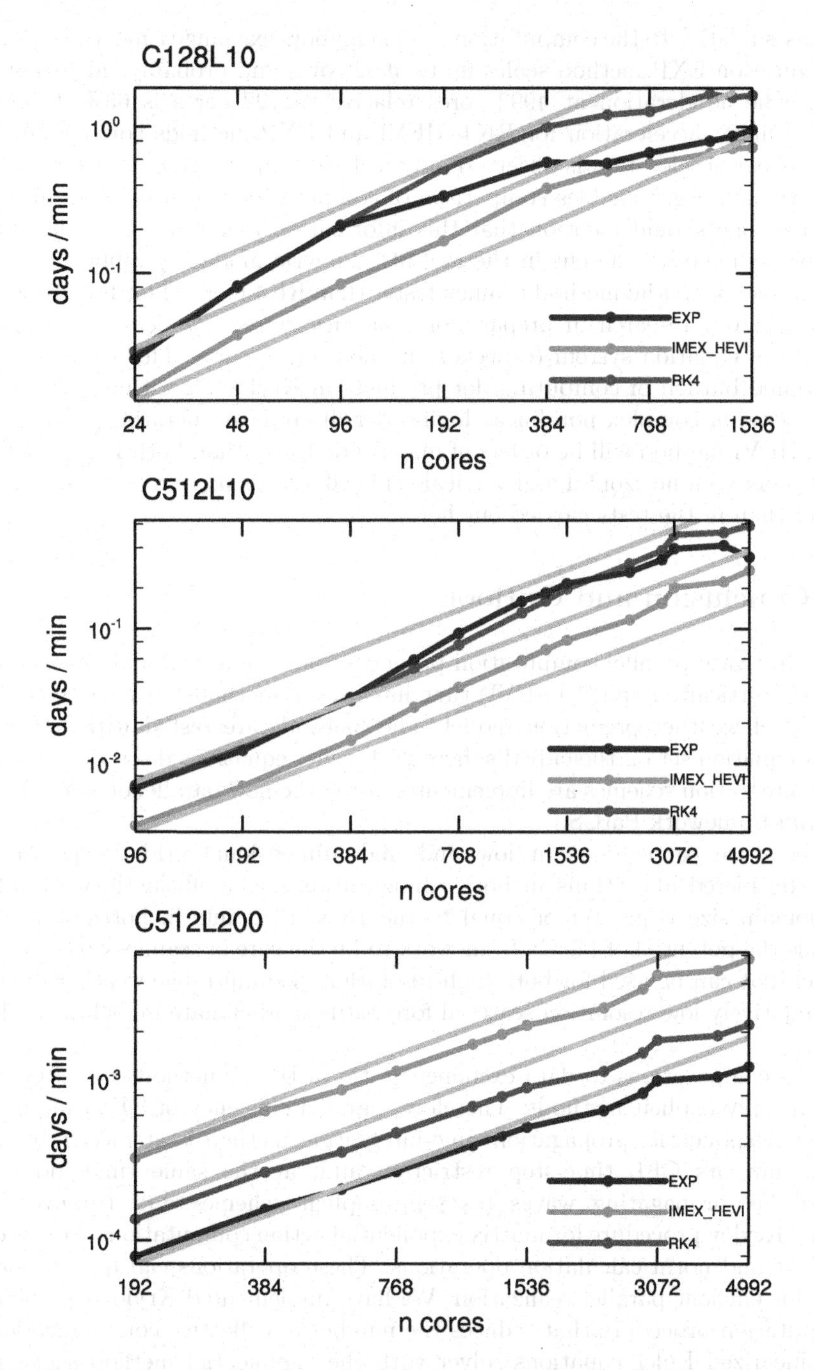

Fig. 3. Computation speed in forecasted days per minute of wall time versus number of used processor cores for three time-stepping schemes in three grid configurations. Gray lines show linear parallel acceleration.

increases relative to the computations and neighbors exchange time. In C512L200 configuration EXP method scales up to 4992 cores and probably this is not the limit. The acceleration at 4992 cores relative to 192 cores is 63%, 67% and 53% of linear acceleration for RK4, HEVI and EXP methods correspondingly. Superlinear acceleration is observed only in L10 configuration.

Although Fig. 3 enables comparison of computation speed between different schemes, one should mention that this information is not reliable. The things can be completely different in the real atmospheric modeling applications. For instance, exponential method is much faster than RK4 method for full non-linear problems (our research in preparation), because it requires less computations of RHS of equation system (especially its non-linear part). This outweighs the additional burden of computing dot products in Krylov algorithm from Sect. 3 in the case of complex non-linear high-order discretized operator. At the same time, HEVI method will be orders of magnitude faster than both RK4 and EXP if ratio between horizontal and vertical grid cell sizes is more realistic (i.e. much bigger than in the tests carried out here).

8 Conclusion and Outlook

We investigate parallel computation properties of exponential and horizontally-explicit vertically-implicit (HEVI) time integration methods for next generation numerical weather prediction model. The methods are tested with linearized Euler equation set on the cubed-sphere grid. Euler equations discretizations and time-integration schemes are implemented using the new parallel object-oriented Fortran framework ParCS.

Tests are carried out in low and high dimensional grid configurations. The considered algorithms in both configurations scale efficiently while MPI-subdomain size is greater or equal to the 16×16 points in horizontal. That reveals the potential of ParCS framework to be the core of seamless atmospheric model that can be used for both high-resolution medium-range weather forecast and relatively low-resolution seasonal forecasting and climate modeling applications.

The computations to data-exchanges ratio in HEVI method is greater than that in fully explicit methods. Therefore, parallel efficiency of HEVI method is higher. Exponential propagation time-integration method is attractive because it circumvents CFL time-step restriction and, at the same time, does not retard fast-propagating waves (as semi-implicit schemes do). However, the robust Krylov procedure for matrix exponential action computation involves dot-product and norm calculation operations. These operations can be the bottleneck for efficient parallel realization. We have implemented Krylov-exponential computation procedure that reduces the number of collective communications.

Linearized Euler equations solver with the exponential method scales efficiently up to the maximum available number of processor cores in the most realistic test configuration. We consider its efficiency as good, however it is slightly worse than the efficiency of explicit and HEVI methods. This result encourages further research in the application of exponential methods to the atmospheric equations. The primary task here is to couple the exponential method with implicit treatment of vertically propagating sound waves. Given the vertical sound-waves CFL of order of 1000 (typical for real atmospheric models) this is needed to use time step size and Krylov subspace dimension that will allow practical simulations. At the same time, gravity waves that are of greater meteorological importance than sound waves won't be retarded.

The work was carried out at G.I. Marchuk Institute of Numerical Mathematics of Russian Academy of Sciences with support of Russian Science Foundation grant 19-71-00160.

References

1. Adams, S., et al.: LFRic: meeting the challenges of scalability and performance portability in weather and climate models. J. Parallel Distrib. Comput. **132**, 383–396 (2019). https://doi.org/10.1016/j.jpdc.2019.02.007. http://www.sciencedirect.com/science/article/pii/S0743731518305306
2. Arakawa, A., Lamb, V.: Computational Design of the Basic Dynamical Processes of the UCLA General Circulation Model, vol. 17, pp. 173–265. Academic Press, New York (1977)
3. Ascher, U.M., Ruuth, S.J., Spiteri, R.J.: Implicit-explicit Runge-Kutta methods for time-dependent partial differential equations. Appl. Numer. Math. **25**(2–3), 151–167 (1997). https://doi.org/10.1016/S0168-9274(97)00056-1
4. Bauer, P., Thorpe, A., Brunet, G.: The quiet revolution of numerical weather prediction. Nature **535**, 47–55 (2015). https://doi.org/10.1038/nature14956
5. Benacchio, T., Klein, R.: A semi-implicit compressible model for atmospheric flows with seamless access to soundproof and hydrostatic dynamics. Mon. Weather Rev. **147**(11), 4221–4240 (2019). https://doi.org/10.1175/MWR-D-19-0073.1
6. Beylkin, G., Keiser, J.M., Vozovoi, L.: A new class of time discretization schemes for the solution of nonlinear PDEs. J. Comput. Phys. **147**(2), 362–387 (1998). https://doi.org/10.1006/jcph.1998.6093. http://www.sciencedirect.com/science/article/pii/S0021999198960934
7. Deconinck, W., et al.: Atlas: a library for numerical weather prediction and climate modelling. Comput. Phys. Commun. **220**, 188–204 (2017). https://doi.org/10.1016/j.cpc.2017.07.006. http://www.sciencedirect.com/science/article/pii/S0010465517302138
8. Gardner, D.J., Guerra, J.E., Hamon, F.P., Reynolds, D.R., Ullrich, P.A., Woodward, C.S.: Implicit-explicit (IMEX) Runge-Kutta methods for non-hydrostatic atmospheric models. Geosci. Model Dev. **11**(4), 1497–1515 (2018). https://doi.org/10.5194/gmd-11-1497-2018. https://www.geosci-model-dev.net/11/1497/2018/
9. Gaudreault, S., Pudykiewicz, J.A.: An efficient exponential time integration method for the numerical solution of the shallow water equations on the sphere. J. Comput. Phys. **322**, 827–848 (2016). https://doi.org/10.1016/j.jcp.2016.07.012. http://www.sciencedirect.com/science/article/pii/S0021999116302911
10. Hochbruck, M., Ostermann, A.: Exponential integrators. Acta Numerica **19**, 209–286 (2010). https://doi.org/10.1017/S0962492910000048
11. Holton, J.R.: An Introduction to Dynamic Meteorology, International Geophysics Series, 4th edn., vol. 88. Elsevier Academic Press (2004)

12. Kennedy, C.A., Carpenter, M.H.: Additive Runge-Kutta schemes for convection-diffusion-reaction equations. Appl. Numer. Math. **44**(1), 139–181 (2003). https://doi.org/10.1016/S0168-9274(02)00138-1. http://www.sciencedirect.com/science/article/pii/S0168927402001381
13. Koskela, A.: Approximating the matrix exponential of an advection-diffusion operator using the incomplete orthogonalization method. In: Abdulle, A., Deparis, S., Kressner, D., Nobile, F., Picasso, M. (eds.) Numerical Mathematics and Advanced Applications - ENUMATH 2013. LNCSE, vol. 103, pp. 345–353. Springer, Cham (2015). https://doi.org/10.1007/978-3-319-10705-9_34
14. Luan, V.T., Pudykiewicz, J.A., Reynolds, D.R.: Further development of efficient and accurate time integration schemes for meteorological models. J. Comput. Phys. **376**, 817–837 (2019). https://doi.org/10.1016/j.jcp.2018.10.018. http://www.sciencedirect.com/science/article/pii/S0021999118306818
15. Melvin, T., Benacchio, T., Shipway, B., Wood, N., Thuburn, J., Cotter, C.: A mixed finite-element, finite-volume, semi-implicit discretization for atmospheric dynamics: cartesian geometry. Q. J. R. Meteorol. Soc. **145**(724), 2835–2853 (2019). https://doi.org/10.1002/qj.3501
16. Mengaldo, G., Wyszogrodzki, A., Diamantakis, M., Lock, S.-J., Giraldo, F.X., Wedi, N.P.: Current and emerging time-integration strategies in global numerical weather and climate prediction. Arch. Comput. Meth. Eng. **26**(3), 663–684 (2018). https://doi.org/10.1007/s11831-018-9261-8
17. Niesen, J., Wright, W.: Algorithm 919: a krylov subspace algorithm for evaluating the ϕ-functions appearing in exponential integrators. ACM Trans. Math. Softw. TOMS **38**, 1–19 (2012). https://doi.org/10.1145/2168773.2168781
18. Rančić, M., Purser, R.J., Mesinger, F.: A global shallow-water model using an expanded spherical cube: gnomonic versus conformal coordinates. Q. J. R. Meteorol. Soc. **122**(532), 959–982 (1996). https://doi.org/10.1002/qj.49712253209
19. Sadourny, R.: Conservative finite-difference approximations of the primitive equations on quasi-uniform spherical grids. Mon. Weather Rev. **100**(2), 136–144 (1972). https://doi.org/10.1175/1520-0493(1972)100⟨0136:CFAOTP⟩2.3.CO;2
20. Thuburn, J., Woollings, T.: Vertical discretizations for compressible Euler equation atmospheric models giving optimal representation of normal modes. J. Comput. Phys. **203**, 386–404 (2005). https://doi.org/10.1016/j.jcp.2004.08.018
21. Ullrich, P.A., Jablonowski, C., Kent, J., Lauritzen, P.H., Nair, R.D., Taylor, M.A.: Dynamical core model intercomparison project (DCMIP) test case document (2012). http://earthsystemcog.org/site_media/docs/DCMIP-TestCaseDocument_v1.7.pdf

Parallel Multilevel Linear Solver Within INMOST Platform

Kirill Terekhov(✉)

Marchuk Institute of Numerical Mathematics of the Russian Academy of Sciences, Moscow, Russia
kirill.terehov@gmail.com

Abstract. The work is dedicated to domain-decomposition parallel iterative linear system solver. An algebraic multilevel preconditioner is used on a subdomain. The cornerstone of the method is the dual-threshold second-order Crout incomplete LU factorization. The Crout version of LU factorization permits condition estimation for inverse factors $\|L^{-1}\|$ and $\|U^{-1}\|$. The factorization of k-th row and column of the matrix is postponed upon the growth of the estimated condition. Following factorization, the Schur complement is computed for the postponed part and the factorization continues in the multilevel fashion until the complete matrix is factorized. Before each level factorization begins, the matrix is first permuted by finding maximum transversal and reverse Cuthill-McKee permutations and then rescaled into I-dominant matrix. The performance of the method on a coupled multiphysics problem of blood coagulation is demonstrated.

Keywords: Multilevel method · Incomplete LU · Schur complement

1 Introduction

There is a growing demand to model coupled multi-physics problems. Such problems produce complex linear systems and require *black-box* solution methods. The primary target is the development of a linear solver, applicable but not limited to coupled blood coagulation model [4,22]. The model couples fluid flow in porous media described by Navier-Stokes-Darcy equations with nine blood components and a stiff cascade of reactions.

The method proposed in this article is built on top of a combination of state-of-the-art techniques. These techniques target improved incomplete factorization robustness: Kaporin [9] proposed second-order accurate incomplete factorization, Bollhöfer [2] proposed condition estimation for inverse factors and with Saad developed the theory for robust multilevel factorization [3]. In [3] it is found that straightforward efficient calculation of Schur complement is of low accuracy. In the present article, the accuracy is recovered using second-order accurate

This work was supported by the Russian Science Foundation through the grant 19-71-10094.

V. Voevodin and S. Sobolev (Eds.): RuSCDays 2020, CCIS 1331, pp. 297–309, 2020.
https://doi.org/10.1007/978-3-030-64616-5_26

incomplete factorization. A combination of these approaches together with static pivoting [6,14] and fill-in reduction [5] leads to efficient and robust multilevel incomplete factorization method.

This work is devoted to a parallel linear system solver from INMOST platform. INMOST is an open-source, flexible and efficient numerical modelling framework that provides application developers all the tools required for the development of parallel multi-physics models [18–21,23]. It interfaces many solvers such as Trilinos [8], PETSc [1], SuperLU [13].

2 INMOST Linear Solvers

The linear solvers implemented in INMOST platform are based on the combination of the following components:

- *Iterative method.* Preconditioned Krylov solver for non-symmetric matrices.
- *Parallelization.* Domain decomposition.
- *Preconditioning.* Incomplete LU-factorization.

The iterative solver is the preconditioned biconjugate gradient stabilized method BiCGStab(ℓ) [16]. Parallelization of the BiCGStab(ℓ) is straightforward as it only requires to accumulate sums of scalar products computed on each processor and synchronize vector elements after matrix-vector multiplication.

Parallel implementation of the preconditioner is based on the combination of an incomplete LU factorization and the restricted Additive Schwartz method with user-specified overlapping parameter. Overlapping of a local matrix and a local vector requires their extension by data from adjacent processors. To construct the overlap, the sparsity pattern is analyzed for the global column indices that lay outside of the local processor and the local matrix is augmented with the rows from remote processors as illustrated in Fig. 1. The procedure of matrix extension is repeated as specified by the user. No overlap corresponds to the block Jacobi method.

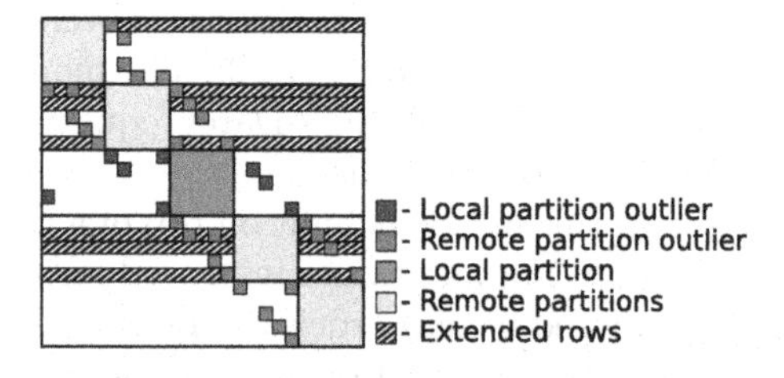

Fig. 1. Sparse matrix extension procedure.

Each processor performs incomplete LU factorization on the local extended matrix. During factorization, all the column indices that fall outside of extended matrix partition are ignored. In BiCGStab(ℓ) iterations, after solution step with the preconditioner, the restriction to the extended preconditioned vector is applied.

There are several choices for incomplete LU factorization algorithm within INMOST platform. Further, the multilevel method is discussed. The multilevel incomplete LU factorization is based on the following components:

- *Factorization.* Crout LU factorization with adaptive dropping.
- *Preprocessing.* Non-symmetric maximum transversal permutation, symmetric fill-in reduction permutation, matrix rescaling
- *Pivoting.* Delayed multi-level factorization.

3 Preprocessing

Before the construction of the preconditioner, the matrix is preprocessed by reordering and rescaling in three steps, that are summarized as follows:

- Non-symmetric permutation for the static choice of pivoting sequence.
- Symmetric permutation for fill-in reduction.
- Rescaling to improve the condition number and dropping strategy.

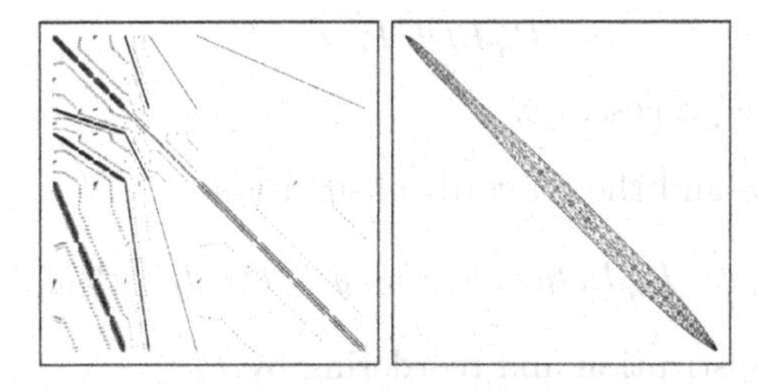

Fig. 2. Original matrix (left). After reordering with weighted reverse Cuthill-Mckee method (right). The matrix corresponds to the blood coagulation model. The non-zeros of the matrix are magnified for visibility.

For static pivoting, the matrix is permuted to maximize the diagonal product [14]. The method rescales the matrix into I-dominant matrix, which has unit values on its diagonal and off-diagonal entries with absolute values not exceeding 1. Efficient implementation of static pivoting for sparse matrices is based on the Dijkstra algorithm [6] utilizing binary heap and linked lists in dense arrays. Such permutation is known to reduce the necessity for pivoting. However, small diagonal values or badly conditioned factors may still be encountered during factorization. To this end, the multi-level factorization with inverse factor condition estimation is used.

The weighted reverse Cuthill-Mckee [5] reordering illustrated in Fig. 2 is used to reduce the number of non-zeros in the triangular factors. The weight corresponding to the sum of absolute values of off-diagonal elements is assigned to each graph vertex and is used instead of vertex degree during computation of reordering. Usually, the construction of reverse Cuthill-Mckee reordering is much cheaper although the fill-in reduction is much better with the node dissection algorithm from METIS [11].

Once the permutations are over, the matrix is rescaled to improve the I-dominance property of the matrix. The I-dominant matrix rescaling is the by-product of the algorithm that finds the maximal transversal and it is further improved iteratively [14].

The incomplete LU factorization is performed for preprocessed matrix $\tilde{A}$ rather than for the original matrix A:

$$\tilde{A} = D_L \hat{A} D_R = D_L P_R A P_C D_R = \tilde{L}\tilde{D}\tilde{U} + e,$$

where e is the factorization error, D_L and D_R are left and right diagonal rescaling matrices, P_R and P_C are permutation matrices for rows and columns, respectively, $\tilde{L}$ and $\tilde{U}$ correspond to the triangular factors, $\hat{A} = P_R A P_C$ is the permuted matrix.

After the factorization is completed, the incomplete factors are rescaled:

$$\hat{L} = D_L^{-1}\tilde{L}D_L, \quad \hat{D} = D_L^{-1}\tilde{D}D_R^{-1}, \quad \hat{U} = D_R\tilde{U}D_R^{-1}.$$

Applying the permuted preconditioner to a vector b requires the solution of the system:

$$P_R^T \hat{L}\hat{D}\hat{U} P_C^T x = b,$$

which is a two-stage process, *i.e.*

- reordering by P_R and the forward substitution:

$$P_R^T \hat{L}\hat{D} y = b \implies y = \hat{D}^{-1}\hat{L}^{-1} P_R b,$$

- the backward substitution and reordering by P_C:

$$\hat{U} P_C^T x = y \implies x = P_C \hat{U}^{-1} y.$$

4 Factorization

The Crout incomplete LDU factorization method [12] is applied to the preprocessed matrix, local to each processor. In the Crout LDU method, one has to access simultaneously both rows and columns of matrix A as well as rows and columns of L and U factors. Access pattern during elimination is illustrated in Fig. 3. Matrix L is constructed and stored by columns whereas matrix U is constructed and stored by rows. This requires additional data structures for traversal of L by rows and of A and U by columns. It requires indices to be ordered. The data structure for each matrix uses three additional arrays of indices:

- I_1 corresponds to the current row (column) position in *ia* array,
- I_2 contains a linked list of next nonzero row (column) positions,

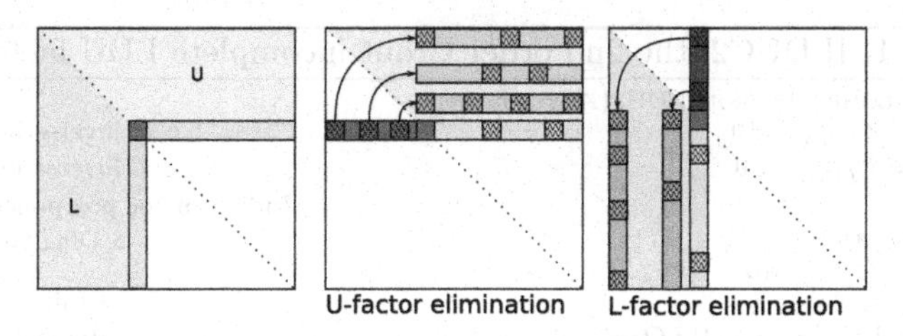

Fig. 3. Access pattern during elimination for a row of factor U and a column of factor L in the Crout LU algorithm.

- I_3 points to the first non-zero for each column (row).

The first nonzero entry of the first column is located in position $I_1[I_2[I_3[1]]]$ of *ia* array of the corresponding matrix, the next nonzero (if present) is located at $I_1[I_2[I_2[I_3[1]]]]$ and so on. If we have to consider the next column, we shift by one the positions in I_1 array for all rows of the current column and update the linked list I_2 and starting positions I_3 according to the new column positions in these rows. The process is illustrated in Fig. 4.

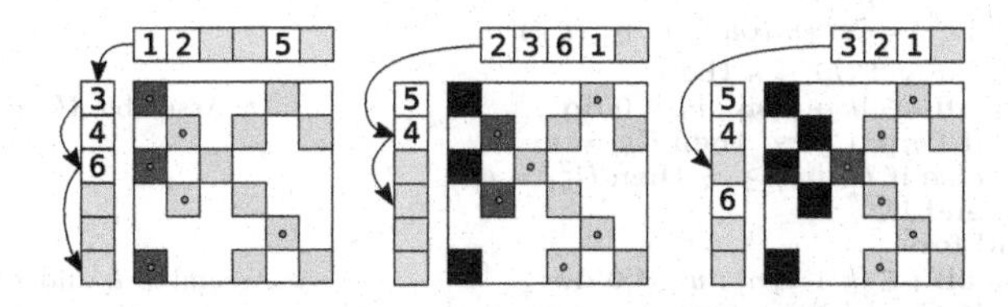

Fig. 4. Illustration for transposed traversal of three columns for matrix stored by rows. Blue squares correspond to already considered non-zeros, yellow squares correspond to yet to be considered nonzeros, red squares correspond to the elements in the currently considered column. Green dots correspond to positions in I_1 array. Arrays I_2 and I_3 are displayed to the left and top of the matrix, respectively. Gray squares in arrays I_2 and I_3 are NaN values indicating the end of the list. (Color figure online)

The elimination process involves the addition of multiple sparse vectors with the coefficient. For this operation, an ordered dense linked list data structure is used. The structure requires two arrays of indices and values of the size that covers all unknowns of local overlapped matrices. It enables fast addition of sparse vectors with ordered indices. The data structure is illustrated in Fig. 5.

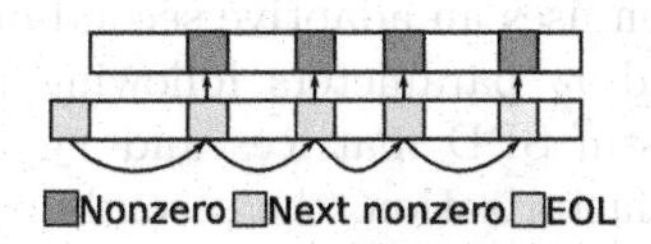

Fig. 5. Illustration for ordered dense linked list data structure.

Algorithm 1. ILDUC2: the 2nd order Crout incomplete LDU factorization.

1: **function** IncompleteCroutLDU2(A, τ_1, τ_2, κ) ▷ Input
2: $\eta_{U[1:n]} = \zeta_{U[1:n]} = 0$ ▷ U inverse norm estimators
3: $\eta_{L[1:n]} = \zeta_{L[1:n]} = 0$ ▷ L inverse norm estimators
4: $Q_{[1:n]} = 1$ ▷ Indicator for postponed factorization
5: $D = \text{diag}(A)$ ▷ Diagonal initialization
6: **for all** $k \in [1, n]$ **do** ▷ Main loop
7: $v_{[k:n]} = A_{k,[k,n]}$ ▷ Initialize by k-th row
8: **for all** $i \in [1 : k-1] : Q_i \neq 0, L_{ki} \neq 0$ **do** ▷ Iteration over row of L
9: $v_{[k:n]} = v_{[k:n]} - L_{ki} D_{ii} U_{i,[k:n]}$ ▷ Addition of sparse vectors
10: $v_{[k:n]} = v_{[k:n]} - L_{ki} D_{ii} U^2_{i,[k:n]}$ ▷ Addition of sparse vectors
11: **end for**
12: **for all** $i \in [1 : k-1] : Q_i \neq 0, L^2_{ki} \neq 0$ **do** ▷ Iteration over row of L^2
13: $v_{[k:n]} = v_{[k:n]} - L^2_{ki} D_{ii} U_{i,[k:n]}$ ▷ Addition of sparse vectors
14: **end for**
15: $v_{[k:n]} = v_{[k:n]} / D_{kk}$
16: $[C_U, c^U_\eta, c^U_\zeta]$ =Estimator(k, v, η_U, ζ_U) ▷ See algorithm 2
17: $w_{[k:n]} = A_{[k,n],k}$ ▷ Initialize by k-th column
18: **for all** $i \in [1 : k-1] : Q_i \neq 0, U_{ik} \neq 0$ **do** ▷ Iteration over column of U
19: $w_{[k:n]} = w_{[k:n]} - U_{ik} D_{ii} L_{[k:n],i}$ ▷ Addition of sparse vectors
20: $w_{[k:n]} = w_{[k:n]} - U_{ik} D_{ii} L^2_{[k:n],i}$ ▷ Addition of sparse vectors
21: **end for**
22: **for all** $i \in [1 : k-1] : Q_i \neq 0, U^2_{ik} \neq 0$ **do** ▷ Iteration over column of U^2
23: $w_{[k:n]} = w_{[k:n]} - U^2_{ik} D_{ii} L_{[k:n],i}$ ▷ Addition of sparse vectors
24: **end for**
25: $w_{[k:n]} = w_{[k:n]} / D_{kk}$
26: $[C_L, c^L_\eta, c^L_\zeta]$ =Estimator(k, w, η_L, ζ_L) ▷ See algorithm 2
27: **if** $C_L \leq \kappa$ and $C_U \leq \kappa$ **then**
28: **for all** $i \in [k+1, n] : v_i \neq 0$ **do** ▷ Assemble U and U^2 k-th rows
29: **if** $C_U |v_i| \geq \tau_1$ **then** $U_{ki} = v_i$
30: **else if** $C_U |v_i| \geq \tau_2$ **then** $U^2_{ki} = v_i$
31: **end if**
32: **end for**
33: **for all** $i \in [k+1, n] : w_i \neq 0$ **do** ▷ Assemble L and L^2 k-th columns
34: **if** $C_L |w_i| \geq \tau_1$ **then** $L_{ik} = w_i$
35: **else if** $C_L |w_i| \geq \tau_2$ **then** $L^2_{ik} = w_i$
36: **end if**
37: **end for**
38: **for all** $i \in [k+1 : n] : v_i \neq 0, w_i \neq 0$ **do**
39: $D_{ii} = D_{ii} - D_{kk} v_i w_i$ ▷ Update diagonal
40: **end for**
41: $\eta_{U[k+1:n]} = \eta_{U[k+1:n]} + c^U_\eta v_{[k+1:n]}$ ▷ Update estimator η_U
42: $\zeta_{U[k+1:n]} = \zeta_{U[k+1:n]} + c^U_\zeta v_{[k+1:n]}$ ▷ Update estimator ζ_U
43: $\eta_{L[k+1:n]} = \eta_{L[k+1:n]} + c^L_\eta w_{[k+1:n]}$ ▷ Update estimator η_L
44: $\zeta_{L[k+1:n]} = \zeta_{L[k+1:n]} + c^L_\zeta w_{[k+1:n]}$ ▷ Update estimator ζ_L
45: **else**
46: $Q_k = 0$ ▷ Postpone k-th row and column factorization
47: **end if**
48: **end for**
49: **return** $[L, D, U, L^2, U^2, Q]$ ▷ Output
50: **end function**

Incomplete factorization uses an adaptive second-order dual-threshold dropping strategy with τ_1 and τ_2 parameters following [9], with $\tau_2 = O(\tau_1^2)$ or $1 \gg \tau_1 \gg \tau_2 > 0$. For stiff SPD matrices and $\tau_2 = \tau_1^2$ in comparison with the conventional $ILU(\tau)$ factorization, the second-order factorization provides the preconditioner quality comparable to $ILU(\tau_2)$ and the preconditioner density comparable to $ILU(\tau_1)$. The second-order factorization method computes two versions of factored matrices L, L^2 and U, U^2. Matrices L and U contain elements whose absolute values exceed the first threshold τ_1 and L^2 and U^2 contain elements whose absolute values exceed the second threshold τ_2. During factorization, elimination is performed based on combinations of all factors except the product of L^2 and U^2. Once the factorization is completed, L^2 and U^2 factors are abandoned and the iterations proceed with the preconditioner involving only factors L and U which speeds up significantly the iterative convergence.

The Crout version of LU factorization enables incremental condition estimation of the inverse factors $C_L = \|L^{-1}\|$ and $C_U = \|U^{-1}\|$. It is possible due to the availability of the entire row and column of factors on each step. The condition estimation cost is the solution procedure with two right hand side vectors per factor performed along with the factorization. Following condition estimation, both thresholds τ_1 and τ_2 are adapted accordingly [2].

Adaptation of the dropping strategy due to condition estimation increases the density of the preconditioner for the same parameters τ_1 and τ_2. However, the method becomes more robust with much larger values of τ_1 and τ_2. Further, following [3] the factorization of a row and column is postponed to the next level if condition estimation exceeds parameter κ. The subsequent elimination is performed without consideration of the delayed rows and columns.

The steps described above are summarized in Algorithm 1. The condition estimator is outlined in Algorithm 2.

Algorithm 2. The condition norm estimator.

1: **function** ESTIMATOR(v, k, η, ζ) ▷ **Input**
2: $\mu_+ = -\eta_k + 1$
3: $\mu_- = -\eta_k - 1$
4: $s_+ = s_- = 0$
5: **for all** $j \in [k+1:n] : v_j \neq 0$ **do**
6: $s_+ = s_+ + |\eta_j + v_j\mu_+|$
7: $s_- = s_- + |\eta_j + v_j\mu_-|$
8: **end for**
9: $c_\eta = \mu_-$
10: **if** $s_+ > s_-$ **then** $c_\eta = \mu_+$ **end if**
11: $C_1 = \max(|\mu_+|, |\mu_-|)$
12: $\mu_+ = -\zeta_k + 1$
13: $\mu_- = -\zeta_k - 1$
14: $n_+ = n_- = 0$
15: **for all** $j \in [k+1:n] : v_j \neq 0$ **do**
16: $v_+ = |\zeta_j + v_j\mu_+|$
17: $v_- = |\zeta_j + v_j\mu_-|$
18: **if** $v_+ > \max(2|\zeta_j|, 1/2)$ **then** $n_+ = n_+ + 1$ **end if**
19: **if** $v_- > \max(2|\zeta_j|, 1/2)$ **then** $n_- = n_- + 1$ **end if**
20: **if** $|\zeta_j| > \max(2v_+, 1/2)$ **then** $n_+ = n_+ - 1$ **end if**
21: **if** $|\zeta_j| > \max(2v_-, 1/2)$ **then** $n_- = n_- - 1$ **end if**
22: **end for**
23: $c_\zeta = \mu_-$
24: **if** $n_+ > n_-$ **then** $c_\zeta = \mu_+$ **end if**
25: $C_2 = \max(|\mu_+|, |\mu_-|)$
26: **return** $[\max(C_1, C_2), c_\eta, c_\zeta]$ ▷ **Output**
27: **end function**

5 Multilevel Method

Before the incomplete LDU factorization, the preprocessed matrix $\tilde{A}$ is reordered symmetrically to place all the postponed elements to the end. Thus we obtain

$$P_Q \tilde{A} P_Q^T = \begin{bmatrix} B & F \\ E & C \end{bmatrix}, \tag{1}$$

where P_Q denotes the permutation matrix and

$$B = (L + L^2 - I)D(U + U^2 - I) + e \tag{2}$$

is the obtained reordered incomplete LDU factorization with rejected error e.

The approximate Schur complement S is computed for the postponed part:

$$S = C - E(U + U^2 - I)^{-1} D^{-1} (L + L^2 - I)^{-1} F. \tag{3}$$

According to [3], (3) corresponds to S-version of Schur complement computation. Given that entries below τ_2/κ are dropped, the accuracy of the Schur complement is $O(\kappa\tau_2)$. This suggests to define $\tau_2 \leq \tau_1/\kappa$ to maintain Schur complement accuracy.

Calculation of the Schur complement is performed by Algorithm 3. It includes premature dropping of entries from EU and LF matrices, which is found to be essential for efficiency, however it may negatively affect the accuracy. The factorization proceeds recursively by setting $A = S$ for the next level. The permutation and scaling factors are accumulated. The resulting factors and E, F blocks of each level are unscaled but the permutation is retained. The complete multi-level factorization process is given in Algorithm 4.

The solution process with the multi-level algorithm involves recursive backwards and forward substitution on each level.

Indeed, given the exact factorization of matrix A with block representation:

$$A = \begin{bmatrix} B & F \\ E & C \end{bmatrix} = \begin{bmatrix} I & \\ EB^{-1} & I \end{bmatrix} \begin{bmatrix} B & \\ & S \end{bmatrix} \begin{bmatrix} I & B^{-1}F \\ & I \end{bmatrix}, \tag{4}$$

with $S = C - EB^{-1}F$, the solution of the system

$$A \begin{bmatrix} u \\ y \end{bmatrix} = \begin{bmatrix} f \\ g \end{bmatrix}, \tag{5}$$

reduces to

$$\tilde{f} = B^{-1} f, \tag{6}$$
$$\tilde{g} = g - E\tilde{f}, \tag{7}$$
$$y = S^{-1}\tilde{g}, \tag{8}$$
$$u = \tilde{f} - B^{-1}Fy. \tag{9}$$

Algorithm 3. Schur complement calculation.

```
1: function SCHUR(C, E, U, U^2, D, L^2, L, F, τ_2, s, κ)          ▷ Input
2:   for all k ∈ [n − s, n] do
3:     ε = τ_2 max_{i∈[1,n−s]}(F_{i,k})/κ
4:     v_{[1:n−s]} = F_{[1,n−s],k}                                ▷ Initialize by k-th column
5:     for all i ∈ [1, n − s] : v_i ≠ 0 do
6:       for all j ∈ [i + 1, n − s] : L_{ji} ≠ 0 do               ▷ Forward substitution with L
7:         if v_j ≠ 0 or |v_i L_{ji}| > ε then                     ▷ Premature dropping
8:           v_j = v_j − v_i L_{ji}
9:         end if
10:      end for
11:      for all j ∈ [i + 1, n − s] : L^2_{ji} ≠ 0 do             ▷ Forward substitution with L^2
12:        if v_j ≠ 0 or |v_i L^2_{ji}| > ε then                   ▷ Premature dropping
13:          v_j = v_j − v_i L^2_{ji}
14:        end if
15:      end for
16:    end for
17:    LF_{[1,n−s],k} = v_{[1,n−s]}                               ▷ Memorize column
18:  end for
19:  for all k ∈ [n − s, n] do
20:    ε = τ_2 max_{i∈[1,n−s]}(E_{k,i})/κ
21:    v_{[1:n−s]} = E_{k,[1,n−s]}                                ▷ Initialize by k-th column
22:    for all i ∈ [1, n − s] : v_i ≠ 0 do
23:      for all j ∈ [i + 1, n − s] : U_{ij} ≠ 0 do               ▷ Backwards substitution with U
24:        if v_j ≠ 0 or |v_i U_{ij}| > ε then                     ▷ Premature dropping
25:          v_j = v_j − v_i U_{ij}
26:        end if
27:      end for
28:      for all j ∈ [i + 1, n − s] : U^2_{ij} ≠ 0 do             ▷ Backwards substitution with U^2
29:        if v_j ≠ 0 or |v_i U^2_{ij}| > ε then                   ▷ Premature dropping
30:          v_j = v_j − v_i U^2_{ij}
31:        end if
32:      end for
33:    end for
34:    EU_{k,[1,n−s]} = v_{[1,n−s]}                               ▷ Memorize row
35:  end for
36:  Reassemble LF into row-wise format
37:  for all k ∈ [n − s, n] do                                    ▷ Assemble Schur complement
38:    v_{[n−s,n]} = C_{k,[n−s,n]}                                ▷ Initialize row
39:    for all i ∈ [1, n − s] : EU_{ki} ≠ 0 do
40:      v_{[n−s,n]} = v_{[n−s,n]} − EU_{ki} D^{−1}_{ii} LF_{i,[n−s:n]}
41:    end for
42:    S_{k,[n−s,n]} = v_{[n−s,n]}
43:  end for
44:  return S                                                     ▷ Output
45: end function
```

The solution of systems in (6), (9) with matrix B involves forward and backwards substitution with already computed factors. Here B matrix is not required. Step (8) solves the system (5) recursively with the block matrix S and the right hand side $\tilde{g}$, until the last level is reached. Before the solution procedure, the input right hand side vector is reordered with P_R permutation matrix and the final solution is reordered with P_C permutation matrix.

Algorithm 4. Multi-level factorization.

```
1: function MultilevelIncompleteCroutLDU2(A, τ1, τ2, κ)
2:     l = 1                                                    ▷ Level number
3:     P_R = P_C = I                      ▷ Global reordering of rows and columns
4:     D_L = D_R = I                      ▷ Global rescaling of rows and columns
5:     repeat
6:         [P_C, D_L, D_R] = MaximumTransversal(A)
7:         [P_R, P_C, D_L, D_R] = WeightedReverseCuthillMckee(D_L A P_C D_R)
8:         [D_L, D_R] = ImproveDominance(D_L P_R A P_C D_R)
9:         [L, U, D, L^2, U^2, Q] = IncompleteCroutLDU2(D_L P_R A P_C D_R, τ1, τ2, κ)
10:        s = n − Σ_{k∈[1,n]} Q_i                              ▷ Schur complement size
11:        if s ≠ 0 then
12:            Construct reordering P_Q
13:            Reorder D_L = D_L P_Q^T and D_R = P_Q D_R
14:            Reorder symmetrically L, U, D, L^2, U^2 with P_Q
15:            Accumulate permutations P_R = P_Q P_R, P_C = P_C P_Q^T
16:            for all k ∈ [1, l − 1] do
17:                Reorder rows E_k = P_R E_k
18:                Reorder columns F_k = F_k P_C
19:            end for
20:            [E, F, C] =Blocks(D_L P_R A P_C D_R)                ▷ Split according to 4
21:            A =Schur(C, E, F, U, U^2, D, L^2, L, F, τ2, s, κ)
22:            Store E_l = D_L^{-1} E D_R^{-1} by rows
23:            Store U_l = D_R U D_R^{-1} by rows
24:            Store F_l = D_L^{-1} F D_R^{-1} by columns
25:            Store L_l = D_L^{-1} L D_L by columns
26:            Store D_l = D_L^{-1} D D_R^{-1} as vector
27:            Advance level l = l + 1
28:        end if
29:    until s ≠ 0
30:    return [L_k, D_k, U_k, E_k, F_k, ∀k ∈ [1, l − 1], P_R, P_C, l]
31: end function
```

6 Application

Figure 6 corresponds to numerically reproduced experiment of the blood clotting in microfluidic capillaries at shear rate $\gamma = 25s^{-1}$ [15]. For illustration purposes, the velocity magnitude is demonstrated in the log-scale. From the velocity distribution, it is evident that the flow is obstructed by the clot.

The model has only 20160 cells that result in 262080 unknowns, *i.e.* there are 13 unknowns per cell. A typical matrix pattern, illustrated in Fig. 2 has 11 millions of nonzeros or 42 nonzeros per row on average. The parameters used for factorization are $\tau_1 = 5 \times 10^{-2}$, $\tau_2 = 2.5 \times 10^{-3} = \tau_1^2$ and $\kappa = 5$. A single layer of overlapping was used for Additive Schwartz method and $\ell = 4$ in BiCGStab(ℓ). The clotting process is simulated in 136 time steps that require a total of 348 Newton iterations for convergence, maximum 4 Newton iterations for a step. These numbers are independent of the number of processors. The linear solver allows performing simulation without breakdown. The computations were performed on the cluster of INM RAS.

The parallel performance of the multi-physics model of blood flow coagulation is demonstrated in Table 1. In the table, the solution time breaks into the solver setup phase (multilevel factorization) and the time taken by the iterative method (Krylov solver). The model scales very well with the growth of the number of

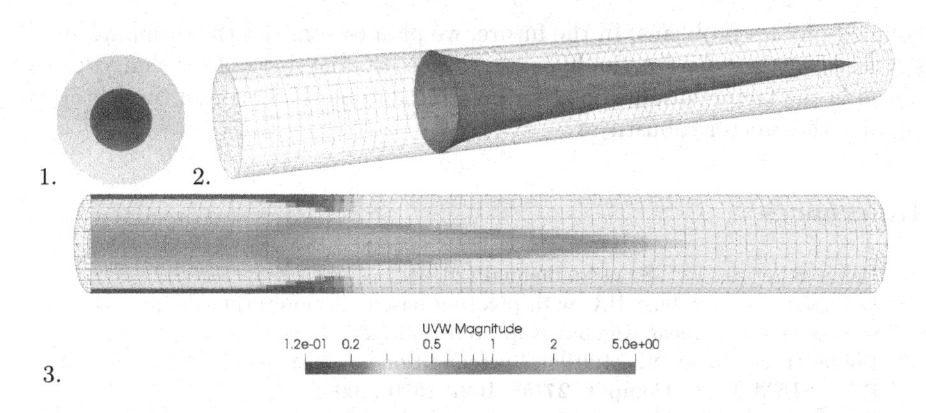

Fig. 6. Simulation of blood clotting at shear rate $\gamma = 25s^{-1}$, the moment of time $t = 60$ s. Isosurface embraces the area with permeability coefficient of one Darcy due to clotting (1,2). Log-scale of the blood velocity field in a middle cutaway of the grid (3).

Table 1. Performance of the multi-physics model of blood flow coagulation.

Processors number	Total time		Assembly time		Solution time		Setup time		Iterations time		Linear iterations
1	30406.1	–	6256.4	–	24137.1	–	21648.6	–	2470.1	–	13372
24	1109.1	27×	415.4	15×	688.7	35×	243.3	88×	444.2	6×	32009
48	548.7	55×	241.1	26×	340.6	71×	80.9	268×	259.2	10×	38773
72	420.5	72×	171.2	37×	247.1	98×	45.5	476×	201.4	12×	41297
96	329.8	92×	130.5	48×	197.4	122×	30.6	707×	166.7	15×	43230
120	282.9	108×	111.2	56×	169.4	142×	22.6	958×	146.7	17×	46604
144	309.7	98×	93.3	67×	214.5	112×	17.9	1210×	196.5	13×	48548
168	248.5	122×	85.0	74×	161.7	149×	14.6	1483×	147.0	17×	50274

processors up to 120. The factorization method speeds up significantly due to the reduction of the local block size. Further, it is evident, that the poor scalability of the iterative scheme due to the larger number of linear iterations required to converge to the solution dominates overall scalability. The Schur complement fill-in is significant on subsequent levels of multi-level factorization. This observation calls for the introduction of additional dropping strategies during the Schur complement assembly phase.

7 Conclusion

This work presents the new functionality of the open-source platform INMOST. This functionality allows the user to take advantage of the multi-level method and greatly facilitate the parallel solution of linear systems arising in complex coupled physics problems. In the future, we plan to consider the dropping strategy for the Schur matrix, the hypergraph partitioning method for shared parallelism in the factorization, the block pivoting and block version of the factorization for the greater robustness.

References

1. Balay, S., et al.: PETSc users manual (2019)
2. Bollhöfer, M.: A robust ILU with pivoting based on monitoring the growth of the inverse factors. Linear Algebra Appl. **338**(1-3), 201–218 (2001)
3. Bollhöfer, M., Saad, Y.: Multilevel preconditioners constructed from inverse-based ILUs. SIAM J. Sci. Comput. **27**(5), 1627–1650 (2006)
4. Bouchnita, A., Terekhov, K., Nony, P., Vassilevski, Y., Volpert, V.: A mathematical model to quantify the effects of platelet count, shear rate, and injury size on the initiation of blood coagulation under venous flow conditions. PloS One **15**(7), e0235392 (2020)
5. Cuthill, E., McKee J.: Reducing the bandwidth of sparse symmetric matrices. In: Proceedings of the 1969 24th National Conference (1969)
6. Duff, I.S., Koster, J.: The design and use of algorithms for permuting large entries to the diagonal of sparse matrices. SIAM J. Matrix Anal. Appl. **20**(4), 889–901 (1999)
7. Duff, I.S., Kaya, K., Uçar, B.: Design, implementation, and analysis of maximum transversal algorithms. ACM Trans. Math. Soft. (TOMS) **38**(2), 1–31 (2012)
8. Heroux, M.A., et al.: An overview of the Trilinos project. ACM Trans. Math. Softw. (TOMS) **31**(3), 397–423 (2005)
9. Kaporin, I.E.: High quality preconditioning of a general symmetric positive definite matrix based on its U^TU+ U^TR+ R^TU-decomposition. Numer. Linear Algebra Appl. **5**(6), 483–509 (1998)
10. Kaporin, I.E.: Scaling, reordering, and diagonal pivoting in ILU preconditionings. Russ. J. Numer. Anal. Math. Model. **22**(4), 341–375 (2007)
11. Karypis, G., Vipin, K.: METIS-unstructured graph partitioning and sparse matrix ordering system, version 2.0. (1995)
12. Li, N., Saad, Y., Chow, E.: Crout versions of ILU for general sparse matrices. SIAM J. Sci. Comput. **25**(2), 716–728 (2003)
13. Li, X.S.: An overview of SuperLU: algorithms, implementation, and user interface. ACM Trans. Math. Softw. (TOMS) **31**(3), 302–325 (2005)
14. Olschowka, M., Arnold, N.: A new pivoting strategy for Gaussian elimination. Linear Algebra Appl. **240**, 131–151 (1996)
15. Shen, F., Kastrup, C.J., Liu, Y., Ismagilov, R.F.: Threshold response of initiation of blood coagulation by tissue factor in patterned microfluidic capillaries is controlled by shear rate. Arterioscler. Thromb. Vasc. Biol. **28**(11), 2035–2041 (2008)
16. Sleijpen, G.L.G., Diederik, R.F.: BiCGstab (l) for linear equations involving unsymmetric matrices with complex spectrum. Electron. Trans. Numer. Anal. **1**(11), 2000 (1993)
17. Terekhov, K., Vassilevski, Y.: INMOST parallel platform for mathematical modeling and applications. In: Voevodin, V., Sobolev, S. (eds.) RuSCDays 2018. CCIS, vol. 965, pp. 230–241. Springer, Cham (2019). https://doi.org/10.1007/978-3-030-05807-4_20

18. Terekhov, K.: Parallel dynamic mesh adaptation within INMOST platform. In: Voevodin, V., Sobolev, S. (eds.) RuSCDays 2019. CCIS, vol. 1129, pp. 313–326. Springer, Cham (2019). https://doi.org/10.1007/978-3-030-36592-9_26
19. Voevodin, V., Sobolev, S. (eds.): RuSCDays 2018. CCIS, vol. 965. Springer, Cham (2019). https://doi.org/10.1007/978-3-030-05807-4
20. Terekhov, K., Vassilevski, Y.: Mesh modification and adaptation within INMOST programming platform. In: Garanzha, V.A., Kamenski, L., Si, H. (eds.) Numerical Geometry, Grid Generation and Scientific Computing. LNCSE, vol. 131, pp. 243–255. Springer, Cham (2019). https://doi.org/10.1007/978-3-030-23436-2_18
21. Vassilevski, Yu.V., Konshin, I.N., Kopytov, G.V., Terekhov, K.M.: INMOST - programming platform and graphical environment for development of parallel numerical models on general grids, no. 144. Moscow University Press (2013). (in Russian)
22. Vassilevski, Yu., Terekhov, K., Nikitin, K., Kapyrin, I.: Parallel Finite Volume Computation on General Meshes, p. 184. Springer, Heidelberg (2020)
23. INMOST - a toolkit for distributed mathematical modeling. http://www.inmost.org/. Accessed 10 Mar 2019

Predictive Quantum-Chemical Design of Molecules of High-Energy Heterocyclic Compounds

Vadim Volokhov[1], Tatyana Zyubina[1], Alexander Volokhov[1], Elena Amosova[1], Dmitry Varlamov[1(✉)], David Lempert[1], and Leonid Yanovskiy[1,2]

[1] Institute of Problems of Chemical Physics of RAS, Chernogolovka, Russia
{vvm,zyubin,vav,aes,dima,lempert}@icp.ac.ru
[2] Central Institute of Aviation Motors, Moscow, Russia
Yanovskiy@ciam.ru

Abstract. Rapidly developing new technologies, especially in the field of modern aircraft, stimulate great interest in creating high-energy materials for various purposes. Recently, modern computer technologies have been playing an increasingly important role in creating new materials with determined properties. Focus of this work is on computer design of new compounds that have not yet been synthesized – high-enthalpy derivatives of heterocycles, such as tetrazine, furazan, furoxan, triazole, etc. using quantum chemical calculation methods. For experimentally studied substances $\mathbf{C_2N_6O_4}$, $\mathbf{C_2N_6O_3}$, $\mathbf{C_2N_8O_4}$, the calculated values of enthalpy $\Delta_{f}{}^{0}_{298}$ (g) are 8–15% higher than the experimental values, which is significantly less than the spread of experimental values for these compounds. The enthalpy of formation of the studied gaseous molecules was calculated using the atomization method. The simulation was performed within the GAUSSIAN 09 software package using the B3LYP hybrid density functional with the basis 6-311+G(2d,p) and the combination of methods G4 and G4(MP2). Tasks have high computational complexity and the calculation time it takes from several hours to months on multi-node supercomputer configurations.

Keywords: Predictive quantum-chemical design · Supercomputer *ab initio* calculations · High-enthalpy substances · Enthalpy of formation · G4 · Gaussian applied package

1 Introduction

Promising fuels for new-generation aircraft that meet the increased requirements for the energy content of their components are currently being developed in advanced countries. One of the modern approaches to create such new fuels is a computer design of molecules of new substances that have not yet been synthesized, but for various reasons are promising for creating components of new fuels. The main parameter defining the energy content of the substance is enthalpy of formation $\Delta_f H^\circ$ in the state of matter in which the test substance is intended for use. The standard enthalpy of formation of a

V. Voevodin and S. Sobolev (Eds.): RuSCDays 2020, CCIS 1331, pp. 310–319, 2020.
https://doi.org/10.1007/978-3-030-64616-5_27

compound is the change of energy during the formation of 1 mol of the substance from its constituent elements, with all substances in their standard states. Physicochemical characteristics (specific impulse of fuels, combustion and detonation rates) in a wide range of values depend approximately linearly on the change in $\Delta_f H^\circ$, which distinguishes this value as the most significant for evaluating energy properties of the substances. In order to evaluate most reliably the prospects of a particular compound in energy compositions for various purposes, it is necessary to have as accurate $\Delta_f H^\circ$ values of the components as possible, especially those present in the composition as the main substance. The most accurate $\Delta_f H^\circ$ values (along with experimental ones) are obtained in quantum chemical calculations based on ab initio approaches [1]. In addition, and most importantly, quantum-chemical methods allow the computer design of promising compounds that have not yet been synthesized. Quantum-chemical methods make it possible to calculate with high accuracy thermochemical parameters of the compounds, and to study in detail their dependence on the structure of the molecule. The study of patterns in the dependence of energy characteristics of a substance on intramolecular parameters requires significant computational resources and is a difficult task. This is due to the very strong dependence of the structure of the molecule and its energy properties on the atomic composition, for example. In this work, we have studied structurally similar substances (N-heterocycles) that are promising for certain categories of use. Table 1 shows values of sublimation energy and enthalpy for three knows substances (**$C_2N_6O_4$, $C_2N_6O_3$ и $C_2N_8O_4$**) in gaseous and crystalline states according to the references. It can be seen that the spread of the $\Delta_f H^{o}_{298}$ (g) values is quite large (10–30%): 20% ($C_2N_6O_4$), 30% ($C_2N_6O_3$) и 10% ($C_2N_8O_4$). In this regard, the question arises of further refinement of these values, for example, by quantum-chemical methods.

Table 1. Values of sublimation energy ($\Delta_{sub} H^{o}_{298,}$ kJ/kg) and enthalpy for **$C_2N_6O_4$** (structure **1** in Table 2), **$C_2N_6O_3$** (structure **2** in Table 2) and **$C_2N_8O_4$** (structure **3** in Table 2) in *gaseous* ($\Delta_f H^{o}_{298}$ (g), kJ/kg) and *crystalline* states ($\Delta_f H^{o}_{298}$ (cr), kJ/kg) according to the references.

Formula	$\Delta_f H^o 298$ (g), kJ/kg	$\Delta_{sub} H^o 298$, kJ/kg	$\Delta_f H^o 298$ (cr), kJ/kg
$C_2N_6O_4$	3051.818 ± 16.72 [2] 3678.4 [3]	331.892 ± 83.6 [2]	2719.926 ± 83.6 [2] 2675.2 [3]
$C_2N_6O_3$	3031.336 ± 20.9 [2] 4321 [4] 4311 [5]	268.356 ± 83.6 [2] 264.176 ± 14.212 [6]	2762.98 ± 87.78 [2] 2813.976 ± 40.128 [5]
$C_2N_8O_4$	3976.016 ± 25.08 [2] 4316 [7, 8]	359.898 ± 83.6 [2]	3616.118 ± 87.78 [2]

In this work, we calculated the structure, formation enthalpy, and IR spectra of the following gaseous molecules: (1) **$C_2N_6O_4$** (*Dioxetrazotrazofuroxan*, 5,6-(3,4-Furoxano-3)-1,2,3,4-tetrazine-1,3-dioxide, FTDOO), (2) **$C_2N_6O_3$** (*Furazantetrazin-dioxide*, FTDO, 5,6-(3,4-Furazano)-1,2,3,4-tetrazine-1,3-dioxide), (3) **$C_2N_8O_4$** (*Tetrazino-tetrazine tetraoxide*, TTTO, Dietetrazine-1,2,3,4-tetraoxide DTTO).

2 Calculation Method

The enthalpy of formation of the studied gaseous molecules was calculated by the atomization method. The methodology and basic approaches to performing calculations are described in detail in our previous work [9]. The simulation was performed within the GAUSSIAN 09 software package [10] using the well-proven B3LYP hybrid density functional [11, 12] with the basis of 6-311+G(2d,p) and combined methods G4 and G4(MP2) [1, 13]

3 Formation Enthalpies of Gaseous Substances

The calculation results are presented in Table 2 which shows structures of the studies gaseous molecules and values of formation enthalpy (kcal/mol, kJ/mol and kJ/kg) obtained on different calculation levels. In general, the values given in Table 2 show insignificant differences (within 1–3%) in the results obtain at the G4 and G4(MP2) calculation levels. The results obtained at the B3LYP/6-311+G(2d,p) calculation level differ from the most accurate ones obtained at the G4 calculation level by 2–11%.

Table 2. Structures of the studied molecules and enthalpy values (in kcal/mol, kJ/mol and kJ/kg) obtained on different calculation levels.

№	Formula and mol. weight	Structure	Enthalpy (kcal/mol)	Enthalpy (kJ/mol)	Enthalpy (kJ/kg)	Calculation level
1	**$C_2N_6O_4$ 172.060**		180.7	756.4	4396.1	B3LYP/6-311+G(2d,p)
			178.7	748.0	4347.3	G4(MP2)
			174.8	731.7	4252.6	G4
2	**$C_2N_6O_3$ 156.061**		180.3	754.7	4835.9	B3LYP/6-311+G(2d,p)
			176.3	738.0	4728.9	G4(MP2)
			173.2	725.0	4645. 6	G4
3	**$C_2N_8O_4$ 200.074**		231.3	967.9	4837.7	B3LYP/6-311+G(2d,p)
			231.8	970.0	4848.2	G4(MP2)
			227.4	951.4	4755.2	G4

For the experimentally studied substances $\mathbf{C_2N_6O_4}$ (structure 1), $\mathbf{C_2N_6O_3}$ (structure 2), $\mathbf{C_2N_8O_4}$ (structure 3), the calculated enthalpy values $\Delta_f H^\circ_{298}$ (g) are 8–15% higher than the experimental ones (15, 8, 10%, respectively), which is significantly less than the spread of the experimental values for these compounds (Table 1).

4 Structural Parameters and IR Spectra of Absorption

Figure 1 and Table 3 show the IR spectra of absorption, frequencies and intensities of vibrations of the studied molecules. The structural parameters and atomic displacements for the most intense vibrations are shown in Figs. 2, 3 and 4.

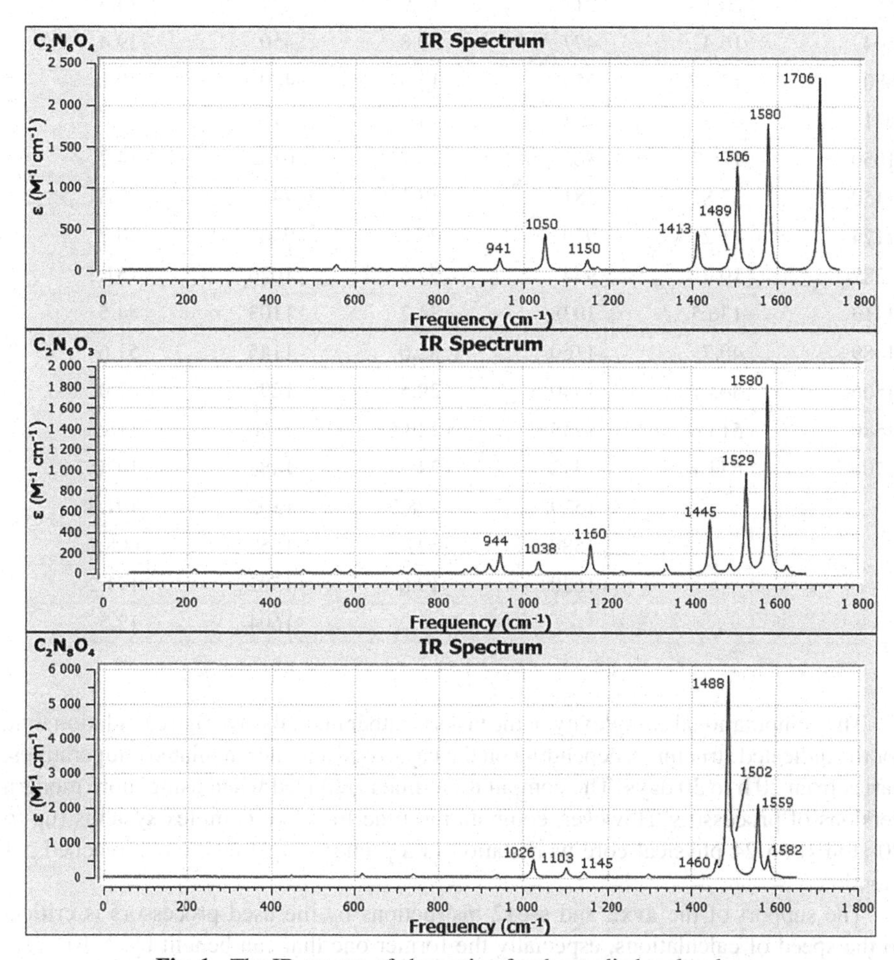

Fig. 1. The IR spectra of absorption for the studied molecules.

5 Details of Computational Calculations

A number of computational configurations (Table 4) based on various Intel Xeon processors, provided pools of computational cores, RAM and disk memory, GPU availability, and versions of the Gaussian application package (https://gaussian.com) were used during the quantum chemical simulation.

Table 3. Frequencies and intensities of vibrations.

Frequency (cm^{-1})	Intensity (km/mol)	Frequency (cm^{-1})	Intensity (km/mol)	Frequency (cm^{-1})	Intensity (km/mol)
$C_2N_6O_4$		$C_2N_6O_3$		$C_2N_8O_4$	
555	20.1	217	10.9	227	14.3
804	16.4	477	10.8	450	19.4
880	13.4	552	12.5	489	10.1
941	41.9	738	14.4	618	20.6
1050	126.7	862	11.7	618	14.2
1150	37.8	881	17.1	741	30.5
1174	11.2	919	25.2	936	20.7
1282	10.7	944	57.8	1026	151.2
1413	136.5	1038	34.2	1103	84.5
1489	40.7	1160	82.6	1145	51.6
1506	364.7	1340	29.5	1299	27.0
1580	513.6	1445	149.5	1460	118.6
1706	674.1	1492	24.5	1488	1670.7
		1529	278.9	1502	191.8
		1580	527.3	1559	608.7
		1628	19.6	1582	171.2
				1604	12.5

The computational complexity of the tasks is rather high; the average calculation time for the indicated structures, depending on the basis used and the calculated temperatures, varies from 20 h to 30 days. The computation time is reduced when using more modern versions of processors. However, computation time for more complex systems (up to 20 atoms) on 24 physical core workstation in a pseudo-single-task mode reached a 1 month.

The support of the **avx2** and **sse42** instructions by the used processors is critical to the speed of calculations, especially the former one that can benefit by 8–10 times on some tasks using the processors with a close clock rate [14]. Unfortunately, at the time of this writing we were not able to accurately assess the possibilities of using GPU accelerators. It is highly desirable to perform calculations on SSD disks or high-speed SAS disks with a large amount of allocated disk memory, since the package creates giant intermediate files up to 2 Tb during the calculations. It can take up to 35–50 min to record them on an SSD disk and, of course, significantly longer on SATA/NFS arrays.

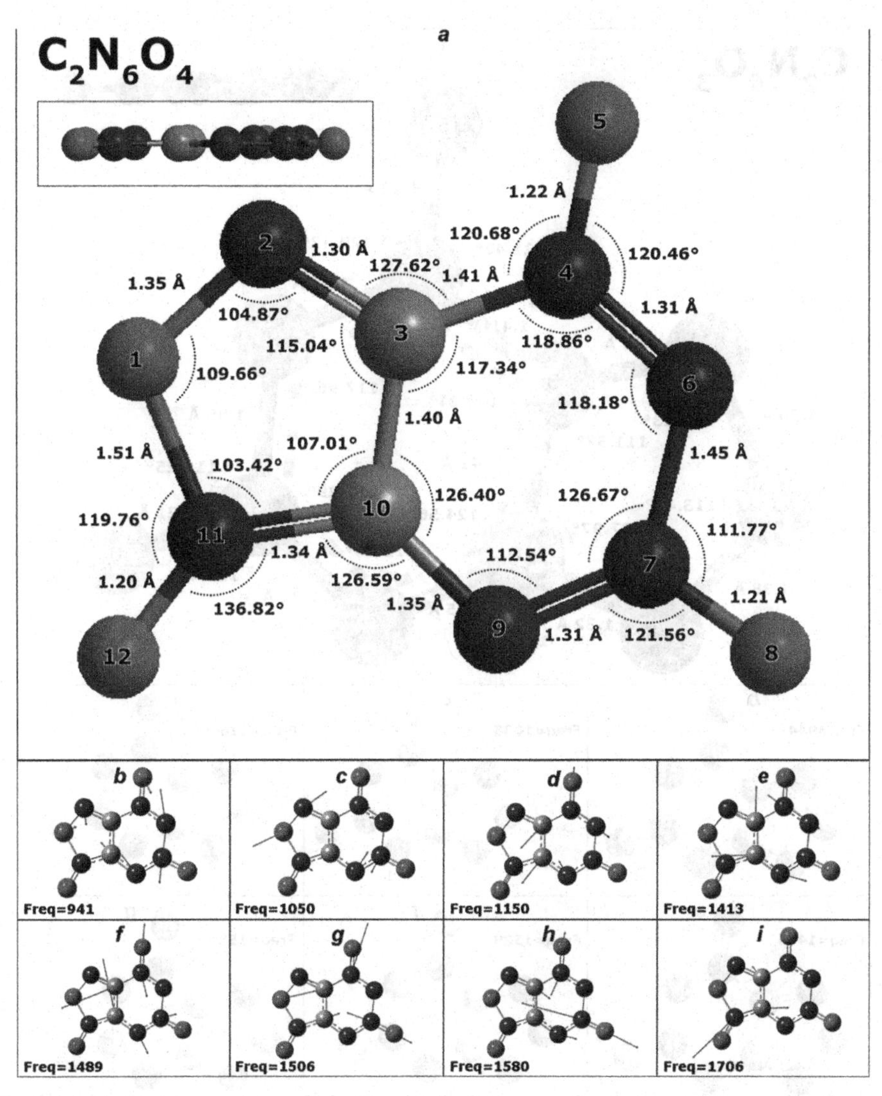

Fig. 2. **a** Structural parameters. **b–i.** Atomic displacements for the most intense vibrations of $C_2N_6O_4$ (structure **1**)

The different publications report that the speed of calculations is greatly influenced by the availability of the latest versions of the Gaussian package (as compared to g9 installed on the IPCP cluster), which fully realize the hardware capabilities of new series of processors, giving acceleration of calculations for most of the used bases up to 7–8 times. The authors did not intend to analyze in detail the degree of parallelization of the performed calculations (despite the fact that the Gaussian package usually uses its own "Linda" parallelization software). However, steady acceleration on pools up to 12 cores had been observed, while further this effect reduced. It also depends on the amount of allocated memory task (but it should be not less than 4 GB per physical core).

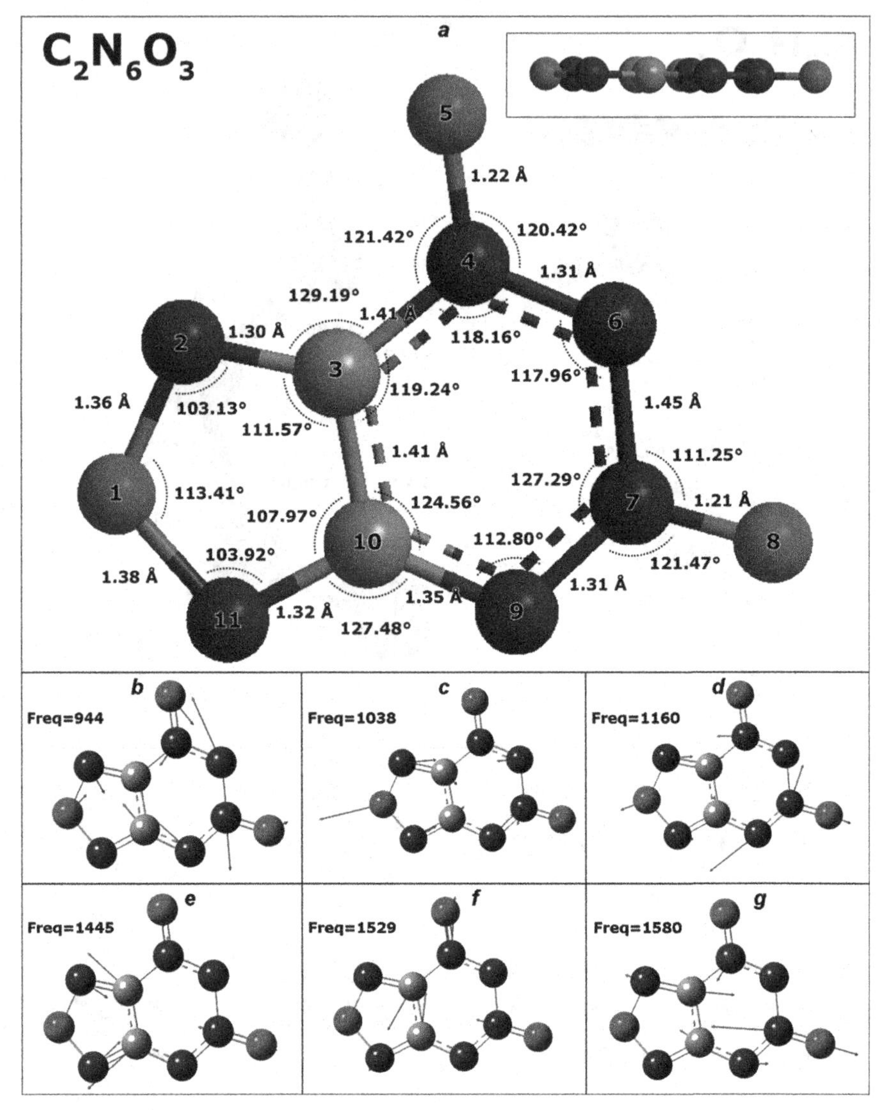

Fig. 3. **a** Structural parameters. **b–g.** Atomic displacements for the most intense vibrations of $C_2N_6O_3$ (structure **2**)

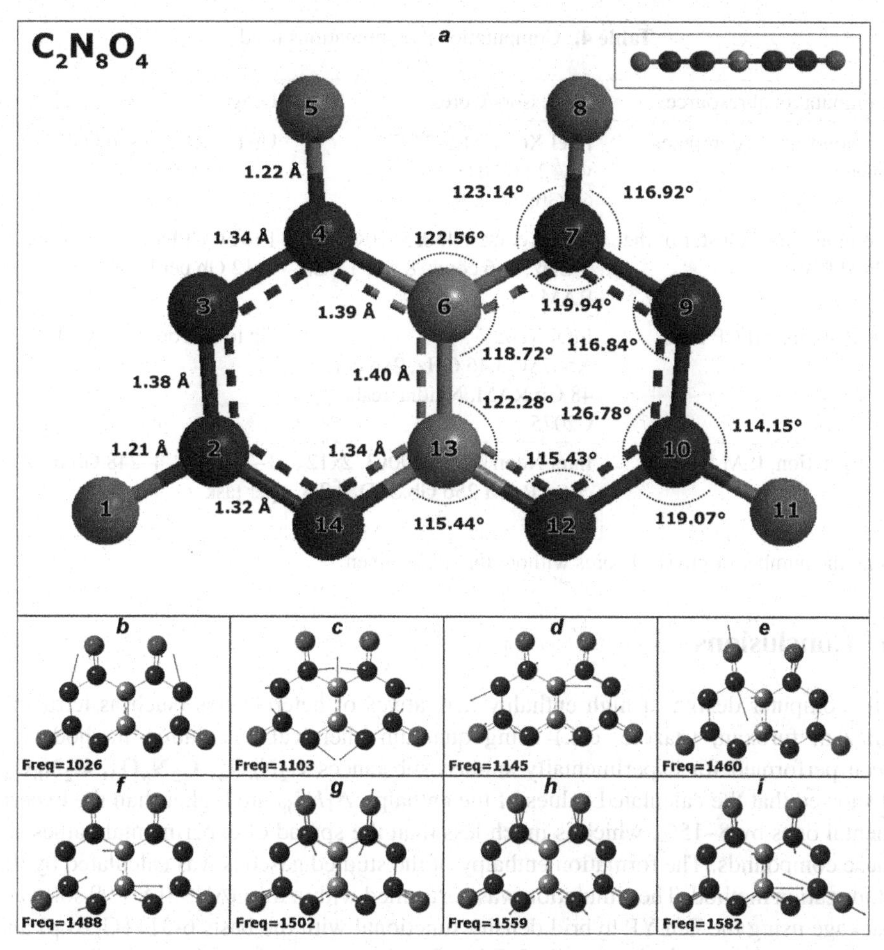

Fig. 4. **a** Structural parameters. **b–i.** Atomic displacements for the most intense vibrations of $C_2N_8O_4$ (structure **3**)

Table 4. Computational combinations used

Computational resource	Processors/Cores/	Usage
Lomonosov-2 «compute» queue	Intel Xeon® E5-2697 v3@2.60 GHz, 14 cores, 64 Gb; Tesla K40s	Up to 104 cores, 64 Gb
Computational cluster of the IPCP RAS:	Intel Xeon® 5450l 5670@ 3 GHz, 4–6 cores, 8 and 12 Gb RAM	1 to 50 CPU (up 200 cores), 1–12 Gb per node
Workstations IPCP RAS	Intel Xeon® X5675@3.46 GHz, 2x6 cores, 48 Gb RAM, Nvidia Tesla C2075	Up to 12 cores, +1 GPU
Workstation, IEM RAS	Intel Xeon® E5-2690v3, 2x12 cores,RAM 256 Gb,SSD, 4 Tb HDD	1–24 cores, 4–248 Gb RAM per task

note: the number of physical cores without threads is given

6 Conclusions

The computer design of high enthalpy derivatives of heterocycles (such as tetrazine, furazan, furoxan, triazole, etc.) using quantum chemical calculation methods has been performed for experimentally studied substances $\mathbf{C_2N_6O_4}$, $\mathbf{C_2N_6O_3}$, $\mathbf{C_2N_8O_4}$. It showed that the calculated values of the enthalpy $\Delta_f H^\circ_{298}$ are higher than the experimental ones by 8–15%, which is much less than the spread of experimental values for these compounds. The formation enthalpy of the studied gaseous was calculated by the atomization method. The simulation was performed within the GAUSSIAN 09 software package using the B3LYP hybrid density functional with the basis 6-311+G(2d,p) and the combined methods G4 and G4(MP2). The obtained results make it possible to use the method presented in this work to calculate and predict formation enthalpy of high-energy compounds that have not yet been obtained experimentally. The IR spectra of absorption, structural parameters and atomic displacements for the most intense vibrations of the studied high-energy compounds $\mathbf{C_2N_6O_4}$, $\mathbf{C_2N_6O_3}$ and $\mathbf{C_2N_8O_4}$ have been obtained for the first time.

Acknowledgements. The work was performed using the equipment of the Center for Collective Use of Super High Performance Computing Resources of the Lomonosov Moscow State University [15, 16] (project "Enthalpy-2065") and the Computing Center of the Institute of Problems of Chemical Physics of the Russian Academy of Sciences.

(This work was supported by RFBR according to the research project No. 20-07-00319a, and by Government of the Russian Federation according to the contract AAAA-A19-119120690042-9 and AAAA-A19-119061890019-5).

Funding. Quantum chemical simulation of high energy gaseous molecules was funded by RFBR according to the research project No. 20-07-00319a. Calculations of IR-spectra were performed in accordance with the state task, state registration No. AAAA-A19-119061890019-5 and AAAA-A19-119120690042-9.

References

1. Curtiss, L.A., Redfern, P.C., Raghavachari, K.: Gaussian-4 theory. J. Chem. Phys. **126**, 084108 (2007). https://doi.org/10.1063/1.2436888
2. Suntsova, M.A.: Prediction of the enthalpies of formation of new nitrogen-containing high-energy compounds based on quantum chemical calculations. PhD thesis for the Degree of Candidate of Chemical Sciences, Moscow (2016). (in Russian)
3. Rybakov, N.A., Tsaplin, A.I.: Defining the specific impulse of mixed solid propellants. Vestnik the Samara State Aerosp. Univ. **1**(21), 161–165 (2010). (in Russian)
4. Simonenko, V.N., et al.: Combustion of model compositions based on furazanotetrazine dioxide and dinitrodiazapentane. I. Binary systems. Combust. Explos. Shock Waves **50**(3), 306–314 (2014). https://doi.org/10.1134/S0010508214030083
5. Pepekin, V.I., Matyushin, Yu, N., Gubina, T.V.: Enthalpy of formation and explosive properties of 5,6-(3,4-furazano)-1,2,3,4-tetrazine-1,3-dioxide. Rus. J. Phys. Chem. B. **5**, 97–100 (2011). https://doi.org/10.1134/s1990793111020102
6. Kiselev, V.G., Gritsan, N.P., Zarko, V.E., Kalmykov, P.I., Shandakov, V.A.: Multilevel quantum chemical calculation of the enthalpy of formation of [1,2,5]oxadiazolo[3,4-e][1,2,3,4]-tetrazine-4,6-di-N-dioxide. Combust. Explos. Shock Waves. **43**, 562–566 (2007). https://doi.org/10.1007/s10573-007-0074-6
7. Lempert, D.B., Dorofeenko, E.M., Soglasnova, S.I.: The energy potential of some high-enthalpy N-oxides as oxidizers. Omsk Sci. Bull. Ser. «Aviation-Rocket and Power Engineering» **2**(3), 58–62 (2018). https://doi.org/10.25206/2588-0373-2018-2-3-58-62
8. Politzer, P., Lana, P., Murray, J.S.: Computational characterization of two Di-1,2,3,4-tetrazine tetraoxides, DTTO and iso-DTTO, as potential energetic compounds. Central Eur. J. Energ. Mater. **10**(1), 37–52 (2013)
9. Volokhov, V.M., et al.: Quantum chemical simulation of hydrocarbon compounds with high enthalpy of formation. Russ. J. Phys. Chem. B (2020). (in print)
10. Frisch, M.J., et al.: Gaussian 09, Revision B.01. Gaussian, Inc., Wallingford (2010)
11. Becke, A.D.: Density-functional thermochemistry. III. The role of exact exchange. J. Chem. Phys. **98**(4), 5648 (1993). https://doi.org/10.1063/1.464906
12. Johnson, B.J., Gill, P.M.W., Pople, J.A.: The performance of a family of density functional methods. J. Chem. Phys. **98**(7), 5612 (1993). https://doi.org/10.1063/1.464906
13. Curtiss, L.A., Redfern, P.C., Raghavachari, K.: Gn theory. Comput. Mol. Sci. **1**, 810–825 (2011). https://doi.org/10.1002/wcms.59
14. Grigorenko, B., Mironov, V., Polyakov, I., Nemukhin, A.: Benchmarking quantum chemistry methods in calculations of electronic excitations. Supercomput. Front. Innov. **5**(4), 62–66 (2019). https://doi.org/10.14529/jsfi180405
15. Voevodin, V.V., et al.: Supercomputer Lomonosov-2: large scale, deep monitoring and fine analytics for the user community. Supercomput. Front. Innov. **6**(2), 4–11 (2019). https://doi.org/10.14529/jsfi190201
16. Voevodin, V.V., et al.: Practice of Lomonosov supercomputer. Otkrytye sistemy [Open Systems] **7**, 36–39 (2012). [in Russian]

Simulations in Problems of Ultrasonic Tomographic Testing of Flat Objects on a Supercomputer

Sergey Romanov(✉)

Moscow Center of Fundamental and Applied Mathematics, Moscow, Russia
romanov60@gmail.com

Abstract. This paper presents a computer simulation study on a problem of ultrasonic tomographic imaging of welded joints in flat metal objects. The developed algorithms and supercomputer software for reconstructing sound speed images and identifying defects in welded joints were tested on model problems. A specific feature of welded joint inspection is that in most cases the inspected object is accessible only from a single side. This study investigates a tomographic scheme in which a flat object is sounded from a single side by transducer arrays, and reflections from the flat bottom of the object are taken into account. Various schemes of tomographic imaging are investigated and compared for the cases in which the position of the bottom of the object is known, partially known or unknown. The results of this study showed that taking into account reflections from the bottom is of fundamental importance for high-quality image reconstruction. This method allows us to greatly increase the angular range of sounding and to register the waves transmitted through the object in order to achieve high-precision sound speed image reconstruction. The computations were carried out on "Lomonosov" supercomputer at Lomonosov Moscow State University. The developed software was optimized on a supercomputer for varying number of MPI processes per node. The optimal setup yielded more than a two-fold performance increase.

Keywords: Supercomputer simulations · High-performance scientific computing · Inverse problems · Ultrasound tomography · Nondestructive testing

1 Introduction

This paper is concerned with an important task of developing the methods of nondestructive tomographic imaging of solids. Algorithms and supercomputer software employing the methods of ultrasonic tomography for reconstructing sound speed images and identifying defects of welded joints in flat metal objects have been developed. The problem of tomographic image reconstruction is posed as a coefficient inverse problem for a scalar wave equation. An approximate solution to this problem is found via minimizing the residual functional. The algorithms of solving the inverse problem are based on the possibility to compute the gradient of the residual functional explicitly. This is a breakthrough result in the field of solving coefficient inverse problems of wave tomography.

V. Voevodin and S. Sobolev (Eds.): RuSCDays 2020, CCIS 1331, pp. 320–331, 2020.
https://doi.org/10.1007/978-3-030-64616-5_28

To date, wave tomography imaging methods are making a start in medicine, seismic studies, and industrial nondestructive testing (NDT) applications [11, 13, 27, 32, 37, 39–41]. In the field of topological imaging, the main efforts are aimed at detecting the boundaries of inhomogeneous inclusions using reflected radiation [1, 8, 29, 34]. Synthetic aperture (focused synthesis) imaging methods employing reflected waves are widely used [2, 24, 25, 30, 38]. However, ultrasonic imaging methods that employ only reflected waves can provide high-resolution images of the boundaries of inhomogeneities inside the inspected object, but cannot determine the wave velocity inside an inhomogeneous object [18].

In preceding works of the author, the inverse problem of ultrasound tomography was formulated, iterative gradient-based solution methods were developed, numerical simulations in application to the problem of tomographic diagnostics of soft tissues in medicine were performed on a supercomputer. In the present work, an attempt is made to apply the previously obtained results to the problem of NDT of metal samples. This task is nontrivial because many factors are fundamentally different, such as the medium of wave propagation, sounding frequencies, and tomographic schemes.

In authors' previous papers [4, 5, 36], methods and algorithms for solving 2D ultrasound tomography problems for NDT of solids were developed. Multi-angle sounding schemes employed in these works assume that the inspected object is accessible from multiple sides. Unlike the previous works, this study is concerned with objects that are accessible only from a single side. In previous works, an important result was obtained on experimental verification of adequacy of the 2D scalar wave model in the proposed scheme of the experiment. A test bench which implements a multi-angle sounding scheme was developed for studies on ultrasonic tomography in application to NDT. The experiments have confirmed the effectiveness of the employed methods and algorithms for obtaining high-resolution tomographic images.

Inspection of welded joints in flat metal objects is one of the most important problems in NDT. In this case, the inspected object is usually accessible only from a single side. We investigate a tomographic scheme in which the object is sounded from a single side using multi-element ultrasonic transducer arrays. The transducer elements sequentially emit short sounding pulses, and the waves reflected from the bottom of the inspected object are then received by all the transducer elements. Taking into account the reflections from the bottom significantly increases the number of sounding angles and allows the waves transmitted through the object to be registered.

Various schemes of tomographic imaging are investigated and compared for the cases in which the position of the bottom of the object is known, partially known or unknown. Tomographic schemes differ in precision of determining the velocity profile inside the tested samples. Reconstructing the wave velocity is of paramount importance in nondestructive testing applications, because it makes characterization of the internal structure of the object possible. The research on ultrasonic tomographic imaging of welded joints was carried out using numerical simulations with parameters corresponding to those of real physical experiments with metal samples.

In the numerical experiments, the tomographic scheme contained several dozen positions of ultrasound emitters and detectors. The employed iterative methods of solving inverse problems require performing numerical simulations of wave propagation in inhomogeneous media tens of thousands of times. Thus, solving inverse problems of ultrasonic tomographic diagnostics is not possible without high-performance supercomputers [3, 21, 23]. Reconstruction of sound speed images was performed on a CPU partition of "Lomonosov" supercomputer at Lomonosov Moscow State University [42]. Performance of the developed software on a supercomputer was measured for varying number of MPI processes per computing node. An optimal setup has accelerated the computations by more than 2 times.

2 Formulations of the Direct and the Inverse Problems

We take into consideration only the waves described by the scalar wave equation. The scalar wave model is widespread in NDT due to its simplicity. This model allows us to compute the scalar wave field $u(r, t)$ from given initial data using the equation

$$c(r)u_{tt}(r, t) - \Delta u(r, t) = \delta(r - r_0)g(t), \tag{1}$$

$$u(r, t = 0) = u_t(r, t = 0) = 0. \tag{2}$$

Here $c(r) = 1/v^2(r)$ and $v(r)$ is the longitudinal wave velocity in the medium; $r \in R^2$; Δ is the Laplace operator with respect to r; δ is the Dirac delta-function which defines the position of the point source at r_0. The initial pulse emitted from the source is described by function $g(t)$. Equation (1) accurately describes wave diffraction and refraction effects.

The numerical experiment on ultrasound tomographic diagnostics of flat objects is set up as follows. We consider a simple two-dimensional problem (Fig. 1), in which flat objects 1 and 2 of the same material are separated by insert 3 of different material, such as a welded joint. Linear multi-element transducer array A is placed on the top surface of the plates. The transducer elements can both emit and receive ultrasound waves. Number 4 in Fig. 1 denotes the boundary between a low-speed medium 5 (air) and objects 1, 2 and 3. Due to a large difference in acoustic impedance at boundary 4, the waves emitted by the elements of transducer array A are reflected from boundary 4 and registered by the transducer array A.

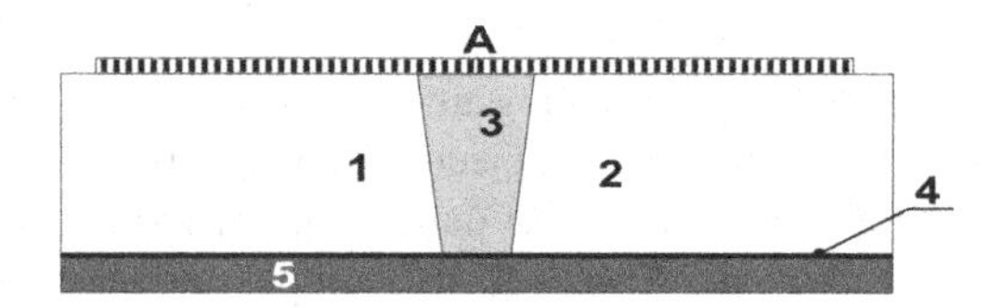

Fig. 1. The scheme of the tomographic experiment.

The following data registration scheme was employed in the numerical experiment. The elements of transducer array A emit short sounding pulses sequentially one by one, while all the elements of A act as receivers simultaneously. The sounding waves propagating through regions 1, 2, and 3 are reflected once from the boundary 4 and registered by the receiving elements of the transducer array. For simplicity, in this formulation we do not take into account the waves reflected from boundary 4 multiple times and the waves that pass between emitters and receivers without being reflected from boundary 4. These waves can be cut off by limiting the time interval for which the data are registered by the detectors.

In the direct problem, the objective is to obtain the data $U^{ij}(t)$ registered by the receiving transducer elements, where j is the index of the emitting element ($j = 1,\ldots,M$), i is the index of the receiving element ($i = 1,\ldots,M$), and M is the number of the transducer array elements. A constant velocity of longitudinal wave propagation $v(r)$ is specified for each of the regions 1, 2, 3 and 5. The wave velocities in regions 1 and 2 are equal and differ from the velocity in region 3 by a few percent. The wave velocity in region 5 is several times lower than the velocity in regions 1, 2, 3.

The wave field $u(r, t)$ computed via Eqs. (1)–(2) for the j-th emitting element at the position of the i-th receiving element are denoted as $u^{ij}(t)$. Thus, the wave field must satisfy the following equation for all emitting and receiving elements

$$u^{ij}(t) = U^{ij}(t) \tag{3}$$

The Eqs. (1)–(3) defines the problem. The objective in the inverse problem of ultrasound tomography under the scalar wave model is to reconstruct the unknown wave velocity $v(r)$ in region 3 according to Eqs. (1)–(3) using a measured wave field $U^{ij}(t)$ at the receiving elements and a known initial pulse $g(t)$. The positions of the emitters and detectors are known, but the geometry of region 3 is unknown.

The longitudinal wave velocity $v(r)$ in region 3 and the exact placement of region 3 are both unknown. The velocity $v(r)$ in regions 1 and 2 is a known constant $v(r) = v_0$, the velocity $v(r)$ in region 5 is also known and constant. We consider multiple variants of the placement of boundary 4. In the first formulation, we assume that the position of boundary is known, for example if we know the thickness of metal plates 1, 2 and 3. In the second formulation, the position of boundary 4 is unknown, and in the third formulation the position of boundary 4 is known only partially.

The problem of reconstructing the wave velocity is a nonlinear coefficient inverse problem [12, 16, 17, 26, 28, 31]. In the considered formulation, the object being inspected is not accessible from all directions. Emitters and detectors can be placed only on one of the sides of the object. Such conditions are common in ultrasound diagnostics. A typical approach in this case is to record the waves reflected from inhomogeneities [7], but this way we can determine only the boundaries of inhomogeneities. The presence of a reflecting flat bottom is a specific feature of the problem considered. In addition to the waves reflected from the boundaries of inhomogeneities, the detectors can register the waves passed through the inhomogeneities and reflected from the bottom surface. The presence of a reflective flat bottom surface adds a promising possibility for tomographic imaging.

The solution method is based on minimizing the residual functional $\Phi(c)$ with respect to $c(r)$. The residual functional represents the difference between measured data $U^{ij}(t)$ and the wave field computed at the detectors using a given $c(r)$ [14, 19, 22]

$$\Phi(c) = \sum\nolimits_{j=1}^{M} \sum\nolimits_{i=1}^{M} \frac{1}{2} \int_0^T \left(u^{ij}(t; c) - U^{ij}(t)\right)^2 dt. \tag{4}$$

Here, $u^{ij}(t; c)$ are the values of the wave field computed via solving the direct problem (1)–(2) with a given $c(r)$. For multiple ultrasound emitters in the setup, the residual functional is the sum of the residual values over emitters $j = 1,\ldots,M$. For each emitter index j, the residual value is integrated over time interval $(0,T)$ and summed over detectors $i = 1,\ldots,M$, which receive the signal from j-th emitter. Mathematically, the inverse problem is posed as a problem of finding an unknown function $\bar{c}(r)$ that minimizes the residual functional (4) $\bar{c}(r) : \min_{c(r)} \Phi(c) = \Phi(\bar{c})$. The solution $\bar{c}(r)$ obtained is the approximate solution of the problem.

Gradient methods have proven to be effective for minimizing the residual functional $\Phi(c)$. The algorithms developed in this study are based on explicit computation of the gradient of the $\Phi(c)$ [15, 17, 20]. Various iterative algorithms for minimizing the $\Phi(c)$ can be constructed using an explicit representation for the gradient.

As well as most inverse problems, the inverse problems of ultrasound tomography are ill-posed. Iterative solution algorithms with the given stopping rule have regularizing properties. Numerical simulations showed that if the scattered wave field is measured over all the sides of the object, the detector array pitch is smaller than the wavelength, and the object is sounded from multiple sides, then the reconstructed image is resistant to measurement errors of ~5–10%. Completeness of the angular range of sounding is of great importance in wave tomography. The issues of uniqueness of a solution to the inverse problem are discussed in the work [33] for various configurations of sources, receivers, and sounding frequencies.

3 Numerical Methods and the Parallel Implementation

To compute the gradient of the residual functional, it is necessary to compute the wave field $u(r, t)$ for a given speed of sound $v(r)$. To this end, we use finite-difference method in the time domain (FDTD). A regular discrete grid with a step of h and a time step of τ is set up over the range of spatial coordinates (x, y) and time t. Time step τ must satisfy the Courant stability condition. The Laplacian in Eq. (1) is approximated using a fourth-order finite difference scheme [35].

In [35], the method of mathematical modeling was used to analyze the influence of the finite difference grid step on the numerical dispersion of the propagating pulse and, consequently, on the quality of tomographic image reconstruction. It was shown that for finite difference schemes of the 2nd-order of accuracy, 10–15 grid points per central wavelength are not sufficient if the distance between the source and receiver exceeds the wavelength by a factor of 40 or more. This ratio is typical for the tomographic problem considered. Halving the grid step significantly reduced the numerical dispersion. It was also shown that using a 4th-order finite difference scheme resulted in a sufficiently small numerical dispersion at 10–15 grid points per wavelength.

Numerical simulations were performed in a rectangular region, which consists of regions 1, 2, 3 and 5 (Fig. 1). Non-reflecting boundary conditions were applied at the boundary of the rectangular region [9, 10]. In this study, we used 2nd-order approximate non-reflecting boundary conditions $\frac{\partial^2 u}{\partial x \partial t} - \frac{1 \partial^2 u}{v \partial t^2} + \frac{v \partial^2 u}{2 \partial y^2} = 0$. The time interval of the numerical simulation was limited to the interval T during which the sounding pulse reflected once from the boundary 4 reached the detectors. Thus, multiple reflections from the upper and the lower boundary were not taken into account. The waves passing from emitters to detectors along the upper boundary without being reflected from the bottom were also cut off by limiting the data registration time.

The volume of experimental data amounted to approximately one gigabyte; the number of unknowns in the inverse problem reaches 160000. A specific feature of the iterative methods employed is that a direct problem of wave propagation in inhomogeneous medium must be solved three times for each of approximately 100 iterations and for each of 50 emitters. Such computational workload is impossible to tackle without high-performance supercomputers [23, 35]. The numerical algorithms were implemented in C++ software designed for high-performance computing systems. MPI interface was used for inter-process data exchange. The computations were performed on Lomonosov supercomputer of Lomonosov Moscow University Supercomputing Center [42] equipped with two Intel Xeon 5570 Nehalem quad-core CPUs with 12 GB of RAM per node and QDR InfiniBand interconnecting network.

During the process of solving inverse problem, the computations for each emitter we carried out on a single CPU core; thus, 50 CPU cores were used for a 50-emitter setup. This parallelizing method has proven effective for the task, providing a nearly 50-fold acceleration. The most computationally expensive part of the algorithm is simulating the wave propagation according to Eq. (1) using a finite-difference scheme. This problem is data-intensive, and the performance depends heavily on the cache size and the memory throughput. In order to optimize the performance of the algorithm, the tests were performed on a supercomputer and the performance was measured for varying number of MPI processes per computing node. The number of processes was specified using "sbatch –ntasks-per-node = N" command, which launches N parallel processes on each node.

Table 1 shows the computation time in seconds measured for 20 iterations of the gradient descent method. The number of MPI processes per node varied from 1 to 8. The tests were performed for three different finite difference grid sizes: $x \times y \times t = 400 \times 200 \times 1000$, $400 \times 300 \times 1000$ and $800 \times 400 \times 2000$ points. The amount of computations and the volume of data transferred for the 2nd and 3rd grid sizes are larger than those for the 1st grid size by factors of approximately 1.5 and 8, respectively. Rows $\boldsymbol{T}_1$, $\boldsymbol{T}_2$, $\boldsymbol{T}_3$ show the computation time in seconds for the three grid sizes. Rows $\boldsymbol{T}_2/\boldsymbol{T}_1$ and $\boldsymbol{T}_3/\boldsymbol{T}_1$ show the respective quotients. The tests showed that the computation times are roughly proportional to the volumes of data involved. For all grid sizes, a performance increase is observed as the number of MPI processes per node decreases. Figure 2 plots the computation time versus the number of MPI processes for three grid sizes. The computation time for the $800 \times 400 \times 2000$ grid is scaled down by a factor of 8 in order to present all the cases in a single plot.

In all test cases, more than a two-fold performance increase was observed when the number of MPI processes per node was decreased from eight to one. Such execution profile is typical for data-intensive tasks, in which the performance is limited by the number of memory access channels. These results show that the problem considered can be solved more efficiently by computer systems with multiple memory access channels

Table 1. Computation time for 20 iterations versus the number of MPI processes.

ntasks-per-node	1	2	3	4	5	6	7	8
T_3, time (s) (grid size 800 × 400 × 2000)	508	586	692	780	870	1104	1244	1308
T_3/T_1	8.47	9.45	10.18	9.75	9.90	9.52	10.45	9.48
T_1, time (s) (grid size 400 × 200 × 1000)	60	62	68	80	98	116	119	138
T_2/T_1	1.48	1.56	1.71	1.68	1.60	1.65	1.65	1.46
T_2, time (s) (grid size 400 × 300 × 1000)	89	97	116	134	157	191	196	202

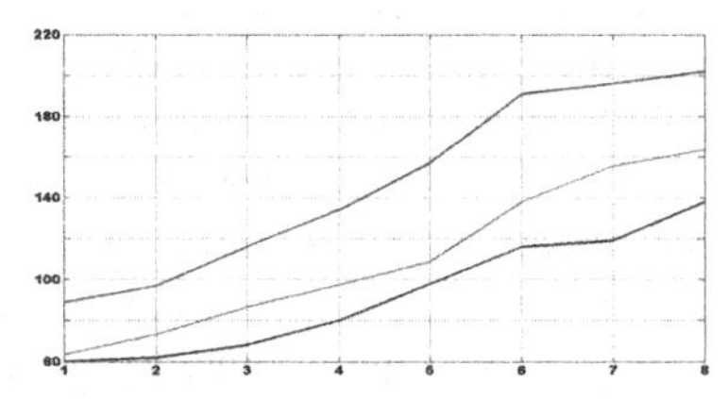

Fig. 2. Computation time versus number of MPI processes per node for 3 different grid sizes.

and lightweight computing nodes such as IBM Blue Gene systems, or systems with fast memory such as GPU.

4 Numerical Simulation Results

The numerical simulations were carried out on a model 2D inverse problem of non-destructive tomographic ultrasonic imaging of welded joints in metal samples. Figure 1 shows the tomographic scheme of the simulations [6]. The inverse problem is to determine the distribution of the longitudinal wave velocity over region 3, the exact position of which is unknown, using the measurements of ultrasonic waves taken by a linear transducer array placed on the upper boundary. In the numerical experiment, each transducer element sequentially emits a sounding pulse, which propagates through the inspected object. The sounding wave reflected from boundary 4 is then measured by all the elements of the transducer array.

In this study, we consider three formulations of the inverse problem with different assumptions for the boundary 4: in the first variant, the position of boundary 4 is known; in the second variant, the position of boundary 4 is unknown; and in the third variant, the position of boundary 4 is known only partially. Naturally, the best case is if the position of boundary 4 is known. However, this is not always the case in practice. For example, in welded pipes the thickness of the welded sheets is known, but the lower boundary of the welded joint itself is unknown. These assumptions led to the formulation with a partially known boundary.

In the numerical simulations, first we solve the direct problem of wave propagation in the simulated inhomogeneous medium shown in Fig. 3a and Fig. 4a for each emitting element. A scalar wave model and a 4th-order finite difference method for Eq. (1) were used. Second-order transparency condition (5) was applied at the boundaries of the computational domain. The wave field at the detectors was stored and used as simulated experimental data for solving the inverse problem. The sounding pulses in the simulation were short wide-band pulses identical for each emitter.

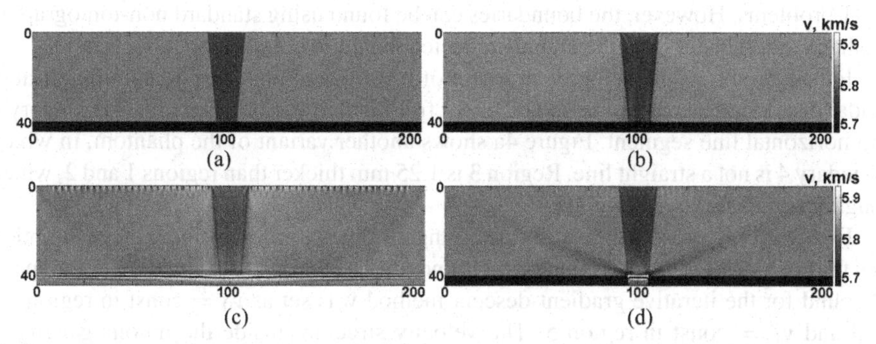

Fig. 3. Numerical simulations for a straight boundary 4: (a) – the phantom, (b) – the reconstruction with a known position of boundary 4, (c) – the reconstruction with unknown position of boundary 4, (d) – the reconstruction with the position of boundary 4 partially known.

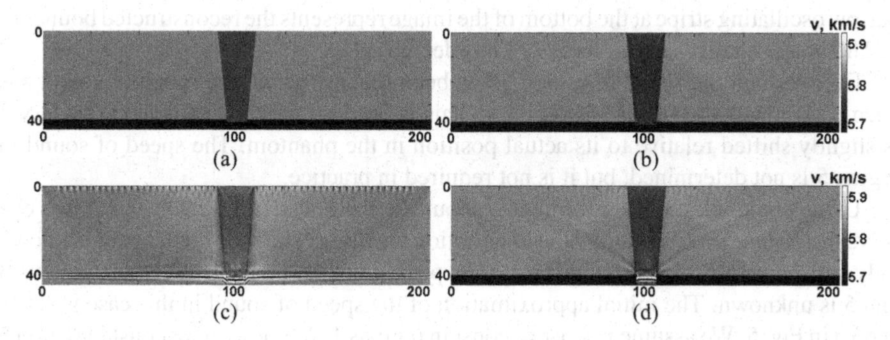

Fig. 4. Numerical simulations: (a) – the exact image (phantom), (b) – the image reconstructed assuming a known lower boundary, (c) – the image reconstructed assuming the lower boundary is unknown, (d) – the image reconstructed assuming a partially known lower boundary.

The following parameter values were assumed in the simulations. The wave velocity in regions 1 and 2 was $v_0 = 5.9$ mm/μs, the wave propagation velocity in region 3 was 5.7 mm/μs, which differs from v_0 by approximately 3.4%. The wave propagation velocity in region 5 amounted to $v_1 = 1.0$ mm/us, which differs from v_0 almost by 6 times. The thickness of regions 1, 2, and 3 was approximately 45 mm. The width of region 3 was 19 mm at the top and 10 mm at the bottom. The central wavelength of the sounding pulse was $\lambda = 2.62$ mm, which corresponds to the central frequency of 2.25 MHz. The size of the computational domain was 200 × 50 mm, and the size of the finite difference grid was 800×200 points. In the simulations, 50 transducer elements were located on the upper side of the computational domain, as shown in Fig. 1.

The difference in the speed of sound in different areas of welded joints can be very small and amount to 2–5%. This leads to the need to determine the speed of sound with at least 1% accuracy. Should the error be greater, precise characterization of the material would be impossible, and incorrect conclusions about the composition and properties of the inspected area would be made. Of course, imaging only the boundaries of regions (without determining the speed of sound inside the regions) is also of interest in NDT problems. However, the boundaries can be found using standard non-tomographic methods that register only the signals reflected from the boundaries.

Figure 3a shows the velocity structure of a simulated phantom containing a trapezoidal insert with a speed of sound different from that in surrounding areas. Boundary 4 is a horizontal line segment. Figure 4a shows another variant of the phantom, in which boundary 4 is not a straight line. Region 3 is 1.25 mm thicker than regions 1 and 2, which roughly simulates a welded joint.

Figures 3b, 4b present the results of solving the inverse problem under the assumption that the position of reflective boundary 4 is known. The initial approximation of the speed of sound for the iterative gradient-descent method was set as $v_0 = \text{const}$ in regions 1, 2, 3 and $v_1 = \text{const}$ in region 5. The velocity structure inside the inhomogeneity is reconstructed with high precision in this case.

Figures 3c, 4c show the results of solving the inverse problem under the assumption that the position of the reflecting boundary 4 is unknown. The initial approximation of the speed of sound in this case was set to $v_0 = \text{const}$ over the entire computational domain. A narrow oscillating stripe at the bottom of the image represents the reconstructed boundary 4. The image quality in regions 1, 2, 3 has decreased.

Even though the shape of region 3 has been reconstructed, the speed of sound has been determined with a significant error. This is because the reconstructed boundary 4 is slightly shifted relative to its actual position in the phantom. The speed of sound in region 5 is not determined, but it is not required in practice.

Using some additional information about the thickness of layers 1 and 2, we can solve the inverse problem under the assumption that the position of boundary 4 is known only between region 5 and regions 1, 2. The position of the boundary between regions 3 and 5 is unknown. The initial approximation of the speed of sound in this case is set as shown in Fig. 5. We assume $v = v_0 = \text{const}$ in regions 1, 2, 3; $v = v_1 = \text{const}$ in region 5 under the known segments of boundary 4; $v = v_0 = \text{const}$ in region 5 under the unknown segments of boundary 4. Figures 3d, 4d present the results of solving the inverse problem in a formulation where the position of the reflective boundary 4 is partially known. The quality of sound speed reconstruction in regions 1, 2, 3 has improved significantly. Some artifacts remain near the area where the position of boundary 4 is unknown.

The computations were carried out on 50 CPU cores of the supercomputer. Each node performed the computations for one ultrasound emitter. The number of iterations amounted to 100–150 for all considered formulations of the inverse problem. The computation time amounted to approximately one hour.

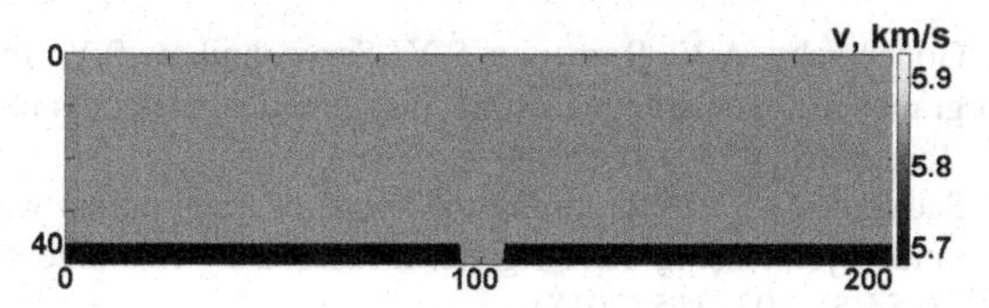

Fig. 5. Initial approximation in the case with the position of boundary 4 partially known.

5 Conclusion

In this article, tomographic methods of imaging flat objects using transducer arrays are proposed. The methods assume that the inspected objects are accessible only from a single side. In this case, the use of waves reflected from the bottom of the inspected object presents a promising approach. By means of numerical simulation on a supercomputer, various imaging schemes are investigated and compared for the cases in which the position of the bottom of the object may be known, unknown or partially known. The results demonstrated that the highest quality of sound speed image reconstruction is obtained if the position of the bottom is fully known. If the position of the bottom is unknown, the shapes of objects can be reconstructed, but the speed of sound is determined with significant errors. Additional information about the position of the reflective bottom can be used to improve the image quality.

The proposed numerical algorithms can be efficiently parallelized on multi-CPU supercomputers. The computations were performed on 50 CPU cores. The developed software was tested on a supercomputer for varying number of MPI processes per node. The optimal setup yielded more than a two-fold increase in performance.

Acknowledgement. The work is carried out according to the research program of Moscow Center of Fundamental and Applied Mathematics. The research is carried out using the equipment of the shared research facilities of HPC computing resources at Lomonosov Moscow State University.

References

1. Bachmann, E., Jacob, X., Rodriguez, S., Gibiat, V.: Three–dimensional and real–time two–dimensional topological imaging using parallel computing. J. Acoust. Soc. Am. **138**(3), 1796 (2015)
2. Bazulin, E.G.: Comparison of systems for ultrasonic nondestructive testing using antenna arrays or phased antenna arrays. Russ. J. Nondestr. Test. **49**(7), 404–423 (2013). https://doi.org/10.1134/S1061830913070024
3. Bazulin, E.G., Goncharsky, A.V., Romanov, S.Y., Seryozhnikov, S.Y.: Parallel CPU- and GPU-algorithms for inverse problems in nondestructive testing. Lobachevskii J. Math. **39**(4), 486–493 (2018). https://doi.org/10.1134/S1995080218040030
4. Bazulin, E., Goncharsky, A., Romanov, S.: Solving Inverse Problems of Ultrasound Tomography in a Nondestructive Testing on a Supercomputer. In: Voevodin, V., Sobolev, S. (eds.) RuSCDays 2019. CCIS, vol. 1129, pp. 392–402. Springer, Cham (2019). https://doi.org/10.1007/978-3-030-36592-9_32

5. Bazulin, E.G., Goncharsky, A.V., Romanov, S.Y., Seryozhnikov, S.Y.: Inverse problems of ultrasonic tomography in nondestructive testing: mathematical methods and experiment. Russ. J. Nondestruct. Test. **55**(6), 453–462 (2019)
6. Bazulin, E.G., Sadykov, M.S.: Determining the speed of longitudinal waves in anisotropic homogeneous welded joint using echo signals measured by two antenna arrays. Russ. J. Nondestruct. Test. **54**(5), 303–315 (2018)
7. Blitz, J., Simpson, G.: Ultrasonic Methods of Non–Destructive Testing. Springer, London (1995)
8. Dominguez, N., Gibiat, V.: Non–destructive imaging using the time domain topological energy. Ultrasonics **50**(3), 367–372 (2010)
9. Engquist, B., Majda, A.: Absorbing boundary conditions for the numerical simulation of waves. Math. Comput. **31**, 629 (1977)
10. Givoli, D., Keller, J.B.: Non-reflecting boundary conditions for elastic waves. Wave Motion **12**(3), 261–279 (1990)
11. Goncharsky, A.V., Kubyshkin, V.A., Romanov, S.Y., Seryozhnikov, S.Y.: Inverse problems of experimental data interpretation in 3D ultrasound tomography. Numer. Methods Programm. **20**, 254–269 (2019)
12. Goncharsky, A.V., Romanov, S.Y.: Two approaches to the solution of coefficient inverse problems for wave equations. Comput. Math. Math. Phys. **52**, 245–251 (2012)
13. Goncharsky, A.V., Romanov, S.Y.: Supercomputer technologies in inverse problems of ultrasound tomography. Inverse Probl. **29**(7), 075004 (2013). https://doi.org/10.1088/0266-5611/29/7/075004
14. Goncharsky, A.V., Romanov, S.Y.: Inverse problems of ultrasound tomography in models with attenuation. Phys. Med. Biol. **59**(8), 1979–2004 (2014). https://doi.org/10.1088/0031-9155/59/8/1979
15. Goncharsky, A.V., Romanov, S.Y.: Iterative methods for solving inverse problems of ultrasonic tomography. Numer. Methods Programm. **16**, 464–475 (2015)
16. Goncharsky, A.V., Romanov, S.Y.: Iterative methods for solving coefficient inverse problems of wave tomography in models with attenuation. Inverse Probl. **33**(2), 025003 (2017). https://doi.org/10.1088/1361-6420/33/2/025003
17. Goncharsky, A.V., Romanov, S.Y.: A method of solving the coefficient inverse problems of wave tomography. Comput. Math. Appl. **77**, 967–980 (2019). https://doi.org/10.1016/j.camwa.2018.10.033
18. Goncharsky, A.V., Romanov, S.Y., Seryozhnikov, S.Y.: Problems of limited-data wave tomography. Numer. Methods Programm. **15**, 274–285 (2014)
19. Goncharsky, A., Romanov, S., Seryozhnikov, S.: Inverse problems of 3D ultrasonic tomography with complete and incomplete range data. Wave Motion **51**(3), 389–404 (2014). https://doi.org/10.1016/j.wavemoti.2013.10.001
20. Goncharsky, A.V., Romanov, S.Y., Seryozhnikov, S.Y.: Low–frequency three–dimensional ultrasonic tomography. Doklady Phys. **61**(5), 211–214 (2016). https://doi.org/10.1134/s1028335816050086
21. Goncharsky, A., Romanov, S., Seryozhnikov, S.: Supercomputer technologies in tomographic imaging applications. Supercomput. Front. Innov. **3**, 41–66 (2016)
22. Goncharsky, A., Romanov, S., Seryozhnikov, S.: A computer simulation study of soft tissue characterization using low-frequency ultrasonic tomography. Ultrasonics **67**, 136–150 (2016)
23. Goncharsky, A.V., Romanov, S.Y., Seryozhnikov, S.Y.: Comparison of the capabilities of GPU clusters and general-purpose supercomputers for solving 3D inverse problems of ultrasound tomography. J. Parallel Distrib. Comput. **133**, 77–92 (2019)
24. Hall, T.E., Doctor, S.R., Reid, L.D., Littlield, R.J., Gilber, R.W.: Implementation of real–time ultrasonic SAFT system for inspection of nuclear reactor components. Acoust. Imag. **15**, 253–266 (1987)

25. Jensen, J.A., Nikolov, S.I., Gammelmark, K.L., Pedersen, M.H.: Synthetic aperture ultrasound imaging. Ultrasonics **44**, 5–15 (2006)
26. Klibanov, M.V., Kolesov, A.E.: Convexification of a 3-D coefficient inverse scattering problem. Comput. Math. Appl. **77**(6), 1681–1702 (2019)
27. Klibanov, M.V., Kolesov, A.E., Nguyen, D.-L.: Convexification method for an inverse scattering problem and its performance for experimental backscatter data for buried targets. SIAM J. Imag. Sci. **12**(1), 576–603 (2019)
28. Klibanov, M.V., Li, J., Zhang, W.: Convexification for the inversion of a time dependent wave front in a heterogeneous medium. SIAM J. Appl. Math. **79**(5), 1722–1747 (2019)
29. Lubeigt, E., Mensah, S., Rakotonarivo, S., Chaix, J.-F., Baquè, F., Gobillot, G.: Topological imaging in bounded elastic media. Ultrasonics **76**, 145–153 (2017)
30. Metwally, K., et al.: Weld inspection by focused adjoint method. Ultrasonics **83**, 80–87 (2018)
31. Natterer, F.: Possibilities and limitations of time domain wave equation imaging. In: AMS: Tomography and Inverse Transport Theory, vol. 559, pp. 151–162. American Mathematical Society (2011). https://doi.org/10.1090/conm/559
32. Pratt, R.G.: Seismic waveform inversion in the frequency domain, Part 1: Theory and verification in a physical scale model. Geophysics **64**, 888–901 (1999)
33. Ramm, A.G.: Multidimensional inverse scattering problems. Wiley, New York (1992)
34. Rodriguez, S., Deschamps, M., Castaings, M., Ducasse, E.: Guided wave topological imaging of isotropic plates. Ultrasonics **54**, 1880–1890 (2014)
35. Romanov, S.: Optimization of numerical algorithms for solving inverse problems of ultrasonic tomography on a supercomputer. In: Voevodin, V., Sobolev, S. (eds.) Supercomputing. RuSCDays 2017. Communications in Computer and Information Science, vol. 793, pp. 67–79. Springer, Cham (2017)
36. Romanov, S.Y.: Supercomputer simulations of nondestructive tomographic imaging with rotating transducers. Supercomput. Front. Innov. **5**(3), 98–102 (2018). https://doi.org/10.14529/jsfi180318
37. Ruiter, N.V., Zapf, M., Hopp, T., Gemmeke, H., van Dongen K.W.A.: USCT data challenge. In: Duric N., Heyde B. (eds.) Medical Imaging 2017: Ultrasonic Imaging and Tomography. Proceedings of SPIE vol. 10139 (SPIE, Bellingham, WA, 2017) 101391 N
38. Schmitz, V., Chakhlov, S., Müller, W.: Experiences with synthetic aperture focusing in the field. Ultrasonics **38**, 731–738 (2000)
39. Seidl, R., Rank, E.: Iterative time reversal based flaw identification. Comput. Math. Appl. **72**, 879–892 (2016)
40. Vinard, N., Martiartu, N.K., Boehm, C., Balic, I.J., Fichtner, A.: Optimized transducer configuration for ultrasound waveform tomography in breast cancer detection. In: Duric N., Heyde B. (eds.) Medical Imaging 2018: Ultrasonic Imaging and Tomography. Proceedings of SPIE vol. 10580 (SPIE, Bellingham, WA, 2018) 105800I
41. Virieux, J., Operto, S.: An overview of full-waveform inversion in exploration geophysics. Geophysics **74**, WCC1–WCC26 (2009)
42. Voevodin, Vl., et al.: Supercomputer Lomonosov-2: large scale, deep monitoring and fine analytics for the user community. Supercomput. Front. Innov. **6**(2), 4–11 (2019)

Supercomputer Implementation of a High Resolution Coupled Ice-Ocean Model for Forecasting the State of the Arctic Ocean

Leonid Kalnitskii[1,2](✉), Maxim Kaurkin[2], Konstantin Ushakov[1,2], and Rashit Ibrayev[1,2]

[1] Marchuk Institute of Numerical Mathematics, Russian Academy of Sciences, Moscow, Russia
leoni.yurevic@gmail.com, ushakovkv@mail.ru, ibrayev@mail.ru
[2] Shirshov Institute of Oceanology, Russian Academy of Sciences, Moscow, Russia
sherema@yandex.ru

Abstract. The paper describes the construction of the coupled ocean-ice Global model INMIO-CICE-CMF2.0 for predicting the state of water and ice in the Arctic Ocean with high spatial resolution (0.1°). The 3D ocean model INMIO is developed at the Institute of Numerical Mathematics and Institute of Oceanology, Russian Academy of Sciences. The sea-ice model CICE (Community Ice CodE) is developed Los Alamos National Laboratory. The models are fully coupled at each four-time steps using own software named Compact Modeling Framework (CMF ver.2.0). Outputs are the surface variables sea level and ice conditions (concentration, thickness, velocity, convergence, strength, etc.) and 3-dimensional maps of current, temperature and salinity. The main aim of the research is developed the algorithm to find the optimal processor configuration of the coupled ocean-ice model for increase performance and optimizing computer resources usage. This is nontrivial problem because these two models are not uniformed and are used for numerical experiments with thousands of processors cores on Supercomputers with shared memory using MPI technology. The theoretical and practical aspects of the problem are discussed.

Keywords: Arctic Ocean · Coupled models · Sea ice · Ocean · High resolution · Parallel computations

1 Introduction

The processes taking place in the Arctic Ocean are one of the most important elements of the general circulation of the World Ocean. The Earth's climate change has a strong influence on the state of the waters of the Arctic Ocean, in particular, it provokes an accelerated reduction in ice cover in the Arctic Ocean, and especially in the so-called Atlantic region, including the Barents Sea and the western part of the Nansen Basin. It is here that in recent years there has been a noticeable reduction in the area of sea ice not only in the summer season, but also at the peak of its seasonal distribution [1].

V. Voevodin and S. Sobolev (Eds.): RuSCDays 2020, CCIS 1331, pp. 332–340, 2020.
https://doi.org/10.1007/978-3-030-64616-5_29

All this causes great interest in the study of the circulation of water and sea ice in the Arctic Ocean. A large number of works have been devoted to numerical studies of the circulation of the Arctic basin, including those carried out as part of the AOMIP and FAMOS projects [2]. Among the works of Russian scientists, it is worth noting the studies conducted at the AARI [3]; at the INM RAS, these are works using the sigma model of general ocean circulation [4] and calculations of ice dynamics [5, 6], at the ICM&MG SB RAS [7, 8]. In such works, first of all, attention was paid to studies of the climate and seasonal variability of the Arctic Ocean. The main question was to bring the quality of ocean and sea ice models to a level that allows us to describe several basic features of the circulation of water in the Arctic Ocean.

The high-resolution model of the ocean plays an important role in reliably modeling ocean dynamics. An explicit reproduction of vortex dynamics is possible starting from scales comparable to the Rossby radius, which is several kilometers in the Arctic. The need to use high spatial resolution in modeling the World Ocean and the Arctic is revealed in the works [9–12].

However, with the increase in spatial and temporal resolution and with the complexity of the models used, the computational cost of the calculation also grows. Besides the issue of optimizing each component individually, there is the issue of speeding up their joint use. The main aim of the research is developed the algorithm to find the optimal processor configuration of the coupled ocean-ice model for increase performance and optimizing computer resources usage. This is nontrivial problem because these two models are not uniformed and are used for numerical experiments with thousands of processors cores on Supercomputers with shared memory using MPI technology.

2 Model Components

This work is a step in the development of the eddy-resolving model of the World Ocean [11]. Of particular interest from the point of view of modeling is the polar region of the World Ocean, since sea ice has a significant impact on the processes taking place here. However, at present, numerical modeling of the global ocean together with sea ice is difficult due to limited computing resources. According to our estimates, the integration of the World Ocean 1/10 $\times$ 1/10 $\times$ 49 model (horizontal resolution 0.1°, 49 vertical levels) based on the coupled ice-ocean model requires 13,500 processors hours per a model year on a supercomputer like MVS-10P MP Petastream (JSCC RAS).

Our coupled model consists of INMIO ocean dynamics model [11, 13] and CICE sea ice model [14]. The ocean dynamics model INMIO has jointly developed at the Institute of Computational Mathematics of the Russian Academy of Sciences. G.I. Marchuk and Institute of Oceanology RAS. P.P. Shirshov for the study of marine and oceanic hydrodynamic processes. The model is based on the complete system of equations of the three-dimensional dynamics of the ocean in the Boussinesq and hydrostatic approximations. In this work, we used such configuration of the ocean model to successfully reproduce the main elements of the circulation of the Arctic Ocean. The time step for baroclinic ocean processes is 3 min.

The sea ice model CICE5.1 is used in our simulation. Sea ice is considered in this model in terms of a function of the concentration distribution over thickness categories.

The configuration we use includes 5 categories of ice and one of snow, elastic-viscous-plastic rheology for dynamics parameterization, upwind transport scheme. We apply 7 layers mushy thermodynamics to update ice temperature as well as salinity and enthalpy too. Thermodynamic time step is 4 min; for every thermodynamic step, 240 dynamic subcycles are necessary.

Atmospheric forcing is defined by cyclically repeated intra-annual fluctuations in atmospheric parameters, precipitation, and solar radiation, taken from the CNYFv2 database in accordance with the CORE-I protocol [15]. Wind speed, air temperature and specific humidity are set at a height of 10 m and are transmitted to the ice/ocean system every 6 h.

The INMIO4.1 ocean dynamic model and the CICE5.1 sea ice model operate in parallel mode on computers with distributed memory. Collaboration and exchange of boundary conditions are also carried out in parallel mode by the Compact Modeling Platform CMF2.0, which interpolate model fields on the corresponding grids when transferring data between components, make the intramodel communications (halo exchanges on tripolar and latitude-longitude grids) and work with the file system [16]. The ocean model transmits the ice models the temperature and salinity on the ocean surface, horizontal velocities and the freezing–melting potential of the upper computational cell; the ice model send summarized in ice categories concentration, horizontal friction stress and fluxes of mass, salt, heat and penetrating short-wave radiation. Flux values are multiplied by the ice concentration inside the ice model before being transferred to the ocean to avoid the interpolation error that occurred in cells with intense ice formation processes on an ice edge. Thus, the concentration of ice in the ocean component is used as a mask for ocean-atmosphere fluxes. Note that such an interpolation error appeared only when switching to high resolution and was not present at a coarser resolution.

3 Model Optimization

Ocean and ice model components use similar B-type grids with 3600×400 horizontal points. To save the instantaneous state of the ocean, 4 three-dimensional (two velocity components, temperature and salinity) and 8 two-dimensional arrays (sea surface level, barotropic velocity, etc.) are written to restart file, which takes about 1 GB storage space. During the calculations, approximately 90 three-dimensional arrays are used to store the current, past, and future states, grid arrays, flows, etc., so more than 25 GB of memory is required for the ocean component.

In this work we use the ice model with 5 thickness categories and 7 layers mushy thermodynamics, and thus the size of restart file is roughly 2 GB. Totally CICE model requires about 15 GB of memory to run.

Ice-ocean exchange occurs every 12 min, that corresponds to the 3 and 4 steps of the ice and ocean models, respectively. We chose this exchange rate based on the following considerations. On the one hand, as was shown in [17], the exchange period should not exceed 1 h. However, we also observed instability at a half-hour interval between exchanges. On the other hand, an increase in frequency leads to exchange costs rise.

The issue of computational performance is quite significant for us because of the high computational cost of the task. The effectiveness of a coupled model depends both on the performance of each model component individually and on the organization of joint work. One possible solution is to precisely distribute the processor cores among the components. To find the optimal ratio, we carried out a series of short starts, which differ in the number of core processors used per component. We take the wall-clock time as a marker value.

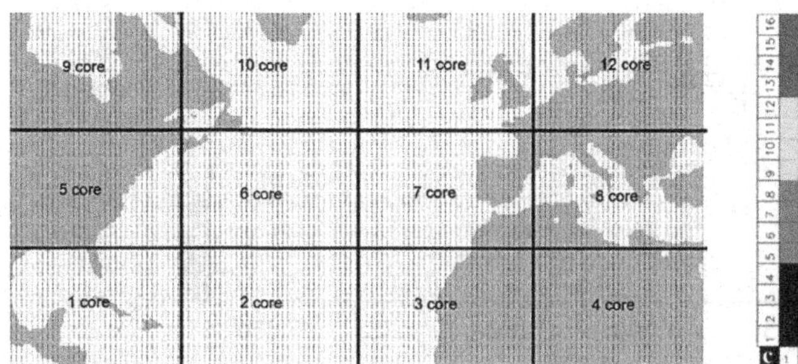

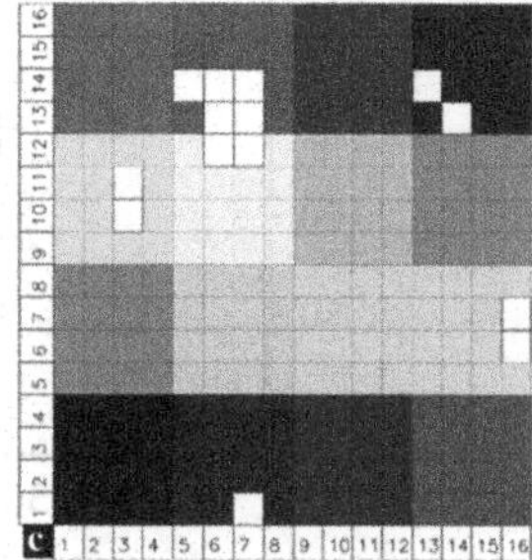

Fig. 1. Two-dimensional processor decomposition of the region in the INMIO (on the left) and CICE. (on the right, from [14]) models. Each subdomain processes a separate processor core.

The two-dimensional decomposition of the INMIO model is used in our experiments (Fig. 1). As was shown earlier [18], a significant fraction of the operating time of the ocean model is made by halo exchanges between domains in a shallow water block. Therefore, to reduce the amount of data sent in this process, we choose domain decomposition of the model grid at which the perimeter of domains is minimal, i.e. the shape of the domains is close to square. The CICE model is divided into domains in the similar way. There is no such obvious imbalance problem that can be observed in the case of a square decomposition of a global model, because in the region under consideration, ice is present in many domains. It is worth noting that in this problem formulation, the loading of the computational cores is nonuniform, not least because ice cover is seasonal.

For each possible core distribution among the components, a launch of 4 model days was carried out for the middle of each seasons. Figure 2 shows the color of the dependence of the execution time on the core number of each component. After that, the dependencies of the walltime on the number of cores allocated to one component were constructed and analyzed with a fixed number of cores from the other. Then each dependence on a double logarithmic scale was approximated by the least squares method to two intersecting straight lines (Fig. 3). We assumed that the time for performing calculations within each component does not depend on the speed of performing calculations on the second component. However, a faster component is forced to wait for a slower execution. A linear scaling was expected with a small number of cores allocated for the varying component. In contrast, this component should be executed naturally faster with a large number of cores than the other, so a further increase component cores does not affect the overall execution time. In other words, the execution time depends on the time of the slowest component, and we assumed that the scalability of each component individually is linear and exchanges between components do not have a significant impact on runtime. From this supposition, the following equation describes the relationship between runtime (T)

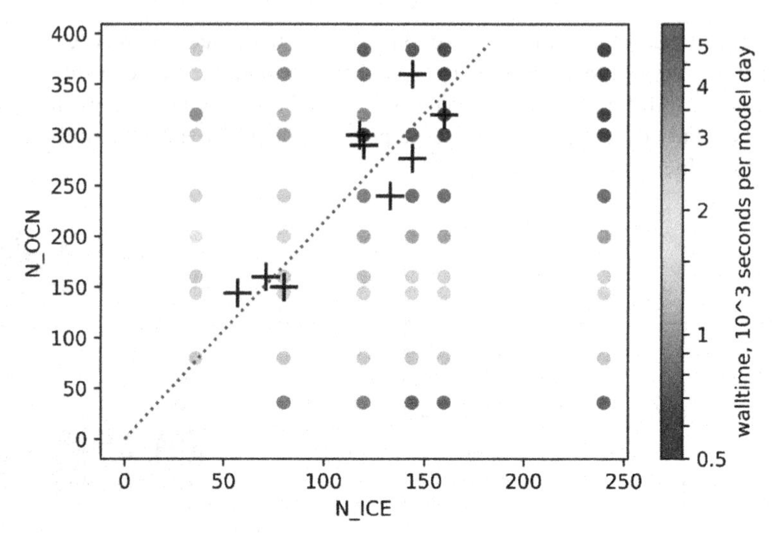

Fig. 2. Wall-clock time for one model day (color) as a function of core number per ice (N_ICE, horizontal axis) and ocean (N_OCN, vertical axis) component. Optimal core distribution is marked with "+".

and the number of processors per each components (N_1, N_2): $T = max\left(\frac{C_1}{N_1}; \frac{C_2}{N_2}\right)$, where C_1 and C_2 are the computational costs of each components. Thus, the break point (Fig. 3) corresponds to the optimal distribution of cores between the components, for which there is no timeout in one component relative to another.

The procedure described above was applied separately for each fixed number of processor cores per ice (N_ICE) and ocean (N_OCN) component. The optimal points obtained in this way are marked with "+" in the Fig. 2. Note that not with each such dependence it was possible to find the break point by the algorithm described above; in some cases, the dependence remains linear everywhere, in some it was impossible to do this with high reliability, due to lack of data or large error. The line marked with dots in Fig. 2 corresponds to the ratio between the components approximately $N_{ICE} : N_{OCN} =$ 2:1.

Another possible way to determine this ratio is to consider the dependence of the computational cost of one component on the core ratio between the components. Under the assumptions described earlier, we obtain the following theoretical dependence: $T \cdot N_1 = max(C_1; \alpha \cdot C_2)$, where $\alpha = N_1{:}N_2$. Thus, a constant dependence is obtained with a ratio less than optimal. But with larger values a linear dependence is obtained on a double logarithmic scale. In accordance with this approach, the resulting optimal ratio is close to the value obtained above (Fig. 4). The constant value corresponds to the computational cost of the launch of the ocean component in the stand alone experiment without ice component. The spread of values can be caused by factors unaccounted for in this experiment, such as the decomposition of each component and the data exchange between them.

To validate the model, we performed a numerical experiment to reproduce the intra-annual variability of the state of ice and the ocean in the Arctic Ocean. The experiment

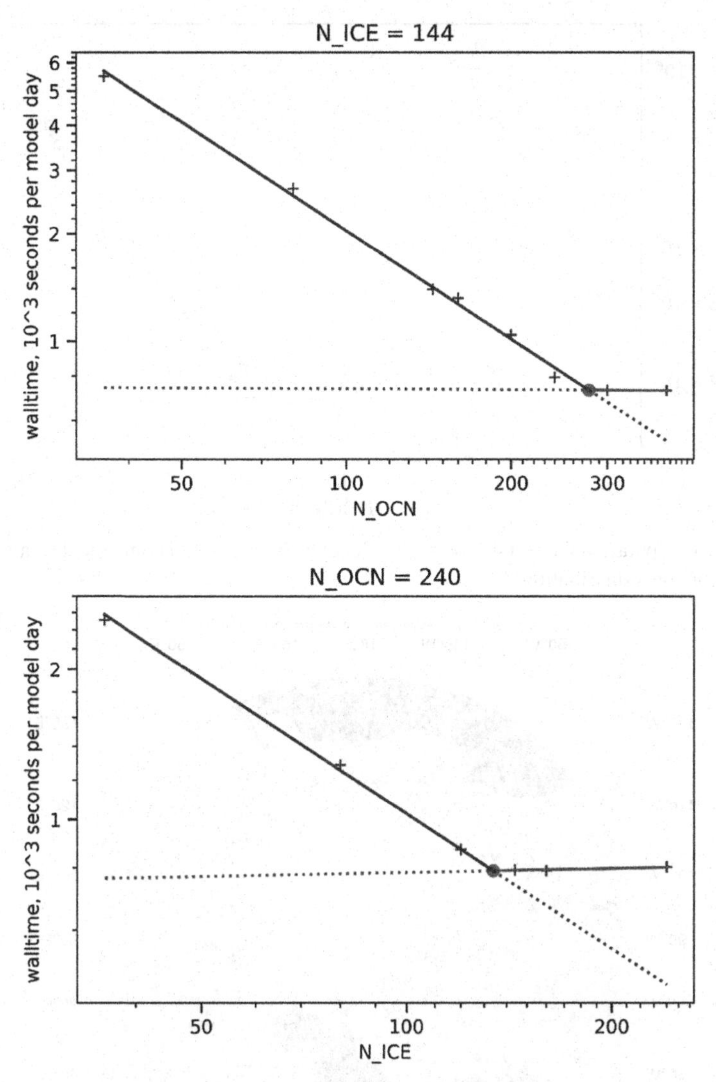

Fig. 3. Wall-clock time per model day as a function of the processor cores of one component (N_OCN for INMIO at the top, N_ICE for CICE is below) with another fixed.

was conducted in accordance with the CORE-I protocol for 5 model years. Based on the study, we used 384 and 180 cores for the ocean and ice components, respectively. The fields of the state of ice, the integrated characteristics of the currents, the fields of salinity and ocean temperature were analyzed and compared with observations. The Fig. 5 shows the field of ice concentration at the time of greatest distribution. The modeling of the interseasonal variability of the ice cover in the Arctic showed good agreement with the observational data and results of other models. Detailed model validation results are published in [19].

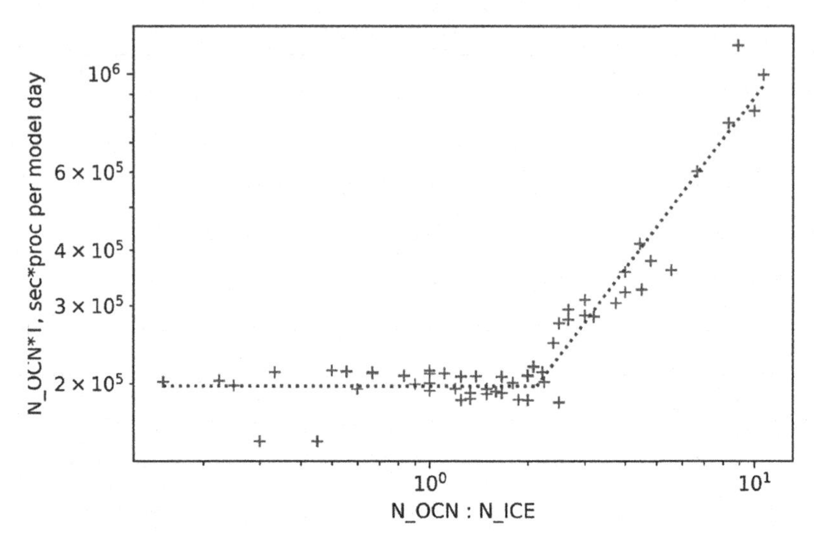

Fig. 4. The computational cost (s*proc per model day) for the ocean component as a function of the processor cores distribution.

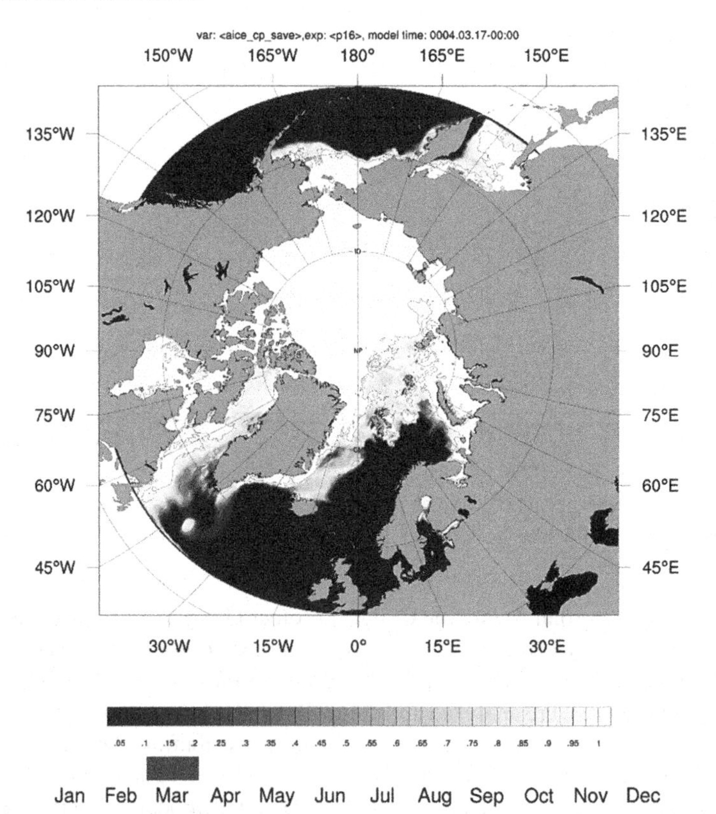

Fig. 5. Sea ice concentration field in March after 5 years model integration.

4 Conclusion

In the research we present an algorithm for optimal loading computer resources in the case of coupling two parallel program codes, each of which runs on hundreds of cores. This problem arose in development of a coupled high-resolution ice-ocean model for the study of the Arctic basin. The models of the ocean dynamics INMIO [11, 13] and of the sea ice CICE [14] are well scalable. It occurs that distribution of available number of cores between two components of coupled model in nontrivial, if one needs to get maximum performance.

The practical experiments shown that usage of optimal processors configuration (OCN_1 = 384 and ICE_1 = 180 cores) can accelerate computations on 30% for ocean-ice model compare to non-optimal configuration (OCN_2 = 280 and ICE_2 = 280 cores) on the same processors pool (560 cores).

Acknowledgements. The research was supported by the Russian Science Foundation (project no. 19-77-00104) and performed at the Shirshov Institute of Oceanology, Russian Academy of Sciences, Moscow, Russia. The research was carried out using the high-performance computing resources at the Joint Supercomputer Center of the RAS (www.jscc.ru) and of Marine Hydrophysical Institute of the RAS.

References

1. Ivanov, V., et al.: Arctic ocean heat impact on regional ice decay: A suggested positive feedback. J. Phys. Oceanogr. No 46. C., 1437–1456 (2016). https://doi.org/10.1175/jpo-d-15-0144.1
2. Proshutinsky, A., Aksenov, Y., Kinney, J.C., et al.: Recent advances in Arctic Ocean studies employing models from the Arctic Ocean model intercomparison project. Oceanogr. **24**(3), 102–113 (2011)
3. Kulakov, M.Y., Makshtas, A.P., Shutilin, S.V.: AARI-IOCM: Combined model of water circulation and ices in the Arctic Ocean. Probl. Arkt. Antarkt. **2**, 6–18 (2012)
4. Moshonkin, S.N., Bagno, A.V., Gusev, A.V., Diansky, N.A., Zalesny, V.B., Alekseev, G.V.: Numeracal simulation of the North Atlantic-Arctic Ocean-Bering Sea circulation in the 20th Century. Russian J. Numer. Anal. Math. Modell. **26**(2), 161–178 (2011). https://doi.org/10.1515/RJNAMM.2011.009
5. Yakovlev, N.G.: Reproduction of the large-scale state of water and sea ice in the Arctic ocean in 1948–2002: Part I. Numerical model. Izv. Atmos. Ocean. Phys. **45**(3), 357–371 (2009)
6. Yakovlev, N.G.: Reproduction of the large-scale state of water and sea ice in the Arctic Ocean from 1948 to 2002: Part II. The state of ice and snow cover. Izv. Atmos. Ocean. Phys. **45**(4), 478–494 (2009)
7. Golubeva, E.N., Platov, G.A., Iakshina, D.F.: Numerical simulations of the current state of waters and sea ice in the Arctic Ocean. Ice Snow **55**(2), 81–92 (2015). (in Russian)
8. Golubeva, E.N., Platov, G.A.: On improving the simulation of Atlantic water circulation in the Arctic Ocean. J. Geophys. Res., **112** (2007). https://doi.org/10.1029/2006jc003734
9. Chen, C., et al.: Circulation in the arctic ocean: Results from a high-resolution coupled ice-sea nested global-fvcom and arctic-fvcom system. Prog. Oceanogr. **141**, 60–80 (2016). https://doi.org/10.1016/j.pocean.2015.12.002

10. Maltrud, M.E., McClean, J.L.: An eddy resolving global 1/10 ocean simulation. Ocean Model. **8**, 31–54 (2005)
11. Ibrayev, R.A., Khabeev, R.N., Ushakov, K.V.: Eddy-resolving 1/10 model of the World Ocean. Izv. Atmos. Ocean. Phys. **48**(1), 37–46 (2012). https://doi.org/10.1134/S0001433812010045
12. Ushakov, K.V., Ibrayev, R.A.: Assessment of mean world ocean meridional heat transport characteristics by a high-resolution model. Russian J. Earth Sci. **18**(1), 1–14 (2018)
13. INMIO Ocean model. http://model.ocean.ru
14. Hunke, E.C., Lipscomb, W.H., Turner, A.K., Jeffery, N., Elliott, S.: CICE: The Los Alamos Sea Ice Model Documentation and Software User's Manual Version 5.1.—2015. URL: http://oceans11.lanl.gov/trac/CICE/attachment/wiki/WikiStart/cicedoc.pdf?format=raw
15. Griffies, S.M., Biastoch, A., Bning, C., et al.: Coordinated ocean-ice reference experiments (COREs). Ocean Model. **26**(1–2), 1–46 (2009). https://doi.org/10.1016/j.ocemod.2008.08.007
16. Kalmykov, V.V., Ibrayev, R.A., Kaurkin, M.N., Ushakov, K.V.: Compact Modeling Framework v3.0 for high-resolution global ocean-ice-atmosphere models. Geosci. Model Dev., **11**, 3983–3997 (2018). https://doi.org/10.5194/gmd-11-3983-2018
17. Fadeev, R.Y., Ushakov, K.V., Tolstykh, M.A., Ibrayev, R.A.: Design and development of the SLAV-INMIO-CICE coupled model for seasonal prediction and climate research. Russ. J. Numer. Anal. Math. Modell. **33**(6), 333–340 (2018). https://doi.org/10.1515/rnam-2018-0028
18. Ushakov, K.V., Ibrayev, R.A., Kaurkin, M.N.: INMIO high resolution global ocean model as a benchmark for different supercomputers. Russian supercomputer days, 23–24 September 2019, Moscow, MAKS Press, pp. 208–209 (2018)
19. Kalnitskii, L.Y., Kaurkn, M.N., Ushakov, K.V., Ibrayev, R.A.: Seasonal variability of water and sea ice circulation in the high-resolution model of the Arctic Ocean. Izv. Atmos. Ocean. Phys. **56**, 522–533 (2020)

Supercomputer Modeling of the Hydrodynamics of Shallow Water with Salt and Heat Transport

Alexander Sukhinov[1], Alexander Chistyakov[1,2], Vladimir Litvinov[1,3(✉)], Asya Atayan[1,2], Alla Nikitina[2,4,5], and Alena Filina[2,5]

[1] Don State Technical University, Rostov-on-Don, Russia
sukhinov@gmail.com, cheese_05@mail.ru, litvinovvn@rambler.ru, atayan24@mail.ru
[2] Scientific and Technological University "Sirius", Sochi, Russia
nikitina.vm@gmail.com, j.a.s.s.y@mail.ru
[3] Azov-Black Sea Engineering Institute of Don State Agrarian University, Zernograd, Russia
[4] Southern Federal University, Rostov-on-Don, Russia
[5] Supercomputers and Neurocomputers Research Center, Taganrog, Russia

Abstract. The paper covers the investigation of hydrodynamics processes in shallow water taking into account the salt and temperature transport on supercomputer. Proposed model is based on the equations of motion (Navier-Stokes), the continuity equation for the case of variable density, as well as the equations of transport of salts and heat. The complex geometry of the coastline and bottom surface of natural reservoirs involves solving a number of problems related to the selection of parameters and the construction of the computational grid used for the numerical implementation of a discrete analogue of the mathematical model of hydrodynamic processes. Improving the accuracy of calculations, fulfilling the conditions of convergence and stability of the numerical solution of the problem involves crushing the step of the computational grid, which leads to the need to use supercomputer computing systems to obtain calculation results for a given period of time. The developed numerical methods and algorithms formed the basis of a software package tested on a supercomputer cluster using OpenMP and MPI technologies. Algorithms are being developed that adapt individual elements of the software package for the NVIDIA CUDA architecture and cloud computing.

Keywords: Shallow water · Salt and heat transport · Hydrodynamics · Parallel algorithm · Software

1 Introduction

The implementation of large-scale engineering projects affecting the ecological state of water requires prompt forecasting in order to prevent irreparable

V. Voevodin and S. Sobolev (Eds.): RuSCDays 2020, CCIS 1331, pp. 341–352, 2020.
https://doi.org/10.1007/978-3-030-64616-5_30

impact on the environment. In the event of an emergency in the water area of the reservoir, it is necessary to forecast hydrodynamic processes in an accelerated time scale, which is almost impossible without the use of modern high-performance computing systems and parallel algorithms. The solution of such problems requires a complex of interconnected mathematical models, including the three-dimensional mathematical model of hydrodynamics developed by the authors, taking into account the transport of salts and heat.

Among the works of Russian scientists devoted to the research and forecast of hydrodynamics of shallow water, the following researches are represented by such authors as Marchuk G.I. [1], Matishov G.G. [2], Belotserkovskii O.M.[3], Yakyshev E.V., Ilyichev V.G., etc.

Existing software for simulation hydrodynamical processes (SALMO, MARS-3D, CHTDM, CARDINAL, PHOENICS, ECOINTEGRATOR, etc.) do not have the necessary accuracy for simulation vortex structures of currents, do not fully utilize computing resources, are not conservative, do not take into account the complex shape of bottom and shore topography, evaporation, river flows, salinity, temperature, and other factors, and are also unstable at significant depth differences and changes in the density of water environment.

2 Problem Statement

The developed model for calculation three-dimensional fields of water velocity vector, temperature and salinity is based on the mathematical hydrodynamic model of shallow waters, which takes into account the heat and salts transport [3,4]:

- the Navier-Stokes motion equation

$$u_t' + uu_x' + vu_y' + wu_z' = -\frac{1}{\rho}P_x' + (\mu u_x')_x' + \left(\mu u_y'\right)_y' + (\nu u_z')_z' + \\ + 2\Omega(v\sin\theta - w\cos\theta), \quad (1)$$

$$v_t' + uv_x' + vv_y' + wv_z' = -\frac{1}{\rho}P_y' + (\mu v_x')_x' + \left(\mu v_y'\right)_y' + (\nu v_z')_z' - 2\Omega u\sin\theta, \quad (2)$$

$$w_t' + uw_x' + vw_y' + ww_z' = -\frac{1}{\rho}P_z' + (\mu w_x')_x' + \left(\mu w_y'\right)_y' + (\nu w_z')_z' + \\ + 2\Omega u\cos\theta + g\left(\rho_0/\rho - 1\right), \quad (3)$$

- the continuity equation in the case of variable density

$$\rho_t' + (\rho u)_x' + (\rho v)_y' + (\rho w)_z' = 0, \quad (4)$$

– the heat transport equation:

$$T_t' + uT_x' + vT_y' + wT_z' = \left(\mu T_x'\right)_x' + \left(\mu T_y'\right)_y' + \left(\nu T_z'\right)_z' + f_T, \quad (5)$$

– the salt transport equation:

$$S_t' + uS_x' + vS_y' + wS_z' = \left(\mu S_x'\right)_x' + \left(\mu S_y'\right)_y' + \left(\nu S_z'\right)_z' + f_S, \quad (6)$$

where $\mathbf{V} = (u, v, w)$ are velocity vector components; P is the full hydrodynamic pressure; ρ is the water density; μ, ν are horizontal and vertical components of the turbulent exchange coefficient; $\boldsymbol{\Omega} = \Omega\,(\cos v \cdot \mathbf{j} + \sin v \cdot \mathbf{k})$ is the angular velocity of the Earth rotation; θ is the latitude of region; g is the gravity acceleration; f_T, f_S are sources of heat and salt (located on the border of region); T is the water temperature; S is the water salinity.

Two components are conditionally distinguished from the total hydrodynamic pressure: the pressure of the liquid column and the hydrodynamic part [5]:

$$P\,(x, y, z, t) = p\,(x, y, z, t) + \rho_0 g z, \quad (7)$$

where p is the hydrostatic pressure of the unperturbed fluid; ρ_0 is the fresh water density under normal conditions.

State equation for density:

$$\rho = \tilde{\rho} + \rho_0, \quad (8)$$

where ρ_0 is the fresh water density under normal conditions; $\tilde{\rho}$ is determined by the equation recommended by UNESCO:

$$\begin{aligned}\tilde{\rho} = \tilde{\rho}_w &+ (8.24493 \cdot 10^{-1} - 4.0899 \cdot 10^{-3}T + 7.6438 \cdot 10^{-5}T^2 \\ &- 8.2467 \cdot 10^{-7}T^3 + 5.3875 \cdot 10^{-9}T^4)S + (-5.72466 \cdot 10^{-3} \\ &+ 1.0227 \cdot 10^{-4}T - 1.6546 \cdot 10^{-6}T^2)S^{3/2} + 4.8314 \cdot 10^{-4}S^2, \quad (9)\end{aligned}$$

where $\tilde{\rho}_w$ is the fresh water density, specified by a polynomial [6]:

$$\begin{aligned}\tilde{\rho}_w = &\ 999.842594 + 6.793952 \cdot 10^{-2}T - 9.095290 \cdot 10^{-3}T^2 \\ &+ 1.001685 \cdot 10^{-4}T^3 - 1.120083 \cdot 10^{-6}T^4 + 6.536332 \cdot 10^{-9}T^5. \quad (10)\end{aligned}$$

Equation 9 is applicable for salinity in the range 0–42 ‰ and temperature from −2 to 40 °C.

The system of Eqs. (1)–(6) is considered under the following boundary conditions [7,8]:

– at the input:

$$\mathbf{V} = \mathbf{V}_1, P_{\mathbf{n}}' = 0, T = T_1, S = S_1, \quad (11)$$

– the bottom boundary:

$$\rho_v \mu \left(\mathbf{V}_{\boldsymbol{\tau}}\right)'_{\mathbf{n}} = -\boldsymbol{\tau}, \mathbf{V}_{\mathbf{n}} = 0, P_{\mathbf{n}}{}' = 0, T_{\mathbf{n}}{}' = 0, S_{\mathbf{n}}{}' = 0, f_T = 0, f_S = 0, \tag{12}$$

– the lateral boundary:

$$\left(\mathbf{V}_{\boldsymbol{\tau}}\right)'_{\mathbf{n}} = 0, \mathbf{V}_{\mathbf{n}}{}' = 0, P_{\mathbf{n}}{}' = 0, T_{\mathbf{n}}{}' = 0, S_{\mathbf{n}}{}' = 0, f_T = 0, f_S = 0, \tag{13}$$

– the upper boundary:

$$\rho_v \mu \left(\mathbf{V}_{\boldsymbol{\tau}}\right)'_{\mathbf{n}} = -\boldsymbol{\tau}, w = -\omega - P'_t/\rho g, P'_n = 0,$$
$$T_{\mathbf{n}}{}' = 0, S_{\mathbf{n}}{}' = 0, f_T = k\left(T_a - T\right), f_S = \frac{\omega}{h_z - \omega} S, \tag{14}$$

– at the output (Kerch Strait):

$$\mathbf{V}_{\mathbf{n}}{}' = 0, P_{\mathbf{n}}{}' = 0, T_{\mathbf{n}}{}' = 0, S_{\mathbf{n}}{}' = 0, f_T = 0, f_S = 0, \tag{15}$$

where ω is the liquid evaporation rate, equal to 606 m^3/s; $\mathbf{n}$ is outer normal vector to the boundary of computational domain; $\mathbf{V}_n, \mathbf{V}_{\boldsymbol{\tau}}$ are normal and tangential components of the velocity vector; $\boldsymbol{\tau} = \{\tau_x, \tau_y, \tau_z,\}$ is the tangential stress vector; ρ is the water density; ρ_v is the sediment density; T_a is the atmospheric temperature, k is the heat transport coefficient between the atmosphere and water environment.

Components of tangential stress for free surface: $\boldsymbol{\tau} = \rho_a C d_s \left|\mathbf{V}\right| \mathbf{V}$, where $\mathbf{V}$ is the wind velocity vector relative to water; ρ_a is the atmosphere density; Cd_s is the dimensionless coefficient of surface resistance, which depends on the wind velocity, is considered in the range 0.0016–0.0032 [9].

Tangential stress for the bottom $\boldsymbol{\tau} = \rho C d_b \left|\mathbf{V}\right| \mathbf{V}$, where $Cd_b = gK^2/h^{1/3}$; K is a group roughness coefficient in the Manning formula, equal to 0.025; $h = H + \eta$ is the water depth; H is the depth of undisturbed surface; η is the height of free surface relative to the geoid (sea level).

The system of Eqs. (1)–(6) is considered under the following initial conditions:

$$\mathbf{V} = \mathbf{V_0}, T = T_0, S = S_0, \tag{16}$$

where $\mathbf{V_0}, T_0$ and S_0 are predefined functions.

3 Solution Method of Hydrodynamic Problem

According to the pressure correction method, the original model of hydrodynamics is divided into three subproblems [10–13]. The first subproblem is represented by the diffusion-convection-reaction equation, used for calculation the components of velocity vector field on the intermediate layer in time:

$$\frac{\tilde{u}-u}{\tau}+u\overline{u}'_x+v\overline{u}'_y+w\overline{u}'_z=(\mu\overline{u}'_x)'_x+(\mu\overline{u}'_y)'_y+(\nu\overline{u}'_z)'_z+2\Omega(v\sin\theta-w\cos\theta),$$
$$\frac{\tilde{v}-v}{\tau}+u\overline{v}'_x+v\overline{v}'_y+w\overline{v}'_z=(\mu\overline{v}'_x)'_x+(\mu\overline{v}'_y)'_y+(\nu\overline{v}'_z)'_z-2\Omega u\sin\theta,$$
$$\frac{\tilde{w}-w}{\tau}+u\overline{w}'_x+v\overline{w}'_y+w\overline{w}'_z=(\mu\overline{w}'_x)'_x+(\mu\overline{w}'_y)'_y+(\nu\overline{w}'_z)'_z+$$
$$+2\Omega u\cos\theta+g\left(\frac{\rho_0}{\rho}-1\right), \quad (17)$$

where τ is the step along the time coordinate; u,v,w are the components of the velocity vector field on the previous layer in time; $\tilde{u}$ is the value of the velocity field on the intermediate layer in time;

Note that the term $g(\rho_0/\rho-1)$ describes the buoyancy (the Archimedes' Power). Numerous experiments on simulation the water environment transport in shallow waters such as the Azov Sea have shown that this term makes a minor contribution to the solution of the problem and can be ignored. Schemes with weights are used to approximate the diffusion-convection-reaction equation in time. Here $\overline{u}=\sigma\tilde{u}+(1-\sigma)u$, $\sigma\in[0,1]$ is the weight of scheme.

The calculation of the pressure distribution (the second subproblem) is based on the Poisson equation:

$$P''_{xx}+P''_{yy}+P''_{zz}=\frac{\breve{\rho}-\rho}{\tau^2}+\frac{(\breve{\rho}\tilde{u})'_x}{\tau}+\frac{(\breve{\rho}\tilde{v})'_y}{\tau}+\frac{(\breve{\rho}\tilde{w})'_z}{\tau}. \quad (18)$$

The value of the velocity field at the upper boundary (water surface) is defined as $w=-\omega-P'_t/\rho g$. As an initial approximation for this problem, a simplified hydrostatic model of the water environment transport was used, which significantly reduces the calculation time. The third subproblem allows us to use explicit formulas to determine the velocity distribution on the next layer in time:

$$\frac{\breve{u}-\tilde{u}}{\tau}=-\frac{1}{\breve{\rho}}P'_x,\ \frac{\breve{v}-\tilde{v}}{\tau}=-\frac{1}{\breve{\rho}}P'_y,\ \frac{\breve{w}-\tilde{w}}{\tau}=-\frac{1}{\breve{\rho}}P'_z, \quad (19)$$

where $\breve{u},\breve{v},\breve{w}$ are the components of the velocity field on the current layer in time.

The computational domain is inscribed in a parallelepiped. A uniform grid is introduced for software implementation of the three-dimensional mathematical model of hydrodynamics. $o_{i,j,k}$ is the "fullness" of the cell (i,j,k). The degree of cell occupancy is defined by the pressure of water column at the cell bottom. In general, the degree of cell occupancy is calculated based on the expression [5]:

$$o_{i,j,k}=\frac{P_{i,j,k}+P_{i-1,j,k}+P_{i,j-1,k}+P_{i-1,j-1,k}}{4\rho g h_z}, \quad (20)$$

where h_z is the sampling step in the vertical direction.

The approximation of calculation problem the velocity field of water environment by spatial variables is based on the balance method taking into account the filling coefficients of control domains.

4 Parallel Implementation

For numerical implementation of the developed model, parallel algorithms oriented on a multiprocessor computer system (MCS) and NVIDIA Tesla K80 graphics accelerator were developed.

Parallel Implementation on Multiprocessor Computer System. We describe parallel algorithm with various types of domain decomposition for solving the problem (1)–(16) on MCS. Parallel algorithms for the modified alternating triangular method (MATM) were implemented on MCS with technical parameters: the peak performance is 18.8 TFlops; 8 computational racks; the computational field of MCS is based on the HP BladeSystem c-class infrastructure with integrated communication modules, power supply and cooling systems; 512 single-type 16-core HP ProLiant BL685c Blade servers are used as computational nodes, each of which is equipped with four 4-core AMD Opteron 8356 2.3 GHz processors and 32 GB RAM; the total number of computational nodes is 2048; the total amount of RAM is 4 TB.

The *k-means* method was used for geometric partition of computational domain for the uniform loading of MCS calculators (processors).

Theoretical estimates of acceleration and efficiency of the developed parallel algorithm [4]:

$$E^t = S^t/p = \chi/\left(1 + \left(\sqrt{p} - 1\right)\left(\frac{36}{50N_z} + \frac{4p}{50t_0}\left[t_n\left(\frac{1}{N_x} + \frac{1}{N_y}\right) + \frac{t_x\sqrt{p}}{N_xN_y}\right]\right)\right),$$

where χ is the ratio of the number of computing nodes to the total number of nodes (computing and fictitious); p is the total number of processors; t_0 is the execution of an arithmetic operation; t_x is the response times (latency); N_x, N_y, N_z are number of nodes in the spatial directions.

The estimation for comparing the efficiency values of algorithms (E_1 is efficiency of the standard algorithm [4]; E_2 is the efficiency of algorithm based on *k-means*) has the form:

$$\delta = \sqrt{\sum_{k=1}^{n}\left(E_{(2)k} - E_{(1)k}\right)^2} \Big/ \sqrt{\sum_{k=1}^{n} E_{(2)k}^2}. \tag{21}$$

We obtained that the efficiency is increased on 10–20% using of algorithm on the basis of *k-means* method for solving problem in the form (1)–(16) compared to the standard algorithm. Increasing of computing nodes from 1 to 128 led the execution time of the *k-means* algorithm decreased from 6.073 to 0.117 s. The maximum acceleration was achieved using of 128 nodes and amounted to 51.933. The efficiency was 0.406 [4].

Parallel Implementation on Graphic Accelerator. Analysis of the CUDA architecture characteristics showed the algorithms for numerical implementation of the developed mathematical model of hydrodynamical processes could be used for creation high-performance information systems.

For numerical implementation of proposed interrelated mathematical model of hydrodynamics, we developed parallel algorithms which will be adapted for hybrid computer systems using the NVIDIA CUDA architecture. Switching to the NVIDIA CUDA architecture will reduce the calculation time from 8–48 h to 2–3 h, depending on the calculation grid size and time step value [14].

Modified Storage Format of the Sparse Matrix with a Repeating Sequence of Elements. The solution of hydrodynamics problem (1)–(16) by the finite difference method (FDM) on uniform grids leads to the necessary operate with sparse matrices, elements of which are a repeating sequence for internal nodes. In the case of high-dimensional problems, this leads to inefficient memory consumption. Using the CSR (Compressed Sparse Rows) matrix storage format avoids the need to store their null elements. However, all nonzero elements, including many repeating, are stored in the corresponding array. This disadvantage is not critical at using computing systems with shared memory. However, this can adversely affect performance at data transferring between nodes in heterogeneous and distributed computing systems. We performed the modification of the CSR format to improve the efficiency of data storage with a repeating sequence of CSR1S elements for modeling continuous biological kinetics processes by the finite difference method. In this case, it is enough to change them in an array that preserves a repeating sequence to change the differential operator, instead of repeatedly finding and replacing values of non-zero elements in an array. The memory capacity for the CSR format was estimated as:

$$P_{csr} = N_{nz}B_{nz} + (N_{nz} + R + 1)B_{idx},$$

for the CSR1S format:

$$P_{csr1s} = B_{nz}[N_{nz}(k_i + 1) - N_{seq}(k_i k_r R + k_r R + 1) - k_i(k_r R - R - 1)],$$

where R is the number of matrix rows; R_{seq} is the number of matrix rows that contain a repeating sequence of elements; N_{nz} is the number of non-zero matrix elements; N_{seq} is the number of elements in a repeating sequence; B_{nz} is the memory capacity to store one non-zero element; B_{idx} is the memory capacity to store one index; $k_r = R_{seq}/R$, $k_i = B_{idx}/B_{nz}$.

Efficient function libraries were developed for solution of grid equations at discretization of model problem (1)–(16) in CSR format on GPUs using CUDA technology. The developed algorithm uses a CSR1S modified data storage format with further conversion to CSR format to solve the resulting system of linear algebraic equations (SLAE) on graphics accelerator using NVIDIA CUDA technology. This raises the problem to develop an algorithm of matrix conversion from CSR1S format to CSR format in minimal time. A number of computational experiments with fivefold repetition and fixation of the average value of the calculation time was performed. The time to solve a 5-diagonal SLAE increased linearly from 26 to 55 s when the dimension was increased from $5 \cdot 10^7$ to 10^8 [14].

5 Software Description and Results of Numerical Experiments

The software complex (SC) "Azov3D" was developed for solution of hydrodynamics and biological kinetics problems with implementation on MCS and graphic accelerator. The SC includes the following modules: control module, oceanological and meteorological databases, application library for solving hydrobiology grid problems, integration with various geoinformation systems (GIS), Global Resource Database (GRID) for geotagging and access to satellite data collection systems, NCEP/NCAR Reanalysis database.

The input data for the hydrodynamic model (1)–(16) are three-dimensional fields of the speeds of the water flow of the Sea of Azov, temperature and salinity. At the initial time, the liquid is at rest. The initial values of the salinity and temperature fields of the Sea of Azov were set using the climate atlas [2] and the Unified State System of Information on the Situation in the World Ocean (ESIMO) [15]. In addition, due to the incompleteness of these literature sources, the salinity and temperature fields of the Sea of Azov were reconstructed from cartographic information based on calculations by the method developed by the authors, for which the Laplace equation was solved on the basis of higher approximation schemes. The data of the ADCP probe were used to verify the adequacy of the hydrodynamic model when calculating the field of the water flow velocity and parameterizing the coefficient of vertical turbulent exchange.

To obtain the salinity function, we can use the diffusion equation solution, which is reduced to the solution of the Laplace equation for long time intervals. However, the salinity function obtained in this way may not have the sufficient smoothness degree at points where the field values are given. Therefore, we use the equation applied to obtain the schemes of high order of accuracy for the

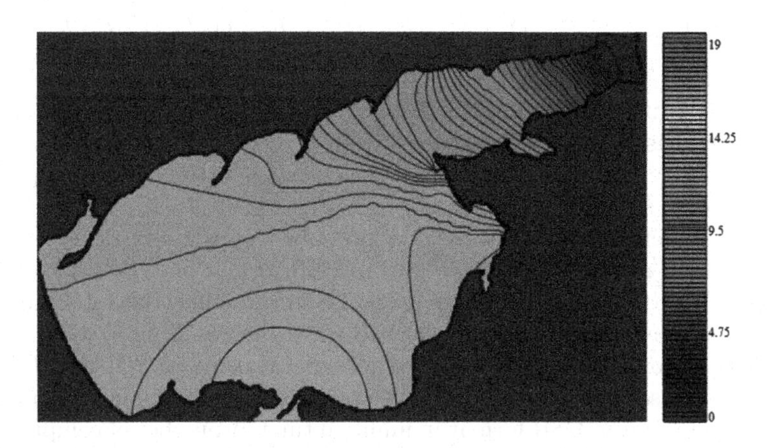

Fig. 1. Restored salt field of the Azov Sea

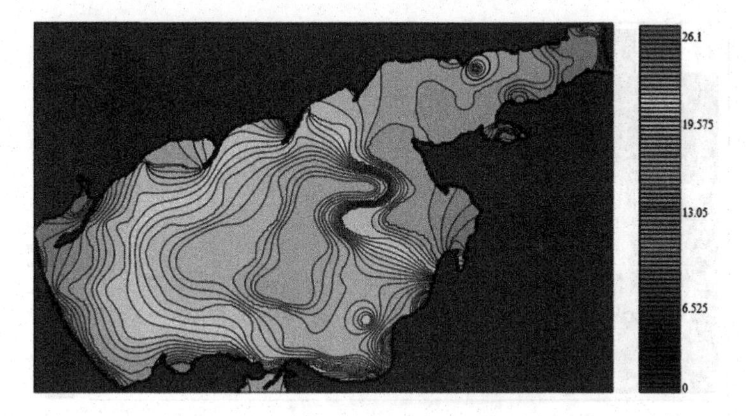

Fig. 2. Restored temperature field of the Azov Sea

Laplace equation:

$$\Delta\varphi - \frac{h^2}{12}\Delta^2\varphi = 0, \varphi \in \{S, T\}, \tag{22}$$

where h is the average distance between the given values of the function φ isolines of the computational domain.

The salinity and temperature isolines were obtained using the recognition algorithm at solving the problem of hydrological information processing. Maps of the salinity and temperature of the Azov Sea were obtained using the interpolation algorithm and by overlaying the boundaries of region (Fig. 1, 2).

As a result of the research, a software was developed to more accurately describe hydrodynamic processes, salts and heat transport in shallow waters, such as the Azov Sea, with complex spatial structures of currents in the conditions of reducing freshwater flow of the Don river, increasing the flow of highly saline waters of Sivash lake and filtering the waters of salt lakes in the North-East of Crimea. Software implementation of mathematical models takes into account the Coriolis force, wind currents and bottom friction, turbulent exchange, evaporation, river flows, as well as the complex geometry of the bottom and coastline. The computational domain corresponds to the physical dimensions of the Azov Sea: the length is 355 km; the width is 233 km; the horizontal step is 1000 m. The time interval is 30 days. On a grid of 350 × 250 × 45 in size and with an estimated interval of one day (time step 500 s), the time for executing a computational experiment on a personal computer is from 8 h to two days, when using a multiprocessor computing system - from 10 min to an hour.

A three-dimensional mathematical model describing hydrophysical processes in shallow waters was developed to reconstruct the ecological disaster. Results of numerical simulation of water environment transport in the Azov Sea water area based on the "Azov3d" software complex are given in Fig. 3.

Complex expedition measurements of water environment parameters of the Azov Sea and Taganrog Bay water areas were performed to update the databases of long-term observations of water environment situation. Based on water area

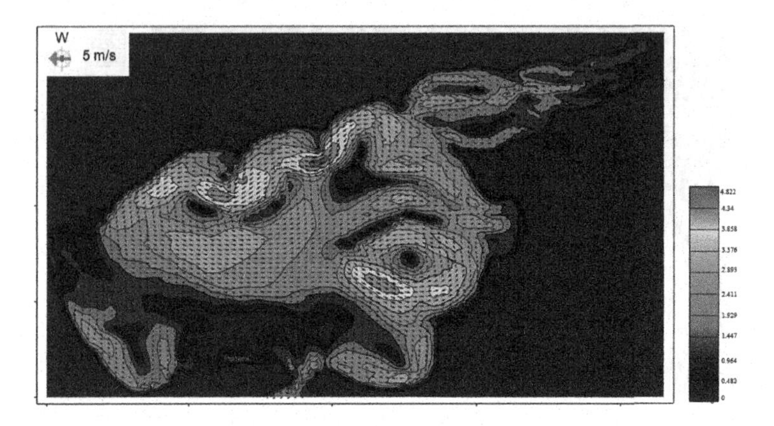

Fig. 3. Results of mathematical modeling the water environment transport (barotropic flows)

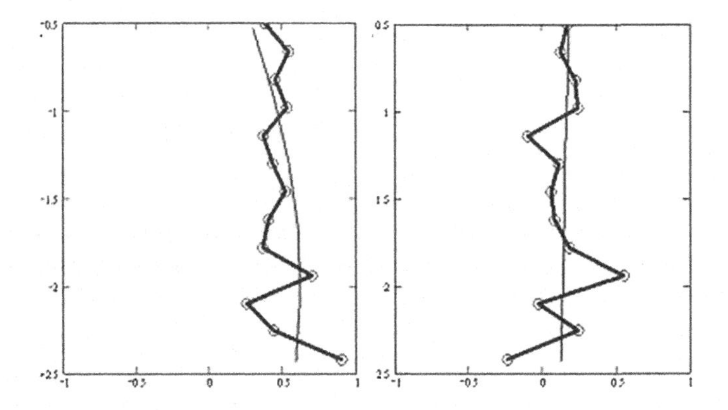

Fig. 4. Profiles of the horizontal component of the velocity vector, directed from West to East on the left, and from North to South – on the right

monitoring, the three-dimensional mathematical models of water environment transport were developed to predict possible development scenarios of the Azov Sea ecosystem, to avoid the occurrence of anaerobic infection areas and to take timely measures to localize them. Profiles of horizontal components of the velocity vector are given in Fig. 4 (thin line are results of numerical experiments; thick line – natural).

The instantaneous values of the velocity vector were measured using the acoustic "ADCP Workhorse 600 Sentinel" Profiler. Using this device, you can get the vertical profile of the three-dimensional velocity vector (128 values with a minimum step of 10 cm along the vertical coordinate). The authors' team was performed the expedition measurements of water environment parameters in the Azov Sea. Based on expedition results, there is a region with overseas phenomena, the appearance of which is seasonal. Results of full-scale measurements of the temperature, salinity and dissolved oxygen profile obtained using

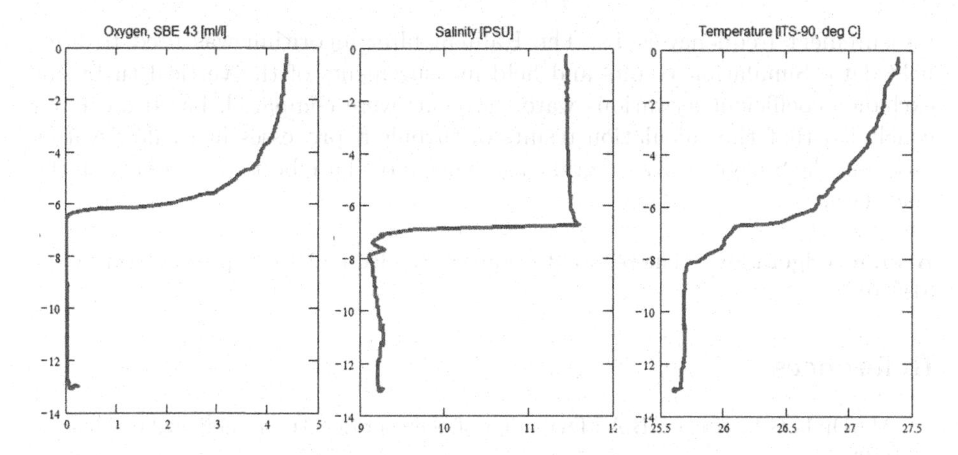

Fig. 5. The dissolved oxygen profile [ml/l] (on the left); the temperature [C] (on the right); the salinity [PSU] (at the center)

the hydrophysical probe "Sea Bird Electronics 19 Plus" at one of the expedition stations are given in Fig. 5, on which the oxygen value at depth of more than 6.5 m is close to zero. The largest water area was subjected to anoxia phenomena at this station.

The measurements were performed vertically, starting from the near sensitivity zone of the ADCP probe to the bottom. The profiler operates on the doppler effect: it transmit the acoustic signal at fixed frequency and receive the signal reflected on the inhomogeneities of water environment in the thickness of water column located under the emitter (starting from the sensitivity zone). The influence of changes in the values of the vertical turbulent exchange coefficient on the content of dissolved oxygen in the bottom layer of shallow water was researched. The values of the vertical turbulent exchange coefficient are close to zero approximately at a depth of 3 m and below. This means the reduced vertical turbulent exchange in this area and explains the hypoxia phenomenon in the bottom layer of the Central-Eastern part of the Azov Sea.

6 Conclusions

The proposed 3D hydrodynamic model of shallow waters allows us to obtain three-dimensional fields of the water flow velocity vector, pressure, sea water density, salinity, and temperature. The geometry of water bottom has a great influence on the flow fields in hydrodynamics models of shallow water. Results of expedition measurements were used at developing hydrodynamic models of shallow waters. Measurements of water flow velocities were performed and based on the ADCP probe, which measures instantaneous values of the vertical profile of the velocity vector. The following parameters were specified at device setting: the vertical step is 10 cm; the number of vertical measurements is 128; the measurement frequency is 1 s. The Kalman filter algorithm was used to filter field data. Simulation results and field measurements of the vertical turbulent exchange coefficient at various water horizons were compared; based on it, we concluded that the calculation results of turbulent processes in shallow waters based on the Smagorinsky subgrid turbulence model are best consistent with the field data.

Acknowledgements. The reported study was funded by RFBR, project number 19-31-51017.

References

1. Marchuk, G.I., Kagan, B.A.: Dynamics of oceans tides. Hydrometeoizdat, Moscow (1983)
2. Matishov, G., et al.: Climatic Atlas of the Sea of Azov. NOAA Atlas NESDIS 59, p. 103. U.S. Government Printing Office, Washington, D.C. (2006)
3. Belotserkovskii, O.M.: Turbulence: New Approaches. Nauka, Moskow (2003)
4. Gushchin, V., Sukhinov, A., Nikitina, A., Chistyakov, A., Semenyakina, A.: A model of transport and transformation of biogenic elements in the coastal system and its numerical implementation. Comput. Math. Math. Phys. **58**, 1316–1333 (2018). https://doi.org/10.1134/s0965542518080092
5. Sukhinov, A., Chistyakov, A., Shishenya, A., Timofeeva, E.: Predictive modeling of coastal hydrophysical processes in multiple-processor systems based on explicit schemes. Math. Models Comput. Simul. **10**, 648–658 (2018). https://doi.org/10.1134/s2070048218050125
6. Sukhinov, A., Khachunts, D., Chistyakov, A.: A mathematical model of pollutant propagation in near-ground atmospheric layer of a coastal region and its software implementation. Comput. Math. Math. Phys. **55**, 1216–1231 (2015). https://doi.org/10.1134/s096554251507012x
7. Marchesiello, P., Mc.Williams, J.C., Shchepetkin, A.: Open boundary conditions for long-term integration of regional oceanic models. Oceanic Model. J. **3**, 1–20 (2001)
8. Wolzinger, N.E., Cleany, K.A., Pelinovsky, E.N.: Long-wave dynamics of the coastal zone. Hydrometeoizdat, Moskow (1989)
9. Androsov, A.A., Wolzinger, N.E.: The straits of the world ocean: a General approach to modelling. Nauka, Saint Petersburg (2005)
10. Tran, J.K.: A predator-prey functional response incorporating indirect interference and depletion. Verh. Internat. Verein. Limnol. **30**, 302–305 (2008)
11. Alekseenko, E., Roux, B., Sukhinov, A.I., Kotarba, R., Fougere, D.: Coastal hydrodynamics in a windy lagoon. Comput. Fluids **77**, 24–35 (2013)
12. Samarskii, A.A., Vabishchevich, P.N.: Numerical methods for solving convection-diffusion problems. URSS, Moscow (2009)
13. Konovalov, A.N.: Method of rapid descent with adaptive alternately-triangular recondition. Differ. Equ. **40**(7), 953–963 (2004)
14. Nikitina, A., et al.: The computational structure development for modeling hydrological shallow water processes on a graphics accelerator. In: AIP Conference Proceeding, vol. 2188, p. 050025 (2019). https://doi.org/10.1063/1.5138452
15. Unified State Ocean Oceans Information System (ESIMO). http://hmc.meteorf.ru/sea/azov/sst/sst_azov.html

Supercomputer Simulations in Development of 3D Ultrasonic Tomography Devices

Alexander Goncharsky and Sergey Seryozhnikov[(✉)]

Moscow Center of Fundamental and Applied Mathematics, Moscow, Russia
goncharOsrcc.msu.ru, s2110sj@gmail.com

Abstract. This study aims to determine the optimal characteristics of ultrasound tomographic scanners for differential breast cancer diagnosis. Numerical simulations of various tomographic schemes were performed on a supercomputer. The parameters of simulations were matched to those of physical experiments. One of the most important problems in designing a tomographic scanner is the choice of a tomographic examination scheme. The layer-by-layer ("2.5D") scheme is the most widely used in medical and industrial tomography. In this scheme, the inverse problem is solved for each 2D plane separately. A fully 3D tomographic scheme is an alternative variant, in which the acoustic properties of the object are reconstructed as 3D images. This study shows that the layered 2.5D scheme has limited capabilities and cannot be used for precise tissue characterization in a general case. The paper presents a comparison of 2.5D and 3D image reconstruction methods in terms of vertical and horizontal resolution, computational complexity of the methods and technical parameters of tomographic scanners. The inverse problem of tomographic image reconstruction is posed as a coefficient inverse problem for the wave equation. The reconstruction algorithms are designed for GPU clusters.

Keywords: Ultrasound tomography · Coefficient inverse problem · Medical imaging · Gpu cluster

1 Introduction

Differential diagnosis of breast cancer at early stages is one of the most pressing issues in modern medicine. Conventional ultrasound technology provides poor image quality, while other existing technologies, such as X-ray, magnetic resonance imaging (MRI), positron-emission tomography (PET) employ hazardous radiation or chemical contrast agents which are unsafe for regular checkups. At present, imaging devices based on ultrasonic tomography are being developed in the USA, Germany, Russia [1–3]. These works are currently at the stage of experiments and prototypes. Ultrasonic tomography is both safe and informative, however, wave propagation physics are much more complicated than X-ray

V. Voevodin and S. Sobolev (Eds.): RuSCDays 2020, CCIS 1331, pp. 353–364, 2020.
https://doi.org/10.1007/978-3-030-64616-5_31

propagation. Ample computing resources are required in order to reconstruct tomographic images under precise mathematical models that account for wave diffraction, refraction, and multiple scattering effects.

The choice of a tomographic imaging scheme is one of the most important choices that must be made in order to design a tomography device. Currently, the "2.5D" layer-by-layer imaging scheme is the most widely used in medical and industrial tomography [1,2,4]. Image reconstruction is performed separately for each 2D section of the object. This approach originates from X-ray tomography, however, ultrasonic wave propagation is always three-dimensional; thus, a 2D model imposes severe limitations on the imaging method. Fully 3D imaging methods are more adequate to reality, however, these methods are much more computationally expensive [5,6]. Only the latest computing technologies made it possible to conduct experiments on 3D wave tomography.

The inverse problem of wave tomography is a nonlinear ill-posed coefficient inverse problem [7–9]. The number of unknowns reaches 10^7. The method of solving this problem is based on explicit computation of the gradient of the residual functional between the measured and simulated wavefields [5,10]. In this study, 2D and 3D tomographic schemes are investigated by means of numerical simulation. Spatial resolution and image reconstruction precision are assessed. Computational complexity of imaging methods and technical parameters of imaging devices are estimated. The parameters of numerical simulations were matched to those of physical experiments conducted at the Lomonosov Moscow State University [11–13].

The results showed that a high spatial resolution 1.5 mm can be achieved even for low-contrast objects such as soft tissues. Sound speed images were reconstructed with high precision. 2D methods provide precise image reconstruction only in some special cases. The image reconstruction algorithms are designed for GPU clusters [14]. The developed software was tested on various GPU devices. The computations were carried out on "Lomonosov-2" supercomputer at Lomonosov Moscow State university [15].

2 The Inverse Problem of Wave Tomography and the Solution Method

The scheme of a tomographic examination is shown in Fig. 1. The object being imaged occupies region G. Region L is filled with water with known properties. The detectors are located on circle Γ. The objective is to reconstruct the acoustic properties in region G using the waves scattered by the object and registered by the detectors. This formulation also applies to the 3D case, in which the surface Γ represents a cylindrical or some other set of detectors surrounding the object.

The inverse problem of wave tomography is posed as a coefficient inverse problem, in which the unknowns are the speed of sound and the absorption factor at each point of the object [7–9]. A scalar wave model based on a second-order hyperbolic differential equation (1) was used to simulate the wave propagation.

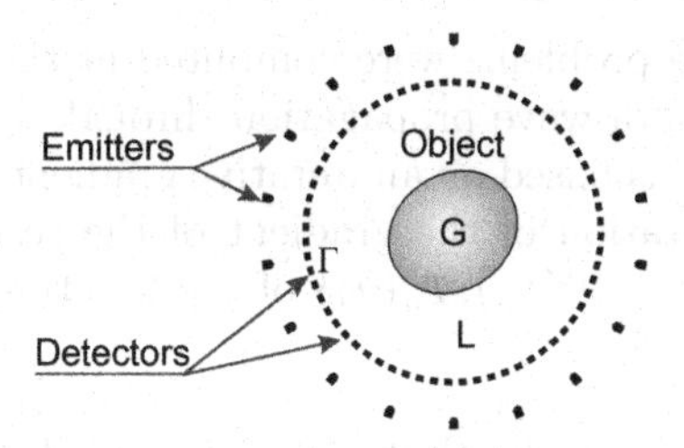

Fig. 1. The scheme of a tomographic examination

$$c(\boldsymbol{r})u_{tt}(\boldsymbol{r},t) + a(\boldsymbol{r})u_t(\boldsymbol{r},t) - \Delta u(\boldsymbol{r},t) = 0; \tag{1}$$

$$u(\boldsymbol{r},t)|_{t=0} = F_0(\boldsymbol{r}), \;\; u_t(\boldsymbol{r},t)|_{t=0} = F_1(\boldsymbol{r}). \tag{2}$$

This model accounts for diffraction, refraction, multiple scattering and absorption of ultrasound waves. Here, $u(\boldsymbol{r},t)$ is the acoustic pressure; $c(\boldsymbol{r}) = 1/v^2(\boldsymbol{r})$, where $v(\boldsymbol{r})$ is the speed of sound; $a(\boldsymbol{r})$ is the absorption factor; $\boldsymbol{r}$ is a point in space, and Δ is the Laplacian with respect to $\boldsymbol{r}$. Non-reflecting boundary condition [16] in the form $\partial u/\partial \boldsymbol{n} = -c^{-0.5}\partial u/\partial t$ were applied at the boundary of the computational domain. Here, $\boldsymbol{n}$ is a vector pointing towards the ultrasound emitter. A first-order approximation was used for the boundary conditions. The remaining reflected waves were attenuated by a sponge-type absorbing layer. The initial conditions $F_0(\boldsymbol{r})$ and $F_1(\boldsymbol{r})$ represent the wavefield at the initial time of the numerical simulation, which was computed as a spherical wave radiating from a point source: $u(\boldsymbol{r},t) = f(vt - |\boldsymbol{r}-\boldsymbol{r_0}|)/|\boldsymbol{r}-\boldsymbol{r_0}|$, where $f(x)$ is the sounding waveform and $\boldsymbol{r_0}$ is the emitter position.

Both 2D and 3D inverse problems are considered in this study. The 2D inverse problems were solved using the data obtained in the 3D simulations. The initial conditions for the 2D simulations were obtained via the time-reversal method [17] by solving Eq. (1) in reverse time with a boundary condition $u(\boldsymbol{s},t)|_{s\in\Gamma} = U_0(\boldsymbol{s},t)$, where $U_0(\boldsymbol{s},t)$ are the data computed at surface Γ via the 3D wave simulation.

The objective is to determine the coefficients $c(\boldsymbol{r})$ and $a(\boldsymbol{r})$ of the wave equation using the measurements of the acoustic pressure at surface Γ. This inverse problem is non-linear and ill-posed. The methods for solving ill-posed inverse problems were developed in [18–20]. We formulate the inverse problem as a problem of minimizing the residual functional

$$\Phi(u(c,a)) = \frac{1}{2}\int_0^T\int_S (u(\boldsymbol{s},t) - U(\boldsymbol{s},t))^2 \,\mathrm{d}\boldsymbol{s}\,\mathrm{d}t \tag{3}$$

with respect to its argument (c,a). Here $U(\boldsymbol{s},t)$ are the data measured at surface Γ for the time period $(0,T)$, $u(\boldsymbol{s},t)$ is the solution of the direct problem (1)–(2) for given $c(\boldsymbol{r}) = 1/v^2(\boldsymbol{r})$ and $a(\boldsymbol{r})$. In this study, "measured" wave fields for both 2D and 3D inverse problems were computed at the detector positions via 3D numerical simulation of wave propagation through a 3D object.

The solution method is based on an iterative gradient-descent algorithm that involves explicit computation of the gradient of the residual functional [5,10]. The gradient $\Phi'(u(c,a)) = \{\Phi'_c(u), \Phi'_a(u)\}$ of the functional (3) has the form:

$$\Phi'_c(u(c)) = \int_0^T w_t(\boldsymbol{r},t)u_t(\boldsymbol{r},t)\,\mathrm{d}t, \quad \Phi'_a(u(a)) = \int_0^T w_t(\boldsymbol{r},t)u(\boldsymbol{r},t)\,\mathrm{d}t. \tag{4}$$

Here, $u(\boldsymbol{r},t)$ is the solution of the direct problem (1)–(2), and $w(\boldsymbol{r},t)$ is the solution of the "conjugate" problem with given $c(\boldsymbol{r})$, $a(\boldsymbol{r})$, and $u(\boldsymbol{r},t)$:

$$c(\boldsymbol{r})w_{tt}(\boldsymbol{r},t) - a(\boldsymbol{r})w_t(\boldsymbol{r},t) - \Delta w(\boldsymbol{r},t) = E(\boldsymbol{r},t); \tag{5}$$

$$w(\boldsymbol{r},t=T) = 0, \quad w_t(\boldsymbol{r},t=T) = 0; \tag{6}$$

$$E(\boldsymbol{r},t) = \begin{cases} u(\boldsymbol{r},t) - U(\boldsymbol{r},t), & \text{where } \boldsymbol{r} \in \Gamma \text{ and } U(\boldsymbol{r},t) \text{ is known;} \\ 0, & \text{otherwise.} \end{cases} \tag{7}$$

Non-reflecting boundary condition [16] is applied at the boundary of the computational domain. Computing the gradient (4) involves solving direct (1)–(2) and "conjugate" (5)–(7) problems.

The residual functional is minimized using an iterative gradient-descent algorithm [21]. In order for the gradient method to converge to the global minimum of the functional, a multi-stage approach [12] is employed, in which the bandwidth of the signal is gradually increased, starting with a sufficiently large wavelength. This method has regularizing properties and stops when the value of the residual functional becomes equal to the error of the input data [19].

Finite-difference time-domain method (FDTD) on uniform grids was used to solve Eqs. (1)–(2) and (5)–(7). The following finite difference scheme was used to approximate Eq. (1):

$$c_{ij}\frac{u_{ij}^{k+1} - 2u_{ij}^{k} + u_{ij}^{k-1}}{\tau^2} + a_{ij}\frac{u_{ij}^{k+1} - u_{ij}^{k-1}}{2\tau} - \frac{\boldsymbol{L}_{ij}^{k}}{h^2} = 0. \tag{8}$$

This scheme is second-order accurate in time and fourth-order accurate in space. Here, $u_{ij}^k = u(x_i, y_j, t_k)$ are the values of $u(\boldsymbol{r},t)$ at point (i,j) at the time step k; c_{ij} and a_{ij} are the values of $c(\boldsymbol{r})$ and $a(\boldsymbol{r})$ at point (i,j). A similar scheme is used to solve the Eqs. (5)–(7) for $w(\boldsymbol{r},t)$ in reverse time.

The 2D discrete Laplacian $\boldsymbol{L}_{ij}^k$ was computed using a fourth-order optimized finite difference scheme [22]. The 3D Laplacian was computed using the same scheme replicated along the X-Y, X-Z and Y-Z planes. This is not an optimal 3D scheme, but it has the same numerical dispersion in the X-Y plane as the 2D scheme. Thus, no numerical errors were introduced into the 2D simulations when the data from the 3D simulations were used as input.

3 Numerical Simulation Results

The aim of this study is to assess the capabilities of 2D and 3D tomographic imaging schemes and to identify the factors that affect the image quality. The test object (phantom) in numerical simulations is a digital representation of a real test sample used in physical experiments on ultrasonic imaging [11,12], so the simulation results are applicable to physical experiments.

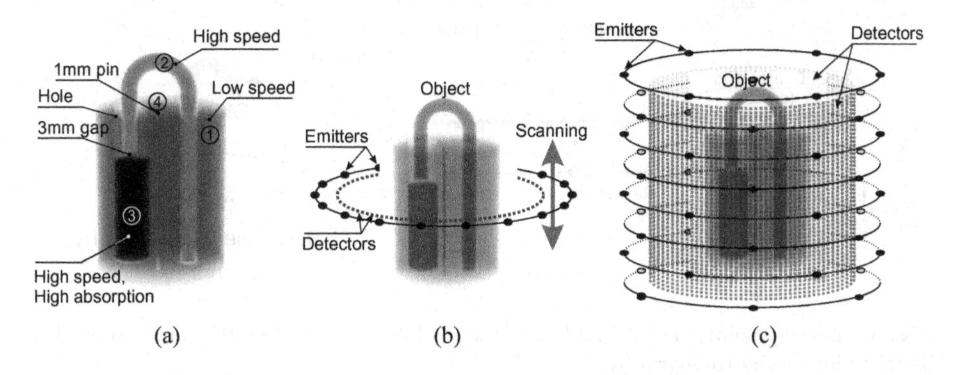

Fig. 2. The test phantom (a), the 2D imaging scheme (b), the 3D imaging scheme (c)

Figure 2a shows the phantom used in numerical simulations. The object contains a low-speed cylinder (1), $v = 1.44$ km·s^{-1}, a high-speed rod (2), $v = 1.8$ km·s^{-1}, curved into a loop in order to achieve strong refraction, an absorbing cylinder (3), $v = 1.92$ km·s^{-1}, $a = 0.2$, and a 1 mm-thick pin (4). There is 3 mm gap between rods 2 and 3.

Figure 2b shows the 2D tomographic imaging scheme. The emitters (18 in total) and detectors are located in a single plane and can move vertically relative to the object. For each vertical position, a two-dimensional coefficient inverse problem is solved and a cross-section of the object in the imaging plane is reconstructed. Figure 2c shows the 3D imaging scheme. The emitters and detectors are located on cylindrical surfaces around the object. There are 48 emitters in total. The detectors are located in multiple circles, representing a vertical transducer array that rotates around the object. The acoustic properties of the object are reconstructed as 3D images.

The 2D scheme is exact if the object is cylindrical and is fully defined by its cross-section in the imaging plane. However, real objects aren't necessarily cylindrical, so various image artifacts will be observed. Vertical resolution of the 2D scheme is expected to be poor (Fig. 3a). Artifacts are expected to appear where the waves join or leave the imaging plane. Figure 3b shows examples of such cases. A wave passing through a low-speed area arrives at the detectors at the same time as the wave reflected from some off-plane object. A wave leaving the imaging plane may appear as being absorbed. In the 3D imaging scheme with a rotating transducer array the view is limited—some waves escape through the open ends of the cylindrical surface. The image quality is expected to degrade near the edges. In this study, we perform the simulations in order to assess these effects and to determine the limits of applicability of various tomographic schemes.

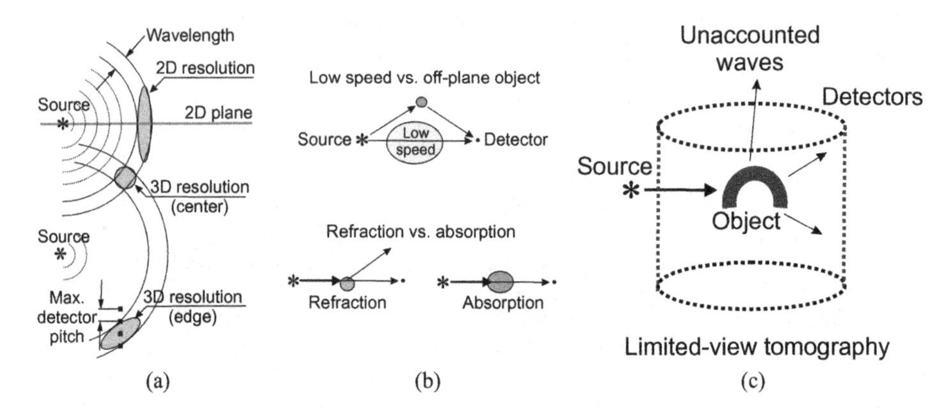

Fig. 3. Spatial resolution of 2D and 3D methods (a), ambiguity in 2D tomography (b), limited view in 3D tomography (c)

The parameters of the simulation were set as follows: reconstructed volume – 120 × 120 mm (cylindrical); total volume simulated, including emitters – 256 × 256 × 256 mm; height of the detector array – 140 mm, vertical pitch – 1.5 mm, rotation radius – 90 mm; the 2D image stack was computed with a vertical step of 1.5 mm. The sounding signal bandwidth amounted to 600 kHz with a central frequency of 400 kHz. The finite difference grid step h was set 0.3 mm in order to accommodate the signal bandwidth with a reasonably low numerical dispersion. The time step τ was set to 0.1 μs to satisfy the Courant stability condition $\sqrt{3}v\tau < h$ for the explicit FDTD method.

Figure 4 shows the exact and reconstructed sound speed images in the vertical cross-section through the loop structure. In the 2D image (a), the speed of sound inside the loop at z = 5 mm is completely incorrect. A region of low speed is displayed where the rod is actually located, and a "shadow" extends for ≈10 mm above the loop. The top end of the low-speed cylinder is irregular. T3 mm gap at z = 42 mm is visible, however, the sound speed in the gap is averaged and appears higher than actual in 10 mm range above the gap. The 3D image (b) is correctly reconstructed with a spatial resolution of ≈1.5 mm, except for the lowest part of the image where the number of emitters and detectors surrounding the material decreases significantly.

Figure 5 shows the exact and reconstructed absorption factor in the vertical cross-section through the loop structure. The absorption factor is reconstructed less precisely than the speed of sound. The speed of sound is a coefficient of the second derivative of the acoustic pressure in Eq. 1, and the absorption factor is a coefficient of the first derivative.

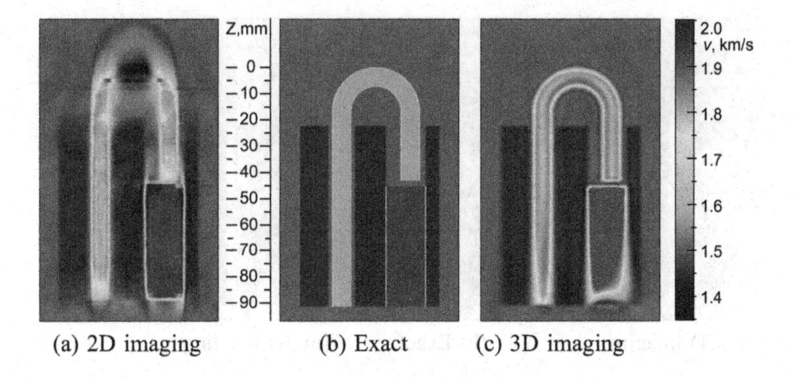

(a) 2D imaging (b) Exact (c) 3D imaging

Fig. 4. Exact and reconstructed sound speed images, vertical cross-section

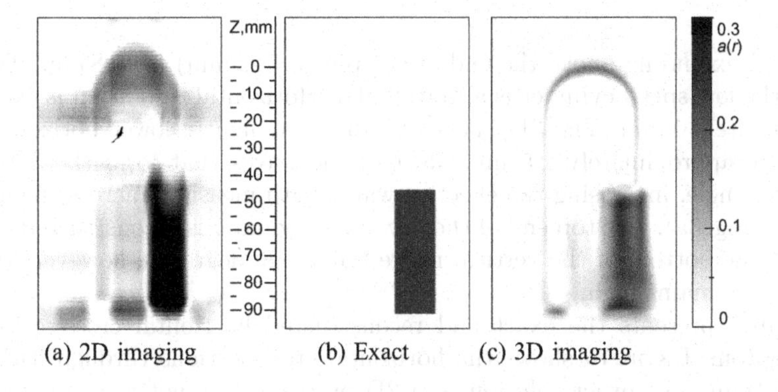

(a) 2D imaging (b) Exact (c) 3D imaging

Fig. 5. Exact and reconstructed absorption factor, vertical cross-section

The 2D image (a) shows high absorption not only where an absorbing material is present, but also in the place of the loop structure (z≈5 mm). The material of the curved rod is transparent to ultrasound, but it bends the waves out of the imaging plane. This effect is registered as heightened absorption factor in 2D images. If some additional information about the absorption factor of the material is available, the reconstructed absorption image can show the areas where the 2D model is not applicable due to refraction.

The 3D image (c) shows a better defined absorbing cylinder image and also some absorption along the loop structure. This is the case of a limited-view tomography, in which some of the scattered waves miss the detectors. However, this effect in the 3D scheme is much less than in the 2D scheme. The sound speed image is still reconstructed with high precision. In medical imaging for breast cancer diagnosis, the detectors cannot be placed at the top boundary and some of the waves will not be registered. Thus, the effect of a limited view must be taken into account.

Figure 6 shows the exact and reconstructed sound speed images in the vertical cross-section through 1 mm pin. The 2D image (a) contains a low-speed

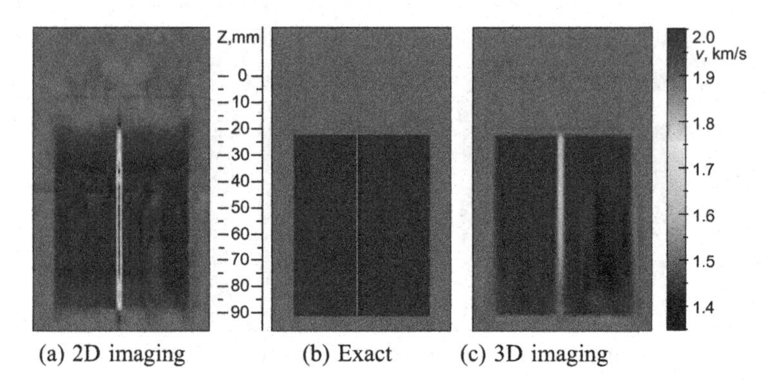

(a) 2D imaging (b) Exact (c) 3D imaging

Fig. 6. Exact and reconstructed sound speed, vertical section through 1 mm pin

"shadow" extending above the end of the pin (z = 22 mm) for ≈8 mm. The top end of the low-speed cylinder is also irregular. Horizontal resolution is excellent, approximately 1 mm. The 3D image (b) shows a slightly lower horizontal resolution of approximately 1.5 mm. Most of the waves that form the 3D image fall at an angle, increasing the effective wavelength measured across the imaged features (Fig. 3a). The top end of the low-speed cylinder is reconstructed equally well. At the bottom of the vertical range the image degrades, however, vertical resolution remains high.

Figure 7 presents the exact and reconstructed horizontal cross-sections of the phantom. Figure 7a shows the horizontal cross-sections through 3 mm gap at z = 42 mm. The exact solution to a 2D problem does not exist in this case, because the object changes in the vertical direction. The sound speed in the gap is averaged, and the image quality degrades. The 3D method reconstructs the speed of sound in the gap correctly, with some minor fluctuations.

Figure 7b shows the horizontal cross-sections through the cylindrical part of the phantom. Both 2D and 3D imaging methods show good image quality. For cylindrical objects, the 2D method covers all possible sounding angles, thus providing the highest image quality.

Figure 7c shows the horizontal cross-sections through the refractive loop of the phantom. The 2D image is completely incorrect. The 3D image is slightly blurred. This area is positioned near the top of the detector array and some of the refracted waves do not reach the detectors, as shown in Fig. 4c. The effective resolution decreases in this case. Except for the aforementioned cases where the reconstruction is incorrect, the sound speed inside the materials is reconstructed with high precision, up to 30 $m{\cdot}s^{-1}$.

A major drawback of the 2D scheme is the dependence of the image quality on the object being imaged, which leads to inability to precisely characterize the material of the object at a given point. The 2D scheme can be used to detect small inclusions in an otherwise homogeneous medium, or to inspect some cylindrical objects. Vertical resolution of the 2D scheme can become as poor as 10 mm, or

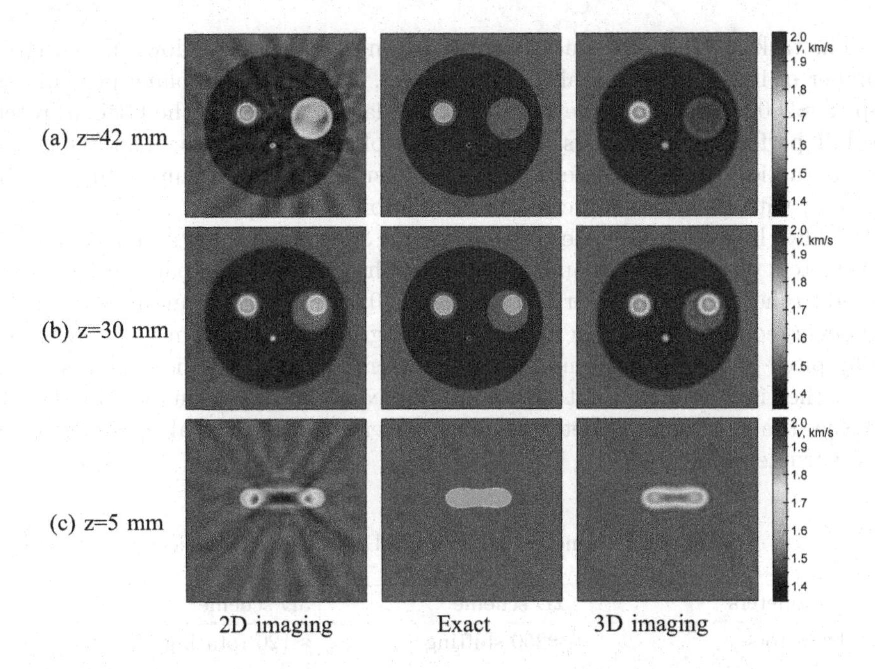

Fig. 7. Exact and reconstructed speed of sound, horizontal cross-sections

the image can be completely incorrect. In medical imaging, soft tissues can take almost any form; thus, the 3D tomography scheme shows the most potential.

4 Computational Complexity of the Algorithms and Technical Parameters of Tomography Devices

The GPU algorithms for solving 2D and 3D inverse problems of wave tomography were described in previous works [11,12]. The 2D and 3D schemes are computationally very similar—the memory access pattern is almost the same and the number of emitters in the 2D scheme acts as the third image dimension [23]. The total volume of the data (image size multiplied by the number of emitters) serves as a good estimate of the computation time.

Currently, graphics processors are the most efficient devices for FDTD numerical methods. The computations were performed on “Lomonosov-2” supercomputer at Lomonosov Moscow State University, equipped with NVidia Tesla K40M GPUs [15]. Small-scale tests were performed on personal computers with AMD Radeon RX580 and AMD Radeon VII graphics cards. Radeon RX580 had outperformed Tesla K40M by a factor of 1.3, and Radeon VII equipped with High Bandwidth Memory (HBM2 with 1 TB/s throughput) had outperformed Tesla K40M by a factor of 3. The FDTD algorithms are data-intensive, and the memory throughput of the device is the primary factor that determines the performance.

The task of 2D image stack reconstruction can be broken down into a large number of independent parallel subprocesses, one per image plane per emitter (up to ≈2000 in total). However, a workload large enough for the GPU to reach its full performance consists of 2–3 image planes per device. The 3D image reconstruction can be parallelized up to one subprocess per emitter (up to ≈50 in total) with no additional cost per simulation time step.

Table 1 lists the properties of the 2D and 3D tomographic schemes and the parameters of possible 2D and 3D tomography devices. The parameters correspond to the setup with a vertical range of 150 mm, vertical scanning step of the 2D device equal to 1.5 mm, a finite difference grid step of 0.3 mm, and a detector array pitch of 1.5 mm. The number of emitters and detectors is chosen so that its further increase does not noticeably increase the image quality. In the 3D scheme, the emitters and detectors must cover a larger vertical range than the reconstructed image.

Table 1. Parameters of 2D and 3D tomography devices

Parameters	2D scheme	3D scheme
Detectors	≈250 shifting	≈120 rotating
Emitters	≈20	≈50
Best-case resolution	1 mm	1.5 mm
Vertical resolution	3–10 mm	1.5 mm
Image quality	Depends on the object	Depends on the geometry
Tissue characterization	No	Yes
Image planes	100	500
Total data volume	2 Gigapixels	40 Gigapixels
Highly parallel tasks	2000	50
Minimum GPU memory	1 GB	8 GB
System RAM (per GPU)	2 GB	16 GB
Resolution factors	N^3 time, N^2 memory	N^4 time, N^3 memory
Total time 0.3 mm grid	5 GPU-hours	100 GPU-hours

GPU-hours are calculated assuming the GPUs are equipped with HBM or GDDR6 memory. In the 3D case, the computational cost rises as a fourth power to the image resolution. Thus, a half-resolution 3D image can be computed in ≈1/16 time—almost as fast as a 2D image stack. The iterative gradient method provides high flexibility in image resolution and quality vs. computational cost [12]. At first, a low-resolution image is reconstructed using a coarse grid. This image is used as an initial approximation at subsequent stages on higher-resolution grids. Thus, the resolution increases gradually and the process can be stopped when the image quality is sufficient to perform the diagnostics required. Assuming a cluster of 25 GPU devices, with modern hardware a 0.7×

resolution (≈2 mm) can be reached in one hour. Such GPU cluster can be used as a computing device in a tomographic diagnostic facility.

5 Conclusion

In this study, cylindrical 2D and 3D tomographic imaging schemes were investigated via numerical simulations. Horizontal and vertical resolution of 2D and 3D schemes were assessed. Various physical factors that affect the image quality in 2D and 3D cases were considered.

The developed 3D imaging methods reconstruct sound speed images with high precision. The 2D imaging scheme can provide high quality only in specific cases. The image reconstruction algorithms are designed for GPU-supercomputers. The parameters of possible ultrasonic tomography devices are estimated. The results of this study can be used in physical experiments on ultrasonic tomography and in development of wave tomography devices.

Acknowledgements. The work is carried out according to the research program of Moscow Center of Fundamental and Applied Mathematics. The research is carried out using the equipment of the shared research facilities of HPC computing resources at Lomonosov Moscow State University.

References

1. Sak, M., et al.: Using speed of sound imaging to characterize breast density. Ultrasound Med. Biol. **43**(1), 91–103 (2017). https://doi.org/10.1016/j.ultrasmedbio.2016.08.021
2. Wiskin, J., et al.: Three-dimensional nonlinear inverse scattering: quantitative transmission algorithms, refraction corrected reflection, scanner design, and clinical results. J. Acoust. Soc. Am. **133**(5), 3229 (2013). https://doi.org/10.1121/1.4805138
3. Birk, M., Dapp, R., Ruiter, N.V., Becker, J.: GPU-based iterative transmission reconstruction in 3D ultrasound computer tomography. J. Parallel Distrib. Comput. **74**, 1730–1743 (2014). https://doi.org/10.1016/j.jpdc.2013.09.007
4. Burov, V.A., Zotov, D.I., Rumyantseva, O.D.: Reconstruction of the sound velocity and absorption spatial distributions in soft biological tissue phantoms from experimental ultrasound tomography data. Acoust. Phys. **61**(2), 231–248 (2015). https://doi.org/10.1134/s1063771015020013
5. Goncharsky, A., Romanov, S., Seryozhnikov, S.: A computer simulation study of soft tissue characterization using low-frequency ultrasonic tomography. Ultrasonics **67**, 136–150 (2016). https://doi.org/10.1016/j.ultras.2016.01.008
6. Goncharsky, A.V., Romanov, S.Y., Seryozhnikov, S.Y.: Low-frequency 3D ultrasound tomography: dual-frequency method. Numer. Methods Program. **19**(4), 479–495 (2018). https://doi.org/10.26089/NumMet.v19r443
7. Natterer, F.: Sonic imaging. In: Handbook of Mathematical Methods in Imaging, pp. 1–23. Springer, Heidelberg (2014). https://doi.org/10.1007/978-3-642-27795-5_37-2

8. Klibanov, M.V., Timonov, A.A.: Carleman Estimates for Coefficient Inverse Problems and Numerical Applications. Walter de Gruyter GmbH (2004). https://doi.org/10.1515/9783110915549
9. Goncharsky, A.V., Romanov, S.Y.: Iterative methods for solving coefficient inverse problems of wave tomography in models with attenuation. Inverse Prob. **33**(2), 025003 (2017). https://doi.org/10.1088/1361-6420/33/2/025003
10. Goncharsky, A.V., Romanov, S.Y.: Supercomputer technologies in inverse problems of ultrasound tomography. Inverse Prob. **29**(7), 075004 (2013). https://doi.org/10.1088/0266-5611/29/7/075004
11. Goncharsky, A., Seryozhnikov, S.: Supercomputer technology for ultrasound tomographic image reconstruction: mathematical methods and experimental results. In: Voevodin, V., Sobolev, S. (eds.) RuSCDays 2018. CCIS, vol. 965, pp. 401–413. Springer, Cham (2019). https://doi.org/10.1007/978-3-030-05807-4_34
12. Goncharsky, A., Seryozhnikov, S.: Three-dimensional ultrasound tomography: mathematical methods and experimental results. In: Voevodin, V., Sobolev, S. (eds.) RuSCDays 2019. CCIS, vol. 1129, pp. 463–474. Springer, Cham (2019). https://doi.org/10.1007/978-3-030-36592-9_38
13. Goncharsky, A.V., Kubyshkin, V.A., Romanov, S.Y., Seryozhnikov, S.Y.: Inverse problems of experimental data interpretation in 3D ultrasound tomography. Numer. Methods Program. **20**, 254–269 (2019). https://doi.org/10.26089/NumMet.v20r323
14. Goncharsky, A.V., Seryozhnikov, S.Y.: The architecture of specialized GPU clusters used for solving the inverse problems of 3D low-frequency ultrasonic tomography. In: Voevodin, V., Sobolev S. (eds.) Supercomputing. RuSCDays 2017. Communications in Computer and Information Science, vol. 793, pp. 363–395. Springer, Cham (2017). https://doi.org/10.1007/978-3-319-71255-0_29
15. Voevodin, V., et al.: Supercomputer Lomonosov-2: large scale, deep monitoring and fine analytics for the user community. Supercomput. Front. Innov. **6**(2), 4–11 (2019). https://doi.org/10.14529/jsfi190201
16. Engquist, B., Majda, A.: Absorbing boundary conditions for the numerical simulation of waves. Math. Comput. **31**(139), 629 (1977). https://doi.org/10.1090/s0025-5718-1977-0436612-4
17. Fink, M.: Time reversal in acoustics. Contemp. Phys. **37**(2), 95–109 (1996). https://doi.org/10.1080/00107519608230338
18. Tikhonov, A.N.: Solution of incorrectly formulated problems and the regularization method. Soviet Math. Dokl. **4**, 1035–1038 (1963)
19. Bakushinsky, A., Goncharsky, A.: Ill-Posed Problems: Theory and Applications. Springer, Heidelberg (1994). https://doi.org/10.1007/978-94-011-1026-6
20. Tikhonov, A.N., Goncharsky, A.V., Stepanov, V.V., Yagola, A.G.: Numerical Methods for the Solution of Ill-Posed Problems. Springer, Dordrecht (1995). https://doi.org/10.1007/978-94-015-8480-7
21. Bakushinsky, A., Goncharsky, A.: Iterative Methods for Solving Ill-Posed Problems. Nauka, Moscow (1989)
22. Hamilton, B., Bilbao, S.: Fourth-order and optimised finite difference schemes for the 2-D wave equation. In: Proceedings of the 16th International Conference on Digital Audio Effects (DAFx-13), pp. 363–395. Springer, Heidelberg (2013). https://doi.org/10.1121/1.4800308
23. Goncharsky, A.V., Romanov, S.Y., Seryozhnikov, S.Y.: Inverse problems of layer-by-layer ultrasonic tomography with the data measured on a cylindrical surface. Numer. Methods Program. **18**(3), 267–276 (2017)

The Numerical Simulation of Radial Age Gradients in Spiral Galaxies

Igor Kulikov(✉), Igor Chernykh, Dmitry Karavaev, Victor Protasov, Vladislav Nenashev, and Vladimir Prigarin

Institute of Computational Mathematics and Mathematical Geophysics SB RAS, Novosibirsk, Russia
kulikov@ssd.sscc.ru, chernykh@parbz.sscc.ru, kda@opg.sscc.ru, inc_13@mail.ru, arni.12@mail.ru, vovkaprigarin@gmail.com

Abstract. In this paper, will present the new results of mathematical modeling of the two-arms galaxy formation and radial age gradients in spirals. The numerical model include self-gravity hydrodynamics equation for gas component of galaxy and collisionless Boltzmann equation for stellar component. To numerical model include important sub-grid physics: star-formation, supernova feedback, stellar wind, cooling and heating function, and non-equilibrium chemistry to ion helium hydride.

Keywords: Parallel computing · Computational astrophysics · Numerical methods

1 Introduction

Isolated galaxies are important due to the fact that they least affected from interaction over the past billion years and their morphology is associated with the evolution of gravitational instability [1]. Thus, the study of instabilities evolution allows us to explain all their diversity [2]. An important property of spiral galaxies is the radial age gradient parameter, which shows the age of stars [3].

Using the developed two-phase hydrodynamic model of galaxies, we shall model the radial age gradient parameter in disk galaxies. To do this, we simulate the evolution of spiral arms of disk galaxies taking into account the star formation process and sufficiently developed sub-grid physics. The second section describes in detail the numerical model of galaxies. The third chapter is devoted to a brief description of parallel implementation and the study of scalability. The fourth chapter presents the results of mathematical modeling. The fifth chapter concludes.

2 Numerical Model

The model presented here is a major extension of an initial gas-dynamic model [4] by adding a model of the collisionless component [5] and a minimum model of

V. Voevodin and S. Sobolev (Eds.): RuSCDays 2020, CCIS 1331, pp. 365–374, 2020.
https://doi.org/10.1007/978-3-030-64616-5_32

subgrid physics [6] and taking into account the requirements of modern simulation methodology [7]. To describe the interstellar medium (ISM) hydrodynamics we will use the following notation: ρ is the gas mixture density, p is the gas pressure, ρ_i is the density of the i-th species, s_i is the formation rate of the i-th species, $\mathbf{u}$ is the velocity vector, ρE is the total mechanical gas energy, Φ is the gravity potential, γ is the adiabatic index, $\mathcal{S}$ is the supernovae formation rate, $\mathcal{D}$ is the star formation rate, Λ is a cooling function, and Γ is a heating function. To describe the star component and the dark matter (collisionless component) we use the following notation: n is the density of the collisionless component, $\mathbf{v}$ is the velocity of the collisionless component, nW_{ij} is the density of the total mechanical energy of the collisionless component, Π_{ij} is the tensor of velocity dispersion of the collisionless component, and G is the gravity constant.

To describe the gas components, we will use the system of single-speed component gravitational hydrodynamics equations:

$$\frac{\partial \rho}{\partial t} + \nabla \cdot (\rho \mathbf{u}) = \mathcal{S} - \mathcal{D}, \tag{1}$$

$$\frac{\partial \rho_i}{\partial t} + \nabla \cdot (\rho_i \mathbf{u}) = s_i + \mathcal{S}\frac{\rho_i}{\rho} - \mathcal{D}\frac{\rho_i}{\rho}, \tag{2}$$

$$\frac{\partial \rho \mathbf{u}}{\partial t} + \nabla \cdot (\rho \mathbf{u}\mathbf{u}) = -\nabla p - \rho \nabla(\Phi) + \mathbf{v}\mathcal{S} - \mathbf{u}\mathcal{D}, \tag{3}$$

$$\frac{\partial \rho E}{\partial t} + \nabla \cdot (\rho E \mathbf{u}) = -\nabla \cdot (p\mathbf{v}) - (\rho \nabla(\Phi), \mathbf{u}) - \Lambda + \Gamma + \rho^{\gamma}\frac{\mathcal{S}}{\rho} - \rho^{\gamma}\frac{\mathcal{D}}{\rho}. \tag{4}$$

$$\rho E = \frac{1}{2}\rho \mathbf{u}^2 + \frac{p}{\gamma - 1}. \tag{5}$$

To describe the collisionless components, we will use the system of equations for the first moments of the Boltzmann collisionless equation:

$$\frac{\partial n}{\partial t} + \nabla \cdot (n\mathbf{v}) = \mathcal{D} - \mathcal{S}, \tag{6}$$

$$\frac{\partial n\mathbf{v}}{\partial t} + \nabla \cdot (n\mathbf{v}\mathbf{v}) = -\nabla \Pi - n\nabla(\Phi) + \mathbf{u}\mathcal{D} - \mathbf{v}\mathcal{S}, \tag{7}$$

$$\frac{\partial nW_{ij}}{\partial t} + \nabla \cdot (nW_{ij}\mathbf{v}) = -\nabla \cdot (v_i \Pi_j + v_j \Pi_i) - (n\nabla(\Phi), \mathbf{v}) + \rho^{\gamma}\frac{\mathcal{D}}{\rho} - \rho^{\gamma}\frac{\mathcal{S}}{\rho}. \tag{8}$$

$$nW_{ij} = n \times v_i \times v_j + \Pi_{ij}. \tag{9}$$

The following Poisson equation is formulated for the total density of the ISM and the collisionless component:

$$\Delta \Phi = 4\pi G\left(\rho + n\right). \tag{10}$$

We will use the following necessary condition for star formation formulated as in [8]:

$$T < 10^4 K, \qquad \nabla \cdot \mathbf{u} < 0, \qquad \rho > 1.64 M_{\odot} pc^{-3}. \tag{11}$$

Then the star formation rate can be written as follows:

$$\mathcal{D} = \mathcal{C}\rho^{3/2}\sqrt{\frac{32G}{3\pi}}, \tag{12}$$

where $\mathcal{C} = 0.034$ is the coefficient of star formation efficiency.

The supernova formation rate [9] can be described by the equation

$$\mathcal{S} = \beta\mathcal{C}n^{3/2}\sqrt{\frac{32G}{3\pi}}, \tag{13}$$

where $\beta = 0.1$ is the coefficient of explosion of young stars. In supernovae feedback we will distinguish two classes of supernovae: core-collapse supernovae (ccSN) and thermonuclear supernovae (SNeIa), as well as new stars with stellar wind. The total contribution of supernovae and new stars can be formulated as

$$\mathcal{S} = \mathcal{S}_{ccSN} + \mathcal{S}_{SNeIa} + \mathcal{S}_{*}. \tag{14}$$

This differentiation is important for taking into account the explosion energy typical for supernovae and new stars. To determine the fraction of massive stars, we will use the initial mass function [10] and the stellar lifetimes function [11] with allowance for nucleosynthes [12]. From the total number of stars, we will exclude the binary systems producing type Ia supernovae [13]. The mass fraction $\mathcal{R}_{ccSN,SNeIa,*}$ and the star formation rate are determined for each of the star types by the simple equation

$$\mathcal{S} = \beta\mathcal{C}n^{3/2}\sqrt{\frac{32G}{3\pi}} \times (\mathcal{R}_{ccSN} + \mathcal{R}_{SNeIa} + \mathcal{R}_{*}). \tag{15}$$

The parameters of [7] are averaged under the assumption that the type Ia supernovae have a limit, namely, the Chandrasekhar mass; core-collapse supernovae are possible with a mass greater than $8M_{\odot}$. As a result we have

$$\mathcal{R}_{ccSN} = 0.15, \quad \mathcal{R}_{SNeIa} = 0.63, \quad \mathcal{R}_{*} = 0.22. \tag{16}$$

These values comply with a statistics of supernova stars which has been conducted since 1855. Note that supernova explosions also contribute to gas heating:

$$\Gamma = \Gamma_{51} \times \beta\mathcal{C}n^{3/2}\sqrt{\frac{32G}{3\pi}} \times (\mathcal{R}_{ccSN} + \mathcal{R}_{SNeIa}),$$

where $\Gamma_{51} = 10^{51}$ Erg.

The cooling function will be considered in two temperature regimes:

1. Low-temperature cooling. At low temperatures such metals as O, C, N, Si, and Fe are ionized due to collision. The collision frequency and the corresponding cooling function can be found in [14].
2. High-temperature cooling. At high temperatures such metals as C, N, O, Ne, Mg, Si, and Fe are subject to a radiation process. The cooling function can be found in [15].

The heating process, in addition to supernova explosions, consists of ionization of hydrogen and helium atoms [16] and also some photoelectric effects on the dust [17].

The chemical reactions network are considered up to the helium hydride ion, which is currently observed and is a basis for further chemical networks [18].

1. Ionization of hydrogen by cosmic rays [19]:

$$H + c.r. \rightarrow H^+ + e$$

2. Collisional ionization of helium [20]:

$$He + e \rightarrow He^+ + 2e$$

3. Dissociative recombination of hydrogen [21]:

$$H^+ + e \rightarrow H + \gamma$$

4. Collisional ionization of hydrogen [22]:

$$H + e \rightarrow H^+ + 2e$$

5. Collision ionization of hydrogen with radiation [23]:

$$H + e \longrightarrow H^- + \gamma$$

6. Dissociative hydrogen electron attachment [20]:

$$H^- + H \longrightarrow H_2 + e$$

7. Association of hydrogen with radiation [24]:

$$H + H^+ \longrightarrow H_2^+ + \gamma$$

8. Recombination of hydrogen [25]:

$$H_2^+ + H \longrightarrow H_2 + H^+$$

9. Recombination of molecular hydrogen [22]:

$$H_2 + H^+ \longrightarrow H_2^+ + H$$

10. Collisional dissociation of molecular hydrogen [26]:

$$H_2 + e \longrightarrow 2H + e$$

11. Collisional hydrogen detachment of electrons (I) [20]:

$$H^- + e \longrightarrow H + 2e$$

12. Collisional hydrogen detachment of electrons (II) [20]:

$$H^- + H \longrightarrow 2H + e$$

13. Neutralization of hydrogen [27]:

$$H^- + H^+ \longrightarrow 2H + \gamma$$

14. Collisional association of hydrogen [24]

$$H^- + H^+ \longrightarrow H_2^+ + e$$

15. Molecular neutralization of molecular hydrogen [27]:

$$H_2^+ + H^- \longrightarrow H + H_2$$

16. Dissociative recombination of molecular hydrogen [28]:

$$H_2^+ + e \rightarrow 2H + \gamma$$

17. Collisional hydrogen molecular association [29]:

$$3H \longrightarrow H_2 + H$$

18. Collisional molecular hydrogen dissociation (I) [30]:

$$H_2 + H \longrightarrow 3H$$

19. Collisional molecular hydrogen dissociation (II) [29]:

$$H_2 + H_2 \rightarrow 2H + H_2$$

20. Molecular hydrogen formation on grain [31]:

$$H + H + grain \rightarrow H_2 + grain$$

21. Molecular hydrogen photodissociation [32]

$$H_2 + \gamma \rightarrow 2H$$

22. Recombination of hydrogen on grains [21]:

$$H^+ + e + grain \rightarrow H + grain$$

23. Collisional helium recombination with radiation [33]:

$$He^+ + e \longrightarrow He + \gamma$$

24. Collisional helium ion ionization [22]:

$$He^+ + e \longrightarrow He^{++} + 2e$$

25. Collisional molecular helium dissociation [34]:

$$He^{++} + e \longrightarrow He^{+} + \gamma$$

26. Radiative association of the helium ion and hydrogen [35]:

$$He^{+} + H \rightarrow HeH^{+} + h\nu$$

27. Dissociative recombination of the helium hydride ion [36]

$$HeH^{+} + e \rightarrow He + H$$

28. Collisional recombination of the helium hydride ion [37]:

$$HeH^{+} + H \rightarrow He + H_2^{+}$$

We will calculate the effective adiabatic index from the equation [32,38]

$$\gamma = \frac{5n_H + 5n_{He} + 5n_e + 7n_{H_2}}{3n_H + 3n_{He} + 3n_e + 5n_{H_2}}, \tag{17}$$

where n_H is the concentration of atomic hydrogen, n_{He} is atomic helium, n_e are electrons, and n_{H_2} is molecular hydrogen.

3 Parallel Numerical Method

The hydrodynamic equations are solved by the HLL method [39]. To increase the order of accuracy of the method we can use a piecewise-parabolic representation of the solution [40]. The Poisson equation is solved by a method based on the fast Fourier transform. The ChemPAK package is used to solve the chemical-kinetic equations [41]. The subgrid processes and gravity are taken into account with an explicit Euler scheme.

Here we present three indicators of the parallel code implementation: weak scalability, which shows the fraction of network interactions in the total calculation time; strong scalability, or the code acceleration per a single Intel Xeon Phi which was used in the computational experiments, as well as the performance achieved in the calculations by code vectorization. The computational experiments were performed on the NKS-1P cluster of the Siberian Supercomputer Center.

3.1 Weak Scalability

To assess weak scalability of the code, a $256p \times 256 \times 256$ computational grid was used. All logical cores for each of the nodes were used, where p is the number of the nodes. Thus, a subdomain size of 256^3 is for each accelerator. To study scalability, the time of performance of the numerical method was measured in seconds at various numbers of nodes being used. It has been shown that a 97% scalability was obtained when using 16 nodes, which is a rather good result. Note that only 3% of the time is used for the network interactions. Earlier we obtained similar values on RSC clusters [42].

3.2 Strong Scalability and Performance

To analyze the acceleration and performance of the code, a $1056 \times 100 \times 100$ computational grid, as well as the OpenMP parameters and data alignment described earlier were used. The results of the computational experiments have shown that the maximum acceleration per a single Intel Xeon Phi 7290 is a factor of 42 when using 64 logical threads, with a performance of 302 GFLOPS. With vtune/advisor tools on a 256^3 grid, a performance of 190 GFLOPS and arithmetic intensity density of 0.3 FLOP/byte with a 100% mask utilization and a real memory capacity of 573 GB/s have been obtained.

4 The Numerical Simulation of Radial Age Gradients

The initial configuration of the galaxy consists of a self-gravitating gas disk immersed in a stationary Halo of dark matter (DM), which is described by the equations of hydrodynamics. DM is described by the equations for the first moments of the collisionless Boltzmann equation, with the profile:

$$\rho_{DM} = \frac{\rho_0}{1 + (r/r_0)^2}, \tag{18}$$

where $\rho_0 = 1.97 M_\odot$ pc^{-3} is the central density of dark matter, $r_0 = 1.6$ kpc is the characteristic length of halo and $r_h = 78.55$ kpc is the halo radius. The mass of dark matter is chosen equal to $5 \times 10^{10} \sim M_\odot$. The procedure for constructing a rotating configuration of equilibrium is given in [43]. Simulation of spiral arms evolution of galaxy is given in [40]. The Fig. 1 shows the approximation of the

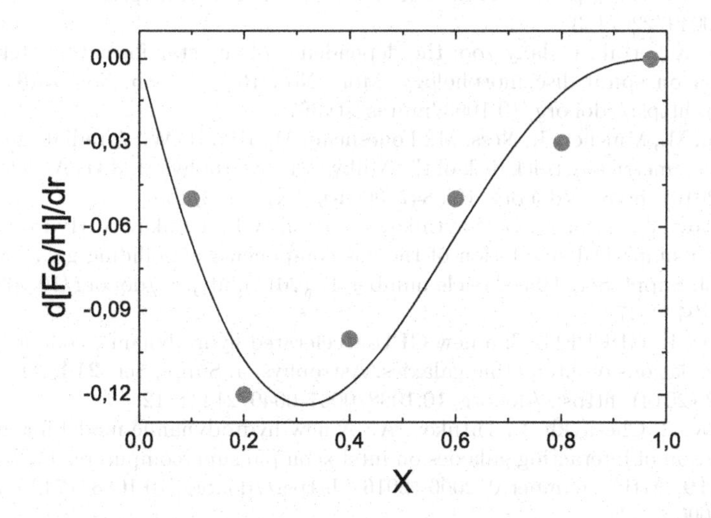

Fig. 1. The numerical profile of radial age gradients in spiral galaxies: simulation (point), approximation (line).

Radial Age Gradients obtained during the simulation. Such graph has analytical equation with approximation:

$$\frac{d[Fe/H]}{dr} = A \times \left(2x^3 - 3x^2 + 1\right) \times \left(x^2 - x\right), \tag{19}$$

where $x = r/R$ and $A \approx 3/4$. In general, such a graph corresponds to observations.

5 Conclusion

The numerical model of galaxies was described. The model of galaxies include self-gravity hydrodynamics equation for gas component of galaxy and collisionless Boltzmann equation for stellar component. To numerical model include important sub-grid physics: star-formation, supernova feedback, stellar wind, cooling and heating function, and non-equilibrium chemistry to ion helium hydride. The new results of mathematical modeling of the two-arms galaxy formation and radial age gradients in spirals was described.

Acknowledgements. The authors was supported by the Russian Foundation for Basic Research (project no. 18-01-00166) and ICMMG SB RAS project 0315-2019-0009.

References

1. Khim, H., et al.: Demographics of isolated galaxies along the hubble sequence. Astrophys. J. Suppl. Ser. **220**, Article number 3 (2015). https://doi.org/10.1088/0067-0049/220/1/3
2. Willet, K., et al.: Galaxy zoo: the dependence of the star formation-stellar mass relation on spiral disc morphology. Mon. Not. Roy. Astron. Soc. **449**, 820–827 (2015). https://doi.org/10.1093/mnras/stv307
3. Martig, M., Minchev, I., Ness, M., Fouesneau, M., Rix, H.-W.: A radial age gradient in the geometrically thick disk of the Milky Way. Astrophys. J. **831**, Article number 139 (2016). https://doi.org/10.3847/0004-637X/831/2/139
4. Vshivkov, V., Lazareva G., Snytnikov A., Kulikov I., Tutukov A. Hydrodynamical code for numerical simulation of the gas components of colliding galaxies. Astrophys. J. Suppl. Ser. **194**, Article number 47 (2011). https://doi.org/10.1088/0067-0049/194/2/47
5. Kulikov, I.: GPUPEGAS: a new GPU-accelerated hydrodynamic code for numerical simulations of interacting galaxies. Astrophys. J. Suppl. Ser. **214**, Article number 12 (2014). https://doi.org/10.1088/0067-0049/214/1/12
6. Kulikov, I., Chernykh, I., Tutukov, A.: A new hydrodynamic model for numerical simulation of interacting galaxies on intel xeon phi supercomputers. J. Phys. Conf. Ser. **719**, Article number 012006 (2016). https://doi.org/10.1088/1742-6596/719/1/012006
7. Vorobyov, E., Recchi, S., Hensler, G.: Stellar hydrodynamical modeling of dwarf galaxies: simulation methodology, tests, and first results. Astron. Astrophys. **579**, Article number A9 (2015). https://doi.org/10.1051/0004-6361/201425587

8. Katz, N., Weinberg, D., Hernquist, L.: Cosmological simulations with TreeSPH. Astrophys. J. Suppl. Ser. **105**, 19–35 (1996). https://doi.org/10.1086/192305
9. Springel, V., Hernquist, L.: Cosmological smoothed particle hydrodynamics simulations: a hybrid multiphase model for star formation. Mon. Not. Roy. Astron. Soc. **339**, 289–311 (2003). https://doi.org/10.1046/j.1365-8711.2003.06206.x
10. Kroupa, P.: On the variation of the initial mass function. Mon. Not. Roy. Astron. Soc. **322**, 231–246 (2001). https://doi.org/10.1046/j.1365-8711.2001.04022.x
11. Padovani, P., Matteucci, F.: Stellar mass loss in elliptical galaxies and the fueling of active galactic nuclei. Astrophys. J. **416**, 26–35 (1993). https://doi.org/10.1086/173212
12. Van den Hoek, L.B., Groenewegen, M.A.T.: New theoretical yields of intermediate mass stars. Astron. Astrophys. Suppl. Ser. **123**, 305–328 (1997). https://doi.org/10.1051/aas:1997162
13. Matteucci, F., Recchi, S.: On the typical timescale for the chemical enrichment from type IA supernovae in galaxies. Astrophys. J. **558**, 351–358 (2001). https://doi.org/10.1086/322472
14. Dalgarno, A., McCray, R.A.: Heating and ionization of HI regions. Ann. Rev. Astron. Astrophys. **10**, 375–426 (1972). https://doi.org/10.1146/annurev.aa.10.090172.002111
15. Boehringer, H., Hensler, G.: Metallicity-dependence of radiative cooling in optically thin, hot plasmas. Astron. Astrophys. **215**, 147–149 (1989)
16. Van Dishoeck, E.F., Black, J.H.: Comprehensive models of diffuse interstellar clouds - physical conditions and molecular abundances. Astrophys. J. Suppl. Ser. **62**, 109–145 (1986)
17. Bakes, E.L.O., Tielens, A.G.G.M.: The photoelectric heating mechanism for very small graphitic grains and polycyclic aromatic hydrocarbons. Astrophys. J. **427**, 822–838 (1994). https://doi.org/10.1086/174188
18. Gusten, R., et al.: Astrophysical detection of the helium hydride ion HeH^+. Nature. **568**, 357–359 (2019). https://doi.org/10.1038/s41586-019-1090-x
19. Lepp, S.: The cosmic-ray ionization rate. astrochemistry of cosmic phenomena. In: Proceedings of the 150th Symposium of the International Astronomical Union, held at Campos do Jordao, Sao Paulo, Brazil, 5–9 August 1991, pp. 471–475 (1992). https://doi.org/10.1073/pnas.0602117103
20. Janev, R.K., Langer, W.D., Evans, K.J., Post, D.E.J.: Elementary processes in hydrogen-helium plasmas. In: Cross Sections and Reaction Rate Coefficients. Springer Series on Atomic, Optical, and Plasma Physics, vol. 4, p. 326 (1987). https://doi.org/10.1007/978-3-642-71935-6
21. Ferland, G.J., Peterson, B.M., Horne, K., Welsh, W.F., Nahar, S.N.: Anisotropic line emission and the geometry of the broad-line region in active galactic nuclei. Astrophys. J. **387**, 95–108 (1992). https://doi.org/10.1086/171063
22. Abel, T., Anninos, P., Zhang, Y., Norman, M.: Modeling primordial gas in numerical cosmology. New Astron. **2**, 181–207 (1997). https://doi.org/10.1016/S1384-1076(97)00010-9
23. Wishart, A.W.: The bound-free photo-detachment cross-section of H^-. Mon. Not. Roy. Astron. Soc. **187**, 59P–60P (1979). https://doi.org/10.1088/0022-3700/12/21/009
24. Shapiro, P., Kang, H.: Hydrogen molecules and the radiative cooling of pregalactic shocks. Astrophys. J. **318**, 32–65 (1987)
25. Karpas, Z., Anicich, V., Huntress Jr., W.T.: An ion cyclotron resonance study of reactions of ions with hydrogen atoms. J. Chem. Phys. **70**, 2877–2881 (1979)

26. Donahue, M., Shull, J.M.: New photoionization models of intergalactic clouds. Astrophys. J. **383**, 511–523 (1991). https://doi.org/10.1086/170809
27. Dalgarno, A., Lepp, S., Vardya, M.S., Tarafdar, S.P.: Astrochemistry, p. 109 (1987)
28. Schneider, I.F., Dulieu, O., Giusti-Suzor, A., Roueff, E.: Dissociate recombination of H2(+) molecular ions in hydrogen plasmas between 20 K and 4000 K. Astrophys. J. **424**, 983–987 (1994). https://doi.org/10.1086/173948
29. Grassi, T., et al.: KROME a package to embed chemistry in astrophysical simulations. Mon. Not. Roy. Astron. Soc. **439**, 2386–2419 (2014). https://doi.org/10.1093/mnras/stu114
30. Dove, J., Mandy, M.: The rate of dissociation of molecular hydrogen by hydrogen atoms at very low densities. Astrophys. J. **311**, L93–L96 (1986). https://doi.org/10.1086/184805
31. Hollenbach, D., McKee, C.F.: Molecule formation and infrared emission in fast interstellar shocks: I physical processes. Astrophys. J. Suppl. Ser. **41**, 555–592 (1979). https://doi.org/10.1086/190631
32. Glover, S., Mac Low, M.: Simulating the formation of molecular clouds: I slow formation by gravitational collapse from static initial conditions. Astrophys. J. Suppl. Ser. **169**, 239–268 (2007). https://doi.org/10.1086/512238
33. Cen, R.: A hydrodynamic approach to cosmology: methodology. Astrophys. J. Suppl. Ser. **78**, 341–364 (1992). https://doi.org/10.1086/191630
34. Osterbrock, D.E.: Astrophysics of Gaseous Nebulae, p. 263 (1974)
35. Zygelman, B., Dalgarno, A.: The radiative association of He^+ and H. Astrophys. J. **365**, 239–240 (1990)
36. Stromholm, C., et al.: Dissociative recombination and dissociative excitation of $4HeH^+$: absolute cross sections and mechanisms. Phys. Rev. A. **54**, 3086–3094 (1996). https://doi.org/10.1103/PhysRevA.54.3086
37. Bovino, S., Tacconi, M., Gianturco, F.A., Galli, D.: Ion chemistry in the early universe. Revisiting the role of HeH+ with new quantum calculations. Astron. Astrophys. **529**, Article number A140 (2011). https://doi.org/10.1051/0004-6361/201116740
38. Glover, S., Mac Low, M.: Simulating the formation of molecular clouds: II rapid formation from turbulent initial conditions. Astrophys. J. **659**, 1317–1337 (2007). https://doi.org/10.1086/512227
39. Kulikov, I.M., Chernykh, I.G., Glinskiy, B.M., Protasov, V.A.: An efficient optimization of Hll method for the second generation of intel xeon phi processor. Lobachevskii J. Math. **39**(4), 543–551 (2018). https://doi.org/10.1134/S1995080218040091
40. Kulikov, I., Vorobyov, E.: Using the PPML approach for constructing a low-dissipation, operator-splitting scheme for numerical simulations of hydrodynamic flows. J. Comput. Phys. **317**, 318–346 (2016). https://doi.org/10.1016/j.jcp.2016.04.057
41. Chernykh, I., Stoyanovskaya, O., Zasypkina, O.: ChemPAK software package as an environment for kinetics scheme evaluation. Chem. Prod. Process Model. **4**, Article number 3 (2009). https://doi.org/10.2202/1934-2659.1288
42. Kulikov, I., Chernykh, I. Glinskiy, B., Weins, D., Shmelev, A.: Astrophysics simulation on RSC massively parallel architecture. In: Proceedings - 2015 IEEE/ACM 15th International Symposium on Cluster, Cloud, and Grid Computing, CCGrid 2015, pp. 1131–1134 (2015). https://doi.org/10.1109/CCGrid.2015.102
43. Vorobyov, E., Recchi, S., Hensler, G.: Self-gravitating equilibrium models of dwarf galaxies and the minimum mass for star formation. Astron. Astrophys. **543**, Article number A129 (2012). https://doi.org/10.1051/0004-6361/201219113

Towards High Performance Relativistic Electronic Structure Modelling: The EXP-T Program Package

Alexander V. Oleynichenko[1,2(✉)], Andréi Zaitsevskii[1,2], and Ephraim Eliav[3]

[1] Petersburg Nuclear Physics Institute named by B.P. Konstantinov of NRC "Kurchatov Institute", Gatchina, Russia
alexvoleynichenko@gmail.com, zaitsevskii_av@pnpi.nrcki.ru
[2] Department of Chemistry, Lomonosov Moscow State University, Moscow, Russia
[3] School of Chemistry, Tel Aviv University, Tel Aviv, Israel
ephraim@tauex.tau.ac.il

Abstract. Modern challenges arising in the fields of theoretical and experimental physics require new powerful tools for high-precision electronic structure modelling; one of the most perspective tools is the relativistic Fock space coupled cluster method (FS-RCC). Here we present a new extensible implementation of the FS-RCC method designed for modern parallel computers. The underlying theoretical model, algorithms and data structures are discussed. The performance and scaling features of the implementation are analyzed. The software developed allows to achieve a completely new level of accuracy for prediction of properties of atoms and molecules containing heavy and superheavy nuclei.

Keywords: Relativistic coupled cluster method · High performance computing · Excited electronic states · Heavy element compounds

1 Introduction

Nowadays first-principle based electronic structure modelling is widely recognized as a powerful tool for solving both fundamental and applied problems in physics and chemistry [1]. A bulk of modern experiments in fundamental physics employing atomic and molecular systems seem to be hardly implementable or even senseless without theoretical predictions and assessments; some recent and the most striking examples are experiments for the electron electric dipole moment search [2], design of laser-coolable molecular systems [3] and spectroscopy of short-lived radioactive atoms and molecules [4,5]. Probably the most intriguing applications of quantum chemical modelling to fundamental problems are associated with molecules containing heavy and superheavy elements [6]; these applications require theoretical predictions to be accurate enough to be useful. For example, recent spectroscopic investigations of short-lived radioactive systems (No, Lr, RaF) required predicted excitation energies to be accurate

V. Voevodin and S. Sobolev (Eds.): RuSCDays 2020, CCIS 1331, pp. 375–386, 2020.
https://doi.org/10.1007/978-3-030-64616-5_33

up to 200–500 cm^{-1} in order to plan spectroscopic experiment, reduce its cost crucially and decode experimentally observed spectrum. Such an outstanding accuracy is unreachable without careful treatment of the so-called relativistic effects, completely changing even the qualitative picture of electronic states and properties [7].

One of the most promising electronic structure models suitable for solution of such problems is the relativistic coupled cluster (RCC) theory [8] and its extensions to excited electronic states [9,10]. Despite such advantages of these methods as correct physical behaviour, conceptual simplicity and controllable accuracy, rather severe drawbacks are to be mentioned. The most important ones are the restricted scope of applicability (not all types of electronic states are accessible at the moment) and high computational cost, at least N^6 (N is a system size parameter), for compact systems where no advantages can be taken from localization techniques [11]. The former obstacle seems to be surmountable at least for systems with three open shells (unpaired electrons); further theoretical developments are required to overcome limitations of currently used models.

A crucial step towards high precision relativistic modelling of molecular systems was made in the frames of the DIRAC project [12]. Within this project, the wide variety of relativistic electronic structure models was developed and implemented as the modern and rather efficient program package. However, the design of RCC codes implemented there seems to be not flexible enough to be able to construct the new more extended generation of coupled cluster models, e.g. models with inclusion of triple excitation and/or more than two open shells. Important requirements for the modern computer implementation of RCC-like models are (a) subroutines should be organized into well-tested elementary blocks which allow working with operators of arbitrary excitation rank; (b) algorithms should be highly scalable and parallelizable with the possible lowest time complexity.

In this paper we discuss the general strategy of building the high performance relativistic coupled cluster code and report first benchmarks of the newly developed EXP-T program package implementing the considered concepts and algorithms.

2 General Considerations

2.1 Relativistic Fock Space Coupled Cluster Method

The Fock space (FS) RCC computational scheme implies the conversion of the relativistic many-electron Hamiltonian into the second quantized form

$$H = \sum_{pq} h_{pq} \{a_p^\dagger a_q\} + \frac{1}{4} \sum_{pqrs} V_{pqrs} \{a_p^\dagger a_q^\dagger a_s a_r\} \tag{1}$$

where $a_p^\dagger$, a_q denote creation/destruction operators associated with one-electron functions (molecular spinors) and curly braces mark normal ordering with respect to some closed-shell Fermi vacuum determinant; coefficients h_{pq} , V_{pqrs}

are molecular integrals in the basis of these spinors. Molecular spinors are normally generated by solving Hartree–Fock-like equations for the vacuum determinant. The conventional FS-RCC version [9,10,13] is based on defining complete model spaces *via* the choice of "active" (valence) spinors and constructing the normal-ordered exponential wave operator,

$$\Omega = \{\exp(T)\}, \qquad T = \sum_{pq\ldots rs\ldots} t_{pq\ldots rs\ldots}\{a_p^\dagger a_q^\dagger \ldots a_s a_r\} \tag{2}$$

where $t_{pq\ldots rs\ldots}$ are cluster amplitudes and the summation is normally restricted to single and double excitation operators (RCCSD) or additionally triple excitations (RCCSDT). The wave operator should reconstruct the target many-electron wavefunctions from their model-space projections. Electronic state energies and model-space parts of the corresponding wavefunctions are obtained as eigenvalues and eigenvectors of the effective Hamiltonian $H^{\mathrm{eff}} = \left(\overline{H\,\Omega}\right)_{Cl}$, where the subscript $_{Cl}$ marks the closed (model-space) part of an operator and the overbar denotes its connected part.

Cluster amplitudes should satisfy the equations

$$t_{pq\ldots rs\ldots} = \frac{1}{D_{pq\ldots rs\ldots}}\left(\overline{V\,\Omega} - \overline{\Omega\,\left(\overline{V\,\Omega}\right)_{Cl}}\right)_{pq\ldots rs\ldots}, \quad V = H - H_0. \tag{3}$$

The subscripts $_{pq\ldots rs\ldots}$ in the r.h.s. indicate that the excitation $\{a_p^\dagger a_q^\dagger \ldots a_s a_r\}$ is considered. H_0 is the Hartree–Fock operator for the Fermi vacuum state and the energy denominators $D_{pq\ldots rs\ldots}$ are the negatives of the differences of H_0 eigenvalues associated with the excitation.

It is convenient to partition the cluster operator T according to the number of valence holes (n_h) and valence particles (n_p) to be destroyed (i.e., related to (n_h,n_p) sectors of the Fock space):

$$T = \sum_{n_h n_p} T^{(n_h n_p)} \tag{4}$$

To describe the electronic states in the (N_h,N_p) sector of the Fock space, one needs to determine only $T^{(n_h n_p)}$ with $n_h \leq N_h$ and $n_p \leq N_p$. Therefore the system of coupled Eqs. (3) is split into subsystems, which can be solved consecutively.

The straightforward application of the complete-model-space FS-RCC method to molecular excited state calculations is severely restricted by unavoidable (at least for certain ranges of nuclear configurations) numerical instabilities of the solutions of Eq. (3) caused by intruder states [14]. The presence of intruder states normally manifests itself as the appearance of small or positive $D_{pq\ldots rs\ldots}$ values in Eq. (3). In Reference [15] we modified the conventional FS-RCC Eq. (3) *via* introduction of adjustable shifts of ill-defined (nearly zero or positive) denominators. This stratagem enables one to obtain stable solutions of amplitude equations in problematic situations. Strongly affecting only highly excited approximate eigenstates, it enables one to achieve an accurate description of low-lying excited states [15,16]. Moreover, results can fe further rectified by extrapolation to the zero-shift limit [17].

2.2 Algorithm Design

The scheme of solving the working Eq. (3) of the FS-RCC method can be formalized using the flowchart shown on Fig. 1. Note that steps I-V are performed consecutively for each Fock space sector from the vacuum through the target one.

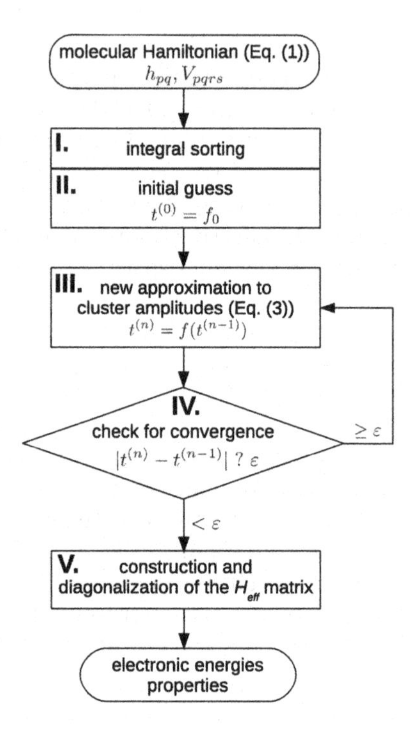

Fig. 1. Flowchart of the FS-RCC method. Steps I–V are performed consecutively for each Fock space sector.

We proceed to the analysis of the time complexity of FS-RCC calculations. The usual measure of the system's size in quantum chemistry is the number of one-electron basis functions N, roughly proportional to the number of atoms in the simulated system. Integral sorting (step I) requires N^4 operations; steps II – IV require at least N^6 or even N^8 (for models involving triple excitations in Eq. (2)) operations. Steps I, II and IV are much cheaper than step III. Finally, the time complexity of step V is completely determined by the dimension of the subspace of active spinors and the number of active quasiparticles. There are numerous well-established and highly efficient algorithms for matrix diagonalization and the step V is the cheapest step of the FS-RCC calculation (at least for sectors with no more than three open shells). Further we will focus on the solution of amplitude equations, since this step is dominating at the RCCSD and higher levels of theory.

Working equations of all RCC models are most conveniently formulated within the language of Goldstone diagrams [18]. Cluster amplitudes can be represented as sums of dozens or even hundreds diagrams. It is worth noting that almost all diagrams for non-trivial FS sectors can be obtained from diagrams for the conventional single-reference CC method simply by "turning down" open lines [19]; this fact greatly simplifies the validation of FS-RCC codes.

Fortunately, all the Goldstone diagrams in FS-RCC amplitude Eq. (3) can be processed in a rather similar way. Consider, for example, one of the simplest diagrams contributing to the $T_2^{(0h,1p)}$ cluster operator amplitudes (see Fig. 2). Its algebraic expression is

$$- P(ab) \sum_{jc} t_{xjcb} V_{ciaj} \quad \text{for all } x, i, a, b \tag{5}$$

where t are cluster amplitudes and $P(ab)$ is a permutation operator. Here we use the widely accepted naming convention for the spinor indices [18]: $i, j, ...$ enumerates holes, $a, b, ...$ – particles, x – active particles. Indices to be contracted over are given in boldface.

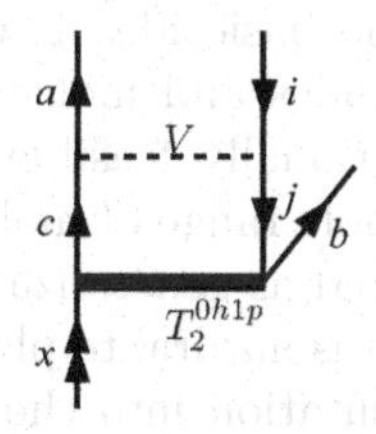

Fig. 2. Goldstone diagram for the contribution (5) to the $T_2^{(0h,1p)}$ operator amplitudes in the (0h,1p) sector.

The straightforward summation is possible, but requires hand-coded loops and non-contiguous memory access, thus resulting in a confusing algorithm and low performance. It is preferable to split the evaluation of this expression into the sequence of elementary operations in such a manner that tensor contraction (5) can be performed as a simple matrix multiplication; in this case multidimensional arrays (which are commonly called tensors in many-body theories) are considered as rectangular supermatrices. These elementary operations will be:

1. $t_{xjcb} \rightarrow t'_{xbjc}$ $\quad \forall\, x, j, c, b \quad N^4$
2. $V_{ciaj} \rightarrow V'_{iajc}$ $\quad \forall\, c, i, a, j \quad N^4$
3. $\sum_{jc} t'_{xb;jc} V'_{ia;jc} \rightarrow \Delta t'_{xb;ia}$ $\quad \forall\, x, i, a, b \quad N^6$
4. $\Delta t'_{xbia} \rightarrow \Delta t_{xiab}$ $\quad \forall\, x, i, a, b \quad N^4$

The most time consuming operation 3 (tensor contraction) can be performed as a matrix multiplication thus allowing to use any high-performance linear algebra package. Additional tensor transpositions (1, 2, 4) are now required for most diagrams. Such tensor transpositions in general case can hardly be implemented in a cache-efficient manner (except of purely 2D matrix transposition-like cases). However, for actual problems these three additional transpositions are necessarily cheaper than the tensor contraction step. This approach is sometimes referred as the Transpose-Transpose-`GEMM`-Transpose (TTGT) approach [20]. Furthermore, the two other important advantages of such a decomposition into elementary operations are to be mentioned here:

(1) only these operations are to be implemented for arbitrary rank tensors; the code for all CC models can be in principle obtained in an automated manner. This ensures flexibility and extensibility of the code written in this elementary building blocks paradigm;
(2) these elementary operations are perfectly suitable for parallel execution.

2.3 Symmetry Handling and Data Structures

Below a brief discussion of data structures optimally compatible with the algorithms described above and ensuring efficient and well-scaling parallel implementation on heterogeneous architectures is presented (in fact, all modern supercomputers are of this type). The basic idea is to choose some partitioning of all the data (e.g. cluster amplitudes and molecular integrals) to be processed into blocks. The most computationally feasible way of such a partitioning is determined by division of the whole range of molecular spinors into subsets; the resulting tensors can be considered as generalizations of block matrices. In case of additional spatial symmetry, it is natural to place the spinors which transform via the same irreducible representation into the same subset, thus allowing to get rid of matrix elements which are *a priori* zero due to symmetry reasons. This approach is known as the direct product decomposition (DPD) technique and is widely used in both non-relativistic [21] and relativistic [22] frameworks. Note that in the general case complex arithmetic is required in relativistic electronic structure calculations; however, for some double groups (e.g. C_{2v}^*, C_{2h}^*, D_2^*, D_{2h}^*, $C_{\infty v}^*$, $D_{\infty h}^*$) real arithmetic can be used [23], resulting in a great reduction of computational effort and memory requirements.

Thus all tensors can be represented as lists of blocks (Fig. 3); all algorithms of elementary operations described in 2.2 are expressed in terms of these blocks (for example, see Fig. 4 for the tensor contraction algorithm). Blocks of one tensor are independent and can be stored on different nodes of the distributed memory system thus reducing memory requirements and allowing treatment of really large systems.

The node-level parallelism (OpenMP [24] or GPGPU) arises naturally for elementary operations with blocks. Current computer implementation of FS-RCC reported here uses highly optimized MKL [25] and CUBLAS [26] libraries to perform parallel contractions on CPU and GPU, respectively. It should be noted that sizes of blocks for high symmetry point groups ($C_{\infty v}^*$ represented by C_{32} and $D_{\infty h}^*$ represented by C_{16h}) can differ by orders of magnitude. This gives rise to a considerable imbalance: strong scaling (wrt number of threads) will be

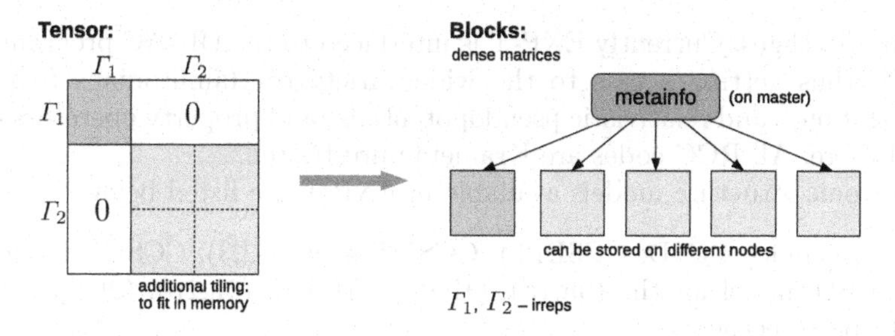

Fig. 3. Partitioning of tensors (cluster amplitudes or molecular integrals) into smaller blocks. The figure shows the case of a rank 2 tensor; generalization to the case of rank 4 and 6 tensors is straightforward.

Data: tensors A, B
Result: tensor $C = \sum_{k_1,\ldots,k_K} A_{i_1,\ldots,i_M;k_1,\ldots,k_K},\ B_{j_1,\ldots,j_N;k_1,\ldots,k_K}$

```
foreach block_c in blocks(C) do
    foreach block_a in blocks(A) do
        load block_a from disk if needed;
        foreach block_b in blocks(B) do
            load block_b from disk if needed;
            calculate dimensions of supermatrices;
            block_c += gemm(block_a, block_b)
        end
    end
    store block_c to disk if needed;
end
```

Fig. 4. Contraction of two tensors stored as lists of dense blocks. Loops can be swapped to ensure that the outer loop runs over the tensor stored on disk (to avoid redundant disk operations).

efficient for large blocks and inefficient for relatively small ones, leading to the degradation of overall scalability of the program. This issue is not addressed here, but the obvious solution can be based on the dynamic selection of optimal number of threads guided by the runtime profiling performed at the first iteration.

3 Implementation and Benchmark

3.1 The EXP-T Program System

Considerations discussed in Sect. 2.2 and 2.3 were implemented in the newly developed electronic structure package EXP-T (named after the formula (2) of the RCC Ansatz).

Parameters of the relativistic Hamiltonian (1), i.e. sets of molecular spinors and molecular integrals, have to be imported from third party electronic structure packages. Currently EXP-T is interfaced to the DIRAC program package [12], thus getting access to the wide variety of Hamiltonians (e.g. four-component ones and relativistic pseudopotentials) and property operators implemented there. All RCC codes are Kramers unrestricted.

Electronic structure models available in EXP-T are listed below:

- single-reference CCSD, CCSD(T), CCSDT-n (n=1,2,3), CCSDT models;
- FS-CCSD model for the (0h,1p), (1h,0p), (1h,1p), (0h,2p), (2h,0p), (0h,3p) Fock space sectors;
- FS-CCSDT-n (n=1,2,3) and FS-CCSDT models are implemented for the (0h,1p), (0h,2p) and (0h,3p) Fock space sectors.

At present only single-point energy calculations are implemented for all the models listed above. The FS-RCC models for the (0h,3p) FS sector were developed to deal with electronic states dominated by determinants with three open shells and was implemented earlier only by Kaldor and coworkers with application to very small non-relativistic atomic systems [27]; the features of these models will be described in our future papers. The corresponding code is to be considered as experimental to the moment.

Some features recently proposed by us which greatly extend the scope of applicability of the FS-RCC method are also included:

- "dynamic" energy denominators shifts and subsequent Padé extrapolation to the zero-shift limit as a solution of the intruder-state problem [15,17];
- finite-field transition property calculations [28];
- decoupling of spin-orbit-coupled states by projection and extraction of SO coupling matrix elements [15].

EXP-T currently supports parallel calculations on shared-memory computers via the OpenMP and CUDA technologies.

EXP-T is written in the C99 programming language and hence can be compiled using the most common development tools available on most platforms. EXP-T is currently oriented to Unix-like operating systems.

3.2 Performance Evaluation

To assess the performance features of the newly developed FS-RCC implementation, the series of the FS-RCCSD calculations of the KCs alkali-metal molecular dimer were done. In order to test the efficiency of the blocking scheme employed we performed the calculations with full account for the point-group symmetry and with artificially lowered symmetries. The size of the problem (112 active spinors, 374 spinors overall) is large enough to demonstrate tendencies in scaling features and relative computational cost of different stages of the routine FS-RCC calculation. Electronic states of KCs are formed by the following Fock space scheme:

$$\mathrm{KCs}^{2+}(0\mathrm{h},0\mathrm{p}) \rightarrow \mathrm{KCs}^{+}(0\mathrm{h},1\mathrm{p}) \rightarrow \mathrm{KCs}^{0}(0\mathrm{h},2\mathrm{p})$$

Results of wall time measurements for the RCCSD(0h,0p) calculation of KCs^{2+} in different point groups are presented in Table 1. The speed-up is rising from C_1^* to C_s^* and from C_{2v}^* to $C_{\infty v}^*$ (represented actually by C_{32}) and comes from the reduction of tensor sizes; the speed-up going from C_s^* to C_{2v}^* is only due to the use of real arithmetic instead of complex one since both groups have the same number of fermionic irreducible representations (two) (see also [29]). It should be mentioned that the C_1^* point group represents the most general case and probably will be the most demanded in future applications of the FS-RCC method to polyatomic molecules.

Table 1. Wall clock time (in seconds on the Intel(R) Xeon(R) E5-2680 v4 CPU) for RCCSD calculation in different point groups. The test calculation concerns the KCs^{2+} ion in the molecular pseudospinor basis (16 electrons, 374 functions). R – real group, C – complex group.

Point group		Total time	Integral sorting	Tensor contractions	Tensor trans-positions	Time per iteration
C_1^*	(C)	105694	19913	37375	2122	2218
C_s^*	(C)	25135	11198	12507	1049	786
C_{2v}^*	(R)	11489	5525	5307	471	324
$C_{\infty v}^*$	(R)	4113	2442	1451	137	87

Furthermore, relative computational costs of stages of the FS-RCC calculation are nearly constant for different point groups. Integral sorting being the N^4 operation requires considerable computational time, but its contribution will decrease for larger problems. Moreover, the tensor contractions/ tensor transpositions ratio is high enough to completely justify the use of the decomposition (TTGT) approach to evaluation of the general tensor contractions presented in Sect. 2.2. This ratio will be much higher for larger problems due to the ratio of time scaling of these tensor operations (N^{4-6} *vs* N^{6-8}).

Nearly 90% of time used for tensor contractions is spent for evaluation of the single term in RCC equations which involves two-electron integrals with four indices of virtual spinors. This tensor (commonly denoted as $\langle pp|pp\rangle$) is typically an order of magnitude larger than the other tensors containing integrals and has to be stored on harddrive in most real-life applications; this tensor is to be read from disk at each CC iteration (see Fig. 4), resulting in limited parallelizabitity of the code (Fig. 5). The possible remedy is to store some part of the $\langle pp|pp\rangle$ tensor in RAM.

The better scaling for the (0h,2p) Fock space sector than for the (0h,1p) sector is due to the amount of data processed in the former case is much bigger: in the example considered the maximum block size is of order $1 \cdot 10^6$ elements for the (0h,1p) sector and $5 \cdot 10^6$ for the (0h,2p) sector. Thus the percentage of computations that can be performed in parallel is considerably higher for the latter case resulting in better scaling (due to the Amdahl's law). Another problem which can lead to worse scaling is the fact that for highly symmetrical

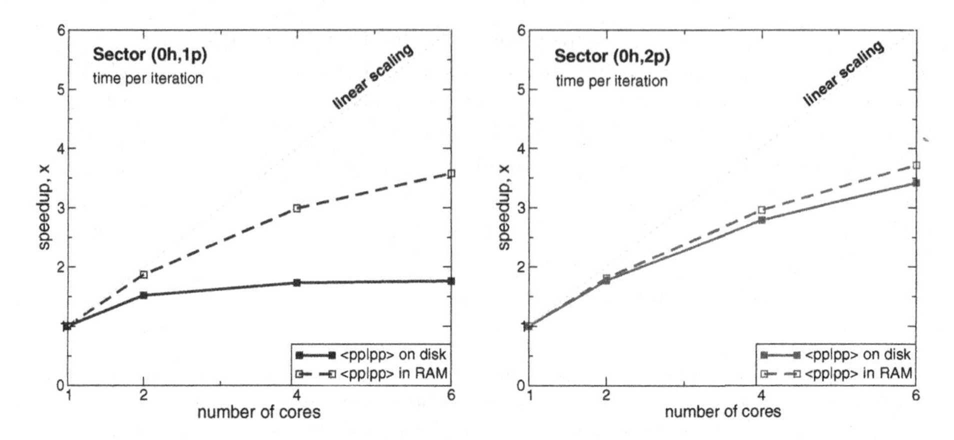

Fig. 5. Scaling of the FS-CCSD calculation in different Fock space sectors with respect to the number of OpenMP threads (for the time per iteration on the Intel(R) Core(TM) i7-4930K CPU). The test calculation concerns the KCs molecule in the molecular pseudospinor basis (16 electrons, 374 functions, 112 active spinors).

point groups like $C^*_{\infty v}$ sizes of blocks can be very small. This results in overheads for thread creating being much larger than the time used for actual calculation and hence in degradation of the overall scaling. Such a situation is observed for the (0h,2p) sector (Fig. 5, right plot). The latter problem can be solved by choosing optimal number of threads for such small blocks.

4 Conclusions and Prospects

A new implementation of the Fock space relativistic coupled cluster method designed for modern parallel systems is presented; underlying method, algorithms and approaches to data handling are discussed. Scaling with respect to the number of OpenMP threads currently is not ideal (in the example presented no more than 4x times faster on 6 CPU cores), but the ways of possible improvements are rather clear. However, conceptual limitations due to the necessity of the usage of harddrive to store molecular integrals can be overcame in future versions of the code by employing the MPI parallelization model.

The future work on the EXP-T program system will address not only improvements of the computational scheme, but also development of new relativistic coupled cluster models aimed at expanding the field of applicability of the FS-RCC method and achieving a principally new level of accuracy for prediction of properties of molecules containing heavy and superheavy nuclei.

Acknowledgements. Authors are grateful to T. A. Isaev, S. V. Kozlov, L. V. Skripnikov, A. V. Stolyarov and L. Visscher for fruitful discussions. This work has been carried out using computing resources of the federal collective usage centre Complex for Simulation and Data Processing for Mega-science Facilities at NRC "Kurchatov Institute", http://ckp.nrcki.ru/, and computers of Quantum Chemistry Lab at NRC "Kurchatov Institute" – PNPI.

The research was supported by the Russian Science Foundation (Grant No. 20-13-00225).

References

1. Dykstra, C., et al. (eds.): Theory and Applications of Computational Chemistry. The First Forty Years, 1st edn. Elsevier Science, Amsterdam (2005). https://doi.org/10.1021/ja059883q
2. Petrov, A.N., Skripnikov, L.V., Titov, A.V.: Zeeman interaction in $^3\Delta_1$ state of HfF^+ to search for the electron electric dipole moment. Phys. Rev. A **96**, 022508 (2017). https://doi.org/10.1103/PhysRevA.96.022508
3. Ivanov, M.V., Bangerter, F.H., Krylov, A.I.: Towards a rational design of laser-coolable molecules: insights from equation-of-motion coupled-cluster calculations. Phys. Chem. Chem. Phys. **21**, 19447–19457 (2019). https://doi.org/10.1039/c9cp03914g
4. Laatiaoui, M., et al.: Atom-at-a-time laser resonance ionization spectroscopy of nobelium. Nature **538**, 495–498 (2016). https://doi.org/10.1038/nature19345
5. Ruiz, R.F.G., et al.: Spectroscopy of short-lived radioactive molecules: a sensitive laboratory for new physics. arXiv preprint arXiv:1910.13416 (2019)
6. Eliav, E., Fritzsche, S., Kaldor, U.: Electronic structure theory of the superheavy elements. Nucl. Phys. A **944**, 518–550 (2015). https://doi.org/10.1016/j.nuclphysa.2015.06.017
7. Dyall, K.G., Faegri Jr., K.: Introduction to Relativistic Quantum Chemistry, 1st edn. Oxford University Press, Oxford (2007)
8. Visscher, L., Lee, T.J., Dyall, K.G.: Formulation and implementation of a relativistic unrestricted coupled-cluster method including noniterative connected triples. J. Chem. Phys. **105**(19), 8769–8776 (1996). https://doi.org/10.1063/1.472655
9. Eliav, E., Kaldor, U., Hess, B.A.: The relativistic Fock-space coupled-cluster method for molecules: CdH and its ions. J. Chem. Phys. **108**, 3409–3415 (1998). https://doi.org/10.1063/1.475740
10. Visscher, L., Eliav, E., Kaldor, U.: Formulation and implementation of the relativistic Fock-space coupled cluster method for molecules. J. Chem. Phys. **115**(21), 9720–9726 (2001). https://doi.org/10.1063/1.1415746
11. Saitow, M., Becker, U., Riplinger, C., Valeev, E.F., Neese, F.: A new near-linear scaling, efficient and accurate, open-shell domain-based local pair natural orbital coupled cluster singles and doubles theory. J. Chem. Phys. **146**(16), 164105 (2017). https://doi.org/10.1063/1.4981521
12. Gomes, A.S.P., Saue, T., Visscher, L., Jensen, H.J.A., Bast, R., et al.: DIRAC, a relativistic ab initio electronic structure program (2016). http://www.diracprogram.org
13. Kaldor, U.: The Fock space coupled cluster method: theory and application. Theor. Chim. Acta **80**, 427–439 (1991). https://doi.org/10.1007/bf01119664
14. Evangelisti, S., Daudey, J.P., Malrieu, J.P.: Qualitative intruder-state problems in effective Hamiltonian theory and their solution through intermediate Hamiltonians. Phys. Rev. A **35**, 4930–4941 (1987). https://doi.org/10.1103/physreva.35.4930

15. Zaitsevskii, A., Mosyagin, N.S., Stolyarov, A.V., Eliav, E.: Approximate relativistic coupled-cluster calculations on heavy alkali-metal diatomics: application to the spin-orbit-coupled $A^1\Sigma^+$ and $b^3\Pi$ states of RbCs and Cs_2. Phys. Rev. A **96**(2), 022516 (2017). https://doi.org/10.1103/physreva.96.022516
16. Kozlov, S.V., Bormotova, E.A., Medvedev, A.A., Pazyuk, E.A., Stolyarov, A.V., Zaitsevskii, A.: A first principles study of the spin-orbit coupling effect in LiM (M = Na, K, Rb, Cs) molecules. Phys. Chem. Chem. Phys. **22**, 2295–2306 (2020). https://doi.org/10.1039/c9cp06421d
17. Zaitsevskii, A., Eliav, E.: Padé extrapolated effective Hamiltonians in the Fock space relativistic coupled cluster method. Int. J. Quantum Chem. **118**, e25772 (2018). https://doi.org/10.1002/qua.25772
18. Shavitt, I., Bartlett, R.J.: Many Body Methods in Chemistry and Physics. Cambridge University Press, Cambridge (2009). https://doi.org/10.1017/cbo9780511596834
19. Kaldor, U.: Open-shell coupled-cluster method: electron affinities of Li and Na. J. Comput. Chem. **8**, 448–453 (1987). https://doi.org/10.1002/jcc.540080423
20. Matthews, D.A.: High-performance tensor contraction without transposition. SIAM J. Sci. Comput. **40**, C1–C24 (2018). https://doi.org/10.1137/16m108968x
21. Stanton, J.F., Gauss, J., Watts, J.D., Bartlett, R.J.: A direct product decomposition approach for symmetry exploitation in many-body methods. I. Energy calculations. J. Chem. Phys. **94**, 4334–4345 (1991). https://doi.org/10.1063/1.460620
22. Shee, A., Visscher, L., Saue, T.: Analytic one-electron properties at the 4-component relativistic coupled cluster level with inclusion of spin-orbit coupling. J. Chem. Phys. **145**, 184107 (2016). https://doi.org/10.1063/1.4966643
23. Saue, T., Jensen, H.J.A.: Quaternion symmetry in relativistic molecular calculations: the Dirac-Hartree-Fock method. J. Chem. Phys. **111**, 6211–6222 (1999). https://doi.org/10.1063/1.479958
24. Dagum, L., Menon, R.: OpenMP: an industry standard API for shared-memory programming. IEEE Comput. Sci. Eng. **5**, 46–55 (1998). https://doi.org/10.1109/99.660313
25. Intel(R) Math Kernel Library Version 2018.0.1
26. https://developer.nvidia.com/cublas
27. Hughes, S.R., Kaldor, U.: The coupled-cluster method in high sectors of the Fock space. Int. J. Quantum Chem. **55**, 127–132 (1995). https://doi.org/10.1002/qua.560550207
28. Zaitsevskii, A.V., Skripnikov, L.V., Kudrin, A.V., Oleinichenko, A.V., Eliav, E., Stolyarov, A.V.: Electronic transition dipole moments in relativistic coupled-cluster theory: the finite-field method. Opt. Spectrosc. **124**, 451–456 (2018). https://doi.org/10.1134/s0030400x18040215
29. Visscher, L.: On the construction of double group molecular symmetry functions. Chem. Phys. Lett. **253**, 20–26 (1996). https://doi.org/10.1016/0009-2614(96)00234-5

Transient Halo in Thin Cloud Layers: Numerical Modeling

Yaroslaw Ilyushin[1,2](✉)

[1] Physical Faculty, Moscow State University, Moscow, Russia
ilyushin@phys.msu.ru
[2] Kotel'nikov Institute of Radio Engineering and Electronics, Russian Academy of Sciences, Moscow, Russia

Abstract. In this paper we investigate time-dependent backscattering halo of a pulsed light beam in a layer of scattering medium. We simulate the polarized radiative transfer in the layer numerically with the three-dimensional upwind-difference scheme. This immediately reveals the dynamic halo effect which we investigate.

Finally, we validate our results against Monte-Carlo radiative transfer simulations and analyze time-dependent structure of the light field using the simulation results.

Keywords: Radiative transfer · Halo · Lidar · Remote sensing

1 Introduction

Radiative transfer (RT) theory [1] is a well established approximated theory for evaluation of energetic parameters of radiative fields in random media. It is widely applied both in radiative heat transfer problems [2] and for analysis and interpretation of remote sensing data [3–7].

Correct modeling of radiation fields requires proper accounting for polarization of the radiation and its changes in scattering events. Disregarding of the polarization significantly simplifies all the calculations, although causes an error up to tens per cent [8]. On the other hand, scalar RT computations are more than ten times faster, due to the scalar phase scattering function instead of 4×4 phase scattering matrix, and four times less memory consuming, because of single unknown scalar function instead of vector of four Stokes parameters. For this reason, scalar approach is still used for qualitative assessments [9], as well as for modeling of complicated fields with singularities [10,11]. Besides, newly posed problems are typically first investigated in scalar approximation.

Among known published papers on transient radiative transfer problems, contrary, there are not so many studies where polarization is accounted for (see [12–16] and references therein). Generally, scalar approach is largely prevalent in studies of nonstationary radiative transfer in scattering media [10,17,18], as well as in the theory of the correlation functions of random wave fields [19–25].

V. Voevodin and S. Sobolev (Eds.): RuSCDays 2020, CCIS 1331, pp. 387–398, 2020.
https://doi.org/10.1007/978-3-030-64616-5_34

However, polarization state of the photon contains valuable information about previous life history of this photon in the medium. Taking the polarization into account not only quantitatively improves the simulation accuracy, but makes it possible to qualitatively analyze physical mechanisms of the formation of the radiative field. Time-resolved distribution of intensity and polarization helps to separate photons with different path length.

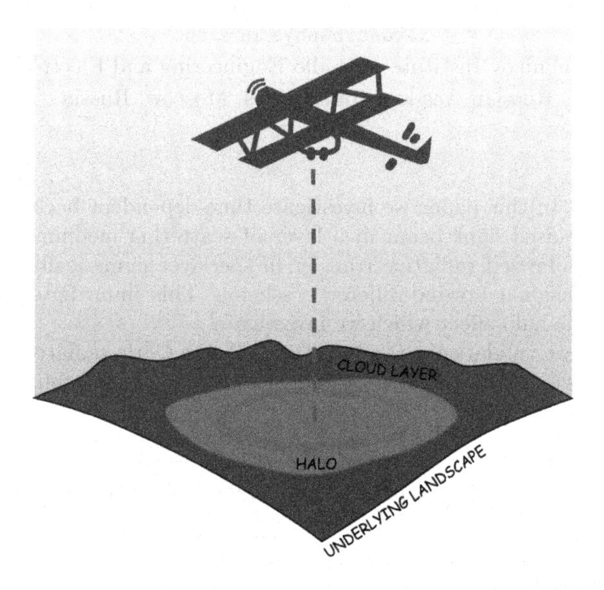

Fig. 1. (Color online) Schematic view of the experimental geometry of lidar sounding of cloud layers.

One of the most complicated problems of the radiative transfer theory is a field of a point directed (PD) source (searchlight, pencil beam etc.) [9,10,26–28]. PD source, on the one hand, is a fundamental elementary radiation source in the RT theory. Every source function of the radiative transfer equation (RTE) can be represented as a linear superposition of the sources of PD type. On the other hand, narrow collimated beams of radiation, in particular laser beams, are widely applied in remote sensing of natural environment[29], geodesy and navigation [30], wireless communication[31], biomedical diagnostics, therapy and surgery [32], various technological processes [33,34] etc. Due to that, many studies are published on the subject of propagation of continuous (e.g. [9,28,35]) and pulsed (e.g.[17,18]) beams of radiation. Most of these studies adopt the scalar approach to the field calculation, i.e. polarization is ignored there.

One of the effects theoretically predicted in the scalar approximation is a backscattering halo, i.e. ring maxima of intensity, centered around the entry point of the beam into the medium [9,28,35]. In the papers [9,28] criteria of the static effect manifestation in a semi-infinite medium are derived. Nonetheless, static backscatering halo has not yet been observed in any nature or laboratory experiment.

Authors of [36] numerically solved the vector radiative transfer equation (VRTE) with a Monte-Carlo algorithm and showed that backscattering halo in slabs of finite thickness ($\mu_s h = 25$) can be formed dynamically. For media with high scattering anisotropy $g = 0.8 \ldots 0.95$ this corresponds to thickness comparable to the transport length l_{tr} ($h = 1.25 \ldots 5l_{tr}$). Nevertheless, authors analyze their results in terms of semi-infinite medium model, i.e. disregard the effect possibly caused by finite slab thickness.

However, time-dependent backscattering halo has been observed in a diffuse reflection of laser radiation from cloud layers with the so-called imaging lidars [37,38] (Fig. 1). Numerical modeling without polarization [17,18] theoretically confirmed these observations.

The objective of this paper is thorough simulation of scattering of polarized pulsed laser beams in cloud layers and formation of dynamically backscattering halos, observed in experiments with imaging lidars. Motivation of the study is development of theoretical model of the halo effect and analysis of conditions of its observation. To do that, direct numerical simulations of the effect have been performed with the two completely different approaches (finite-difference scheme and Monte-Carlo approach) and then cross-validated.

The rest of the paper is organized as follows. In the Sect. 2 the theory of polarized radiative transfer (RT), used in the study, is briefly recalled. In the Sect. 3 results of polarized RT numerical simulations are presented and analyzed. The polarized ray transport model is then validated against scalar RT simulations (Sect. 4). In the Sect. 5 all the results of the study are finally summarized and concluding statements are made.

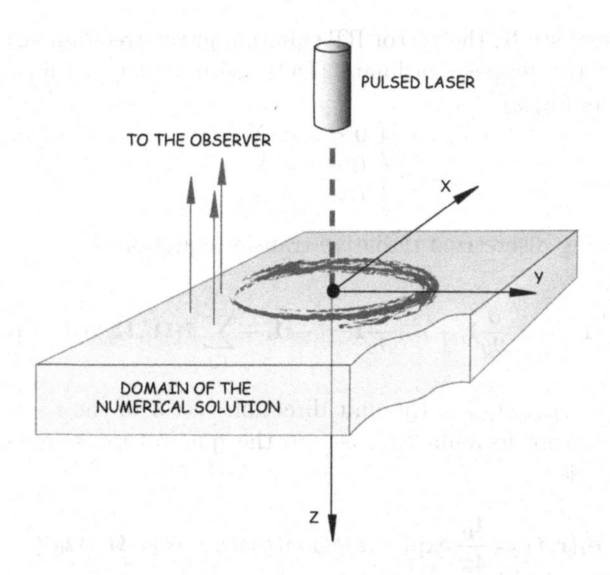

Fig. 2. (Color online) Schematic view of the numerical RT simulation.

2 Polarized Radiative Transfer in Scattering Medium

Spatial and angular distribution of the intensity and polarization of the radiation in a scattering medium obeys the vector radiative transfer equation (VRTE)

$$\frac{1}{c}\frac{\partial \mathbf{I}(\mathbf{r},\boldsymbol{\Omega})}{\partial t}+(\boldsymbol{\Omega}\cdot\nabla)\mathbf{I}(\mathbf{r},\boldsymbol{\Omega})=-\varepsilon\mathbf{I}(\mathbf{r},\boldsymbol{\Omega})+\frac{1}{4\pi}\int\limits_{4\pi}\hat{x}(\boldsymbol{\Omega},\boldsymbol{\Omega}')\mathbf{I}(\mathbf{r},\boldsymbol{\Omega}')d\boldsymbol{\Omega}'+\mathbf{e}(\mathbf{r},\boldsymbol{\Omega},t)\,, \quad (1)$$

where $\boldsymbol{\Omega} = (\mu_x, \mu_y, \mu_z)$ – unit vector of an arbitrary direction, $\mathbf{I}(\mathbf{r}, \boldsymbol{\Omega}) = \{I, Q, U, V\}$ – vector of the Stokes parameters of the polarized radiation, ε – volume extinction coefficient, $\hat{x}(\boldsymbol{\Omega}, \boldsymbol{\Omega}')$ – scattering (Mueller) matrix of the medium, $\mathbf{e}(\mathbf{r}, \boldsymbol{\Omega}, t)$ – source function.

Regarding the limits of applicability of the VRTE (1), we approximate the pulsed light beam by a polarized point directed source

$$\mathbf{I}(\mathbf{r},\boldsymbol{\Omega},t)=\mathbf{I_0}\exp(-\varepsilon z)\delta(\boldsymbol{\Omega})\delta(x)\delta(y)\delta(z-ct)\,, \quad (2)$$

where $\delta(\cdot)$ is the Dirac delta function, $\mathbf{I_0}$ is a polarization state of the incident beam. Thus corresponding source function of the VRTE is

$$\mathbf{e}(\mathbf{r},\boldsymbol{\Omega},t)=\frac{\mathbf{I_0}}{4\pi}\exp(-\varepsilon z)\delta(x)\delta(y)\delta(z-ct)\hat{x}(\boldsymbol{\Omega},\boldsymbol{\Omega}_0)\,, \quad (3)$$

$\boldsymbol{\Omega}_0 = (0, 0, 1)$ is the beam incidence direction, which is parallel to z axis. Simulation of the spatial and angular distributions of the radiation intensity and polarization in the medium is performed by numerical solving of the Eq. (1) in a given medium domain with the proper boundary conditions on the medium boundaries.

In the present study, the vector RT equation in the tree-dimensional medium was solved by the discrete ordinate (DO) method [39] within a rectangular domain (see the Fig. 2)

$$\begin{cases} 0 < x < X \\ 0 < y < Y \\ 0 < z < Z \end{cases}. \quad (4)$$

Corresponding discretized radiative transfer equation is

$$\frac{1}{c}\frac{\partial}{\partial t}\mathbf{I}_i+\mu_{xi}\frac{\partial}{\partial x}\mathbf{I}_i+\mu_{yi}\frac{\partial}{\partial y}\mathbf{I}_i+\mu_{zi}\frac{\partial}{\partial z}\mathbf{I}_i=-\varepsilon\mathbf{I}_i+\sum_{l,j}\hat{x}(\boldsymbol{\Omega}_i,\boldsymbol{\Omega}_j)a_j\mathbf{I}_j+\mathbf{e}_i(\mathbf{r},t)\,, \quad (5)$$

where $\boldsymbol{\Omega}_i = \{\mu_{xi}, \mu_{yi}, \mu_{zi}\}$ is the unit direction vector of the i – th node of the spherical quadrature formula [40], a_j are the quadrature weights. The source function $\mathbf{e}_i(\mathbf{r}, t)$ is

$$\mathbf{e}_i(\mathbf{r},t)=\frac{\mathbf{I_0}}{4\pi}\exp(-\varepsilon z)\delta(x)\delta(y)\delta(z-ct)\hat{x}(\boldsymbol{\Omega}_i,\boldsymbol{\Omega}_0))\,. \quad (6)$$

Boundary conditions of zero incoming radiance has been set on the boundaries of the medium domain

$$\mathbf{I}_i \equiv \mathbf{I}(\boldsymbol{\Omega}_i) = 0 \text{ when } \boldsymbol{\Omega}_i \cdot \mathbf{n} > 0\,, \tag{7}$$

where n is a unit vector of the inner normal to the medium boundary.

Scattering integral on the sphere was calculated with the Gaussian quadrature formula for a sphere of 29th order of accuracy with 302 nodes [41].

Scattering matrix was calculated for the wave length $\lambda = 532$ nm for the C3 cloud model [42]. Sizes of the water droplets in the cloud obey the modified gamma-distribution

$$n(r) = ar^{\alpha} \exp\left(-br^{\gamma}\right)\,, \quad 0 \leq r \leq \infty\,, \tag{8}$$

where n is the droplets concentration in cm^{-3}, r – is a droplet radius in μm, $a = 5.5556$, $b = 1/3$, $\alpha = 8$, $\gamma = 3$. Scattering matrix of the individual droplets were calculated with the T-matrix algorithm [43] for different values of the radii and then averaged according to the statistical distribution (8). For the wave length $\lambda = 532$ nm, this model yields volume scattering coefficient $\mu_s = 0.0029$ m^{-1} and scattering anisotropy parameter $g = 0.83$.

The Eq. (5) has been solved with the upwind-difference scheme [44]. Computer code for the VRTE solution has been developed since 2011 for the thermal microwave RT simulations and used in [45–47]. The code has been validated against tabulated benchmark RTE solutions [1] and cross-validated with other RT codes [48]. For regularization of the strong anisotropy of phase scattering functions with large g parameters, scattering matrices in forward directions $\hat{x}(\boldsymbol{\Omega}_i, \boldsymbol{\Omega}_i)$ were properly renormalized, i.e. the scattering phase functions were effectively roughened according to the spherical quadrature formula used for the simulation.

3 Results of the Finite-Difference Numerical Solution of the VRTE

The finite difference scheme with the upwind-differences [44] have been implemented with the C++ language. The previously developed polarized RT code has been extensively used before for the thermal RT simulations [45–47,49] and thoroughly tested and validated. Since the single simulation could fit the memory available at the single node, the calculations were parallelized with the OpenMP standard. If necessary, the simulation can be distributed over several nodes. For this purpose, the rectangular domain can be divided into several sub-domains with the corresponding transfer of the boundary values between neighboring processes.

Backscattered radiation patterns produced by a circularly polarized pulsed laser beam at the cloud boundary at the time $t = 7\,\mu\mathrm{s}$ after the pulse incidence on the cloud boundary, are shown in the Fig. 2 (upper row). Cloud model – C3 [42], size of the rectangular cloud model – 5000×5000 m. Cloud thickness varied from

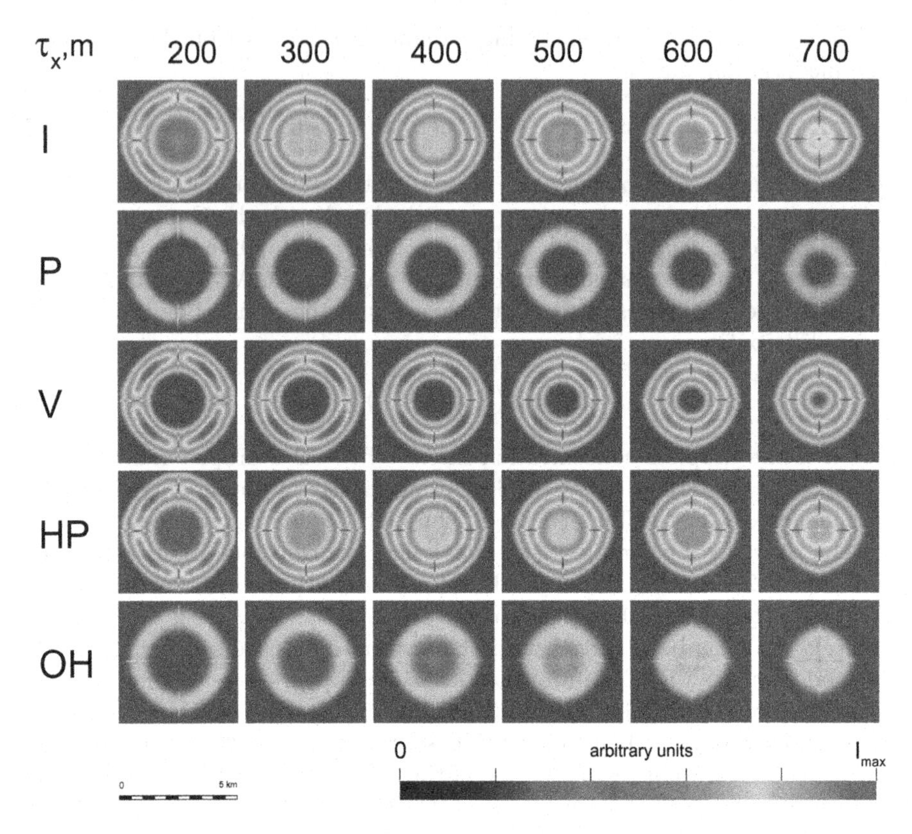

Fig. 3. (Color online) Time-dependent halo in a cloud layer. Circularly polarized incident beam ($U_0 = Q_0 = 0$, $V_0 = 0$). First row – intensity I, second row – polarized part of the intensity $P = \sqrt{U^2 + V^2 + Q^2}$, third row – the fourth Stokes parameter V, fourth and fifth rows – helicity preserving and oppose helicity channels $HP = (I+V)/2$ and $OH = (I-V)/2$, respectively. Size of the shown area is 5000×5000 m.

200 to 700 m. For the wave length $\lambda = 532$, nm the chosen layer thicknesses 200 m approximately correspond to $h \approx 0.1\, l_{tr} \ldots 0.35\, l_{tr}$. The domain was discretized in $128 \times 128 \times N_z$ nodes, where $N_z = 3h/100 + 1$. When the 29th order Gaussian quadrature with 302 nodes was used in DO calculations, such numerical scheme consumed from 2 to 8 Gb RAM for two copies of the solution (the previous and the next values in the iteration). The time step of the numerical scheme was $\Delta t = 2.38 \cdot 10^{-8}$ s, so the source function (3) was approximated numerically with corresponding temporal and spatial resolution. There is an inhomogeneity caused by intrinsic artifacts of the used DO numerical scheme (maxima correspond to the nodes of the quadrature formula lying in the cloud layer plane). Intensity of the polarized component of the radiation $P = \sqrt{Q^2 + U^2 + V^2}$ is shown in the 2nd row of the Fig. 2. One can see that typical radiation pattern consists of bright polarized ring (the so-called "halo") and its less bright and almost not polarized

interior. With the increase of the cloud thickness, the contrast of the halo in the intensity and polarization gradually degrades. This leads us to the two-component model of the scattered field, consisting of the diffuse (background) and ballistic (halo) parts (Fig. 3).

4 Monte-Carlo Simulations of the Transient Radiative Transfer in Layers

To prove the results of the previous sections, we have to make sure that the effect under consideration is not a numerical artifact, and to investigate the impact of the slab thickness on the observed dynamic picture. To show that the time-dependent halo, i.e. outwardly traveling wave from the point source is mesh independent, we use for validation meshless simulation algorithm (Monte-Carlo technique). We also have to investigate how the halo effect depends on the slab thickness. We use scalar RT simulations, because possible errors caused by neglect of polarization (up to tens of percent at worst, and 10% as a typical value [8,11]) probably is roughly on the same order of the calculation error of the finite-difference scheme with the cell sizes selected here.

Monte-Carlo (MC) calculations are usually not too memory consuming, so that we used MPI standard for parallelizing the processes. Each MC process occupies its own processing core, so that single cluster node could support many independent MC simulations with their own unique input parameters.

Besides, in 3D RT problems MC algorithms are somewhat more effective, because the statistical error vanishes as $N^{-1/2}$, where N is the number of randomly scattered photons in the simulation. On the other hand, approximation error of the explicit upwindvscheme is $O(\Delta t) + O(\Delta x)$ [39,50]. In the 3D scheme N is inversely proportional to $\Delta x \Delta y \Delta z$, which makes the approximation error to be $O(N^{-1/3})$. Nevertheless, previously developed finite-difference [45,46] and MC [9,28] codes have been used for polarized and unpolarized RT simulations, respectively.

Numerical accuracy, potentially achievable in these simulations, was limited by the available memory and CPU performance, as well as the length of the pseudo random number generator sequence. However, the codes used here have been extensively tested in previous studies, and the convergence of the result with the increase of the computing time and memory has been firmly established.

Transient radiative transfer equation without account for the polarization is

$$\frac{\partial I}{c\partial t} + (\boldsymbol{\Omega} \cdot \nabla) I = -\varepsilon I + \frac{\Lambda \varepsilon}{4\pi} \oint I(\mathbf{r}, \boldsymbol{\Omega}') x(\boldsymbol{\Omega}, \boldsymbol{\Omega}') d\boldsymbol{\Omega}' + e(\mathbf{r}, \boldsymbol{\Omega}), \tag{9}$$

where $\varepsilon = \mu_a + \mu_s$ – is a volume extinction coefficient, μ_a and μ_s are the volume absorption and scattering coefficients, respectively, $\Lambda \equiv \mu_s/\varepsilon$ is the single scattering albedo, I is the radiance, $x(\boldsymbol{\Omega}, \boldsymbol{\Omega}')$ – phase scattering function of the medium, $e(\mathbf{r})$ – the source function. The Monte-Carlo (MC) algorithm of simple local estimate, used before in [9,28], has also been applied in this study. The number of photons in each run of the algorithm was 10^9. The same configuration with PD source (Fig. 2) was used for MC simulations, with infinite horizontal dimensions of the layer.

We adopted for the calculation the widely used Henyey-Greenstein model phase scattering function [51]

$$x(\mathbf{\Omega}, \mathbf{\Omega}') = \frac{1 - g^2}{(1 + g^2 - 2g\cos\theta)^{3/2}}, \tag{10}$$

where g is the scattering asymmetry parameter, $\cos\theta = \mathbf{\Omega} \cdot \mathbf{\Omega}'$ is the scattering angle cosine. Results of the simulations are shown in the Fig. 4. Scattered radiance is plotted against the dimensionless radius $\rho' = \varepsilon\rho$ at different times $t' = \varepsilon ct$.

One can see that the time-dependent halo gets less and less visible with the increase of the optical thickness of the medium and practically disappears in layers thick enough.

Analysis of the numerical results presented in these figures shows strong dependence of the dynamic halo effect on the scattering layer thickness. In the Fig. 4, left panel it a) time-dependent distribution of the observed intensity of the light field in the slab with the optical thickness $\tau = 1$, which for the asymmetry scattering parameter $g = 0.5$ corresponds to the real physical thickness of the slab $h = 0.5\, l_{tr}$. Annular maximum of the intensity is manifested in the range of the dimensionless normalized radius $\rho' = 0 \ldots 6.5$.

In the slab with the thickness $h = 2\, l_{tr}$, the intensity at every time moment decreases with the radius and any notable manifestations of the halo effect are

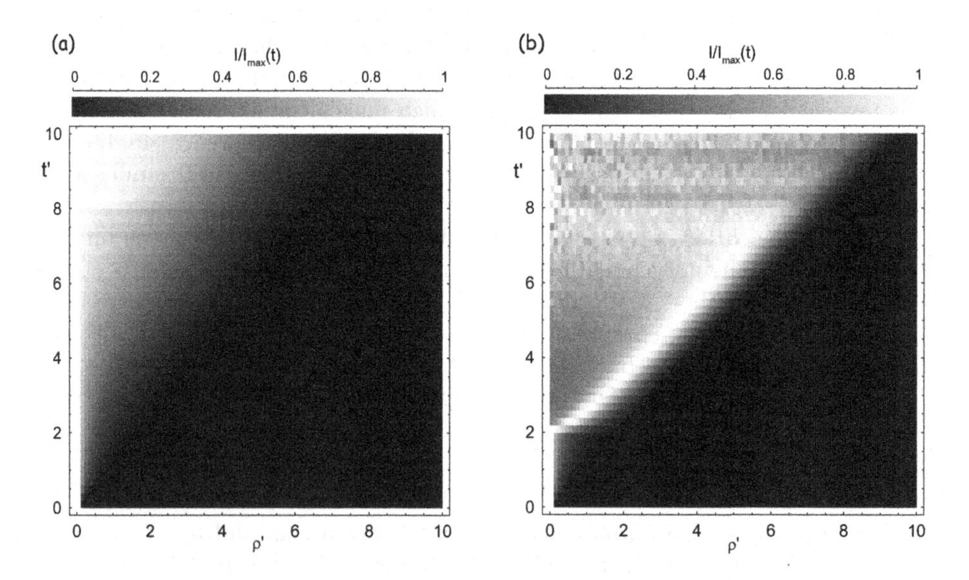

Fig. 4. Time-dependent backscattering halo in a cloud layer. Monte-Carlo numerical simulation. Scattering asymmetry parameter $g = 0.5$, layer optical depth $\varepsilon h = 1.0$ (a), $\varepsilon h = 4.0$ (b). Normalized times εct are shown by the numerical labels near each curve.

absent (the Fig. 4, right panel it b)). To the moment $t' = 10$ diffuse field, similar to Gaussian distribution, is practically formed.

Similar numerical results can also be found in [17,18]. Observations of the effect reported in several papers Measurements [37,38] provide experimental confirmation of the results obtained.

5 Conclusions and Remarks

In the study presented here, effect of time-dependent backscattering halo in layers of scattering media is studied both analytically and numerically. Transient distributions of the intensity and polarization of the light field, revealed here with the numerical simulations, provide meaningful insight into the mechanisms of formation of the scattered radiation pattern and evolution history of photons forming it.

Finally, proposed theory is validated against Monte-Carlo numerical solution of the scalar radiative transfer equation, and satisfactory agreement between the theory and numerical results is shown.

Manifestation of the effect is an indication that the medium is thin enough, and thus can help to constrain its thickness immediately on the basis of the observation data.

The research is carried out using the equipment of the shared research facilities of HPC computing resources at Lomonosov Moscow State University [52]. Support from Russian Science Foundation with the grant 17-77-20087 is kindly acknowledged.

Author thanks all the reviewers of this paper for valuable comments and constructive suggestions.

This study is also partially supported by Russian Fundamental Research Fund with the grants 13-02-12065 ofi-m "Development of new tecniques and means of the satellite microwave remote sensing of the atmospheric precipitation from space" and 15-02-05476 "Development of new techniques and means of meteorological radar sounding of atmospheric precipitation in the millimeter wave band".

References

1. Chandrasekhar, S.: Radiative Transfer. Courier Corporation, North Chelmsford (1960)
2. Ozisik, M.N.: Radiative Transfer and Interactions with Conduction and Convection. Wiley, New York (1973)
3. Evtushenko, A., Zagorin, G., Kutuza, B., Sobachkin, A., Hornbostel, A., Schroth, A.: Determination of the stokes vector of the microwave radiation emitted and scattered by the atmosphere with precipitation. Izvestiya - Atmos. Ocean Phys. **38**(4), 470–476 (2002)
4. Kutuza, B.G.: Spatial and temporal fluctuations of atmospheric microwave emission. Radio Sci. **38**(3) (2003). MAR12/1-MAR12/7

5. Ilyushin, Y.A.: Martian northern polar cap: layering and possible implications for radar sounding. Planet. Space Sci. **52**(13), 1195–1207 (2004)
6. Ilyushin, Y., Seu, R., Phillips, R.: Subsurface radar sounding of the martian polar cap: radiative transfer approach. Planet. Space Sci. **53**(14–15), 1427–1436 (2005)
7. Ilyushin, Y.A.: Radiative transfer in layered media: application to the radar sounding of martian polar ices. II. Planet. Space Sci. **55**(1–2), 100–112 (2007)
8. Emde, C., Mayer, B.: Errors induced by the neglect of polarization in radiance calculations for three-dimensional cloudy atmospheres. J. Quant. Spectros. Radiat. Transf. **218**, 151–160 (2018)
9. Ilyushin, Y.A.: Backscattering halo from the beam in the scattering medium with highly forward peaked phase function: is it feasible? J. Opt. Soc. Am. A **29**(9), 1986–1991 (2012)
10. Ilyushin, Y.A., Budak, V.P.: Narrow beams in scattering media: the advanced small-angle approximation. J. Opt. Soc. Am. A **28**(7), 1358–1363 (2011)
11. Ilyushin, Y.A.: Coherent backscattering enhancement in highly anisotropically scattering media: Numerical solution. J. Quant. Spectrosc. Radiat. Transfer **113**(5), 348–354 (2012)
12. Ilyushin, Y., Budak, V.: Analysis of the propagation of the femtosecond laser pulse in the scattering medium. Comput. Phys. Commun. **182**(4), 940–945 (2011)
13. Yi, H.L., Ben, X., Tan, H.: Transient radiative transfer in a scattering slab considering polarization. Opt. Express **21**(22), 26693–26713 (2013)
14. Zhang, Y., Yao, F.J., Xie, M., Yi, H.L.: Analysis of polarized pulse propagation through one-dimensional scattering medium. J. Quant. Spectrosc. Radiat. Trans. **197**(SI), 141–153 (2017)
15. Wang, C.H., Feng, Y.Y., Zhang, Y., Yi, H.L., Tan, H.P.: Transient/time-dependent radiative transfer in a two-dimensional scattering medium considering the polarization effect. Opt. Express **25**(13), 14621–14634 (2017)
16. Wang, C.H., Yi, H.L., Tan, H.P.: Transient polarized radiative transfer analysis in a scattering medium by a discontinuous finite element method. Opt. Express **25**(7), 7418–7442 (2017)
17. Prigarin, S.M.: Monte Carlo simulation of the effects caused by multiple scattering of ground-based and spaceborne lidar pulses in clouds. Atmos. Oceanic Opt. **30**(1), 79–83 (2017)
18. Prigarin, S., Aleshina, T.: Monte Carlo simulation of ring-shaped returns for CCD lidar systems. Russ. J. Numer. Anal. Math. Model. **30**(4), 251–257 (2015)
19. Vologdin, A., Prikhod'ko, L.: The autocorrelation function of the plane wave phase in the case of oblique sounding of a randomly inhomogeneous planar stratified medium. Radiotekhnika i Elektronika **49**(10), 1218–1221 (2004)
20. Vologdin, A., Vlasova, O., Prikhod'ko, L.: Fluctuations of the group path and the group-delay time of waves obliquely reflected by a plane-layered medium. J. Commun. Technol. Electron. **52**(10), 1100–1103 (2007)
21. Ilyushin, Y.: Impact of the plasma fluctuations in the martian ionosphere on the performance of the synthetic aperture ground-penetrating radar. Planet. Space Sci. **57**(12), 1458–1466 (2009)
22. Ilyushin, Y.: Influence of the ionospheric plasma density fluctuations on the subsurface sounding of the martian soil by a synthetic aperture radar. Radiophys. Quantum Electron. **52**(5–6), 332–340 (2009)
23. Vologdin, A., Prikhod'Ko, L., Shirokov, I.: Fluctuations of the wave amplitude level in a plane-layered medium with random irregularities. J. Commun. Technol. Electron. **55**(8), 870–875 (2010)

24. Ilyushin, Y.: Influence of anisotropic fluctuations of the ionosphere plasma density on deep radio sounding by a ultra wide band radar with synthesized aperture. Cosm. Res. **48**(2), 157–164 (2010)
25. Ilyushin, Y.: Subsurface radar location of the putative ocean on Ganymede: numerical simulation of the surface terrain impact. Planet. Space Sci. **92**, 121–126 (2014)
26. Ilyushin, Y.A., Budak, V.P.: Narrow-beam propagation in a two-dimensional scattering medium. J. Opt. Soc. Am. A **28**(2), 76–81 (2011)
27. Ilyushin, Y.A.: Propagation of a collimated beam in the refractive scattering medium. Radiophys. Quantum Electron. **55**(10), 648–653 (2013)
28. Ilyushin, Y.A.: Backscattering effects in media with strongly elongated scattering indicatrices. Radiophys. Quantum Electron. **60**(4), 323–331 (2017)
29. Eloranta, E.W.: Practical model for the calculation of multiply scattered lidar returns. Appl. Opt. **37**(12), 2464–2472 (1998)
30. Kaloshin, G.A., et al.: Potential capabilities of aircraft laser landing systems. Appl. Opt. **55**(30), 8556–8563 (2016)
31. Giuliano, G., Laycock, L., Rowe, D., Kelly, A.E.: Solar rejection in laser based underwater communication systems. Opt. Express **25**(26), 33066–33077 (2017)
32. Tuchin, V.V.: Light scattering study of tissues. Phys. Usp. **40**(5), 495–515 (1997)
33. Shalupaev, S.V., Maksimenko, A.V., Myshkovets, V.N., Nikityuk, Y.V.: Laser cutting of ceramic materials with a metallized surface. J. Opt. Technol. **68**(10), 758 (2001)
34. Fox, M.D.T., French, P., Peters, C., Hand, D.P., Jones, J.D.C.: Applications of optical sensing for laser cutting and drilling. Appl. Opt. **41**(24), 4988–4995 (2002)
35. Kim, A.D., Moscoso, M.: Backscattering of beams by forward-peaked scattering media. Opt. Lett. **29**(1), 74–76 (2004)
36. Phillips, K., Xu, M., Gayen, S., Alfano, R.: Time-resolved ring structure of circularly polarized beams backscattered from forward scattering media. Opt. Express **13**(20), 7954–7969 (2005)
37. Cahalan, R., McGill, M., Kolasinski, J., Varnai, T., Yetzer, K.: THOR - cloud thickness from offbeam lidar returns. J. Atmos. Oceanic Technol. **22**(6), 605–627 (2005)
38. Polonsky, I., Love, S., Davis, A.: Wide-angle imaging lidar deployment at the ARM southern great plains site: intercomparison of cloud property retrievals. J. Atmos. Oceanic Technol. **22**(6), 628–648 (2005)
39. Richtmyer, R., Morton, K.: Difference Methods for Solving Boundary-Value Problems (1972)
40. Penttila, A., Lumme, K.: Optimal cubature on the sphere and other orientation averaging schemes. J. Quant. Spectrosc. Radiat. Transfer **112**(11), 1741–1746 (2011). Electromagnetic and Light Scattering by Nonspherical Particles XII
41. Lebedev, V.: Quadrature formulas for a sphere of the 25–29th order of accuracy. Sib. Mat. Zh. **18**(1), 132–142 (1977)
42. Deirmendjian, D.: Electromagnetic Scattering on Spherical Polydispersions. R (Rand Corporation). American Elsevier Pub. Co. (1969)
43. Moroz, A.: Improvement of Mishchenko's T-matrix code for absorbing particles. Appl. Opt. **44**(17), 3604–3609 (2005)
44. Richtmyer, R.D., Morton, K.W.: Difference Methods for Initial-Value Problems. Interscience Publishers, New York (1967)
45. Ilyushin, Y.A., Kutuza, B.G.: Influence of a spatial structure of precipitates on polarization characteristics of the outgoing microwave radiation of the atmosphere. Izvestiya - Atmos. Ocean Phys. **52**(1), 74–81 (2016)

46. Ilyushin, Y.A., Kutuza, B.G.: Microwave band radiative transfer in the rain medium: Implications for radar sounding and radiometry. In: 2017 Progress In Electromagnetics Research Symposium - Spring (PIERS), pp. 1430–1437 (2017)
47. Ilyushin, Y., Kutuza, B., Sprenger, A., Merzlikin, V.: Intensity and polarization of thermal radiation of three-dimensional rain cells in the microwave band. In: AIP Conference Proceedings, vol. 1810 (2017)
48. Kokhanovsky, A., et al.: Benchmark results in vector atmospheric radiative transfer. J. Quant. Spectrosc. Radiat. Transfer **111**(12–13), 1931–1946 (2010)
49. Ilyushin, Y., Kutuza, B.: Microwave radiometry of atmospheric precipitation: radiative transfer simulations with parallel supercomputers. Commun. Comput. Inf. Sci. **965**, 254–265 (2019)
50. Kalitkin, N.N.: Numerical Methods. Nauka, Moscow (1971)
51. Henyey, L.G., Greenstein, J.L.: Diffuse radiation in the galaxy. Astrophys. J. **93**, 70–83 (1941)
52. Sadovnichy, V., Tikhonravov, A., Voevodin, V., Opanasenko, V.: "Lomonosov": supercomputing at Moscow state university. In: Contemporary High Performance Computing: From Petascale toward Exascale, pp. 283–307. Chapman & Hall/CRC Computational Science, Boca Raton (2013)

HPC, BigData, AI: Architectures, Technologies, Tools

Analysis of Key Research Trends in High-Performance Computing Using Topic Modeling Technique

Yuri Zelenkov(✉)

National Research University Higher School of Economics, Moscow, Russian Federation
yzelenkov@hse.ru

Abstract. The intellectual structure of scientific discipline consists of a set of interacting topics. The evolution of these topics is the subject of special attention because it reflects the actual interest of researchers and stakeholders. This paper analyzes issues of High-Performance Computing (HPC) on the base of the formal topic modeling technique. Analyzing the abstracts of 7661 publications referenced in Web of Science in 2005–2019, we identified seven topics that concern different aspects of HPC science. The central theme is the *Large Scale Applications* focused on practical and scientific problems solved using HPC. It is closely linked with *Parallel Algorithms* that should effectively exploit the thousands of processing cores, *Parallel Software* for heterogeneous distributed systems, and *Interconnected systems* that study the integration of HPC facilities in systems of larger size. These topics are relatively stable both in terms of popularity (number of publications) and impact (number of citations). The single topic, which popularity and impact continuously grow in the last 15 years, is *Energy efficiency* since power consumption is a critical issue of exascale systems. We also found that the topic of *Heterogeneous systems* dedicated mainly to GPU usage declines after the peak of interest in 2010–2015. The results obtained shed light on the structure of HPC science and supplement the known publications that declare research direction towards exascale performance.

Keywords: High-performance computing · Data-driven analysis · Literature review

1 Introduction

High-performance computing (HPC) plays a significant role in the development of many scientific disciplines. The performance of these systems, measured in floating-point operations per second (FLOPS), has increased substantially over the last decades [1]. In 2008, the first HPC system reached petascale performance (10^{15} FLOPS); the next major milestone is a computer system that should be capable of at least one exaflop (10^{18} FLOPS). China announced the exascale supercomputer to be operational by the end of 2020, the USA aims to build an exascale system in 2021, and the European Union is targeting 2022/2023 [1].

V. Voevodin and S. Sobolev (Eds.): RuSCDays 2020, CCIS 1331, pp. 401–412, 2020.
https://doi.org/10.1007/978-3-030-64616-5_35

As these efforts play an essential role in global development and competing, governments of different countries and industry support these research directions using various incentives. Many government and public organizations publish policy papers declaring research directions (see, for example, [2, 3]); however, a real vast stream of scientific publications sometimes looks fragmented and inconsistent. Thus, the research community needs to detect significant, implicit knowledge associations hidden in fragmented knowledge areas by analyzing existing scientific literature [4]. This activity should not only help in the development of new cross-disciplinary research but also help stakeholders to assess progress in the area of interest.

The intellectual structure of scientific discipline looks like a set of interacting and evolving topics. Analysis of the dynamics of those topics, i.e., changes in the amount and citation of the publications, also can provide additional knowledge on the field because it reflects a shift in the actual interest of researchers and stakeholders [5]. Besides, to understand the structure of a scientific discipline, it is crucial to study the collaborations of individual researchers and research organizations and analyze the productivity of journals, which is the subject of bibliometric studies.

However, despite the importance of HPC research as a driver of other scientific disciplines, there is a relatively small amount of papers that review the overall intellectual structure of this area and its relationship with application fields. Authors of work [6] analyze the evolution of Scientific Communications Networks in HPC and try to predict the challenges of the future uses of this type of advanced services. In work [7], the authors study the impact of the government investment in HPC on the national innovation system and analyze the country specificity of HPC usage in application domains.

Authors of work [1], which is most close to our study, analyze the landscape of exascale research. For this, they review the top challenges identified in landmark research and use data-driven literature analysis to detect the most relevant topics. According to [1–3], top current declared challenges in HPC area are as follows:

- Energy Efficiency: 20 MW is established as a reasonable power limit for exascale systems [2], so, the development of energy-efficient technology is needed to reach this goal.
- Interconnect Technology: both vertical (intra-node) and horizontal (inter-node).
- Memory technology to improve capacity, bandwidth, resilience, and energy-efficiency.
- Scalable Systems Software, e.g., operating systems, runtime and monitor systems.
- Programming Systems, i.e., new methods to provide fine-grained concurrency, locality, and resilience.
- Data Management for handling massive amounts of data.
- Parallel Algorithms that should improve scalability (e.g., reduce communication, avoid synchronization) since explicit parallelism might be the only solution to increase overall system performance [2].
- Algorithms that should optimize the ensembles of many small runs.
- Resilience and Correctness of computations.
- Scientific Productivity, i.e., tools to productively utilize exascale systems.

Analyzing 2017 works published in 2007–2019, authors of [1] detected 25 resear trends that they combined in eight research themes: Energy/Power, Fault Tolerance/Resilience, Data Storage, Network Interconnection, Computer Architecture, System Software, Parallel Programming, and Scientific Computing.

Note that there is no exact overlapping between research topics declared in policy papers and detected in the real flow of scientific publications.

As we highlighted above, yet another issue that is crucial for understanding the evolution of the research area is the dynamics of the topics, i.e., change in publications number and citations over time. Work [1] does not address this issue. Our study aims to bridge this gap.

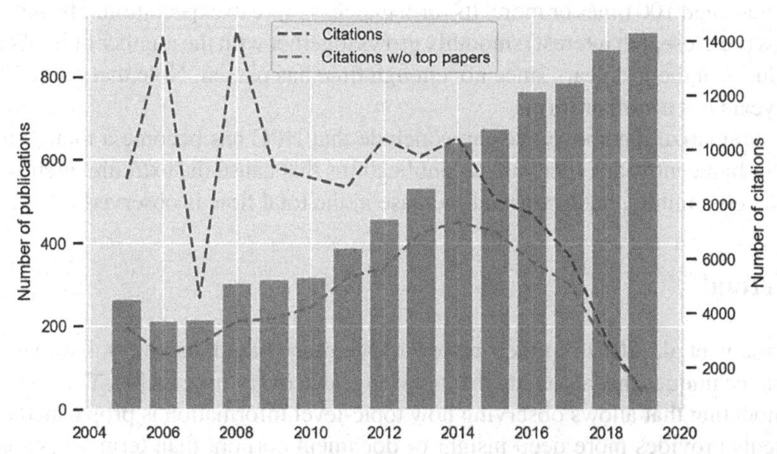

Fig. 1. The distribution of publications (bars) and the number of citations (dashed lines) by year. The red dashed line corresponds to the total number of citations of articles published in a particular year. The green line shows the number of citations without papers cited more than 100 times. (Color figure online)

2 Data

We used the Web of Science database as a source of data and selected publications that conform to the following criteria:

- The paper topic includes at least one term from the list [supercomputing, high-performance computing, supercomputers], that is, the term may appear in the title of the article, annotations, or the list of keywords.
- The language of publication is English.
- Publication type is Article, i.e., we selected just journal papers and excluded conferences proceedings. On the one hand, this made it possible to choose the most mature works; on the other, it limited the data set to reasonable limits.
- The period is 2005–2019, as our analysis showed that a sharp increase in the amount of the publication begins in 2005.

The search query formed according to presented rules yielded 7796 results. After clearing the data, i.e., deleting entries with an empty field AB (abstract), PY (year of publication), etc., we obtained 7661 articles. Figure 1 shows the distribution of publications and the number of citations by years.

As can be seen from Fig. 1, there is a significant variation in the number of citations. It is because, in some years, some researchers published articles that significantly impacted their research area and therefore were cited an enormous amount of times. For example, at the moment of our research, the number of citations of Stamatakis (2006) [8] is 10897, which is more than the total number of citations of other works published in 2006. Yet another work of Stamatakis et al. (2008) [9] was cited 5051 times that deforms the curve in 2008. The green dashed line in Fig. 1 presents the number of citations without papers that cited 100 times or more. Its form matches more to expectation. The number of citations (i.e., research interest) smoothly grows together with the number of publications and reduces in the last years since not enough time has passed. Note that the red line in recent years has a similar form.

To summarize the above, we can conclude that HPC has become a mature area of research. In recent years, there are no publications that cause the extreme interest of the scientific community, but a constant increase in the total flow is observed.

3 Method

Thilakaratne et al. (2019), in their review of literature-based discovery, list main computation techniques that automate the knowledge discovery process [4]. They noted that topic modeling that allows observing how topic-level information is propagated among documents provides more deep insight of document corpora than term-level analysis. However, the topic modeling is still relatively rarely used in literature analysis [4, 10].

Among the recent work, authors of [5] applied topic analysis to the knowledge management area. They pay special attention to the topic dynamics, i.e., how a number of publications and citations regarding each topic changes in time. It allows shedding light on the shift in research interest and identifying critical trends of the present time.

More interestingly, work [1] uses thematic modeling to analyze the exascale research landscape. However, the authors of [1] do not go beyond the identification of the topics and do not analyze their dynamics. Thus, we will use the methodology proposed in [5], expanded in accordance with our goal.

A topic is a set of words often co-occur in texts related to a given subject area. Probabilistic topic modeling bases upon the idea that documents are mixtures of topics, where a topic is a probability distribution over terms [11]. Let there is a finite set of topics T, which is not known. Each use of the term w in document d is associated with some topic $t \in T$. Thus, a collection of documents is considered as a set of triples (d, w, t) selected randomly and independently from the distribution defined on a finite set $D \times W \times T$. Documents $d \in D$ and the terms $w \in W$ are observable variables. The topics $t \in T$ are latent variables whose values must be defined.

The topic model automatically detects latent topics by the observed frequencies of words in the documents $p(w|d) = \sum_{t \in T} p(t|d)p(w|t)$. So, input of algorithm is a matrix $D \times W$, which cells contain counts of the word w in document d.

To prepare matrix $D \times W$, we used abstracts of 7661 papers downloaded from the Web of Science database, as described in the previous section. According to the [12], differences between abstract and full-text data are more apparent within small document collections. Therefore, we have selected abstracts as an object of analysis.

According to the general text mining technique, abstracts were tokenized, and the terms obtained were converted to standard form. Next, words that belong to an extended stop word list were deleted. The extended stop-word list includes standard English stop-words and corpus-specific words that appear in less than 5% and more than 75% of documents. We also created bigrams to join terms often co-occurred beside. As a result, we got a sparse matrix with dimensions of 7661×266, only 9.4% of the cells of which contain values greater than zero.

To compute the topics, we used Latent Dirichlet Allocation (LDA) algorithm that is based on additional assumption that the distribution Θ of documents θ_d and distribution Φ of topics φ_t are spawn by a Dirichlet distributions [13]. To build the model, one should define a number of topics $|T|$; the LDA algorithm computes distributions Θ and Φ. As a result, each topic is presented by the weighted list of words; the weight of word corresponds to its importance in the topic definition. The weighted list of topics presents each document; the weight of the topic corresponds to its significance in the document.

Determining the number of topics is a critical issue in topic analysis; many authors use various kinds of grid search optimizing a specific metric [1, 5]. We used more advanced techniques, namely, Bayesian optimization [14]. Such an approach allows us to optimize simultaneously not only the number of the topics and also parameters of Θ and Φ distributions and other parameters of the algorithm. Optimization target is a perplexity that can be computed as

$$P(D) = \exp\left[\frac{1}{2}\sum_{d \in D}\sum_{w \in d} n_{dw} \ln p(w|d)\right].$$

The perplexity of collection D is a measure of the language quality and often used in computational linguistics. In our case, language is the distribution of words in documents $p(w|d)$. The less perplexity, then more uneven this distribution [5].

When the optimal number of topics and corresponding topic distribution for each document are found, we can study topic dynamics. Let θ_{dt} is the weight of topic t in document d $(0 \leq \theta_{dt} \leq 1)$. So, the overall popularity of topic across all documents can be defined as [5]

$$\hat{\theta}_t = \frac{1}{|D|}\sum_{d \in D} \theta_{dt} \tag{1}$$

To measure the topic popularity in a particular year y, it is enough to set $D = D_y$ in Eq. (1), where D_y is the set of all papers in year y.

Let C_d is the number of citations of document d and $C = \sum_{d \in D} C_d$. An impact of the topic can be defined as follows [5]

$$\hat{i}_t = \frac{1}{C}\sum_{d \in D} \theta_{dt} C_d \tag{2}$$

By analogy, to obtain the topic impact in the particular year, one should set $D = D_y$ in Eq. (2).

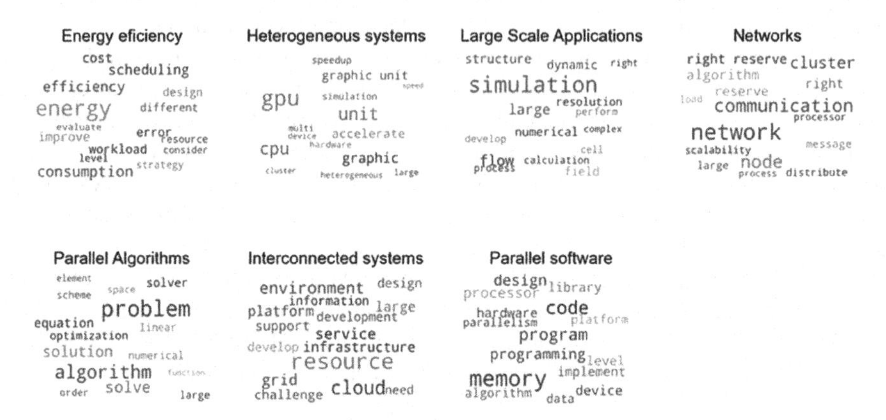

Fig. 2. Visualization of topic model using word clouds. Each word cloud represents one detected topic where the size of words indicates the relevance of each word to that particular topic.

4 Results and Discussion

Performing all preprocessing operation described in the previous Section and 100 iterations of Bayesian optimization of the LDA model, we found that the optimal number of topics is 7, and the corresponding value of perplexity is 212.3.

Analyzing the dominant terms (Fig. 2), we can conclude that each topic represents some coherent area of research. Moreover, the weights of topics in the document are either large (i.e., the topic is strongly related) or near zero (i.e., the topic is unrelated). So, to assign the labels, we analyzed the term distributions and most representative papers for each topic. To select the most representative papers, we sorted the publications by the topic weight and next by the number of citations, both in descending order. Figure 2 presents the labels assigned, and Table 1 lists the topics description. We also provide in Table 1, where it is possible, the reference on the paper that reviews the corresponding research area.

Table 1 also presents the values of popularity and impact for each topic computed according to Eq. (1) and (2) respectively. Please note that the sum of both popularity and impact is 1, so the values presented can be considered as a share of a particular topic in the total flow of HPC research, i.e. its total weight. Thus, 25.2% of all publications dedicate to *Large Scale HPC Applications*, and this topic produces 31% of citations. Next topic both in terms of popularity and impact is the *Interconnected systems*.

Note that the topics determined in our research are only partially consistent with the research directions declared in policy paper [2] and the results of [1] (cf. Introduction). Both papers don't consider the flow of applied research that is most significant according to our results.

Table 1. Topics of HPC research.

Topic	Description	Popularity (Eq. 1)	Impact (Eq. 2)
Energy fficiency	Research efforts focusing exclusively on power and energy efficiency models and techniques [15]	0.125	0.110
Heterogenous systems	Graphics Processing Units (GPUs) and integration with core CPU [16]	0.046	0.048
Large scale applications	Research flow focusing on practical and scientific problems solved using HPC resources. It joins all areas of application: bioinformatics, flow and combustion simulation, climate, etc.	0.252	0.310
Networks	All issues regarding communication in parallel systems, publications about both hardware, and software solutions [17]	0.097	0.090
Parallel algorithms	Since the modern HPC systems offer concurrency with thousands of cores, this research direction dedicates to new algorithms that can effectively exploit multi-core facilities	0.119	0.109
Interconnected system	Systems of systems that connect large HPC facilities from the Grid to HPC Clouds [18]	0.188	0.183
Parallel software	This topic deals with software prototyping, code design, and implementation for parallel, heterogeneous, and distributed platforms	0.173	0.150

The next issue that should be considered is the topic collaboration, i.e. the topics co-occurrence. Let θ_{di} and θ_{dj} are the weights of the topics i and j, respectively, in the document d. So, we can define the topics co-occurrence in this document as a product $\theta_{di}\theta_{dj}$. The maximal value of the co-occurrence of two topics in one document is 0.25 when $\theta_{di} = \theta_{dj} = 0.5$. From this, the maximal possible value of the topics collaboration in the document corpus is 0.25*N, where N is the number of documents. In our case, this value is 1915.25.

So, the topic collaboration in the document corpus can be computed as

$$c_{ij} = \sum_{d \in D} \theta_{di}\theta_{dj}.$$

Figure 3 presents these data. As we can see, the topic of *Large Scale Applications* has relatively strong links with *Parallel Algorithms, Interconnected Systems, Parallel Software,* and, in less extent, *Energy Efficiency*. From this, we can conclude that research in the development of new algorithms, software and systems is carried out, as a rule, in the context of HPC applications. Together, these topics form the core of HPC science. We can estimate the share of efforts that come together in this core, summing up the popularity of the topics listed. The overall popularity of these topics is 86%. We also note that the topic of *Heterogeneous systems*, primarily associated with the use of GPUs, is in relative isolation from other areas.

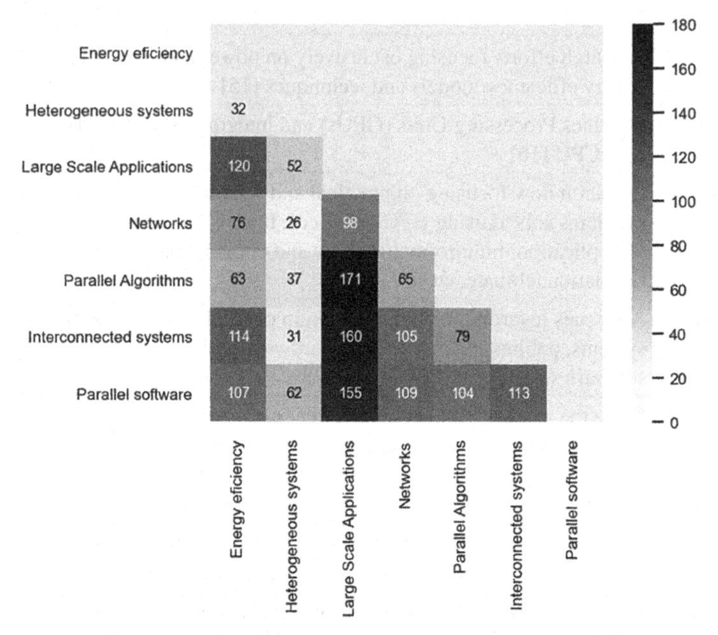

Fig. 3. A topic co-occurrence.

Figure 4 and Fig. 5 present the dynamics of the topics in terms of popularity (Eq. 1) and impact (Eq. 2), respectively. Popularity is the derivative of the number of publications, and the impact is computed using the number of citations. These data shed light on the interests of the HPC community regarding each topic.

The only area of research whose popularity and impact has been steadily growing over the past 15 years is *Energy Efficiency*. In fact, these issues come to the fore when designing exascale systems. The performance of the best system in the last top 500 list (November 2019) is about 0.2 exaflops with a power consumption of 10 MW. As we noted above, the declared power limit of the exascale system is 20 MW [2], therefore, energy consumption per unit of performance should be reduced by at least 2.5 times. This is a very complex engineering task that affects the design of all components: processors, internal and external memory, interconnect and algorithms.

As follows from the data, the direction of *Heterogeneous Systems* associated with GPUs (and other co-processors like Xeon Phi) and their integration with the central processor and other elements, firstly, is relatively isolated from other topics, and secondly, attracts relatively less attention of researchers. A peak in the number of publications and citations in this area was observed from 2010 to 2015. It is because equipment manufacturers proposed new devices, and heterogeneous systems were considered as one of the possible paths to exaflop performance. But the integration problems were so high that, after 2015, research interest in this area is gradually declining. In 2019 Intel closed

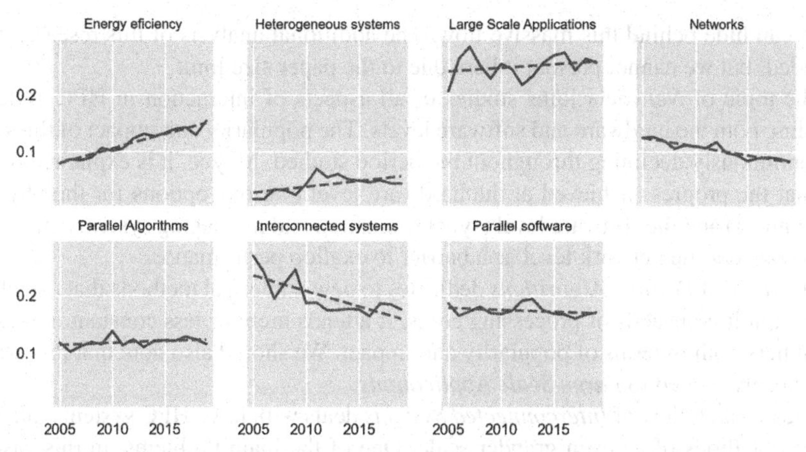

Fig. 4. The dynamics of the topics' popularity (solid line), according to Eq. (1). The dashed line presents the trend.

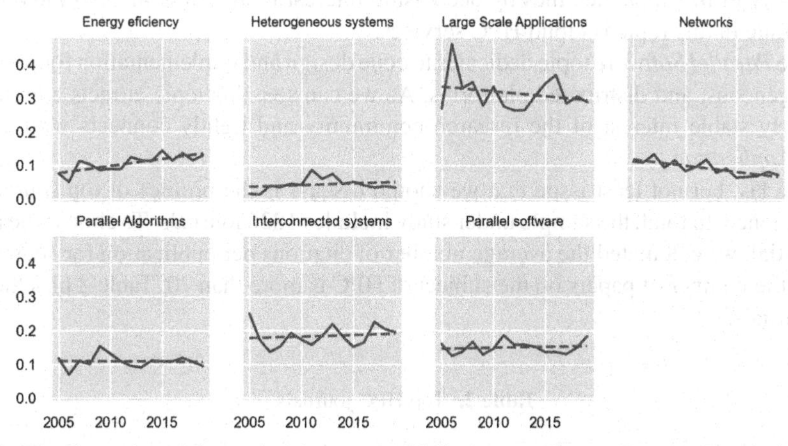

Fig. 5. The dynamics of the topics' impact (solid line), according to Eq. (2). The dashed line presents the trend.

the Xeon Phi project. However, most systems at the top of the Top 500 list are heterogeneous and use various computational accelerators. Also note that *Energy Efficiency* and *Heterogeneous Systems* have a low co-occurrence coefficient, which contradicts the main feature of GPUs, which is a high ratio of performance per watt.

Large Scale Applications is the central topic that consolidates research in other HPC areas. It is the most important direction both in terms of popularity and impact. Significant fluctuations of both metrics of this topic can be explained by the fact that most top-cited papers (e.g. [8, 9]) relate to this area. Nevertheless, one can trace the trend towards an increase in the number of publications in this area, while at the same time a slight decrease in their influence. Note that it is an 'easy-to-note' fact. The majority of the works in the leading conferences and journals present HPC applications. So, the value of "popularity" in this sense can mislead for new emerging trends recognition, which can hide behind this massive flow. The additional analysis of this research flow is needed, but we cannot present it here due to the paper size limit.

The topic of *Networks* joins studies of all aspects of interaction in HPC systems, including both the hardware and software levels. The popularity and impact of this topic are continuously declining throughout the period studied. Maybe, it is explained by the fact that the progress achieved at the hardware level (various options for the physical interconnect) and the software level (synchronization and messaging systems) allows us not to consider the network level as a barrier to exaflop performance.

The topic of *Parallel Algorithms* dedicates to new numerical methods that can effectively exploit thousands of processing cores. It attends more or less constant interest of researchers both in terms of popularity and impact. We should also note that this area is most closely related to *Large Scale Applications.*

The research flow of *Interconnected Systems* deals with large HPC systems integrating into facilities of an even grander scale. One of the main problems, in this case, is the optimal control of the system, which throughput significantly varies at the different levels of the hierarchy, and the elements may belong to different owners. In general, the popularity of this topic declines by decreasing interest in grid systems. Nowadays, the main issue in this topic is cloud HPC services.

The *Parallel Software* topic dedicates to code design and implementation for parallel, heterogeneous, and distributed platforms. As we can see, this topic attracts a high and relatively stable interest of the research community and tightly connects with *Large Scale Application.*

The last but not least issue that we should discuss is the profiles of top sources in HPC science. In total, the sample under study includes 1826 journals. To choose the most influential, we calculated the average number of citations per publication for sources in which the number of papers on the subject of HPC is more than 70. Table 2 lists the top ten journals.

Table 2. Top HPC journals.

Journal	Pubs	Cit	Avg citations
Computer physics communications	185	6816	36.8
Future generation computer systems	134	4290	32.0
Journal of computational physics	74	1596	21.6
IEEE transactions on parallel and distributed systems	154	2220	14.4
Parallel computing	148	2046	13.8
Journal of parallel and distributed computing	140	1689	12.1
SIAM journal on scientific computing	76	803	10.6
International journal of high performance computing applications	188	1814	9.6
Concurrency and computation – practice & experience	277	1941	7.0
Journal of supercomputing	200	889	4.4

We can identify the profile of the source as a weighted mix of topics that is the normalized sum of topics of all publications. Note that the sum of the topics weights for each journal also is 1. Figure 6 shows the distribution of topics weights in top journals for the entire period of research. To assess the dynamics of interests, in Fig. 7, we present the publication profile of these journals for 2019. These data may be useful for researchers choosing a source for publishing their results. As we can see, the leading topic of most journals is *Parallel Software*. In the last time, *Energy Efficiency* moves to the fore, and the subject of *Networks* declines that are consistent with data above.

	Energy eficiency	Heterogeneous systems	Large Scale Applications	Networks	Parallel Algorithms	Interconnected systems	Parallel software
COMPUTER PHYSICS COMMUNICATIONS	0.037	0.055	0.341	0.123	0.140	0.050	0.154
FUTURE GENERATION COMPUTER SYSTEMS-THE INTERNATIONAL JOURNAL OF ESCIENCE	0.160	0.016	0.114	0.219	0.044	0.370	0.077
JOURNAL OF COMPUTATIONAL PHYSICS	0.036	0.052	0.408	0.090	0.328	0.017	0.068
IEEE TRANSACTIONS ON PARALLEL AND DISTRIBUTED SYSTEMS	0.244	0.040	0.057	0.146	0.084	0.140	0.289
PARALLEL COMPUTING	0.091	0.040	0.164	0.248	0.095	0.096	0.245
JOURNAL OF PARALLEL AND DISTRIBUTED COMPUTING	0.193	0.045	0.130	0.285	0.062	0.109	0.176
SIAM JOURNAL ON SCIENTIFIC COMPUTING	0.061	0.019	0.170	0.086	0.475	0.034	0.156
INTERNATIONAL JOURNAL OF HIGH PERFORMANCE COMPUTING APPLICATIONS	0.146	0.053	0.131	0.111	0.118	0.165	0.277
CONCURRENCY AND COMPUTATION-PRACTICE & EXPERIENCE	0.163	0.060	0.090	0.106	0.081	0.243	0.257
JOURNAL OF SUPERCOMPUTING	0.203	0.058	0.081	0.143	0.107	0.168	0.240

Fig. 6. Profiles of top journals in 2005–2019.

	Energy eficiency	Heterogeneous systems	Large Scale Applications	Networks	Parallel Algorithms	Interconnected systems	Parallel software
COMPUTER PHYSICS COMMUNICATIONS	0.029	0.055	0.394	0.080	0.221	0.023	0.197
FUTURE GENERATION COMPUTER SYSTEMS-THE INTERNATIONAL JOURNAL OF ESCIENCE	0.119	0.012	0.122	0.225	0.027	0.401	0.094
JOURNAL OF COMPUTATIONAL PHYSICS	0.003	0.003	0.583	0.024	0.355	0.031	0.003
IEEE TRANSACTIONS ON PARALLEL AND DISTRIBUTED SYSTEMS	0.263	0.087	0.090	0.059	0.081	0.098	0.323
PARALLEL COMPUTING	0.076	0.048	0.147	0.206	0.148	0.058	0.316
JOURNAL OF PARALLEL AND DISTRIBUTED COMPUTING	0.335	0.095	0.145	0.187	0.066	0.051	0.121
SIAM JOURNAL ON SCIENTIFIC COMPUTING	0.113	0.003	0.055	0.091	0.639	0.013	0.086
INTERNATIONAL JOURNAL OF HIGH PERFORMANCE COMPUTING APPLICATIONS	0.184	0.035	0.186	0.089	0.164	0.068	0.284
CONCURRENCY AND COMPUTATION-PRACTICE & EXPERIENCE	0.236	0.040	0.076	0.090	0.084	0.240	0.234
JOURNAL OF SUPERCOMPUTING	0.203	0.064	0.070	0.079	0.111	0.196	0.277

Fig. 7. Profiles of top journals in 2019

5 Conclusion

Using the topic modeling technique, we analyzed the intellectual structure of HPC science. Note that this approach is devoid of subjectivity since based on the formal test analysis. Thus, it allows identifying real clusters of research interests that are partially consistent with direction declared in policy papers.

References

1. Heldens, S., Hijma, P., VanWerkhoven, B., Maassen, J., Belloum, A.S.Z., Van Nieuwpoort, R.V.: The landscape of exascale research: a data-driven literature analysis. ACM Comput. Surv. **53**(2), Article 23 (2020)
2. Lucas, R. et al.: Top Ten Exascale Research Challenges. Technical Report. U.S. Department of Energy, Office of Science. DEO ASCAC Subcommittee Report (2014)

3. European Exascale Software Initiative. http://www.eesi-project.eu/ressources/documentationAccessed 04 May 2020
4. Thilakaratne, M., Falkner, K., Atapattu, T.: A systematic review on literature-based discovery: general overview, methodology, and statistical analysis. ACM Comput. Surv. **52**(6), Article 129 (2019)
5. Zelenkov, Y.: The topics dynamics in knowledge management research. In: Uden, L., Ting, I.-H., Corchado, J.M. (eds.) KMO 2019. CCIS, vol. 1027, pp. 324–335. Springer, Cham (2019). https://doi.org/10.1007/978-3-030-21451-7_28
6. Fernández-González, Á., Rosillo, R., Miguel-Dávila, J.Á., Matellán, V.: Historical review and future challenges in supercomputing and networks of scientific communication. J. Supercomput. **71**(12), 4476–4503 (2015). https://doi.org/10.1007/s11227-015-1544-3
7. Zelenkov, Y.A., Sharsheeva, J.A.: Impact of the investment in supercomputers on national innovation system and country's development. In: Sokolinsky, L., Zymbler, M. (eds.) PCT 2017. CCIS, vol. 753, pp. 42–57. Springer, Cham (2017). https://doi.org/10.1007/978-3-319-67035-5_4
8. Stamatakis, A.: RAxML-VI-HPC: maximum likelihood-based phylogenetic analyses with thousands of taxa and mixed models. Bioinformatics **22**(21), 2688–2690 (2006)
9. Stamatakis, A., Hoover, P., Rougemont, J.: A rapid bootstrap algorithm for the raxml web servers. Syst. Biol. **57**(5), 758–771 (2008)
10. Sebastian, Y., Siew, E., Orimaye, S.: Learning the heterogeneous bibliographic information network for literature-based discovery. Knowl.-Based Syst. **115**, 66–79 (2017)
11. Steyvers, M., Griffiths, T.: Probabilistic topic models. In: Landauer, T., McNamara, D., Dennis, S., Kintsch, W. (eds.), Latent Semantic Analysis: A Road to Meaning. pp. 424–440. Laurence Erlbaum (2007)
12. Syed, S., Spruit, M.: Full-text or abstract? Examining topic coherence scores using latent Dirichlet allocation. In: 2017 IEEE International Conference on Data Science and Advanced Analytics (DSAA), pp. 165–174 (2017)
13. Blei, D.M., Ng, A.Y., Jordan, M.I.: Latent Dirichlet allocation. J. Mach. Learn. Res. **3**, 993–1022 (2003)
14. Mockus, J.: Bayesian Approach to Global Optimization: Theory and Applications. Springer, Heidelberg (2012)
15. O'Brien, K., Pietri, I., Reddy, R., Lastovetsky, A., Sakellariou, R.: A survey of power and energy predictive models in HPC systems and applications. ACM Comput. Surv. **50**(3), Article 37 (2017)
16. Bridges, R., Imam, N., Mintz, T.: Understanding GPU power: A survey of profiling, modeling, and simulation methods. ACM Comput. Surv. **49**(3), Article 41 (2016)
17. Rico-Gallego, J., Díaz-Martín, J., Manumachu, R., Lastovetsky. A.: A survey of communication performance models for high-performance computing. ACM Comput. Surv. **51**(6), Article 126 (2019)
18. Netto, M., Calheiros, R., Rodrigues, E., Cunha, R., Buyya, R.: HPC cloud for scientific and business applications: taxonomy, vision, and research challenges. ACM Comput. Surv. **51**(1), Article 8 (2018)

Core Algorithms of Sparse 3D Mipmapping Visualization Technology

Stepan Orlov^(✉), Alexey Kuzin, and Alexey Zhuravlev

Peter the Great St.Petersburg Polytechnic University, St. Petersburg, Russia
majorsteve@mail.ru, kuzin_aleksei@mail.ru, zhurus@mail.ru

Abstract. The paper presents algorithms implemented in the core components of sparse 3D mipmapping technology, which is a basement for animated visualization of scalar time-dependent fields defined on large (up to 10^{10} nodes) meshes and resulting from CFD simulations. The basic idea of sparse 3D mipmapping is the interpolation of original data on octree meshes and further visualization of fields on subtrees of limited depth using volume rendering algorithms. The paper focuses on data structures and algorithms delivering input for visualization, and presents the results of performance testing.

Keywords: Scientific visualization · Big data · Octree · Volume rendering

1 Introduction

Today, the amount of data that needs to be analyzed often exceeds the processing capabilities of traditional systems [1,2]. This makes it necessary to replace the use of traditional data processing algorithms with specialized tools.

A problem of this kind arises when visualizing a large amount of engineering data. In [3] it was shown that data reading can take a significant fraction of the total processing time of a visualization request. This necessitates the development of a specialized approach to reading and organizing data. Approaches to data preprocessing in distributed environments are discussed in [4].

The goal of the project authors are working on is to provide the ability to visualize scalar fields on large meshes that may contain up to 10^{10} nodes. Those fields typically depend on time, so the visualization naturally implies the animation of fields over the time dimension. That means the visualization system must be able to render tens frames per second, with each frame containing the data from a certain time layer and that frame data is constituted by field values at all nodes of the large mesh. Moreover, the hardware part of the visualization system would consist of a small number of nodes equipped with modern multicore CPUs and GPUs. Preliminary estimations hint that an attempt to visualize the data directly will give a solution with unacceptably poor performance. Therefore we focused on an approach that would allow to operate on small subsets of the data

V. Voevodin and S. Sobolev (Eds.): RuSCDays 2020, CCIS 1331, pp. 413–424, 2020.
https://doi.org/10.1007/978-3-030-64616-5_36

and provide a parameter for trading-off the performance of visualization against its quality. To achieve this goal, we have designed and implemented a number of data structures and algorithms. Part of those are applied at the *preprocessing* stage, which means that the original dataset has to be transformed to another one, which in turn is suitable for visualization (further the *visualization dataset*). Other algorithms are applied at the visualization stage.

2 Preprocessing Pipeline

We assume that the original dataset is distributed across nodes of a computer system (further referred to as *data host*), which is not necessarily the visualization system (further referred to as VS). Each node of the data host has local access to a subset of data, containing all fields at all time layers at a specific part of the computational mesh, further referred to as *subdomain* (subdomains may result, for example, from applying a domain decomposition algorithm [5]). Data host nodes run the preprocessing pipeline resulting in the generation of the visualization dataset. The assembled visualization dataset is stored locally on the VS.

The crucial part of the visualization dataset is the *octree* [6] capturing the spatial density of the original mesh. In our system, octree nodes store no specific data, and the octree is only used to construct a hierarchy of sparse grids on which scalar fields are defined in the visualization dataset. The preprocessing involves five stages, which are (1) the calculation of the bounding box containing the entire mesh; (2) octree generation at each subdomain; (3) the interpolation of scalar fields at each subdomain; (4) merging all octrees into one; and (5) merging scalar fields such that they are defined at sparse grids induced by the global octree. Figure 1 illustrates stages 1, 2, and 4 of that process.[1]

2.1 Octree Generation

The location of the octree in space is defined by specifying the size and center position for its root node. Each node of an octree may have up to eight child nodes, one per octant of Cartesian coordinate system centered at node center.

To generate the octree, one has first to specify its center and size. Therefore, the first preprocessing stage is the *bounding box calculation*. Data host nodes compute bounding boxes of their mesh parts and send those to the VS, which in turn computes the common bounding box. Once the center $\mathbf{R}_c$ of bounding box and maximal length L of its edges are known, octree nodes are generated.

In fact, each node of the data host generates its own octree, further called *octree component*, but all of them share the same root location ($\mathbf{R}_c$ and L) in order to make it possible to further merge all components into one.

To build an octree component, one needs a data structure representing an octree and suitable for fast insertion of new nodes. For that purpose we use a

[1] For clarity, most of the figures show examples for quadtrees, rather than octrees.

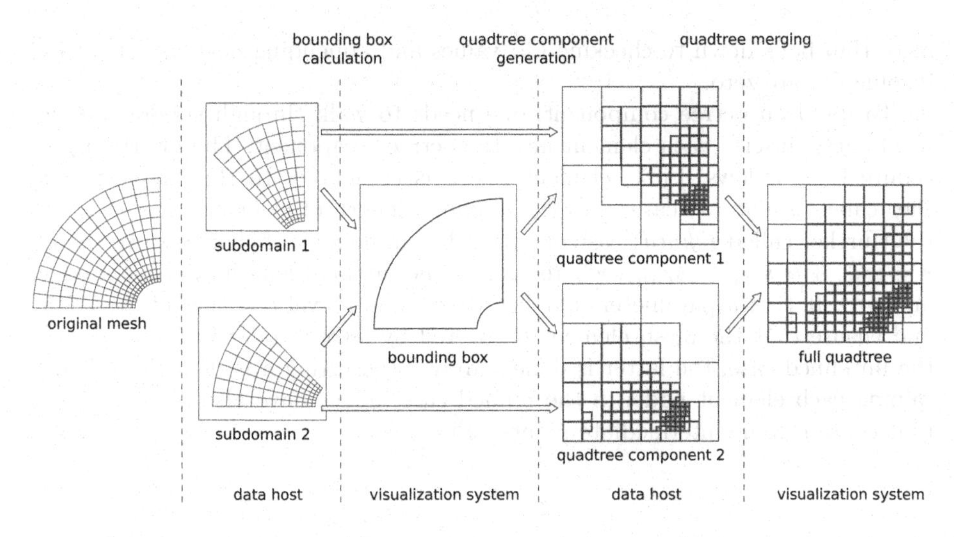

Fig. 1. Octree or quadtree generation stages.

structure containing root location ($\mathbf{R}_c$, L) and an array of elements, C. Array index corresponds to a node of the tree. Each i-th element C_i of C ($i \geq 0$) is an array of 8 numbers C_{ij} ($0 \leq j < 8$), which are the indices of child nodes in the same array, or zeros if there is no child at specified place. The j index identifies each 1/8 cubic part of the node, where child nodes may be located; it is a number whose binary digits make a $2 \times 2 \times 2$ Morton code [7]. Initially, the array contains only one element C_0 for the root, and all $C_{0j} = 0$. An example of child node array C for a quadtree is shown in Fig. 2.

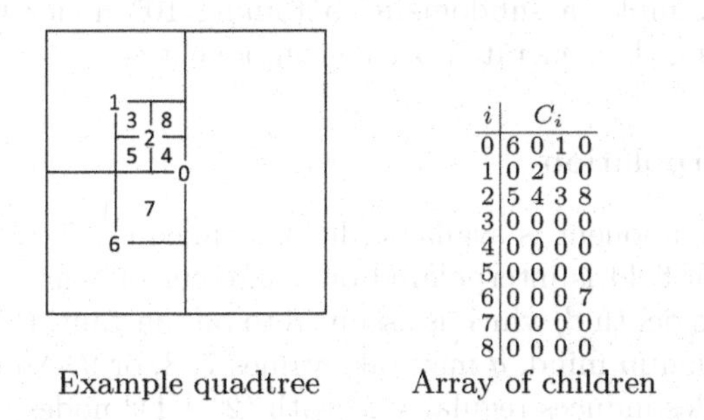

i	C_i
0	6 0 1 0
1	0 2 0 0
2	5 4 3 8
3	0 0 0 0
4	0 0 0 0
5	0 0 0 0
6	0 0 0 7
7	0 0 0 0
8	0 0 0 0

Fig. 2. Example of a quadtree and an array of its child nodes.

A node of an octree can be identified by its distance from the root, l,—the *level*, and a 3D vector $\mathbf{x}$ of integer coordinates of the node in the index space that determines the *location* of the node among its siblings; each coordinate varies in the range $[0, 2^l - 1]$. The complexity of finding/inserting node with given $l, \mathbf{x}$ is proportional to l for the octree data structure described above; the insertion algorithm boils down to checking C_{ij} values and appending new elements to C if some C_{ij} are zero.

To build an octree component, one needs to walk through subdomain elements and "insert" each element into the octree component. The insertion procedure is as follows. First, element size s is estimated. In the simplest (further called *unrefined*) case, it is the maximal length of element edge, s_{max}. In the (further called *refined*) case of noticeable anisotropy of elements, we choose $s = s_{min} + \alpha(s_{max} - s_{min})$, where s_{min} is the minimal length of element edge, and $\alpha \in [0, 1]$ is a parameter. Further we compute level $l = \lfloor \log_2(L/s) + 0.5 \rfloor$ and ensure that the mesh element is covered by level l nodes in the octree. In the unrefined case, the latter is done simply by inserting a node at level l containing each element node; in the refined case, nodes have also to be inserted that contain some intermediate points—Fig. 3.

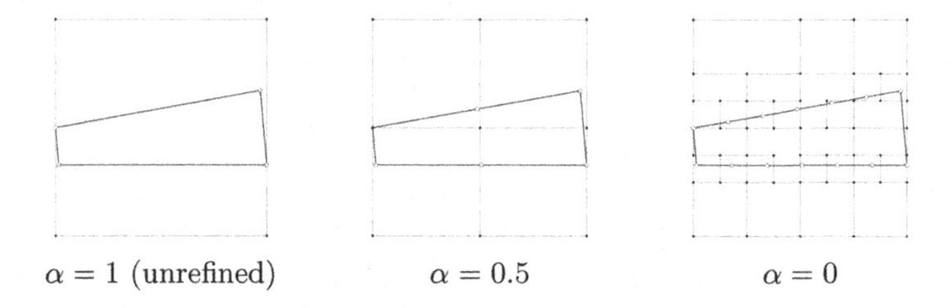

Fig. 3. Quadtree generation for unrefined and refined elements.

The size of each subdomain is supposed to be small enough such that the corresponding octree component fits into operating memory of the process generating it. For example, a subdomain containing 10^8 nodes typically requires 1–10 Gb for storing the generated octree component.

2.2 Field Interpolation

Once an octree component is available, fields defined on the subdomain can be interpolated. Each field is interpolated on a number of sparse grids induced by subtrees of limited depth d, which is assumed to be constant. Practically, keeping further visualization in mind, d may take values 7, 8, or 9. An octree of depth d containing all nodes induces regular grid with $(2^d + 1)^3$ nodes (nodes of the grid are further called *vertices*, to avoid confusion with octree nodes); but in general, induced grids are sparse. By definition, the grid induced by an octree is an array of vertices of all cubes corresponding to octree nodes. Vertices and their ordinal numbers in the grid can be obtained by walking octree nodes recursively.

Each octree defines a set of subtrees of depth at most d. The first subtree has the same root as the octree itself, and contains all octree nodes at distance no more than d from the root (in other words, nodes at levels $0, \ldots, d$). Let us denote D_0 the set of level d nodes that are not leaves of the original octree, and introduce the operation $p(a)$ that maps octree node a onto its parent node. If D_0 is not empty, consider all subtrees of depth d with roots at level 1 nodes belonging to the set $P_1 \equiv p^{d-1}(D_0)$. Those subtrees also may have non-leaves at their maximal depth d, or $d+1$ with respect to the octree root. In that case, the set of such nodes is denoted by D_1, and $P_2 \equiv p^{d-1}(D_1)$ is the set of roots of next subtrees of depth d. This process continues until D_k is empty at some k. In this way we represent an octree as the union of subtrees of depth d. Let P_0 be the set containing one element—the root of the octree. Then the set of subtrees just built can be identified by d and sets of their roots $P_0, P_1, \ldots P_k$. Notice that the subscript of P is the level at which subtree roots reside in the original octree.

Figure 4 illustrates subtrees of depth 4 of a quad tree and grids induced by those subtrees.

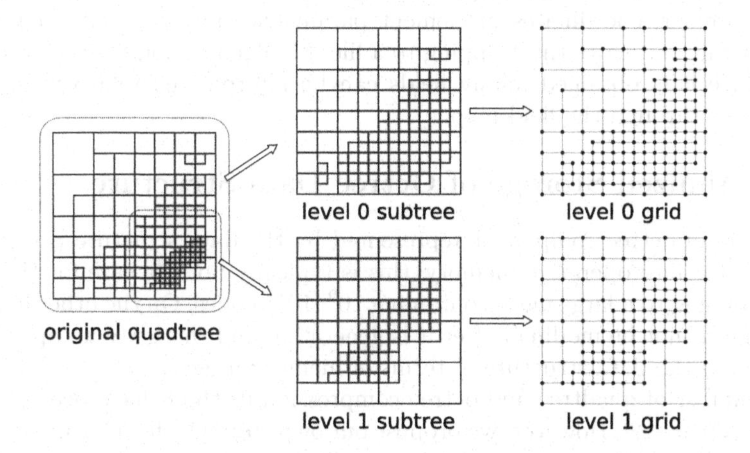

Fig. 4. Subtrees of depth 4 of a quadtree, and induced grids.

The octree, number d and sets $P_0, \ldots, P_k$ of subtree root identifiers constitute *metadata* used further during the field interpolation preprocessing stage, as well as at visualization time. In addition, metadata stores an index for each subtree; the index is used as the position of a scalar field defined on the subtree in a file, in the case each field at each time layer is stored in a separate file. Metadata can correspond to a subdomain or to the entire mesh.

At the interpolation stage, scalar fields defined on a subdomain are interpolated at vertices of the grids induced by subtrees of depth d. The order of subtree processing is $P_k, \ldots, P_1, P_0$. This is reasonable because vertices with even index coordinates of each grid belong to the parent grid, so values interpolated at those vertices are simply copied to parent grid.

When a field value needs to be interpolated at a grid vertex, an element of subdomain mesh has to be found that contains that vertex. However, it is not possible to quickly find such an element in the general case. Therefore, instead of iterating through vertices of the grid, we iterate through elements of the subdomain. For each element, its bounding box is computed, and then all grid vertices belonging to the bounding box are iterated. For each of those vertices, its coordinates in the element parameter space are computed. Currently we support linear simplex elements (2D triangles and 3D tetrahedra) and brick elements (2D bilinear quads and 3D trilinear hexahedra). For a simplex element, parametric coordinates are computed by solving 2 (in 2D) or 3 (in 3D) linear equations. For a brick element, we use Newton's method to solve 2 or 3 nonlinear equations, with zero initial approximation; for reasonable element shapes, the Newton method converges in 2–3 iterations. Once the parametric space coordinates are found, it is known whether a vertex belongs to an element or not, and a field interpolation can be done in the former case.

Since many fields at many time layers need to be interpolated on the same grids, we first compute the mapping from each element to the set of pairs (grid vertex number, coordinates in element parametric space) for all vertices contained by it, and write the mapping to a file. Map generation takes considerable time; once it is obtained, many fields can be interpolated relatively fast in a single pass through the field map.

3 Optimized Storage of Octree Data Structure

Although an octree component represented by the data structure presented in Sect. 2.1 fits into operative memory, this is typically not the case for the octree corresponding to a large mesh containing 10^9–10^{10} nodes. On the other hand, the octree needs not be modified after it is generated, and this gives an opportunity to optimize the data structure in terms of memory usage.

A number of quadtree and octree compression methods have been proposed in literature [8–12]. However, we propose our own storage scheme, which is partly conditioned by different octree definitions: traditionally, an octree node either is a leaf or has all eight children, and leaves store some data. In our case, a node may have any subset of is children, and no specific data is stored at any nodes. Due to the absence of specific data at nodes, a more efficient storage format can be proposed than existing ones.

The basic idea is as follows. Let us go back to Fig. 2 and observe that elements C_i could be reordered such that nonzero C_{ij} grow monotonously as i and j increase—see Fig. 5.

Once an octree component is generated, the reordering of its children array can be done in a single pass, with an algorithm linear in the number of octree nodes. But now it is obvious that we don't need to store an integer value per node, because one-bit value c_{ij} is enough ($c_{ij} = 0$ if $C_{ij} = 0$ and $c_{ij} = 1$ if $C_{ij} \neq 0$). Original values of C_{ij} can easily be restored during the consecutive reading. So the storage of an octree in this form is 32/64 times more memory efficient than in the case C_{ij} are stored as 32/64-bit integers. In addition, all trailing zeros can be omitted, which further reduces the storage size up to 8 times (that is the case for full octree).

At first glance it might seem that the above compact octree storage scheme is unusable for searching nodes or performing a recursive octree walk, because

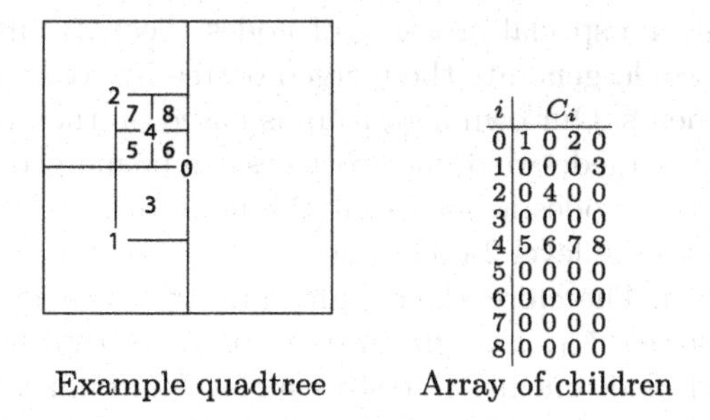

Fig. 5. Reordering of children array for a quadtree.

values C_{ij} are necessary for that. On the other hand, C_{ij} are by construction partial sums of c_{ij}:

$$C_{ij} = M_i + \sum_{\mu=0}^{j} c_{i\mu}, \quad M_i \equiv \sum_{\lambda=0}^{i-1} \sum_{\mu=0}^{7} c_{\lambda\mu}. \tag{1}$$

Direct calculation of the first term M_i is inefficient, because the average number of operations required is proportional to the octree size. But this problem is easily solved by introducing additional "milestone" indexing: assume that during the reading of c_{ij} we computed M_i for $i = sq$, where $s = 0, 1, 2, \ldots$, and q is a positive integer constant. Then a more efficient formula can be proposed to compute C_{ij}:

$$C_{ij} = M_{i_0} + \sum_{\lambda=i_0}^{i-1} \sum_{\mu=0}^{7} c_{\lambda\mu} + \sum_{\mu=0}^{j} c_{i\mu}, \quad i_0 \equiv i - i \bmod q. \tag{2}$$

Here the computation of each term takes a limited number of operations. Practically, it is useful to choose $q = 2^n$ with an integer n, e.g., 3–5, because in this case i_0 is computed using only bitwise binary operations. The smaller n we take, the larger will be the memory required to store M_i, and the faster is the calculation of C_{ij} using formula (2); on the other hand, by increasing n it is possible to use quite small amount of storage for milestones while still gaining efficiency from (2). For example, in the case of octree, at $n = 3$ the amount of memory for milestones is the same as for c_{ij}, and (2) works quite fast. Notice also that internal sums can be computed using precomputed tables. Importantly, the compressed octree with search and recursive walk enabled typically fits into 1–20 Gb of operative memory even for meshes with 10^{10} nodes.

4 Merging Subdomain Data

Once metadata and interpolated fields are computed for subdomains, the visualization dataset has to be assembled. The first step is to merge all octree components into one. Several algorithms exist for that [9,13,14], but we didn't find any that would employ the special ordering of nodes (Sect. 3) allowing compressed storage, and that would generate the merged octree by consecutively writing its children array elements. Our own algorithm is based on the breadth-first traversal of octrees being merged; two or more octrees can be merged at a time. Clearly, our ordering of octree nodes is such that the nodes of level 1 immediately follow the root, then go the level 2 nodes, and so on—the nodes are ordered in the array C level by level. Therefore, the breadth-first traversal algorithm in our case walks through nodes $C_0, C_1, \ldots$, right in that order; yet the algorithm is nontrivial, because for each node one has to restore its level and index coordinates. With the help of milestone indices M_i and a small stack-like data structure, a traversal algorithm can be implemented that takes $O(N \ln N)$ operations, N being the number of nodes. Once this is done, octrees to be merged are traversed simultaneously, and the merged octree is written consecutively. Pseudocode illustrating the algorithm for two octrees is shown in Fig. 6—it is quite trivial, apart from the problem of finding node levels and index coordinates X_i^1 and X_j^2, which is done as outlined above.

1: $i \leftarrow 0, \quad j \leftarrow 0$
2: **while** $i < N_1$ and $j < N_2$ **do**
3: **if** $X_i^1 < X_j^2$ **then**
4: append C_i^1 to C
5: $i \leftarrow i + 1$
6: **else if** $X_i^1 > X_j^2$ **then**
7: append C_j^2 to C
8: $j \leftarrow j + 1$
9: **else**
10: append bitwise-or (C_i^1, C_j^2) to C
11: $i \leftarrow i + 1$
12: $j \leftarrow j + 1$
13: **end if**
14: **end while**
15: Append all remaining C_i^1 or C_j^2 to C

N_1 and N_2 are total number of nodes in the input octrees; C^1 and C^2 are children array of the input octrees, in the compressed format; C is the compressed children array of the output octree. X_i^1 and X_j^2 are node levels and index coordinates in first and second octrees respectively: $X = (l, \mathbf{x})$. Comparisons on lines 3, 6 are lexicographical; l compares first, then compares $\mathbf{x}$. Vectors $\mathbf{x}$ of index coordinates are compared in a way that reflects the ordering of nodes within a level, as described below.

Fig. 6. Algorithm for merging two octrees.

To establish the comparison rule for index coordinates of two nodes, let us consider one vector of index coordinates $\mathbf{x} = \{x_1, x_2, x_3\}$ of level-l node a and denote j-th binary digit of i-th coordinate by x_{ij}, such that $x_i = \sum_{j=0}^{l-1} 2^j x_{ij}$. Now notice that $\{x_{1j}, x_{2j}, x_{3j}\}$ is the Morton code determining the position of node $p^j(a)$ among the children of its parent node $p^{j+1}(a)$. Since by construction the ordinal number of $p^j(a)$ among its siblings is $o_j = x_{1j} + 2x_{2j} + 4x_{3j}$, we should compare index coordinate vectors $\mathbf{x}^1$ and $\mathbf{x}^2$ of two l-level nodes by first computing numbers o_j^1 and o_j^2, $j = 0, \ldots l-1$, and then compare the two series of numbers lexicographically, in the order $o_{l-1}, \ldots o_0$.

Many octree components can be merged, e.g., pairwise in parallel, similar to a parallel reduce operation. Importantly, there is no need to use the much more memory-expensive data structure required for octree component generation.

The next step is to generate subtrees of depth d for the merged octree and then interpolate fields on sparse grids induced by those subtrees. For each subtree, a subdomain has a contribution into the field if its metadata has a record about subtree with the same root; the correspondence between vertices of the sparse grids is established through index coordinates of the vertices.

5 Dense Field Interpolation

Fields interpolated on sparse grids, like those shown in Fig. 4 on the right, but 3D, are subject to further visualization using volume rendering algorithms; those algorithms employ 3D textures, which means that the fields have to be interpolated at regular dense grids whose spatial step size equals the edge length of smallest subtree cubes. Currently we have a CPU implementation of the algorithm that recursively computes field values at vertices of the dense grid by first processing edges, then faces, and then cubes, each time placing average value in the middle. The algorithm can be illustrated by Fig. 7, in which computed vertices are black, and influencing vertices are marked with crosses. Dense field interpolation algorithm has crucial influence on the performance of the animation, therefore it is important to provide its GPU implementation, which is planned.

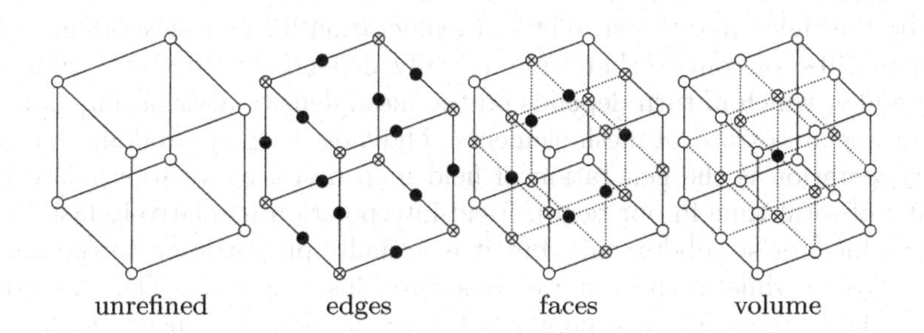

Fig. 7. Dense field interpolation.

6 Performance of Algorithms

The presented dataset preprocessing algorithms, except subdomain merging, have been implemented in C++ and tested in terms of performance and result validity. The measurement of algorithm performance data presented below has been done on a desktop system with Intel(R) Core(TM) i7-8700 CPU, 16 Gb

RAM, a regular hard disk drive, and NVIDIA GeForce 1060 GPU with 6 Gb RAM. Preliminary testing on an NVIDIA Tesla V100 system gave similar results.

Figure 8 shows performance data for test dataset containing $N = 10^3$–10^8 nodes (a typical subdomain is expected to have up to 10^8 nodes). Preprocessing time of largest subdomain is 500 s, and the algorithms are sequential. As long as different subdomains can be preprocessed independently, we expect linear scalability of subdomain preprocessing with respect to CPU count, provided the I/O system does not create a bottleneck.

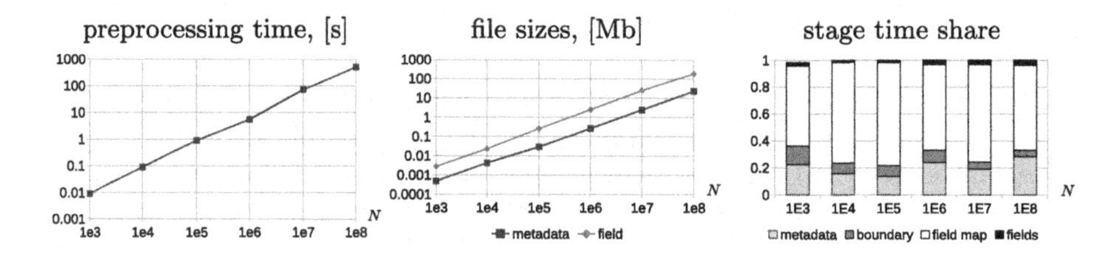

Fig. 8. Performance of preprocessing algorithms.

The size of visualization dataset is reasonable. Importantly, the metadata fits into RAM for quite large dataset due to the use of octree storage format proposed in Sect. 3. In any case, size vs quality trade-off is possible when necessary by limiting the total depth of the octree.

Contribution of each preprocessing stage into its overall time is shown in Fig. 8 on the right. In our test, metadata generation for largest subdomain (10^8 nodes) is 140 s; of course, that time actually depends on the total number of octree nodes, which in turn depends on the mesh density distribution and especially on the anisotropy of mesh elements. The most lengthy subdomain preprocessing operation is the generation of field map necessary to interpolate fields (60–80% of total time in our tests). Field interpolation is relatively fast (in our test, 18 s for largest subdomain), but it is actually proportional to the number of time layers, while in the test we considered just one level. The "boundary" stage in the histograms corresponds to the calculation of a special scalar field, whose isosurface approximates domain boundary. Its contribution into the total preprocessing time is 5–15%.

The visualization system authors are currently working on has been partially implemented. At the moment of writing, it is able to visualize animated fields defined on a sparse grid induced by one fixed-depth subtree, using 3D textures and volume rendering algorithms. An example of isosurface visualization is shown in Fig. 9.

The results of visualization performance measurements are shown in Fig. 10. For grids induced by depth 7 subtrees, the total time of rendering one frame, including the reading of data from a file and dense field interpolation, is 44 ms, which corresponds to 23 frames per second (fps). For depth 8 subtrees, that

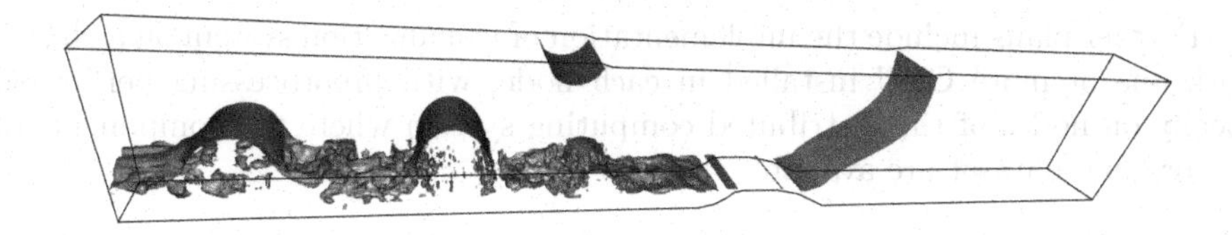

Fig. 9. An example of isosurface visualization.

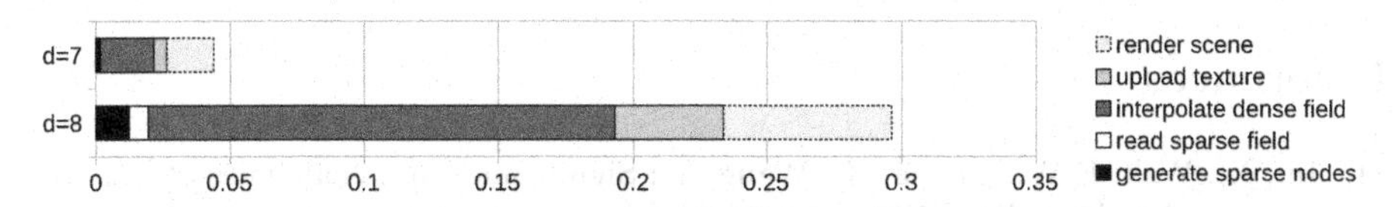

Fig. 10. Performance of animated field rendering.

time is 300 ms (3 fps). As can be seen from histograms, the most lengthy operation during the rendering is the interpolation of dense field, followed by texture upload to GPU: $20 + 5$ ms for $d = 7$ and $170 + 40$ ms for $d = 8$. All other operations last 19 ms and 82 ms for $d = 7$ and $d = 8$ respectively, which corresponds to 52 and 12 fps. Obviously, the CPU implementation of dense field interpolation creates performance bottleneck and therefore the algorithm needs to be reimplemented on GPU. This would significantly reduce dense field interpolation time; besides, only sparse field would be uploaded to the GPU, and the resulting dense field, generated on the GPU, would be bound to texture using CUDA-OpenGL interoperability API. In other words, we expect much better visualization performance after having the algorithm implemented on GPU (about 50 fps for $d = 7$ and about 10 fps for $d = 8$).

7 Conclusions

The paper describes the sparse 3D mipmapping technology for visualization of time-dependent scalar fields on large meshes containing up to 10^{10} nodes, and its core algorithms used at the preprocessing stage. An efficient data format is proposed for visualization dataset. Subdomain processing algorithms have been proposed and implemented, and efficient algorithms for merging subdomain contributions into the visualization dataset are proposed. Tests show good performance of preprocessing.

The visualization of the preprocessed dataset has been partially implemented and its performance is tested. Existing implementation reaches 22 fps for animated field visualization with depth 7 subtrees, and 3 fps with depth 8 subtrees. Reimplementing dense field interpolation on GPU promises to have better performance, estimated as 50 fps with depth 7 subtrees and 10 fps with depth 8 subtrees. Frame rates are expected to grow with the evolution of graphics hardware. Therefore, we conclude that our approach to large dataset visualization is feasible.

Further plans include the implementation of visualization system on a cluster with one or more GPU installed in each node, with preprocessing performed locally on nodes of the distributed computing system where the components of the original dataset are available.

Acknowledgements. Supported by Russian Science Foundation (Grant No. 18-11-00245).

References

1. Jin, X., Wah, B.W., Cheng, X., Wang, Y.: Significance and challenges of big data research. Big Data Res. **2**(2), 59–64 (2015)
2. Bikakis, N.: Big data visualization tools. In: Sakr, S., Zomaya, A. (eds.) Encyclopedia of Big Data Technologies. Springer, Cham (2018). https://doi.org/10.1007/978-3-319-63962-8_109-1
3. Childs, H., et al.: A contract based system for large data visualization. In: VIS 2005, IEEE Visualization, pp. 191–198 (2005). https://doi.org/10.1109/VISUAL.2005.1532795
4. García, S., Ramírez-Gallego, S., Luengo, J., Benítez, J.M., Herrera, F.: Big data preprocessing: methods and prospects. Big Data Anal. **1**, 9 (2016). https://doi.org/10.1186/s41044-016-0014-0
5. Karypis, G., Kumar, V.: MeTis: Unstructured Graph Partitioning and Sparse Matrix Ordering System, Version 4.0. University of Minnesota, Minneapolis, MN (2009). http://www.cs.umn.edu/~metis
6. Jackins, C.L., Tanimoto, S.L.: Oct-trees and their use in representing three-dimensional objects. Comput. Graph. Image Process. **14**(3), 249–270 (1980)
7. Morton, G.M.: A Computer Oriented Geodetic Data Base and a New Technique in File Sequencing. IBM Ltd. (1966)
8. Gargantini, I.: An effective way to represent quadtrees. Commun. ACM **25**(12), 905–910 (1982)
9. Oliver, M.A., Wiseman, N.E.: Operations on quadtree encoded images. Comput. J. **26**(1), 83–91 (1983)
10. Woodwark, J.: Compressed quad trees. Comput. J. **27**(3), 83–91 (1984)
11. Choi, M.G., Ju, E., Chang, J.W., Lee, J., Kim, Y.J.: Linkless octree using multi-level perfect hashing. Comput. Graph. Forum **28**(7), 1773–1780 (2009)
12. Lefebvre, S., Hoppe, H.: Perfect spatial hashing. ACM Trans. Graph. **25**(3), 579–588 (2006)
13. Pham, T.T., Kim, Y.H., Ko, S.L.: Development of a software for effective cutting simulation using advanced octree algorithm. In: 2007 International Conference on Computational Science and its Applications (ICCSA 2007), Kuala Lampur, pp. 324–334 (2007). https://doi.org/10.1109/ICCSA.2007.19
14. Jessup, J., Givigi, S.N., Beaulieu, A.: Merging of octree based 3D occupancy grid maps. In: 2014 IEEE International Systems Conference Proceedings, Ottawa, ON, pp. 371–377 (2014). https://doi.org/10.1109/SysCon.2014.6819283

Describing HPC System Architecture for Understanding Its Capabilities

Dmitry Nikitenko[1,2(✉)], Alexander Antonov[1,2], Artem Zheltkov[1], and Vladimir Voevodin[1,2]

[1] Lomonosov Moscow State University, Moscow, Russian Federation
{dan,asa,voevodin}@parallel.ru, artzlt@mail.ru
[2] Moscow Center of Fundamental and Applied Mathematics, Moscow, Russian Federation

Abstract. There is a variety of known HPC ratings nowadays which represent machine capability for solving a fixed problem, based on a certain algorithm, but these ratings represent a top of the iceberg, and as a rule, one can't compare application tuning features even for the selected system, and the details of system architecture are not usually described precisely. At the same time lots of efforts are made to describe diverse algorithm features formally, AlgoWiki is one of the most notable recent projects. The idea of Algo500 is joining precise description of computer system with detailed formal descriptions of algorithms using implementation performance data, and building an engine over such joint base to allow various queries, thus giving means of building user-defined ratings regarding selected method, algorithms and/or computer platform features. This paper gives an overview of Algo500 design and some use cases.

Keywords: High-performance computing · HPC ratings · Co-design · Algo500 · AlgoWiki · Open encyclopedia of algorithms · Top50 HPC rating

1 Background

To assess the performance of supercomputer systems, it is generally accepted to use various benchmarks. The most famous of them at the moment are: Linpack [1] (HPL [2], as the most famous implementation), HPCG [3], Graph500 [4], NAS Parallel Benchmarks [5], etc. The main drawback for a real user is that each of these benchmarks tests a supercomputer only in terms of the efficiency of only one implementation of one algorithm and says little about how good the computer will be when executing other programs. Even all well-known benchmarks taken together can cover only a small part of the set of all computational algorithms and their implementations, therefore they provide only quite abstract and far from real practice information.

On the other hand, the number of algorithms is constantly growing. Properties of them are described in detail in the framework of the AlgoWiki Open

V. Voevodin and S. Sobolev (Eds.): RuSCDays 2020, CCIS 1331, pp. 425–435, 2020.
https://doi.org/10.1007/978-3-030-64616-5_37

encyclopedia of parallel algorithmic features [6]. For many algorithms, a number of their implementations for various types of supercomputer systems are given, and dynamic characteristics of the execution of these implementations are investigated. The idea arises [7] that each of these implementations can serve as the basis for a benchmark, the results of which on various platforms can form a new rating of supercomputer systems. This rating will characterize computers in terms of the efficiency of this particular implementation of a particular algorithm. But when there will be many such ratings, the user will be able to choose the algorithms that are closest to the practical problems that they are solving, compare the effectiveness of various algorithmic approaches, choose the most effective implementations of these algorithms on supercomputer systems available to him, and solve many more problems for which it was not previously possible to solve.

Of course, the formation of a rating system for many implementations of various algorithms described in the framework of AlgoWiki is an extremely difficult and time-consuming task. Its complete solution is possible only through the efforts of the entire computing community. Within the framework of the described project, tools are being developed that allow us to approach the solution of this problem, as well as methods for using the generated rating system to solve a variety of practical problems are being developed.

2 Introducing Algo500

The idea of Algo500 [8] is as a joint methodology of describing algorithm properties, based on AlgoWiki encyclopedia, and platform-related properties, based on advanced Top50-like [9] detailed machine description, enriched with query engine that would allow building preset and custom ratings based on various combinations of parameters.

3 Algo500 Project Design

Algo500 design is shown in Fig. 1. It consists of four general blocks:

1. AlgoWiki—Description of Algorithms;
2. PerfData—Repository and Performance Results;
3. CompZoo—Description of Computer Systems;
4. Algo500 Engine.

The following subsections describe these blocks and functional interaction of the components.

3.1 AlgoWiki—Description of Algorithms

AlgoWiki was originally created as an encyclopedia of parallel algorithmic features. A universal structure consisting of two parts was proposed for this description [10]. The first part describes machine-independent properties that do not

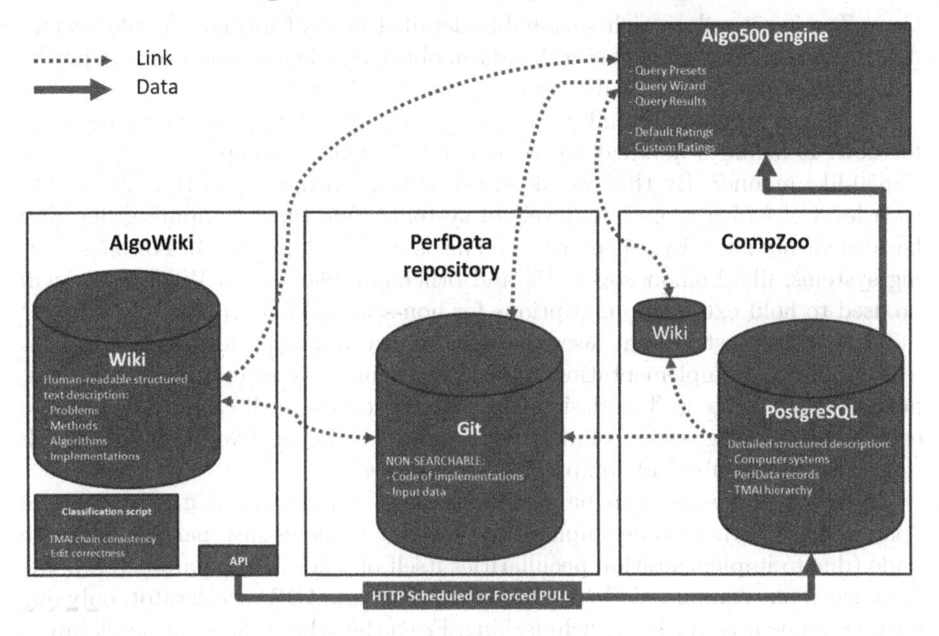

Fig. 1. Algo500 project concept design.

change when switching from one target platform to another. The second part describes the dynamic properties of the implementations of algorithms [11]. Dynamic properties already depend significantly on the features of the software and hardware platform, therefore, they characterize not only the algorithm itself and its software implementation, but also their correspondence to the target computer.

Subsequently, from simply describing the properties of the algorithms, a transition was made to the description of the problems to be solved. For this purpose, the chains Problem–Method–Algorithm–Implementation (PMAI) [12] corresponding to the main stages of solving computational problems on supercomputer systems began to be built and described. In order to be able to analyze the effectiveness of various options for solving problems, this chain must be closed to descriptions of computer systems, resulting in an extended chain Problem–Method–Algorithm–Implementation–Computer (PMAIC). In the future, fixing the various stages of such chains, it will be possible to select the most effective solutions for the remaining stages, obtaining optimal solutions to a variety of emerging tasks.

Besides AlgoWiki, there is a script mentioned on the scheme that is aimed at keeping an eye on editing AlgoWiki substances to avoid broken PMAI chains. Secondly, it will actually serve a base for API for export such PMAI data for the needs of external services like Algo500 engine.

3.2 CompZoo—Description of Computer Systems

CompZoo is actually a rich searchable detailed base of formal real-life system descriptions and their benchmarking data, obtained when running certain implementations from PMAI chains, described by AlgoWiki.

This base is generally built on SQL database, we plan to use PostgreSQL as for now. It includes detailed structured descriptions of computer systems in a Top50-like manner. By that we mean describing machines as a tree-like graph, each level of which represents levels of system, node type, compute group and basic elements like CPU, memory etc. This is described in [13, 14]. For outstanding systems, like Lomonosov-2 [15] and other notable ones, a Wiki engine will be used to hold extended descriptions for non-searchable purposes.

One of the most general cases that can be met among performance results, is a situation when implementation is being run using only several compute nodes, not the entire system. The design of CompZoo covers such cases well, because each benchmarking result is linked with an entity of type "Node group", which represents some subset of computer system nodes.

However, there also can be situations (that happen much more rarely) of "partial runs" when some implementation run is using just part of compute node (due to implementation peculiarities itself or some of run parameters). For instance, if one compute node contains two CPUs and GPU accelerator, only one CPU could be used during benchmarking. From the other side, some accelerators may operate without including CPU into the computations.

CompZoo result description is able to reflect these specialties of implementation runs. It is made by using special "Partial runs" entity, where for each partial run it is specified which components of nodes were used (such subsets of components are represented by entity of type "Component group", which is very similar to "Node group" entity described above). If there are no "Partial runs" records, corresponding to the given implementation run, it is considered to have used the whole compute node, it was run at (because this is the most general case).

The set of possible characteristics belonging to one vertex depends on the type of this vertex (see Table 1).

It should be noted, that we fill the value of each characteristic belonging to vertex only if we have reliable information about it. Hence, the situations, when only a small part of possible characteristics is present, are completely regular. CompZoo is designed with the assumption that not all specifications of systems may be provided.

The database also includes benchmarking data with all details that could be needed for queries. For example, not only TEPS for a graph algorithms, but also a graph type, size, and so on together with specified compiling options and used resources.

This database also holds a copy of PMAI hierarchy from AlgoWiki to build query results without need of every time pulling such data from a Wiki engine.

Table 1. The set of possible vertex characteristics

Whole system	*CPU and GPU characteristics*
Rpeak, TFLOPs	Vendor (e.g., Intel or AMD)
Rmax, TFLOPs	Model (e.g. Xeon Gold 6126)
Task manager	Family (e.g., Intel Skylake)
Power, GW	Base clock frequency, MHz
Compute node	Max clock frequency, MHz
OS (e.g., Red Hat 7.5)	Peak performance, GFLOPS
OS family (e.g., Linux)	Power, W
OS distribution (e.g., CentOS)	Direct memory access capability
Interconnect	Memory bandwidth, GB/s
Vendor (e.g., Mellanox)	Manufacturing process, nm
Interconnect model (e.g., Nvlink 1.0)	Number of cache levels
Interconnect family (e.g., Infiniband)	Number of micro clusters
Number of channels	*CPU-specific attributes*
Bandwidth, Gbit/s	Number of cores
Latency, us	Hyperthreading capability
Topology (e.g., fat tree)	Threads per core
GPU-specific attributes	L1 cache size
Number of streaming multiprocessors	L2 cache size
Number micro cores	L3 cache size
Number of tensor cores	L4 cache size
Inbound memory type (e.g., HBM2)	Vectorization technology (e.g., AVX-2)
Inbound memory size	Vector registers width, bits
Inbound memory clock frequency	CPU architecture (e.g., x86-64)
LLC capacity, MB	Bus speed, GT/s
LLC bandwidth, GB/s	

3.3 PerfData—Repository and Performance Results

The PerfData repository contains a part of benchmarking data which is not going to be used as a query parameter. It contains such information like source codes of the specified implementation, input data if needed, and author credentials.

GitHub seems to be a good base for that. Note, that the hierarchical structure of PMAI chains should not be duplicated into the repository, because it will cause sufficient difficulties if any changes would be applied to it.

3.4 Algo500 Engine

Algo500 engine is the heart of the project, which drives all components together. The main goal of this engine is to provide techniques to build and execute queries over the whole diversity of algorithms and computer system peculiarities.

The most unified functional block of the engine is the **Query Wizard**. At the first stage it will be implemented as a dynamical web form, but the final version could provide means of visual constructor.

Based on that, we introduce **Query Presets**, which are actually use cases of Query Wizard usage with some fixed parameters. These presets can be too weak to pretend being ratings, but they will help us linking a certain branch or leaf of PMAI chain, or part of hardware description, with the benchmarking results, corresponding to it. For example, one of the presets could be like "all the benchmarking results in FLOPS which correspond to implementations, run on systems with x86 architecture, having more, then one core used".

The correct and clear displaying of **Query Results** is needless for the tool, if we want it to be useful. The main idea, which we're going to keep to, is to allow going into details on any result observing stage, saying nothing about user-friendly interface.

Every **Default Rating** is a subset of Query Presets. As it was mentioned, not every query can be considered as rating. For example, we can build rating of architectures or systems, solving specified algorithm, but it obvious, that it is not always possible to compare system capabilities by the results of any arbitrary set of implementations. Top500, Graph500 are good examples of ratings. Nevertheless, if a user builds a query that is strong enough to become a rating, it will be possible to save it as a **Custom Rating**, and it will be available after moderation. Of course, speaking about ratings, we do not mean their regular revisions, so we won't have a number of time-related lists, corresponding to one rating, at least for the observed future. We could do it later, saving timestamps of benchmarking data and system description.

3.5 Components Interaction

Let us discuss briefly main use cases of Algo500, and data distribution between stated above blocks. At present the design supports implementation of two major scenarios of Algo500 complex functioning, and two rare ones.

Algo500 user can choose between several modes, roughly, they are grouped into two clusters—fixed (previously created and approved) ratings and preset queries, and query wizards. In both cases the lists of results will provide interactive general performance data and information about the used hardware platform. Most of records in these lists will be clickable and will point to AlgoWiki and CompWiki pages with full algorithmic and hardware descriptions, as well as links to repository with source codes.

Any AlgoWiki user would find two add-ons on every page of AlgoWiki: a performance block, and a PerfData submission invitation. The performance block can be different, depending of the problem, method, algorithm, or implementation page is visited. For example, we plan to show top performance results in such a block for implementation (every result contains a link to a page with details in Algo500 engine, link to the used platform, and to the repository location, which holds the corresponding code). For algorithm page, such block reasonably contains best results for various implementations. The method page will have a top of productive algorithms, and the problem page—a top of methods.

Repository visitor, who came across dome implementation will have a direct link to AlgoWiki implementation page to get access to the whole PMAI chain and all corresponding services.

CompWiki reader will find PerfData query shortcuts for this certain system, and a full computer system description, of course.

4 Describing Real Computer Systems

Let us demonstrate the application of the CompZoo system description model by the example of the following supercomputer systems: the Top500 ranking leader from 2018 to the present, the "Summit" supercomputer, and one of the most productive Russian supercomputers "Lomonosov-2" of Moscow State University. Below are the hierarchy schemes for the description of these systems in the form of a graph whose vertices are certain parts of the system. Each of the vertices in this case contains data that are characteristics directly inherent in the corresponding part of the computing system.

4.1 "Summit"

The proposed description model allows you to reflect the features of the architecture of computing systems, including at the level of the structure of the computing node. For example, the description of the "Summit" supercomputer is illustrated by the fact that each node of this computing system, on which there are 2 IBM Power9 processors and 6 NVIDIA Volta GV100 graphics accelerators, is logically divided into 2 connected components, each of which consists of 1 processor and 3 accelerators. Such connectivity components correspond to vertices of the "Compute group" type.

The description of the "Summit" supercomputer within CompZoo, the schematic representation of which is given in Fig. 2, contains information that this system consists of 4608 computing nodes connected by the Mellanox EDR Infiniband interconnect with the Non-blocking fat tree topology and 100 G bandwidth. At the same time, each computing node contains 2 logical groups of components that are connected at the CPU level via an X-Bus interconnect with a bandwidth of 64 Gb/s. Each of these groups consists of 1 processor, 3 graphics accelerators and two memory modules (256 Gb DDR4 and 1600 Gb NVRAM), see Fig. 3. It is also reflected in the description that within one group the processor connects to Nvlink accelerators with a bandwidth of 50 Gb/s, and that the accelerators are interconnected with the same characteristics). For the processor and accelerators,

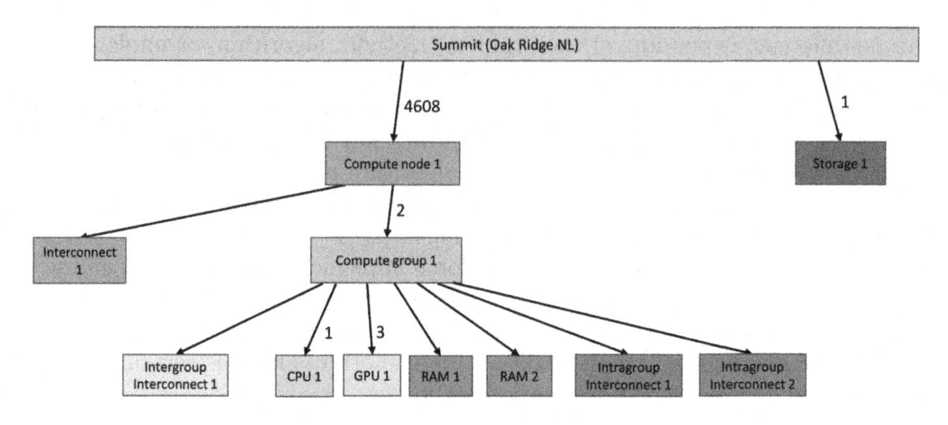

Fig. 2. "Summit" description schema.

a set of their characteristics is set, such as model, clock frequency, manufacturing process, number of cores/multiprocessors, cache sizes at different levels, etc. In addition, the description contains information that the storage system consists of 250 Pb storage with the GPFS file system and a speed of 2.5 Tb/s for reading and writing.

4.2 "Lomonosov-2"

In the "Lomonosov-2" system, there is no logical separation of components into groups within a node. In such a trivial case, each computing node consists of a single vertex "Compute group", which contains all the components located on the node.

However, "Lomonosov-2", unlike "Summit", contains heterogeneous computing nodes, which is reflected in the CompZoo description in the form of two groups of nodes containing 1536 and 160 computing nodes, see Fig. 4. We also know more about "Lomonosov-2" more reliable information about the data storage system and the interconnect between the storage and computing nodes, therefore these aspects of the description are presented in more detail: the fact that the storage system consists of two storages (operational on 94 Tb with the Lustre file system and archived at 445 Tb).

As for the components of computing nodes, nodes of the same type (Compute node 1) contain one Intel Xeon processor E5-2697v3, one NVIDIA Tesla K40M graphics accelerator and 64 Gb DDR4 memory; nodes of another type (Compute node 2) are equipped with one Intel Xeon Gold 6126 processor, two NVIDIA Tesla V100 accelerators and 96 Gb DDR4 memory. For the processor and accelerators are also set more detailed characteristics listed in the description of the "Summit".

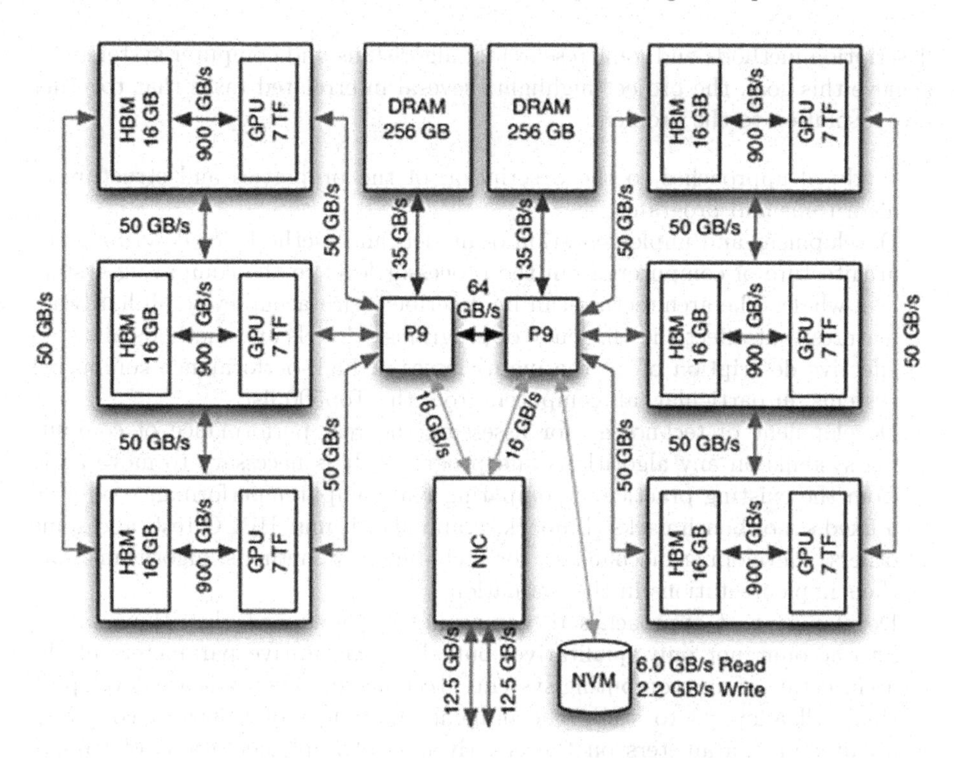

Fig. 3. "Summit" node architecture (https://www.researchgate.net/figure/The-architecture-of-a-computational-node-on-Summit_fig3_337110568).

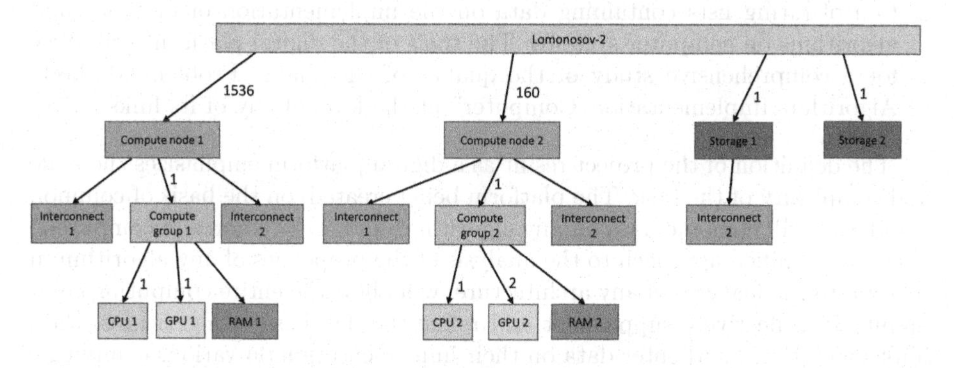

Fig. 4. "Lomonosov-2" description schema.

5 Conclusions and Future Work

The main goal of the project is the development of a set of models, methods and software for a joint analysis of the properties of algorithms and architecture of computer systems. The task is large-scale, has a large number of applications in various fields and requires an integrated approach to the development of both description methods and joint research of algorithms and computer systems. To achieve this goal, the project highlights several interrelated tasks that together give a solution to the problem:

- Study of approaches to the description of the properties and structure of algorithms and programs.
- Development and implementation of models and methods for describing the architecture of computers from the processor level to the computing system as a whole. The architecture can be described with accuracy to all hardware features that affect the efficiency of programs. Models and methods allow an effective description of all the main classes of high-performance computing systems, in particular, all computers from the Top500 list.
- Development of technology for assessing the real performance of computing systems on any algorithms and programs. It is necessary to move away from the existing practice of comparing real computer performance only on a fixed set of benchmarks (Linpack, graph algorithms, HPCG test, and some others), developing mechanisms for including any methods, algorithms, and their implementations in the evaluation.
- Development of approaches to comparing methods and algorithms taking into account not only qualitative, but also quantitative parameters of the architecture of any computing systems. Technologies and tools are developing that will allow us to study the influence of ranges of values of computer architecture parameters on the effectiveness of implementations of various methods and algorithms.
- Design and creation of a scalable digital platform Algo500 for the formation of rating lists containing data on the implementation of methods and algorithms on computer systems. The tools of the digital platform will allow for a comprehensive study of the quality of the chain "Problem–Method–Algorithm–Implementation–Computer" at the level of any of its links.

The definition of the project result as a digital platform emphasizes the scale and complexity of the task. The platform being created, on the basis of common standards, will combine data on any algorithms and architectures of computers, allow for a unified approach to the analysis of the properties of any algorithm in relation to the features of any architecture, will allow the entire computing community to collectively supplement and refine the database of algorithms, their implementations, and enter data on their implementation on various computing systems, generate, upon request, arbitrary individual ranking lists.

Acknowledgements. The results were obtained in Lomonosov Moscow State University with the financial support of the Russian Science Foundation (agreement N 20-11-20194). The research is carried out using the equipment of the shared research facilities of HPC computing resources at Lomonosov Moscow State University [15].

References

1. Dongarra, J., Luszczek, P., Petitet, A.: The LINPACK benchmark: past, present and future. Concurr. Comput. Pract. Exp. **15**, 803–820 (2003). https://doi.org/10.1002/cpe.728
2. Petitet, A., Whaley, R.C., Dongarra, J., Cleary, A.: HPL – A Portable Implementation of the High-Performance Linpack Benchmark for Distributed-Memory Computers. Innovative Computing Laboratory, September 2000. http://www.netlib.org/hpl/
3. HPCG Benchmark. http://hpcg-benchmark.org/
4. Graph 500. http://graph500.org/
5. NAS Parallel Benchmarks. https://www.nas.nasa.gov/publications/npb.html
6. Open Encyclopedia of Parallel Algorithmic Features. https://algowiki-project.org
7. Antonov, A., Dongarra, J., Voevodin, V.: AlgoWiki project as an extension of the Top500 methodology. Supercomput. Front. Innov. **5**(1), 4–10 (2018). https://doi.org/10.14529/jsfi180101
8. Antonov, A.S., Nikitenko, D.A., Voevodin, V.V.: Algo500—a new approach to the joint analysis of algorithms and computers. Lobachevskii J. Math. **41**(8), 1435–1443 (2020). https://doi.org/10.1134/S1995080220080041
9. Nikitenko, D., Zheltkov, A.: The Top50 list vivification in the evolution of HPC rankings. In: Sokolinsky, L., Zymbler, M. (eds.) PCT 2017. CCIS, vol. 753, pp. 14–26. Springer, Cham (2017). https://doi.org/10.1007/978-3-319-67035-5_2
10. Voevodin, V., Antonov, A., Dongarra, J.: AlgoWiki: an open encyclopedia of parallel algorithmic features. Supercomput. Front. Innov. **2**(1), 4–18 (2015). https://doi.org/10.14529/jsfi150101
11. Antonov, A., Voevodin, V., Voevodin, V., Teplov, A.: A study of the dynamic characteristics of software implementation as an essential part for a universal description of algorithm properties. In: 24th Euromicro International Conference on Parallel, Distributed, and Network-Based Processing Proceedings, 17–19 February, pp. 359–363 (2016). https://doi.org/10.1109/PDP.2016.24
12. Antonov, A., Frolov, A., Konshin, I., Voevodin, V.: Hierarchical domain representation in the AlgoWiki encyclopedia: from problems to implementations. In: Sokolinsky, L., Zymbler, M. (eds.) PCT 2018. CCIS, vol. 910, pp. 3–15. Springer, Cham (2018). https://doi.org/10.1007/978-3-319-99673-8_1
13. Zheltkov, A.: Development of methods for constructing ratings of computing systems based on implementations of various algorithms. In: Proceedings of Russian Supercomputing Days 2019, 23–24 September, pp. 192–199 (2019). (in Russian)
14. Zheltkov, A., Nikitenko, D.: An approach for analyzing influence factors on performance of implementations of standard algorithms. In: 2019 International Multi-Conference on Engineering, Computer and Information Sciences (SIBIRCON), Novosibirsk, Russia, pp. 0828–0832 (2019). https://doi.org/10.1109/SIBIRCON48586.2019.8958070
15. Voevodin, V., et al.: Supercomputer Lomonosov-2: large scale, deep monitoring and fine analytics for the user community. Supercomput. Front. Innov. **6**(2), 4–11 (2019). https://doi.org/10.14529/jsfi190201

LLVM Based Parallelization of C Programs for GPU

Nikita Kataev(✉)

Keldysh Institute of Applied Mathematics RAS, Moscow, Russia
kaniandr@gmail.com

Abstract. The paper proposes an approach to semi-automatic program parallelization in SAPFOR (System FOR Automated Parallelization). SAPFOR proposes opportunities to perform user-guided source-to-source program transformations and to reveal implicit parallelism in sequential programs. The LLVM compiler infrastructure is used to examine a program and Clang is used to perform source-to-source program transformation. This paper highlights benefits of IR-level (Intermediate Representation) program analysis which allows us to apply low-level program transformations to investigate properties of the original program. To exploit program parallelism SAPFOR relies on DVMH which is a directive-based programming model. We use subset of C-DVMH language which allows us to run parallel program on GPU as well on multiprocessors. Evaluation of presented approach has been performed using the C version of the NAS Parallel Benchmarks.

Keywords: Program analysis · Program transformation · Semi-automatic parallelization · SAPFOR · DVM · GPU · LLVM

1 Introduction

Today's parallel hardware platforms are usually heterogeneous and they are not only equipped with multi-core processors but also provide accelerators. In order to fully utilize the available resources, the developers have to update existing software which relies on sequential programming models. To start this time-consuming effort the developers choose from a large set of available approaches to parallel programming.

Low-level data parallel programming models (CUDA, OpenCL) allow us to achieve the best performance but at the same time require the greatest effort. Pragma based models (OpenMP, OpenACC, DVMH [1,2]) simplify programming and increase software maintainability while still providing high performance. DSLs [3–5] and corresponding compilers automate the development of high-performance parallel programs in a given domain. And finally, general purpose libraries enable exploiting heterogeneous platforms through the use of standard high-level programming languages.

V. Voevodin and S. Sobolev (Eds.): RuSCDays 2020, CCIS 1331, pp. 436–448, 2020.
https://doi.org/10.1007/978-3-030-64616-5_38

However, any of these diverse approaches still requires expert knowledge. In this situation, the development of user assistance tools can significantly reduce the cost of parallel programming. The most desirable tools are automatic parallelizing compilers which return a fully parallelized source code for a given sequential one [6–8]. Unfortunately, a generated code of such compilers may not be optimal. In this case, successful program optimization may require preliminary manual transformation of a sequential source code or even further manual optimization of a generated one. Other tools are applicable only on some stages of parallelization [9,10]. For example, they assume that parallelism will be exploited in a manual way, while static or dynamic dependence analysis or source code profiling could be done automatically. Some of them make suggestions how to improve performance of already written parallel program.

Alternative solution is to follow an implicit parallel programming methodology [11,12]. This implies that the programmer is aware that the program must be well-formed for automatic parallelization. Thus, he should be able to increase algorithm-level parallelism, still relies on expressivity of standard sequential programming languages. It is also possible to guide the compiler by the hints which emphasis high-level program properties which are essential for parallelization.

This paper is devoted to the System FOR Automated Parallelization (SAPFOR) [13] which combines approaches mentioned above to automate development of parallel programs.

SAPFOR relies on an implicitly parallel programming model. This means that the system includes an automatic parallelizing compiler and it does not require the user to parallel program explicitly. The system implements both static and dynamic analysis techniques which complement each other. Thus, the static analysis reduces overheads of program evaluation at runtime [14]. In general, dynamic analysis tools are input sensitive and the application of static analysis techniques also reduces the number of analysis results that the user must control. SAPFOR implements LLVM [15] based static analysis. The paper [16] shows how some kind of IR-level (Intermediate Representation) transformations are used in SAPFOR to increase the quality of program analysis. It also argues that despite the use of a low-level program transformations, the analysis report is closely related to the original high-level source code.

The system also provides the user with a set of automatically performed source-to-source transformations (inline expansion, dead code elimination, expression propagation and other) that he can apply to the original sequential program. Guided by the analysis report, the programmer may choose some of them for automatic execution.

Unlike traditional compilers SAPFOR performs source-to-source parallelization and produces a parallel version according to high-level DVMH parallel programming model [1,2]. It was designed to create parallel programs of scientific-technical calculations for heterogeneous computational clusters. C-DVMH is a directive based programming language. The programmer can annotate a C source code to highlight regions of code that should be executed in parallel. Thus, application of DVMH model improves parallel program maintainability and the developer also can optimize it if necessary. Using DVMH extremely simplifies the development of an automatic parallelizing compiler since it is not necessary to use various parallel programming models to generate programs for heterogeneous parallel platforms. Moreover, DVMH runtime system controls low-level data transfer and synchronization and it makes some optimizations of data transfer between CPU and accelerators. Hence SAPFOR should not insert low-level specifications of these operations in a source code.

The rest of the paper is organized as follows. Section 2 presents our approach to automatic parallelization of well-formed sequential programs. It also focuses on the application of lower-level transformations to increase the quality of program analysis. Section 3 outlines the implementation details of the interaction of the higher level transform passes, which perform program parallelization, and lower level analysis passes. Section 4 is devoted to semi-automatic parallelization of the C version of the NAS Parallel Benchmarks [20]. It summarizes the necessary source-to-source transformations and presents the performance results. Section 5 discusses the related work and finally Sect. 6 concludes this paper.

2 Automatic Parallelization

In this paper we consider parallelization for compute devices with shared memory. This means that a parallel program can be run on multi-core CPU or accelerator. For this purpose DVMH model requires that three kinds of annotations be inserted into the source code:

- specifications of the loops which can be executed in parallel, as well as specifications of private and reduction variables,
- specification of the compute regions which can be executed on the accelerators, each region may enclose one or more parallel loops,
- high-level specifications of data transfer between a memory of CPU and a memory of accelerator (actualization directives).

All specifications are in the form of directives. Each directive may contain number of clauses. To make sure the insertion of these directives is permissible, the automatic parallelizing compiler must investigate the properties of the code section to be parallelized. There are two main groups of these properties. Firstly, these are the properties of the program variables and corresponding memory locations. Secondly, these are properties associated with the control flow of the program.

The first group of properties includes loop-carried data dependencies, spurious dependencies, input, output and local data for compute regions. The second group includes summary information on control flow of each function and loop. It is necessary to identify whether function calls have side effect (including I/O operations), whether a program may terminate inside a function call, whether a function captures a pointer (i.e. it makes any copies of the pointer that outlive the function call). It is also necessary to make sure that there is no recursion leading to nested compute regions, including due to indirect function calls.

The paper [16] discusses the implementation of static analysis in SAPFOR. It introduces a novel data structure which is called a source-level alias tree. The source-level alias tree depicts the structure of accessed memory and it allows us to apply transform passes to improve the quality of the source program analysis. This means that SAPFOR analyzes transformed program and propagate its properties to the original source code. Moreover, it is possible to make property-sensitive transformations, i.e. to make some transformation to analyze one kind of properties and another transformation to analyze another kind of property.

In that way, the source-level alias tree is suitable to examine the first group of properties which are necessary for automatic parallelization. If transform passes do not cross bounds of the analyzed code section (function or loop) it is safe to investigate the second group of properties.

After the properties of the program have been explored, SAPFOR looks up for sections of code that can be executed in parallel. At this point, the original source code is processed.

In the first step, we use a depth-first ordering [17] to traverse strongly connected components of a call graph. To avoid explicit recursion, strongly connected components with a single function are considered only. As a result, we prevent the appearance of nested compute regions which are prohibited in the DVMH model.

In the next step, the body of the visited function is processed. Depth-first search of a loop tree allows us to find the outermost loops which can be parallelized. For each loop the following constraints are examined:

1. Safety of control flow. That means the absence of function calls which have side effects, the absence of multiple exits outside the loop body, the absence of I/O operations inside the loop body, the absence of indirect calls to user-defined functions.
2. Safety of memory accesses. That means the absence of loop-carried data dependencies and captured pointers. If a pointer references a privatizable variable and if this pointer is captured, then, after variable privatization, the relation between the pointer and the variable will be lost. Spurious dependencies such as private and reduction variables are allowed.
3. Direction of data usage. An input data is intended to have the newest values at the beginning of the loop. An output data is updated in the loop and is live [17] at any exit from the loop. A local data are updated in the loop, but the values of corresponding memory locations are not used outside the loop. This property will be useful for data transfer optimizations at runtime if this loop is supposed to be executed on GPU. The corresponding clauses will follow a compute region which surrounds this loop.
4. Canonical loop form according to the OpenMP [18] standard. DVMH as well as OpenMP disallows parallelization of a loop that does not have canonical loop form. Source-level alias tree is useful to ensure that loop boundaries and step are loop invariant expressions.
5. The ability to express properties of memory locations with DVMH directives. A source-level alias tree allows us to represent any memory location in the

program. However, directive based programming models have some restrictions on variables listed in clauses. For example, it is not possible to privatize memory that is allocated in the heap. The other case is accesses to a global memory inside the loop body. If there is a call which accesses this global memory, the corresponding variable cannot be placed in *private* or *reduction* clauses. Each new item is allocated for the corresponding variable listed in these constructs. Hence these items are not associated with global memory which is accessed in callees.

6. The ability to collapse iteration spaces of nested loops into one larger iteration space. The corresponding specification in the *parallel* directive is similar to collapse clause in OpenMP [18]. It increases the amount of computation, which is especially important for running a program on accelerators.

If all constraints are satisfied, the current loop will be parallelized and corresponding DVMH directives will be created. Some constraints can be relaxed to allow parallel execution on CPU even if the utilization of GPU is not possible. For instance, calculation of data usage is not necessary in this case and data transfer specifications can be omitted in a parallel code.

In the last step, we optimize the placement of data transfer specifications in a source code. We use postorder traversal [17] to prepare data for accelerator as early as possible and to request data from the accelerator as late as possible. The call graph and loop trees are traversed. Neighboring regions at the same level of a loop tree are joined. Even if regions cannot be joined compiler tries to insert actualization directives before the first region and after the last one to avoid data transfer between computations. For loops which are not parallelized we try to move actualization directives outside the loop body and for functions we move data transfer outside the callees.

3 Implementation Details

We have implemented proposed approach to automatic parallelization in SAPFOR. Low-level LLVM IR is used to perform program analysis and source-to-source transformations rely on Clang AST (Abstract Syntax Tree). LLVM 7.1.0 is currently supported.

As stated in the previous section, we need to have a transformed representation of the program in the form of a LLVM IR as well as its original representation. Transformed representation is suitable to analyze memory locations and to build summary information on control flow while the parallelization pass also investigates original LLVM IR. Maintaining the correspondence between original LLVM IR and Clang AST is necessary for source-to-source transformations.

For this purpose, a separate thread (analysis server) can be started inside the automatic compiler (Fig. 1). It clones the original LLVM IR and then it performs analysis and transform passes according to [16].

The client thread, which is responsible for program parallelization, requests the necessary information from the server. A source-level alias tree is used to

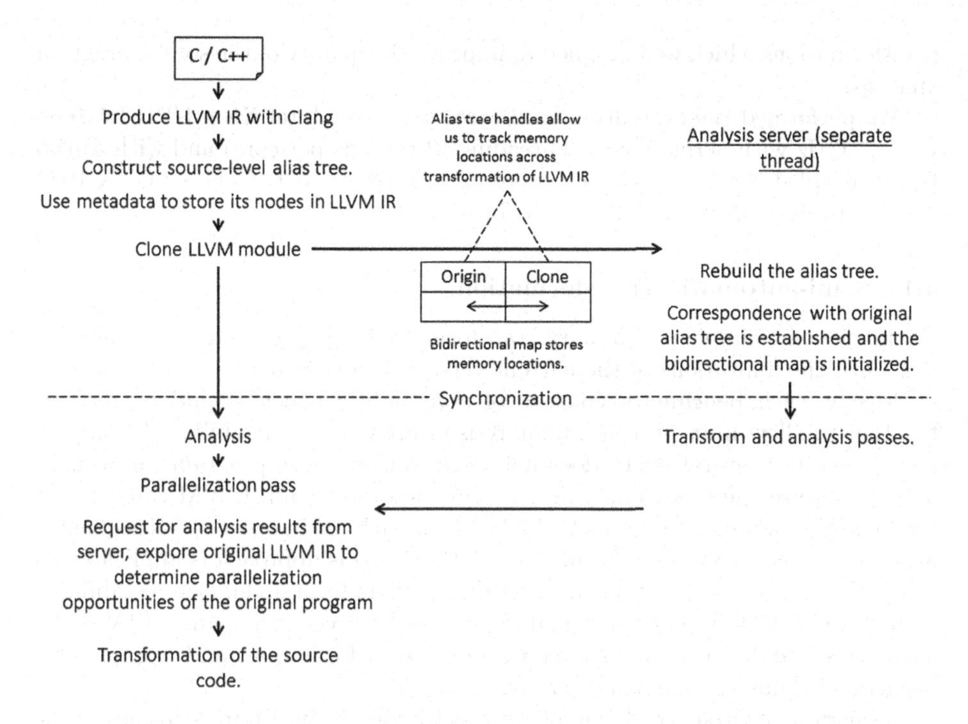

Fig. 1. Implementation scheme of the automatic parallelizing compiler in SAPFOR

synchronize the analysis results. An instance of the alias tree will be built both on the client side and on the server side. As noted in [16] for the alias tree, special handlers are implemented that allow the compiler to track memory locations across rebuilding of the alias tree after IR transformation. These handlers are also used to establish correspondence between objects on the server and on the client.

A separate LLVM pass has been created to start analysis server. Built-in support for threads, which is included in C++11, is used. It is possible to create multiple analysis servers at the same time in order to examine transform-sensitive properties. A separate pass is also implemented to transform original program.

4 Evaluation

The applicability of the implemented approach was examined on the C versions of the NAS Parallel Benchmarks (NPB) [19]. In this section we consider in detail the semi-automatic parallelization of three benchmarks: EP (Embarrassingly Parallel), BT (Block Tri-diagonal solver) and CG (Conjugate Gradient). Each of these programs has features that affect the complexity of its analysis and further parallelization. This section highlights SAPFOR capabilities which are helpful to overcome these issues. We also summarize source-to-source program transformations which were applied to improve the quality of the source program analysis.

We performed the evaluation on 6-cores processor Intel Xeon CPU E5-1660 v2, 3.70 GHz with active Hyper Threading (2 threads per core) and with Turbo Boost disabled. GPU experiments were performed on GPU GeForce GTX 1660 Ti. We use Intel Compiler 19.0.2.187 for all tests.

4.1 Semi-automatic Transformation

We made some simple preliminary transformations of benchmarks manually to deal with the limitations of the current version of SAPFOR.

Firstly, each benchmark consists of several files which should be merged together to allow inter-procedural analysis of programs. The ability of Clang to merge together several ASTs does not work well for large programs, especially ones which use the C standard library. On the other hand, LLVM comprises a tool which allows us to obtain an LLVM IR for each file in order to subsequently generate a single LLVM IR for all files. Although this approach is applicable to analyze programs, at the moment, SAPFOR suffers from inability to establish a correspondence between the original higher level sources and a single LLVM IR. Thus, this inability prevents source-to-source transformations and it implies us to merge all sources except header files manually.

Secondly, the presence of macros drastically complicates source-to-source program transformation. For instance, macro with the same name may have different meanings in distinct program regions. Moreover, changes of macro definition after program transformation may lead to changes in the transform region which are unobvious for the user. We replaced definitions of all integer constants with enumerations and definitions of floating point constants with const-qualified variables. However, the wide usage of the preprocessor in real-world applications [21] does not allow us to rely on the absence of macros in programs of scientific-technical calculations as well. In the future versions of SAPFOR the user will be able to force the transformations across presence of macros in the transform region.

The time of the analysis of the merged files, as well as the size of each benchmark in the number of lines of code are given in Table 1. The original versions comprise source files with mentioned above preliminary transformations performed on them. The transformed versions were obtained after manual and user-guided automatic transformations. And finally, the parallel C-DVMH versions were obtained after automatic parallelization.

The increase in code size of the transformed versions is primarily due to the inline expansion. This is the most important transformation which significantly reduce the complexity of program analysis. At this moment SAPFOR implements some kinds of inter-procedural analysis known as classical methods of interprocedural summary dataflow analysis [22]. Unfortunately, as mentioned in [22] this summary information is too coarse to prove the absence of data dependencies. Moreover, in the presence of pointers which are essential for every C program it is necessary to ensure that the callee does not make any copies

Table 1. The analysis time(s) of the NAS Parallel Benchmarks (NPB)

Benchmark	Original		Transformed		Parallel
	Lines	Time (s)	Lines	Time (s)	Lines
BT	3488	46.15	8805	2213.09	8954
CG	1283	0.59	1460	0.67	1515
EP	623	0.2	908	0.4	947

of the pointer that outlive the callee itself. The inline expansion allows SAPFOR to perform pairwise comparison of array accesses to determine whether dependences exist between two subscripted references [23] to this array in the loop nest. In case of inline expansion the loop nest may grow notably causing significant rise of analysis time (up to 37 min for BT).

Call to a function with arguments of pointer type (this is an CG case) is another reason for inline expansion because invocation context should be analyzed to investigate whether or not two pointers can point to the same object. The usage of *restrict* keyword in a source code is another way to overcome this issue.

Along with the choice of transformations, the user can also control the analysis options. As presented in [24] array delinearization is an important technique which is implemented in SAPFOR. It significantly reduces the complexity of data dependence analysis since it allows SAPFOR to perform the pairwise comparison of subscript expressions which calculate the addresses of accessed elements [24]. Unfortunately, variable dimension sizes and loop bounds may prohibit such comparison because C language does not ensure that subscript expression is in bounds value of an array dimension. That is why we introduce an analysis option which allows the user to force data dependence analysis and to assume that subscript expression is in bounds value.

Although SAPFOR is not able to reveal privatizable arrays in a static way, it uses dynamic analysis [14] which is helpful in case of EP and BT programs.

The EP benchmark has two features that require a manual program transformation. Firstly, it uses a reduction array, the use of which in C-DVMH is currently not supported. This array consists of 10 elements and we manually replaced it with 10 scalar variables. In this case, SAPFOR was able to automatically detect the presence of reduction operations and it generated the corresponding DVMH directives. Secondly, this benchmark uses a privatizable array of a very large size; this prevents the execution of the program on GPU. In order to eliminate this array, we manually fused two adjacent loops into a single loop (the first loop initializes this array and the second loop accesses the calculated values) and added a re-calculation of the required elements (two neighboring array elements) at each iteration of the new loop. As a result, the array was replaced with two scalar variables.

Each of benchmarks uses time measurement functions which access global variables to store execution time. If the calls of these functions were placed inside the loop body, this loop could not be parallelized. Whether the control flow reaches these calls depends on the program input, so static analysis techniques are not able to parallelize this loop. We removed unreachable calls manually if SAPFOR detected the likelihood of a side effect which prevents parallelization.

4.2 Performance Results

Table 2 shows the execution times of parallel versions obtained after automatic parallelization of the transformed benchmarks. The benchmarks have been also parallelized using OpenMP and OpenCL manually [19]. The results of corresponding launches are also given.

It can be noted that the sequential transformed version of the BT benchmark is better optimized than the original one. It is the result of inline expansion which provides the compiler with more optimization opportunities. On the other hand, the elimination of the private array in the EP benchmark increases the amount of computations, and as a result, slows down the sequential program.

A significant advantage of the OpenCL version of the CG benchmark compared to the DVMH version is caused by the use of shared memory on GPU. In addition, in the OpenCL version, the developers performed the vectorization of some inner loops. At the same time, DVMH and OpenCL versions of EP benchmark have similar performance, and on the BT benchmark, the DVMH program significantly outperforms OpenCL version. As to the maintainability, the OpenCL program is dramatically inferior to DVMH program because it is very different from the original sequential program.

Table 2. The execution time(s) of the NAS Parallel Benchmarks (NPB)

Benchmark		Sequential		SAPFOR		Manual	
				DVMH		OpenMP	OpenCL
Name	Class	Original	Transformed	CPU	GPU	CPU	GPU
BT	A	39.71	38.75	8.21	9.65	8.3	21.29
	B	169.72	161.49	34.3	34.83	35.71	77.15
	C	720.86	696.74	145.72	127.16	149.85	356.70
CG	A	0.82	0.83	0.33	0.42	0.23	0.07
	B	75.99	75.63	15.2	10.74	14.63	2.0
	C	213.11	222.67	40.14	46.67	39.05	6.45
EP	A	15.83	18.56	1.77	0.53	1.64	0.4
	B	63.2	74.27	7.07	1.49	6.55	1.42
	C	252.94	297.07	28.28	5.35	26.04	5.05

5 Related Work

There is a large number of studies dedicated to the automation of parallel programming. In this section we will consider some of them.

Polly [6] focuses on loop transformations to optimize data-locality and to exploit OpenMP level parallelism as well as to vectorize loops. It relies on a polyhedral model to optimize the program. Polly-ACC [7] extends Polly to bring accelerator support to generated parallel programs. Although the polyhedral model has a high potential for detecting parallelism in the program, it imposes significant restrictions on the source code that can be processed. The low level of LLVM IR, which is used to analyze and transform programs, does not allow the programmer to update or even view the generated code. Moreover, polyhedral based transformations implemented in a source-to-source way as in Plutto [25] produces the code which is significantly different from the original one and it can be quite difficult for the user to maintain it.

The application of the static analysis in these tools limits the possibilities of parallelization. In some cases this analysis does not allow one to estimate the sizes of the array dimensions and of the loop boundaries. Thus, the absence of such information may lead to a conservative assumption of the presence of data dependencies. Unlike Polly and the Polly-ACC, the Apollo [26] optimizer, which also relies on the polyhedral model, applies speculative optimizations at run time. However, Apollo does not allow parallelization of programs for GPU.

DiscoPop [8] relies on dynamic profiling information to reveal task graph which can be transformed with both loop-level and task-level parallelism. Clang is used to perform source-to-source transformation and to obtain parallel code using Intel Threading Building Blocks (TBB). It does not imply a significant transformation of the original program to increase the available parallelism. For each task in the task graph the corresponding code section is taken from the source code. Then it is wrapped up in a separate function which becomes a node in a flow graph of a parallel program.

The authors present the results of the automatic parallelization of programs from NAS Parallel Benchmarks (loop-level parallelism was exploited). Despite the use of dynamic analysis techniques, the performance of the parallel code is quite low and significantly inferior to the manually parallelized programs. Investigation of task-level parallelism is mainly suitable for programs with an unchangeable flow graph, such as application of different filters in image processing. In the case of repeated rebuilding of the flow graph, the overhead is very high.

6 Conclusion

The paper proposes an approach to the automation of parallel programming which follows an implicit parallel programming methodology. This approach was implemented in SAPFOR which includes an automatic parallelizing compiler. SAPFOR also provides source-to-source transformation techniques that allow the user to bring the sequential program to a well-formed version. SAPFOR relies on DVMH directive based programming model to exploit loop-level parallelism for multi-core processors and accelerators. Based on the proposed approach we perform semi-automatic parallelization of some applications from the C version of the NAS Parallel Benchmarks. The paper shows that automatically generated parallel versions have the similar performance to the manually parallelized ones.

This paper advocates the use of low-level program transformations, which are invisible to the programmer, to increase the quality of the analysis of the original program. We propose a novel approach that enables property sensitive transformations. This means that the internal representation of the program can be transformed to the most suitable form for program analysis.

The evaluation results show that SAPFOR is still suffering from an insufficient level of inter-procedural analysis. Future work will focus on this issue. We also intend to increase the number of supported C-DVMH constructions to distribute the data and the computations between several accelerators and nodes of the heterogeneous cluster.

References

1. Konovalov, N.A., Krukov, V.A., Mikhajlov, S.N., Pogrebtsov, A.A.: Fortan DVM: a language for portable parallel program development. Program. Comput. Softw. **21**(1), 35–38 (1995)
2. Bakhtin, V.A., Klinov, M.S., Krukov, V.A., Podderugina, N.V., Pritula, M.N., Sazanov, Yu.L.: Extension of the DVM-model of parallel programming for clusters with heterogeneous nodes. Bull. South Ural State Univ. Ser. Math. Model. Program. Comput. Softw. **18(277)**(12), 82–92 (2012). (in Russian)
3. Ragan-Kelley, J., Barnes, C., Adams, A., Paris, S., Durand, F., Amarasinghe, S.P.: Halide: a language and compiler for optimizing parallelism, locality, and recomputation in image processing pipelines. In: Proceedings of the 34th ACM SIGPLAN Conference on Programming Language Design and Implementation, PLDI 2013, pp. 519–530 (2013)
4. Beaugnon, U., et al.: VOBLA: a vehicle for optimized basic linear algebra. In: Proceedings of the 2014 SIGPLAN/SIGBED Conference on Languages, Compilers and Tools for Embedded Systems, LCTES 2014, New York, NY, USA, pp. 115–124 (2014)
5. Zhang, Y., Yang, M., Baghdadi, R., Kamil, S., Shun, J., Amarasinghe, S.: Graphit: a high-performance graph DSL. In: Proceedings of the ACM on Programming Languages, vol. 2, no. OOPSLA, pp. 121:1–121:30 (2018)
6. Grosser, T., Groesslinger, A., Lengauer, C.: Polly-performing polyhedral optimizations on a low-level intermediate representation. Parallel Process. Lett. **22**(04), 1250010 (2012)
7. Grosser, T., Hoefler, T.: Polly-ACC transparent compilation to heterogeneous hardware. In: ICS 2016: Proceedings of the 2016 International Conference on Supercomputing, June 2016, pp. 1–13 (2016). https://doi.org/10.1145/2925426.2926286
8. Zhao, B., Li, Z., Jannesari, A., Wolf, F., Wu, W.: Dependence-based code transformation for coarse-grained parallelism. In: Proceedings of the International Workshop on Code Optimisation for Multi and Many Cores, San Francisco, CA, USA, pp. 1:1–1:10. ACM, February 2015

9. Kim, M., Kim, H., Luk, C.-K.: Prospector: a dynamic data-dependence profiler to help parallel programming. In: 2nd USENIX Workshop on Hot Topics in Parallelism (HotPar 2010) (2010)
10. Garcia, S., Jeon, D., Louie, C., Taylor, M.B.: Kremlin: rethinking and rebooting gprof for the multicore age. ACM SIGPLAN Not. (2011). https://doi.org/10.1145/1993316.1993553
11. Hwu, W.-M., et al.: Implicitly parallel programming models for thousand-core microprocessors. In: Proceedings of the 44th Annual Design Automation Conference (DAC 2007), pp. 754–759. ACM, New York (2007). https://doi.org/10.1145/1278480.1278669
12. Vandierendonck, H., Rul, S., De Bosschere, K.: The Paralax infrastructure: automatic parallelization with a helping hand. In: 2010 19th International Conference on Parallel Architectures and Compilation Techniques (PACT). IEEE (2010)
13. Klinov, M.S., Krukov, V.A.: Automatic parallelization of Fortran programs. Mapping to cluster. In: Vestnik of Lobachevsky University of Nizhni Novgorod, no. 2, pp. 128–134. Nizhni Novgorod State University Press (2009). (in Russian)
14. Kataev, N., Smirnov, A., Zhukov, A.: Dynamic data-dependence analysis in SAPFOR. In: CEUR Workshop Proceedings, vol. 2543, pp. 199–208 (2020)
15. Lattner, C., Adve, V.: LLVM: a compilation framework for lifelong program analysis & transformation. In: Proceedings of the 2004 International Symposium on Code Generation and Optimization (CGO 2004), Palo Alto, California (2004)
16. Kataev, N.: Application of the LLVM compiler infrastructure to the program analysis in SAPFOR. In: Voevodin, V., Sobolev, S. (eds.) RuSCDays 2018. CCIS, vol. 965, pp. 487–499. Springer, Cham (2019). https://doi.org/10.1007/978-3-030-05807-4_41
17. Aho, A.V., Lam, M.S., Sethi, R., Ullman, J.: Compilers: Principles, Techniques, and Tools, 2nd edn. Addison Wesley, Boston (2006). p. 1038, Chap. 9
18. OpenMP Application Programming Interface. https://www.openmp.org/wp-content/uploads/OpenMP-API-Specification-5.0.pdf. Accessed 14 Apr 2020
19. Seo, S., Jo, G., Lee, J.: Performance characterization of the NAS parallel benchmarks in OpenCL. In: 2011 IEEE International Symposium on Workload Characterization (IISWC), pp. 137–148 (2011)
20. NAS Parallel Benchmarks. https://www.nas.nasa.gov/publications/npb.html. Accessed 14 Apr 2020
21. Ernst, M.D., Badros, G.J., Notkin, D.: An empirical analysis of C preprocessor use. IEEE Trans. Software Eng. **28**(12), 1146–1170 (2002). https://doi.org/10.1109/TSE.2002.1158288
22. Havlak, P., Kennedy, K.: An implementation of interprocedural bounded regular section analysis. IEEE Trans. Parallel Distrib. Syst. **2**(3), 350–360 (1991). https://doi.org/10.1109/71.86110
23. Goff, G., Kennedy, K., Tseng, C.-W.: Practical dependence testing. In: Proceedings of the ACM SIGPLAN 1991 Conference on Programming Language Design and Implementation (PLDI 1991), pp. 15–29. ACM, New York (1991). https://doi.org/10.1145/113446.113448
24. Kataev, N., Vasilkin, V.: Reconstruction of multi-dimensional arrays in SAPFOR. In: CEUR Workshop Proceedings, vol. 2543, pp. 209–218 (2020)

25. Bondhugula, U., Hartono, A., Ramanujam, J., Sadayappan, P.: A practical automatic polyhedral parallelizer and locality optimizer. SIGPLAN Not. **43**(6), 101–113 (2008)
26. Caamano, J.M.M., Sukumaran-Rajam, A., Baloian, A., Selva, M., Clauss, P.: APOLLO: automatic speculative polyhedral loop optimizer. In: 7th International Workshop on Polyhedral Compilation Techniques (IMPACT), Stockholm, Sweden, January 2017 (2017)

Research of Hardware Implementations Efficiency of Sorting Algorithms Created by Using Xilinx's High-Level Synthesis Tool

Alexander Antonov(✉), Denis Besedin, and Alexey Filippov

Peter the Great St. Petersburg Polytechnic University, Saint-Petersburg, Russia
{antonov,filippov}@eda-lab.ftk.spbstu.ru, 1310nero@mail.ru

Abstract. The article describes results of our research of hardware implementation efficiency of sorting algorithms created by using of Xilinx's High-Level Synthesis tools, the Vivado HLS package, and FPGAs. The term efficiency, used in the research, defined as a function of the performance, estimated in time for sorting a random generated array, and hardware "cost", estimated in utilized FPGA resources. In the research, a simulation modeling and a comparative analysis was carried out for the wide range of sorting algorithms implemented on a universal processor and on the Xilinx's FPGAs. The research results prove that hardware implementations of the sorting algorithms, synthesized by Xilinx HLS tool, do not always provide higher performance comparing with the implementations of the same algorithms on a universal processor. The article shows that the hardware implementation of the Merge sort algorithm, created by Xilinx's HLS tool, can speed up, comparing with software implementation, the process of sorting arrays of small and medium size.

Keywords: Hardware acceleration · Sorting algorithms · High-level synthesis · Reconfigurable hardware accelerator · FPGA

1 Introduction

There are rapidly growing demands on high-performance computing for Big Data analysis, Data Mining and Management tasks. A modern trend in the development of high performance computing systems is an attempt to accelerate them through the introduction of Distributed Reconfigurable Heterogeneous High-Performance Computing systems (DRH HPC) [1].

Modern DRH HPC consists of [2, 3]: multiprocessor units (MPUs), Single Instruction Multiple Data (SIMD) accelerators, commonly known as General Purpose Graphics Processing Units (GPGPU), and Reconfigurable Hardware Accelerators (RHA), based on Field Programmable Gate Arrays (FPGAs) [4, 5]. FPGA is an Integrated Circuit (IC) that can change its internal structure in accordance with the particular task. A modern FPGA consists of programmable logic cells (LCELL), which can perform any functions of logic or memory, and a hierarchy of programmable matrices, which can connect nearly all LCELL in FPGA together to implement complex logic and memory functions. The

V. Voevodin and S. Sobolev (Eds.): RuSCDays 2020, CCIS 1331, pp. 449–460, 2020.
https://doi.org/10.1007/978-3-030-64616-5_39

modern FPGAs contain not only LCELL and the programmable matrices, but also digital signal processing units (DSP), integrated memory units (BRAM), high-throughput memory units (HBM), hardware-implemented controllers for external DDR4 memory and multi gigabits transceivers for external PCIe interfaces and 10-100G Ethernet ports. State-of-art FPGAs can be configured "on the fly"; it means that ones can be configured during the normal operation of the device. Some modern FPGAs support partial configuration and reconfiguration via PCIe and Ethernet. Partial reconfiguration of the FPGA means that a part of the FPGA can be reconfigured, while the rest of the FPGA continues to solve the current task [6].

DRH HPC system allows you to create highly specialized computational "pipes" for solving the particular tasks. The computational "pipe" can consist of just Reconfigurable Hardware Accelerator, a reconfigurable FPGA based accelerator, or, for solving a complex task, can include MPUs, SIMD accelerators and Reconfigurable Hardware Accelerators working together. A huge advantage for the performance of such DRH HPC systems is the ability to create and reconfigure the computational "pipes" on the fly [7], in accordance with the particular task, and so, to satisfy to one of the most important criteria for high-performance computing systems: performance and energy efficiency [8].

The traditional procedure for developing a Reconfigurable Hardware Accelerators, hardware solution for FPGA, commonly uses Hardware Description Languages (HDL), such as VHDL, Verilog HDL, System Verilog. This approach is very time-consuming and requires painstaking work both at the development stage and at the debugging stage, since it is necessary to operate with logical functions and assemble complex systems from small pieces, between which it is also necessary to establish the correct connections and check them during debugging [9].

The modern approach is a development flow, which uses High-Level Synthesis (HLS) tools provided by leading manufacturers of FPGA programmable logic, such as Xilinx [9] and Intel PSG [10], and some other companies engaged in the development of HLS tools, for example, Mentor Graphics [11].

Modern HLS tools allow not only to synthesize a hardware solution, often called an synthesized implementation, for an algorithm described by high-level programming languages such as C or C++, but also to simulate the source code and the implementation for verifying the correctness with respect to the expected results. Modeling simulation of the source code and the synthesized implementation use a common test, often referred as self-checking testbench, described in C or C++.

The goal of the research described in this paper is to find some sorting algorithms (at least one) which, when they are implemented in hardware by using Xilinx's HLS tools, improves performance of sorting tasks, in comparison with software implementations, implementation based on universal MPU, of the same algorithms.

There are some related works dealing with sorting networks implemented on FPGA. The most relevant articles are [12–14]. The article [12] describes Bitonic sort algorithm implementation flow for Xilinx FPGA and hardware aspects for FPGA based implementations, such as static and dynamic approaches for FPGA configuration. All estimations are for very small arrays with up to 600 elements. The article [13] highlights a research dealing with synthesis of recursive Binary-tree algorithm and suggests hardware-oriented parallel implementations of recursive sorting algorithms. The article demonstrate a "pure" hardware approach for building sorting networks without using C\C++ descriptions and HLS tools. The article [14] highlights created framework for building sorting network architectures consisting of basic sorting primitives synthesized by HLS. Results provided for performance and hardware "cost" such algorithms for very small arrays with up to 16384 elements.

The contributions of the paper are: optimized hardware implementation of the wide range of sort algorithms by using HLS; evaluation and comparison of performance and hardware "cost" for arrays with small (up to 10 000 elements), medium (up to 100.000 elements) and large (up to 1.000.000) number of elements.

2 The Research Objects

When choosing algorithms for research, libraries of already implemented algorithms such as MatLab and IP library from Xilinx were considered. MatLab library contains many commonly used sorting algorithms described in .m language. Among them are comb sorting, merge, heap, quick sort, insertion, bubble, bucket, cocktail, choice, shell sorting algorithms [15]. IP library from Xilinx contains merging, inserting and bitonic sorting algorithms described in C++ [16].

Summarizing the both lists of the algorithms, we distinguished those types of algorithms, in accordance with the classification in Fig. 1, that were objects for the research. Thus, the following types of algorithms were distinguished: exchanging algorithms, selection algorithm, insertion sorting, merging algorithm, counting algorithms and unique algorithms that use non-standard sorting approaches.

We picked at least one algorithm of each type to conclude about the most successful type for hardware implementation.

From this list, we chose for our research the following algorithms: Gnome sort, Comb sort, Merge sort, Heap sort, Binary tree, Bucket sort, Counting sort and Bitonic sort. During the research, it is necessary to consider, that Xilinx's HLS tool has limited capabilities to implement a recursive algorithms. Therefore, non-recursive modifications for the recursive algorithms were developed during the research.

Comb sorting and gnome sorting belong to the class of exchange sorting algorithms, they, as all representatives of this class, are simple to implement, but considered the slowest when implemented on universal processors. Those have high computational complexity $O(n^2)$, do not require additional memory, with memory costs equals to (O(1)) [17, 18].

Merge sorting is an algorithm whose sorting principle differs significantly from exchange sorting algorithms, but it is also simple for implementation. This sorting algorithm, comparing to the exchange sorting algorithms, is faster when executed on universal processors, since it has less computational complexity O(n*log(n)), but have the significantly higher memory costs O(n) [19].

Heap sorting algorithm has a very good combination of performance and memory costs: O(n*log(n)) and O(1) [20].

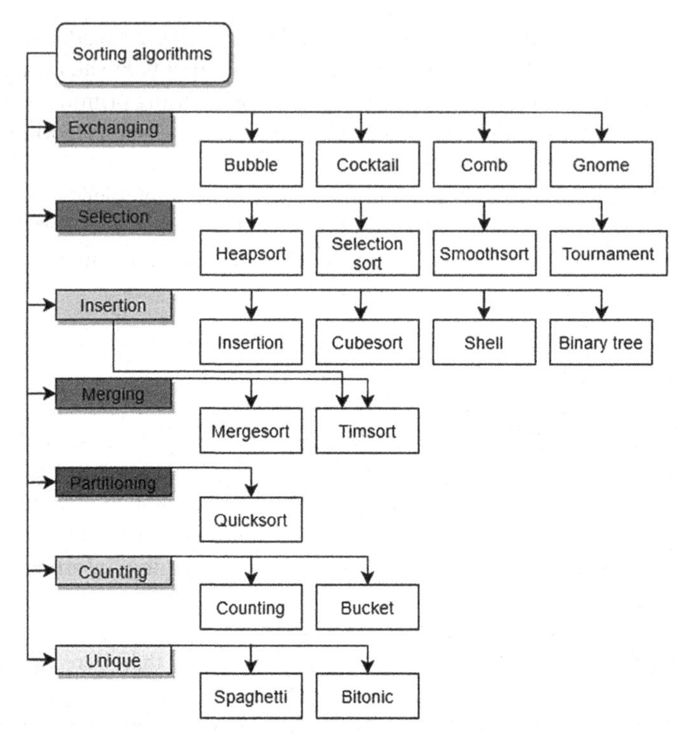

Fig. 1. Simplified classification of sorting algorithms.

The binary tree sorting algorithm is an insertion sorting algorithm and it has the ability to process stream data [21]. The standard implementation of this algorithm is recursive; therefore, to synthesize a hardware implementation by using Xilinx's HLS tool, its non-recursive modification was developed. The computational complexity of the algorithm is estimated as O(n*log (n)), and the additional memory cost is O(n) [21].

The bucket and counting sorting algorithms have a similar idea of preparing a single array, which counts the number of specific numbers [Bucket sort] [Counting sort]. This leads to significant usage restrictions, namely: sorting only integer arrays or data that can be represented as integers, and additional memory overhead, which completely depend on a range of numbers in the original array. The computational complexity of the Bucket algorithm is estimated as O (n2), and of the Counting algorithms is estimated as O(n) [22, 23].

The Bitonic sorting algorithm has very good internal parallelism [20]. It makes the algorithm a very promising. The commonly known implementations of the algorithm are recursive; therefore, to synthesize a hardware implementation by using Xilinx's HLS tool, its non-recursive modification was developed. The computational complexity of the Bitonic algorithm is estimated as O (log2 (n)), and the memory cost is O (n * log2 (n)) [24].

3 The Research Methods

During the research, we used two research methods:

- Simulation modeling. It means a simulation of each selected algorithm on the different computing architectures: on a universal processor and on synthesized FPGA based hardware, i.e. on Reconfigurable Hardware Accelerator.
- Comparative analysis. It means comparing efficiency of software implementation and a synthesized FPGA based hardware implementation of the same algorithm.

Since the architectures of a universal processor and a reconfigurable FPGA are different, the term efficiency, used in our research, was defined as a function of:

- Performance
  - Software implementation performance, estimated as a time, which is necessary to sort a randomly generated array by using a universal processor;
  - Hardware implementation performance, estimated as a time, which is necessary to sort a randomly generated array by using a synthesized hardware implementation;
- Hardware "cost", estimated in utilized FPGA resources for a particular implementation.

The hardware "cost" was used for comparing between different synthesized hardware implementations.

To analyze the hardware implementation efficiency created by using HLS tools, it is necessary for the same C or C++ description of an algorithm, using the same self-checking testbench:

- To simulate and analyze efficiency of a software implementation, based on universal processors;
- To simulate and analyze efficiency of a hardware implementation, based on reconfigurable hardware solutions generated by HLS tool;
- To do a comparative analysis of the efficiency.

During the simulation modeling procedure, we used the following set of software and hardware staff:

- For software implementation:
  - IDE – JetBrains CLion.
  - Multi core processor Intel core i7-4710HQ, 2.50 GHz c 12G Bytes DDR3 RAM.
- For the HLS synthesis flow and hardware implementation:
  - IDE – Vivado HLS [9].
  - FPGA – XCVU125-flvc2104-3.

The JetBrain CLion development environment was chosen because it provides a cross-platform development environment for C and C++ development, as well as the ability to work with all the popular compilers for the pointed languages: GCC, MinGW, Clang, Cygwin. For the current research, these features are sufficient, and mean the opportunity for a wide range of readers to repeat our results.

Vivado HLS synthesis environment was selected because it allows:

- To synthesize, optimize and implement in hardware the sorting algorithm, described in C or C++ language;
- To provide an estimation of the performance for a particular hardware implementation;
- To evaluate a hardware "cost" for a particular hardware implementation.

To determine the performance of a synthesized hardware solution, Vivado HLS provides the minimum possible clock period and Initiation Interval (II). II is the number of clock cycles before it is possible to accept a new array for processing by hardware. Using this data, we estimated the hardware solution performance by multiplying the obtained clock period on II.

3.1 The Research Procedure

The research procedure included the following steps:

1. Creating a project that is used for both software and hardware solutions for all chosen algorithms and all array size.
2. Creating of source files of the sorting algorithms in C language and their modification if necessary (switching to non-recursive implementations, replacing dynamic memory allocation with static memory etc.).
3. Creating a single self-checking testbench, that is used to verify the functionality of both a software implementation and the synthesized hardware implementation.
4. Modeling the sorting algorithm on a universal processor:

    a. Testing a software solution for the correctness.
    b. Estimating the performance of the software implementation of the sorting algorithm for the array of a given size.

5. Synthesizing, optimizing and implementing in hardware the sorting algorithm, described in C language, by using HLS tool:

    a. Testing the source code of the algorithm by using the HLS tool based on the self-checking testbench for a given array size.
    b. The iterative process of optimization the hardware implementation efficiency in order to improve the performance and reduce the hardware "cost". There is a limitation for the hardware "cost", which is imposed from above by available internal resources of the target FPGA.

        (1) Synthesizing of a hardware implementation for the sorting algorithm

(2) Evaluating the performance of the synthesized hardware solution
(3) Estimating the hardware "costs" of the synthesized hardware solution.
(4) Comparative analysis of the performance and the hardware "cost" between synthesized hardware implementations.

c. Final modeling simulation of the optimized hardware implementation by using HLS tool based on the self-checking testbench for a given array size.

6. Comparative analysis of software and hardware implementations for the sorting algorithm.

For this research, we used the following sizes of the input arrays: 1.000, 10.000, 100.000, and 1.000.000. All numbers have an integer positive type, i.e. unsigned integer 32 bits. For the counting sorting algorithms all numbers had an upper limit of their values, which was 100.000. It should be noted that in the case with a lower limit, a decrease in hardware "cost" and an increase in the performance are expected.

During the iterative process of optimizing a synthesized hardware solution, the following optimization directives were applied in attempts to improve the device performance and reduce the hardware "cost":

- Optimizing source code (e.g. change divide by 2 operation by shift operation or make inner loops "perfect" (there are no actions in outer loop)). It allows HLS to use different resources to perform necessary actions and can increase performance and decrease hardware "cost".
- Directives optimizing an interface for the input unsorted array and the resulting array of sorted data. It can increase read and write values speed (e.g. using Dual-Port RAM), also it can allow us to use streaming data reading (e.g. using FIFO).
- Directives for splitting input, output and internal arrays into multiple memory blocks to increase parallel read and write capabilities.
- Directives for unroll of the execution of cycles. The unroll procedure leads to the parallel execution of several iterations of a given cycle, but it requires an increase of hardware "cost". This directives and directives for splitting arrays allow us to perform multiple actions at once with high degree of parallelism.
- Directives for pipelining cycles in the source code. The pipelining reduces the number of cycles spent on each iteration of the loop by adding possibility to start next iteration before the previous one ends.
- Directives for data flow control. The directives allow to have hardware implemented intermediate data buffers between loops and functions, which allows better pipelining of the entire device.

4 The Research Results and Analysis

The research results are presented in Table 1, on Fig. 2 and Fig. 3.

Table 1 highlights the hardware "cost" for all investigated sorting algorithms. All the results are for the optimized hardware implementations. Table 1 contains the following columns:

Table 1. Hardware costs estimation after optimization

Algorithm	Array size	LCELL	BRAM	DSP	Interface
Gnome	1.000	290	0	0	ap_memory
	10.000	290	0	0	
	100.000	290	0	0	
	1.000.000	290	0	0	
Comb	1.000	688	0	0	ap_memory
	10.000	741	0	0	
	100.000	798	0	0	
	1.000.000	813	0	0	
Merge	1.000	2.590	18	0	ap_memory
	10.000	2.510	231	0	
	100.000	2.600	2.830	0	
	1.000.000	2.650	33.600	0	
Heap	1.000	870	0	0	ap_memory
	10.000	870	0	0	
	100.000	870	0	0	
	1.000.000	870	0	0	
Binary tree	1.000	1.040	5	0	ap_fifo
	10.000	1.060	10	0	
	100.000	1.040	96	0	
	1.000.000	1.040	902	0	
Bucket	1.000	688	179	0	ap_memory
	10.000	688	179	0	
	100.000	688	179	0	
	1.000.000	688	179	0	
Counting	1.000	454	179	0	ap_memory
	10.000	462	179	0	
	100.000	470	179	0	
	1.000.000	490	179	0	
Bitonic	1.000	366.000	34	44	ap_memory
	10.000	365.000	34	44	
	100.000	364.000	46	44	
	1.000.000	364.000	63	44	

- Array size – a number of elements in the input and output arrays;
- LCELL – the number of FPGA logical cells, which is necessary to implement the particular sorting algorithm after hardware optimization procedure;
- BRAM – the number of FPGA integrated memory units, which is necessary to implement the particular sorting algorithm;
- DSP – the number of FPGA internal digital signal processing units, which is necessary to implement the particular sorting algorithm;
- Interface – type of an interface used for the input and the output arrays.

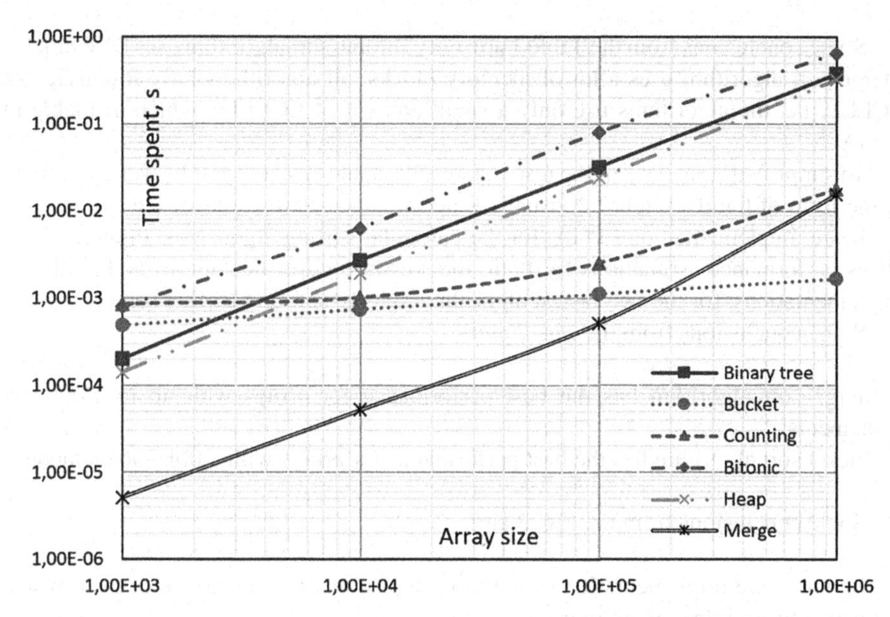

Fig. 2. Performance of the synthesized hardware

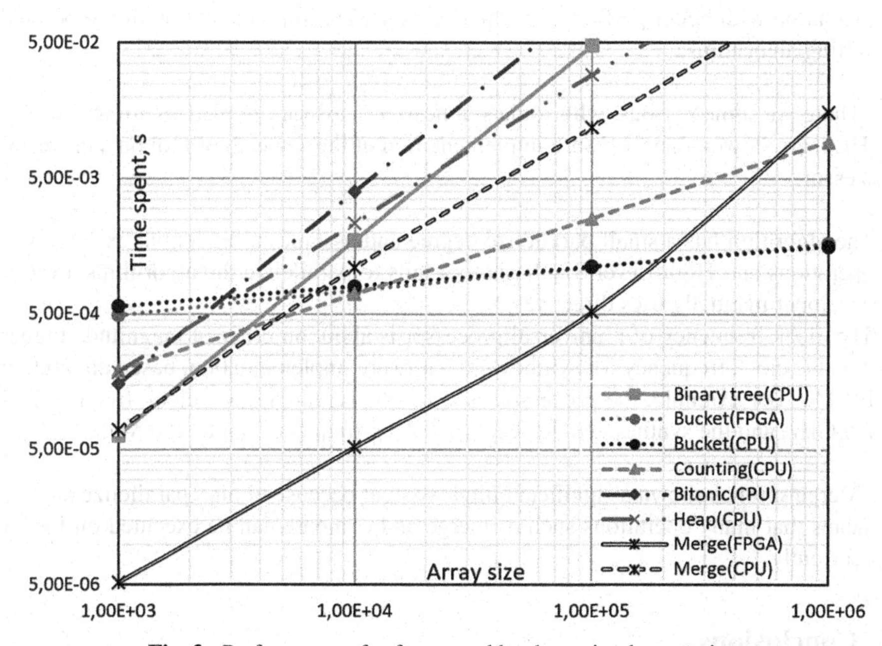

Fig. 3. Performance of software and hardware implementations

Some conclusions from the Table 1 are: only Bitonic sort algorithm uses DSP blocks; Merge sort algorithm uses a lot of memory blocks; all algorithms have low usage of LCELL, i.e. the algorithms use only a small amount of LCELLs, which available in FPGAs.

To simplify the analysis of the research results, some research results were presented in the form of figures. Figure 2 visualizes performance estimations of the synthesized hardware, implemented on FPGA, for the particular sorting algorithms. Figure 3 visualizes performance estimations for both: the software implementations, and hardware implementations, for the two fastest algorithms, Merge sort and Bucket sort.

Some conclusions from the Fig. 2 are:

- Merge sort algorithm has the best performance for arrays with up to 1.00E+05 elements;
- Bucket sort algorithm has the best performance for arrays with 1.00E+06 elements.

Some conclusions from the Fig. 3 are:

- The hardware implementation of the most algorithms is comparable or even worse than software implementation;
- The only Merge sort and Bucket sort algorithms have the synthesized hardware implementation with better performance than any software implementation of the pointed sorting algorithms;

There are some reasons, why some synthesized hardware implementations, created by HLS, are slower, than software implementation of the same algorithm on a universal processor:

- The algorithm has a small parallelize degree and assumes a sequential execution of operations and High-Level Synthesis tool fails to parallelize the algorithms, even if developer manually uses directives.
- The clock frequency of a universal processor is about an order of magnitude higher than the clock frequency for synthesized hardware implementation, based on modern FPGAs. For example, the processor used for the research has a clock frequency of 2.5 GHz, and the synthesized device, has a clock frequency about 250 MHz.

Merge sort has shown a significant improvement because of high parallelize degree. It means that many operations such as merge and compare can be executed on FPGA concurrently.

5 Conclusions

A hardware implementation of a sorting algorithm, synthesized by Xilinx HLS tool, does not always provide higher performance comparing with an implementation of the same algorithm on a universal processor. A positive performance gap depends on internal properties of an algorithm and efforts of optimization conducted.

In accordance with the research the Merge sort is the algorithm that can be recommended to speed up the process of sorting arrays for small (up to 10.000 elements) and medium (up to 100.000 elements) arrays. Bucket sort can be recommended to use for sorting arrays with 1.000.000 and more elements.

A further research should be connected with: further optimization of the investigated algorithms for reducing the hardware "cost" and improving the performance; finding some sorting algorithms (at least one) with better hardware performance, in comparison with software implementations, for the large arrays (with 1.00E+06 elements and more).

Acknowledgements. The reported study was funded by RFBR, project number 18-29-03250.

References

1. Antonov, A., Zaborovskij, V., Kalyaev, I.: The architecture of a reconfigurable heterogeneous distributed supercomputer system for solving the problems of intelligent data processing in the era of digital transformation of the economy. Cybersecur. Issues **33**(5), 2–11 (2019). https://doi.org/10.21681/2311-3456-2019-5-02-11
2. Mantovani, F., Calore, E.: Performance and power analysis of HPC workloads on heterogeneous multi-node clusters. Low Power Electron. Appl. **8**(2), 13–27 (2018). https://doi.org/10.3390/jlpea8020013
3. Usman Ashraf, M., Alburaei Eassa, F., Ahmad Albeshri, A., Algarni, A.: Performance and power efficient massive parallel computational model for HPC heterogeneous exascale systems. IEEE Access **6**, 23095–23107 (2018). https://doi.org/10.1109/ACCESS.2018.2823299
4. Kobayashi, R., Oobata, Y., Fujita, N., Yamaguchi, Y., Boku, T.: OpenCL-ready high speed FPGA network for reconfigurable high performance computing, In: Proceedings of International Conference on High Performance Computing in Asia-Pacific, HPC Asia, pp. 192–201 (2018). https://doi.org/10.1145/3149457.3149479
5. Antonov, A., Zaborovskij, V., Kisilev, I.: Specialized reconfigurable computers in network-centric supercomputer systems. High Availab. Syst. **14**(3), 57–62 (2018). https://doi.org/10.18127/j20729472-201803-09
6. Dongarra, J., Gottlieb, S., Kramer, W.: Race to exascale. Comput. Sci. Eng. **21**(1), 4–5 (2019). https://doi.org/10.1109/MCSE.2018.2882574
7. Haidar, A., Jagode, H., Vaccaro, P., YarKhan, A., Tomov, S., Dongarra, J.: Investigating power capping toward energy-efficient scientific applications. Concurr. Comput. Pract. Exp. 1–14 (2018). https://doi.org/10.1002/cpe.4485
8. Le Fèvre, V., Herault, T., Robert, Y., Bouteiller, A., Hori, A., Bosilca, J.G., Dongarra, J.: Comparing the performance of rigid, moldable and grid-shaped applications on failure-prone HPC platforms. Parallel Comput. **85**, 1–12 (2019). https://doi.org/10.1016/j.parco.2019.02.002
9. IDE Vivado HLS. https://www.xilinx.com/video/hardware/vivado-hls-tool-overview.html. Accessed 19 April 2020
10. Intel HLS compiler. https://www.intel.com/content/www/us/en/software/programmable/quartus-prime/hls-compiler.html?wapkw=HLS. Accessed 19 April 2020
11. Catapult HLS. https://www.mentor.com/hls-lp/catapult-high-level-synthesis/. Accessed 19 April 2020
12. Angermeier, J., Sibirko, E., Wanka, R., Teich, J.: Bitonic sorting on dynamically reconfigurable architectures. In: 2011 IEEE International Symposium on Parallel and Distributed Processing Workshops and Phd Forum, Shanghai, pp. 314–317, (2011). https://doi.org/10.1109/ipdps.2011.164

13. Mihhailov, D., Sklyarov, V., Skliarova I., Sudnitson, A.: Parallel FPGA-based implementation of recursive sorting algorithms, In: 2010 International Conference on Reconfigurable Computing and FPGAs, Quintana Roo, pp. 121–126, (2010). https://doi.org/10.1109/reconfig.2010.30
14. Janarbek M.I., et al.: Resolve: generation of high-performance sorting architectures from high-level synthesis. In: Proceedings of the 2016 ACM/SIGDA International Symposium on Field-Programmable Gate Arrays (FPGA '16), pp. 195–204, New York, NY, USA (2016). https://doi.org/10.1145/2847263.2847268
15. Sorting Methods. https://www.mathworks.com/matlabcentral/fileexchange/45125-sorting-methods?focused=3805900&tab=function. Accessed 19 April 2020
16. Vitis_Libraries. https://github.com/Xilinx/Vitis_Libraries. Accessed 19 April 2020
17. Comb sort. https://www.geeksforgeeks.org/comb-sort/. Accessed 19 April 2020
18. Gnome sort. http://rosettacode.org/wiki/Sorting_algorithms/Gnome_sort. Accessed 19 April 2020
19. Merge sort. http://rosettacode.org/wiki/Sorting_algorithms/Merge_sort. Accessed 19 April 2020
20. Heap sort. http://rosettacode.org/wiki/Sorting_algorithms/Heapsort. Accessed 19 April 2020
21. Binary tree. https://www.geeksforgeeks.org/counting-sort/. Accessed 19 April 2020
22. Bucket sort, Web: https://en.wikipedia.org/wiki/Bucket_sort. Accessed 19 April 2020
23. Counting sort. https://www.geeksforgeeks.org/counting-sort/. Accessed 19 April 2020
24. Bictonic sorter. https://en.wikipedia.org/wiki/Bitonic_sorter. Accessed 19 April 2020

Set Classification in Set@l Language for Architecture-Independent Programming of High-Performance Computer Systems

Ilya Levin[1], Alexey Dordopulo[2], Ivan Pisarenko[2](✉), and Andrey Melnikov[3]

[1] Academy for Engineering and Technology, Institute of Computer Technologies and Information Security, Southern Federal University, Taganrog, Russia
iilevin@sfedu.ru

[2] Supercomputers and Neurocomputers Research Center, Taganrog, Russia
{dordopulo,pisarenko}@superevm.ru

[3] "InformInvestGroup" CJSC, Moscow, Russia
ak@iigroup.ru

Abstract. Traditional programming languages for parallel computer systems do not separate the description of an algorithm from the details of its hardware implementation efficiently. As a result, the porting of the same algorithm between different computational architectures and configurations requires a significant code modification. To reduce the time and complexity of porting, we proposed an architecture-independent Set@l programming language based on the aspect-oriented programming paradigm and set-theoretical code view. In contrast to conventional tools for parallel programming, Set@l operates by sets, subsets, attributes and relations between them. Various aspects of realization transform an architecture-independent source code according to the certain architecture and configuration of a computer system. Set@l provides the porting of parallel applications without the changing of a source code, the modification of an information graph or a computing structure with regard to computer system's features or user's preferences, and the indefinite description of calculations. This paper treats the essential issue of set classification in the Set@l programming language. We propose three basic criteria of set typing: by the parallelism of collection's elements, by the definiteness, and by the processing method. We yield examples of programs in Set@l that demonstrate the techniques and specificities of various set types' usage.

Keywords: Architecture-independent parallel programming · Set@l programming language · Set classification · Types by parallelism · By definiteness and by processing method

1 Introduction

The design of heterogeneous, hybrid and reconfigurable computer systems, which contain processors as well as other types of computing devices [1], is an urgent research

V. Voevodin and S. Sobolev (Eds.): RuSCDays 2020, CCIS 1331, pp. 461–472, 2020.
https://doi.org/10.1007/978-3-030-64616-5_40

direction in the field of modern supercomputer engineering. The variety of architectures used in hybrid computer systems and lack of efficient methods and tools for architecture-independent parallel programming considerably complicate the process of software porting. In the conventional programming languages, the algorithm for the solution of a computational problem and features of its implementation are described by indivisible code fragments. As a result, the change of parallelizing method caused by the porting of a program between computer systems with different architectures or, sometimes, configurations requires the development of a new code. Traditional approaches to the solution of the inter-architecture porting problem have significant disadvantages: they are based on the highly specialized translation algorithms (e.g. the Pyfagor language of functional programming [2], NORMA [3], Paralaxis-III [4]) or fix the procedural parallelization model (e.g. the OpenCL (Open Computing Language) standard [5], GPH (Glasgow Parallel Haskell) [6]).

The high-level programming language COLAMO (Common-Oriented Language for Architecture of Multi-Objects) [7] solves the majority of problems related to the programming of FPGA-based reconfigurable computer systems. In COLAMO, the parallelization of an algorithm is described implicitly by the declaration of access types for arrays and indexing of their elements. However, COLAMO is aimed at the structural and procedural organization of calculations, and the porting of a parallel application in COLAMO to computer systems with different architectures seems to be very problematic.

To deal with the aforementioned issue, we propose an architecture-independent Set@l programming language that develops the essential ideas of COLAMO and the set-theory-based programming language SETL (SET Language) [8]. Set@l is based on a paradigm of aspect-oriented programming (AOP) [9] and describes an algorithm and its implementation features as separate program modules. A program in the Set@l language represents the information structure of a computational problem as sets, subsets and relations between them. The decomposition and typing of collections define different variants of parallelization and other aspects of the algorithm implementation. In contrast to other set-theory-based programming languages (e.g. SETL), Set@l classifies sets by various criteria and operates with indefinite collections according to the alternative set theory (AST) of P. Vopenka [10].

This paper concentrates on the features and techniques of set classification by parallelism, definiteness and processing method in Set@l. Various types of collections are applied for the efficient architecture- and resource-independent description of algorithms.

2 Set Classification by Parallelism in Set@l

For the implicit declaration of the algorithm parallelization, the classification of collections according to the parallelism of their elements during processing is introduced into the Set@l programming language. The basic parallelism types of collections and format of their description in Set@l are given in Table 1.

If the method for the parallelizing of collection's elements is clearly defined, the following types by parallelism are used: "tuple" (`seq` – sequential processing), "pipeline

Table 1. Key types of collections classified by parallelism and formats of their description in Set@l

Type of collection	Processing type	Symbolic notation	Format of description
Tuple	Sequential	[1, 2, ..., p]	`seq (1…p)`
Pipeline tuple	Pipeline	$\langle 1, 2, \ldots, p \rangle$	`pipe(1…p)`
Set	Parallel-independent	{1, 2, ..., p}	`par(1…p)`
Set of processing by iterations	Parallel-dependent	$\overrightarrow{\{} 1, 2, \ldots, p \overrightarrow{\}}$	`conc(1…p)`
Implicit collection	Type is defined in other aspect	[[1, 2, ..., p]]	`imp(1…p)`

tuple" (**`pipe`** – pipeline processing), "set" (**`par`** – parallel-independent processing) and "set of processing by iterations" (**`conc`** – parallel-dependent processing).

However, in some aspects it is impossible to specify the exact type of collections by parallelism because any information about the architecture of computer system is not available. In this case, a special type **`imp`** (implicit or undefined) is applied. The typing of **`imp`** collections is defined in the architectural aspect by means of the following syntax structure:

```
type(<name of collection>)='<type of collection>';
```

Each collection of the Set@l program has the only attribute of parallelism, but one can change it during the passing from one abstraction level to another.

The proposed classification of collections by parallelism provides the architectural independence of the source code in Set@l and allows to describe various parallelizing methods for an algorithm as an entire program. To switch between implementations, user activates the corresponding processing method, architectural and configuration aspects, while the source code of an algorithm remains unchanged.

Lower-upper (LU) decomposition is a well-known computational algorithm for the solution of linear equation systems, which represents a $n \times n$ matrix as the product of a lower and an upper triangular matrices. The source code of the LU-decomposition program in the Set@l language is shown in Fig. 1. It declares the problem's information graph `G` and describes its operational vertices in terms of sets, attributes and relations between them.

The previously represented Set@l program of the Gaussian elimination [11] used the partition and typing of the sets of row (`I`), column (`J`) and iteration (`K`) numbers for the specification of the algorithm parallelism. By contrast, the aforementioned code operates with special collection `G` that includes operational vertices of the information graph. This approach allows to consider the membership relations between sets and their elements in the traditional mathematical sense.

Within the source code, the parallelism type and partition of set `G` are unknown. Therefore, the special parallelism type `imp` (implicit) is applied.

```
int(n); set(a);          // size of matrix and matrix;
set(I,J,K);              // sets of row, column and iteration numbers;
I=1...n;   J=1...n;      K=1...n-1;
{set,graph,imp}(G);      // information graph;
attribute LU_it(k,s1,s2):                        // attribute of LU iteration;
  operand(element(k),set(s1),set(s2)); // operand types;
  (forall i in s1 and j in s2 | i<=k or (i>k and j<k)):
    a(k+1,i,j)=a(k,i,j);                        // transit elements;
  end(forall);
  (forall i in s1 | i>k):                       // recalculated elements;
    a(k+1,i,k)=a(k,i,k)/a(k,k,k);
    (forall j in s2 | j>k):
      a(k+1,i,j)=a(k,i,j)-a(k+1,i,k)*a(k,k,j);
    end(forall);
  end(forall);
end(LU_it);
(forall k in K):                                // graph description;
  G(k)=LU_it(k,I,J);                            // attribute assignment;
end(forall);
```

Fig. 1. The source code of the LU-decomposition program in Set@l

The aspect of processing method for the LU-decomposition program in the Set@l language is given in Fig. 2. This component of the program determines the partitions of the initial information graph G into subgraphs described by subsets with unknown type of parallelism.

```
(forall k in K):                          // decomposition by rows;
  R(k,i)=(G(k,i,j)|j in J);               // i-th row at k-th iteration;
  RB(k,p)=imp(R(k,i)|i in ((p-1)*s+1...p*s)); //p-th row subset;
  GR(k)=imp(RB(k,p)|p in (1...N)); // row-built subgraph;
end(forall);
(forall k in K):                          // decomposition by columns;
  C(k,j)=(G(k,i,j)|i in I);               // j-th column at k-th iteration;
  CB(k,q)=imp(C(k,j)|j in ((q-1)*c+1...q*c));      // column subset;
  GC(k)=imp(CB(k,q)|q in (1...M)); // column-built subgraph;
end(forall);
// Decomposition by iterations:
It(k)=(G(k,i,j)|i in I and j in J);            // k-th iteration;
IB(l)=imp(It(k)|k in ((l-1)*ni+1...l*ni));    // iteration subset;
G=imp(IB(l)|l in (1...T));                     // iteration-built graph;
```

Fig. 2. The aspect of processing method for the LU-decomposition program in Set@l

The aspect of processing method defines the partition of collection G into the row (`RB(k,p)`), column (`CB(k,q)`) and iteration (`IB(l)`) blocks with implicit type of parallelism. Each block consists of rows `R(k,i)`, columns `C(k,j)` or iterations `It(k)`. These partitions allows to describe the processing methods by rows, columns, cells and iterations [8] applied for the parallelizing of the linear algebra algorithms. The values of decomposition parameters `s`, `N`, `c`, `M`, `ni` and `T` depend on the architecture of a computer system and are declared in the aspect of configuration.

The aspect of architecture for the LU-decomposition program in Set@l specifies the parallelism types of the basic set G and its subsets `GR(k)`, `GC(k)`, `RB(k,p)`, `CB(k,q)` and `IB(l)`. The code of the architectural aspect for the reconfigurable architecture of computer system is given in left column of Fig. 3. According to the code, the algorithm of LU-decomposition is parallelized by iterations in the case of implementation on a reconfigurable computer system. In the architectural aspect, the characteristics of set partitions are calculated using the parameters of configuration (R and R0) and the size of processing matrix (n).

If the algorithm of LU-decomposition is realized on a multiprocessor computer system, the architectural aspect is given as in right column of Fig. 3. In contrast to a reconfigurable computer system, the implementation of the LU-decomposition algorithm on a multiprocessor supercomputer assumes the parallelization by cells. Analogously to the previous version of the architectural aspect, the decomposition of sets depends on the configuration parameters (q1 and q2) and the matrix size (n).

// Rows:	
s=n; N=1; **type**(RB(k,p))='**pipe**'; **type**(GR(k))='**pipe**';	s=q1; N=n/s; **type**(RB(k,p))='**par**'; **type**(GR(k))='**seq**';
// Columns:	
c=n; M=1; **type**(CB(k,q))='**pipe**'; **type**(GC(k))='**pipe**';	c=q2; M=n/c; **type**(CB(k,q))='**par**'; **type**(GC(k))='**seq**';
// Iterations:	
ni=**floor**(R/R0); T=(n-1)/ni; **type**(IB(l))='**conc**'; **type**(G)='**pipe**';	ni=n-1; T=1; **type**(IB(l))='**seq**'; **type**(G)='**seq**';

Fig. 3. The aspect of architecture for the LU-decomposition program in Set@l designed for the reconfigurable (left column) and multiprocessor (right column) architectures of computer system

The aspect of computer system's configuration substitutes the specific values of configuration parameters R (available computing resource) and R0 (resource of basic subgraph) for the reconfigurable architecture or q1 and q2 (numbers of processors by rows and by columns) for the multiprocessor architecture into other modules of the program and completes the forming of graph set G.

Owing to the parallelism typing of collections in the Set@l programming language, it is possible to describe different variants of the algorithm parallelizing in a single aspect-oriented program. To switch the method of implementation during the inter-architecture porting, one should change the architectural aspect of a program, but the source code remains unchanged. In contrast to other languages for parallel programming, Set@l does not mix the description of an algorithm (mathematical aspect) with the details of its implementation on a computer system (practical aspect). This feature allows for more efficient development and porting of parallel programs for hybrid computer systems with FPGAs: the algorithm developer describes the problem solution in general set-theoretical form, and the circuit designer creates architectural and configurational aspects.

3 Set Classification by Definiteness in Set@l

If aspects do not modify an algorithm during its architectural adaptation, the solution of a computational problem can be described within the Cantor-Bolzano set theory [12].

However, the functionality of aspects is not limited to the parallelization of algorithms. In some cases, it is reasonable to modify an algorithm according to the architectural features of the computer system used for calculations. Then some collections are indefinite and are not sets; so, it is impossible to describe them using the concepts of the traditional set theory.

The architecture-independent Set@l programming language for high-performance computer systems describes various implementations of an algorithm in a unified aspect-oriented program. For this purpose, the classification of collections by the definiteness of their elements is introduced into Set@l. In the Set@l programming language, indefinite collections are described by the special mathematical objects (classes and semisets). The concepts of a class and semiset were proposed by P. Vopenka within the AST [10].

The type "set" (**set**, { } in symbolic notation) describes a sharply defined and definite collection of certain objects. For a set, we always exactly know if one or another object belongs to it or not. In Set@l, a set can be specified using the direct enumeration of its elements or by means of the relation calculus. In fact, plenty of naturally organized collections are not sets, because their elements are not clearly defined. The AST analyses the phenomenon of indefinite collections with the help of special mathematical objects (classes and semisets).

The type "semiset" (**sm**, { ?) indicates a collection whose indefiniteness is a fundamental characteristic and can not be eliminated by the aspects of a program. The relation of inclusion usually connects a semiset with some sharply defined set. Therefore, to specify a semiset as an object of the Set@l program, it is necessary to form a suitable superset and declare the appropriate relation of inclusion.

A class (**cls**, ? ¿) is the most common and multipurpose type of collections in the Set@l programming language. If the type and structure of a collection are not sharply defined on the current level of abstraction, it is declared as a class and is used in code analogously to standard sets. Owing to the extension of the class definition in the aspects of a program, it is possible to specialize the type and partition of the collection during translation. The application of classes provides the unification of objects' names in all units of an aspect-oriented program in Set@l. The indefiniteness of collections by parallelism denoted by **imp** attribute (see Table 1) can also be described with the help of classes. To specify a collection as a class in the Set@l program, one has to assign **cls** attribute to this collection and give its possible attributes. The collections of rows, columns and iterations *R*, *C*, and *It* (see Fig. 1) are the examples of classes because in the source code they are considered as objects with unknown types and decompositions.

Using classes, sets and semisets, one can describe various modifications methods for an algorithm in a unified aspect-oriented Set@l program. The application features of AST objects in Set@l programs are considered in the example of the Jacobi iterative method given below.

The basic parallelization techniques for the computer-aided solution of linear equation systems by the Jacobi iterative method are shown in Fig. 4.

In the case shown in Fig. 4-a, each iteration of the Jacobi algorithm contains the following operations: the calculation of the column-vector of unknown variables (block *C*); the verification of the termination condition (block *V*) given by $err(k) \leq \delta$, where *err* is the residual; *k* is the number of iteration; δ is the fixed value of tolerance. If

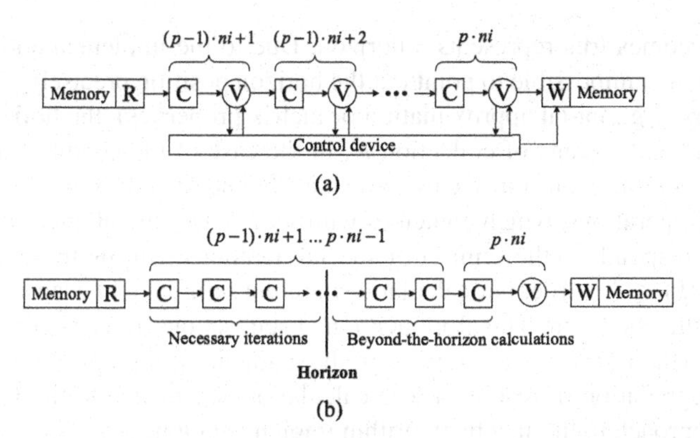

Fig. 4. Parallel implementations of the Jacobi algorithm for the solution of linear equation systems: with verification during each iteration (a) and with one verification after several computational iterations (b)

the condition is true, the control device transfers data via the untapped blocks *C* and *V* and saves the result into a specially allocated area of distributed memory. In practice, the considered approach is efficient, but not for all computational architectures. Each verification block *V* performs the resource-intensive and time-consuming operation of *err(k)* calculation. The hardware resource required for the implementation of *V* block is comparable with the *C* one, and time costs are equivalent too.

For the increase in hardware usage efficiency and reduction of time costs, it is reasonable to modify the Jacobi algorithm in case of the reconfigurable architecture. This modification assumes single verification of the termination condition in a computing structure (see Fig. 4-b) with *ni* iterations. If the condition is fulfilled before the operation of verification (in iterations with numbers from $(p–1)\cdot ni + 1$ to $p\cdot ni–1$), further iterations will not worsen the calculation results. At the same time, the algorithm's modification provides the reduction of time costs: hardware resource freed from *V* blocks can be used for the placement of additional *C* blocks. The quantity of *C* blocks in the computing structure is defined by *ni* parameter. It is worth noting that the both considered variants of the calculations' organization are suitable for multiprocessor computer systems as well as for reconfigurable ones.

In the case of verification during each iteration (see Fig. 4-a), collection *K* of the algorithm iteration numbers is a typical set because its first, intermediate and last elements are sharply defined. The number of the last iteration I_m is unknown in advance, but it is explicitly determined by the termination condition $err(k) \leq \delta$.

If there is one verification in each cadr (see Fig. 4-b), the Cantor-Bolzano set theory provides the description of only one special case K^* corresponding to the fulfillment of the termination condition at the iteration with $T\cdot ni$ number, where *T* is the number of the last iteration block. Otherwise, collections *K* and K^* describe different mathematical objects: set K^* contains not only necessary but also excessive iterations, and it is impossible to specify the exact location of fulfillment point for termination condition $err(k) \leq \delta$. In terms of the AST, the collection of iterations *K* is a semiset, i.e. a class that has the subset relation $K \subseteq K^*$ with set K^*. The iteration at which the termination condition

$err(k) \leq \delta$ becomes true represents a horizon. Due to the implementation features of the algorithm, it is impossible to point out the horizon position precisely. Depending on various factors (e.g., initial approximation or matrix properties), the horizon can move and form different variants of collection K. In the case of the condition fulfillment at iteration $T \cdot ni$, the horizon transforms to a sharp boundary, and semiset K becomes a definite set of operations, which coincides with set K^*. In general, the set difference of K^* and K corresponds to the semiset of special iterations. During these iterations, the condition $err(k) \leq \delta$ is true, but calculations are not terminated because it is impossible to check the condition. The aforementioned semiset describes beyond-the-horizon calculations, which are not necessary from the mathematical point of view, but do not lead to the degradation of results. These calculations occur due to the features of the considered approach to the Jacobi algorithm implementation.

The source code of the Jacobi program in the Set@l language declares the attributes of calculation and verification operations `C` and `V` and utilizes them for the description of full information graph `G`, which is a class within the source code shown in Fig. 5.

```
set(I,J); cls(K); // sets of row and column numbers,class of iteration numbers;
{cls,graph}(G);   // information graph;
set(a,b);         // matrix and vector-column;

attribute C(k):                    // calculation attribute;
   operand(element(k));         // operand type;
   k=1 -> x(k,*)=x_init(*); // initial approximation;
   (forall i in I):
      x(k+1,i)=(b(i)-sum(a(i,j)*x(k,j)|j in J and
                 j!=i))/a(i,i);      // recalculation;
   end(forall);
end(C);

attribute V(k):                 // verification attribute;
   operand(element(k));         // operand type;
   err(k)=max(abs(x(k+1,i)-x(k,i))|i in I);
   err(k)<=delta  ->  x_res(*)=x(k+1,*);
end(V);

(forall k in K):                // graph definition;
   G(k)={conc(C,V)}(k);         // subgraph description;
end(forall);
```

Fig. 5. The source code of the Jacobi program in the Set@l language

Collection `K` of iteration numbers and graph `G` can be sets (Fig. 4-a) as well as semisets (Fig. 4-b). When we develop the source code of the program, the implementation details are still unknown. Therefore, we can not define the types of collections `K` and `G` unambiguously, and they are marked as classes. Within the source code, we consider only one variant of implementation with verification at each iteration because it corresponds to the mathematical sense of the Jacobi algorithm.

The fragment of the processing method aspect that describes two variants of the Jacobi algorithm implementation is given in Fig. 6. At first, we declare the possible types of collection `K` (a clearly defined set or an indefinite semiset). Then, set `K*` is specified, and its partition is suitable for both reconfigurable (`ni` > `1`) and multiprocessor (`ni` = `1`) computer systems. If computer system has the multiprocessor architecture, aspects choose the implementation variant with verification during each iteration (see Fig. 4-a). In this case, `K` is a set, and we describe its elements using reference set `K*`. For the reconfigurable architecture, it is reasonable to verify the termination condition after several computational iterations (see Fig. 4-b). In this case, K is a semiset, and we declare its subset relation with set K and modify the subgraph of iteration G(k) according to its number.

```
typing(K)='set' or 'sm';      // possible types of K;
K*={set,imp}(IT(k)|IT(k)={set,imp}((k-1)*ni+1 ... k*ni)
                   and (k=1 or err((k-1)*ni)>delta) );
type(K)='set';                // verification during each iteration;
K=(k|k in K* and (k=1 or err(k-1)>delta));
type(K)='sm'; // one verification after ni computational iterations;
K=sub(K*);    // subset relation;
G(k)=(P|mod(k,ni)=0 -> P={conc(C,V)}(k) and mod(k,ni)!=0 -> P=C(k));
```

Fig. 6. The fragment of the processing method aspect that describes two variants of the Jacobi algorithm implementation

Owing to indefinite collections in the Set@l programming language, one can change the implementation method of an algorithm by means of aspect linking without the modification of the architecture-independent source code. For example, in COLAMO the considered realizations of the Jacobi algorithm are described by separated programs with different cadr constructions.

4 Set Classification by Processing Method in Set@l

To describe the modification of an information graph according to the architecture or configuration of a computer system, the classification of sets by processing method is introduced into Set@l. Consider the implementation of information graphs with associative operations (e.g. the summation of integer array elements). Traditional topologies of such graphs [13] are shown in Fig. 7. The "Head/Tail" topology (Fig. 7-a) assumes the sequential execution of associative operation f on elements a_i of input set A. In contrast, the "Half-Splitting" topology (Fig. 7-b) is characterized by the parallel processing of input data. Both structures has the same quantity of operational vertices, but the latency of the "Head/Tail" variant is higher than the latency of the "Half-Splitting" topology.

The program code in the Set@l language that describe the "Head/Tail" and "Half-Splitting" topologies (see Fig. 7) is given in Fig. 8. The information graphs are formed recursively using the basic operation `f(<input 1>,<input 2>,<output>)`. In both cases we define topologies in the direction from output vertex to input vertices. Basic attributes `Head(A)` and `Tail(A)` allocate the first element of ordered set `A` and all elements of `A` except the first one, respectively. Attribute `d2` splits set `A` into subsets `A1` and `A2` with the same cardinality.

In terms of the Set@l programming language, it is reasonable to introduce additional set classification by the processing method in order to specify the "Head/Tail" and "Half-Splitting" principles (see Fig. 7 and 8) in compact set-theoretical form. If attributes `H/T` ("Head/Tail") and `DIV2` ("Half-Splitting") are once declared in a program, user can change the topology of information graph with associative operations by means of the

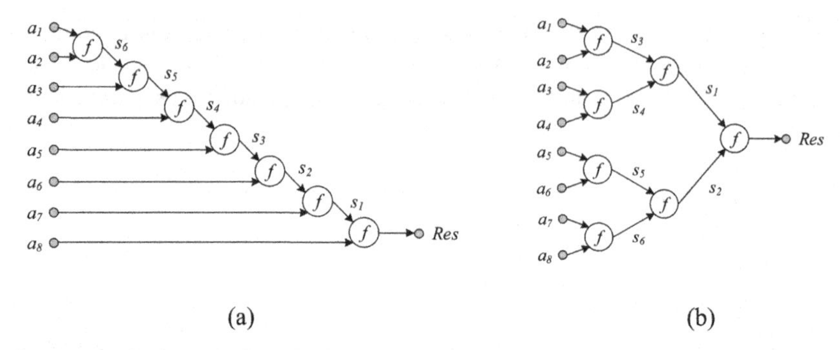

Fig. 7. The standard topologies of information graphs with associative operations: "Head/Tail » (a) and "Half-Splitting" (b)

```
attribute [Lf(A,Res)|Lf=Rec(f),type(A)='H/T']:
    operand(set(A),element(Res));
    element(s);
    Lf(A,Res)=break[card(Tail(A))=1: f(Head(Tail(A)),Head(A),Res)],
                union[Lf(Tail(A),s),f(s,Head(A),Res)];
end(Lf);
attribute [Pf(A,Res)|Pf=Rec(f),type(A)='DIV2')]:
    operand(set(A),element(Res));
    d2(A,A1,A2);
    element(s1,s2);
    Pf(A,Res)=break[card(A1)=1 and card(A2)=1: f(Head(A1),Head(A2),Res)],
                union[Pf(A1,s1),Pf(A2,s2),f(s1,s2,Res)];
end(Pf);
```

Fig. 8. The recursive description of the "Head/Tail" and "Half-Splitting" principles in Set@l

simple type changing, as it is shown in Fig. 9. Moreover, the proposed approach offers new possibilities for the description of combined topologies with parallel and sequential fragments of calculations.

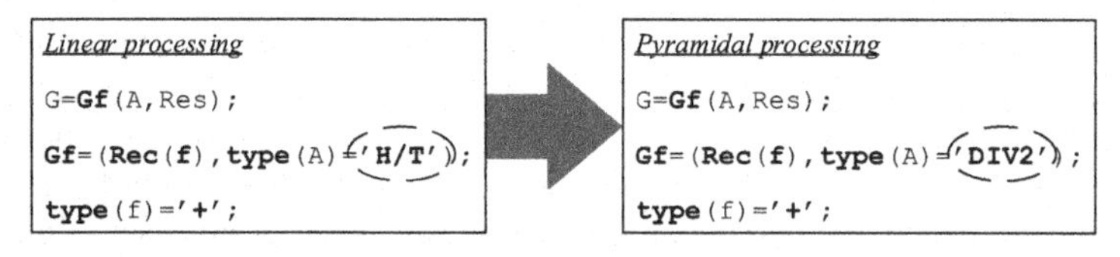

Fig. 9. The modification of processing method requires only the change in type of set A

In fact, the convenient topologies in Fig. 7 are the limiting cases: they are efficient for the maximal (Fig. 7-b) and minimal (Fig. 7-a) amounts of available hardware resource of a computer system. When the hardware resource is higher than the resource of one vertex and is lower than the resource of the whole information graph, we can modify the topology and achieve high specific performance of computing structure after the performance reduction [14]. The modified topology of information graph with associative operations shown in Fig. 10 contains isomorphic subgraphs DIV2 adapted for the free computing resource and one sequential block H/T that is reduced to one operational vertex. Using the processing method types, it is possible to describe the aforementioned topology as follows: $G = [\{a_1 \ldots a_k\}_{\mathrm{DIV2}}, \{a_{k+1} \ldots a_{2k}\}_{\mathrm{DIV2}}, \ldots]_{\mathrm{H/T}}$.

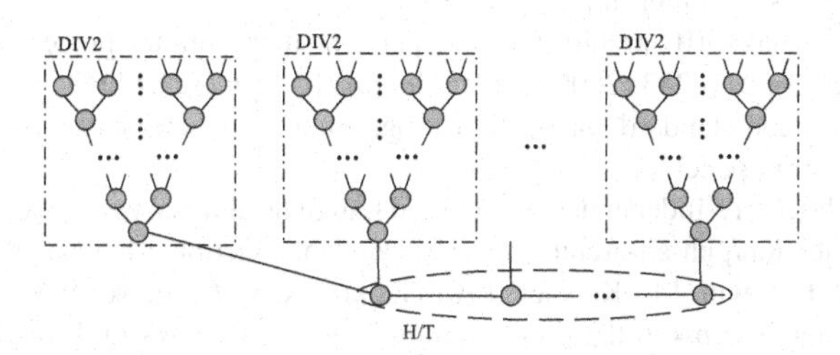

Fig. 10. The modified topology of information graph with associative operations

The classification of sets by processing method allows for the graph modification according to the architecture and configuration of a computer system. This feature of the Set@l programming language is important for the creation of efficient architecture- and resource-independent parallel programs.

5 Conclusions

The Set@l programming language applies the AOP paradigm, AST and relation calculus in order to represent the algorithm for the problem solution as the architecture-independent source code and to separate it from the description of implementation features, which are defined in aspects by means of the partition and classification of sets. Using the aspects of processing method, architecture and configuration, one can provide the efficient porting of parallel applications between computer systems with different architectures and adapt programs to various modifications of hardware resources.

Application of the Set@l language gives fundamentally new possibilities for efficient porting of software to various architectures of computer systems, including heterogeneous and reconfigurable ones, without the change of a source code. This paper summarizes the development of set classification principles in Set@l: the distinctive features of the set types are thoroughly considered, and the fundamental syntax elements and code examples are given. Currently, we are working at the development of the translator prototype for the Set@l programming language.

Acknowledgments. The reported study was funded by the Russian Foundation for Basic Research, project number 20-07-00545.

References

1. Mittal, S., Vetter, J.: A survey of CPU-GPU heterogeneous computing techniques. ACM Comput. Surv. **47**(4), art. no. 69 (2015). https://doi.org/10.1145/2788396
2. Legalov, A.I.: Functional language for creation of architecture-independent parallel programs. Comput. Technol. **10**(1), 71–89 (2005) (in Russian)
3. NORMA: Declarative language for the specification of problems with computational character (in Russian). https://parallel.ru/tech/norma
4. Braunl, T.: Parallaxis-III: architecture-independent data parallel processing. IEEE Trans. Softw. Eng. **26**(3), 227–243 (2000). https://doi.org/10.1109/32.842949
5. OpenCL: The open standard for parallel programming of heterogeneous systems. https://www.khronos.org/opencl/
6. GPH: An Architecture-Independent Functional Language. https://www.microsoft.com/en-us/research/publication/gph-an-architecture-independent-functional-language/
7. Dordopulo, A.I., Levin, I.I., Kalyaev, I.A., Gudkov, V.A., Gulenok, A.A.: Programming of hybrid computer systems in the programming language COLAMO. Izvestiya SFedU. Eng. Sci. **11**, 39–54 (2016) (in Russian). https://doi.org/10.18522/2311-3103-2016-11-3954
8. Dewar, R.: SETL and the evolution of programming. In: Davis M., Schonberg E. (eds.) From Linear Operators to Computational Biology: Essays in Memory of Jacob T. Schwartz. Springer-Verlag, London (2013). https://doi.org/10.1007/978-1-4471-4282-9_4
9. Dessi, M.: Spring 2.5 Aspect-Oriented Programming. Packt Publishing Ltd., Birmingham (2009)
10. Vopenka, P.: Introduction to Mathematics in Alternative Set Theory. Alfa, Bratislava (1989). (in Czech)
11. Levin, I.I., Dordopulo, A.I., Pisarenko, I.V., Melnikov, A.K.: Approach to architecture-independent programming of computer systems in aspect-oriented Set@l language. Izvestiya SFedU. Eng. Sci. **3**, 46–58 (2018) (in Russian). https://doi.org/10.23683/2311-3103-2018-3-46-58
12. Haggarty, R.: Descrete Mathematics for Computing. Pearson Education, Harlow (2002)
13. Karepova, E.D.: The Fundamentals of Multithread and Parallel Programming. Siberian Federal University Publishing, Krasnoyarsk (2016). (in Russian)
14. Levin, I.I., Dordopulo, A.I.: On the problem of automatic development of parallel applications for reconfigurable computer systems. Comput. Technol. **25**(1), 66–81 (2020) (in Russian). https://doi.org/10.25743/ict.2020.25.1.005

Shared Memory Based MPI Broadcast Algorithms for NUMA Systems

Mikhail Kurnosov(✉) and Elizaveta Tokmasheva

Siberian State University of Telecommunications and Information Sciences, Novosibirsk, Russia
{mkurnosov,eliz_tokmasheva}@sibguti.ru

Abstract. `MPI_Bcast` collective communication operation is used by many scientific applications and tend to limit overall parallel application scalability. This paper investigates the design and optimization of broadcast operation for NUMA nodes with GNU/Linux. We describe algorithms for `MPI_Bcast` that take advantage of NUMA-specific placement of queues in a shared memory for message transferring. On a Xeon Nehalem and Xeon Broadwell servers, our implementation achieves on average 20–60% speedup over algorithms of Open MPI `coll/sm` and MVAPICH.

Keywords: MPI · Broadcast · Collectives · NUMA

1 Introduction

High-performance computing systems are growing intensively in two directions: compute node counts and number of cores per node. Many of the supercomputers are built on multi-processor nodes with non-uniform memory architecture (NUMA), it becomes increasingly important for MPI to leverage shared memory for intra-node communication.

Broadcast is an important communication operation in HPC. For a significant number of parallel algorithms and packages of supercomputer simulation, the performance (execution time) of broadcast operation is critical. The MPI standard defines an `MPI_Bcast` routine for single source non-personalized broadcast operation, in which data available at a root process is sent to all other processes. On shared memory systems broadcast can reduce the number of memory transfers with multiple consumers accessing a shared buffer. The most used double-copy (copy-in/copy-out) algorithms involve a shared buffer space used by local processes to exchange messages. The root process copies the content of the message into the shared buffer before the receiver reads from it.

In this paper, we investigate the problem of message broadcasting from the root process to other processes over shared memory of a NUMA machine with GNU/Linux operating system.

This work is supported by Russian Foundation for Basic Research (project 18-07-00624).

V. Voevodin and S. Sobolev (Eds.): RuSCDays 2020, CCIS 1331, pp. 473–485, 2020.
https://doi.org/10.1007/978-3-030-64616-5_41

Main contributions of this paper include: (1) NUMA-aware algorithms for MPI_Bcast operation are based on k-ary, k-nomial, chain and flat notification trees. In contrast to other works our algorithms explicitly allocate memory for queues from local NUMA nodes even with active linux page cache readahead subsystem; (2) Optimal values of the size s of buffer and length s of the queue what takes no more than b bytes and provides minimum algorithm time. On NUMA machines with Xeon Nehalem and Xeon Broadwell processors, our implementation based on Open MPI achieves on average 20–60% speedup over algorithms of Open MPI coll/sm and MVAPICH (mv2_shm_bcast).

The paper is organized as follows. Section 2 discusses related work. Section 3 presents an overview of our approach and describes the shared-memory MPI_Bcast for NUMA system implemented within the Open MPI. Analyses and experimental results are presented in Sect. 4. Section 5 summarizes and concludes.

2 Related Work

Modern MPI implementations optimize intra-node collective communication in two different ways: (1) using intra-node point to point communication and minimizing inter-node interactions [1,2,7–9,11,12,15]; (2) allocating a shared memory region that can be used for the communication across processes in the same node [4–6,10,13,14]. The main part of shared memory based MPI_Bcast algorithms are based on two step procedure [1–6]. At communicator creation time a set of queues is formed in a shared memory region and a message is transferred over queues at each call of MPI_Bcast. The root process copies fragments of the message into the shared queue and the non-root reads from it. This approach is called copy-in/copy-out (CICO, double-copy) and is widely used in practice because it provides portability, and does not require additional libraries and additional permissions from the operating system. Scalability of CICO algorithms are limited by double copying of fragments and waiting for the readiness of the data in the queue.

Along with the CICO method in many MPI implementations a zero-copy approach is used. Zero-copy algorithms perform one copying of each fragment without using of an intermediate buffer. They use special possibilities of operating system to copy of a data from address space of one process into another. Well known examples are KNEM [13], XPMEM and linux Cross Memory Attach. In [6,13,14] a process distance-aware adaptive collective communication framework based on KNEM is proposed. Kernel-assisted collective algorithms do not use intermediate queues in a shared memory segment. This paper addresses problems of CICO algorithms with queues in a shared memory region.

In MVAPICH [3] processes create a shared memory segment with a cyclic queue of $w = 128$ slots for each process. Each slot contains a buffer to store a fragment of $f = 8192$ bytes and an operation number psn. The root process uses flat tree and psn to notify other processes about data readiness. If the queue is full, the root process waits on the barrier until all processes have finished copying. The total size of the shared memory segment is $O(pwf)$, and an each process requires an $O(pw)$ bytes of memory.

In the paper [4] authors proposed to use p cyclic queues in a shared memory. The queue includes w buffers and is divided into $q = 2$ sets (banks) of memory with each set having several buffers. When the last buffer in the set is used, a non-blocking barrier is initiated. Multiple sets are used to allow the non-blocking barrier to complete while another set is in use reducing the synchronization costs. The root process uses a complete k-ary tree for message transferring and notifications. Algorithm is implemented in `coll/sm` component of the Open MPI. The total size of the shared memory segment is $O(pwf)$ and an each process requires $O(w + pk)$ bytes of memory.

In [5] authors use a single queue divided into $w = 4$ buffers and two synchronization flags per process. An each buffer occupies $f = 8192$ bytes of memory. One of the synchronization flags is used when a process copies its data to the shared buffer, to notify that new data is available. The other flag is used when a process copies the data out of the shared buffer, to signal that it has read the contents of the buffer and the buffer can be reused. A broadcast is implemented using a release followed by a gather step. During the release step, the parent copies the message into the shared queue and updates the children's release flag. Child processes wait on the shared release flag and copy out the data from the buffer. After the release step, in the gather step the children processes signal the parent that they have completed copying the data. Authors use k-ary and k-nomial trees for notifications. The size of the shared memory segment is $O(p + wf)$.

Algorithms in MVAPICH, Open MPI and in [5] allocate memory pages for queues without explicit binding to local NUMA nodes. This can lead to allocating of memory pages for queues from a NUMA node of the master-process which created shared-segment. As a consequence, the amount of inter-socket exchanges can increase. Our approach takes advantage of NUMA-specific placement of queues in a shared memory and tries to minimize a volume of inter-socket traffic.

3 Bcast Algorithms

The developed algorithms include two stages. At communicator creation time they form a set of queues in a shared memory region for inter-process communications. After that, on each call of `MPI_Bcast` a message is transferred from the root process over its queue to others processes.

3.1 Shared Memory Segment Structure

At MPI communicator creation time (including `MPI_COMM_WORLD`) all processes form a shared memory segment. The POSIX-compatible system call `mmap` is used for this purpose. Process 0 allocates memory in shared region and other processes attach it to its address space. The size of the allocated segment and individual blocks is a multiple of a memory page. The reason is that NUMA memory binding is controlled by linux kernel at the level of memory pages.

Each of the first q memory pages contains two shared counters *shm_op* and *shm_nreaders* (by default $q = 2$). They are used to synchronize access of processes to shared queues during the `MPI_Bcast` operation. The addresses of the counters are aligned to a cache line boundary to reduce possible false sharing. Further, the shared memory region contains for each of p processes a cyclic queue of s buffers ($shm\_queue[rank][s]$) and an array of s control blocks ($shm\_controls[rank][s]$). Each buffer has f bytes length and occupies minimum number of memory pages. A size of control block is one page length. By default we use $s = 8$ and $f = 8192$ bytes. Figure 1 shows an example of shared memory segment structure for $p = 8$ processes running on two NUMA nodes (two quad core processors). In general, the size of a shared segment depends linearly on the number p of processes and queue length s and occupies $O(qw + ps(f + w))$ bytes of memory, where w is the memory page size. In practice, the queue's length s and buffer size f should be chosen taking into account the available memory size. For example, at $p = 64$ processes and $s = 1024$, $f = 8192$ the memory segment will occupy 384 MB.

After calling `mmap`, each process initializes areas of the segment with its data structures: it zeroes control blocks and the first byte of each page of all queue buffers. This ensures that physical memory pages are allocated from its local NUMA node (using the *first touch policy* of the linux kernel). Memory pages with shared counters *shm_op* and *shm_nreaders* are initialized by the process 0. Overall initialization time linearly depends on the number p of processes, queue length s and the number q of sets.

According to default linux memory policy, the first access to any address *addr* on the segment will allocate a physical memory page from the local NUMA node of the process and a certain number of pages for the following addresses will be allocated from the same NUMA node. This is done by page caching subsystem (linux page cache readahead) which speculatively sequentially reading memory-mapped file (shared region) into the page cache. Default behavior of readahead subsystem may cause incorrect allocation of memory pages for queues and control blocks of processes $1, 2, \ldots, p-1$ from NUMA node of the process 0 (it performs a first modification of the shared region). Algorithms of Open MPI `coll/sm` and MVAPICH ignore NUMA topology – pages for shared data structures are allocated from NUMA node of process 0. This increases `MPI_Bcast` operation time due to the increased number of accesses to remote NUMA nodes. In our algorithms to establish correct allocation of memory pages from NUMA nodes we temporarily disable sequential readahead immediately after `mmap` by calling `madvise(seg, segsize, MADV_RANDOM)`. This ensures correct allocation of memory pages for queues and control blocks from local NUMA nodes.

Control blocks are used by the root process to notify other processes about data readiness in the queue. The root copies the fragment i of a message to his queue $shm\_queue[root][i]$ and notifies processes *rank* by writing fragment size to their control blocks $shm\_controls[rank][i]$. Each non-root process spin waits

Process owner	Data Block	Content	Size	NUMA-node
0	*shm_op*	1	4KB	0
	shm_nreaders	0		
0	*shm_op*	1	4KB	0
	shm_nreaders	0		
0	*shm_controls*[*s*]		8 · 4KB	0
	shm_fragments[*s*]		8 · 8KB	
1	*shm_controls*[*s*]		8 · 4KB	1
	shm_fragments[*s*]		8 · 8KB	
2	*shm_controls*[*s*]		8 · 4KB	0
	shm_fragments[*s*]		8 · 8KB	
...				
7	*shm_controls*[*s*]		8 · 4KB	1
	shm_fragments[*s*]		8 · 8KB	

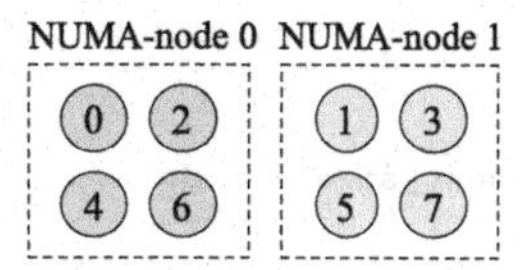

Fig. 1. Shared memory segment structure: $p = 8$ processes on two NUMA nodes; memory page size $w = 4\,\text{KB}$; queue length $s = 8$, number of sets per queue $q = 2$, buffer size $f = 8\,\text{KB}$ (total segment size is 776 KB).

on its own control block until the value becomes positive. We have implemented four algorithms using various trees to propagate notification from the root process to others: completed k-ary tree, k-nomial tree, chain tree and flat (linear) tree (Fig. 2).

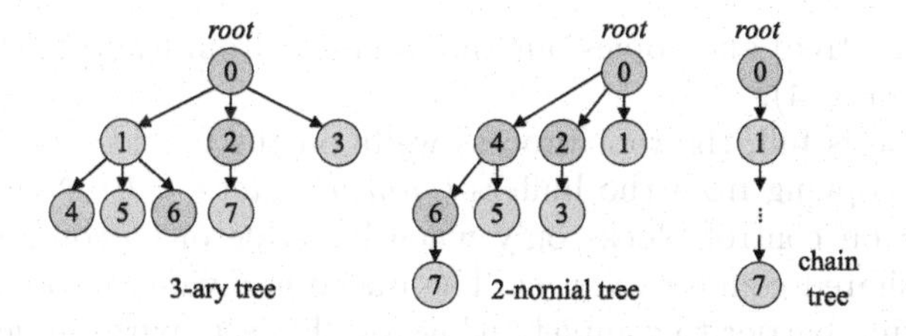

Fig. 2. Notification trees: $p = 8, root = 0$.

At communicator creation time the root process of `MPI_Bcast` operation is unknown. For this reason each process generates a fragment of a tree to all p possible values of root. A process stores information only about his parent and children nodes, it requires $O(p)$ bytes per process.

3.2 MPI_Bcast

The root process implements a pipelined message transferring. It divides the message into $\lceil m/f \rceil$ fragments and copies them through the queue in a shared memory. The root copies the current fragment *index* into the next available buffer in the queue $shm\_queue[root][index]$ and notifies children processes in the tree – updates their control blocks $shm\_controls[rank][index]$ with the current fragment size (Fig. 3). Non-root process *rank* waits on its control block until the value becomes positive, then notifies its children processes (propagates the

```
while sent_size < m do
  set = op % q;  op++
  wait_for(shm_nreaders[set] == 0)
  shm_nreaders[set] = p - 1
  shm_op[set] = op - 1
  i = set * (s / q);
  while i < (set + 1) * (s / q) and
        sent_size < m do
    // Copy to the queue
    frag = get_next_frag(m, sent_size)
    copy(frag, shm_queue[root][i],
         frag_size)
    write_memory_barrier()
    // Notify children
    for each child in children[root] do
      shm_controls[child][i] = frag_size
    end for
    sent_size += frag_size
    i++
  end while
end while
```

Fig. 3. Root process.

```
while sent_size < m do
  set = op % q
  wait_for(shm_op[set] == op)
  op++
  i = set * (s / q);
  while i < (set + 1) * (s / q) and
        sent_size < m do
    // Wait for a data
    wait_for(shm_controls[rank][i] > 0)
    frag_size = shm_controls[rank][i]
    shm_controls[rank][i] = 0
    // Notify children
    for each child in children[root] do
      shm_controls[child][i] = frag_size
    end for
    // Copy data from the queue
    frag = get_next_frag(m, sent_size)
    copy(shm_queue[root][i], frag,
         frag_size)
    sent_size += frag_size
    i++
  end while
  write_memory_barrier()
  atomic_dec(shm_nreaders[set])
end while
```

Fig. 4. Non-root process.

signal down the tree) and copies out the fragment from the root's queue to the output buffer (Fig. 4).

If the queue is full, the root process waits on the barrier until all processes have finished copying from the buffers ($shm\_nreaders = 0$). Non-root process starts to wait on control blocks only when its value of *op* counter is equal to the value of shared counter *shm_op*. The queue is divided into q sets to allow the non-blocking barrier to complete while another set (part of queue) is in use, reducing the synchronization costs [4]. For example, in the case of 12 fragments and the queue of $s = 8$ buffers is divided into $q = 2$ sets, the root process fills the first four buffers (the first set) and without blocking starts to copying data into next four buffers (the second set).

Proposed algorithms are implemented within the Open MPI code base (v4.0.x) as a separate collective component. `Wait_for` operation is implemented by spin waiting with periodic calling of the Open MPI's progress engine. The correctness of the page allocation from NUMA nodes is checked by the `move_pages()` linux system call.

4 Analysis of Algorithms

4.1 Theoretical Analysis

In general, the algorithm execution time is determined by the time of leaf processes in the notification tree. Figure 5 shows time diagrams for the root and leaf process in a flat tree. Let us consider three important cases.

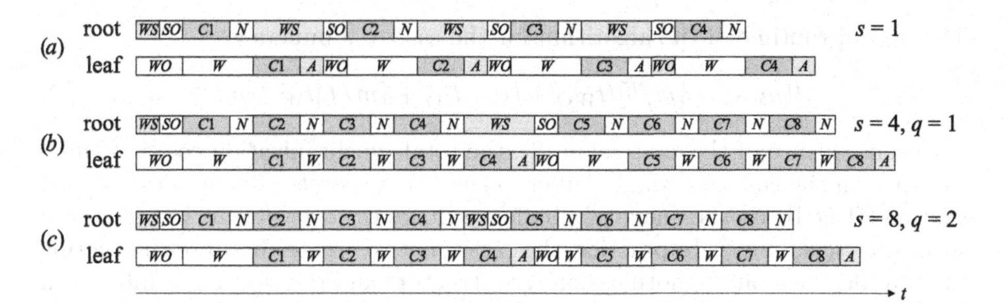

Fig. 5. Time diagrams of the algorithm (WS – waiting for a set in the root, SO – setting the *shm_op* counter, Ck – copying of the fragment k, N – notifying the child process, WO – waiting for a set in the leaf, W – waiting for a notification from the root, A – atomic decrement of *shm_nreaders* counter): *a*) single buffer queue ($s = 1$), $m = 4f$; *b*) queue of $s = 4$ buffers and one set ($q = 1$), $m = 8f$; *c*) queue of $s = 8$ buffers and two sets ($q = 2$), $m = 8f$.

A Single Buffer Queue. In the case of single buffer queue ($s = 1$) a pipelined message transmission is not possible (Fig. 5a). The root and leaf processes perform $\lceil m/f \rceil$ copies of fragments over the single shared buffer. On the first step the root process waits for the readiness of the set (WS) because a buffer may be occupied by non-root processes that complete the previous call of `MPI_Bcast`. After that, the root notifies child about beginning of operation (SO) and starts a loop of fragments copying. A leaf process waits for the readiness of the set (WO) for t_{WO} time and starts copying fragments from the root's queue. The leaf process receives notification for the readiness of the first buffer no sooner then the root copies it ($C1$) for a t_C time. To receive a notification from the root process (W), t_W units of time are required, it depends on the notification tree structure and the number of processes p. Thus, the copying of the first fragment by the leaf process is finished at the time $t_{WO} + t_C + t_W + t_C$. After copying, the leaf notifies the root with an atomic operation (A) for a t_A time about releasing of the set. The root process begins to re-fill the buffers. The time t_C of copying the fragment is mt, where t is the time for reading/writing one byte. Thus, the overall runtime of the algorithm is

$$t(m) = \lceil m/f \rceil (t_{WO} + t_A) + \lceil m/f \rceil t_W + 2mt. \quad (1)$$

A Queue of s Buffers. If the message size is larger than the buffer size, then the message is transferred in a pipeline mode ($m > f, s > 1$, Fig. 5b). The presence of s buffers allows the root process to copy s fragments to the queue without waiting. Copying of the fragment k by the root is performed simultaneously with copying of fragment $k-1$ by the leaf (child) process. After filling all s buffers the root process waits for the completion of copying by all processes, which requires at least $t_C + t_A$ time units (second step *WS*, Fig. 5b). A total number of barrier synchronizations (WO) is $\lceil m/b \rceil$, where $b = fs$ is the total queue size in bytes.

The overall runtime of the algorithm for the case of s buffers is

$$t(m,s) = \lceil m/b \rceil (t_{WO} + t_C + t_A) + \lceil m/f \rceil t_W + mt. \quad (2)$$

As a consequence of the expression (2), the total synchronization cost is s times less than in the case of a single buffer queue (2). Theoretically, at zero costs on waiting (WO, W, A) the queue of s buffers reduces the overall time by a factor of two: $t(m)/t(m,s) < 2$. In practice, the ratio may be greater because the waiting time t_W depends on the notification tree structure and the runtime is influenced by the process placement and copying from local/remote NUMA nodes.

A Queue Divided into q Sets. When the last buffer in the queue is used $(m \geq b)$, a barrier is initiated (WS) and the root process waits for the time $t_C + t_A$ for notifications from the child processes. The waiting cost can be reduced by time t_C if we split s buffers into q sets (Fig. 5c). This allows the root process to start filling the buffers of the next set while the child processes finish copying the fragments from the buffers of the previous set. The runtime of the algorithm:

$$t(m,s,q) = q\lceil m/b \rceil (t_{WO} + t_A) + t_C + \lceil m/f \rceil t_W + mt. \quad (3)$$

4.2 Experimental Results

Experiments were conducted on Intel Xeon Broadwell and Intel Xeon Nehalem dual-processor servers. Intel Xeon Broadwell server has two Intel Xeon E5-2620 v4 processor sockets (8 cores, HyperThreading disabled, L1 cache 32 KB, L2 256 KB, L3 20 MB) and 64 GB of RAM (2 NUMA nodes); linux 4.18.0-80.11.2.el8_0.x86_64, gcc 8.2.1. Intel Xeon Nehalem has two Intel Xeon E5620 processor sockets (4 cores, HyperThreading disabled, L1 256 KB cache, L2 1 MB, L3 12 MB) and 24 GB of RAM (2 NUMA nodes); linux kernel 4.16.3-301.x86_64, gcc 8.2.1.

The performance measurements were taken using Intel MPI Benchmarks (IMB 2019 Update 2). We run one rank per core. For all the figures we use time of the slowest process (`t_max`). An each experiment was run 5 times, discard the slowest and fastest runs, and we average the other 3. The IMB were run with parameters:

```
IMB-MPI1 Bcast -off_cache 20,64 -iter 5000,250 -msglog 6:24
      -sync 1 -imb_barrier 1 -root_shift 0 -zero_size 0
```

In our evaluation we used MVAPICH 2.3.2 and the Open MPI 4.0.x. Both libraries are built with optimizations (`CFLAGS=-O3 CXXFLAGS=-O3`). Figure 6 presents the performance comparison of different queue lengths s on Xeon Broadwell and Xeon Haswell servers. It shows normalized time to the single buffer queue $(s = 1)$. From the formula $t(m,s)$ it follows that the algorithm execution time depends linearly on the message size m, inversely proportional to the queue length s and its size $b = sf$ in bytes. If the messages size m is less then the buffer length f then pipelining is not possible. In Fig. 6 such situation is presented for messages less then buffer size (8 KB).

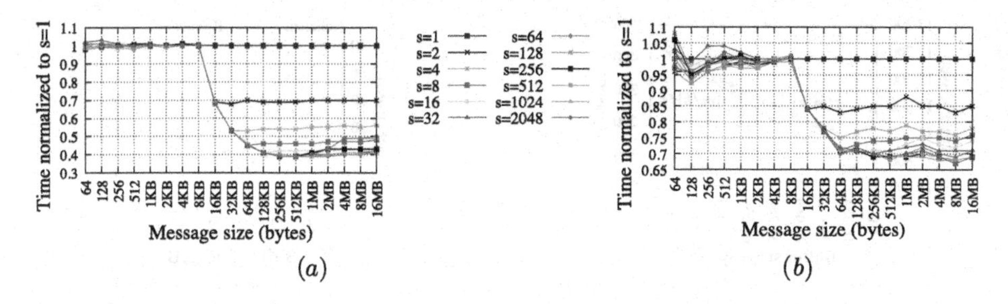

Fig. 6. Latencies of `MPI_Bcast` for different queue lengths s (time normalized to $s = 1$; one set $q = 1$, buffer size $f = 8192$, $p = 8$, one NUMA node, flat notification tree): *a*) Intel Xeon Broadwell server; *b*) Intel Xeon Haswell server.

For messages larger then 8 KB the root process fills s/q buffers without waiting of non-root processes. For example, queue of two buffers ($s = 2$) allows to transfer messages up to 16 KB without blocking of the root and it reduces the runtime by 30% relative to $s = 1$. Similarly, the queue of four buffer reduces the runtime by 40–46%. As noted above, synchronization costs do not allow to reach speedup by a factor of two. Also, the significant size of the shared memory segment limits the use of long queues. Experiments have shown that queues of 32–64 buffers provide good performance. Our results show that a binary or ternary notification trees provide, in most cases, the best performance.

4.3 Optimizing Queue Parameters

Let us find the size s of buffer and length s of the queue, what takes no more than b bytes and provides minimum algorithm time. For example, for a given MPI application it is necessary to determine the optimal configuration of the queue, which fits in 1% of the memory per core. We denote $t_C = tf$ and assume that m is divided by f without remainder. Let find the optimal value of s:

$$t(m, s) = \lceil m/b \rceil (t_{WO} + t_A) + \lceil m/b \rceil ft + m/f \cdot t_W + mt, \quad (4)$$

$$\frac{\partial t}{\partial f} = -mt_W/f^2 + \lceil m/b \rceil t = 0, \quad (5)$$

$$f^* = \sqrt{m/\lceil m/b \rceil \cdot t_W/t} \approx \sqrt{b \cdot t_W/t}, \qquad s^* = b/f^* = \sqrt{b \cdot t/t_W}. \quad (6)$$

In $t(m, s)$ two terms depend on f as f increases, the time $\lceil m/b \rceil ft$ also increases linearly, but the total time $\lceil m/f \rceil t_W$ decreases inversely with f. Figure 7 illustrates minimum point of $t(m, s)$ – intersection point of $\lceil m/b \rceil ft$ and $\lceil m/f \rceil t_W$. Figure 8 shows `MPI_Bcast` latency on Nehalem and Broadwell servers as the function of a fragment size f. The minimum time has been reached at the buffer sizes of 8 KB and 12 KB bytes, which corresponds to the obtained f^* and s^*.

Considering that $t_W > t$, it is practical to use buffers of size $f \geq \sqrt{b}$, rounded up to the nearest multiple of a page size. Consider the choice of the queue parameters for different cases.

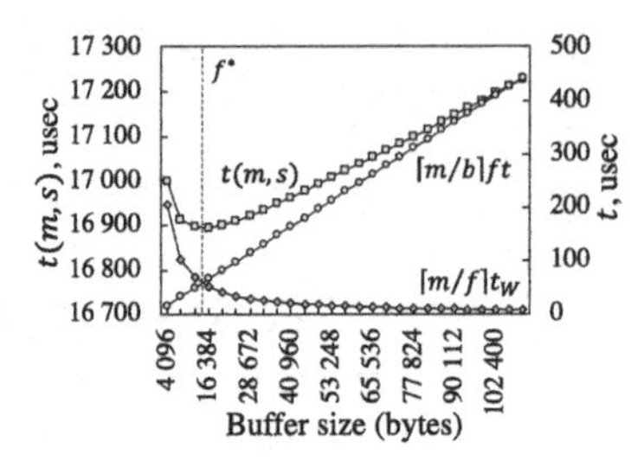

Fig. 7. Algorithm runtime (model, left axis) and terms $\lceil m/b \rceil ft$, $\lceil m/f \rceil t_W$ (right axis): $m = 16\,\text{MB}$, $b = 4\,\text{MB}$, $s = b/f, q = 1, t = 10^{-9}$ s, $t_{WO} = 100t$, $t_A = 10t$, $t_W = 50t$).

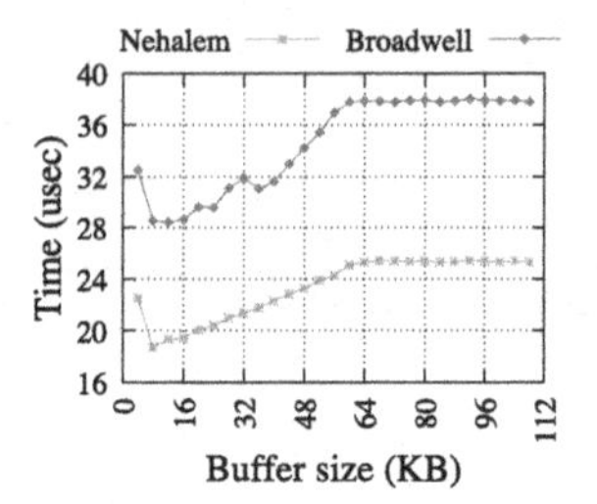

Fig. 8. Latency of `MPI_Bcast`: $m = 64\,\text{KB}$, $b = 4\,\text{MB}$, $s = b/f$, $q = 1$, chain tree, Nehalem ($p = 8$), Broadwell ($p = 16$).

1. The message size m is known or the upper bound for it (for example after application profiling). The best choice is to get f and s such that $f < m \le fs$. Let be $f = \sqrt{m}$ and rounds up obtained f to the nearest multiply of a page size; $s = m/f$.
2. The buffer size f is given, we need to find the queue length s. Let m_{max} denote the upper bound of message size, then $s = \lceil m_{max}/f \rceil$.
3. The queue length s is given, we need to get the buffer size f. Let $b = m_{max}$ and apply (6): $f = \sqrt{m_{max}}$.

4.4 NUMA-Aware Queues Placement

Much of the algorithms time is spent copying from the input buffer into root's queue. For this reason it is important to store the input buffer and root's queue on a same NUMA node. In our algorithms we temporarily disable sequential readahead by calling `madvise(seg, segsize, MADV_RANDOM)`.

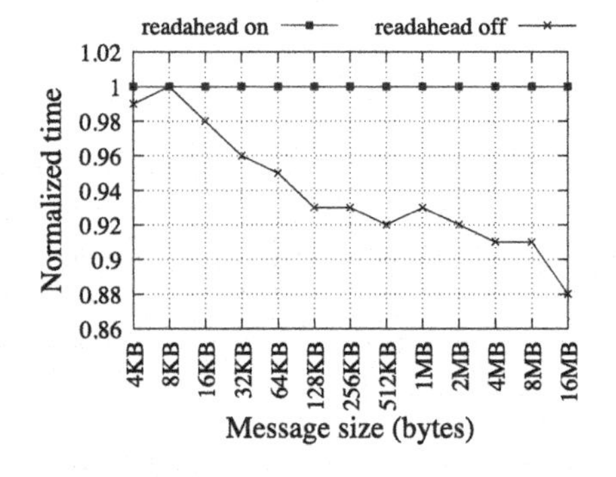

Fig. 9. `MPI_Bcast` normalized time: $s = 64$, $f = 8192$, $q = 1$, binary tree, $p = 16$, two NUMA nodes Xeon Broadwell, IMB `-root_shift on`.

This ensures correct allocation of memory pages from local NUMA nodes within the first touch policy.

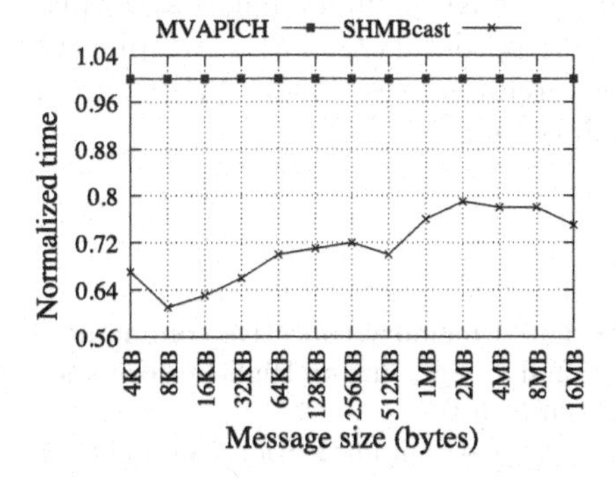

Fig. 10. Normalized time of the developed algorithm (SHMBcast) and MVAPICH ($f = 8192, s = 128$): $s = 64$, $f = 8192, q = 1$, binary tree, $p = 8$, two NUMA nodes of Intel Xeon Nehalem, IMB `-root_shift on`.

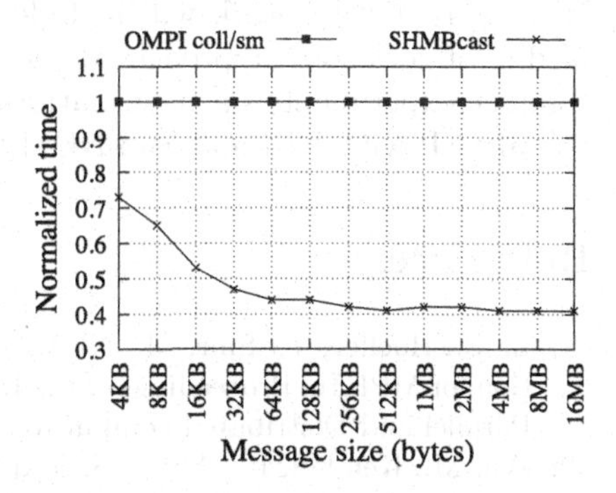

Fig. 11. Normalized time of the developed algorithm (SHMBcast) and Open MPI `coll/sm` ($f = 8192, s = 8$): $s = 64$, $f = 8192$, $q = 1$, binary tree, $p = 16$, two NUMA nodes of Intel Xeon Broadwell, IMB `-root_shift on`.

Figure 9 shows the proposed algorithm's runtime in the "readahead on" mode and without it ("readahead off", `madvise(MADV_RANDOM)`). To estimate overhead due to remote NUMA node access we have run IMB with `-root_shift on` option to cyclically change root on each iteration of measurements. Clearly, as the message size increases, the time for copying fragments to remote NUMA node also increases. Our approach with explicit placement of queues on NUMA nodes (readahead off) allows to reduce inter-socket communications. Similarly, MVAPICH allocates memory for the queues without taking into account a topology of NUMA node (Fig. 10). The Open MPI `coll/sm` algorithm implements a partial binding of buffers to the NUMA nodes, but it is influenced by the readahead subsystem and allocates significant amounts of pages from the NUMA node of the process 0 (Fig. 11). Our algorithms SHMBcast achieve on average 20–40% speedup over Open MPI `coll/sm` and 20–60% over MVAPICH.

5 Conclusion

In this paper we have examined the benefits of NUMA-aware placing of shared queues for optimizing `MPI_Bcast` operation. Proposed algorithms use k-ary, k-nomial, chain and flat trees to propagate notifications from the root process to others. On a Xeon Nehalem and Xeon Broadwell servers, our implementation achieves on average 20–60% speedup over algorithms of Open MPI `coll/sm` and MVAPICH. We find that a binary or ternary trees provides, in most cases, the best performance.

The same approach could be used to optimize other algorithms of collective operations. Future work will include the use of huge memory pages and optimizing of zero-copy approach to the MPI derived datatypes. Also, we plan to conduct experiments on new platforms and architectures (AMD EPYC, Intel Skylake-SP with UPI and Sub-NUMA Clusters, ARMv8).

References

1. Li, S., Hoefler, T., Snir, M.: NUMA-aware shared memory collective communication for MPI. In: Proceedings of the International Symposium on High-Performance Parallel and Distributed computing, pp. 85–96 (2013)
2. Wu, M., Kendall, R., Aluru, S.: Exploring collective communications on a cluster of SMPs. In: Proceeedings of the HPCAsia, pp. 114–117 (2004)
3. MVAPICH: MPI over InfiniBand, Omni-Path, Ethernet/iWARP, and RoCE. http://mvapich.cse.ohio-state.edu/
4. Graham, R.L., Shipman, G.: MPI support for multi-core architectures: optimized shared memory collectives. In: Lastovetsky, A., Kechadi, T., Dongarra, J. (eds.) EuroPVM/MPI 2008. LNCS, vol. 5205, pp. 130–140. Springer, Heidelberg (2008). https://doi.org/10.1007/978-3-540-87475-1_21
5. Jain, S., et al.: Framework for scalable intra-node collective operations using shared memory. In: Proceedings of the International Conference for High Performance Computing, Networking, Storage, and Analysis (SC 2018), pp. 374–385 (2018)
6. Ma, T., Herault, T., Bosilca, G., Dongarra, J.J.: Process distance-aware adaptive MPI collective communications. In: Proceedings of the 2011 IEEE International Conference on Cluster Computing, pp. 196–204 (2011)
7. Bienz, A., Olson, L., Gropp, W.: Node-aware improvements to allreduce. In: Proceedings of ExaMPI 2019: Workshop on Exascale MPI (SC 2019), pp. 19–28 (2019)
8. Li, S., Hoefler, T., Hu, C., Snir, M.: Improved MPI collectives for MPI processes in shared address spaces. Cluster Comput. **17**(4), 1139–1155 (2014). https://doi.org/10.1007/s10586-014-0361-4
9. Graham, R., et al.: Cheetah: a framework for scalable hierarchical collective operations. In: Proceedings of IEEE/ACM International Symposium on Cluster, Cloud and Grid Computing, pp. 73–83 (2011)
10. Chakraborty, S., Subramoni, H., Panda, D.K.: Contention-aware kernel-assisted MPI collectives for multi-/many-core systems. In: Proceedings of IEEE International Conference on Cluster Computing, pp. 13–24 (2017)
11. Luo, X., Wu, W., Bosilca, G., Patinyasakdikul, T., Wang, L., Dongarra, J.J.: ADAPT: an event-based adaptive collective communication framework. In: Proceedings of International Symposium on High-Performance Parallel and Distributed Computing, pp. 118–130 (2018)
12. Träff, J.L., Rougier, A.: MPI collectives and datatypes for hierarchical all-to-all communication. In: Proceedings of EuroMPI/ASIA, pp. 27–32 (2014)
13. Goglin, B., Moreaud, S.: KNEM: a generic and scalable kernel-assisted intranode MPI communication framework. J. Parallel Distrib. Comput. **73**(2), 176–188 (2013)

14. Ma, T., Bosilca, G., Bouteiller, A., Dongarra, J.: HierKNEM: an adaptive framework for kernel-assisted and topology-aware collective communications on many-core clusters. In: Proceedings of Parallel and Distributed Processing Symposium, pp. 970–982 (2012)
15. Tu, B., Zou, M., Zhan, J., Zhao, X., Fan, J.: Multi-core aware optimization for MPI collectives. In: Proceedings of International Conference on Cluster Computing, pp. 322–325 (2008)

Similarity Mining of Message Passing Delays in Supercomputer Networks Based on Peak and Step Detection

Alexey Salnikov, Artur Begaev(✉), and Archil Maysuradze

Lomonosov Moscow State University, Moscow, Russia
{salnikov,maysuradze}@cs.msu.ru, akagitsunesan@gmail.com

Abstract. The problem of analyzing the delays arising in the transmission of data in a multiprocessor computing system is considered. This problem is actual due the need to create parallel programs that effectively utilize the resources of a cluster computing system. Also, the analysis of delays in the communication environment is helpful for system administrators. They can use it for automated troubleshooting and fine-tuning of computing cluster system software. We propose a special approach of analysis of delays in supercomputer communications. There the clustering algorithm is proposed. This algorithm allows smoothing out outliers and significantly reduce the amount of stored data of delay values by the way of finding the nodes with similar behavior of delays. Algorithm is based on methods of correlations analysis and K-means clustering algorithm. The research was conducted using K60 supercomputer of Keldysh Institute of Applied Mathematics and supercomputers of Moscow State University.

Keywords: Benchmarks · MPI · Cluster analysis · Outlier detection · Correlation analysis · Communication delay · Computer cluster

1 Introduction

Modern supercomputers are designed to make possible solve a many types of numerical tasks: mathematical modeling, processing large amounts of data, etc. Calculations are performed in parallel on a large number of computational nodes which constitute a computational cluster. During implementation of parallel application it is necessary to form strategy of distribution the calculations over the supercomputer nodes. Data transfer is organized using MPI (Message Passing Interface). During these transfers delays occur. Supercomputer nodes are connected using network cards, switches and wires in one system. The equipment that forms the network of a supercomputer is called a switching environment. The node that initiates the message transfer is called 'sender', and the node which receives this message is called 'receiver'.

The execution time of the parallel program can be estimated using knowledge about algorithm implementation, but at the same time there are indirect factors

V. Voevodin and S. Sobolev (Eds.): RuSCDays 2020, CCIS 1331, pp. 486–499, 2020.
https://doi.org/10.1007/978-3-030-64616-5_42

that also significantly affect the execution time of program. One such factor is the variability of delay in transmission messages between supercomputer nodes. This variability arises due to the fact that messages between nodes are not sent directly, but through other nodes or some switches. The magnitude of the delay is affected by the structure and state of the switching environment in supercomputer, settings of this environment, moments of message transfer starts between nodes and other applications distribution among the supercomputer nodes [1]. There a lot of unconsidered factors which can indirectly influent on execution time of parallel program and its calculations. But some statistical prediction of the behavior delays between nodes may provide a performance boost of parallel program, when it is working in an environment of a specific supercomputer [2]. Analyzing delays also can help to find out an undocumented features of data partitioning during the transmission of messages, failures of the network components of the supercomputer, or their incorrect configuration.

Various testing and benchmarking methods are used for studying the state of supercomputer interconnect. One of the benchmarking approaches consists in measuring the time of message transmission from node to node varying the message size. Using the results of these benchmarks it is possible to analyze the state of the communication environment – its settings and organization. In this article following features of the delay behavior are considered:

- Step – change in delays after which, delays keep the same with the message length increase.
- Peak – significant change of delay, but it goes back for the next size of message length.

It was noticed that the benchmarking applications often collect large amount of data. For example, when it works over N nodes where performs K measurements for each pair sender - receiver, it turns out the amount of data dimensions equals $N^2 \times K$. The obtained data can be represented in the form of a three-dimensional space where the dimensions are: the index of the node initialized the message transmission, the index of the node received the message, and index corresponding to the message length. So the amount of data needs to stored and analyzed is huge. It is needed to simplify analysis of such data containing a lot of information.

It was noticed that pairs of sender - receiver nodes can be broken down into set of groups. One group contain set of such pairs, where the delay between nodes is similar. Thus, such grouping allows to store less amount of data and perform simpler analysis of communicational environment. For acquiring these groups K-Means clustering algorithm is used.

Some measured delays for pairs of nodes from supercomputer Lomonosov-2 [3] are presented in Fig. 1. We see several pairs of delays between nodes have a similar nature. It is clear that the character of delay behavior during message passing between nodes $(55, 46)$, $(69, 10)$, $(76, 61)$ is similar. Steps in delays when sending messages between these pairs of nodes occur at approximately the same message lengths. Thus, the similarities in behavior an be find by clustering.

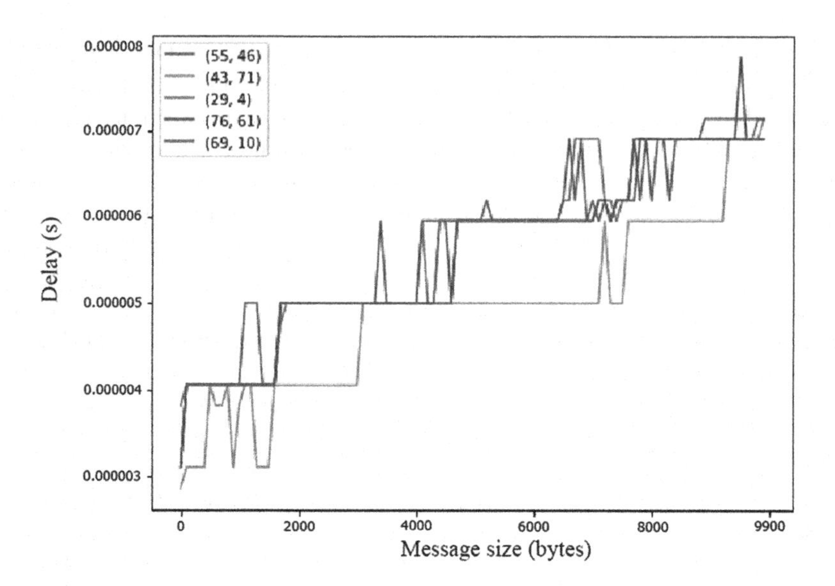

Fig. 1. Plots of delays between medians 5 random nodes for K60 supercomputer

Rare "peaks" in data are useless for characterizing the behavior of delays, as their occurrence is random. The sizes of messages during those transmissions "peaks" occurred are not robust, they differ from pair to pair. So, "peaks" cannot be used as property characterizing the behavior of delay. Thus "steps" characterize delay behavior between a pair of nodes.

2 Related Work

There are not a lot of works performing delay analysis of supercomputer interconnect. Meanwhile the author consider this task as necessary and challenging. It becomes difficult to make any statements about supercomputer network status due increasing number of components of computational cluster. This is also the reason why semi-automatic tools for delay analysis are needed.

In one of the previous work [4] the authors developed a toolkit system, which collects delays by method mentioned earlier – varying message length. It was hard to analyze collected data by older tools because they required a lot of manual work. Proposed paper extends this toolkit by special tools for analysis.

The similar analysis of delays as in this paper is performed in [5]. The authors of this article investigated another type of delays. They proposed to send a message with fixed length several times periodically. The obtained measurements were clustered. By analyzing the clustering result with special graphical tool they try to estimate the bottleneck in supercomputer interconnect network. At the same time the researchers didn't take into account the fact that depending on size of transmitted data the congestion algorithm and routing algorithm in supercomputer network can be switched. So, it is necessary to use varying length messages for benchmarking.

3 Clustbench Project Components Description

The software package Clustbench [4,6] has been developed with the participation of the authors of this paper. The package includes several programs:

- `network_tests2`—an application that measures the amount of delays when sending messages of different lengths between the supercomputer nodes. Also it collects statistical information about delays: median, average, minimum value and standard deviation. The implementation of messaging occurs through the MPI software interface;
- `network_viewer_qt_v2`—application for visualisation. The results for it are obtained by the program `network_tests2`.

Benchmarking application `network_tests2` supports multiple modes for obtaining delay values from supercomputer interconnections. Each mode emulates a potentially implemented scheme for transferring messages between nodes of a supercomputer in a parallel program. In our work we use following modes of application:

- `one_to_one` mode blocks the process of transferring data from one node to another, nodes exchange messages sequentially;
- `all_to_all` mode makes processes start to send messages to each other at the same time;
- `async_one_to_one` mode does not block the data transfer process from one node to another;

It is possible to specify a range of message lengths for which benchmarking is performed. With purpose to obtain more accurate collection of statistics on delays, program allows to specify the number of iterations. The results of this application will be used to obtain information about the behavior of delays between nodes of a supercomputer, depending from the length of the transmitted messages.

Application `network_viewer_qt_v2` contains several modes of data visualization which allows to perform a manual delay analysis based on the results of rendering delays. There is the mode of rendering in the form of matrices. Each matrix corresponds to a specific message length for which values were measured delays in the transmission of this message between pairs of nodes sender - receiver. Selection by self-similarity groups also occurs manually, which is not quickly resolved.

4 Analysis of Delays

4.1 Peak and Step Detection

As it was mentioned above it is necessary to make semi-automatic peak and step detections, because it is difficult to detect them manually. Following methods allow such operations.

Set of vectors with medians of delay for each sender - receiver pair is constructed. For sender node i and receiver node j one vector looks like $M_{i,j} = m_1, ..., m_L$, where L is the number of message lengths for which delays has been measured. In this paper medians are considered because this statistic is less affected by peaks. With a large amount of peaks further analysis is impossible, since their impact on median behavior will be high and proposed analysis can work incorrectly.

The calculation of the Pearson correlation [7] with the near-model is used for detection of peaks and steps. Model is represented as a single vector $y^{(k)} = (0, ..., 1, ..., 0)$, of length L, where the "1" is located at the position k. This position corresponds to the length of the message where it is assumed that there was a step. A step is represented as an outlier in a series of variations in delay values Δm, where $\Delta m_l = m_{l+1} - m_l$, the boundary values are assumed as follows: $\Delta m_0 = \Delta m_L$, $\Delta m_{L-1} = \Delta m_{L-2}$.

The calculated correlation coefficients serve as an indicator of the behavior of delays, so the dependence between the delay values for the lengths are corresponding to specific indices are used in the terms of the near-model. Changing the position of the "1" in the model series and calculating the coefficients correlation for all lengths of messages, we find out where the step occurred, and where the peak.

To search for peaks, it is required to calculate correlation coefficients between all values of delays m_k, contained in a series m, and series of the form $y^{(k)}$. To search for steps, it is required to calculate the correlation coefficients between all values of the delays Δm_k, contained in the series Δm, and the series of the form $y^{(k)}$.

Since structure of each vector $y^{(k)}$ always remains similar, some of the coefficients in the Pearson correlation formula can be shrunk to constants. This means that the formulas for calculating the correlation coefficients for the m and Δm series will take the following form:

$$corr^p{}_k = \frac{m_k - \overline{m}}{\sqrt{\sigma^2{}_{y^{(k)}}}\sqrt{\sigma^2{}_m}}, k \in [0, l-1] \tag{1}$$

$$\sqrt{\sigma^2{}_{y^{(k)}}} = \frac{n-1}{n^2}, \quad \sqrt{\sigma^2{}_m} = \overline{m^2} - \overline{m}^2 \tag{2}$$

$$corr^j{}_k = \frac{\Delta m_k - \overline{\Delta m}}{\sqrt{\sigma^2{}_{y^{(k)}}}\sqrt{\sigma^2{}_{\Delta m}}}, k \in [0, l-1] \tag{3}$$

$$\sqrt{\sigma^2{}_{y^{(k)}}} = \frac{n-1}{n^2}, \quad \sqrt{\sigma^2{}_{\Delta m}} = \overline{\Delta m^2} - \overline{\Delta m}^2 \tag{4}$$

Where $corr^p{}_k$ are the correlation coefficients corresponding to the peak indicator in a number of medians of delays m in the length of the message corresponding to the index k, and $corr^j{}_k$ – correspond to peak correlation coefficients in a series of changes medians of delays Δm, respectively, steps in a number of medians m in length messages to the appropriate index k. $\sqrt{\sigma^2{}_{y^{(k)}}}$, $\sqrt{\sigma^2{}_m}$ $\sqrt{\sigma^2{}_{\Delta m}}$ are the standard deviations of the series $y^{(k)}$, m, Δm.

Before calculating the indicators (1), (3), input data should be preprocessesed. Elimination of some peaks can be done using median filtering. Applying the methods of correlation analysis is impossible to non-stationary series. Clarification is following, values of the indicators are strongly influenced by the increase in the value of delay with increasing in message size. This increase is called a linear trend. Therefore when we are looking for peaks, we must first subtract linear trend value from a number of medians of delays. Note that this is not required for step search, because the Δm series is stationary.

4.2 Revealing the Similarities of Delays Behavior and Visualization

Calculated indicators are useful for revealing the whole structure of delays behavior for each pair of nodes sender - receiver. Analysis of indicators requires the selection of threshold values $threshold^{(p)}$ and $threshold^{(j)}$ to determine the lengths of messages that are sent when between nodes, peaks and steps occur. After the selection of these values is required evaluate the quality of the data for further clustering. Analysis approach peak indicators and steps are common—they require to build identical graphs and histograms for analysis.

The choice of thresholds is made empirically. It is based on data and calculated indicators. For the selection of these values, we design a special program that allows us to make a prediction on how to choose threshold values. The threshold below 0.2 is considered as very weak correlation and shouldn't be taken into account at all [9].

For accurate determination of threshold values the obtained values should be considered more in detail. For the purpose of further analysis the indicators distribution histogram should be plotted.

After calculating these thresholds the peaks and steps location can be revealed. Looking at the number of peaks in whole dataset it can be told is it useful for further analysis. If the number of peaks is too high, the obtaining "sender" - "receiver" pairs with similar behavior of delays is imprecise. Allowed number of peaks is a parameter of presented algorithm. The authors suggest that their total number shouldn't be higher 5% of total number of collected delays for appropriate precision. So the supercomputer testing should be conducted again, retrieved data should be analysed the same way. If outliers occurred again for the same nodes and same message lengths the administrators should be told that there is malfunctioning components in the network of supercomputer.

If obtained dataset is fine, calculated steps will completely represent the structure of delays behavior. The most common class algorithms for revealing object with similar structure or characteristics is clustering. For our task we used well-known algorithm K-Means [11]. This algorithm calculates the centroid of each cluster, which generalizes all values of delays between pairs sender - receiver, which are appeared to be located in the same cluster. This centroid stores the message sizes, where steps appeared and average delays for all message sizes. The tuning the K number is done by using the SSE metric [10]. The essence of this metric is to calculate how much the elements that are in the cluster differ from the center of this cluster. Its calculation is made according to the formula:

$$SSE = \sum_{i=1}^{K} \sum_{x \in C_i} d^2(\mu_i, x)$$

Where C_i is a cluster, μ_i is the center of mass of the cluster, x is the values of the delay indicators of the pairs sender - receiver.

Verification of similarity of median delays between pairs of nodes in one cluster is performed as following. For each cluster median delays between some random pairs sender - receiver each tested message length are plotted. After that the calculated average delay for this cluster is drawn. By looking at the behavior of average and plotted median delays it can be conducted about the quality of performed clustering.

Special 3D visualizing tool was developed. This tools draws acquired clusters in 3D space, corresponding to analysed space. Each axis represents following: one of the axis is the index of sender node, another one is the index of receiver node and the last one corresponds to message length. Developed tool has 2 modes: to draw steps found in clusters (steps are calculated from centroid of cluster) or draw chosen single cluster. Visualizing lets analyze the state of supercomputer interconnect. The whole analytic pipeline is shown in Fig. 2.

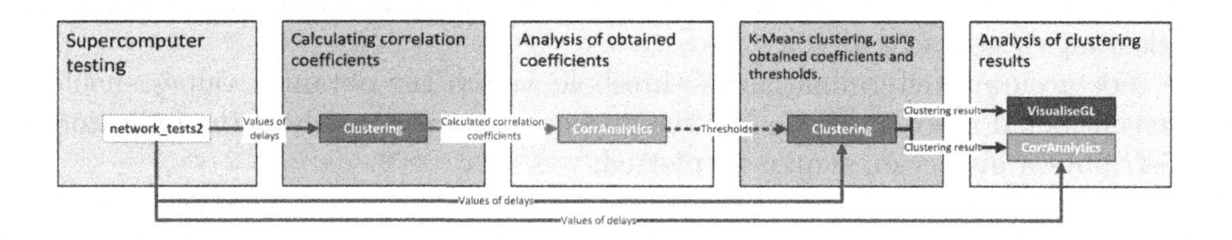

Fig. 2. Analytic pipeline

5 Results

K60 supercomputer [8] is located at Keldysh Institute of Applied Mathematics. The authors and system administrators of this computer were very interesting in detailed information on delays structure in communications and methods of analysis of this information. The dataset with information about delays was collected by `network_tests2`. Testing K60 was conducted with parameters:

- Initial message length – 0 bytes;
- End message length – 10000 bytes;
- Step of length change – 100 bytes;
- Number of iterations for collecting statistics – 100;
- Testing mode – `one_to_one`.

K60 consists of 78 computational nodes connected by switches, network cards and fiber optic cables Mellanox Infiniband 56G. Testing requires all nodes to be idle, so it was done during the prevention, while no other were running programs on the supercomputer.

After gaining dataset the needed pre-processing was done the indicators were calculated.

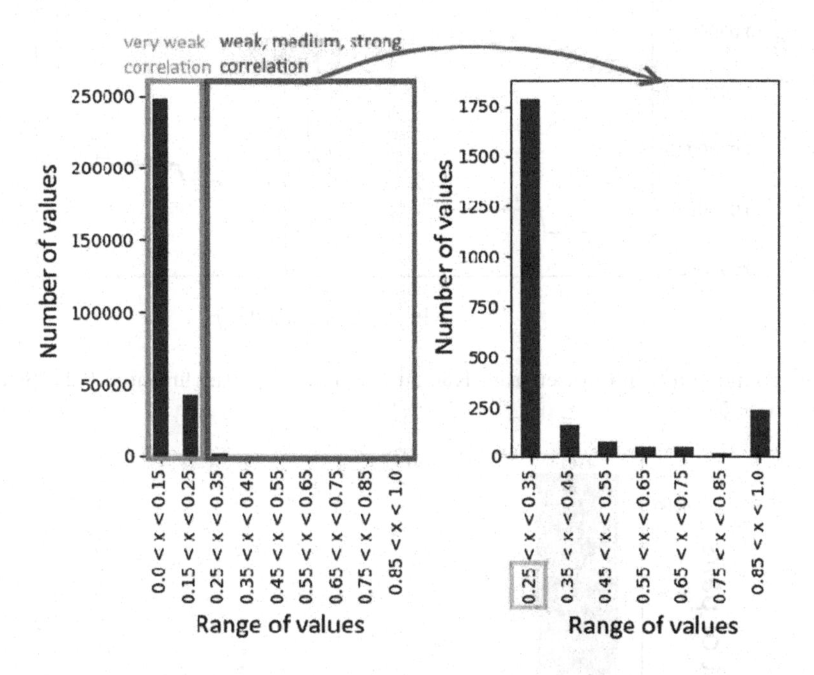

Fig. 3. Distribution of calculated indicators values for K60

In Fig. 3 the threshold value for peaks is 0.25. This value allows to determine the number of peaks in the whole dataset. The distribution for steps looks similar for peaks, so the threshold is 0.25 again.

In Fig. 4 one can see that the threshold value was picked up correctly and peaks are detected in the sets of delays.

Looking at the peaks distribution plot in Fig. 5, found in dataset collected from K60 supercomputer, it can be seen, that the number of peaks is not very big. Most part of delay sets have zero or one peak, which will be corrected by median filtering. According to this fact, it can be claimed, that the dataset is good and further delay analysis can be conducted.

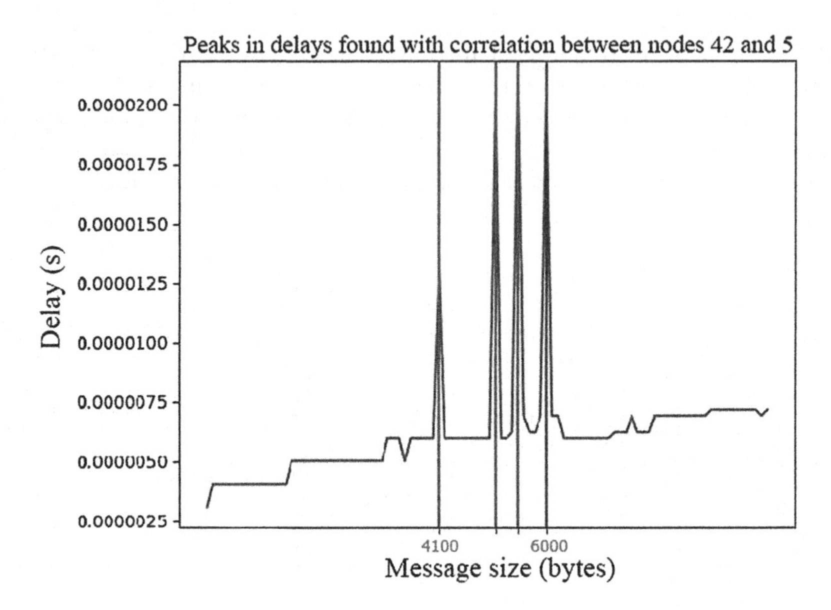

Fig. 4. Found peaks in dataset from K60 supercomputer with threshold 0.25. Sender: 42, receiver: 5.

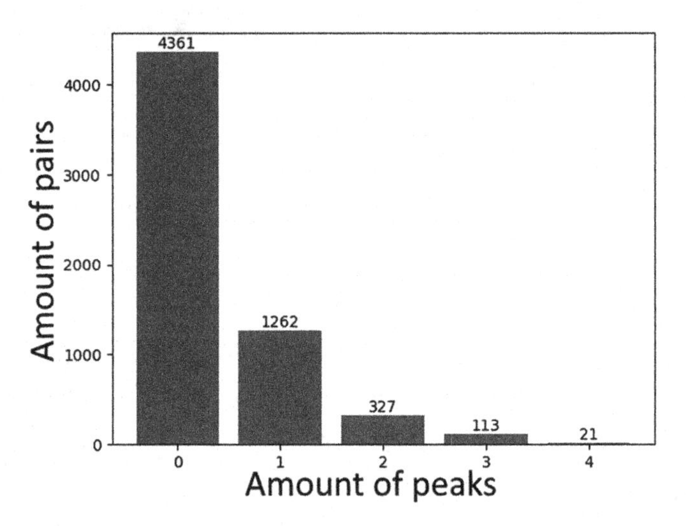

Fig. 5. Distribution of peaks number in each pair sender - receiver in dataset from K60 supercomputer

Exploring the plot in Fig. 6 of steps distribution by the message length, it can be seen that there are lengths of messages, where the number of steps is relatively high. From this Figure it can concluded, that there is pattern of steps distribution, so there are pairs sender - receiver, which have similarities in delay behavior. So, the further clustering can be conducted.

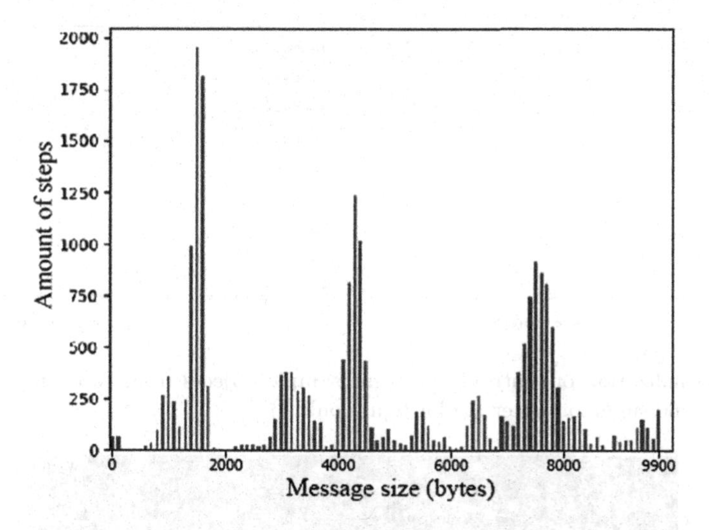

Fig. 6. Steps distribution in dataset from K60 supercomputer

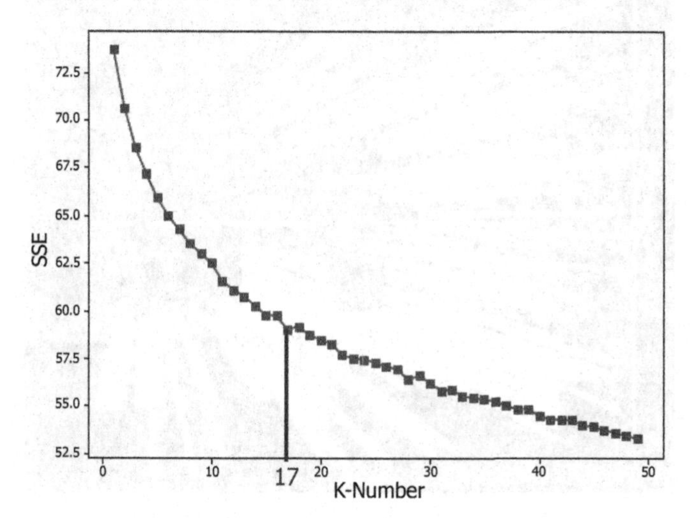

Fig. 7. SSE metric for K-Means

For K-Means algorithm the K number was chosen equal to 17 for K60 supercomputer. This number in this paper was selected empirically by looking at Fig. 7. But in the future works the proper research of techniques of tuning this parameter should be performed. One of the ideas is to use a priori information about supercomputer network architecture, such as topology or "cable journal" – special data, containing information about interconnect.

After completing K-Means clustering for K60 centroids and some sample objects from two random clusters were plotted. The red line on the plots is centroid the paler ones are sample objects. As it can be seen at Fig. 8, the

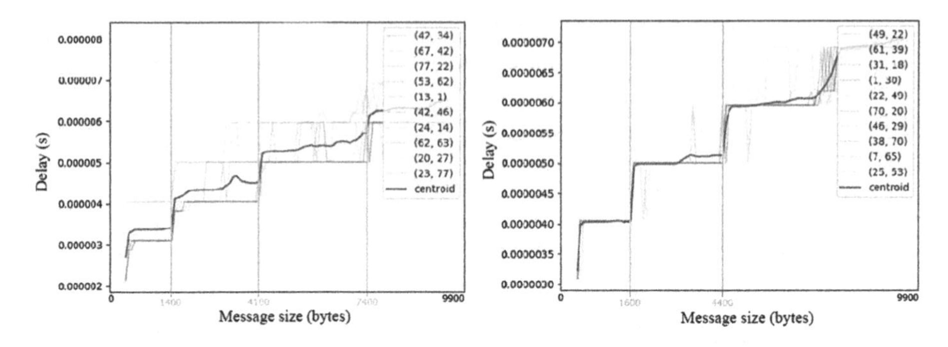

Fig. 8. Visualization of centroid and some sample objects from two random clusters found for K60 supercomputer (Color figure online)

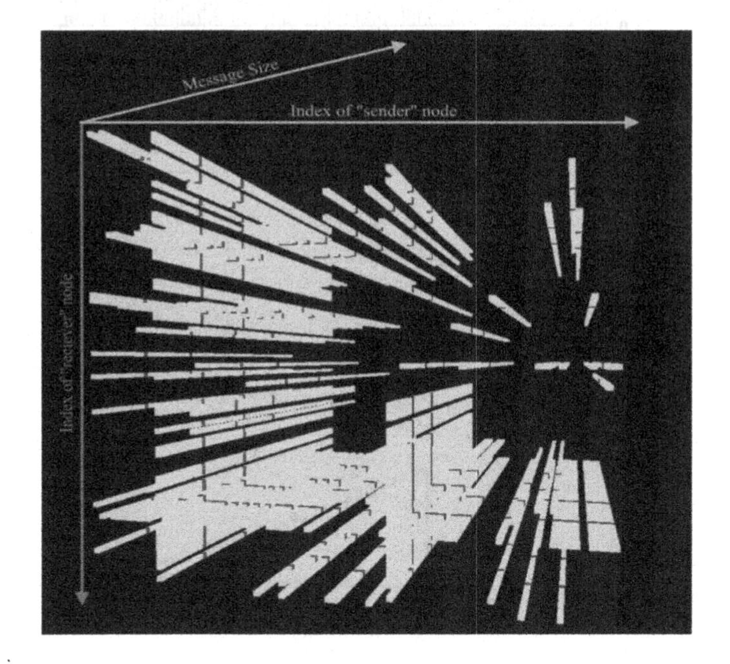

Fig. 9. One of the clusters found on K60

centroid behavior of clusters and behavior of samples are similar. Applying to this fact, the clustering was made correctly and desired results were reached.

The 3D representation of one of the clusters found for K60 supercomputer in Fig. 9 did not provide any valuable information. Any of found clusters do not have any recognizable patterns, so it is difficult to make any statements about logic hidden behind the shape of cluster.

In the 3D plot in Fig. 10, the common steps for acquired clusters provided interesting information, concluded in the fact, that all steps are combined in four planes. So, it can be concluded, that the interconnect of K60 supercomputer is

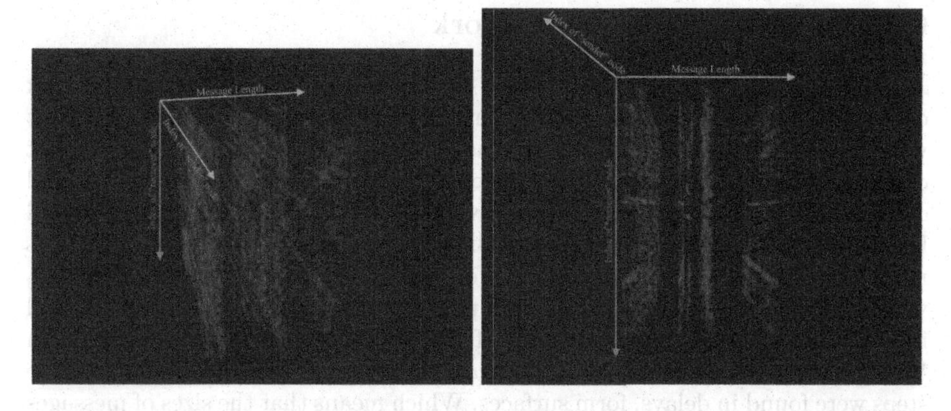

Fig. 10. 3D visualization of all clusters centroid steps for K60 supercomputer

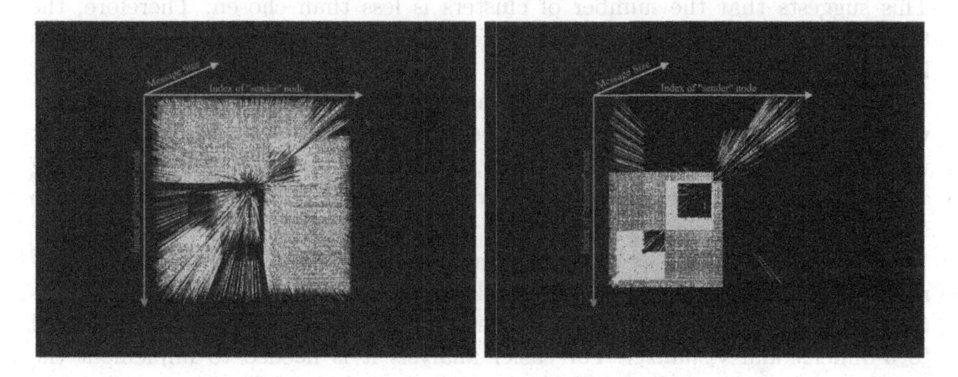

Fig. 11. 3D visualization of all clusters centroid steps

homogeneous and it works correctly. Possibly, but for supercomputer malfunctioning this pattern would haven't been found. Supercomputer is difficult system to tune and the authors were not allowed to make any interventions in network settings or connection schemes.

Besides the K60 supercomputer, the Lomonosov supercomputer [12] was also tested. The `network_tests2` was launched with same parameters as for K60, but the characteristics of Lomonosov is different—it has 1280 computational nodes, connected by Mellanox Infiniband 40G. The same clustering algorithm was conducted for collected dataset.

The Fig. 11 shows that communication environment in Lomonosov is heterogeneous. There are several types of switches. So, unlike K60, we see different steps on same messages length.

6 Conclusion and Future Work

A little number of peaks was detected in the data [13], meaning the network components of K60 supercomputer environments are working properly and they were configured correctly. Clustering was performed, which reduced the amount of stored data in 48 times. The `network_tests2` testing tool helped to find out the malfunctioning optical wire of K60 supercomputer. This tool needs full bandwidth of supercomputer network, but because of wire the bandwidths was reduced, so, it led to program crash. For each cluster, features were detected, which describe the behavior of delays in pairs of nodes source - receiver, collected in one cluster. Based on the results of the three-dimensional visualization in Fig. 10, it can be concluded that the parallelepipeds, displaying the places where steps were found in delays, form surfaces. Which means that the sizes of messages at which the steps occur are either close to each other, or they are the same. This suggests that the number of clusters is less than chosen. Therefore, the topology (i.e. the forwarding scheme) of message delivery supercomputer K60 is homogeneous.

Without developed tools, such an analysis is difficult, as all data volumes would have to be analyzed manually. Despite the fact that the amount of peak was small as a result testing of the K60, there is a possibility that when testing another supercomputer, their number will be significant. Thus, it is required to analyze peaks more deeply.

This analysis allows us to identify faulty components of the switching environment of the supercomputer and quickly eliminate the causes of faults by system administrators. To increase the clarity of the display grid should be added to 3-dimensional visualizer. For easier analysis it is needed to implement the possibility to click on the element of a cluster to build different plots for clusters, without additional applications. For more convenient use of a 3-dimensional visualizer, it is required to implement an user interface.

Also, as a result of clustering, it was not possible to detect the features of delay behavior for large message sizes. This happened due to the fact that the delay value does not change dramatically with the increase in the message size, as during a step, but grows more smoothly. Some techniques for choosing the K-parameter in K-means clustering algorithm for users of developed tool should be proposed.

Also, the behavior of delays collected during the K60 supercomputer testing in `all_to_all` mode is not so sharp, but much more smoother. It is required to refine the algorithm for analyzing such behavior of delays, which would take into account the nature of increasing or decreasing delays with increasing length of the message being sent.

References

1. Meyer, H., Sancho, J.C., Mdrakovic, M., et al.: Optical packet switching in HPC. An analysis of applications performance. Future Gener. Comput. Syst. **82**, 606–616 (2018)

2. Faizian, P., Mollah, M.A., Tong, Z., et al.: A comparative study of SDN and adaptive routing on dragonfly networks. In: Proceedings of the International Conference for High Performance Computing, Networking, Storage and Analysis, pp. 51–62. ACM, New York (2017)
3. Top500: Lomonosov 2 - T-Platform A-Class Cluster. https://www.top500.org/system/178444. Accessed 16 May 2020
4. Salnikov, A.N., Andreev, D.Y., Lebedev, R.D.: Toolkit for analyzing the communication environment characteristics of a computational cluster based on MPI standard functions. Moscow Univ. Comput. Math. Cybern. **36**, 41–49 (2012). https://doi.org/10.3103/S0278641912010074
5. Fujiware, T., Li, K., Mubarak, M., Ross, C., et al.: A visual analytics system for optimizing the performance of large-scale networks in supercomputing systems. Vis. Inform. **2**(1), 98–110 (2018)
6. Clustbench community, Clustbench: HPC cluster benchmarking toolkit. https://github.com/clustbench. Accessed 12 Mar 2020
7. Cohen, J., Cohen, P., West, S.G., Aiken, L.S.: Applied Multiple Regression/Correlation Analysis for the Behavioral Sciences, p. 736. Lawrence Erlbaum Associates, New Jersey (2003)
8. Keldysh Institute of Applied Mathematics, K60 Supercomputer specifications. http://www.kiam.ru/MVS/resourses/k60.html. Accessed 15 Mar 2020
9. Puth, M.-T., Neuhauser, M., Ruxton, G.D.: Effective use of Pearson's product-moment correlation coefficient. Anim. Behav. **93**, 183–189 (2014)
10. Du, W., Lin, H., Sun, J., et al.: Combining statistical information and distance computation for K-Means initialization. In: 2016 12th International Conference on Semantics, Knowledge and Grids (SKG), IEEE, pp. 97–102 (2016)
11. Everitt, B.S., Landau, S., Leese, M., Stahl, D.: Cluster Analysis, p. 330. Wiley, Chichester (2011)
12. Sadovnichy, V., Tikhonravov, A., Voevodin, V., Opanasenko, V.: Lomonosov: supercomputing at Moscow State University. In: Contemporary High Performance Computing: From Petascale toward Exascale (Chapman and Hall/CRC Computational Science), pp. 283–307. CRC Press, Boca Raton (2013)
13. Begaev, A., Salnikov, A.: network_tests2 benchmark results from Lomonosov, Lomonosov-2, K60, Juropa supercomputers. Mendeley Data, V2. https://doi.org/10.17632/bffc8w295s.3

SoftGrader: Automated Solution Checking System

Alexander Sysoyev(✉), Mikhail Krivonosov, Denis Prytov, and Anton Shtanyuk

Lobachevsky State University of Nizhni Novgorod, Nizhni Novgorod, Russia
{alexander.sysoyev,krivonosov}@itmm.unn.ru, prytovdd@gmail.com, ashtanyuk@gmail.com

Abstract. One of the most effective ways to enhance students' learning in programming disciplines is carrying out practical assignments covering all key topics on their own. Ideally, a student deals with an individual set of tasks. While the relevance of such an approach is self-evident, its practical implementation is no easy matter. The main difficulty encountered by teachers is providing a high-quality assessment of the solutions developed by students on a variety of problems. Therefore, it is extremely important to automate the evaluation process. The present article is devoted to the description of SoftGrader – automated checking system, developed in UNN, which allows performing mass testing of student program correctness, both sequential and parallel programs.

Keywords: Automated checking · Programming training · An individual set of tasks · The correctness of programs · Mass testing

1 Introduction

The best-known example of the use of automated program checking systems is Olympiads, especially those held online. Some of such systems are freely available and can be used for automating program checking, including parallel ones. But due to the original purpose of such software, the functionality for training courses is redundant and usually is absent.

Therefore, the Institute of Information Technologies, Mathematics and Mechanics of Lobachevsky State University of Nizhni Novgorod is developing its own system of automated solution checking SoftGrader, which was initially developed considering the requirements on course maintenance, including parallel programming.

The article is structured in the following way: Sect. 2 gives an overview of popular systems of automated problem checking used in various Olympiads and competitions; Sect. 3 gives the key features of the proposed system; Sect. 4 discusses the interaction with the system from the viewpoint of the course participant; Sect. 5 – from the viewpoint of the teacher; Sect. 6 presents information on the use of the system in the educational process; in the conclusion the further improvements of the system are described.

V. Voevodin and S. Sobolev (Eds.): RuSCDays 2020, CCIS 1331, pp. 500–510, 2020.
https://doi.org/10.1007/978-3-030-64616-5_43

2 Overview of Automated Problem Checking Systems

2.1 PC^2 System

PC^2 (Programming Contest Control) [1] is a system developed at California University, Sacramento, USA. The International Collegiate Programming Contest (ICPC) uses it. A detailed description of the system, including documentation for administrators, judges and teams of participants, as well as permissions and download links are available at https://pc2.ecs.csus.edu/. The system has been under development since 1988, the current version 9.6.0 is dated January 4, 2019.

PC^2 consists of client and server parts written in Java. Deployment of the system is performed on a local server and requires manipulations with configuration files described in the administrator's manual. To work with PC^2 you can use two shells: a graphical one and a management console.

PC2 supports various compilers and interpreters. Administrator can expand the list of compilers manually. The list of supported interpreters is not managed by the administrator, it includes PERL, PHP, Ruby, Python, and shell.

2.2 Yandex.Contest

Yandex.Contest (https://contest.yandex.ru/) [2] is a service for online checking of assignments, focused primarily on the competition in programming. On the basis of Yandex.Contest the annual championship on programming of Yandex company, and also various olympiads, including, recently, the All-Russia school olympiad on computer science passes.

Yandex.Contest supports more than twenty programming languages and allows using different schemes of competitions determined by the tournament organizer. They also should prepare and place assignments in the system (task definition, test sets, evaluation criteria). Solutions are checked automatically according to the set of tests. The participants submit their solutions to the testing system, and the system provides the result. The Yandex company states that "the service is capable of simultaneously processing terabytes of data, so it can easily withstand the load of more than a thousand participants".

Access to the Yandex.Contest for competition organizers can be granted on request.

2.3 The ejudge System

The ejudge [3] system is designed for any type of event that requires automatic program checking, including training courses. Installation manuals, user guides, ways of functionality extending (the system is available in source codes), and download links are available at https://ejudge.ru. The system development has been underway since 2002. The current version "ejudge 3.8.0" is dated February 1, 2020, and provides the following features:

- Web-interface for both users and administrators. Additional administrative features are provided through command-line utilities.
- Support for many tournaments simultaneously.
- Maintaining a user database based on person tournament binding.
- Flexible configuration of scoring systems, tournament time limits (including unlimited).
- Management of user access rights.
- Support for almost all popular programming languages: C, C ++, Java, Pascal, Perl, Python, etc. During the first run, ejudge automatically detects the installed programming environments and performs initial configuration. Extending list of compilers is also possible in an already running system.

3 Key Features of SoftGrader System

Section 1 presents selected systems for automated program checking. In general, the available systems can be classified according to several most significant parameters:

- Requirement of local deployment or online access mode. Most modern systems either have a web-interface or, like Yandex, are a cloud service.
- Support only for olympiads/competitions or the possibility of learning process conducting. Only a small number of systems declare assistance in training courses arranging.
- Testing only sequential programs or parallel ones as well. In olympiads the main criteria of evaluating programs are passing all the tests, satisfying the limitations on the execution time and (sometimes) the memory consumption. Automatic estimation of acceleration and scalability of parallel programs is also essential as criteria for student solution assessment.

At the time of the review of the automated checking systems (2012), there is none that satisfied all the requirements. As a result, the decision was made to develop a proprietary system (SoftGrader), initially focused on the needs of the educational process in the university.

User interaction with the SoftGrader system [4] takes place via web-interface. To start with, one should create an account in the system (http://softgrader.itmm.unn.ru/). Once logged in, a user can choose one of two global education sections: online courses or university courses. Inside the section, there is a list of available courses, as well as courses that are already in the progress.

Further actions depend on the category to which the user belongs: course participant (student) or teacher. Section 3 describes the workflow from a student's point of view, and Sect. 4 shows the differences that are specific to the teacher workflow.

4 Course Participant Workflow

Access to a specific course in the educational section requires course registration (online courses has an open registration, university courses – may require teacher confirmation).

After registration and navigation to the selected course, the user sees the main page of the course (Fig. 1) with the following tabs.

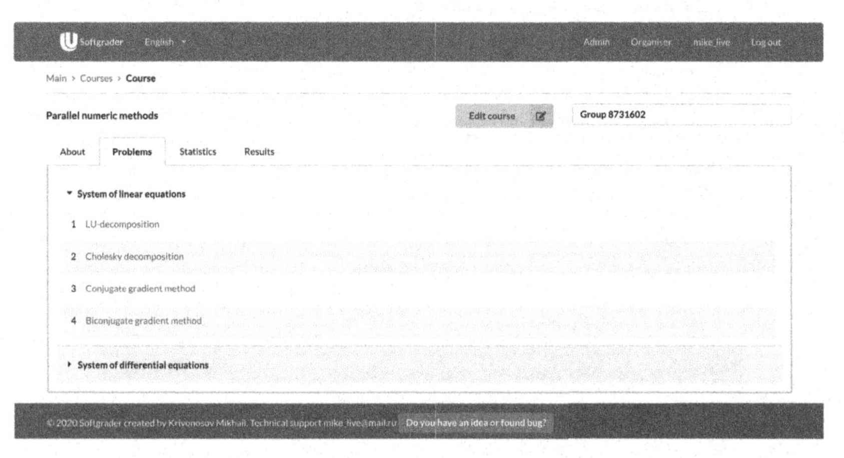

Fig. 1. Course page "Parallel numeric methods"

1. Tab "About". Contains a brief description of the course: goals and objectives, requirements for students, course content, skills that can be acquired by the student after completion of all tasks of the course, as well as a list of recommended literature.
2. Tab "Problems". Contains a table with a list of all course problems included in the automated checking system. The list of tasks can be grouped into different topics.
3. The "Statistics" tab. Provides summary information on tasks, such as the number of students who solved this task, the minimum and maximum scores for the solutions, as well as the number of attempts required by users to solve this problem.
4. Tab "Results". Contains a standings table of all course participants with their grades and results corresponding to course problems.

On the "Problems" tab, a student can select one of the tasks to be solved and will be taken to a page with detailed information on it (Fig. 2). The page contains the problem statement, the requirements and limits, the compiler selection field, the field of picking the file with the source code of the solution, and so on. Clicking the "Submit" button sends the solution code for checking.

The "My Submissions" tab allows a participant to get information about all their attempts, the received grade and verdicts. The "Standings" tab contains a list of all students who have solved this task and their best score on this task (Fig. 3) and is intended primarily for teachers.

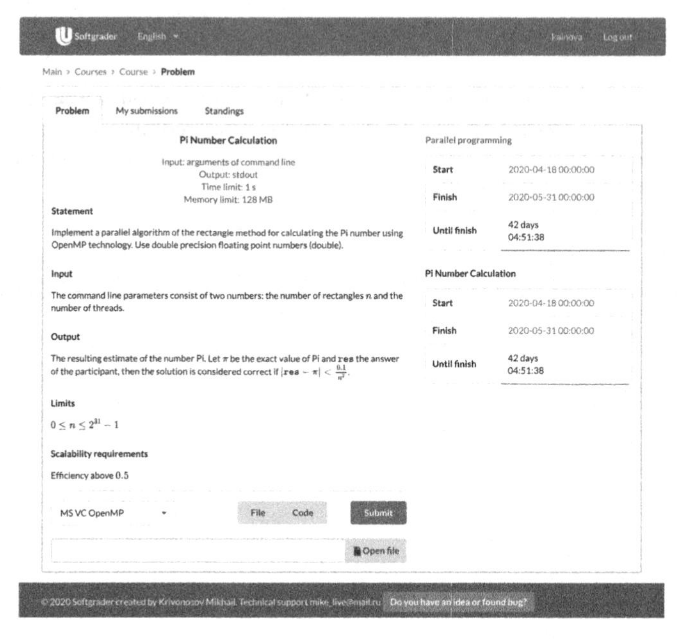

Fig. 2. Page of the "Pi Number Calculation" problem

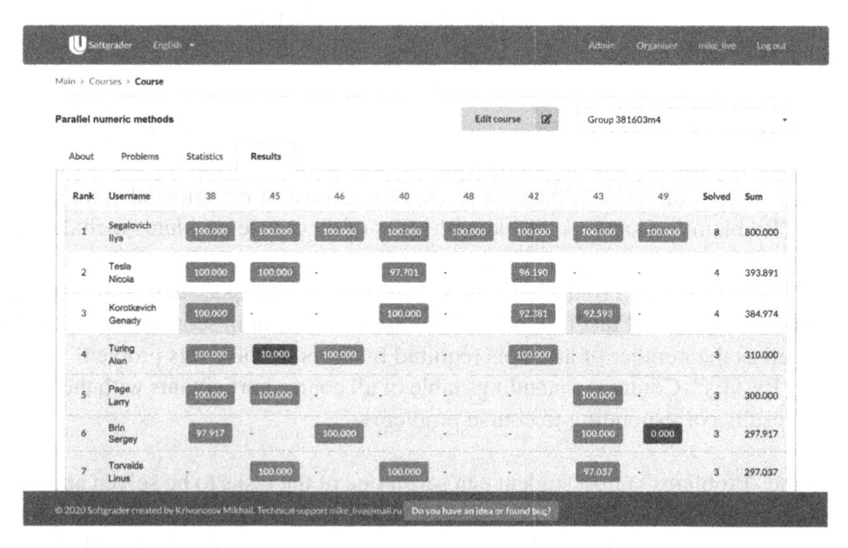

Fig. 3. The "Results" tab on the course page "Parallel numeric methods"

5 Teacher Workflow

The applying of an automated checking system in the training process requires substantial preparatory work. Thus, in order to add a new problem into the system a teacher (or his assistants) is required to perform the following preliminary steps:

1. formulate a detailed and precise problem statement;
2. implement a reference sequential version of the problem solution in the programming language from the list of supported, placing the whole code in one file;
3. implement a reference parallel version of the problem solution (if necessary) in C + + language, placing the whole code in one file;
4. develop a test generator program;
5. generate 20-100 tests with different sizes of the task input data;
6. develop a program-checker which determines the correctness of the obtained result by input data, output data and a known correct answer;
7. launch solutions from points 2 and 3 on the prepared set of tests;
8. run a "checker" on the received input data sets and answer sets;
9. achieve correctness of the obtained results by analyzing the "checker" answers, the source data and answers of the reference implementation, correcting the reference implementation if necessary.

Besides, for each task the teacher formulates requirements to efficiency of parallel implementation, which will be checked by the system of automated checking. The teacher should also form a scheme of assigning points for each task taking into account the number of tests passed and the efficiency indicators achieved.

As it can be seen from the above description, the preparatory work for the inclusion of tasks into the SoftGrader system is quite large, but these efforts will quickly pay off in the future use of the system, significantly reducing the time spent on checking the solutions prepared by the students.

Adding of the prepared task into the system is carried out through the web-interface. To create the task in the selected course it is necessary to press the button "Edit course" (Fig. 1). On the opened page (Fig. 4) press the button "Add new problem".

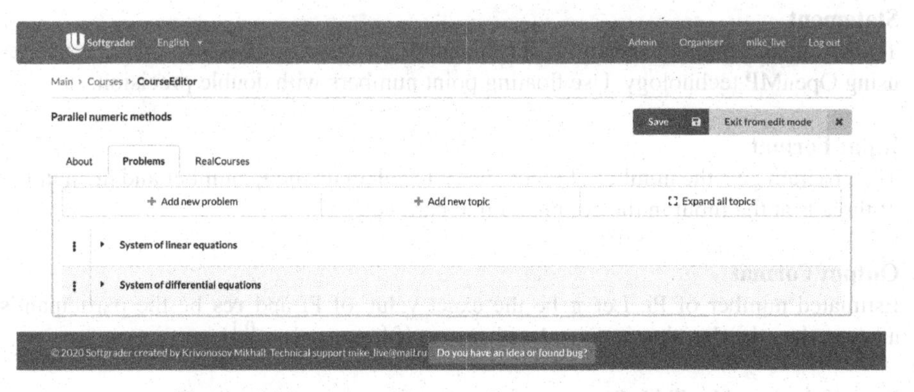

Fig. 4. Editing page of the "Parallel programming" course

There are three types of tasks (see Fig. 5): those that require development of an OpenMP-program, a regular sequential program or a program with a client-server architecture. At present, the first two variants are implemented.

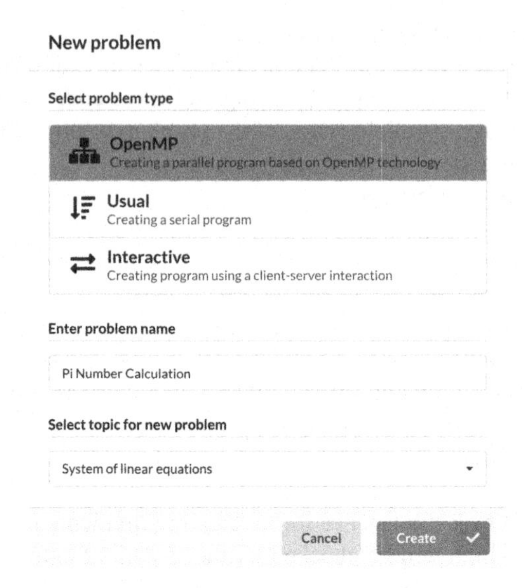

Fig. 5. The window of a new problem creating

After selecting the task type, entering the name, selecting the topic, in which the task will be added, and pressing the "Create" button, the page for editing the task (Fig. 6) will open, where you can enter its description in the provided typical sections or form your own rubric. You can form a task condition in both HTML and LaTeX format.

Let's consider the typical structure of description on the example of the task "Calculation of pi number".

Statement
To implement a parallel algorithm of calculating pi number by the method of rectangles using OpenMP technology. Use floating point numbers with double precision.

Input Format
The program gets the number of rectangles n to calculate the Pi number and the number of threads at the input in the command line parameters.

Output Format
Estimated number of Pi. Let π be the exact value of Pi and **res** be the participant's answer, then the decision is considered correct if $|res - \pi| < \frac{0.1}{n^2}$.

Restrictions on the Task Size

$$0 \leq n \leq 2^{31} - 1.$$

Requirements for Scalability
Efficiency above 0.5.

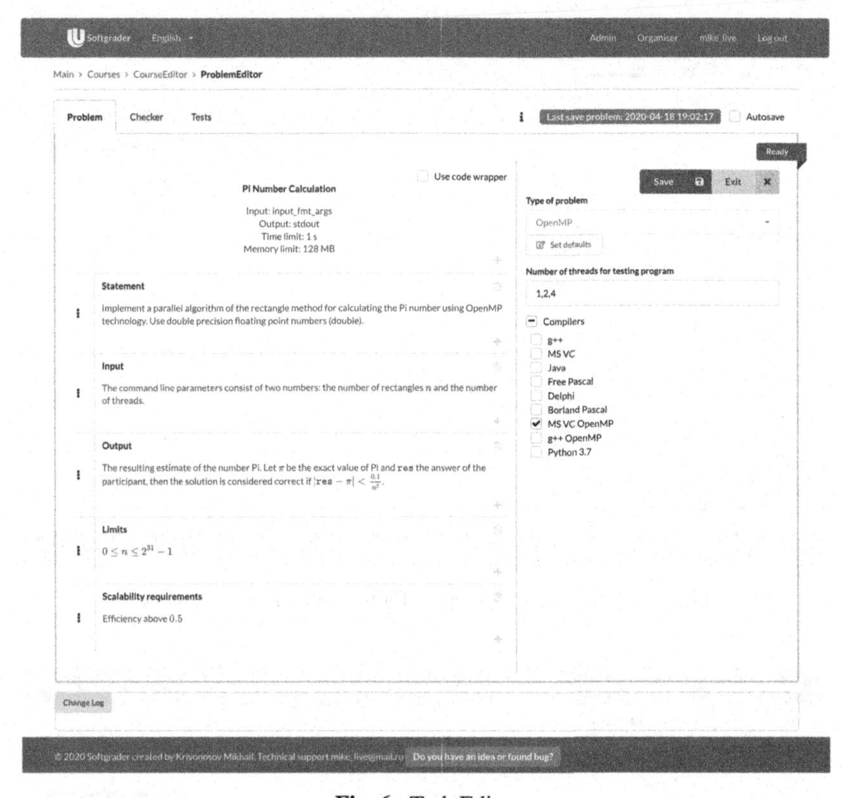

Fig. 6. Task Editor

The list of available compilers (with OpenMP support in this case) is also configured for the task. Input and output parameters are passed in the form of parameters of a function to be developed by the participant. The running time (100 s) and memory limits (255 megabytes) are set.

For tasks requiring development of sequential programs, there is obviously no need for a section with scalability requirements.

The remaining steps to enter the task into the system are performed on the same page (Fig. 6) on the other tabs.

On the tab "Code wrapper" it is necessary to load the code or write it directly in the editor. The code should perform the loading of the file with the input data, calls the function of the specified format, which should be prepared by the student, entering to it the read data, save the result of the functions in the resulting file. To facilitate the process of preparing a wrapper, the system provides a preparation of code, which the teacher can edit directly in the web-interface (Fig. 7).

On the "Checker" tab you need to upload a file with a program that will check the decisions of participants. "Checker" should be written in a certain format and return provided verdicts as a result of its work, so the system provides the generated template (Fig. 8), in which all the preparatory work is done, and the teacher only needs to implement the checking code.

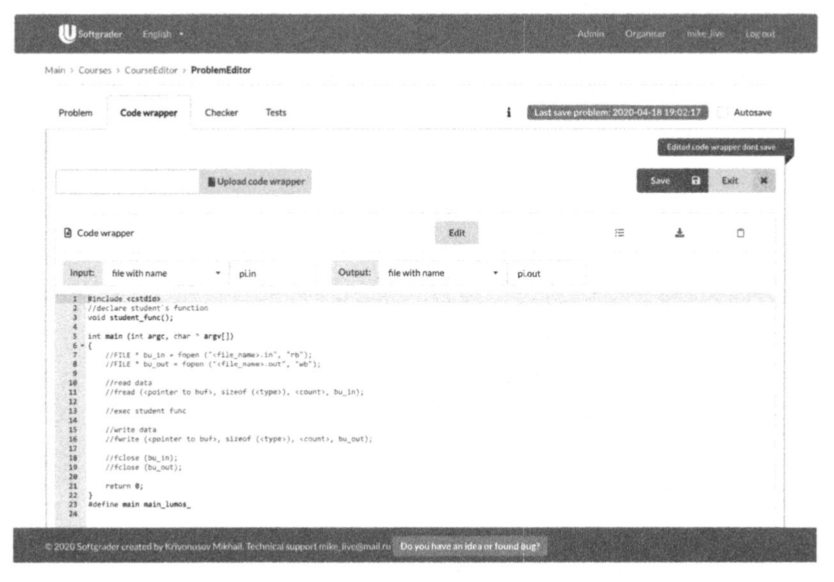

Fig. 7. Editor of the "wrapper" for the problem solving code

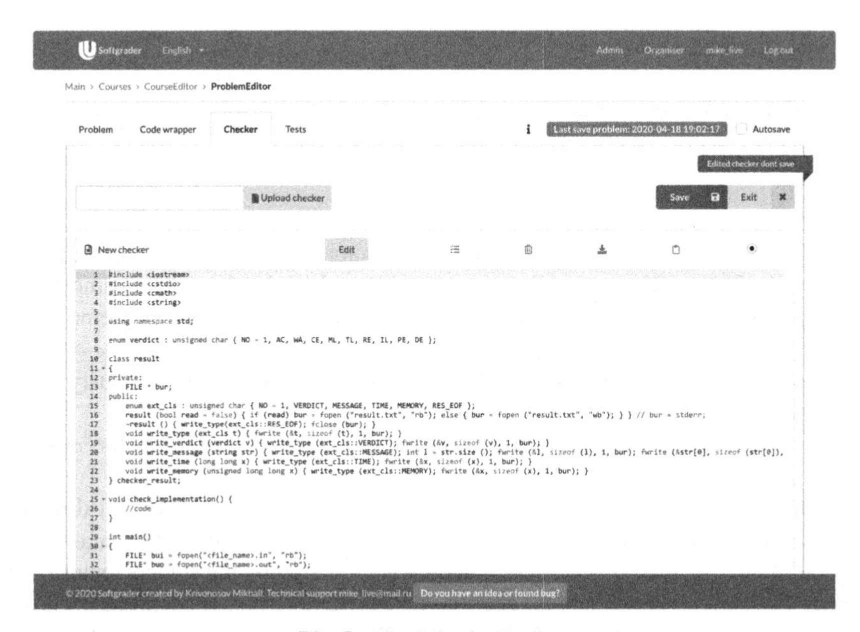

Fig. 8. The "checker" editor

The simplest situation with tests: prepared files are loaded on the same tab. Note only that each test should be formed as two files: a file with input data, a file with the result.

6 Using in the Learning Process

SoftGrader system has been used in the educational process of the Institute of Information Technologies, Mathematics and Mechanics of Lobachevsky State University of Nizhni Novgorod since 2017, providing control over the programs developed by students while mastering the disciplines:

- "Parallel Programming" is taught for 3rd year students in the specialties "Applied Mathematics and Informatics", "Fundamental Informatics and Information Technologies" and "Software Engineering". In the academic year 2019-2020, about 150 students study the discipline. About 30 tasks in various sections of computational mathematics were entered into the system already [5–7]. The adding process continues.
- "Parallel numeric methods" is taught for master's students of the 1st year of study in the direction "Applied mathematics and computer science". About 15 undergraduates study the discipline annually. 8 tasks were entered into the system on the main topics addressed in the course [8, 9].

The plans for 2020 include the introduction of SoftGrader system in practical classes on the basic programming disciplines "Programming fundamentals" and "Algorithms and data structures" which are studied at the ITMM Institute of UNN at 1st and 2nd year.

7 Conclusion

The paper describes SoftGrader automated checking system developed at the Institute of Information Technologies, Mathematics and Mechanics of the UNN. The system allows to carry out mass checking of programs developed by students in the process of mastering the disciplines of programmer orientation, primarily related to the field of high-performance computing and parallel programming [10].

Ongoing work on SoftGrader system is carried out in three directions:

1. addition of new courses and expansion of the set of tasks in the existing ones;
2. elaboration of the question on support of automated MPI-program checking;
3. bringing the system to a replicable state, which will allow its deployment in the interested organizations.

References

1. The PC2 Home Page. https://pc2.ecs.csus.edu/
2. Yandex.Contest Service. https://contest.yandex.ru/about/
3. Ejudge System. https://ejudge.ru/
4. SoftGrader system. http://softgrader.itmm.unn.ru/
5. Gergel, V., Sysoyev, A., Barkalov, K., Meyerov, I. et el.: High performance computing. 100 assignments for an extended laboratory workshop. FIZMATLIT (2018). (In Russian)

6. Gergel, V., Liniov, A., Meyerov, I., Sysoyev, A.: NSF/IEEE-TCPP curriculum implementation at the state university of nizhni novgorod. In: Proceedings of the International Parallel and Distributed Processing Symposium, № 6969501, pp. 1079–1084 (2014)
7. Meyerov, I., Bastrakov, S., Sysoyev, A., Gergel, V.: Comprehensive collection of time-consuming problems for intensive training on high performance computing. Commun. Comput. Inform. Sci. **965**, 523–530 (2019)
8. Gergel, V., Barkalov, K., Meyerov, I., Sysoyev, A. et el.: Parallel Numerical Methods and Technologies. UNN Press (2013). (in Russian)
9. Meyerov, I., Bastrakov, S., Barkalov, K., Sysoyev, A., Gergel, V.: Parallel numerical methods course for future scientists and engineers. Commun. Comput. Inform. Sci. **793**, 3–13 (2017)
10. Voevodin, V., Gergel, V., Popova, N.: Challenges of a systematic approach to parallel computing and supercomputing education. In: Hunold, S., et al. (eds.) Euro-Par 2015. LNCS, vol. 9523, pp. 90–101. Springer, Cham (2015). https://doi.org/10.1007/978-3-319-27308-2_8

Students' Favorite Parallel Programming Practices

Igor Konshin[1,2,3,4](✉)

[1] Marchuk Institute of Numerical Mathematics of the Russian Academy of Sciences, Moscow, Russia
igor.konshin@gmail.com
[2] Dorodnicyn Computing Centre of FRC CSC RAS, Moscow, Russia
[3] Moscow Institute of Physics and Technology (State University), Dolgoprudny, Moscow Region, Russia
[4] Sechenov University, Moscow, Russia

Abstract. The paper describes the students' most favorite parallel programming practices when studying a parallel computing course. Students' learning activities are considered as the motivation to study the features of parallel programming. A presentation of parallel computing theory for shared and distributed memory is introduced in a sequence that allows to make students interested in the practical work. There are also derived estimates of the programs' parallel efficiency which can be directly applied to the described learning activities.

Keywords: High performance computing · Parallel programming · Parallel efficiency estimation · Speedup

1 Introduction

The basis of any subject teaching is a sequence of material presentation, keeping up interest to the course by questions to the audience as well as by solving practical tasks, which results are not obvious to students. This is especially crucial while teaching high-performance computing or parallel programming as students are supposed to have already taken courses in programming languages such as C or C++. Sometimes students have already been familiar with OpenMP [1] and/or MPI [2], and for them a nonconventional teaching is particularly important. Furthermore, most of the young people initially consider themselves specialists in contemporary computer technology and its development trends. It is they who are so important to get interested in simple questions that students have not thought about before and which answers could become unexpectable.

There are many courses on high performance computing. For example, the site [3] has 267 courses in parallel programming, 255 courses in parallel computing, and 169 courses in high performance computing. Many of them are very popular, so about 50 thousand students took the "Parallel Programming Course

V. Voevodin and S. Sobolev (Eds.): RuSCDays 2020, CCIS 1331, pp. 511–523, 2020.
https://doi.org/10.1007/978-3-030-64616-5_44

by EPFL" course [4]. Among the courses in Russian, it is worth noting the selection of courses [5], as well as courses from N.I. Lobachevsky Nizhny Novgorod State University (see [6]) which include the practical training as well.

In contrast to the common approaches, the course presented is intended both for those who are faced with the topic of high-performance computing for the first time, and for students who have already studied this topic and are familiar with OpenMP, MPI, and even GPGPU programming. Of course, one can present material by classical books (see, for example, [7]), but it may be interesting for students to present the material from the OpenMP or MPI developer point of view, a kind of "invention" based on the needs of the parallel computer user. The presentation can be carried out from simple and obligatory to more complex and less commonly used. It is useful to immediately set a goal for students—to estimate the parallel efficiency for a given algorithm or program.

With the right selection of practical tasks, they can become not only a means of fastening the material covered and recognizing new possibilities of parallel programming, but also becoming a more exciting lesson than theoretical material in lectures. When performing practical tasks, the creative potential of students is manifested, this can become the basis of their interest in the subject and the further use of parallel computing in their future scientific work.

Additionally, the importance of emphasizing the differences in parallelization approaches for systems with shared and distributed memory can be noted. It is necessary to consider the difference in paradigms, goals and objectives of parallelization, dissimilarity in the methods used.

The offered course differs from the more traditional ones in that it is oriented at various levels of students' preliminary training, a close connection between the theoretical and practical parts of the course, a permanent emphasis on the common features and differences in parallel programming paradigms for systems with shared and distributed memory, in teaching students the ability to estimate the speedup, which he can theoretically and practically get for the selected algorithm on the parallel computing system used.

Section 2 discusses the structure of lectures on OpenMP and the sequence of presentation, Sect. 3 provides the parallel efficiency estimates for OpenMP programs. Section 4 contains intermediate tasks on OpenMP that fasten the material passed. Section 5 describes the sequence of MPI material presentation, while Sect. 6 provides the distributed memory parallel programming paradigm and estimates of the algorithms parallel efficiency. Section 7 contains intermediate test cases for parallel programming, including hybrid MPI and OpenMP parallelism. Section 8 gives a list of final practical tasks in parallel programming. Section 9 sums up the key findings.

2 OpenMP Lecture Structure

It is worth mentioning to students on the main reasoning for the appearance of OpenMP: the need to obtain a parallel version of the serial program on a shared memory computer with minimal code changes and the desire to keep it working when returning to a serial compiler or computer. In this regard, one can propose the following sequence of material presentation:

- The most "harmless" way to interfere with the program code is to use macros. Using the macro _OPENMP, you can decide whether this compiler supports parallelization via OpenMP:

```
#ifdef _OPENMP
    printf("OpenMP is supported\n");
#endif
```

- The next equally "harmless" way for a working program is to analyze the environment variables. For example, on the Linux operating system, in the command shell bash, the number of threads can be set up in the environment variable OMP_NUM_THREADS by the following command:

```
export OMP_NUM_THREADS=n
```

  Naturally, there are other OpenMP-related environment variables.
- The next group of OpenMP parallelization management tools is preprocessor directives. There are directives declaring parallel and sequential program areas; performing synchronization, as well as distributing work across threads. Since the main computational work is always concentrated in loops, the most important and most frequently used directive will be the loop parallelization directive:

```
#pragma omp parallel for
```

  Primarily, one may discuss the data dependency of loop iterations and then proceed to the option that controls the distribution of loop iterations across threads schedule(type[,chunk]). It is worth noting that so far only the most "light" and easily reversible parallelization tools have been considered and it has not even been necessary to include a file with description of OpenMP structures and functions:

```
#include <omp.h>
```

- In conclusion, one may consider the special functions of OpenMP. These functions specify the number of threads, the distribution of loop iterations, the functioning of locks (in the most difficult cases of parallelization). There are only two functions that can not be avoided with creative work on parallel computing. These are functions that can help calculate the parallel efficiency of a program or its individual fragments:

```
int omp_get_max_threads(void);
```

  getting the number of threads used (usually specified in applications by the environment variable), as well as the time measurement function:

```
double omp_get_wtime(void);
```

Mention may be made of the possible necessity for users of the newly invented parallel environment to run sometimes their parallel program on an older compiler that does not support OpenMP. In principle, there are special "stubs" that provide the correct behavior of OpenMP functions for a single thread mode. One of the additional practical tasks may be the development of such a library, supplemented by some script for preprocessing (or just removing from code) OpenMP directives.

To make the program work without changes both with and without OpenMP support, the following somewhat cumbersome code is usually used:

```
#ifdef _OPENMP
#pragma omp parallel for
#endif
    for (i=0; i<n; i++) {...}
```

The following loop parallelization tool may be interesting for students, that reduces the program code and makes it more clear:

```
#ifdef _OPENMP
   #include <omp.h>
   #ifdef _WIN32 /* MS Windows */
      #define OMP_PARALLEL_FOR __pragma(omp parallel for)
   #else /* Linux */
      #define OMP_PARALLEL_FOR _Pragma("omp parallel for")
   #endif
#else /* OpenMP is not supported */
   #define OMP_PARALLEL_FOR
#endif

#define FOR(iii) OMP_PARALLEL_FOR for(iii)

    FOR (i=0; i<n; i++) {...}
```

This macro will be compiled without warning on both Windows and Linux, as well as on compilers that support and do not support OpenMP.

3 Estimation of Parallel Efficiency for OpenMP Program

Right before proceeding to the estimation of parallel efficiency for programs on OpenMP, it is necessary to emphasize that parallelization for shared memory computers involves parallelizing of arithmetic (assuming that the memory access is ideally parallelized by the operating system itself). From this point of view, one can easily obtain an upper bound for speedup (S) and parallel efficiency (E) of the program, assuming that a certain fraction (σ) of computational work is performed sequentially:

$$S(p) = \frac{T_1}{T_p} = \frac{T_1}{\sigma T_1 + \dfrac{(1-\sigma)T_1}{p}} = \frac{p}{1+\sigma(p-1)}, \quad E(p) = \frac{S(p)}{p} = \frac{1}{1+\sigma(p-1)},$$

where p is the number of processes (threads), T_p is the program execution time for p processes, and $0 \leqslant \sigma \leqslant 1$.

The resulting estimate is ideal for parallelization by OpenMP and is called the Amdahl's law, formulated for the first time in the distant 1967, long before inventing OpenMP itself (1997). Students will probably be interested in looking at the paper [8], carefully stored on the Internet.

It may be unexpected for students when their programs, which provide evidently perfect parallelization, in practice show behavior that is far from ideal. Moreover, with an increase in the number of threads used, even a slowdown of the program may occur. The following paradigm can be proposed to explain this situation.

Suppose that several builders (computational threads) are working (performing arithmetic operations) in some not too large room. Due to the hustle (memory access conflicts) in a limited space (cash memory) or the limited speed of the construction resources delivery (RAM speed), delays can occur, up to a slowdown in program execution.

Remark 1. As some compensation for the disappointment brought about by the slowdown in their program, students can be asked to find the number of threads at which the operating time will be minimal.

4 OpenMP Practices

To reinforcement of learning, it is useful to carry out practical tasks that bring up non-obvious issues and allow one to discover new parallelization features. This section contains four tasks, each of which contains from 1 to 4 subjobs:

1. Print out the value of macro `_OPENMP` and determine
   - the supported version number in `yyyymm` format.
2. Print out the return value of the function `omp_get_threads_num()`:
   - in serial
   - and parallel sections.
3. In one program, set the number of threads in three ways:
   - `export OMP_NUM_THREADS=3`
   - `#pragma omp parallel for num_threads(4)`
   - `omp_set_num_threads(2)`
4. Implement matrix–matrix multiplication `dgemm` (see BLAS-3 [9]) and check the program performance depending on:
   - cycle order (6 different options);
   - task dimensions, for example, 1000 and 4000;
   - `schedule`;
   - `chunk`.

5 MPI Lecture Structure

In the further presentation of the material, one can follow a similar MPI "invention" strategy with the help of the audience, leaving unresolved issues for individual research when performing practical tasks.

The First Priority Functions. One can start the presentation with a set of the most necessary functions.

```
#include <mpi.h>
int main (int argc, char **argv) {
    int id, np;
    MPI_Init (&argc, &argv);
    MPI_Comm_rank (MPI_COMM_WORLD, &id);
    MPI_Comm_size (MPI_COMM_WORLD, &np);
    ... some code ...
    MPI_Finalize();
}
```

One can discuss with students what should be done in `MPI_Init` function, which can be indicated by `MPI_COMM_WORLD`, what information should the MPI communicator contain, why `id` and `np` are returned namely as arguments, but not by the function itself, what in this case is returned by MPI functions, is it possible to call MPI functions before invoking `MPI_Init` and after `MPI_Finalize`, why?

Blocking Communications. One can ask students to imagine what arguments the blocking sending function might have:

```
int MPI_Send( void *buf, int count, MPI_Datatype datatype,
              int dest, int msgtag, MPI_Comm comm );
```

The following questions can be suggested: why buffer `buf` is of type `void *`, why instead of the message length in bytes the two parameters `count` and `datatype` are used, what information could be passed in `msgtag` variable?

For arguments to the message receiving function

```
int MPI_Recv( void *buf, int count, MPI_Datatype datatype,
              int source, int msgtag, MPI_Comm comm,
              MPI_Status *status );
```

one can discuss the difference between `count` argument for `MPI_Send` and `MPI_Recv` functions, can `MPI_ANY_SOURCE` and `MPI_ANY_TAG` be used instead of `source` and `msgtag`, respectively, in which cases it can lead to misunderstandings, what information should be contained in `status` structure, how the exact length of the received message can be found out? One may also ask, does the termination of the function `MPI_Send` mean that the message has completely reached its destination? that at least they have already begun to receive it?

Non-blocking Communications. For non-blocking functions `MPI_Isend` and `MPI_Irecv`, the main issue for discussion is differences in the use of the buffer `buf` compared to the blocking version of the respective function.

```
int MPI_Waitall( int count, MPI_Request *requests,
                 MPI_Status *statuses );
```

For the function `MPI_Waitall`, one can find out whether it is allowed to combine requests in the `requests` array to complete operations with both `MPI_Isend` and `MPI_Irecv`?

Persistent Requests for Communication. The main question here is to find out why it may be necessary to combine different send/receive operations in a single MPI call, if non-blocking communication functions already exist? Perhaps the students will be convinced by the possibility of reducing the overhead during the transfer of information between the process and the network controller. Although in practice, unfortunately, the opposite result may turn out, it depends on the MPI implementation.

Coupled Communications. Operation `MPI_Sendrecv`, which organizes the simultaneous reception and sending of a message, deserves a separate conversation. Hardware support for two-way communication can really help reduce the total time of data exchanges.

Collective Communications. Some algorithms may require functions `MPI_Bcast`, `MPI_Gather`, and `MPI_Reduce`, but in any case the function `MPI_Allreduce` is used most often in numerical methods, which is worth paying special attention to:

```
int MPI_Allreduce( void *sbuf, void *rbuf, int count,
        MPI_Datatype datatype, MPI_Op op, MPI_Comm comm );
```

despite the apparent simplicity of its arguments. The main issue is the discussion of the properties that operation `op` is to be satisfied. For sure, a difference in the result when calling this function again with the same arguments may become a surprise. This behavior can happen due to the influence of round-off errors in the non-deterministic order of calculations with data received from the processes. It can be said that for a sufficiently large number of processes (more than 1000) this function can significantly slow down. This effect caused the appearance of hundreds of scientific papers devoted to modifications of the iterative BiCGStab method, in which instead of three calls to `MPI_Allreduce` function, only two calls are made at each iteration, despite the fact that the modified method becomes less stable.

Synchronization and Time Measurement. Why does one need to synchronize processes? The command `time` from the command shell `bash` returns three types of time (real, user, system). What time should the time measurement function `MPI_Wtime` return? What does the letter `W` present in the function name mean ("wall clock" time but not "world" time)? Can this function be called before `MPI_Init` and after `MPI_Finalize`? Why?

What is the most correct way to measure the execution time of a program fragment?

```
double t;
MPI_Barrier( MPI_COMM_WORLD );
t = MPI_Wtime() ;
... some code ...
MPI_Barrier( MPI_COMM_WORLD );
t = MPI_Wtime() - t;
```

Will there be differences in measured time on several processes? What is the sense of time measurement in the above fragment if the second synchronization is removed? If in the latter case the differences in the measured time at different processes turn out to be significant, what relation is it to the parallel efficiency of the program fragment being studied? How can one characterize such a situation (disbalance in processes loading)?

6 Estimation of Algorithm' Parallel Efficiency for MPI

To evaluate the parallel efficiency of the algorithm for distributed memory, in contrast to the harmonious "builders" of Sect. 3, we can propose a different parallel computing paradigm.

Several submarines (MPI processes) float in an ocean, each one contains a sailor (for performing arithmetic calculations) and a radio operator (for sending/receiving messages). Local data (local memory) are nearby, and nothing is known about data on other submarines (only submarine numbers are available). One can work with your own data quite successfully (arithmetic is fast), but requesting (by Morse code) and then receiving data from neighboring submarines is a very long process (communications are relatively slow). In this case, the main delays in work may occur due to the transmission of messages, and they should be taken into account when estimating the parallel efficiency of collaboration.

Remark 2. If parallelization of MPI is used in conjunction with OpenMP, then this means that the submarine has several sailors (computational threads) who share the local data of their submarine (which is shared memory for them).

For message transmission time, a commonly used estimate is applied:

$$T_c = \tau_0 + \tau_c L_c,$$

where τ_0 is the message initialization time, τ_c is the message transmission rate (message transmission time for unit length), and T_c is the message transmission time for length L_c. If the messages are long enough, then one can put $\tau_0 = 0$ and the previous formula will be essentially simplified:

$$T_c = \tau_c L_c.$$

The execution time of arithmetic operations can be expressed by a similar formula $T_a = \tau_a L_a$, where τ_a is the execution time of one arithmetic operation, and L_a is the total number of arithmetic operations.

Then we can introduce two auxiliary values

$$\tau = \tau_c/\tau_a, \qquad L = L_c/L_a,$$

where τ is the characteristic of the "parallelism" resource of the computer used (how many arithmetic operations can be completed during the transfer of one number to another MPI process), while L is the "parallelism" reserve of the algorithm under study (how many numbers should be transferred from another MPI process to perform one arithmetic operation).

If $T(p)$ is the program execution time on p MPI processes, then following [10, 11], one can easily estimate the resulting speedup:

$$\begin{aligned} S = S(p) &= T(1)/T(p) = T_a/(T_a/p + T_c/p) = pT_a/(T_a + T_c) \\ &= p/(1 + T_c/T_a) = p/(1 + (\tau_c L_c)/(\tau_a L_a)) = p/(1 + \tau L). \end{aligned} \quad (1)$$

Estimate of the algorithm' parallel efficiency is even simpler:

$$E = S/p = 1/(1 + \tau L).$$

One should definitely ask students: where in the last estimate the dependence on the number of processes p is hidden.

The resulting estimate can be applied to various linear algebra algorithm [10], mathematical physics problems [11], as well as practical tasks from Sect. 8.

7 Parallel Programming Practices

When doing practical tasks, it is important that the students are interested in the statement of the task itself, so that they know in advance why this is necessary and what question they want to answer with this experiment. It is also very important to combine in one program two models of parallelization via MPI and OpenMP, to contrast their parallelization techniques, as well as the ability to directly compare their performance [12]. Especially informative there will be the fact that in most cases usage of one variant (either MPI or OpenMP) will be less time effective than their hybrid operation.

- Implement parallelization of `daxpy` operation (see BLAS-1 [9]) ($\overrightarrow{y} = \alpha \cdot \overrightarrow{x} + \overrightarrow{y}$) simultaneously using MPI and OpenMP. For how many processes in each of the three possible startup options (OpenMP, MPI, MPI+OpenMP) will the execution time be minimal?
- Implement the calculation of the norm of the vector `dnrm2` (see BLAS-2 [9]) similarly to `daxpy`, and then determine the optimal number of processes for each of the three (OpenMP, MPI, MPI+OpenMP) options (the results will be very different from `daxpy`).
- Implement with MPI in several ways the summation over a single number of type `double` (similar to `MPI_Allreduce`) and compare their performance (see, for example, implementation using `MPI_Send/MPI_Recv`, `MPI_Isend/MPI_Irecv`, `MPI_Gather/MPI_Scatter`, as well as using `MPI_Allreduce` itself).
- Calculate the execution time of one arithmetic operation (in `double`) τ_a, transfer a single number (`double`) τ_c, and find their ratio $\tau = \tau_c/\tau_a$ as the main characteristic of computer parallelism.

- Calculate the initialization time τ_0 for the message from the formula $T_c = \tau_0 + \tau_c L_c$ and compare it with the time τ_c, getting τ_0/τ_c.

In the G.V. Baidin paper [13] on stereotypes of parallel programming, several tests are considered. Some of these tests are reformulated so that students can perform them on their own or using ready-made program fragments.

- Is it efficient to perform communications in the background of computations? Compare the program execution time in the case of blocking and non-blocking communications.
- How does the message transfer rate depend on its length? What does the transmission time of a message tend to when its length decreases (to τ_0)? When its length increases (to τ_c)? Is there a maximum rate somewhere in the middle (actually, it depends on the MPI implementation)?
- What is faster: collective communications (say, `MPI_Bcast`) or consecutive blocking communications (for example, through `MPI_Sendrecv`)?

Remark 3. The most surprising thing in the last three tests is that the answer can be reversed depending on the implementation of MPI and the computer used.

8 Final Practical Task

As a final practice, students can choose one of the following rather serious tasks:

(a) *Research on the "fine" characteristics of communications.* Find the dependence of the communication rate (τ_c) and its initialization time (τ_0) on the following parameters:
  - communication length;
  - communication type (blocking vs. non-blocking);
  - the number of simultaneously running non-blocking communications;
  - the number of processes involved;
  - type of compute nodes (if available).

(b) *2D game "Life"* [14]. Random initial distribution of living cells. Periodic boundary conditions with the closure of a square region in a torus. Stationarity/periodicity control. Investigation of several rules for "alive/dead" cells transitions. Interested in the rules providing the longest variations of the initial state of the field. 2D decomposition of the field by processes. The most brave can try to implement a 3D version of the game with a 3D decomposition by processes (see [15]).

(c) *3D heat equation:* $\partial u/\partial t - a^2 \Delta u = f$. The cold walls of the region have a fixed temperature, a constant heat source is located in the center of the domain. 3D decomposition of the source grid by processes.

(d) *3D diffusion equation:* $\partial c/\partial t = \mathrm{div}(D\, \mathrm{grad}\, c) + f$. 3D decomposition of the source grid by processes.

(e) *Preconditioned conjugate gradient method for the Poisson equation.* 2D grid of nodes, 5-point discretization stencil (sparse matrix with elements: -1 -1 4 -1 -1), matrix is distributed by block rows. A preconditioner without overlap (block Jacobi method) with IC0 factorization (preserving the structure of the original matrix) for its central block of the matrix. The exact solution is $x^* = 1$, the initial guess is $x_0 = 0$, iterating until the norm of the initial residual is reduced by 10^6 times.

(f) *Preconditioned conjugate gradient method for a system with a dense matrix.* A small diagonal dominance of the (symmetric!) matrix, the distribution of the matrix by block rows. Non-overlapping preconditioner (block Jacobi method) with the Cholesky factorization of its central matrix block. The exact solution is $x^* = 1$, the initial guess is $x_0 = 0$, iterating until the norm of the initial residual is reduced by 10^6 times.

(g) *Matrix–matrix multiplication* `dgemm` (see BLAS-3 [9]): $A \cdot B = C$. Dense square matrices with random elements. Row/column distribution of data across processes without duplication of the initial and final distribution of data, for example, A and C – by block rows, B – by block columns.

(h) *Parallel sorting with memory limit.* Parallel implementation of Batcher's odd-even mergesort, bitonic sort, or a similar one. Each process sorts its own part of a random vector, than it sends portions of the vector to other processes and collects received portions which is in its own sorting interval. The total working memory should not exceed 3 total vector lengths, and the full vector in parallel implementation is not collected in one process.

Remark 4. In each of the proposed problems, there is a sufficient parallelism resource to obtain a good parallel efficiency, especially if one take a larger dimension of the problem. For problems (b)–(e) arithmetic costs are proportional to the number of internal nodes, and exchanges – to the number of boundary nodes; (f) arithmetic costs are proportional to N^2, and exchanges – to N; (g) arithmetic costs are proportional to N^3, and exchanges – to N^2; (h) arithmetic costs for sorting are proportional to $N \log N$, and exchanges – to N. Here, N denotes the dimension of the vector.

For the selected task, the following should be done:

- design and write the program code with parallelization via MPI;
- by himself verify the correct operation of the program;
- measure the execution time on 1, 2, 4, 8, 16, 32, 64 processes and calculate the speedup obtained;
- theoretically estimate the speedup using formula (1) by the computational and communication costs of the algorithm and the machine-dependent characteristic τ (the ratio of communication rate to arithmetic execution rate), which is required to be measured by himself for a specific computer and the arithmetic operations used in the program;
- plot speedup graphs and compare theory with practice;
- write a short scientific report.

Remark 5. Instead of choosing one of the presented practices, the student can apply parallel programming skills in his scientific work and present the results of parallel calculations, which reduce the time for his own problem. This would be the most valuable result of studying the course.

Remark 6. As a practical task, multiplication of dense matrices and parallel sorting algorithms are most often chosen. And for those who want to give a lecture, parallel processing in Python, parallel programming with CUDA, or parallel programming C++11 threads are more interesting.

9 Conclusion

The paper describes the sequence of material presentation on parallel programming by OpenMP and MPI for computers with shared and distributed memory, respectively. There are highlighted theoretical issues to be worth for studying as well as the sequence of their presentation. The theoretical estimates of the algorithms and programs parallel efficiency are also derived. The most interesting activities for students on OpenMP, MPI, and the hybrid use of MPI and OpenMP are widely covered by the paper. The programming practices has been divided into two groups: intermediate tasks performed to clarify the features of every parallelization method, and final tasks, where along with their parallel programming skills students should apply creative and analytical abilities in study theoretical and practical parallelism properties of the implemented algorithm.

Acknowledgements. This work has been supported by the RAS Research program No. 26 "Basics of algorithms and software for high performance computing" and RFBR grant No. 18-00-01524.

References

1. The OpenMP API specification for parallel programming. https://www.openmp.org. Accessed 15 Apr 2020
2. MPI: The Message Passing Interface standard. http://www.mcs.anl.gov/research/projects/mpi/. Accessed 15 Apr 2020
3. https://www.coursera.org/. Accessed 15 Apr 2020
4. Parallel Programming Course by EPFL. https://www.coursera.org/learn/parprog1. Accessed 15 Apr 2020
5. Training materials on parallel computing and supercomputers. https://parallel.ru/info/education. Accessed 15 Apr 2020
6. Gergel, V.P.: Multiprocessor computing systems and parallel programming. (in Russian) http://www.hpcc.unn.ru/?doc=98. Accessed 15 Apr 2020
7. Voevodin, V.V., Voevodin, Vl.V.: Parallel Computing. BHV-Petersburg, St. Petersburg (2002). (in Russian)
8. Amdahl, G.M.: Validity of the single-processor approach to achieving large scale computing capabilities. In: AFIPS Conference Proceedings, Atlantic City, N.J., 18–20 April, vol. 30, pp. 483–485. AFIPS Press, Reston, Va. (1967). http://www-inst.eecs.berkeley.edu/~n252/paper/Amdahl.pdf. Accessed 15 Apr 2020

9. BLAS (Basic Linear Algebra Subprograms). http://www.netlib.org/blas/. Accessed 15 Apr 2020
10. Konshin, I.: Parallel computational models to estimate an actual speedup of analyzed algorithm. In: Voevodin, V., Sobolev, S. (eds.) RuSCDays 2016. CCIS, vol. 687, pp. 304–317. Springer, Cham (2016). https://doi.org/10.1007/978-3-319-55669-7_24
11. Konshin, I.: Efficiency estimation for the mathematical physics algorithms for distributed memory computers. In: Voevodin, V., Sobolev, S. (eds.) RuSCDays 2018. CCIS, vol. 965, pp. 63–75. Springer, Cham (2019). https://doi.org/10.1007/978-3-030-05807-4_6
12. Konshin, I.: Efficiency of basic linear algebra operations on parallel computers. In: Voevodin, V., Sobolev, S. (eds.) RuSCDays 2019. CCIS, vol. 1129, pp. 26–38. Springer, Cham (2019). https://doi.org/10.1007/978-3-030-36592-9_3
13. Baidin, G.V.: On some stereotypes of parallel programming. Vopr. Atomn. Nauki Tekhn. Ser. Mat. Model. Fiz. Prots. **1**, 67–75 (2008). (in Russian)
14. Conway's Game of Life. https://en.wikipedia.org/wiki/Game_of_Life. Accessed 15 Apr 2020
15. Bays, C.: Candidates for the game of life in three dimensions. Complex Syst. **1**(3), 373–400 (1987). https://wpmedia.wolfram.com/uploads/sites/13/2018/02/01-3-1.pdf. Accessed 15 Apr 2020

The Algorithms Properties and Structure Study as a Mandatory Element of Modern IT Education

Alexander Antonov[1,2](✉) and Vladimir Voevodin[1,2]

[1] Lomonosov Moscow State University, Moscow, Russia
[2] Moscow Center of Fundamental and Applied Mathematics, Moscow, Russian Federation
{asa,voevodin}@parallel.ru

Abstract. The paper describes the system of mass practical assignments formed in the framework of the course "Supercomputing Simulation and Technologies" at the Faculty of Computational Mathematics and Cybernetics at Lomonosov Moscow State University. The practical assignments were held for four years, from 2016 to 2019, and each year about 200 students of the second year of the magistracy passed through them. These practical assignments are aimed at developing skills in the efficient use of parallel computing systems and knowledge of properties of parallel algorithms. The basics of performing the described assignments is the deep analysis of parallel algorithms and use of the concept of the information structure of algorithms and programs. This concept is extremely important for writing efficient parallel programs, so it should be the cornerstone of modern IT education.

Keywords: High-performance computing education · Structure of parallel algorithms · Information structure · Parallelism resource · Parallel programming · Supercomputers · AlgoWiki

1 Introduction

Understanding the fact that parallelism is an integral part of modern computing devices has long spread throughout the world. Therefore, it is not surprising that the study of parallel technologies has become the basis for the construction of many educational curricula in a variety of educational institutions [1–13].

The study of the algorithms parallel structure and properties has become a key objective of a series of practical assignments that were held as a part of the course "Supercomputing Simulation and Technologies" in 2016–2019. This course is taught to Master's degree students of the Faculty of Computational Mathematics and Cybernetics (CMC) at Lomonosov Moscow State University (MSU) during the final year of education. Annually, about 230 students study this discipline.

V. Voevodin and S. Sobolev (Eds.): RuSCDays 2020, CCIS 1331, pp. 524–535, 2020.
https://doi.org/10.1007/978-3-030-64616-5_45

The course covers the following topics:

- computationally complex tasks in various application fields;
- architectural features of modern processors, new approaches to creating high-performance systems;
- introduction to the analysis of the parallel structure of algorithms;
- scalable algorithms for parallel supercomputer systems: problems and prospects;
- algorithmic approaches for solving problems in various application areas: turbulence, molecular dynamics, climate change, supercomputer drug design etc.;
- additional chapters of MPI, OpenMP and CUDA parallel programming technologies;
- and others.

This course consists of lectures and seminars. Lectures are given on a weekly basis in the fall semester, for four academic hours per week.

The purpose of the practical assignments given to students in the framework of this course is developing skills in the efficient use of parallel computing systems.

2 Formulation of Practical Assignments

The overall goal of the practical assignments was to learn to explore and understand the parallel structure and properties of algorithms. In some cases, students were allowed to work in pairs, which contributes to the development of teamwork skills.

To perform numerical experiments, a wide variety of supercomputer platforms available in the supercomputing center of Moscow State University were provided:

- Lomonosov-2 supercomputer [14],
- Lomonosov supercomputer [15],
- Blue Gene/P supercomputer,
- computing cluster with Angara network,
- systems based on Power8 / Phi (KNL) / GPU (K40, Pascal).

A feature of the verification and acceptance of practical assignments was that in this case it was important not only to give an assessment, but to learn how to analyze and describe the properties of algorithms correctly. This, of course, requires highly qualified teachers and considerable time for multiple iterations in communication with students.

2.1 Practical Assignments on the Study and Description of the Structure and Properties of Algorithms (2016)

Study and describe the structure and properties of the selected algorithm.

Despite the simplicity of the formulation, the assignment is not at all trivial. It is difficult to even explain what it means to describe the structure and properties of algorithms. In our case, the methodological basis of this practical task is an AlgoWiki Open encyclopedia of parallel algorithmic features [16,17].

From the complete universal structure for describing the algorithms of the AlgoWiki project, for the practical assignments it was required to describe the entire first part related to the machine-independent properties of the algorithms. Of the dynamic characteristics of the algorithms described in the second part [18], only the study of the scalability of the algorithm implementation and the compilation of a list of existing implementations were left (Fig. 1).

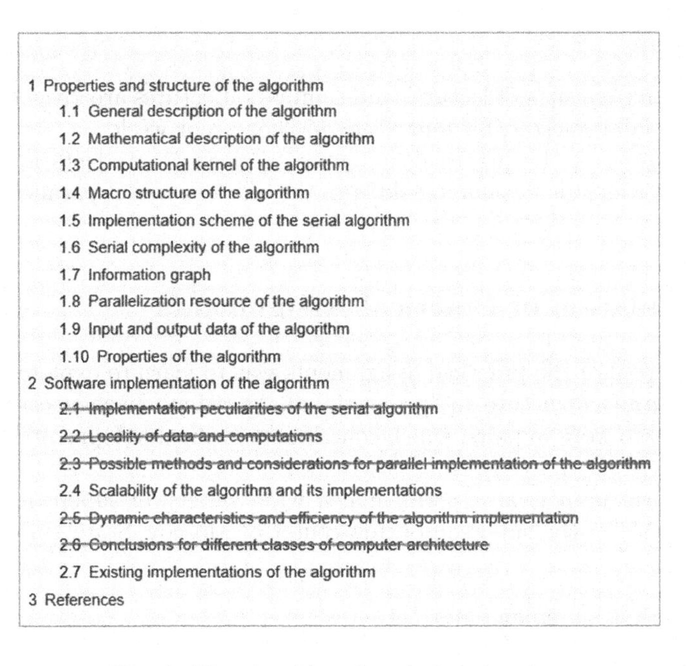
1 Properties and structure of the algorithm
1.1 General description of the algorithm
1.2 Mathematical description of the algorithm
1.3 Computational kernel of the algorithm
1.4 Macro structure of the algorithm
1.5 Implementation scheme of the serial algorithm
1.6 Serial complexity of the algorithm
1.7 Information graph
1.8 Parallelization resource of the algorithm
1.9 Input and output data of the algorithm
1.10 Properties of the algorithm
2 Software implementation of the algorithm
~~2.1 Implementation peculiarities of the serial algorithm~~
~~2.2 Locality of data and computations~~
~~2.3 Possible methods and considerations for parallel implementation of the algorithm~~
2.4 Scalability of the algorithm and its implementations
~~2.5 Dynamic characteristics and efficiency of the algorithm implementation~~
~~2.6 Conclusions for different classes of computer architecture~~
2.7 Existing implementations of the algorithm
3 References

Fig. 1. The algorithm description structure

The students (or pairs) could select one of 30 proposed algorithms for description, particularly:

- Jacobi's method for the symmetric eigenvalue problem,
- Jacobi's method for the singular value decomposition,
- the Lanczos algorithm with full reorthogonalization for the symmetric eigenvalue problem,
- Gram-Schmidt orthogonalization process,
- GMRES (Generalized Minimal RESidual method) as an iterative method for solving systems of linear algebraic equations,
- QR algorithm for solving the algebraic eigenvalue problem,
- Newton's method for solving systems of nonlinear equations,

- fast discrete Fourier transform,
- minimum spanning tree based clustering algorithm,
- *k*-means clustering algorithm,
- and other algorithms.

Each algorithm is accompanied by links to well-known books that explain it. Thus, students have a reliable source of information. At the same time, both a student and a teacher know exactly which algorithm should be described.

Despite the apparent simplicity, this assignment contains many hidden difficulties. It is far from easy sometimes to answer even the most basic questions that arise, for example:

- What does it mean to "distinguish and describe the full resource of algorithm parallelism"?
- How to show possible ways of parallel execution?
- What does it mean to "define and show the information structure of an algorithm"?
- How to show the macrostructure of the algorithm?
- By what signs should the computing core of the algorithm be distinguished?
- and other.

More details about the features of this practical assignments can be found in the paper [19]. As a result of completing the assignment for each of the 30 proposed algorithms, several descriptions were obtained within the AlgoWiki encyclopedia, using the data of which experienced experts could make full-fledged articles on these algorithms.

Table 1 shows the results of the evaluation of student practical assignments. Most students received positive assessments, a fairly high average mark (4.03) also characterizes the overall high quality of work.

Table 1. Final assessments (2016)

Year	2016
Groups (students)	145 (246)
5 (Excellent)	59 (41%)
4 (Good)	36 (25%)
3 (Satisfactory)	48 (33%)
2 (Unsatisfactory)	2 (1%)
Average mark	4.03

2.2 Practical Assignments on Scalability Study (2017–2019)

Studying the scalability of algorithms and their implementations on various computing platforms when changing the size of the task and the number of processors.

Students needed to conduct a series of computational experiments, collect the necessary data, build graphs with dependencies, interpret, and draw conclusions about scalability. In [20] scalability is defined as a property of a parallel application that describes the dependency of changes in the full range of dynamic characteristics for that program on the full range of its startup parameters. In these assignments, students investigated the dependence of one dynamic characteristic (execution time or performance metric) on two parameters—the number of processes involved and the size of the problem to be solved. For graph problems, the size was determined by the number of vertices, and for matrix ones, by the linear size of the matrix. An example of a graph similar to which it was required to obtain can be seen in Fig. 2.

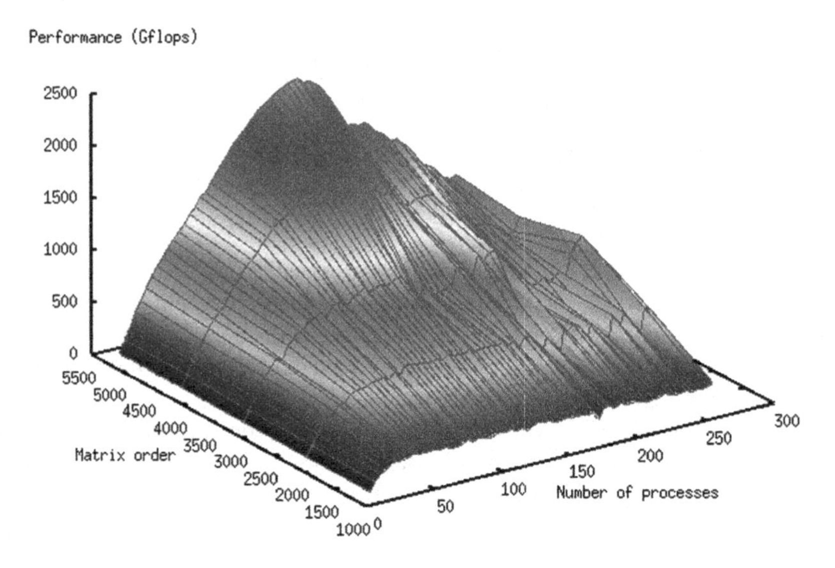

Fig. 2. The scalability of the Cholesky decomposition

In 2017 and 2019, the object of the study were algorithms solving graph problems such as:

- single source shortest path (SSSP),
- breadth-first search (BFS),
- Page Rank,
- construction of the minimum spanning tree (MST),
- finding the strongly connected components (SCC).

For each problem, several algorithms were considered, for example, for the single source shortest path problem, these are the algorithms:

- Bellman-Ford algorithm,
- Dijkstra's algorithm,
- Δ-stepping algorithm.

For each algorithm, up to 5 different ready-made implementations were also provided.

Each student had to experiment with two types of graphs: R-MAT [21] and SSCA2 [22]. Graph generators were provided.

The resulting task for students suggested that for both types of graphs (RMAT and SSCA2) and the selected chain "Problem → Algorithm → Implementation → Computer Platform" [23] was necessary:

- determine the dependence of performance (TEPS) on the number of processors (cores) and graph size;
- find a combination of the number of processors and the size of the graph on which maximum performance is achieved.

Particular attention was required to be paid to large graphs (2^{25} – 2^{28} vertices), since this corresponds to real practical applications, and also in order to avoid the influence of cache memory.

In 2018, the object of the study were linear algebra algorithms:

- SLAU solution,
- eigenvalue calculation,
- eigenvector calculation,
- calculation of singular values,
- LU factorization,
- QR factorization,
- LQ factorization,
- QL factorization,
- RQ factorization,
- and others.

For each algorithm, different ready-made implementations from ScaLAPACK [24] and Magma [25] libraries were provided.

As computing platforms, the SMP nodes of all computer platforms available at the RCC of Moscow State University were considered.

As a result of the practical assignments, students obtained filled tables with the running times of various versions for the chosen algorithm implementations. Some students got the same version of the assignment to complete, which led to an element of competition, which is always useful. Many students began to wonder why their results were worse than others, looked for the reasons for this, and tried to improve something. In the search for the correspondence of the architecture and the implemented algorithm, even ideas of supercomputer co-design [26] were appeared.

This version of the practical assignments requires serious computer resources to get all the necessary performance data for a large number of parallel program runs:

- various sizes of graphs or matrices,
- different numbers of processors,
- various input options, for example, two types of graphs: RMAT and SSCA2,

- for each point on the graph, it is necessary to conduct several experiments in order to remove random fluctuations (artifacts) in performance.

When studying scalability, it is important to draw students' attention to the need to explain all the special points on the graphs: peaks, inflection points, asymptotes, and others. A detailed analysis is complicated, but all the features of the behavior of the algorithm (implementation) must be connected with the properties of the architecture of the computing system.

Table 2 shows the results of the evaluation of student practical assignments. The reason for obtaining very high marks was the need for repeated finishing work in the presence of shortcomings noted by teachers.

Table 2. Final assessments (2017, 2019)

Year	2017	2019
Groups (students)	143 (207)	155
5 (Excellent)	121 (85%)	130 (84%)
4 (Good)	15 (10%)	12 (7.7%)
3 (Satisfactory)	5 (3%)	4 (2.6%)
2 (Unsatisfactory)	2 (1%)	9 (5.8%)
Average mark	4.78	4.7

In 2018, the grading scale for the assignment was more differentiated, including putting down assessments with minuses. The results are shown in Table 3. It can be seen here that a more differentiated approach allows more flexible assessment of the remaining shortcomings of student work.

Table 3. Final assessments (2018)

Students	176
5	25 (14%)
5–	79 (45%)
4	33 (19%)
4–	22 (12.5%)
3	14 (8%)
3–	3 (1.5%)

In general, it is clearly seen that the overall level of assessments for these practical tasks is significantly higher than the level of assessments received in 2016 (see Sect. 2.1). This is due to a noticeably simpler formulation of the task.

The results obtained from these practical assignments are used to populate the database of the project for creating the Algo500 system [27].

2.3 Practical Assignments on the Information Structure of Algorithms and Programs (2019)

Identify and analyze the information structure of the algorithm or a piece of program code.

When preparing the assignment, despite the simple formulation, it was required to find the answer to a number of difficult questions, such as: "How to build the information structure of the algorithm?", "How to depict the information structure of the algorithm?" (the question only at first glance seems simple), "If we know (see) the information structure, how to describe potential parallelism?" etc.

The basis of the analysis of the properties of programs and algorithms is the analysis of the information structure, which is determined by the presence of information dependencies in the analyzed fragment. An informational dependency [28] between two operations exists if the second of them uses what was calculated in the first. Based on the concept of information dependence, an information graph can be constructed using an algorithm or fragment which is an oriented acyclic multigraph whose vertices correspond to the operations of the algorithm, and arcs correspond to information dependencies between them. The information graph is used as a convenient representation of the algorithm in the study of its structure, parallelism resource, as well as other properties.

Our first task was to distribute fragments of the studied program code for more than 200 students. For this, a special parameterized program fragment was constructed.

```
for(i = 1; i <= n; ++i)
   C[i] = C[i+18] * e;
for(i = 1; i <= n; ++i)
   for(j = 1; j <= n; ++j)
      B[i][j] = B[i+16][j+17] + (1-15)*C[i];
for(i = 1; i <= n; ++i)
   for(j = 1; j <= n; ++j){
      for(k = 1; k <= n; ++k)
         A[i][j][k] = A[i][j][k] + 14*A[i+11][j+12][k+13] +
                                 (1-14)*A[i+11][j+12][n];
      A[i][j][n] = A[i][j][n] + 15*B[i][j];
   }
```

A specific variant for each student was given by a set of parameter values $l1, \ldots, l8$. Some of these parameters took only the values 0 or 1, which corresponds to the presence or absence of an information dependence for the corresponding measurement, other parameters could take other values to specify different types of dependencies. In total, the number of variants exceeded the number of students in the course, so each student received an individual option for research.

In addition to constructing an information graph, students were required to investigate a number of properties:

- The number of vertices in the information graph of a fragment (sequential complexity).
- The length of the critical path in the information graph (parallel complexity).
- The width of the level parallel form (with explanations for which specific LPF the value of the width is given).
- Maximum depth of nesting cycles.
- The number of different types of arcs (the type of arcs is determined by the direction vector and length).
- The presence of long arcs (i.e. arcs whose length depends on external parameters).
- The number of regularity areas in the information graph. A region of regularity is a set of vertices of a graph from which arcs of the same type come from (arcs of different types can come from one region of regularity).

After the study, it was necessary to mark parallel loops of the given program fragment using the OpenMP directive `#pragma omp parallel for` [29].

Students were not provided with a ready-made tool for constructing and analyzing an informational graph of a fragment. Therefore, the reports accepted any images of information graphs—both those built in any graphics editor, as well as hand-drawn on a sheet of paper and scanned. The main thing was to correctly display the information structure of the fragment under consideration and determine its properties. Not all students did this right away; some had to redo their reports several times to achieve an acceptable result. Figure 3 shows the information graph of one of the variants, obtained using the AlgoView [30,31] software tool.

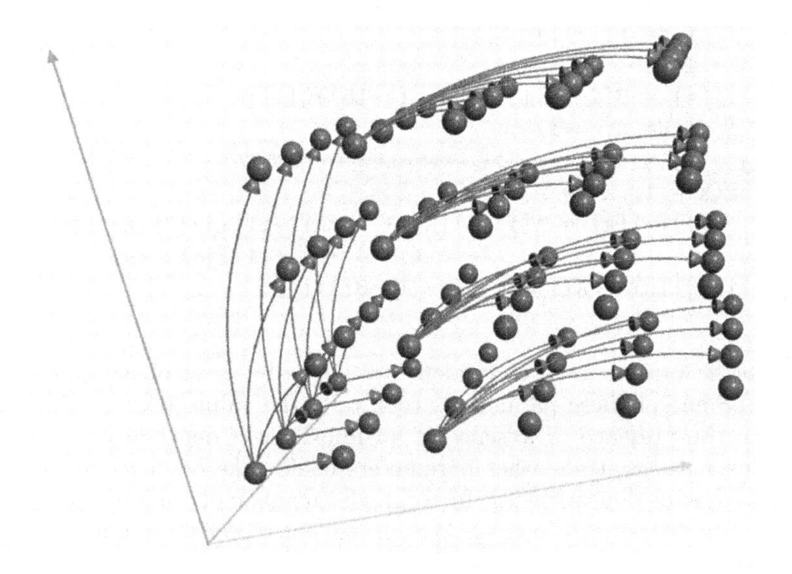

Fig. 3. Information graph of one variant

These assignments were carried out by 183 students, the results are shown in Table 4.

Table 4. Final assessments (2019)

Year	2019
Students	183
5 (Excellent)	168 (91.8%)
4 (Good)	9 (4.9%)
3 (Satisfactory)	3 (1.6%)
2 (Unsatisfactory)	3 (1.6%)
Average mark	4.87

3 Conclusions

Thus, as a part of the course "Supercomputing Simulation and Technologies", the design of a system of assignments aimed at developing skills of efficient use of parallel computing systems is carried out. Different options for practical assignments had varying complexity and were aimed at developing different practical skills. Multiple iterations students-teachers and bringing the report on the practical assignments to a decent state requires students to learn the concepts deeply and thoroughly. The use of modern supercomputer technology prepares students for future work in various areas of the modern science, commerce and industry. The considered system of practical assignments in its current form has proved to be useful, and we are going to develop it further in the framework of this course.

Acknowledgements. The results were obtained in Lomonosov Moscow State University with the financial support of the Russian Science Foundation (agreement N 20-11-20194). The research is carried out using the equipment of the shared research facilities of HPC computing resources at Lomonosov Moscow State University.

References

1. Computer Science Curricula (2013). http://ai.stanford.edu/users/sahami/CS2013
2. NSF/IEEE-TCPP Curriculum Initiative on Parallel and Distributed Computing. http://tcpp.cs.gsu.edu/curriculum/
3. SIAM: Graduate Education for Computational Science and Engineering. SIAM Working Group on CSE Education (2014). http://www.siam.org/students/resources/report.php
4. Future Directions in CSE Education and Research. Report from a Workshop Sponsored by the Society for Industrial and Applied Mathematics (SIAM) and the European Exascale Software Initiative (EESI-2). http://wiki.siam.org/siag-cse/images/siag-cse/f/ff/CSE-report-draft-Mar2015.pdf

5. Prasad, S.K., et al.: NSF/IEEE-TCPP Curriculum Initiative on Parallel and Distributed Computing – Core Topics for Undergraduates, Version I, p. 55 (2012). http://www.cs.gsu.edu/~tcpp/curriculum
6. Voevodin, V., Gergel, V.: Supercomputing education: the third pillar of HPC. Comput. Methods Softw. Dev. New Comput. Technol. **11**(2), 117–122 (2010)
7. Gergel, V., Liniov, A., Meyerov, I., Sysoyev, A.: NSF/IEEE-TCPP curriculum implementation at the state University of Nizhni Novgorod. In: Proceedings of the International Parallel and Distributed Processing Symposium, IPDPS, vol. 6969501, pp. 1079–1084 (2014). https://doi.org/10.1109/IPDPSW.2014.128
8. Voevodin, Vl.V., Gergel, V.P., Popova, N.N.: Challenges of a systematic approach to parallel computing and supercomputing education. In: Lecture Notes in Computer Science, vol. 9523, pp. 90–101 (2015). https://doi.org/10.1007/978-3-319-27308-2_8
9. Rüde, U., Willcox, K., McInnes, L.C., de Sterck, H.: Research and education in computational science and engineering. SIAM Rev. **60**(3), 707–754 (2016). https://doi.org/10.1137/16M1096840
10. Meyerov, I., Bastrakov, S., Barkalov, K., Sysoyev, A., Gergel, V.: Parallel numerical methods course for future scientists and engineers. Commun. Comput. Inf. Sci. **793**, 3–13 (2017). https://doi.org/10.1007/978-3-319-71255-0_1
11. Prasad, S.K., Gupta, A., Rosenberg, A., Sussman, A., Weems, C. (eds.): Topics in Parallel and Distributed Computing. Springer, Cham (2018). https://doi.org/10.1007/978-3-319-93109-8
12. Ghafoor, S., Brown, D.W., Rogers, M.: Integrating parallel computing in introductory programming classes: an experience and lessons learned. In: Heras, D., et al. (eds.) Euro-Par 2017: Parallel Processing Workshops. Euro-Par 2017. Lecture Notes in Computer Science, vol. 10659, pp. 216–226. Springer, Cham (2018). https://doi.org/10.1007/978-3-319-75178-8_18
13. Meyerov, I., Bastrakov, S., Sysoyev, A., Gergel, V.: Comprehensive collection of time-consuming problems for intensive training on high performance computing. Commun. Comput. Inf. Sci. **965**, 523–530 (2019). https://doi.org/10.1007/978-3-030-05807-4_44
14. Voevodin, V., et al.: Supercomputer Lomonosov-2: large scale, deep monitoring and fine analytics for the user community. Supercomput. Front. Innov. **6**(2), 4–11 (2019). https://doi.org/10.14529/jsfi190201
15. Sadovnichy, V., Tikhonravov, A., Voevodin, V., Opanasenko, V.: Lomonosov: supercomputing at Moscow State University. In: Contemporary High Performance Computing: From Petascale toward Exascale, ser, pp. 283–307. Chapman & Hall/CRC Computational Science, Boca Raton (2013)
16. Open Encyclopedia of Parallel Algorithmic Features. http://algowiki-project.org
17. Voevodin, V., Antonov, A., Dongarra, J.: AlgoWiki: an open encyclopedia of parallel algorithmic features. Supercomput. Front. Innov. **1**(2), 4–18 (2015). https://doi.org/10.14529/jsfi150101
18. Antonov, A., Voevodin, Vad., Voevodin, Vl., Teplov, A.: A study of the dynamic characteristics of software implementation as an essential part for a universal description of algorithm properties. In: 24th Euromicro International Conference on Parallel, Distributed, and Network-Based Processing Proceedings, 17th–19th February 2016, pp. 359–363 (2016). https://doi.org/10.1109/PDP.2016.24
19. Antonov, A.S., Voevodin, V.V., Popova, N.N.: Parallel structure of algorithms and training computational technology specialists. J. Phys. Conf. Ser. **1202** (2019). https://doi.org/10.1088/1742-6596/1202/1/012021

20. Antonov, A., Teplov, A.: Generalized approach to scalability analysis of parallel applications. Lect. Notes Comput. Sci. **10049**, 291–304 (2016). https://doi.org/10.1007/978-3-319-49956-7_23
21. Chakrabarti, D., Zhan, Y., Faloutsos, C.: R-MAT: a recursive model for graph mining (2006). http://www.cs.cmu.edu/~christos/PUBLICATIONS/siam04.pdf
22. Bader D.A., Madduri K.: Design and implementation of the HPCS graph analysis benchmark on symmetric multiprocessors. In: Bader, D.A., Parashar, M., Sridhar, V., Prasanna, V.K. (eds) High Performance Computing – HiPC 2005. Lecture Notes in Computer Science, vol. 3769. Springer, Heidelberg (2005). https://doi.org/10.1007/11602569_48
23. Antonov, A., Frolov, A., Konshin, I., Voevodin, V.: Hierarchical domain representation in the AlgoWiki encyclopedia: from problems to implementations. Commun. Comput. Inf. Sci. **910**, 3–15 (2018). https://doi.org/10.1007/978-3-319-99673-8_1
24. ScaLAPACK – Scalable Linear Algebra PACKage. http://www.netlib.org/scalapack/
25. Matrix Algebra on GPU and Multicore Architectures. http://icl.eecs.utk.edu/magma/
26. Dosanjh, S.S., Barrett, R.F., Doerfler, D.W., Hammond, S.D., et al.: Exascale design space exploration and co-design. Fut. Gener. Comput. Syste. **30**, 46–58 (2014). https://doi.org/10.1016/j.future.2013.04.018
27. Antonov, A., Nikitenko, D., Voevodin, V.: Algo500 - a new approach to the joint analysis of algorithms and computers. Lobachevskii J. Math. **8**(41), 1435–1443 (2020). Special issue "Supercomputing Applications, Algorithms and Software Tools"
28. Voevodin, V., Voevodin, Vl.: Parallel Computing, p. 608. BHV-Petersburg, St. Petersburg (2002)
29. The OpenMP API specification for parallel programming. https://www.openmp.org/
30. Antonov, A.S., Volkov, N.I.: An AlgoView web-visualization system for the AlgoWiki project. Commun. Comput. Inf. Sci. **753**, 3–13 (2017). https://doi.org/10.1007/978-3-319-67035-5_1
31. Antonov, A., Volkov, N.: Interactive 3D representation as a method of investigating information graph features. In: Russian Supercomputing Days: Proceedings of the international conference, 24–25 September 2018, pp. 262–273. Moscow State University, Moscow (2018). https://doi.org/10.1007/978-3-030-05807-4_50

Tuning ANNs Hyperparameters and Neural Architecture Search Using HPC

Kamil Khamitov(✉), Nina Popova, Yuri Konkov, and Tony Castillo

Lomonosov Moscow State University, Moscow, Russia
berserq0123@gmail.com, popova@cs.msu.ru, konkov96@gmail.com, cmtony4@gmail.com

Abstract. Rapid development of deep ANNs made the number of hyperparameters constantly grow. As a sequence various aspects of ANNs, such as inference time, efficient resources utilization, losses and even training time were strongly influenced. In general methods of hyperparameters tuning are used for adapting well-known ANN models to new tasks or to tasks in similar areas without pre-training or for synthesis of new particular architectures for particular problems. In this article we compare different types of hyperparameters tuning like CoDeepNEAT, Naive Evolution, Tree-Parzen estimation, structured annealing with MorphNet post-tuning. We apply these methods to particular network architectures for image processing and HRM signal estimation. The process of adaptation this technology to big networks requires a lot of computational resources, so it's necessary to use parallel implementations. It can be done by utilizing HPC with hybrid computational nodes. Also we propose new type of tool based on Microsoft NNI. It is used for tuners comparison, convergence analysis, and runs different tuners in parallel mode on cluster nodes.

Keywords: Hyperparameters tuning · NNI · HPC · Neural architecture search

1 Introduction

Recent studies demonstrate significant increase of ANN's hyperparameters of models currently used in production. Number of BERT's [3] parameters exceeds 110M. It means that even if we use computing nodes with quite powerful video card single Tesla V100 each, it is not enough for this amount of parameters neither for inference nor for training. Nowadays, many researchers solving particular tasks tend to reuse already pre-trained models in order to adapt them to their tasks. But practice shows the difference between general analytically created architectures and particular architectures that best fit to constraints of the specific task [6]. It leads us to the question about Neural Architecture Search and tuning. Obviously, general graphs-evolution approach can not be directly applied to ANN's because of graph constraints and neural network parameters (like

V. Voevodin and S. Sobolev (Eds.): RuSCDays 2020, CCIS 1331, pp. 536–548, 2020.
https://doi.org/10.1007/978-3-030-64616-5_46

dropout rate, learning rate etc.). In such a way we need to consider NAS(Neural Architecture search) problem and refining particular network topology problem as large-scale optimization task with certain constraints. This search methodology and connecting nodes in graphs obliges us to use algorithms with prebuilt constraints to limit search space [6]. Computational power insufficiency of inference devices can be solved via hyperparameters tuning, choosing the appropriate regularizer that is demonstrated in [15].

2 Neural Architecture Search Problems

In this article we mainly focus on neural architecture search problems. We take into consideration two main problems:

- refining existing topology by applying to new particular task,
- synthesizing new topology from scratch (Neural Architecture search).

Topology adaptation to existing hardware is very important question too, because modern DNNs require resources increasing. In this case the process looks like a best model development for the particular task that can be used to tune different sets of hyperparameters and then to buld a "distilled" model which fits the resource limitations on the particular device. In this article we want to cover both steps of a process. The implementation of hyperparameters tuning requires a lot of computational resources and it's significant to provide a possibility to perform such tuning on HPC clusters.

Adaptation Problem. The adaptation problem is considered as refining the existing model $min_\theta L(\theta_n), \theta_n = X(\theta_{n-1}, ...\theta_{n-k})$, where L – loss function, θ – hyperparameters set, θ_0 – initial hyperparameters (initial model) that are used as a core of method, X – the iterative process of new model building, based on previous hyperparameters.

Neural Architecture Search Problem. In the neural architecture search problem we don't have the initial value of hyperparameters. It limits the capabilities of methods that rely on the quality of the initial approximation. The formal process can be described as follows: $L(\theta_n), \theta_n = X(\theta_{n-1}, ...\theta_{n-k}),\ \theta_0 = \mathbf{0}$.

Improving Inference Metrics Modifying Architecture. The post-processing step is the essential step in the process of deploying a particular model to a particular computational device. Usually the computational power of the device which is responsible for inference is significantly lower than the computational power/latency of the device where nets have been trained. So it creates requirements for "distillation" – a process of limiting or changing the dimension of some layers or other hyperparameters or even changing the topology in order to optimize some inference-time metrics. But it's not the only way, because post-processing can even improve loss of other properties, not directly connected to the inference, optimizing the way how layers should map to specific device which should run inference. The post-processing also can be considered as optimization with some kind of regularizer that describes which property of the model should be optimized.

Distributed Hyperparameters Optimization. Since hyperparameters optimization is a computation-heavy problem, the usage of distributed technologies is essential to lift such task. Evolution-based techniques have a natural fit for parallel calculations, and large internal resource of parallelism with parameter-server models, or island-based model. With other methods, even which doesn't imply parallel implementation, benefits of distributed optimization can be obtained with running different tuners at the same time. Implementations of parallel TPE can be found in [2]. To combine approaches, we decide to implement it on top of the existing system for running different optimization tasks. The system we've chosen is NNI (Neural Network Intelligence) [5].

Key Concepts of NNI. NNI [5] is a toolkit that helps users to design and tune machine learning models (e.g., hyperparameters), neural network architectures or complex system's parameters efficiently and automatically. It has several appealing properties: convenience, scalability, flexibility and efficiency. Key concepts of NNI are:

1. **Search Space.** It means the feasible space for tuning the model. For example, the value range of each hyperparameter.
2. **Configuration.** A configuration is an instance from the search space, that is, each hyperparameter has a specific value.
3. **Trial.** Trial is an individual attempt at applying a new configuration (e.g., a set of hyperparameter values, a specific ANN topology). Trial code should be able to run with the provided configuration.
4. **Tuner.** Tuner is AutoML algorithm, which generates new configuration for the next try. New trial will run with this configuration.

For every experiment user needs to define a search space and to update a few lines of code, then leverage NNI built-in Tuner/Assessor and training platforms to search the best hyperparameters and/or neural architecture.

3 Techniques for Tuning Hyperparameters

Naive Evolution. Naive Evolution comes from [4]. It randomly initializes the population based on search space. For each generation, it chooses the best ones and performs some mutation (e.g., changes a hyperparameter value, adds/removes one layer) on them to get next generation. Naive Evolution requires many trials to work, but it's straightforward and can easily be expanded to get new features. DNA of evolving algorithms stores tensor dimensions and learning rate. During the evolution following mutations procedures are applicable:

1. Reset weights.

2. Insert/remove convolution.
3. Change number of channels or filter size
4. Add/remove random skip.

For selection tournament selection processes were used. The particular implementation from [4] also implies that the number of workers performing mutations, measure losses and accuracy of the specific synthesized network should be larger than the quarter of the population size. It allows efficient load distribution, but largely increases computational resources demand.

CoDeepNEAT. The idea of multilevel neuroevolution was further developed in the CoDeepNEAT (Coevolution DeepNEAT) method [6]. This approach uses hybrid evolution, in which general population consists of two disjoint groups. First group is responsible for the network topology (so-called network template). The template defines blocks and layers used in the network, and second group is responsible for layers themselves or blocks of the deep neural network. The network topology also appears as a graph of neurons, as in the original NEAT. During fitness function value calculation, templates and blocks are connected to form a large deep neural network. Like in DeepNEAT [6] it has procedures to link multiple levels of the network and perform reshaping procedures like upsampling or task-dependent downsampling to connect multidimensional layers. Unlike DeepNEAT, the value of fitness function is distributed in the population of modules and templates as an average among all networks containing this module or constructed using the particular template. The performance analysis of parallel implementation CoDeepNEAT was made in [14]. Results are demonstrated the high degree of parallelism and beneficial usage of GPU-RAM NVLink connection, during tuning large number of parameters of CRNN in data-labeling task.

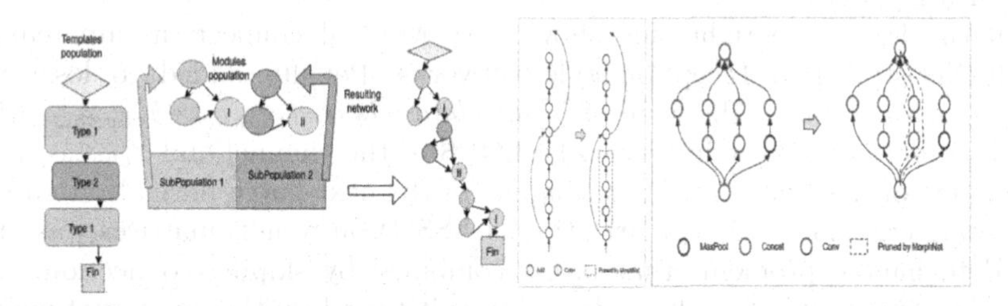

Fig. 1. Evolution scheme of CoDeepNEAT [6] and MorphNet scheme for ResNet-style and Inception-style networks. [15]

An approximate structure of CoDeepNEAT is shown in Fig. 1. CoDeepNEAT's convergence is better than its predecessors, since this method has more possibilities for mutations, and the smallest mutations among the population of templates or modules lead to a significant change in final networks. The applicability of the method is also much higher in comparison to its predecessors.

Tree-Parzen Estimation. The Tree-structured Parzen Estimator (TPE) is a sequential model-based optimization (SMBO) approach. SMBO methods sequentially construct a model to approximate the performance of hyperparameters based on historical measurements. Then it's subsequently choosing new values of hyperparameters to integrate them into this model. The TPE approach models $P(x|y)$ and $P(y)$ where x represents hyperparameters and y the associated evaluate matrix. $P(x|y)$ is modelled by transforming the generative process of hyperparameters, replacing distributions of the configuration prior with non-parametric densities. This optimization approach is described in details in [1].

Simulated Annealing. The simulated annealing is a prevalent technique for optimization. It's widely used in computational simulation as an efficient solver for other large-scale optimization problems. Considering the application to the hyperparameters tuning, it can be used for obtaining good first range points in large search spaces for some non-discrete parameters.

MorphNet Post-processing. MorphNet [15] implements the approach for existing well-known architecture post-processing to adapt topological-based hyperparameters to the particular task configuration on the specific computation device. The method is based on tensors dimensions modification in some types of layers (mostly in convolutional ones) and on convolution sizes optimization to minimize the aimed regularizer loss. The main step can be considered as shrinking and sparsifying convolutions depends on regularizer values. MorphNet supports large scale models "out of the box" and utilizes the iterative approach of learning the new architecture. Instead of trying numerous architectures across a large design space, MorphNet starts with an existing architecture for a similar problem and, in one shot, optimizes it for the task at hand, trying to zero some outputs of the layer using penalties.

In the Fig. 1 the right side shows how residual connections are removed in ResNet-style and Inception-style networks. Penalty added to loss (with weight factor λ) takes the form of $\sum all\_channels|cost_{channel} * Y_{channel}|$ where $cost\_channel$ is the cost in terms of FLOPS of the channel and $Y_{channel}$ is the scaling factor associated with the channel in the next batch normalization layer (prunning criterion). L1 regularization (LASSO) on weight matrices was introduced to handle problems that layers combined by skiping connections must have the same number of channels and uses it to reduce the number of nonzero weights without or with little effect on the performance (e.g. accuracy) of the deep neural network. Whereas MorphNet mainly used for network distillation in this article, we consider it as the post-processing tool, which allows optimizing inference time without significant changes in the loss. The parallel implementation of the MorphNet will be discussed in the further sections.

4 Automatic System Architecture

4.1 System Overview

The following automatic system is proposed for tuning ANNs hyperparameters and neural architecture search on HPC clusters. System consists of three major parts: *Tuners*, *Platform-dependent Executors*, *Post-processing part.* The system's classes diagram is presented in the Fig. 2. During tuning activities user chooses the optimizer, type of task (tuning or NAS) and appropriate metric for the post-processing. The system performs set of launches to optimize large set of hyperparameters and provide a number of network-topologies as output. Typical process of user interaction consists of 4 main stages:

1. Selecting the type of task.
2. Providing data.
3. Selecting particular platform and tuners for Optimization.
4. Obtaining best architectures for the task and comparative metrics of efficiency.

User can optionally do a post-processing of the network or distillation using different kind of tuners. It also can be done via tuning task on HPC cluster as well as other optimizations routine.

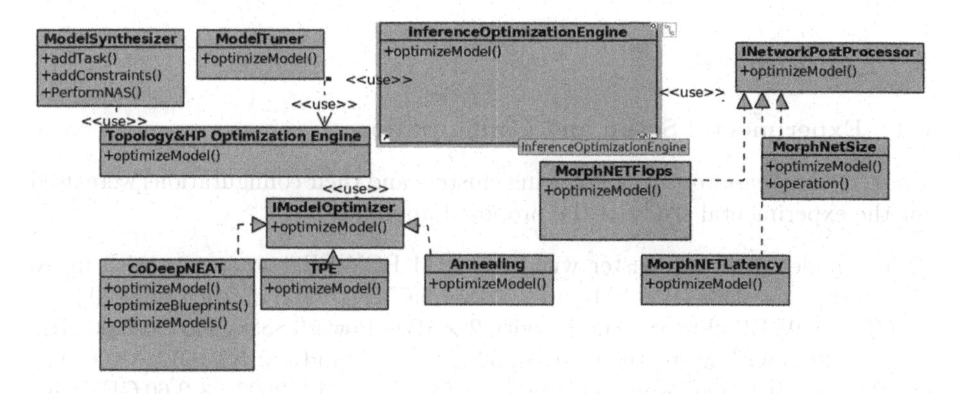

Fig. 2. Modules UML class diagram, that briefly describe system's architecture [5]

4.2 Performing HPO Using Distributed Computing and HPC

Module "SlurmTrainingService" was designed to perform distributed experiments with NNI. Experiments with NNI were performed on a cluster with Slurm Workload Manager, because usually users don't have direct SSH access to computing nodes (before the start of our job), and therefore they can't use available distributed approaches in NNI. Module "SlurmTrainingService" based on

"RemoteTrainingService" (included in NNI) that uses SSH to establish a connection between nodes. "SlurmTrainingService" uses Slurm utilities to get information about allocated nodes and then establish connections between master and compute nodes as well as "RemoteTrainingService". Briefly description of "SlurmTrainingService" is shown on the Fig. 3. The startup script of NNI can be described as follows. User creates the NNI configuration file with parameters of the experiment ("slurm" is set of "trainingServicePlatform" parameter). The task is put into the queue using "sbatch" utility. When resources are allocated for the task, the user will be able to connect to the NNI interface. However "SlurmTrainingService" module makes execution of NNI on cluster more simple, so users still have to put jobs into the Slurm queue and to organise SSH port forwarding to the node with NNI Web server. We also have plans to create user-friendly interface for using NNI on clusters with a variety of workload managers (Slurm, IBM LSF).

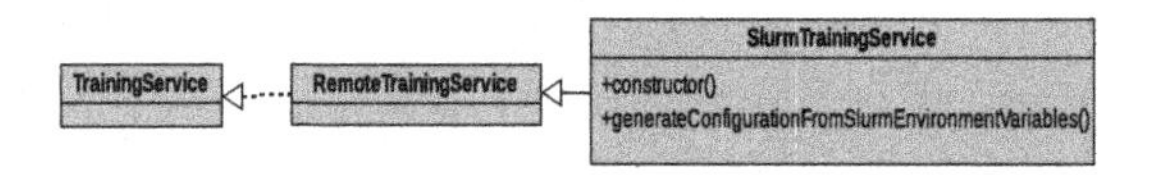

Fig. 3. UML class diagram of SlurmTrainingService, describing system's architecture

5 Experiments

5.1 Experimental Setup and Configuration

Clusters Configuration. The following clusters and their configurations were used for the experimental study of the proposed approach

- 4 × nodes of Polus cluster with 2 × IBM POWER8 processors with up to 160 threads, 256 GB RAM and 2 × Nvidia Tesla P100 on each node [9].
- CTE-POWER cluster 16 nodes with 2 × IBM Power9 8335-GTG @ 3.00 GHz processors with up to 160 threads, 32 GB RAM, and 4 × NVIDIA V100 [11].
- 33 nodes of Lomonosov-2 cluster with Intel Xeon E5-2697 v3 2.60 GHz with up to 28 threads, 64 GB RAM and Tesla K40s on each node [10].

MNIST. For testing the neural architecture search, we've chosen digit's recognition problem on MNIST dataset with appropriate NAS-based tuner (CoDeepNEAT). Due to the small complexity of MNIST classification task, templates of possible neural networks were built without any base network (i.e. without applying transfer learning approach). Initial templates of possible neural networks are directed acyclic graphs with one input block, one output block and intermediate blocks. First layer in the input block and last layer in the output block were fixed to convolutional layer and dense layer respectively. Each block in the template has at most two inputs (in case of two inputs the concatenation layer was used). One of the problems of applying CoDeepNEAT to NAS problem is the alignment of input's dimensions. Since MNIST is a classification task of images, upsampling layers were used to align the output of convolutional and pooling layers. Other CoDeepNEAT's parameters described in Table 1.

Table 1. MNIST CoDeepNEAT's parameters

Parameter	Type	Value
Number of filters in convolutional layers	Random integer	[16, 64]
Kernel size of convolutional layers	Random choice	[1, 3, 5]
Kernel size of pooling layer	Fixed	[2]
Stride of pooling and convolutional layers	Fixed	Preserving input's dimension
Dropout rate	Random float	[0.00001, 0.001]]
Mutation rate of block's/template's population	Fixed	[0.5]
Crossover rate of block's/template's population	Fixed	[0.5]
Template's population size	Fixed	[10]
Block's population size	Fixed	[15]
Number of generations	Fixed	[40]
Number of epochs to train neural network	Fixed	[10]
Batch size	Fixed	[200]

Image Recognition on ImageNET. To estimate post-processing benefits and analyze the scalability and parallel efficiency, we performed a set of experiments on ImageNET dataset. For testing Neural architecture search, we've chosen the image recognition problem on ImageNET dataset and refining Inception architecture [16]. We utilize Horovod as a framework for carrying out experiments in distributed training and inference [17]. It's an open-source distributed training framework for TensorFlow, which makes distributed deep learning fast and easy with minimal modifications to the code. The working process of the step can be described as follows.

- On each step it computes model updates (gradients) from the run

- Average gradients among those multiple copies.
- Update the model.

Horovod averages gradients and communicates these gradients to all nodes that follow the decentralized ring reduction scheme using the Baidu algorithm [18]. The main advantage of Horovod in comparison with distributed Tensorflow is simple code implementation, also it uses a programming paradigm better known by users, because of rich api that was introduced in Tensorflow recently. In Horovod each process has a unique role like a parameter server that receives data to be averaged or a worker role who processes training data, calculates gradients and then sends them to parameter server. Unlike in distributed Tensorflow, the Horovod use role models and ring reduce process, that we described above.

HRM Signal Prediction Problem on UBFC-RPPG. We performed and reconsidered large network for HRM signal estimation using UBFC-RPPG dataset [13]. It consists of two major parts: the convolutional part and recurrent detector. In CoDeepNEAT evolving processes considered these parts of such ANN as an independent part and perform independent evolving, which reduces needed computational power per cluster node. For Naive evolution search space, used for optimizing "base1" convolutional layers, has fixed values of number of convolutional layers (equals to 3) and activation function was fixed to ReLU. Each layer "base1" and "base2" can have kernel size of 3×3 or 5×5. Other hyperparameters of search space are described as follows:

1. First layer can consist of 48 or 64 output channels.
2. Second layer can consist of 32 or 48 output channels.
3. Third layer can consist of 8 or 16 output channels.

Pooling layer on the end of the first group of layers could be AvgPool or MaxPool. Search space, used for optimizing "base2" convolutional layers, could be described as follows:

1. First layer can consist of 64 or 128 output channels.
2. Second and Third layers can consist of 32 or 64 output channels.
3. Fourth and Fifth layers can consist of 8 or 16 output channels.

For the second part of the model, maximum of hidden layers was fixed to 2 and activation function to PReLU. The first hidden layer could consist of only 256 or 512 neurons, the last hidden layer could have 64 or 256 neurons. Learning rate hyperparameter was chosen from a uniform distribution ([0.0001, 0.01]). The dropout rate at the end of "base2" was selected from a uniform distribution ([0.00001, 0.001]). In NEAT-based techniques different scheme was used for two parts of the network, as discussed before. Search space for GRU part was fixed to only two dimensions. Dropout rate of seq_class varies from 0.0001 to 0.00002 in discrete space with step 0.00002. Number of GRU layers size in second GRU block varies from 1 to 5. Probability of NEAT mutation addNode/removeNode was fixed to 0.13 in MLP part. Learning rate has a uniform distribution from 0.001 to 0.0006. Dropout factor has a uniform distribution from 0.1. to 0.75. Convolutional window size has a uniform distribution from 2 to 7.

5.2 Results

HRM Signal Prediction Problem on UBFC-RPPG. Best loss values for different obtained topologies (with Naive Evolution (ITK8C), TPE (CoYE6), Structured Annealing (ozEKA) and CoDeepNEAT) on training subset of UBFC dataset are presented in Table 2.

Table 2. Best loss during training for HRM series prediction on UBFC-RPPG

	Unmodified net	Naive evolution	TPE (C0YE6)	Anneal	CoDeepNEAT
Test loss	0.729	0.568	0.573	0.612	0.592
Train loss	0.538	0.606	0.583	0.576	0.44

Loss dynamics on the test dataset is presented in the Fig. 4

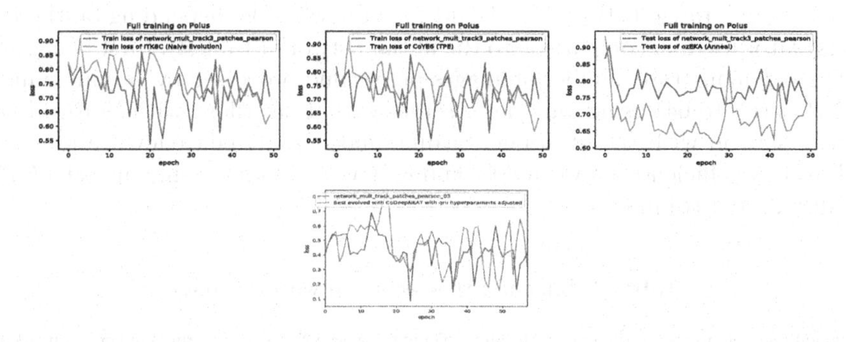

Fig. 4. Loss dynamics of best tuned architectures for HRM data series prediction on UBFC-RPPG

MNIST. During MNIST experiments after 40 epochs of evolution with CoDeepNEAT, we obtain CNN for MNIST with appropriate loss and accuracy. Particular values of loss are presented in Table 3.

Table 3. Comparing accuracy and loss on test data with [12].

DNN (784-800-10) [12] accuracy	DNN (784-800-10) [12] loss	MNIST model accuracy	MNIST model loss
0.91	0.061	0.97	0.094

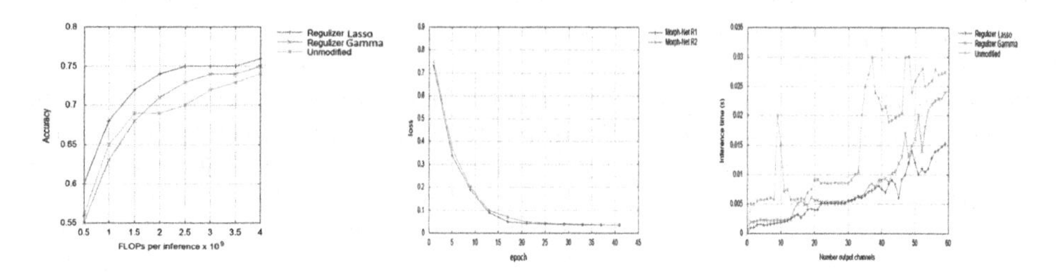

Fig. 5. Evaluation of the loss and accuracy and during the training process, and Inference time for different sizes of regulators

ImageNet. The graph in the Fig. 5 shows the number of FLOPS used to obtain a loss value determination, using the network given initially, as well as 2 modifications using the network created by MorphNet.

Figure 5 shows variations of the number of output channels for a convolutional layer in case we use different regularizers in MorphNet. Inference time decreases in a range between 5.7% and 7% in new model depending on the type of regulator used, in comparison to the initial network (Table 4). First two graphs in Fig. 6 demonstrate number of images per second processed by varying number of GPUs (up to 64) using parameter server with Tensorflow, with Horovod, and its comparison with the ideal one. Scaling and parallel efficiency are presented in Fig. 6. An efficiency between 80% and 87% was obtained using up to 64 GPUs for distributed training.

Table 4. Inference time achieved with test data.

Unmodified net (inference time ms)	MorphNet Gamma	MorphNet Lasso	Gain Gamma	Gain Lasso
132,4	124,9	123,1	5,7 %	7,0 %

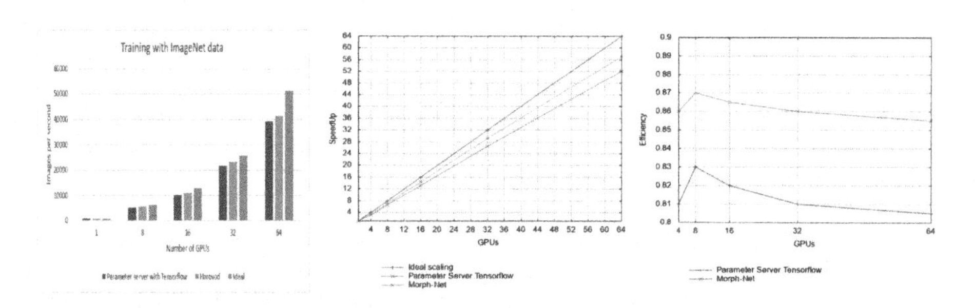

Fig. 6. Comparison of the Tensorflow Parameter server and Horovod and Speedup analysis of MorphNet post-processing ImageNet dataset

6 Conclusion

The approach to analysis and tuning hyperparameters in both problems was tested in distributed clusters environment. Architecture of analized system and processes was tested using HPC-running integration module for NNI package. We also integrated MorphNet as an example of post-tuning refining process, that can use different communication models for suitable cluster network architectures. MorphNet refined version allows us to decrease training time up to 7%. During synthesis problem we've found that CodDeepNEAT's best models can provide comparative accuracy with the best well-known non-synthetic solutions, it has 3% better accuracy than DNN from [12]. And version tuned to MorphNet also demonstrates speedup in inference time. It demonstrates appropriate resources utilization, estimating inference time by varying number of input and output channels using MorphNet significantly helped to find optimal values for it. Parallel implementation of MorphNET and CoDeepNEAT demonstrates proper good scaling and parallel efficiency in both types of nodes resources (GPU and CPU-based). The comparison of MorphNET implementation on ImageNET based tasks shows good strong scaling and efficiency (up to 80% on 64 GPU nodes) of CTE-POWER cluster.

Acknowledgements. Reported study was funded by RFBR according to research project №20-07-01053. This research was carried out using shared research facilities equipment of HPC computing resources at Lomonosov Moscow State University.

References

1. Bergstra, J.S., et al.: Algorithms for hyper-parameter optimization. In: Advances in Neural Information Processing Systems (2011). https://doi.org/10.5555/2986459.2986743
2. Amado, N., Gama, J., Silva, F.: Parallel implementation of decision tree learning algorithms. In: Brazdil, P., Jorge, A. (eds.) EPIA 2001. LNCS (LNAI), vol. 2258, pp. 6–13. Springer, Heidelberg (2001). https://doi.org/10.1007/3-540-45329-6_4
3. Devlin, J., et al.: Bert: pre-training of deep bidirectional transformers for language understanding. arXiv preprint arXiv:1810.04805 (2018)
4. Real, E., et al.: Large-scale evolution of image classifiers. In: Proceedings of the 34th International Conference on Machine Learning, vol. 70. JMLR. org (2017). https://doi.org/10.5555/3305890.3305981
5. Neural Network Intelligence, April 2020. https://github.com/microsoft/nni
6. Miikkulainen, R., Liang, J., Meyerson, E., et al.: Evolving deep neural networks. In: Artificial Intelligence in the Age of Neural Networks and Brain Computing, pp. 293–312. Elseiver (2019). https://doi.org/10.1016/B978-0-12-815480-9.00015-3
7. Gordon, A., et al.: MorphNet: fast & simple resource-constrained structure learning of deep networks. In: Proceedings of the IEEE Conference on Computer Vision and Pattern Recognition (2018). https://doi.org/10.1109/CVPR.2018.00171
8. CoDeepNEAT implementation base, April 2020. https://github.com/sash-a/CoDeepNEAT
9. Polus cluster specifications, April 2020. http://hpc.cs.msu.su/polus

10. Voevodin, V.l., et al.: Supercomputer Lomonosov-2: large scale, deep monitoring and fine analytics for the user community. Supercomput. Front. Innov **6**(2), 4–11 (2019). https://doi.org/10.14529/jsfi190201
11. CTE cluster configuration, April 2020. https://www.bsc.es/user-support/power.php#ToC-systemoverview
12. Simard, P.Y., Steinkraus, D., Platt, J.C.: Best practices for convolutional neural networks applied to visual document analysis. In: Seventh International Conference on Document Analysis and Recognition, 2003. Proceedings, Edinburgh, UK, pp. 958–963 (2003). https://doi.org/10.1109/ICDAR.2003.1227801
13. Bobbia, S., Macwan, R., Benezeth, Y., Mansouri, A., Dubois, J.: Unsupervised skin tissue segmentation for remote photoplethysmography. Pattern Recogn. Lett. (2017). https://doi.org/10.1016/j.patrec.2017.10.017
14. Khamitov, K., Popova, N.: Research of measurements of ANN hyperparameters tuning on HPC clusters with POWER8. In: Russian Supercomputing Days 2019: Proceedings of International Conference, 23–24 September 2019, Moscow, pp. 176–184 (2019)
15. Poon, A., Narayanan, D.: MorphNet: Towards Faster and Smaller Neural Networks. Google AI Perception (2019)
16. Szegedy, C., et al.: Going deeper with convolution (2014)
17. Sergeev, A., Balso, M.D.: Meet Horovod: Uber's Open Source Distributed Deep Learning Framework for TensorFlow. Uber Engineering Blog (2017). https://eng.uber.com/horovod
18. Patarasuk, P., Yuan, X.: Bandwidth optimal all-reduce algorithms for clusters of workstations (2009). https://doi.org/10.1016/j.jpdc.2008.09.002

Distributed and Cloud Computing

Availability-Based Resources Allocation Algorithms in Distributed Computing

Victor Toporkov(✉) and Dmitry Yemelyanov

National Research University "MPEI", Moscow, Russia
{ToporkovVV,YemelyanovDM}@mpei.ru

Abstract. In this work, we introduce resources co-allocation algorithms for parallel jobs execution in distributed computing with non-dedicated and heterogeneous hosts. Complex distributed computing systems often operate under conditions of the resources availability uncertainty. Imprecise estimations of jobs execution runtime, unplanned maintenance works and other global and local events do not allow to consider accurate schedules of the resources utilization. On the other hand, an efficient job-flow execution in compliance with QoS constraints requires reliable mechanisms for advanced resources allocation and reservation. The novelty of the proposed resources allocation approach is in a general procedure efficiently selecting computing nodes according to the resources availability criteria. Special knapsack and greedy algorithms are implemented and compared in a market-based computing simulation model.

Keywords: Distributed computing · Grid · Resource · Uncertainty · Availability · Probability · Job · Allocation · Optimization · Economic scheduling

1 Introduction and Related Works

High-performance distributed computing systems (HPCS), such as Grids, cloud and hybrid infrastructures, provide access to large amounts of resources. These resources are typically required to execute parallel jobs submitted by HPCS users and include computing nodes, data storages, network channels, software, etc. The actual requirements for resources amount and types needed to execute a job are defined in resource requests and specifications provided by users [1–5].

HPCS organization and support bring certain economical expenses: purchase and installation of machinery equipment, power supplies, user support, etc. As a rule, HPCS users and service providers interact in economic terms and the resources are provided for a certain payment.

Economic models [3–5] are used to efficiently solve resource management and job-flow scheduling problems in distributed environments such as cloud computing and utility Grids. Majority of scheduling solutions for distributed environments implement scheduling strategies on a basis of efficiency criteria [1–5]. A metascheduler [2, 5] implements economic policies and criteria optimization based on information from local resource schedules and user jobs. So, a base component for a job-flow scheduling is a

V. Voevodin and S. Sobolev (Eds.): RuSCDays 2020, CCIS 1331, pp. 551–562, 2020.
https://doi.org/10.1007/978-3-030-64616-5_47

procedure for selecting a set of resources required to execute a particular computational job.

Traditional models consider this problem in a deterministic way. Such an approach is sometimes justified by the strict market rules for resources acquisition and utilization during the purchased period of time. Commercial Grids and cloud service providers usually own full control over the resources and may reliably consider their local schedules for some scheduling horizon time [1, 3]. Besides, market-based interactions and QoS constraints compliance require deterministic model for the resources utilization profile. Thus, it is convenient to represent available resources as a set of slots: time intervals when particular nodes are idle and may be used for user jobs execution [4–8].

However general distributed computing systems with non-dedicated resources usually can't rely on deterministic utilization schedules and instead make predictions based on the utilization predictions and probabilities [9–12]. The probabilities of the resources availability and utilization at any given time may originate from jobs execution and completion time uncertainties, local activities of the resource provider, maintenance or numerous failure events. Particular utilization characteristics and patterns usually strongly depend on the resource types. However, according to [9] about 20% of Grid computational nodes exhibit truly random availability intervals.

The scheduling problem in Grid is NP-hard due to its combinatorial nature and many heuristic solutions have been proposed. When scheduling under uncertainties, proactive and reactive approaches are usually distinguished [12]. Proactive algorithms concentrate on the resources utilization predictions and heuristic-based advanced resources allocations and reservations. Reactive algorithms analyze current state of the computing environment and make decisions for jobs migration and rescheduling. Both types of algorithms may be used in a single system to achieve even greater resource usage efficiency.

The resources availability predictions for the considered scheduling interval may be obtained based on the historical data processing, linear regression models or with help of expert and machine learning systems [9–11]. In [10], a set of availability states is defined to model resource behavior and probabilities of the states' transitions. On the other hand, sometimes it is possible to identify distributions of resources utilization and availability intervals [9].

Economic scheduling models are implemented in modern distributed and cloud computing simulators GridSim and CloudSim [13]. They provide reliable tools for resources co-allocation, but consider price constraints on individual nodes and not on a total window allocation cost. However, as we showed in [6], algorithms with a total cost constraint are able to perform the search among a wider set of resources and, thus, increase the overall scheduling efficiency. Algorithms [14, 15] implement knapsack based slot selection optimization according to a specified criterion with a total job execution cost constraint.

In this paper, we propose proactive algorithms for resources selection and co-allocation in heterogeneous market-based computing environments with non-dedicated resources and corresponding availability uncertainties. The uncertainties are modeled as resources availability events and probabilities: a natural way of machine learning and statistical predictions representation. The novelty of the proposed approach consists in a dynamic programming scheme performing resources selection with a total availability criterion maximization. The paper is organized as follows. Section 2 defines availability-based scheduling problem and a general scheme for the reliable resources selection. Then several greedy and knapsack implementations are proposed and considered. Section 3 contains an experiment setup and simulation results obtained for the considered algorithms. Section 4 summarizes the paper and describes further research topics.

2 Resource Selection Algorithm

2.1 Job Execution Under Uncertainties

We consider a set R of heterogeneous computing nodes with different performance p_i and price c_i characteristics.

The probabilities (predictions) of the resources availability and utilization for the whole scheduling interval L are provided as input data. Thus, we model resources utilization profile as an ordered list of utilization events, such as resources *allocation* or *release* events. Jobs execution time uncertainties are modeled as a sequence of *allocation, occupation* (actual execution) and *release* events with the *occupation* probability $P_o = 1$. *Global* resources utilization uncertainties, such as maintenance works or network failures, are modeled as a continuous *occupation* events with $P_o \ll 1$ during the whole considered scheduling interval.

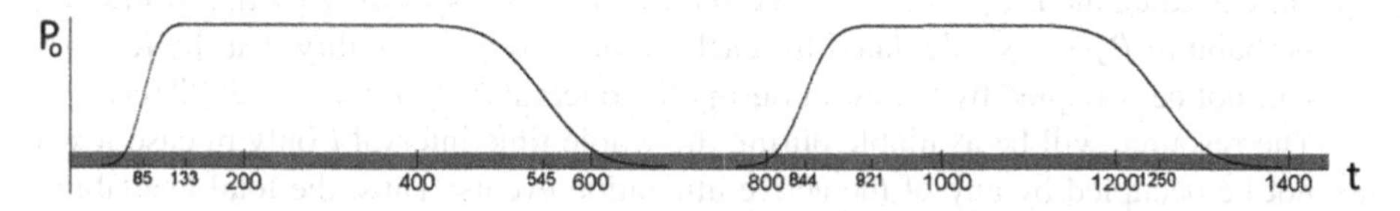

Fig. 1. Example of a resource occupation probability schedule.

Figure 1 shows an example of a single resource occupation probability P_o schedule. With two jobs already assigned to the resource, there are two resources allocation events (with expected times of allocation at 85 and 844 time units), two resources occupation events (starting at 133 and 921 time units) and two resources release events (expected release times are 545 and 1250 time units respectively). Gray translucent bar at the bottom of the diagram represents a sum of global utilization events with a total resource occupation probability $P_o = 0.06$.

Resources allocation and release events are modeled by random variables with a *normal distribution.* Expected allocation and release times are derived from the job's replication and execution time estimations. Corresponding standard deviations depend on the job's features and may be predicted based on user estimations or historical data [9–11]. Hence, in Fig. 1 the resource occupation probability at expected times of allocation and release events are: $P_o(85) = P_o(545) = P_o(844) = P_o(1250) = 0.5$.

In order to execute a parallel job a set of simultaneously idle nodes (a window) should be allocated ensuring user requirements from the resource request. The resource request usually specifies number n of nodes required simultaneously, their minimum applicable performance p, job's total computational volume V and a maximum available resources allocation budget C. The required window length and job execution time estimation are defined based on a window node with the minimum performance. For example, if a window consists of nodes with performances $p \in \{p_i, p_j\}$ and $p_i < p_j$, then the resources should be allocated for a time $T = \frac{V}{p_i}$. In this way V really defines an average computational volume for a single node. The total cost of a window allocation is then calculated as $C_W = \sum_{i=1}^{n} T * c_i$.

These parameters constitute a formal generalization for resource requests common among distributed computing systems and simulators [13–16].

Common allocation and release times for all the window resources ensure the possibility of inter-node communications during the whole job execution. In this way, the occupation and availability probabilities should be estimated for each resource during the considered time interval required to execute the whole job. For this purpose we implement the following procedure to calculate an availability probability P_a of a resource r during the job execution interval l.

1. Retrieve a set of independent utilization events e_i active for the resource r during the time interval l. When a subset of dependent events is active during the interval, then only a single event providing the maximum occupation probability P_o is retrieved. For example from the *allocation-occupation (execution)-release* events chain only the *execution* event is retrieved with $P_o = 1$.
2. For each independent event e_i a maximum occupation probability during the interval l is calculated: $P_o^{max}(e_i) = \max_{t \in l} P_o(e_i, t)$. Corresponding partial availability probability $P_a(e_i)$ is calculated for each event e_i as a probability that the resource will not be occupied by the event during the interval l: $P_a(e_i) = 1 - P_o^{max}(e_i)$.
3. The resource will be available during the whole time interval l only in case it will not be occupied by any of the active utilization events. Thus, the total availability probability for the resource r is a product of all partial availability probabilities calculated for independent events e_i:

$$P_a^r = \prod_i P_a(e_i).$$

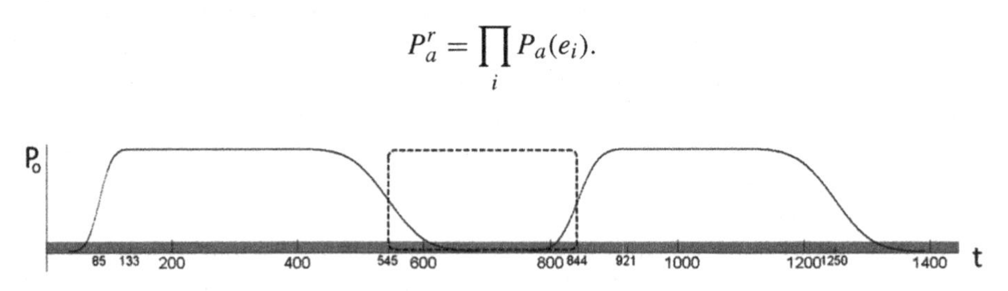

Fig. 2. Example of a resource occupation probability schedule.

For example consider a resource availability probability for an interval $l : [545; 844]$ presented as a dotted rectangle in Fig. 2. Three independent events are active during the interval: 1) resource release event e_1 with the expected release time at 545 time units, 2) resource allocation event e_2 with the expected allocation time at 844 time units, and 3) a global utilization event e_3 with a constant occupation probability $P_o = 0.06$ (related details were provided with a Fig. 1 example). Corresponding partial occupation and availability probabilities are: $P_o^{max}(e_1) = 0.5$, $P_o^{max}(e_2) = 0.5$, $P_o^{max}(e_3) = 0.06$, while $P_a(e_1) = 0.5$, $P_a(e_2) = 0.5$, $P_a(e_3) = 0.94$. So, the total probability of the resource availability during the whole interval l is $P_a^r = 0.235$.

When a set of n resources is required for a job execution for a time T, the total window availability P_a^W during the expected job execution interval can be estimated as a product of availability probabilities of each independent window nodes:

$$P_a^W = \prod_i^n P_a^{r_i}.$$

If any of the window nodes will be actually occupied during the expected job execution interval, the whole parallel job will be postponed or even aborted. Therefore, a common resources allocation problem is a maximization of a total resources availability probability.

Based on the model above we consider the following job resources allocation problem in heterogeneous computing environment with non-dedicated resources and corresponding local and global utilization uncertainties: during a scheduling interval L allocate a set of n nodes with performance $p_i \geq p$ for a time T required to compute V instructions on each node, with common allocation and release times and a restriction C on the total allocation cost. As a target optimization criterion we assume maximization of a whole window availability probability P_a^W.

2.2 General Resources Allocation Procedure

For a general resources allocation procedure and the problem statement presented in Sect. 2.1, we combined core ideas and solutions from algorithms [4, 6, 7, 14, 15]. These approaches function in deterministic models and perform window search procedure based on a list of slots retrieved from a heterogeneous computing environment. Start and finish times of the system slots (i.e. the exact resources allocation and release times) represent key moments when the state of the computing environment changes. Therefore it is convenient to search for suitable resources allocations by scanning through the list of all available slots.

However the main difference of the present model is the presence of *continuous* probability functions of the resources utilization. Thus, there is no way to identify exact time moments when the system has the most favorable utilization conditions for a particular parallel job execution. So, the following general window search algorithm implements a rather computationally expensive *time scan* instead of a slot scan.

1. Initializing variables for the best availability criterion value and corresponding best window: $P_{max} = 0$, $W_{max} = \{\}$.
2. From the computing nodes available during the interval L we select different groups by their performance p_i. For example, group G_k contains nodes with performance $p_i \geq G_k$. Thus, one computing node may be included in several groups. In a general case number of such groups does not exceed the number $|R|$ of available heterogeneous resources.

3. Next is a cycle through all discrete time moments during the scheduling interval $t_k \in L$. All the sub-items represent a cycle body for $k = 1..L$.

   a. Next is a cycle for all retrieved groups G_i starting from the max performance group G_{max}. All the sub-items represent a cycle body for $i = 1..|R|$.

      (1) The resources reservation time required to compute V instructions on nodes with performance $p_i = G_i$ is $T_i = \frac{V}{p_i}$.
      (2) Calculate availability probabilities $P_a^{r_j}$ for each resource r_j in group G_i during the time interval $[t_k; t_k + T_i]$.
      (3) Based on $P_a^{r_j}$ values allocate from group G_i a subset $S_w^{k,i}$ of n nodes with a maximum total availability probability $P_a^{k,i}$ and a total usage cost less then C.
      (4) This subset $S_w^{k,i}$ represents a window with the maximum attainable availability probability at time t_k with nodes of performance $p_j \geq G_i$. If the resulting $P_a^{k,i} > P_{max}$ then reassign $P_{max} = P_a^{k,i}$ and $W_{max} = S_w^{k,i}$.

4. End of the algorithm. After all $k = 1..L$ and $i = 1..|R|$ values are scanned, W_{max} will contain the resulting window with the maximum attainable availability probability P_{max}.

2.3 Near-Optimal Resources Allocation Algorithm

Let us discuss in more details the procedure which allocates an optimal (according to the probability criterion P_a^w) subset $S_w^{k,i}$ of n resources from a group G_i (algorithm step 3a(3)).

We consider the following total resources availability criterion $P_a^w = \prod_j^n P_a^{r_j}$, where $P_a^{r_j} = P_j$ is an availability probability of a single resource r_j.

In this way we can state the following problem of an n - size window subset allocation out of m nodes in a group G_i:

$$P_a^w = x_1 P_1 * x_2 P_2 * \cdots * x_m P_m, \tag{1}$$

with the following restrictions:

$$x_1 c_1 + x_2 c_2 + \cdots + x_m c_m \leq C,$$
$$x_1 + x_2 + \cdots + x_m = n,$$
$$x_j \in \{0, 1\}, j = 1..m,$$

where c_j is total cost required to allocate resource r_j for a time T_i, x_j - is a decision variable determining whether to allocate resource r_j ($x_j = 1$) or not ($x_j = 0$) for the current window.

This problem relates to the class of integer linear programming problems, which imposes obvious limitations on the practical methods to solve it. However, we used 0-1 knapsack problem as a base for our implementation. The classical 0–1 knapsack problem with a total weight C and items-resources with weights c_j and values P_j have a similar formal model (1) except for extra restriction on the number of items required: $x_1 + x_2 + \cdots + x_m = n$. Besides, a special initialization step should be performed to take into account a product-wise form of the target criterion P_a^w. Therefore, we implemented the following dynamic programming recurrent scheme:

$$f_j(c, v) = \max\{f_{j-1}(c, v),\ f_{j-1}(c - c_j, v - 1) * P_j\}, \tag{2}$$

$$j = 1, .., m,\ c = 1, .., C,\ v = 1, .., n,$$

where $f_j(c, v)$ defines the maximum availability probability value for a v-size window allocated from the first j resources of G_i for a budget c. After the forward induction procedure (2) is finished the maximum availability value $P_a^w{}_{max} = f_m(C, n)$. x_j values are then obtained by a backward induction procedure.

An estimated computational complexity of the presented recurrent scheme is $O(m * n * C)$, which is n times harder compared to the original knapsack problem $(O(m * C))$. On the one hand, in practical job resources allocation cases this overhead doesn't look very large as we may assume that $n \ll m$ and $n \ll C$. On the other hand, in a general window search algorithm this subset allocation procedure (2) is called multiple times in cycles for $k = 1..L$ and $i = 1..|R|$ (step 3a(3)). So, the total computational complexity of the general resources allocation procedure presented in 2.2 can be estimated as $O(m^2 * n * C * L)$.

2.4 Greedy Resources Allocation Algorithm

Another approach for a $S_w^{k,i}$ subset allocation (algorithm step 3a(3)) is to use a more computationally efficient greedy approach. We outline four main greedy algorithms to solve the problem (1). The task is to select n out of m nodes from the group G_i, providing maximum total availability probability P_a^w with a constraint on a total allocation cost C.

1. *MaxP* selects first n nodes providing maximum availability probability P_j values. This algorithm doesn't take into account total usage cost limit and may provide infeasible solutions. Nevertheless *MaxP* can be used to determine the best possible availability options and estimate a budget required to obtain them.
2. An opposite approach *MinC* selects first n nodes providing minimum usage cost c_j or an empty list in case of exceeding a total cost limit C. In this way, *MinC* doesn't perform any availability optimization, but always provides feasible solutions when possible. Besides, *MinC* outlines a lower bound on a budget required to obtain a feasible solution.
3. Third option is to use a weight function to regularize nodes in an appropriate manner. *MaxP/C* uses $w_j = P_j/c_j$ as a weight function and selects first n nodes providing maximum w_j values. Such an approach doesn't guarantee feasible solutions for all G_i groups configurations, but nonetheless performs some availability optimization by implementing a compromise solution between *MaxP* and *MaxC*.
4. Finally we consider a joint approach *GreedyJnt* for an efficient greedy-based resources allocation. The algorithm consists of three stages.

a. Obtain *MaxP* solution and return it if the constraint on a total usage cost is met.
b. Else, obtain *MaxP/C* solution and return it if the constraint on a total usage cost is met.
c. Else, obtain *MinC* solution and return it if the constraint on a total usage cost is met.

This combined algorithm is designed to perform the best possible greedy optimization taking into account a restriction on a total resources allocation cost.

Estimated computational complexity for the greedy resources allocation step is $O(m * \log m)$. So the total computational complexity of the general resources allocation procedure presented in Sect. 2.2 can be estimated as $O\left(m^2 * \log m * L\right)$.

3 Simulation Study

3.1 Simulation Environment

An experiment was prepared as follows using a custom distributed environment simulator [6, 14, 15]. For our purpose, it implements a heterogeneous resource domain model: nodes have different usage costs and performance levels. A space-shared resources allocation policy simulates a local queuing system (like in CloudSim [13]) and, thus, each node can process only one task at any given simulation time. The execution cost of each task depends on its execution time which is proportional to the dedicated node's performance level. The execution of a single job requires parallel execution of all its tasks.

Each node supports a list of active global and local job utilization events.

Global uncertainty events represent resources failure or shutdown susceptibility and keep a constant occupation probability during the whole scheduling interval. Global utilization is generated for each resource based on a random variable P_o of occupancy probability with a normal distribution and an expected value $P_o^* = 0$. We use only positive values to determine occupancy probabilities P_o^j for each resource r_j. System-wide *global-load* parameter defines a standard deviation for P_o and is used to set an average global utilization for the whole computing environment. Thus, for example, when *global-load* $= 0.1$, about 68% of resources on average have global occupancy probability $P_o^j < 0.1$.

Local job-based utilization uncertainty is generated based on a preliminary job-flow scheduling simulation. For each resource a list of single-node jobs is generated with random jobs' submit times, lengths, start time and finish time uncertainty estimations. The jobs are ordered by their submit time and are scheduled in advance starting either at the submit time, or after the previous job is finished. During this scheduling, a chain of the resource *allocation*, *occupation* and *release* events is generated for each job. Corresponding expected times and standard deviations are defined by the job length and uncertainty parameters. A total length of jobs generated for each resource is determined by a system wide *job-load* parameter. For example, when *job-load* $= 0.1$, a total length of locally generated jobs constitute nearly 10% of a considered scheduling interval L.

Figures 1 and 2 show a single resource utilization schedule with global and local utilization events generated based on the procedures described above.

3.2 Resources Allocation Under Global Uncertainties

The experiment is set to demonstrate general behavior, features and efficiency of the proposed algorithms in a simplified environment setup with a single job scheduling, common job start time $t_k = 0$, and without local jobs' utilization uncertainties (*job-load* = 0).

During the experiment series we performed resources allocation for a single job requesting $n = 8$ nodes with performance level $p \geq 1$ and a computational volume $V = 200$.

In each experiment a new instance for the computing environment was automatically generated with the following properties. The resource pool includes 80 heterogeneous computational nodes. Each node performance level is given as a uniformly distributed random value in an interval [1, 8]. This configuration provides a sufficient resources diversity level while the difference between the highest and the lowest resource performance levels does not exceed one order.

A specific cost of a node is proportional to its performance level and correspondingly distributed on an interval [6, 12]. Cost and budget parameters were selected so that each job could be executed at least on the cheapest resources. This determines a steady market state appropriate for the consistent scheduling results comparison.

The resources allocation is independently performed by greedy-based algorithms *MaxP*, *MinC*, *GreedyJnt* and a *Knapsack*-based resources allocation approach presented in Sect. 2.3. The comparison is obtained based on different values of a total job execution budget limit C, representing the main market factor for the job scheduling management.

Figure 3 shows resulting availability probability of eight computing nodes allocated by the considered algorithms with different total allocation cost limits and *global-load* = 0.1. 1000 independent experiments were simulated to obtain results for each value of C on Fig. 3. *MaxP* and *MinC* results define an interval between the highest attainable resources availability and no availability optimization at all correspondingly. Both solutions depend mainly on the resource environment configuration, and graphs for *MaxP* and MinC represent straight lines independent of a budget limit C.

On the other hand, both *GreedyJnt* and *Knapsack* algorithms are able to improve window resources availability with increasing budget limit C.

Knapsack -provides better optimization results compared to *GreedyJnt* for all budget limit values $C \in [10500; 23000]$. In our experiment setup $C = 10500$ cost units is (at average) a lower bound required to schedule an eight-nodes parallel job. While with $C > 23000$, *MaxP*, *GreedyJnt* and *Knapsack* are able to select resources with the maximum availability probability in each experiment without breaking the total cost limit. Starting at $C = 10500$ *Knapsack* gradually improves the allocated resources availability and nearly approaches *MaxP* bound at $C > 16000$.

GreedyJnt generally improves resources availability except for values $C \in [11000; 16000]$. The corresponding plateau on *GreedyJnt* graph may be observed on Fig. 3. It may be explained by a ratio of inner *MinC*, *MaxP/C* and *MaxP* algorithms usage (wins) in *GreedyJnt* presented in Fig. 4. With a relatively low $C \in [10500; 12000]$ the

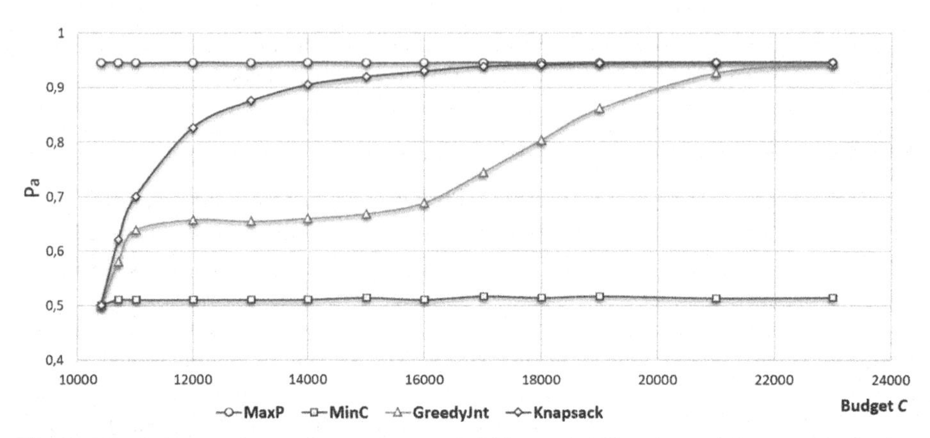

Fig. 3. Simulation results: total resources availability probability versus job execution budget limit C.

budget becomes sufficient to move *GreedyJnt* from *MinC* to *MaxP/C* solution and, thus, improve the resources availability compared to *MinC*. However for $C \in [1200; 15000]$ *GreedyJnt* uses *MaxP/C* in almost all experiments as lacks a budget for *MaxP* allocations. Starting at $C = 16000$ *GreedyJnt* gradually begins to use *MaxP* solutions and continue resource availability improvements up to the *MaxP* bound at $C = 23000$. The corresponding *MaxP* usage ratio on Fig. 4 tends to 1.

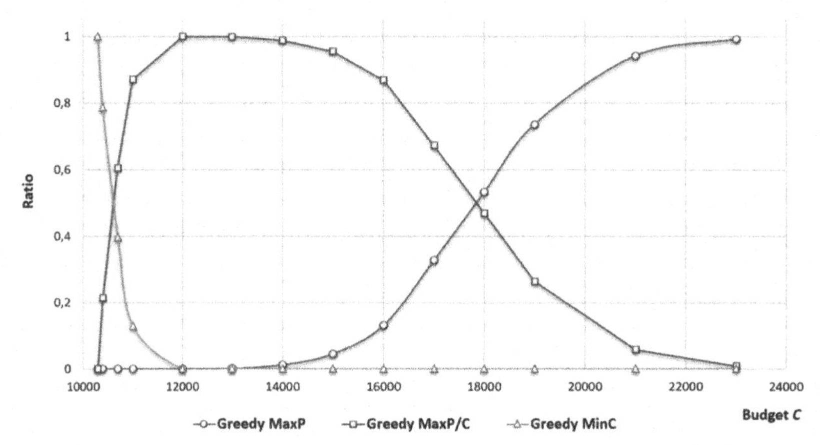

Fig. 4. Simulation results: a ratio of greedy algorithms usage in *GreedyJnt* versus job execution budget limit C.

Average algorithms' processing time obtained on Intel Core i5 system presented in Table 1. *Knapsack* algorithm requires almost 10000 more time to compute a solution compared to greedy algorithms. This figure mainly depends on quite a large budget values $C \in [10000; 23000]$ included in the algorithm's computational complexity. In a simple experiment configuration with only a global resources utilization of 80 available nodes and a common job start time $t_k = 0$ the total *Knapsack* runtime of 0.5 s remains reasonable. However, the presence of continuous probability functions will require a full time scan to consider all possible utilization configurations of the computing environment at each time point. Thus, an important area of development is *Knapsack* algorithm optimization and parallelization.

Table 1. Resources allocation algorithms working time.

Algorithm	*MaxP*	*MinC*	*GreedyJnt*	*Knapsack*
Working time, ms	0,029	*0,033*	*0,063*	*500*

4 Conclusion and Future Work

In this work, we address the problem of dependable resources co-allocation for parallel jobs in distributed computing with non-dedicated resources under global and local utilization uncertainties. For this purpose a general probability-based window allocation algorithm was proposed and considered. A special knapsack-based procedure is implemented to optimize a total resources availability probability. A set of greedy algorithms is implemented as an alternative with a less computational complexity.

We considered several types of global and local job-based resources utilization events. The simulation study proved optimization efficiency of the proposed knapsack and greedy algorithms according to the composite resources *availability* criterion.

As a drawback, the general case algorithm has a relatively high computational complexity. The presence of continuous probability functions requires a full time scan to consider all possible utilization configurations of the computing environment. Therefore we conducted simulation experiment in a simplified environment only under global utilization uncertainties. In our further work, we will refine a general resource co-allocation scheme in order to decrease its computational complexity.

Acknowledgments. This work was partially supported by the Council on Grants of the President of the Russian Federation for State Support of Young Scientists (YPhD- 2979.2019.9), RFBR (grants 18-07-00456 and 18-07-00534).

References

1. Lee, Y.C., Wang, C., Zomaya, A.Y., Zhou, B.B.: Profit-driven scheduling for cloud sevices with data access awareness. J. Parallel Distrib. Comput. **72**(4), 591–602 (2012)
2. Garg, S.K., Konugurthi, P., Buyya, R.: A linear programming-driven genetic algorithm for meta-scheduling on utility grids. Int. J. Parallel Emergent Distrib. Syst. **26**, 493–517 (2011)
3. Buyya, R., Abramson, D., Giddy, J.: Economic models for resource management and scheduling in grid computing. J. Concurr. Comput. Pract. Exp. **5**(14), 1507–1542 (2002)

4. Ernemann, C., Hamscher, V., Yahyapour, R.: Economic scheduling in grid computing. In: Feitelson, D.G., Rudolph, L., Schwiegelshohn, U. (eds.) JSSPP 2002. LNCS, vol. 2537, pp. 128–152. Springer, Heidelberg (2002). https://doi.org/10.1007/3-540-36180-4_8
5. Kurowski, K., Nabrzyski, J., Oleksiak, A., Weglarz, J.: Multicriteria aspects of grid re-source management. In: Nabrzyski, J., Schopf, J.M., Weglarz, J. (eds.) Grid Resource Management. State of the Art and Future Trends, pp. 271–293. Kluwer Academic Publishers, New York (2003)
6. Toporkov, V., Toporkova, A., Bobchenkov, A., Yemelyanov, D.: Resource selection algorithms for economic scheduling in distributed systems. In: Proceedings of International Conference on Computational Science, ICCS 2011, 1–3 June 2011, Singapore, Procedia Computer Science, vol. 4. pp. 2267–2276. Elsevier (2011)
7. Netto, M.A.S., Buyya, R.: A flexible resource co-allocation model based on advance reservations with rescheduling support. In: Technical report, GRIDSTR-2007–17, Grid Computing and Distributed Systems Lab University of Melbourne, Australia (2007)
8. Jackson, D., Snell, Q., Clement, M.: Core algorithms of the Maui scheduler. In: Feitelson, D.G., Rudolph, L. (eds.) JSSPP 2001. LNCS, vol. 2221, pp. 87–102. Springer, Heidelberg (2001). https://doi.org/10.1007/3-540-45540-X_6
9. Javadi, B., Kondo, D., Vincent, J., Anderson, D.: Discovering statistical models of availability in large distributed systems: an empirical study of SETI@home. IEEE Trans. Parallel Distrib. Syst. **22**(11), 1896–1903 (2011)
10. Rood, B., Lewis, M.J.: Grid resource availability prediction-based scheduling and task replication. J. Grid Comput. **7**, 479 (2009)
11. Tchernykh, A., Schwiegelsohn, U., El-ghazali, T., Babenko, M.: Towards understanding uncertainty in cloud computing with risks of confidentiality, integrity, and availability. J. Comput. Sci. **36** (2016)
12. Chaari, T., Chaabane, S., Aissani, N., Trentesaux, D.: Scheduling under uncertainty: survey and research directions. In: 2014 International Conference on Advanced Logistics and Transport (ICALT), pp. 229–234 (2014)
13. Calheiros, R.N., Ranjan, R., Beloglazov, A., De Rose, C.A.F., Buyya, R.: CloudSim: a toolkit for modeling and simulation of cloud computing environments and evaluation of resource provisioning algorithms. J. Soft. Pract. Exp. **41**(1), 23–50 (2011)
14. Toporkov, V., Yemelyanov, D.: Optimization of resources selection for jobs scheduling in heterogeneous distributed computing environments. In: Shi, Y., Fu, H., Tian, Y., Krzhizhanovskaya, V.V., Lees, M.H., Dongarra, J., Sloot, P.M.A. (eds.) ICCS 2018. LNCS, vol. 10861, pp. 574–583. Springer, Cham (2018). https://doi.org/10.1007/978-3-319-93701-4_45
15. Toporkov, V., Yemelyanov, D.: Resources co-allocation optimization algorithms for distributed computing environments. In: ACM International Conference Proceeding Series, 47th International Conference on Parallel Processing, ICPP 2018, Paper 43 (2018)
16. Epema, D., Iosup, D.: Grid computing workloads. J. IEEE Internet Comput. **15**(2), 19–26 (2011)

Development of Experimental Data Processing Workflows Based on Kubernetes Infrastructure and REANA Workflow Management System

Anton Teslyuk[1,2](✉), Sergey Bobkov[1], Alexey Poyda[1,2], Alexander Novikov[1], Vasily Velikhov[1], and Viacheslav Ilyin[1,2]

[1] Russian Research Centre "Kurchatov Institute", Moscow, Russia
{anthony.teslyuk,s.bobkov}@grid.kiae.ru, poyda_aa@nrcki.ru, novikov@wdcb.ru, velikhovve@kiae.ru, ilyin@sinp.msu.ru

[2] Moscow Institute of Physics and Technology, Dolgoprudny, Moscow Region, Russia

Abstract. In this paper we present the design of data processing workflow for scientific experiments, which require complicated multi-step analysis procedure. We test it on datasets from Single Particle Imaging (SPI) experiments. The workflow is based on microservice architecture, Docker containers and Kubernetes platform. For workflow setup and management we use REANA software which is compatible with Kubernetes ochestrator and supports standard Common Workflow Language (CWL) to describe complex computing jobs. Our approach allows easy construction of workflows of diverse architecture for a wide range of applications. It allows integration of heterogeneous software in a uniform way as well as easy modification or replacement of workflow components. In the same time it allows easy scaling of computations in a cloud infrastructure. We show the applicability of the designed scheme and estimate the overhead of the platform middleware.

Keywords: Containers · Kubernetes · Docker · Single particle imaging · Software workflow · REANA

1 Introduction

Modern research facilities generate vast amounts of data during experiments which require powerful computing infrastructure for subsequent data analysis. The need to process big volumes of data efficiently has led to emergence of a new trend in information technology - High Performance Data Analysis (HPDA) [1]. For rapidly developing fields of science such as study of biological objects with single particle imaging (SPI) [2] at European XFEL [3] or cryo-electron microscopy (CryoEM) [4,5], data analysis techniques and software progress rapidly.

This requires computing infrastructure used for data analysis to be flexible, allowing easy changes and updates to software workflows. In the same time it

V. Voevodin and S. Sobolev (Eds.): RuSCDays 2020, CCIS 1331, pp. 563–573, 2020.
https://doi.org/10.1007/978-3-030-64616-5_48

should be scalable, allowing to distribute computations to a large number of computing nodes when needed.

A common way to build scalable and flexible data processing infrastructures is microservice architecture [6]. The core idea behind it is to define utilised software as a set of small loosely coupled components, "microservices", each to be put in its own application environment, usually implemented by the use of container technology such as Docker, LXC or CRI-O. Application environment isolation makes it possible to use heterogeneous software with various requirements to OS-level dependencies and auxiliary software in a uniform way. Microservices management and scaling is done using container orchestration software such as Kubernetes [7], Docker Swarm [8], Rancher [9] and others. The purpose of container orchestrator is to distribute containerized software across computing infrastructure, manage the execution and scaling of containers and set up network communications between them. Data processing in modern experiments often aims to extract a tiny fraction of meaningful information from huge raw datasets collected during experiment. Extraction of such information is usually a many step procedure which requires plenty of distributed computing resources and configuration of numerous parameters due to complexity of applied algorithms and amount of data to be analysed. Evolution of data processing causes further complication of the procedure due to development of new methods and additional processing steps that improves quality of extracted information.

Construction of data processing workflows requires additional software component that is aware of the data analysis procedure as a whole and manage the computations according to the predefined dependency graph of workflow elements. It tracks the availability of the input data for workflow components, starts required containers with proper parameters and controls results of individual workflow elements operation.

Containerized data processing workflows bring two important advantages. They provide the possibility to run complex data analysis tasks on a large-scale computing infrastructure. In the same time, they provide researchers with a standardised tools to reproduce existing results of the data analysis. It requires the platform middleware components to be implemented using standard software tools, which are available in most scientific data centres as well as in popular commercial cloud providers such as Amazon AWS or Google Cloud. This approach will allow easy transfer of application workflows between data centres without additional modifications or setup.

In this paper we present the review of a software platform being under development at Kurchatov Institute. We use it for analysis of SPI experimental data from European XFEL facility and laboratory of Cryo-EM at KI. However, its design allows to adapt it for other scientific applications which require multiplex data processing. In the next sections we describe its main components and estimate the performance overhead caused by the platform middleware.

2 Related Work

A vast number of ways exists to build data analysis pipelines [10]. They differ significantly in aims and realization strategies. For example, in bioinformatics field there is a very popular software Galaxy [11] aimed to provide users with user-friendly web interface for typical data analysis jobs. It provides users with a simple web-based interface to access various command line tools and can integrate multiple programs into a complex pipeline chains. Its core intent is the simplicity of operations for researchers without strong IT-background, and this is the sources of its popularity. In the same time, the most common setup for Galaxy is running on a single server, so normally it is not used for applications which require large-scale computing resources.

A different approach to build data processing pipelines is used when we have large sets of data and homogeneous algorithms and software tools to process the data. This is often the case for machine learning applications when a single model has to be trained using large distributed dataset or a pre-trained model needs to be applied to process large amounts of data for example for data classification tasks. A common scenario for this class of pipelines is to use map-reduce data storage system, e.g. Hadoop [12] with a machine learning framework e.g. Apache Spark MLLib [13] on its top. This approach allows extremely efficient data processing but it not flexible in the choice of data processing tools. One has to use Apache Spark MLLib or similar framework to access the data and develop data analysis algorithms. In the same time users need to have strong programming skills to develop pipelines using this approach.

When it comes to compute-intensive data analysis jobs with general purpose software tools another set of solutions arise. A common way is to use some resource manager software that can run general-purpose computing jobs: Slurm [14] for high performance computing resources, HTCondor [15] for high throughput computing, Kubernetes [7] or Singularity [16]) for cloud-based computing. On top of that, a workload management software is used for management of complex composite applications (data processing pipelines). An example of this paradigm is cryogenic microscopy data acquisition and data processing pipelines developed at SLAC National Accelerator Center [17]. These pipelines use Kubernetes as a resource manager and Apache Airflow [18] as a workload management system. In Apache Airflow python code is used to setup graph of pipeline elements and relationships between them and then Airflow workload manager takes care of execution and management of the pipeline providing users with a web interface to see the progress. Being very flexible and powerful this style of pipeline construction can be obscure for users without python programming skills.

Various pipeline types significantly differ in their features, for our task it is important to have the pipeline scalable, flexibile and simple for end users. Performance is needed to apply sophisticated algorithms to large sets of experimental data. Flexibility is important because data processing tools evolve rapidly and need a simple mechanism to be replaced or upgraded. Also we try to separate IT machinery internals from pipeline logic level letting users to construct workflows using high level abstractions.

3 Components of Software Infrastructure for HPDA

The main components of the software platform under our development include: Docker CE as a container technology [19], Kubernetes as a container orchestrator [7], CEPHfs [20] as distributed storage system and REANA [21] for data processing workflow management. The platform is deployed on a dedicated testbed including four computing nodes each having 2xIntel Xeon X5670 CPU and 24 Gb RAM, Gigabit Ethernet interconnect and 3xNVIDIA Tesla M2050 GPUs. Host operating system is Centos Linux version 7. Below we present deployment details of container middleware of our system, while in the next sections we make a review of existing workflow management systems as well as provide details of our deployment and the use of REANA.

3.1 Docker CE

For container technology we used Docker Community Edition version 19.03.8. To allow use of NVIDIA GPUs inside containers we used nvidia-container-runtime [22], which was set as a default Docker runtime for Kubernetes.

3.2 Kubernetes Orchestrator and CEPHfs File Storage

Container orchestrations across the cluster were perfomed by Kubernetes software version 1.18.0. Nvidia Device Plugin for Kubernetes [23] was used to take advantage of GPU devices. Network communications between containers were established by Flannel Fabric [24] which automatically deploys and maintains a virtual layer 3 network which interconnects all containers in a cluster.

We used CEPHfs version 13.2.8 as a persistent storage for containerized infrastructure. It was integrated with Kubernetes cluster using CEPHfs external storage provisioner [25]. The storage is used in two ways:

1. reservation of individual storage resources of desired size for containerized applications using persistent volume claims mechanism.
2. use of static CEPHfs volumes to share data between containers. Our data model assumes that all input, intermediate and output files are located in a shared storage resource, so we don't transfer huge data volumes through the network.

4 Workflow Management Software

A number of solutions were developed to integrate heterogeneous data processing software in workflows in a convenient way for users. They allow creation of workflows or workflows containing processing steps with defined dependencies using high-level description language. Users got the ability to define and execute workflows as a whole, while inner steps were processed automatically. Initially each workflow management software used its own approach to define workflow and dependencies between processing steps. To unify workflow definitions between software packages high level languages were proposed: Common Workflow Language (CWL) [26] and Workflow Description Language (WDL), which provide users with a simple abstraction level to describe data processing jobs.

Another important feature of workflow management software is the ability to take advantage of underlying computing infrastructure such as a high performance cluster or cloud system. If several steps can be processed independently, significant computing speedup can be achieved. Effective usage of computing resources along with automated execution of complex workflows allow dramatic reduction of processing time.

We analysed a number of existing workflow management software to build our containerised workflows. The list includes:

- Apache Airflow (initially developed inside Airbnb company), one of the most popular open-source workflow management solutions.
- REANA, a platform for reproducible data analysis developed in CERN.
- Luigi a simple Python framework for building software workflows.
- HKube a framework to run distributed computing workflows in Kubernetes infrastructure.
- Galaxy, one of the most widely used platforms for data analysis in biomedical research.

While all the software products mentioned above are suitable to build complex data analysis workflows they differ in certain properties. The features we consider important for our use include: compatibility with Docker containers and Kuberntetes container orchestration, standard workflow definition language, support parallel operations and availability of command line interface and REST API. Comparison of selected packages is presented in Table 1. After initial testing we selected REANA as a workflow manager for our platform as it supports kubernetes cluster with docker containers, CWL workflow definitions and allows full control over workflow execution via REST API.

5 Reana on KI Cluster

REANA is a platform for reusable and reproducible data analysis. It allows researchers to structure their analysis workflows and to run them on remote containerised compute clouds. REANA supports several different workflows systems (CWL, Serial, Yadage) and aimed to use different job execution backends (originally developed for Kubernetes backend it was extended to use SLURM and HTCondor backends) [27].

Table 1. Comparison of workflow management software packages

	Apache Airflow	REANA	Luigi	HKube	Galaxy
Kubernetes cluster	+	+	+	+	+
Docker containers	+	+	+	+	+
Workflow definition language	Python code, CWL with plugin	CWL, Yadage	Python code	JSON	XML, JSON
Parallel workflow step execution	+	+	+	+	+
REST API	Limited	Full	Limited	Full	Full
Command line interface	+	+	+	+	+

The platform consists of several core modules:

- server, which handles user requests using REST API and mediates them to other services. It can create a new workflow, load data, get logs and results, control workflow lifecycle;
- database that stores users, workflows definition and operational status, user data, log files and other system information;
- workflow controller manager, controlling the execution of workflows on job execution backends;
- messages queue service for information exchange between services;
- base and additional workflow execution backends (serial, CWL,Yadage);
- workflow single step execution adapters - jobs (Kubernetes, SLURM, HTCondor);
- command line client or UI service;
- 'traefik' service as a proxy for Kubernetes services.

Users communicate with the platform via HTTP requests using a client application running on the user side.

All modules are implemented as Kubernetes deployments which could be scaled on demand are connected through message queue service. The whole system could be configured via a deployment scheme file with parameter variables. Workflow execution modules can handle data with temporary K8s volumes, CEPHfs volume claims and optional common file system (in the native approach only the CVMFS, developed at CERN).

Important part is a workflow scheme, a description of workflow steps and their relationships combined with input and output data and software dependencies for each single step. REANA supports serial, CWL or YADAGE workflow definition language [28]. This scheme defines the relationship between individual steps. Each step is determined by the Docker container with the software to be executed, the set of input data (that are the result of the previous steps or the workflow input), and the output data (that are input of the next steps or the final result).

To adopt REANA installation for the Kurchatov supercomputer, we had to solve a number of problems:

1. We had to connect CEPHfs as a shared file system because input data occupy volume of several terabytes and it would be ineffective to copy data between steps. In the native approach REANA assumes that workflow input data is pre-copied to a local data storage or taken from a shared file system CVMFS which is not available at our installation. To solve this problem, we have to connect the specified CEPH-volume to each Docker container involved in the execution of the workflow. In addition, we need to control the access rights of connected CEPH-volume to ensure that workflow steps will not break permissions of the authorized user.
2. REANA authorize users by kerberos service [29] or native CERN user authentication mechanism, which was not available in our case. Additionally at the Kurchatov supercomputer the resources allocated to the users with its own user id, which are to be synchronized with REANA ones to execute workflows and access the data. In the future, we plan to use FreeIPA service which supports the Kerberos as well as LDAP directory service, but for now we have implemented a simpler solution:
   (a) create local REANA accounts for users who should have access to the platform;
   (b) associate REANA accounts with Kurchatov supercomputer users (uids) by extending the REANA user table in the database with a field containing the user uid;
   (c) modify the REANA workflow execution code so that all containers and volumes related with a certain workflow use the corresponding user id.
3. Original REANA implementation implies the ‘default’ namespace of K8s for installation, which does not fit every K8s cluster due to security reasons.

Thus, the solution of all three problems required code modifications of the core REANA packages and REANA Docker-images rebuild. We also modified several configuration parameters. Except for these changes, the installation and use of REANA remains the same. To start the workflow one need to perform a number of steps:

1. Use one of the existing workflow schemes from a centralized library or create a new one. Users can also combine several basic single step workflows into complex one. There is an open question how users can share or distribute these schemes between different groups.
2. Modify steps, values and data sources using the UI service or simple text editor at the user side. For example, we can use CWL schemas and specify absolute path for input and output data at shared CEPHfs.

3. Define a REANA workflow from that scheme and load the required input data and code into it by means of the UI service or the client application at the user side.
4. Start the REANA workflow.
5. Check the status of the workflow. If an error occures, user can download and analyze workflow logs.
6. After finish of workflow execution the user can get the results placed at shared file system for consequent workflows or download them locally.

6 Experimental Data and Applied Workflow

The sample data was obtained in the experiment performed at the Atomic Molecular Optics (AMO) instrument [30,31] at the Linac Coherent Light Source (LCLS) at SLAC National Accelerator Laboratory. The details of the experiment setup, raw data collection and sample preparations were reported [32]. The data was deposited in the coherent x-ray imaging data bank id 58 [33].

Steps of the applied workflow were implemented in Python code, computations were optimised with NumPy, SciPy and Numba libraries. The software can take advantage of multicore computing resources and does not use GPU accelerators.

The applied workflow were developed for preprocessing of diffraction images obtained in SPI experiments preformed at XFELs. It includes the following major steps: estimation of center position for diffraction patterns, size filtering and 2D clustering of diffraction images. Some of the major steps were divided into minor steps each packed into its own containerized application and requiring set of input and output files. The applied workflow consists of 12 minor steps. Four most performance-demanding steps were optimised to allow natural data parallelism at workflow level. It means that CWL workflow manager (both REANA and cwltool) can split the set of input data into a number of batches and process each batch independently in separate workflow threads. This regime allows us distribute computations across multiple nodes when using REANA. The other eight steps use the input data as a whole and should be computed within one node.

We investigated the total processing time for two workflow execution modes: processing on a cluster via REANA service and processing on a single node via cwltool. REANA adds additional software layer for synchronization and data migration between cluster nodes and Kubernetes layer for job distribution. Comparison of results on a single node allow to estimate the overhead caused by these additional layers. We also checked how computation time changes when we use REANA and vary the number of working nodes. This will allow us to test the effectiveness of parallelization in multi-node regime.

Another important topic is the location of input, intermediate and output files. By default REANA creates a separate copy of input and output data for every workflow step. We applied a number of optimizations for REANA to replace file copy operations with file move when it is possible. Also, cwltool saves intermediate files in local filesystem which leads to ineffective data copy between local filesystem and CEPHfs. We have tested cwltool in two regimes: with intermediate files in local filesystem and optimized regime with intermediate files in CEPHfs. In the first case, cwltool can not replace all copy operations with move operations. Therefore, cwltool and REANA have to perform the same set of copy operations.

The results of time measurements of different regimes of operations is presented in Table 2.

Table 2. Performance measurements. The time of data processing is calculated for parallelized and un-parallelized steps of the pipeline for single-node and distributed modes of operation. Singe-node operation was tested in default and optimized regimes.

Mode of operation	Parallelized steps	Un-parallelized steps	Total time
Single node (default)	34 m 8s	25 m 7 s	59 m 15 s
Single node (optimized)	13 m 33 s	14m 35s	28 m 8 s
REANA, one node	24 m 30 s	27 m 23s	51 m 53 s
REANA, two nodes	14 m	28 m 53 s	42 m 53 s
REANA, three nodes	13 m 47 s	34 m 2s	47 m 49 s

One can see that optimized mode of singe node operation is about twice faster than the default one. The performance of default single node mode is comparable to single worker REANA configuration. The amount of additional data coping is comparable for REANA and default single node regime. Thus we can roughly estimate the overhead of excessive I/O operations to the half of the total time. It means that it is vitally important to optimize the location of input and intermediate files.

In the same time one can see REANA can distribute computations among two nodes, significantly reducing time of computation for parallelized jobs. We don't see significant acceleration with three nodes setup. We believe that this is because we have too few data to share between large number of workers and the overhead cost of managing the additional containers becomes comparable to the benefit of the additional processing power.

7 Summary

We have presented the design of workflow for scientific data processing. The combination of its three core components: Docker containers, Kubernetes container ochestration and workflow management with REANA brings us features which can be of great significance for a wide range of applications. The key features are:

- High level description language for workflow definition with CWL allows easy construction of complicated and diverse workflows.

- Containerization technology allows to integrate heterogeneous applications with various requirements to application or OS-level environment. Containerized applications become standard building blocks for the flexible design of complex application pipelines. One can easily modify or replace individual blocks of a complex workflow.
- Container orchestration technology allows easy scaling of computations from a single server system to large computing infrastructures. High performance data analysis can be implemented using existing parallelization techniques: natural data parallelism, parallel computing with MPI, GPU-acceleration, Multi-Core parallelization and others which have standard deployment patterns in Kubernetes platform.

We have shown that our workflow can distribute computations across multiple nodes giving significant improvements in performance for operations run in parallel. In the same time it is important to bound the number of workers depending on the amount of data to be processed. Another important point for optimizations is the amount of I/O operations for the data. We show that the distribution of files used in pipeline operation across file systems can critically affect the performance. This is a topic of the further research.

Acknowledgements. This research was partially supported by the Helmholtz Associations Initiative and Networking Fund and the Russian Science Foundation (project No. 18-41-06001, workflow development and deployment done by AT, SB, AN, VI and VV) and by RFBR grant 18-29-23020 (review of existing workflow management software done by AP). The work has been carried out using computing resources provided by NRC Kurchatov institute project "Development of modular platform for scientific data processing and mining" (Project No. 1571).

References

1. Osseyran, A., Giles, M.: Industrial Applications of High-Performance Computing: Best Global Practices, vol. 25. CRC Press, Boca Raton (2015)
2. Gaffney, K., Chapman, H.: Imaging atomic structure and dynamics with ultrafast X-ray scattering. Science **316**(5830), 1444–1448 (2007)
3. Altarelli, M., Brinkmann, R., Chergui, M., Decking, W., Dobson, B., Düsterer, S., Grübel, G., Graeff, W., Graafsma, H., Hajdu, J., et al.: The European X-ray free-electron laser technical design report. DESY **97**(2006), 4 (2006)
4. Callaway, E.: The revolution will not be crystallized: a new method sweeps through structural biology. Nature News **525**(7568), 172 (2015)
5. Danev, R., Yanagisawa, H., Kikkawa, M.: Cryo-electron microscopy methodology: current aspects and future directions. Trends in biochemical sciences (2019)
6. Nadareishvili, I., Mitra, R., McLarty, M., Amundsen, M.: Microservice architecture: aligning principles, practices, and culture. O'Reilly Media, Inc. (2016)
7. Kubernetes. https://kubernetes.io/
8. Docker swarm. https://docs.docker.com/get-started/swarm-deploy/
9. Rancher. https://rancher.com/
10. Computational data analysis workflow systems. https://github.com/common-workflow-language/common-workflow-language/wiki/Existing-Workflow-systems

11. Galaxy project. https://galaxyproject.org
12. Apache hadoop. https://hadoop.apache.org
13. Apache spark mllib. https://spark.apache.org/mllib/
14. Slurm workload manager. https://slurm.schedmd.com/documentation.html
15. Htcondor project. https://research.cs.wisc.edu/htcondor/
16. Singularity project. https://sylabs.io/
17. Data acquisition and data processing pipelines for cryoem at slac. https://github.com/slaclab/cryoem-airflow
18. Apache airflow. https://airflow.apache.org
19. Docker. https://www.docker.com/
20. Ceph. https://docs.ceph.com/docs/master/
21. Reana. http://reanahub.io/
22. Nvc-github. https://github.com/NVIDIA/nvidia-container-runtime
23. Nvidia device plugin for kubernetes. https://github.com/NVIDIA/k8s-device-plugin
24. Flannel for kubernetes. https://github.com/coreos/flannel
25. Cephfs volume provisioner for kubernetes. https://github.com/kubernetes-incubator/external-storage/tree/master/ceph/cephfs
26. Amstutz, P., et al.: Common workflow language, v1. 0 (2016)
27. Maciulaitis, R., et al.: Support for htcondor high-throughput computing workflows in the reana reusable analysis platform. Technical report (2019)
28. Reana documentation. https://reana.readthedocs.io/en/latest/
29. Kohl, J., Neuman, C., et al.: The kerberos network authentication service (v5). Technical report, RFC 1510, September 1993
30. Bozek, J.D.: AMO instrumentation for the LCLS X-ray FEL. Europ. Phys. J. Special Top. **169**(1), 129–132 (2009)
31. Ferguson, K.R., et al.: The atomic, molecular and optical science instrument at the linac coherent light source. J. Synchrotron Radiation **22**(3), 492–497 (2015)
32. Reddy, H.K., et al.: Coherent soft x-ray diffraction imaging of coliphage pr772 at the linac coherent light source. Sci. Data **4**, 170079 (2017)
33. Maia, F.R.: The coherent x-ray imaging data bank. Nat. Methods **9**(9), 854 (2012)

Discrete Event Simulation Model of a Desktop Grid System

Evgeny Ivashko[1,2], Natalia Nikitina[1], and Alexander Rumyantsev[1,2](✉)

[1] Institute of Applied Mathematical Research, Karelian Research Centre of RAS, Petrozavodsk, Russia
{ivashko,nikitina,ar0}@krc.karelia.ru
[2] Petrozavodsk State University, Petrozavodsk, Russia

Abstract. The paper describes a discrete event simulation model of a Desktop Grid system. Firstly, we present a stochastic model of a volunteer computing project. We then employ the event simulation approach based on the generalized semi-Markov processes to develop a discrete event simulation model. Finally, using the simulation model, we describe a performance optimization problem aiming to optimize the project runtime as a function of the task size under performance constraints.

Keywords: Desktop Grid · Volunteer computing · BOINC · Dynamic load balancing · Discrete event simulation · Generalized Semi-Markov Process

1 Introduction

Desktop Grids are important tools for computationally demanding scientific research. They combine the non-dedicated geographically distributed computing resources connected to the central server by the Internet or a local access network. The resources are either provided by individuals and organizations related to the research performed, or donated by the volunteer community. Many of the world's leading research institutions perform Desktop Grids based large-scale computational experiments (Washington University: Rosetta@home, Folding@home; CERN: LHC@home; University of Oxford: Climateprediction.net, and many others).

Distributed computing platforms may be organized in various ways. However, the BOINC middleware, introduced in the early 2000s [1], is currently the standard way of Desktop Grid project implementation. More than 130 projects were performed with this platform, with the toll of 30 active ones of total capacity exceeding 90 PetaFLOPS at present. Delivering scalability and flexibility into distributed computing, BOINC is actively developed and widely used. For this reason, in this work we restrict ourselves to an architecture of Desktop Grid inherent to BOINC, that is, the centralized master-worker computing paradigm. Each worker node communicates with the server independently

V. Voevodin and S. Sobolev (Eds.): RuSCDays 2020, CCIS 1331, pp. 574–585, 2020.
https://doi.org/10.1007/978-3-030-64616-5_49

through HTTP-based RPC interfaces, requesting new tasks and reporting the results. The detailed description of BOINC is given in [1].

The Desktop Grids are best suited for the so-called embarrassingly parallel applications, that is, the applications consisting of a large number of independent tasks. Moreover, due to a relatively slow network connection and highly unpredictable computational time, the *size of each task* is selected in such a way to guarantee relatively lengthy computation. To minimize the server load, both task assignment and result acquisition may be performed *in batches* of various size. All these parameters heavily affect the overall computing time. Finally, due to a long development history, BOINC has a sophisticated configuration procedure with a number of parameters affecting the distributed computing system performance. This motivates the study of a Desktop Grid system with a stochastic model.

This work is aimed at design of a mathematical model for providing the balance between characteristics of a computational project and the server load. In Sect. 2, we present an overview of related papers in the field of Desktop Grid, BOINC and queuing systems. In Sect. 3, we deliver a stochastic model of a volunteer computing project and performance optimization parameters. Along with that, we discuss optimization problems and model extensions. Finally, in Sect. 4, we summarize the reasoning and conclude the paper.

2 Related Works

A large number of configuration parameters and project implementation peculiarities, ranging from task scheduling [2] up to psychological aspects of working with volunteers, affect the overall system performance [3,4]. Below we summarize some recent advances in Desktop Grid performance studies.

Technical limitations of a BOINC server bound the scalability of a project. The exponential back-off embedded connection leveraging mechanism [5,6] may be insufficient or even inappropriate for some projects, and thus the *server load minimization* techniques are actively studied. One of the promising techniques is the data processing outsourcing suggested in [7], which was demonstrated to be effective for data-intensive applications, such as ATLAS@Home.

An *optimal size* of a BOINC task is needed to balance the overhead of serving small tasks and the waste of resources on computing large-sized tasks on unreliable nodes. The paper [8] investigates an optimal BOINC task size for genome-scale data analysis. Grouping the tasks into *batches* is also an important server load minimization technique. The work [9] aims to reach the game-theoretic balance between the server load and the total time of computations in a virtual drug screening project.

Apart from optimizing a stabilized workflow, one also needs to adapt volunteer computing projects to workload volatility, e.g. the community-wide *competitions* such as Formula BOINC (http://formula-boinc.org), or the so-called computational experiments (e.g. the project Gerasim@home [10] issues about 10 experiments a year in addition to the regular project activity). The technical capabilities of BOINC allow to temporarily increase server performance during such periods by parallelizing task-handling daemons and varying workflow parameters such as the maximal number of tasks to send to one client and the maximal number of results to report at once. As a consequence, the workflow characteristics temporarily change.

Finally, a BOINC project can be studied as a queueing system or a queueing network. As such, various BOINC performance optimization problems may be addressed by means of the stochastic modeling. In [11] the optimal *replication and quorum* redundancy parameters of a BOINC server were studied. In [12] a large-scale Desktop Grid project was approximated by Gaussian process, and project computation time was estimated. In [13], the fork-join type models were used to study the distributed computing model. This makes stochastic modeling a promising approach to solve practical tasks of Desktop Grid performance optimization.

3 Stochastic Model of a Volunteer Computing Project

In this Section we develop a stochastic model of a single volunteer computing project, with the aim to formalize reasonable optimization problems related to the project's performance. First, we discuss the typical volunteer computing infrastructure on the basis of the BOINC computing platform. We outline the project performance optimization management parameters available both at the server side and at the volunteer side. We do not discuss the resource concurrency at the volunteer side, however, we still take into account some management parameters available at client side only, and assume these are reachable by project administrators (either directly, or indirectly, by requesting the project users to configure their software the desired way). Finally, we state the mathematical model of a single volunteer computing project, using the framework of a queueing network model. We outline some optimization problems that can be solved using the model suggested at the end of the Section.

3.1 BOINC Infrastructure Overview

A typical volunteer computing project is based on a master-worker type network with a dedicated BOINC server responsible for the project organization, and multiple hosts performing the work assigned by the server. Usually, the project contains a large, yet fixed amount of computational work, which is divided into relatively small, independent pieces (tasks).

The communication and process organization are performed over the Internet using RESTful interface over HTTP in an asynchronous way. The hosts request a server to assign them tasks in an asynchronous manner, and once the computational tasks are available, the hosts keep performing computations without the need to be controlled by a server. Using sophisticated decision making algorithms, the hosts report the results of computations in a timely manner, and can be assigned some new tasks at the time of reporting, or independently, at the

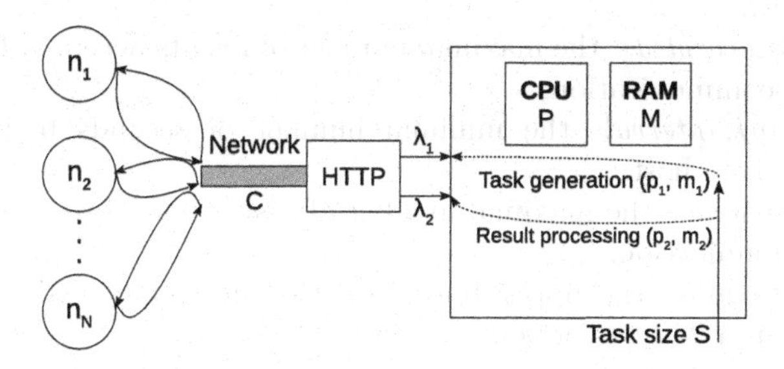

Fig. 1. Queueing model of BOINC

asynchronous request times. Schematically, the structure of a BOINC network is depicted in Fig. 1.

Desktop Grid projects typically possess dynamic changes in the number of hosts. For this reason, the usage of server resources is also dynamic. In the process of project completion, a number of project parameters (system configuration), including task sizes, may be adopted so as to better respond the number and capacity of the hosts.

During the project lifecycle, hosts contact the BOINC server with various types of requests, e.g. (1) sign-up to the project; (2) request of appropriate application files; (3) request of a task (or a group of tasks) with input data; (4) report of a result (or a group of results) with output data; (5) request of altering client's preferences; (6) detach from the project.

In our model, we restrict ourselves to the request flows of types (3) and (4), assuming that requests of other types are sporadic and do not significantly influence the server load. Their rates depend on the average computational time, certain settings on the project's side and on the client's side, and the stochastic characteristics of the computational process. More exactly, there is a fraction of tasks that end up with an error (these have another rate); there are lost tasks (client generates a request of type (3) but never reports the result); finally, the requests of type (3) and (4) are sometimes bundled together.

It is clear from the project description that the network possesses random and unpredictable characteristics. Moreover, there is a clear delayed feedback effect moderated by the amount of work assigned to a host. To leverage this effect, several parameters are available both at the host and at the server side. We discuss these parameters in the following subsection.

3.2 Project Performance Optimization Parameters

The BOINC platform allows to adjust project performance by a number of parameters. Many of them impact client-server communications. In this subsection we restrict ourselves to those that directly control the frequency of client-server communications and the size of transmitted data.

At the server side of BOINC, the following parameters may be set in the project-wide configuration file:

- *max_results_accepted*—the maximal number of results accepted from one host in a single communication;
- *min_sendwork_interval*—the minimal number of seconds between sending tasks to a given host;
- *max_wus_to_send*—the maximal number of tasks to send to a given host in a single communication;
- *daily_result_quota*—the upper bound on the number of tasks to send to a given host in a 24-hour period.

At the client side of BOINC, the following parameters may be set:

- *max_file_xfers*—the maximal number of simultaneous file transfers;
- *max_file_xfers_per_project*—the maximal number of simultaneous file transfers for a single project;
- *max_tasks_reported*—the maximal number of results to report to the server in a single communication.

3.3 Stochastic Model and Optimization Problems

It is clear from the description of a typical BOINC project that a natural model for the volunteer computing can be found in the framework of stochastic simulation. As such, we suggest to model the volunteer computing system as a discrete-event simulation (DES) model based on the so-called Generalized Semi-Markov Processes (GSMP) [14,15].

To define such a model in a simplest case, we need to construct two (multi-dimensional) stochastic processes, say, $X(t)$ being the system *state* at time t (which is in most cases a discrete-state multidimensional process), and $T(t) = (T_1(t), \ldots, T_m(t))$ being the multidimensional clock (timer) process. As such, an event epoch is an expiration of (at least) one of the components of a clock, that is, an *event of type* i occurs at time t, if $T_i(t) = 0$. For simplicity, in what follows we assume that only a *single event occurs at a time*, however, a general GSMP model may incorporate also multi-event epochs.

At each event epoch, a change in the system state may occur, and in general it is assumed that the transition depends on the previous state, as well as on the clock value, in stochastic manner. In most general case, the transition to x_1 at time t given $X(t-) = x_0$, happens with probability $P(T(t), x_0, x_1)$. It remains to note that between the events, the clocks decrease linearly, that is, $T(t + x) = T(t) - x, x < \min(T(t))$. Such a model allows to obtain both point and confidence estimates of various performance measures of interest [16].

Now we construct the required processes. In a general setup, the BOINC system consists of fixed, large number N of heterogeneous BOINC hosts, and a dedicated BOINC server communicating over the Internet. For simplicity, we omit the network delays and network-related issues, assuming immediate request transmission (in general, the time for task/result transmission may be added to the computation time). We note however, that the flexibility of the model allows to incorporate these effects separately.

The project handles the (fixed) amount $S \leq \infty$ of work to be done eventually. This amount is split into small (fixed) pieces of work, $s \ll S$, that are distributed to the hosts at their request times. Note that in the model we omit the replication and quorum parameters, since they are intrinsically included in the parameter S. The atomic amount of work, s, is an important management parameter of the model.

Consider now a BOINC host $i = 1, \dots, N$. Denote $\{X^{(i)}(t), t \geq 0\}$ the associated state process, and $\{T^{(i)}(t), t \geq 0\}$ the clock process of the given host. Define now at time $t \geq 0$:

$X_1^{(i)}(t)$ – the number of incomplete tasks assigned to the host, and
$X_2^{(i)}(t)$ – the number of complete tasks not yet reported to the server,

Now associate the following clock variables:

$T_1^{(i)}(t)$ – the residual time to server connection initiation,
$T_2^{(i)}(t)$ – the residual service time (to complete the task computed at time t),
$T_3^{(i)}(t)$ – the residual server processing time (time to accept the reported result).

It remains to note that the BOINC server in the simplest case can be modeled by a one-dimensional process $X^{(0)}(t)$ being the number of incomplete tasks in the project. The server's clock component, though, is associated with the clients as the time to accept the result, $T_3^{(i)}$.

As such, the processes $\{X^{(0)}(t), \dots, X^{(N)}(t), t \geq 0\}$ constitute a state, and $\{T^{(1)}(t), \dots, T^{(N)}(t), t \geq 0\}$ the clock multidimensional processes of the volunteer computing model.

Now we need to describe the evolution of the stochastic processes involved. In a general case, these could be defined by stochastic recursions. Note that the model allows to define various global and local constraints due to the versatile state transition structure. It is more easy to define these transitions on the basis of events.

Connection Event. If at time t an event of type 1 (connection) occurs, that is, the client (say, i) initiates a connection to the server, first the type of connection is selected (among the available types, say, *request*, *report*, and both) at the connection time using the state variable so as to guarantee reporting only if $X_2^{(i)} > 0$. Based on the type of event, the following changes are made in the system state at time t:

- at the *request* connection, the variable $X_1^{(i)}$ is increased in such a way as to guarantee the per-host instant limit

$$X_1^{(i)}(t) - X_1^{(i)}(t-) \leq L_{(TH)}^{(i)},$$

  where $L_{(TH)}^{(i)}$ is a management parameter, the upper bound for the number of tasks assigned per host in one connection event.

At the same time, the following changes are made in the system clock vector:

- the time to server connection, $T_1^{(i)}$, is initialized e.g. from some distribution, based both on the system state and the system clock (in a simple case, the process $\{T_1^{(i)}(t), t \geq 0\}$ is a renewal process with independent identically distributed intervals);
- if $X_1^{(i)}(t-) = 0$ and $X_1^{(i)}(t) > 0$, the residual service time, $T_2^{(i)}(t)$ is initialized e.g. from the service time distribution of i-th host which *in general depends on the size of a task, s*;
- at the *report* connection, the clock $T_3^{(i)}$ is initialized based on the global condition so as to guarantee the server is not overloaded, e.g.

$$\sum_{j=1}^{N} I(T_3^{(j)} > 0) \leq L_{(C)}^{(0)},$$

where $I(\cdot)$ is the indicator function used for the non-random events, and $L_{(C)}^{(0)}$ is the management parameter, the upper bound for the number of simultaneously processed results at server side (note also that server processing time may depend on the task size, s).

The model flexibility allows to define more sophisticated rules, e.g. the per-host integral limit for the number of tasks requested in some time interval (e.g. day), which though requires one additional clock per host, say, the time to limit renewal, $T_4^{(i)}(t)$; one additional state variable, e.g. the residual number of tasks to be requested, $X_3^{(i)}(t)$, and appropriate treatment of the corresponding clock expiration event (renewal of the limit $X_3^{(i)}(t)$).

Note also that the time to server connection may be initialized as $T_1^{(i)}(t) = \infty$, that is, the host has left the project forever. As such, the model allows to model the dynamics of the BOINC project network.

Service Completion Event. It is more straightforward to define the changes related to service completion. Indeed, the following changes appear in the state vector:

- the number of results ready for reporting, $X_2^{(i)}$, is increased by one (though the model allows also to define e.g. batch service at this point, by allowing more than one result completion per event, which, though, requires to do additional check at the service time initiation epoch),

$$X_2^{(i)}(t) = X_2^{(i)}(t-) + 1,$$

while the number of incomplete tasks is decreased,

$$X_1^{(i)}(t) = X_1^{(i)}(t-) - 1;$$

The following changes are made in the clock vector:

- if $X_1^{(i)}(t) > 0$, the residual service time, $T_2^{(i)}(t)$ is initialized e.g. from the service time distribution of i-th host.

Note that the service completion event may take into account more sophisticated models of host, e.g. the so-called host availability, that is, at the service completion epoch the host may not necessarily start serving a new task, but rather become unavailable for some (e.g. random) interval; this, however, may require to add one more state variable per host (the availability of the host).

Report Processed Event. At the time of expiration of a residual server processing time, $T_3^{(i)}(t-) = 0$, the following changes appear in the system state:

- the variable $X_2^{(i)}$ is decreased if the per-host instant limits are satisfied,

$$X_2^{(i)}(t-) - X_2^{(i)}(t) \leq L_{(RH)}^{(i)},$$

  where $L_{(RH)}^{(i)}$ is a management parameter, the upper bound for the number of results reported per host in one connection event;
- the variable $X^{(0)}$ is decreased by the same amount,

$$X^{(0)}(t) = X^{(0)}(t-) - X_2^{(i)}(t-) + X_2^{(i)}(t);$$

and the following changes in clock variables appear:

- the residual server processing time is initialized as $T_3^{(i)}(t) = \infty$, since the server processing is initialized only at the server connection event of type *report*.

Note that the number of results reported, $X_2^{(i)}(t-) - X_2^{(i)}(t)$, may be selected by the host, e.g. some batch reporting policy may be implemented.

Model Initialization. It remains to define the initial state of the model. First, all tasks are marked as incomplete:

$$X^{(0)}(0) = S/s.$$

For each host, the state is a zero vector initially:

$$X_1^{(i)}(0) = X_2^{(i)}(0) = 0, i = 1, \ldots, N.$$

The timers are initialized in the following way:

$$T_2^{(i)}(0) = T_3^{(i)}(0) = \infty,$$

and $T_1^{(i)}$ is selected so as to initialize the server connections at some (random) per-host times. Note that by selecting the first connection time, $T_1^{(i)}(0)$, it is easy to implement the host dynamics, by interpreting the time $T_1^{(i)}(0)$ as the time at which the host *joins* the project.

3.4 Model Extensions

Apart from possible model extensions discussed inline, we outline the following possibilities that may require some changes in the model definition, though still keep the model within the framework of GSMP:

- Global multidimensional limits may be defined, e.g. the overall memory and disk limitations per server. Indeed, additional per-host state variable, say, $X_4^{(i)}(t) \leq X_2^{(i)}(t)$, being the number of results being processed by the server, allows to define global limits

$$\sum_{j=1}^{N} m_j X_4^{(j)}(t) \leq L_M^{(0)},$$

  where m_j is the amount of memory consumed by the result reported by host j at server side, and $L_M^{(0)}$ is the global server memory limit (the server disk limit may be defined similarly).
- The value s, instead of being a global static parameter, may be defined on a per-host basis (that is, the size s_i of task sent to the host i), or even become a dynamic parameter (in such a way, it becomes a per-host state variable). If such an extension takes place, only the residual service time clock initialization is affected and, possibly, the global limits need to be equipped with this information.
- The host availability model may be implemented. However, to implement a model with task suspend/resume capability, we need to implement a GSMP model with clock speed variation. This is possible, but due to many technical details, we leave this discussion out of the scope of the present paper.

3.5 Optimization Problems

The server controls the task size, s, per-host, $(L_{(TH)}^{(i)}, L_{(RH)}^{(i)})$, and global limit, $L_{(C)}^{(0)}$, in such a way that to meet a certain QoS criterion. Firstly, task size, s, determines the time host needs for a task completion, and consequently, the rates of server's incoming requests flows. Secondly, it determines the amounts of input and output data and, accordingly, the necessary channel capacity, demands of serer processing time and e.g. memory. Thus, the performance of a model depends on the *model configuration*, $(s, \mathbf{L})$, where

$$\mathbf{L} = \left(L_{(TH)}^{(1)}, L_{(RH)}^{(1)}, \ldots, L_{(TH)}^{(N)}, L_{(RH)}^{(N)}, L_{(C)}^{(0)}\right)$$

is the vector of local and global limits.

Using the model parameters, it is easy to define various performance metrics. Let $\{t_i, i \geq 1\}$ be the global event time epochs sequence, and let $\{t_j^{(CR)}, j \geq 1\}$ be the *reporting type connection* attempts subsequence of $\{t_i\}$. Then define recursively the following subsequence of events of *failed reporting attempts due to server capacity exceeded*:

$$t_{j+1}^{(CRF)} = \min \left\{ t_i^{(CR)} > t_j^{(CRF)} : \sum_{j \geq 1, j \neq i} I(T_3^{(j)} > 0) \geq L_{(C)}^{(0)} \right\}, i \geq 1,$$

with obvious initialization of the first member of subsequence.

Take τ as the global modeling time and η be the number of modeling events in time τ, that is, $\tau = t_\eta$. It is easy then to define the estimate of probability of server capacity exceeded, $\hat{P}_C(s, L)$, as the following time-average:

$$\hat{P}_C(s, \mathbf{L}) = \frac{1}{\tau} \sum_{i=1}^{\eta-1} (t_{i+1} - t_i) I \left[t_i = t_j^{(CRF)}, j \geq 1 \right],$$

where in the notation we stress the dependence of the value $\hat{P}_C(s, \mathbf{L})$ on the model configuration. Then it is easy to define the following optimization constraint:

$$\hat{P}_C(s, \mathbf{L}) < \varepsilon, \quad \varepsilon > 0.$$

Note that in a simple case, decreasing the probability $\hat{P}_C(s, \mathbf{L})$ is possible by e.g. lowering the rate of server connections by hosts. At the same time, one needs to receive requests at the highest possible rate, e.g., to provide high results throughput (project efficiency). Such a performance characteristic is important in case $S = \infty$. Let $\{t_j^{(RP,i)}, j \geq 1\}$ be the sequence of report processed events (which is a subsequence of events, $\{t_i, i \geq 1\}$) of the host $i = 1, \dots, N$. At each such an event, the number of reported results is increased by the value

$$\Delta_{i,j} := X_2^{(i)} \left(t_j^{(RP,i)} - \right) - X_2^{(i)} \left(t_j^{(RP,i)} \right).$$

Thus, the time-average throughput of the system may be defined as follows:

$$\theta(s, \mathbf{L}) = \frac{1}{\tau} \sum_{i=1}^{N} \sum_{j=1}^{M_i} \Delta_{i,j} \left[t_j^{(RP,i)} - \min \left(t_i > t_j^{(RP,i)} \right) \right],$$

where $M_i = \max(j \geq 1 : t_j^{(RP,i)} < \tau)$. As such, we may define the following throughput maximization problem:

$$\theta(s, \mathbf{L}) \to \max.$$

Note that more delicate problems may be defined similarly, e.g. the fair throughput maximization,

$$\min_{i=1,\dots,N} \frac{1}{\tau} \sum_{j=1}^{M_i} \Delta_{i,j} \left[t_j^{(RP,i)} - \min \left(t_i > t_j^{(RP,i)} \right) \right] \to \max.$$

Another example of the performance optimization problem is the overall project computing time minimization, i.e.

$$\tau^{(s,\mathbf{L})}(S) \to \min,$$

where $\tau^{(s,\mathbf{L})}(S)$ is the overall computing time defined, if S is finite, as follows:

$$\tau^{(s,\mathbf{L})}(S) = \min\left(t \geq 0 : X^{(0)}(t) = 0\right),$$

and in notation we stress the dependence on system configuration.

Therefore, the model allows to define an optimization problem where one needs to maximize the objective function of project efficiency by the system configuration management, under existing constraints on the QoS and resources usage. In a similar way, we may define various cost optimization problems, e.g. for lost computational capacity due to insufficient server resources, as well as define the largest number of hosts, N that the server with given model configuration is capable of. However, we leave these for future research.

4 Conclusion

Desktop Grid is a powerful high-throughput computing concept highly demanded by scientific computing. Still, it suffers from lack of simulators suitable for studying performance issues. Among the promising is the ComBos project implementing a BOINC simulator in a flexible way [17]. However, the problem of stochastic modeling of the input sequences of a simulator still needs to be solved.

In this paper we present a discrete event simulation model of a Desktop Grid system. First, we state the stochastic model of a volunteer computing project. This model tracks computing nodes requests and use of server's resources. Second, we employ event simulation approach based on the Generalized Semi-Markov Processes to develop a discrete event simulation model. This model runs a stochastic state and clock processes used to generate events. The events of client-server interaction (and use of server's resources) are described.

Finally, using the simulation model, we describe an optimization problem of balance between the project runtime and size of task aiming higher performance. The proposed discrete event simulation model could be used to develop a next generation Desktop Grid simulator.

Acknowledgements. This work was partially supported by the Russian Foundation of Basic Research, projects 18-07-00628, 19-57-45022, 19-07-00303, 18-07-00156, 18-07-00147.

References

1. Anderson, D.P.: BOINC: a platform for volunteer computing. J. Grid Comput. **18**(1), 99–122 (2019). https://doi.org/10.1007/s10723-019-09497-9

2. Ivashko, E., Chernov, I., Nikitina, N.: A survey of desktop grid scheduling. IEEE Trans. Parallel Distrib. Syst. **29**(12), 2882–2895 (2018)
3. Chernov, I., Nikitina, N., Ivashko, E.: Task scheduling in desktop grids: open problems. Open Eng. **7**(1), 343–351 (2017)
4. Mengistu, T.M., Che, D.: Survey and taxonomy of volunteer computing. ACM Comput. Surv. (CSUR) **52**(3), 1–35 (2019)
5. Scheduler RPC timing and retry policies. BOINC Wiki (2020). https://boinc.berkeley.edu/trac/wiki/RpcPolicy. Accessed 12 May 2020
6. BOINC. GitHub repository. Client release 7.16.5 (2020). https://github.com/BOINC/boinc/blob/master/client/work_fetch.cpp. Accessed 12 May 2020
7. Alonso-Monsalve, S., García-Carballeira, F., Calderón, A.: A new volunteer computing model for data-intensive applications. Concurr. Comput. Pract. Exp. **29**(24), e4198 (2017)
8. Bazinet, A.L., Cummings, M.P.: Subdividing long-running, variable-length analyses into short, fixed-length BOINC workunits. J. Grid Comput. **14**(3), 429–441 (2016). https://doi.org/10.1007/s10723-015-9348-5
9. Mazalov, V.V., Nikitina, N.N., Ivashko, E.E.: Task scheduling in a desktop grid to minimize the server load. In: Malyshkin, V. (ed.) PaCT 2015. LNCS, vol. 9251, pp. 273–278. Springer, Cham (2015). https://doi.org/10.1007/978-3-319-21909-7_27
10. Vatutin, E., Zaikin, O., Kochemazov, S., Valyaev, S.: Using volunteer computing to study some features of diagonal Latin squares. Open Eng. **7**(1), 453–460 (2017)
11. Chakravarthy, S.R., Rumyantsev, A.: Efficient redundancy techniques in cloud and desktop grid systems using MAP/G/c-type queues. Open Eng. **8**(1), 17–31 (2018). https://www.degruyter.com/view/j/eng.2018.8.issue-1/eng-2018-0004/eng-2018-0004.xml
12. Morozov, E., Lukashenko, O., Rumyantsev, A., Ivashko, E.: A Gaussian approximation of runtime estimation in a desktop grid project. In: 2017 9th International Congress on Ultra Modern Telecommunications and Control Systems and Workshops (ICUMT), pp. 107–111. IEEE, Munich (2017)
13. Osipov, O.A.: Analysis of fork/join queueing networks with retrials. Vestnik Tomskogo gosudarstvennogo universiteta. Upravlenie, vychislitel'naya tekhnika i informatika (43), 49–55 (2018). http://journals.tsu.ru/informatics/&journal_page=archive&id=1706&article_id=37744. (in Russian)
14. Henderson, S.G., Glynn, P.W.: Regenerative steady-state simulation of discrete-event systems. ACM Trans. Model. Comput. Simul. **11**(4), 313–345 (2001). http://portal.acm.org/citation.cfm?doid=508366.508367
15. Glynn, P.W., Haas, P.J.: On transience and recurrence in irreducible finite-state stochastic systems. ACM Trans. Model. Comput. Simul. **25**(4), 1–19 (2015). http://dl.acm.org/citation.cfm?doid=2774955.2699721
16. Ross, S.: Simulation. Elsevier (2013). https://doi.org/10.1016%2Fc2011-0-04574-x
17. Alonso-Monsalve, S., García-Carballeira, F., Calderón, A.: ComBos: a complete simulator of volunteer computing and desktop grids. Simul. Model. Pract. Theory **77**, 197–211 (2017). http://www.sciencedirect.com/science/article/pii/S1569190X17301028

Enumerating the Orthogonal Diagonal Latin Squares of Small Order for Different Types of Orthogonality

Eduard Vatutin[1(✉)] and Alexey Belyshev[2]

[1] Southwest State University, Kursk, Russia
evatutin@rambler.ru
[2] Internet portal BOINC.ru, Moscow, Russia
alexey-bell@yandex.ru

Abstract. The article describes computational experiments aimed to enumerating the number of orthogonal diagonal Latin squares for general and special types of orthogonality. General type of orthogonality can be verified using Euler-Parker method, corresponding the number of main classes of orthogonal diagonal Latin squares, the number of normalized orthogonal diagonal Latin squares and total number of orthogonal diagonal Latin squares of general type form previously unknown numerical series A330391, A305570 and A305571 (calculated up to order 8) has been added to OEIS. Self-orthogonal (SODLS), doubly self-orthogonal (DSODLS) and extended self-orthogonal diagonal Latin squares (ESODLS) form the set of special types of orthogonality. For each of these types corresponding numerical series was calculated and published in OEIS with numbers A329685, A287761, A287762 (SODLS, up to order 10), A333366, A333367, A333671 (DSODLS, up to order 10) and A309210, A309598, A309599 (ESODLS, up to order 8). Values for orders 1–8 were obtained by analyzing the complete lists of canonical forms of the main classes of orthogonal DLSs obtained by the authors by exhaustive search. Values for order 9 were derived from the SODLS list of order 9 provided by Harry White. Values for order 10 were obtained by analyzing the list of SOLS of order 10, available online (van Vuuren et al.). The values obtained confirm the similar values for SODLS and DSODLS obtained previously by Francis Gaspalou and partially published by Harry White. For some of the obtained numerical values previously unknown mathematical relations are empirically established.

Keywords: Desktop grid · Volunteer computing · BOINC · Combinatorics · Orthogonal diagonal Latin squares · Self-orthogonal diagonal Latin squares · Doubly self-orthogonal diagonal Latin squares · Integer sequences · OEIS · Gerasim@Home

1 Introduction

Latin squares (*abbr.* LS) are a well-known type of combinatorial objects the study of which is devoted to a fairly large number of scientific papers [1,2]. Diagonal

V. Voevodin and S. Sobolev (Eds.): RuSCDays 2020, CCIS 1331, pp. 586–597, 2020.
https://doi.org/10.1007/978-3-030-64616-5_50

Latin squares (*abbr.* DLS) are a special case of LS where in addition to the difference in the values of the elements in the rows and columns, an additional restriction is imposed on the difference in the values of the elements on the diagonals (or, in other words, the sets of elements forming the diagonals must be transversals). Pair of squares A and B is called orthogonal (*abbr.* OLS or ODLS) if all ordered pairs of values (a_{ij}, b_{ij}), $i, j = \overline{1, N}$, where N – order of square, are different. Latin squares are used in such fields of science as experiment planning, error correction codes, cryptography, construction of schedules of a certain kind. They have a close relationship with magic squares.

The classic approach to checking LS/DLS for the presence of OLS/ODLS is the Euler-Parker approach [3] based on getting transversals (diagonal transversals) set [4,5] and then finding a subset of N disjoint transversals. In practice this approach most efficiently works using software implementation of the dancing link algorithm (*abbr.* DLX) [6,7]. The search pace in this case is limited precisely by the DLX algorithm.

There are a set of open questions associated with the OLS/ODLS: problems of enumerative combinatorics, classification of the combinatorial structures (graphs) from the ODLS on the set of orthogonality binary relation [8–10]. The most famous unsolved problem in this area is the problem of finding a triple (or pseudotriples) of mutually orthogonal LS/DLS (*abbr.* MOLS/MODLS) of order 10, which despite the numerous attempts at construction was not found, but it has not been conclusively proven that it does not exist.

2 The Concept of Orthogonality of General and Special Types

In enumerative combinatorics there are a number of problems associated with counting the number of combinatorial objects of a certain type depending on its dimension (for example, the number of permutations with specified properties [11], the number of graph isomorphism classes [12], etc.). Typically, as a result of solving the problem a numerical series is obtained, the N-th value of which determines the number of corresponding objects of dimension N. These numerical series are of fundamental interest and have a number of practical applications. A large number of such series (more than 300,000) is a part of the Online Encyclopedia of Integer Sequences (*abbr.* OEIS) [13].

The team of authors developed a highly efficient generator of DLS of a given order N based on a Brute-Force method and the principle of varying the order of filling the square cells and software implementation based on nested loops (without recursion) and bit arithmetic. Using it within the volunteer distributed computing project Gerasim@Home[1] on BOINC platform [14,15] a direct enumeration of all DLS orders up to $N = 9$ were organized, the search pace for $N = 10$ using this generator amounted to about 6–7 million DLS/s for a single-threaded CPU implementation. It took about 3 months to enumerate all DLSs of

[1] http://gerasim.boinc.ru.

order $N = 9$ (using grid within Gerasim@Home project and, regardless of this, on academician Matrosov computing cluster) [16]. As a result of the calculations, the numerical series "the number of DLS of order N with a fixed first row" and "the number of DLS of order N" were obtained and included in the OEIS under numbers A274171 and A274806.

Similar numerical series for ODLS which are rarer objects than DLS were unknown at the time of the start of scientific research (the number of OLS is known up to order 9 [17]).

As already noted above, the determination of the presence or absence of ODLS for a given DLS is efficiently possible using the Euler-Parker method by reducing the source problem to the exact cover problem and solving it using the DLX algorithm and its software implementation [6,7]. In this case the search pace is limited by the DLX software implementation and amounts to about 1,000–2,000 DLS/s for the order $N = 10$, i.e. 3 orders of magnitude lower than the above pace of the DLS generators (it would be nice to make a direct ODLS generator, however, at present the effective implementation of this idea is not known). When searching for ODLS within the Gerasim@Home project, the effective search pace was additionally increased to 7,000–8,000 DLS/s by using the canonization procedure which reduced to finding symmetrically placed transversals in the LS and setting them to the place of the main and back diagonals by rearranging rows and columns of square [18], combined with a check for the presence of ODLS.

The processing pace can be significantly increased if we get away from the use of transversals and the DLX algorithm. In the general case this is impossible, but in some special cases this idea has a right to life. For example, in the RakeSearch[2] volunteer distributed computing project all possible ODLS pairs of order 9 were found where the square B, orthogonal to the checked square A, was obtained by rearranging the rows of square A (special case of ESOLS). At the same time the search pace is approximately an order of magnitude higher (in the region of 70,000–80,000 DLS/s), which made it possible to discover several dozen new combinatorial structures from DLSs of order 9[3], the ODLS to which were found using the Euler-Parker method during post-processing of row-permutation pairs of ODLS [9]. Another well-known type of ODLS, also formed without the use of transversals, is a self-orthogonal LS/DLS (*abbr.* SOLS/SODLS) [19], where the orthogonal square B of a pair of ODLS is obtained by transposing square A. For example, the square below is SODLS of order 10:

[2] https://rake.boincfast.ru/rakesearch/.

[3] http://evatutin.narod.ru/evatutin_ls_all_structs_n9_eng.pdf.

$$\begin{pmatrix} 0&1&2&3&4&5&6&7&8&9 \\ 5&2&0&9&7&8&1&4&6&3 \\ 9&5&7&1&8&6&4&3&0&2 \\ 7&8&6&4&9&2&5&1&3&0 \\ 8&9&5&0&3&4&2&6&7&1 \\ 3&6&9&5&2&1&7&0&4&8 \\ 4&3&1&7&6&0&8&2&9&5 \\ 6&7&8&2&5&3&0&9&1&4 \\ 2&0&4&6&1&9&3&8&5&7 \\ 1&4&3&8&0&7&9&5&2&6 \end{pmatrix}.$$

SODLS search can be organized at an efficient pace that is several orders of magnitude faster than the Euler-Parker method, both due to the fact that transposing a square and subsequent checking a pair of squares for orthogonality are very fast operations, and due to the possibility of developing a specialized highly efficient ODLS generator of corresponding type based on X-based diagonal fillings, principles of variation of filling cells order and specialized software implementation with nested loops and bit arithmetic [16]. In addition to self-orthogonal squares, some articles also mention doubly self-orthogonal squares (*abbr.* DSOLS/DSODLS) [20], where in addition to the requirement of orthogonality from transposition, an additional requirement is imposed on the presence of an orthogonal square after the transposition of the square from the back diagonal. DSOLS/DSODLS are a special kind (subset) of SOLS/SODLS. An example of such DSODLS of order 9 is given below:

$$\begin{pmatrix} 0&1&2&3&4&5&6&7&8 \\ 2&4&3&0&7&6&8&1&5 \\ 4&6&7&1&8&2&3&5&0 \\ 8&3&5&6&0&7&1&2&4 \\ 7&8&1&4&5&3&0&6&2 \\ 3&7&0&2&1&8&5&4&6 \\ 1&5&4&7&6&0&2&8&3 \\ 5&0&6&8&2&1&4&3&7 \\ 6&2&8&5&3&4&7&0&1 \end{pmatrix}.$$

When working with Latin squares, it is necessary to take into account an isomorphism which allows in some cases to reduce the search space significantly (by several orders of magnitude) and the corresponding computational costs for practical software implementation of algorithms. We will say that DLSs belong to the same main class if they have the same canonical forms (*abbr.* CF) [21] – lexicographically minimal string representations for the corresponding squares within the indicated classes of isomorphism. All DLSs within the main class can be obtained by applying a combination of a number of equivalent transformations [22] and are characterized by identical properties (the number of transversals, the presence of ODLS, the number and composition of generalized symmetries (automorphisms), etc.).

The above definition of self-orthogonality can be extended to the class of ODLS where corresponding DLSs within the pair of orthogonal squares have the same main class. We will call such type of orthogonal squares as ESODLS (Extended SODLS). Their search can also be effectively organized without the use of transversals, and their properties are of theoretical interest (for example, all cliques from ODLS of orders 1–8 with a cardinality of more than 2 and most of the currently known cliques of order 9 includes ESODLS [23]). SODLS is a special case of ESODLS by definition. An example of ESODLS of order 10 is shown below:

$$\begin{pmatrix} 0&1&2&3&4&5&6&7&8&9 \\ 1&2&0&6&7&9&8&3&4&5 \\ 3&6&7&9&8&4&2&5&1&0 \\ 4&0&8&5&2&3&7&1&9&6 \\ 5&9&4&8&3&6&0&2&7&1 \\ 7&8&6&4&0&1&3&9&5&2 \\ 6&4&5&2&1&7&9&0&3&8 \\ 9&5&1&7&6&0&4&8&2&3 \\ 2&3&9&0&5&8&1&4&6&7 \\ 8&7&3&1&9&2&5&6&0&4 \end{pmatrix}.$$

At the start of the study the number of orthogonal DLSs was calculated only for small-order SODLS and DSODLS by Francis Gaspalou[4], and its results (including DLS lists) were either not published at all or were known only in private correspondence, which makes actual the task of independent verification of the relevant values for SODLS and DSODLS, as well as a similar calculation of the unknown number of small order ESODLS and ODLS and publication corresponding lists of CFs and numerical series in OEIS.

3 Enumerating the ODLS of General Type

The search for all ODLS of orders 1–7 using the software developed by the authors was performed on the single PC, for order 8 the organization of the corresponding computational experiment in the Gerasim@Home project was required. As a result of the calculations complete lists of ODLS CFs of orders 1–8 were obtained. Based on these lists, we can obtain both special type ODLS lists (DSODLS/SODLS/ESODLS) by dropping some of the corresponding ODLS CFs with not interesting properties, and expand the corresponding lists on the DLSs with the ordered first row (also called as normalized DLS) and on the general type DLSs. For orders more than 8, obtaining complete lists of CFs of the ODLS is difficult due to their large number and the huge computational costs required for this. The obtained numbers of the ODLS main classes, the numbers of ODLS with a fixed first row and the numbers of general type ODLS of orders 1–8 are given in Tables 1, 2 and 3 (5th column) and were added to the OEIS

[4] http://www.gaspalou.fr/magic-squares/index.htm.

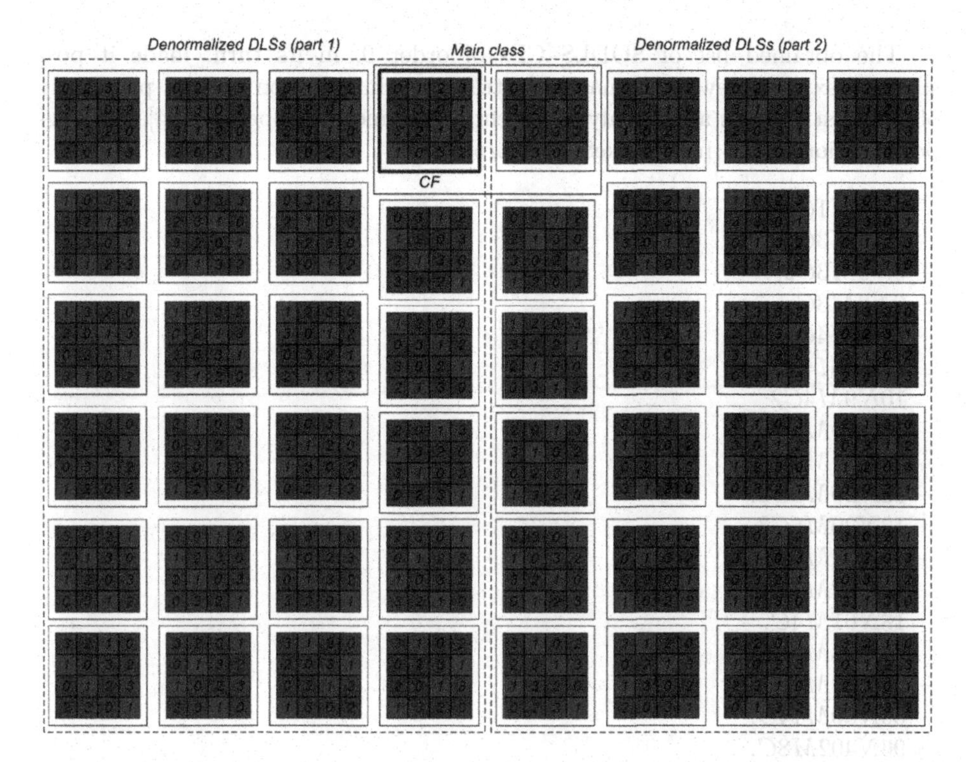

Fig. 1. An example of obtaining isomorphic squares from CF (top middle in a bold frame). A thin line picked out the corresponding main class from 2 normalized by the first row DLSs obtained by applying equivalent transformations to CF. Each of the normalized DLS from the main class forms a subclass of 1 normalized and the remaining $4! - 1 = 24 - 1 = 23$ denormalized DLSs (dashed rectangles on the left and on the right sides). In total, the canonical form corresponds to the isomorphism class from 48 general type DLSs.

under the numbers A330391, A305570 and A305571. Schematically the process of obtaining squares of various types from the CF is shown in Fig. 1.

4 Enumerating the SODLS, DSODLS and ESODLS

Based on the data of the ODLS CFs lists we can determine the number of main classes of SODLS of orders $1 \leq N \leq 8$. Determining the number of SODLS main classes of orders $N > 8$ in this way is not possible due to the lack of similar lists of ODLS CFs. However, the SODLS list of order 9 is known which was provided by Harry White[5] at the request of the authors and includes 224,832 DLSs with an ordered first row. Based on it, it is possible to find corresponding different CFs and calculate their number which turned out to be equal to 470 which is the 9th member of the corresponding numerical series.

[5] http://budshaw.ca/SODLS.html.

The obtained list of SODLS CFs of order 9, in its turn, made it possible to explore new combinatorial structures in addition to the previously obtained structures after analysis of central symmetry properties [24] and row-permutation ODLS from RakeSearch project [9]:

- $4N6M1C$,
- $6N5M3C$,
- $6N5M3C2$,
- $6N8M3C$,
- $8N8M4C$,
- $10N9M4C$,
- $10N9M4C2$,
- $10N19M3C$,
- $10N21M3C$,
- $12N20M6C2$,
- $12N20M6C3$,
- $12N21M2C$,
- $14N22M7C$,
- $18N61M4C$,
- $20N28M9C$,
- $32N42M12C$,
- $36N80M14C$,
- $96N402M8C$,
- $98N470M21C$,
- $98N502M21C$,
- $162N606M37C$,
- $540N1500M11C$,
- $760N944M57C$.

SOLS list of order 10 is also known [25]. By canonizing it, we can get a list of SODLS of order 10 including 30,502 different CFs, which is the 10th member of the investigated numerical series (these lists, identical in composition, were obtained by the co-authors of the article independently from each other using various software implementations, which is an independent confirmation of the correctness of the obtained values). The obtained numerical values form a numerical series (see Table 1, 3rd column) and were added to OEIS under the number A329685.

Having lists of SODLS CFs, by applying equivalent transformations to each CF in their composition, we can get the full main classes from the DLS corresponding to them, sum their cardinalities and get the total number of normalized SODLS, which is the next numerical series (see Table 2, 3rd column) has been added to OEIS with number A287761.

An interesting feature of the obtained series is the fact that all SODLS main classes of order 10 are full-sized: they are formed by 15,360 normalized DLSs each and include 7,680 SODLS and 7,680 ESODLS with orthogonality by transposition from the back diagonal. This feature was identified empirically in the course of a computational experiment where all SODLS main classes were constructed and their cardinalities evaluated. For dimensions $1 \leq N \leq 9$, this property is not fulfilled, and different main classes contain a different number of normalized DLSs, apparently, due to the presence of generalized symmetries (automorphisms) in the corresponding SODLS [26], which reduces the cardinalities of some main classes in comparison with the theoretical maximum (see A299784 and A299787 series in OEIS).

Multiplying the obtained numerical values by $N!$, we can get the total number of SODLS (see Table 3, 3rd column), forming the sequence A287762 that also has been added to OEIS.

Having ready lists of SODLS main classes, normalized SODLS and SODLS of general type we can get the corresponding numerical series for DSODLS by performing an additional check for the presence of an orthogonal square when transposing the corresponding SODLS from the back diagonal. In this case we can get the following numerical series, also not presented in the OEIS at the time of the research (see Tables 1, 2 and 3, 2nd columns in each table). Currently, they have also been tested and added to OEIS under the numbers A333366, A333367 and A333671.

It is easy to notice that DSODLS of order 10 do not exist (previously a similar result was obtained for DSOLS of order 10 [20]). The latter property, in particular, implies a meager set of combinatorial structures formed by SODLS of order 10 compared to, for example, combinatorial structures obtained in the neighborhoods of generalized symmetries [26] or from SODLS of lower orders (for example, of order 9).

During the publication of preliminary information related to the calculations mentioned above, it turned out that similar values for SODLS and DSODLS were obtained earlier by Francis Gaspalou and partially published by Harry White. Thus, the results of calculations performed within the Gerasim@Home project with subsequent post-processing can be an independent confirmation of the correctness of the previously found values.

The calculation of the ESODLS number can be performed similarly to the considered above: firstly, for each of the ODLS CFs of a given order it is necessary to find the corresponding main classes. Then, the resulting main classes may be expanded by a combination of equivalent transformations to normalized ESODLS and then by changing the values of the elements to general type ESODLS.

For order $N = 10$ a set of 33,240 ESODLS CFs obtained in the Gerasim@Home project through a targeted search of ESODLS is currently known (33,238 CFs of them were obtained by searching for cell mapping schemes (*abbr.* CMS) with a canonical loops structure $\{1{:}10, 2{:}45\}$ (where, among other things, all SODLS are included) and 2 more CFs were obtained during the study of the neighborhoods of generalized symmetries, they correspond to CMSs with the loops structure $\{4{:}25\}$ and combinatorial structures of the "loop-4" type [10] including 4 DLSs with the same CF). During the computational experiments that are currently ongoing, other ESODLS CFs are not obtained: they either do not

Table 1. ODLS main classes of different type of orthogonality

ODLS type DLS order N	DSODLS (A333366)	SODLS (A329685)	ESODLS (A309210)	ODLS (A330391)
1	1	1	1	1
2	0	0	0	0
3	0	0	0	0
4	1	1	1	1
5	1	1	1	1
6	0	0	0	0
7	2	2	5	5
8	8	8	23	1,105
9	88	470	?	?
10	0	30,502	$\geq$ 33,240	?

Table 2. Normalized ODLS of different type of orthogonality

ODLS type DLS order N	DSODLS (A333367)	SODLS (A287761)	ESODLS (A309598)	ODLS (A305570)
1	1	1	1	1
2	0	0	0	0
3	0	0	0	0
4	2	2	2	2
5	4	4	4	4
6	0	0	0	0
7	64	64	256	256
8	1,152	1,152	4,608	632,064
9	28,608	224,832	?	?
10	0	234,255,360	$\geq$ 510,566,400	?

exist or are very rare. The resulting value allows us to impose a lower limit on the total number of ESODLS main classes of order 10 which is very close to the exact value. All currently known ESODLS main classes (as well as SODLS) are full-sized, which was established empirically during the calculation of the cardinalities of each of them within the performed computational experiment.

Corresponding numerical series (see Tables 1, 2 and 3, 4th columns) were also added to the OEIS under the numbers A309210, A309598 and A309599.

Table 3. General case ODLS of different type of orthogonality

ODLS type DLS order N	DSODLS (A333671)	SODLS (A287762)	ESODLS (A309599)	ODLS (A305571)
1	1	1	1	1
2	0	0	0	0
3	0	0	0	0
4	48	48	48	48
5	480	480	480	480
6	0	0	0	0
7	322,560	322,560	1,290,240	1,290,240
8	46,448,640	46,448,640	185,794,560	25,484,820,480
9	10,381,271,040	81,587,036,160	?	?
10	0	850,065,850,368,000	$\geq$ 1,852,743,352,320,000	?

All found numerical series turned out to be new, no coincidence with known numerical series for other types of combinatorial objects was revealed. All the found lists of ODLS CFs of general and special types are available online[6].

Taking into consideration the fact that DSODLS $\subseteq$ SODLS $\subseteq$ ESODLS $\subseteq$ ODLS, we can formulate the following relations between the numerical values found:

$$A333366(n) \leq A329685(n) \leq A309210(n) \leq A330391(n),$$
$$A333367(n) \leq A287761(n) \leq A309598(n) \leq A305570(n),$$
$$A333671(n) \leq A287762(n) \leq A309599(n) \leq A305571(n).$$

In addition, there is a number of interesting relations between the numerical values found. All of them are found empirically and currently do not have a theoretical explanation.

1. All ODLSs of orders 1, 4, and 5 are simultaneously DSODLS, SODLS and ESODLS; there are no other types of ODLSs for these orders.
2. All ODLSs up to order 7 are ESODLS.
3. All SODLS up to order 8 are DSODLS.
4. The number of normalized SODLS and general type SODLS of orders 7 and 8 is exactly 4 times less than the corresponding numbers of ESODLS of the same type and the same dimension:

$$A287761(n) \cdot 4 = A309598(n), n = \overline{7, 8},$$
$$A287762(n) \cdot 4 = A309599(n), n = \overline{7, 8}.$$

[6] http://evatutin.narod.ru/evatutin_odls_1_to_8.zip (ODLS CFs),
http://evatutin.narod.ru/evatutin_esodls_1_to_8.zip (ESODLS CFs),
http://evatutin.narod.ru/evatutin_sodls_1_to_10.zip (SODLS CFs),
http://evatutin.narod.ru/evatutin_dsodls_1_to_10.zip (DSODLS CFs).

For the SODLS and ESODLS main classes of orders 7 and 8 similar relations do not revealed. The similar ODLS/SODLS and SODLS/DSODLS values are also fractional.

5 Conclusion

Thus, using volunteer distributed computing, new numerical series were found (for ODLS and ESODLS) and the validity of previously partially known numerical series (for SODLS and DSODLS) was confirmed. All of them added to OEIS, a well-known resource in this field of knowledge. The obtained results prove the efficiency of distributed computing in enumerative combinatoric problems.

The authors thank the user citerra [Russia Team] from the BOINC.ru forum, as well as Harry White and Francis Gaspalou for a number of valuable comments related to the history of studying the properties of SODLS and DSODLS. The author also wish to thank Anna Vayzbina for assistance in preparing the English version of the article.

References

1. Colbourn, C.J., Dinitz, J.H.: Handbook of Combinatorial Designs. Discrete Mathematics and Its Applications, 2nd edn. Chapman & Hall/CRC, Boca Raton (2006)
2. Keedwell, A.D., Denes, J.: Latin Squares and Their Applications. Elsevier, Amsterdam (2015)
3. Parker, E.T.: Orthogonal Latin squares. Proc. Natl. Acad. Sci. USA **45**(6), 859–862 (1959)
4. McKay, B.D., McLeod, J.C., Wanless, I.M.: The number of transversals in a Latin square. Des. Codes Cryptogr. **40**, 269–284 (2006). https://doi.org/10.1007/s10623-006-0012-8
5. Vatutin, E.I., Kochemazov, S.E., Zaikin, O.S., Valyaev, S.Yu.: Enumerating the transversals for diagonal Latin squares of small order. In: CEUR Workshop Proceedings. Proceedings of the Third International Conference BOINC-Based High Performance Computing: Fundamental Research and Development (BOINC: FAST 2017), vol. 1973, pp. 6–14 (2017)
6. Knuth, D.E.: Dancing links (2000). arXiv preprint arXiv:cs/0011047v1
7. Knuth, D.E.: The Art of Computer Programming, Volume 4A: Combinatorial Algorithms. Addison-Wesley Professional, Boston (2013)
8. Vatutin, E.I., Manzuk, M.O., Titov, V.S., Kochemazov, S.E., Belyshev, A.D., Nikitina, N.N.: Orthogonality-based classification of diagonal Latin squares of orders 1–8 (in Russian). High-Perform. Comput. Syst. Technol. **3**(1), 94–100 (2019)
9. Manzyuk, M., Nikitina, N., Vatutin, E.: Start-up and the results of the volunteer computing project RakeSearch. In: Voevodin, V., Sobolev, S. (eds.) RuSCDays 2019. CCIS, vol. 1129, pp. 725–734. Springer, Cham (2019). https://doi.org/10.1007/978-3-030-36592-9_59
10. Vatutin, E.I., Titov, V.S., Zaikin, O.S., Kochemazov, S.E., Manzuk, M.O., Nikitina, N.N.: Orthogonality-based classification of diagonal Latin squares of order 10. In: CEUR Workshop Proceedings. Proceedings of the VIII International Conference "Distributed Computing and Grid-technologies in Science and Education" (GRID 2018), vol. 2267, pp. 282–287 (2018)

11. Simpson, T.: Permutations with unique fixed and reflected points. Ars Combin. **39**, 97–108 (1995)
12. Bona, M.: Handbook of Enumerative Combinatorics. Apple Academic Press Inc., Williston (2015)
13. Sloane, N.J.A.: The on-line encyclopedia of integer sequences (2020). https://oeis.org/
14. Anderson, D.P.: BOINC: a platform for volunteer computing (2019). arXiv preprint arXiv:1903.01699
15. Anderson, D.P.: BOINC: a system for public-resource computing and storage. In: 5th IEEE/ACM International Workshop on Grid Computing, Pittsburgh, USA, pp. 1–7, November 2004
16. Kochemazov, S., Zaikin, O., Vatutin, E., Belyshev, A.: Enumerating diagonal Latin squares of order ip to 9. J. Integer Sequences **23** (2020). Article 20.1.2
17. Egan, J., Wanless, I.M.: Enumeration of MOLS of small order. Math. Comput. **85**, 799–824 (2016)
18. Brown, J.W., Cherry, F., Most, L., Most, M., Parker, E.T., Wallis, W.D.: Completion of the Spectrum of Orthogonal Diagonal Latin Squares. Lecture Notes in Pure and Applied Mathematics, vol. 139, pp. 43–49 (1992)
19. Brayton, R.K., Coppersmith, D., Hoffman, A.J.: Self-orthogonal Latin squares of all orders n! = 2, 3, 6. Bull. Am. Math. Soc. **80**(1), 116–118 (1974)
20. Lu, R., Liu, S., Zhang, J.: Searching for doubly self-orthogonal Latin squares. In: Lee, J. (ed.) CP 2011. LNCS, vol. 6876, pp. 538–545. Springer, Heidelberg (2011). https://doi.org/10.1007/978-3-642-23786-7_41
21. Vatutin, E., Belyshev, A., Kochemazov, S., Zaikin, O., Nikitina, N.: Enumeration of isotopy classes of diagonal Latin squares of small order using volunteer computing. In: Voevodin, V., Sobolev, S. (eds.) RuSCDays 2018. CCIS, vol. 965, pp. 578–586. Springer, Cham (2019). https://doi.org/10.1007/978-3-030-05807-4_49
22. Chebrakov, Yu.V.: Theory of magic matrices (in Russian). Saint-Peterburg (2016)
23. Vatutn, E.I., Nikitina, N.N., Manzuk, M.O., Zaikin, O.S., Belyshev, A.D.: Cliques properties from diagonal Latin squares of small order (in Russian). In: Intellectual and Information Systems (Intellect - 2019), Tula, Russia, pp. 17–23, November 2019
24. Vatutin, E.I., Kochemazov, S.E., Zaikin, O.S., Manzuk, M.O., Nikitina, N.N., Titov, V.S.: Central symmetry properties for diagonal Latin squares. Probl. Inf. Technol. **2**, 3–8 (2019)
25. Burger, A.P., Kidd, M.P., van Vuuren, J.H.: Enumerasie van self-ortogonale latynse vierkante van orde 10. LitNet Akademies (Natuurwetenskappe) **7**(3), 1–22 (2010)
26. Vatutin, E.I., Belyshev, A.D., Zaikin, O.S., Nikitina, N.N., Manzuk, M.O.: Investigating of properties of generalized symmetries in diagonal Latin squares using voluntary distributed computing (in Russian). High-Perform. Comput. Syst. Technol. **3**(2), 39–51 (2019)

Privacy-Preserving Logistic Regression as a Cloud Service Based on Residue Number System

Jorge M. Cortés-Mendoza[1], Andrei Tchernykh[1,2,3](✉), Mikhail Babenko[4], Luis Bernardo Pulido-Gaytán[2], Gleb Radchenko[1], Franck Leprevost[5], Xinheng Wang[6], and Arutyun Avetisyan[3]

[1] South Ural State University, Chelyabinsk, Russia
{kortesmendosak,gleb.radchenko}@susu.ru,
{chernykh,lpulido}@cicese.edu.mx
[2] CICESE Research Center, Ensenada, BC, Mexico
[3] Ivannikov Institute for System Programming, Moscow, Russia
arut@ispras.ru
[4] North-Caucasus Federal University, Stavropol, Russia
mgbabenko@ncfu.ru
[5] University of Luxembourg, Esch-sur-Alzette, Luxembourg
franck.leprevost@uni.lu
[6] Xi'an Jiaotong-Liverpool University, Suzhou, China
xinheng.wang@xjtlu.edu.cn

Abstract. The security of data storage, transmission, and processing is emerging as an important consideration in many data analytics techniques and technologies. For instance, in machine learning, the datasets could contain sensitive information that cannot be protected by traditional encryption approaches. Homomorphic encryption schemes and secure multi-party computation are considered as a solution for privacy protection. In this paper, we propose a homomorphic Logistic Regression based on Residue Number System (LR-RNS) that provides security, parallel processing, scalability, error detection, and correction. We verify it using six known datasets from medicine (diabetes, cancer, drugs, etc.) and genomics. We provide experimental analysis with 30 configurations for each dataset to compare the performance and quality of our solution with the state of the art algorithms. For a fair comparison, we use the same 5-fold cross-validation technique. The results show that LR-RNS demonstrates similar accuracy and performance of the classification algorithm at various thresholds settings but with the reduction of training time from 85.9% to 96.1%.

Keywords: Cloud security · Homomorphic encryption · Residue number system · Logistic regression

1 Introduction

The cloud computing paradigm provides an easy way to store, retrieve, and process data as a part of its services. Machine Learning as a Service (MLaaS) has emerged as a

V. Voevodin and S. Sobolev (Eds.): RuSCDays 2020, CCIS 1331, pp. 598–610, 2020.
https://doi.org/10.1007/978-3-030-64616-5_51

flexible and scalable solution [1–3]. Unfortunately, security and privacy issues still pose significant challenges, especially when data must be decrypted.

Homomorphic Encryption (HE), Fully Homomorphic Encryption (FHE), Somewhat Homomorphic Encryption (SHE), and secure Multi-Party Computation (MPC) are ways to address vulnerabilities of data processing. These cryptosystems allow applying certain mathematical operations directly to the ciphertext and safely delegate the processing of data to an untrusted remote party, it guarantees that the remote party will learn neither the input nor the output of the computation [4].

In the last decade, there is considerable interest in using the Residue Number System (RNS) as a variant of FHE [9]. It is a widely known and studied number theory system. It codes the original number as a tuple of residues with respect to a moduli set that can be processed in parallel. This coding technique is one of the core optimization techniques used in several implementations of HE schemes. The advantages of RNS include security, parallel processing, scalability, error detection, and correction.

The paper focuses on developing RNS Logistic Regression (LR-RNS) for the processing of confidential information in cloud computing. The goal is to enrich the paradigm of Machine Learning as a Service. Our contribution is multifold:

- We propose a logistic regression algorithm with a homomorphic encryption scheme based on a residue number system.
- Training, testing, and prediction process are performed with ciphertexts.
- We conduct comprehensive simulation with six known datasets of different domains in medicine (diabetes, cancer, drugs, etc.) and genomics.
- We show that LR-RNS has similar accuracy and classification performance compared with the state of the art algorithms, with considerable training time decrease.

The rest of the paper is structured as follows. Next section provides information about logistic regression and gradient descent. Section 3 discusses related works to solve our problem. Section 4 presents information about homomorphic encryption. Section 5 describes the characteristics of the logistic regression in RNS. Section 6 presents the experimental results. Finally, we conclude and discuss future work in Sect. 7.

2 Logistic Regression and Gradient Descent

Logistic Regression (LR) is a simple and powerful strategy to solve problems in different domains: detection of prostate cancer, diabetes, myocardial infarction, infant mortality rates, cardiac problems, treatment for drug abuse, genomic, fraudulent transactions detection, etc. [10–15]. It is a statistical method for analyzing information with independent variables. It determines a binary (or dichotomous) outcome (success/failure, yes/no, etc.) using logistic functions to predict a dependence on the data. A dataset with dimension d defined by $X^{(i)} \in \mathbb{R}^d$ and their corresponding labels $Y^{(i)} \in \{0, 1\}$ for $i = 1, 2, \ldots, N$ are used to model a binary dependent variable. The inference of logistic regression is considered within hypotheses $h_\theta(X^{(i)}) = g(\theta^T X^{(i)})$, where the sigmoid function is defined as $g(z) = \frac{1}{1+e^{-z}}$ and $\theta^T X^{(i)} = \theta_0 + \theta_1 X_1^{(i)} + \theta_2 X_2^{(i)} + \ldots + \theta_d X_d^{(i)}$, for $\theta^T = [\theta_0, \theta_1, \ldots, \theta_d]^T$ and $X^{(i)} = \left[1, X_1^{(i)}, X_2^{(i)}, \ldots, X_d^{(i)}\right]^T$.

Logistic regression uses a likelihood function to make inference on parameter θ. Simplifying the likelihood function for the whole data by log yields

$$J(\theta) = -\frac{1}{\mathrm{N}}\sum\nolimits_{i=1}^{N} Y^{(i)}\log\left(h_{\theta}\left(X^{(i)}\right)\right) + \left(1 - Y^{(i)}\right)\log\left(1 - h_{\theta}\left(X^{(i)}\right)\right).$$

Techniques to minimize $J(\theta)$ vary, however, the gradient descent algorithm is a common option.

Gradient Descent (GD) is an optimization algorithm to minimize the objective function $J(\theta)$. At each iteration, it updates the parameters θ in the opposite direction of the gradient of the function. The learning rate α determines the dimension of the steps to reach a (local) minimum. The direction of the slope, created by the objective function, guides the search downhill until to reach a valley.

Batch Gradient Descent (BGD) is the most common version of GD. It updates the values of θ considering the entire training dataset. BGD guarantees to converge to the global minimum for convex error surfaces and local minimum for non-convex surfaces. But it has a slow time of convergence. Big datasets can be intractable due to memory limitations and low access speed.

Stochastic Gradient Descent (SGD) performs an update of θ for each training example. Advantages of SGD include reduced convergence time, exploration of new valleys (potentially with better local minima), and online learning.

In order to estimate the parameter θ, GD with Fixed Hessian Newton method (GD-FHN) applies the Newton-Raphson method to solve the equation numerically which iteratively determines the zeros of a function. The Hessian matrix is the second partial derivative of $J(\theta)$, but its evaluation and inverse are quite expensive. So, GD-FHN simplifies the process by approximating the Hessian matrix.

Momentum is a method to accelerate the direction and reduce the oscillations of SGD. It defines a fraction γ and uses it to update θ. Nesterov Accelerated Gradient (NAG) takes advantage of the momentum term to improve the performance of GD. It provides an approximation of the next position of the parameters with partial updates of θ.

The training phase of logistic regressions focuses on finding values θ^* that minimizes the cost function $J(\theta)$. θ^* values are used to estimate the binary classification of new data. For example, for a given data $X = [1, X_1, X_2, \ldots, X_d] \in \mathbb{R}^{d+1}$, it is possible to guess its binary value $Y \in \{0, 1\}$ by setting:

$$Y = \begin{cases} 1\,\mathit{if}\;\; h_{\theta^*}(\mathrm{X}) \geq \tau \\ 0\,\mathit{if}\;\; h_{\theta^*}(\mathrm{X}) < \tau \end{cases} \tag{1}$$

where τ defines a variable threshold in $0 < \tau < 1$, typically with value equal to 0.5.

3 Related Work

Homomorphic Encryption (HE) is an active research field and has a long list of approaches and improvements [5–8]. Solutions focus on algorithms, arithmetic operations, applications data analytics, approximation functions, Machine Learning (ML) [10–15], Internet of Things (IoT) [22], etc.

Aono et al. [10] propose a homomorphism-aware LR system where the training and predicting data are protected under encryption. The authors study the quality of the additive homomorphic encryption schemes with the public key (Paillier), Learning With Errors (LWE), and ring-LWE (see Sect. 4).

Bonte et al. [11] develop a privacy-preserving logistic regression using SHE. The central server can combine the data of several users to train the binary classification model without learning anything about the underlying information.

Kim et al. [12] present a method to train logistic regression with a SHE. The authors use NAG method to increase the speed of convergence and a packing method to reduce the storage of encryption data.

Cheon et al. [13] define a variant of HE scheme for Arithmetics of Approximate Numbers (HEAAN) based on the RNS representation of polynomials in a software library that implements homomorphic encryption and supports fixed-point arithmetic. The algorithm uses RNS decomposition of cyclotomic polynomials and Number Theoretic Transformation (NTT) conversion.

Cheon et al. [14] propose a variation of GD for logistic regression. Ensemble gradient descend runs several standard GDs on a partial dataset and then takes an average on resulting solutions (from each of the partial datasets). The algorithm reduces the number of iterations to train the model, hence, improving the execution time of logistic regression. However, the errors from approximate computations may disrupt the convergence of the ensemble algorithm in an encrypted state.

Yoo et al. [15] develop an LR for HE using binary approximation. It can represent reals numbers, perform subtraction, division, and exponential functions based on the encrypted bitwise operation.

Table 1 summarizes the main characteristics of the approaches.

Table 1. Main characteristics of HE schemes for logistic regression.

Encryption	Evaluation function	Gradient descent	Metrics (Sect. 6.1)	Library	Datasets	Ref.
Paillier, LWE, Ring-LWE	Taylor series	BGD	F-score, AUC	–	Pima, SPECTF	[10]
Ring-LWE	Taylor series	GD-FHN	ROC, accuracy	NFLlib	iDASH, financial data	[11]
Ring-LWE	Polynomials of degree 3, 5 and 7	NAG	AUC, accuracy	HEAAN	iDASH, lbw, mi, nhanes3, pcs, uis	[12]
Ring-LWE, RNS	A polynomial of degree 7	NAG	AUC, accuracy	HEAAN	Lbw, uis	[13]
Ring-LWE	A polynomial of degree 5	NAG	AUC	HEAAN	MNIST, credit	[14]
–	Logistic function	BGD	AUC	–	NIDDK	[15]

Our work focuses on analyzing the gradient descent algorithm with RNS as a variation of FHE. Security, parallel processing, scalable storage, error detection, and correction of the data are advantages of RNS that can be used to design homomorphic encryption functions for cloud computing.

4 Homomorphic Encryption

Cloud computing provides data protection from theft, leakage, deletion, integrity, etc. at levels of firewalls, penetration testing, obfuscation, tokenization, Virtual Private Networks (VPN), etc. However, the use of third-party services can bring several cybersecurity risks. Using conventional data encryption does not avoid the problem. At some point, the data must be decrypted for processing, for instance, for statistical analysis or training a logistic regression model. At this moment, the vulnerability of the information is high. A possible solution is to use encryption schemes that allow performing operations over the ciphertext.

HE allows performing operations on ciphertexts based on publicly available information without having access to any secret key. An additively homomorphic encryption scheme generates ciphertexts c_1 and c_2 with the encrypted content of the messages m_1 and m_2, respectively. Then, decryption of the ciphertext $c_+ = c_1 + c_2$ produces $m_1 + m_2$. Similarly, a multiplicatively homomorphic encryption scheme produces a ciphertext $c_\times$ that decrypts $m_1 \times m_2$.

HE includes multiple types of encryption schemes and has been developed using different approaches: Partially Homomorphic Encryption (PHE), Somewhat Homomorphic Encryption (SHE), and Full Homomorphic Encryption (FHE). PHE supports only one type of operation, for example, a scheme supports homomorphic addition but not multiplication. PHE is generally more efficient than SHE and FHE. SHE supports additions and multiplications but can perform only a limited number of operations before the error grows too much to maintain the correctness of the evaluation. FHE allows an increased number of operations due to bootstrapping.

Paillier encryption scheme is a probabilistic asymmetric cryptography algorithm. It consists of a public key $pk = n$ used to encrypt plaintext in the interval $\{0, \ldots, n-1\}$. The additive homomorphic encryption defines $PaiEnc_{pk}(m, r) = g^m r^n mod\ n^2$ to encrypt a message m, where $n = p * q$ for primes numbers p and q, a randomly selected $r \in \{0, \ldots, n-1\}$, and integer $g \in \mathbb{Z}^*_{n^2}$. The encryption function $PaiEnc_{pk}(m, r)$ has the additively homomorphic property $PaiEnc_{pk}(m_1, r_1) * PaiEnc_{pk}(m_2, r_2) = PaiEnc_{pk}(m_1 + m_2, r_1 r_2)$.

LWE encryption scheme provides a public key $pk = (A, P, (p, l), (n_{lwe}, s, q))$ in the interval $\mathbb{Z}_p = (-p/2, p/2]$ to represent plaintext in a vector $\mathbb{Z}^l_p$, where p and l are security parameters, and A with P define a matrix concatenation of public matrices $A \in \mathbb{Z}_p^{n_{lwe} \times n_{lwe}}$. For a plaintext $m \in \mathbb{Z}_p^{1 \times l}$, the encryption message m can be generated by: $LweEnc_{pk}(m) = e_1[A|P] + p[e_2|e_3] + [0_{n_{lwe}}|m] \in \mathbb{Z}_q^{1 \times (n_{lwe}+l)}$, in which $e_1 \in \mathbb{Z}_q^{1 \times n_{lwe}}$, $e_2 \in \mathbb{Z}_q^{1 \times n_{lwe}}$, $e_3 \in \mathbb{Z}_q^{1 \times l}$ are Gaussian noise vectors of deviation s. The additive encryption produces ciphertext $m_1 + m_2$ by $LweEnc_{pk}(m_1) + LweEnc_{pk}(m_2)$.

Ring LWE encryption scheme defines a public key $pk = (a, p) \in R_p^2$ in the interval $\mathbb{Z}_p = (-p/2, p/2] \cap \mathbb{Z}$ with a ring $R = \mathbb{Z}[x]/f(x)$ where $f(x) = x^{n_{rlwe}} + 1$ and quotient rings $R_q = R/q, R_p = R/p$. The additive homomorphic encryption generates a ciphertext with the message m by: $RlweEnc_{pk}(m) = (e_1 a + e_2, e_1 p + e_3 + p/qm)$, where $e_1, e_2, e_3 \in R_{(0,s)}$ are noises and $R_{(0,s)}$ stand for polynomials in R with small Gaussian coefficients of mean 0 and deviation s. The addition of the two ciphertexts m_1 and m_2 can be done as $RlweEnc_{pk}(m_1) + RlweEnc_{pk}(m_2)$.

Approximate computation improves the efficiency of the FHE cryptosystem. In the last decade, there has been considerable interest in using RNS in FHE schemes. Different HE cryptosystems are applied for cloud computing: RSA, Paillier, El Gamal, Goldwasser-Micali, Boneh-GohNissim, and Gentry [20].

5 Logistic Regression with Residue Number System

5.1 Residue Number System

Residue Number System (RNS) is a variation of finite ring isomorphism widely known and studied. The RNS represents original numbers as residues over the moduli set. The advantages of this representation include inherently carry-free operations, smaller numbers that encode original numbers, and no-positional system with independent arithmetic units. The use of RNS has gained considerable interest in HE schemes due to some characteristics of the system: security, parallel processing, scalability, error detection, and correction [16, 17].

A moduli set of pairwise co-prime numbers $\{p_1, p_2, \ldots, p_n\}$ defines the representation of the values and the range $P = \prod_{1}^{n} p_i$. An integer number X, where $X \in [0, P-1)$, is defined in RNS as a tuple $(x_1, x_2, ..., x_n)$ where x_i represents the remainder of the division of X by p_i, defined by $x_i = |X|_{p_i}$.

The RNS system also allows performing arithmetic operations with several properties. For instance, given X and Y integer numbers, and their representation in RNS by the tuples $(x_1, x_2, ..., x_n)$ and $(y_1, y_2, ..., y_n)$ then:

$$X \otimes Y = (x_1, x_2, ..., x_n) \otimes (y_1, y_2, ..., y_n) =$$
$$\left(|x_1 \otimes y_1|_{p_1}, |x_2 \otimes y_2|_{p_2}, \ldots, |x_n \otimes y_n|_{p_n}\right) = (z_1, z_2, \ldots, z_n) = Z \quad (2)$$

where $\otimes$ denotes one operation: addition, multiplication, and subtraction; with $z_i = |Z|_{p_i}$, for all $i = 1, n$.

Equation 2 shows that the RNS can be defined as a variant of HE [22]. We can obtain secure data processing and storage since the representation of numbers in RNS can be seen as coding and secret sharing scheme. For every n-tuple, the corresponding integer $X \in [0, P-1)$ can be recovered by means of the Chinese Remainder Theorem (CRT) $X = \left(\sum_{i=1}^{n} x_i P_i b_i\right) mod\ P, \forall i = 1, \ldots, n$, where $P_i = P/p_i$ and b_i is the multiplicative inverse of $P_i\ mod\ p_i$.

Additionally, in order to avoid using real $X_R \in R$ numbers, we use the scaling factor method to represent real values in the integer domain, where $X = \lfloor 2^{powerInt} X_R \rfloor$. In general, operations that require the magnitude of a number are high complexity in the

RNS, for instance, magnitude comparison, sign and overflow detection, division, etc. More efficient algorithms to compute division [18] and comparison of numbers in RNS were proposed in [19].

5.2 Logistic Function Approximation

To implement a logistic regression in RNS, we approximate the standard logistic function $g(z)$ using first-degree polynomials, see Sect. 2. On one hand, a restriction on the degree of a polynomial arises due to restrictions on the number of multiplication operations that can be performed with encrypted text using homomorphic encryption. On the other hand, there is a great computational complexity of the multiplication operation in comparison with the addition in homomorphic encryption.

As a simplest initial approximation, we use the function defined by three lines $y_1(z) = 0$ if $z < -2$, $y_2(z) = \frac{1}{4}(2+z)$ if $-2 \leq z \leq 2$, and $y_3(z) = 1$ if $z > 2$ (see Fig. 1). $y_2(z)$ is the tangent to the reference sigmoid function in point $z = 0$. The maximum approximation error is $\frac{1}{1+e^2} \approx 0.1192$. Since the sigmoid function has the following property of $g(-z) = 1 - g(z)$, its calculation for a negative argument is equal to the calculation for a positive argument.

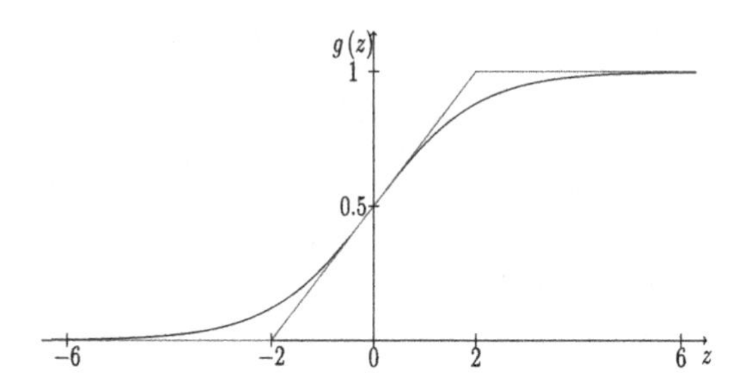

Fig. 1. Approximation function of $g(z)$

5.3 Homomorphic Logistic Regression with Residue Number System

We implemented the Batch Gradient Descendent algorithm with Residual Number System (LR-RNS) to provide security in the training and prediction phases of logistic regression. Both processes can be computed in a third-party infrastructure without worrying about the data leaking. At each iteration of the algorithm, all records in the training set are used to update the values of θ. After processing the training set, all theta values are updated at the same time.

Parallelism is one of the inherent advantages of RNS. The main loop of the algorithm can be done in parallel to reduce the execution time. Assuming that the number of resources is equal to the number of moduli, each resource performs the operation of each moduli for all the elements in the training set. It reduces the execution time of the loop by the number of elements in the moduli set. At the end of the execution, the algorithm returns θ^* to estimate the prediction values of new elements.

Before updating the values of θ in each iteration of the algorithm, a rescale function is used to eliminate the scaling factors. After each multiplication, a scaling factor is accumulated in the result of the operation. We use CRT to partially decode θ with an accumulated scaling factor, then adjust the value and encrypt the information again. The only risk of the data leaking is partial computing of θ.

6 Experimental Results

In this section, we describe the evaluation method, configuration setup, and experimental results. The LR and LR-RNS algorithms were implemented using jdk 1.8.0_221 64-bit and the compute of the metrics was performed by Metrics library (in R).

6.1 Evaluation Method

The efficiency of a classifier is defined by the number of correct and incorrect classification forms of each class. A Confusion Matrix (CM) displays the difference between the true and predicted classes for a set of examples. Accuracy (A) expresses the systematic error to estimate a value. Precision (P) defines the closeness of the measurements between classes. Recall (R) or sensitivity measures the proportion of positives values that are correctly identified. Specificity (S) measures the proportion of negative values (see Eq. 3).

$$A = \frac{T_p + T_n}{T_p + T_n + F_p + F_n}; P = \frac{T_p}{T_p + F_p}; R = \frac{T_p}{T_p + F_n}; S = \frac{T_n}{T_n + F_p} \quad (3)$$

T_p defines the number of elements classified correctly as positive, T_n - correctly classified as negative, F_p incorrectly classified as positive, and F_n - incorrectly classified as negative.

The measures $A, P, R,$ and S depend on the threshold τ(see Eq. 1). It can increase or decrease the classification efficiency. The Receiver Operating Characteristic (ROC) allows plotting multiples values of R and S with respect to different values of τ. ROC curve shows the ability of a binary classifier to discriminate when the threshold is varied. It can be used as a tool to compare and select models. The Area Under the ROC Curve (AUC) is a way to estimate the efficiency of models with different thresholds. AUC is an indicator of the performance comparison of two classifiers. ROC curve, AUC, and A provide good guidance in order to evaluate the proposed algorithm.

6.2 Datasets

We consider six datasets widely used in the literature [10, 12, 13, 21] (see Table 2). They define a series of continuous input variables and two output classes. To normalize the values of the features in the range of [0, 1], we use known min-max normalization, the simplest method using the formula $x' = \frac{x - \min(x)}{\max(x) - \min(x)}$, where x is an original value, and x' is the normalized value.

Table 2. Datasets characteristics and size of sets.

Dataset	Name	N	Features	N-training	N-testing
Lbw	Low Birth Weight study	189	9	151	38
Mi	Myocardial Infarction	1,253	9	1,002	251
Nhanes3	National Health & Nutrition Examination	15,649	15	12,519	3,130
Pcs	Prostate Cancer Study	379	9	303	76
Pima	Indian's diabetes	728	8	582	146
Uis	Umaru Impact Study	575	8	460	115

Low Birth Weight (Lbw) dataset consists of information about births to women in an obstetrics clinic. Myocardial Infarction (Mi) is a heart disease dataset. National Health and Nutrition Examination Survey (Nhanes3) includes a database of human exposomes and phenomes. Pima is the Indian's diabetes dataset. Prostate Cancer Study (Pcs) dataset of patients with and without cancer of prostate. Umaru Impact Study (Uis) dataset stores information about resident treatment for drug abuse.

We use the 5-fold Cross-Validation (CV) technique to compare the performance of our algorithms and consider other similar results in the literature. 5-fold technique divides the dataset randomly in five subsets, four of them are used to train the prediction model, and the last set is used to validate the model.

6.3 Configuration Setup

We perform a series of experiments to find the best setup for LR and LR-RNS. The accuracy is calculated with a threshold of 0.5 and a scaling factor of 32 bits to represent real values with integer values. Our moduli set contains seven pair-wise relatively primes p_i: 18446744073709551629, 18446744073709551653, 18446744073709551667, 18446744073709551697, 18446744073709551709, 18446744073709551757, and 18446744073709551923.

Tables 3 and 4 present AUC and accuracy for LR and LR-RNS, respectively, for Lbw dataset. Each value represents the average of 30 execution with different initial values for θ. The idea is to define a set of values to analyze the behavior of the LR and LR-RNS algorithms. We use six values for $\alpha = \{1.6, 1.1, 0.6, 0.1, 0.06, \text{and } 0.01\}$. The numbers of iteration of the algorithm are 100, 250, 500, 750, and 1,000.

The worst configuration for AUC for both metrics and both algorithms is $\alpha = 0.01$ and 100 iterations. The best configuration for AUC is $\alpha = 1.1$ with 500 iterations for LR and $\alpha = 1.6$ and 100 iterations for LR-RNS. The best accuracy is provided by the configuration $\alpha = 0.6$ with 250 iterations for LR and $\alpha = 1.1$ with 100 iterations for LR-RNS.

Table 3. Average AUC and accuracy for several LR configurations with Lbw.

Iter α	AUC					Accuracy				
	100	250	500	750	1,000	100	250	500	750	1,000
1.6	0.7679	0.7642	0.7629	0.7628	0.7628	0.7518	0.7325	0.7105	0.7105	0.7105
1.1	0.7628	0.7682	**0.7702**	0.7628	0.7628	0.7474	0.7421	0.7149	0.7105	0.7105
0.6	0.7569	0.7665	0.7682	0.7640	0.7694	0.7237	**0.7544**	0.7412	0.7272	0.7114
0.1	0.6122	0.7181	0.7534	0.7589	0.7624	0.6877	0.7000	0.7193	0.7377	0.7456
0.06	0.5693	0.6549	0.7317	0.7502	0.7567	0.6807	0.6886	0.7079	0.7140	0.7237
0.01	0.5120	0.5252	0.5560	0.5859	0.6123	0.5807	0.6465	0.6719	0.6825	0.6886

Table 4. Average AUC and accuracy for several LR-RNS configurations with Lbw.

Iter α	AUC					Accuracy				
	100	250	500	750	1,000	100	250	500	750	1,000
1.6	**0.7591**	0.7537	0.7468	0.7468	0.7468	0.7289	0.7368	0.7368	0.7368	0.7368
1.1	0.7567	0.7573	0.7500	0.7468	0.7468	**0.7377**	0.7368	0.7368	0.7368	0.7368
0.6	0.7553	0.7585	0.7562	0.7530	0.7493	0.7298	0.7298	0.7368	0.7368	0.7368
0.1	0.6159	0.7192	0.7533	0.7554	0.7562	0.6833	0.7035	0.7219	0.7307	0.7316
0.06	0.5703	0.6590	0.7328	0.7499	0.7553	0.6763	0.6904	0.7070	0.7193	0.7298
0.01	0.5116	0.5262	0.5584	0.5876	0.6144	0.5781	0.6439	0.6675	0.6807	0.6825

6.4 Experimental Analysis

Tables 5 and 6 show results for the datasets described in Table 2 and configurations described in Sect. 5.3. We also consider the results of Kim [12] and Aono [10] marked *

Table 5. Average accuracy and AUC for datasets with 5-fold CV after 30 execution.

Name	Accuracy (%)			AUC		
	LR	LR-RNS	Other	LR	LR-RNS	Other
Lbw	75.44	73.77	69.19*	0.7702	**0.7591**	0.689*
Mi	89.31	88.99	91.04*	0.9470	0.9407	**0.958***
Nhanes3	84.99	85.11	79.22*	0.9011	**0.9012**	0.717*
Pcs	71.05	70.35	68.27*	0.7604	**0.7563**	0.740*
Pima	79.93	78.72	80.7**	0.8570	0.8576	**0.8763****
Uis	74.78	74.78	74.44*	0.6383	**0.6343**	0.603*

Table 6. Best accuracy and AUC for datasets with 5-fold CV.

Name	Accuracy (%)			AUC		
	LR	LR-RNS	Other	LR	LR-RNS	Other
Lbw	81.58	78.94	69.19*	0.8141	**0.8077**	0.689*
Mi	90.43	90.43	91.04*	0.9496	0.9496	**0.958***
Nhanes3	85.11	85.94	79.22*	0.9017	**0.9021**	0.717*
Pcs	75	75	68.27*	0.7748	**0.7748**	0.740*
Pima	81.51	80.82	80.7**	0.8579	0.8585	**0.8763****
Uis	76.52	76.52	74.44*	0.6387	**0.6343**	0.603*

and **, respectively, as references because a direct comparison is not fair due to different configurations and parameters.

LR-RNS provides similar results to LR, even with the use of the approximation function. Table 5 shows that for six datasets, the LR-RNS has similar average accuracy and AUC compared with other approaches. Table 6 shows the best solutions for datasets with 5-fold CV. It confirms that LR-RNS provides similar results to RL.

Table 7 presents computing and updating times of theta per iteration for all datasets. Table 8 provides the learning time of the LR-RNS with 1,000 iterations. Training time decrease of LR-RNS is between 85–96%, 3.89%, and 14.12% of time reported in the literature for five of the six datasets.

Table 7. Execution time per iteration of the algorithm (milliseconds).

Name	Processing theta	Updating theta
Lbw	6.6	1.1
Mi	29.4	1.1
Nhanes3	654.1	1.4
Pcs	16.5	1.0
Pima	33.6	0.9
Uis	21.7	1.2

LR-RNS has worse performance in the Nhanes3 dataset increasing time about 49.67%. Reducing the number of iterations to the half, hence, reducing time twice, LR-RNS provides a solution of 1.04% lesser than the value reported in Table 5 for AUC.

Table 8. Leaning time for datasets with 5-fold CV and 1,000 iterations.

Name	Kim [12] (min.)	LR-RNS (min.)	Time decrease (%)
Lbw	3.3	0.128	96.11
Mi	3.6	0.508	85.88
Nhanes3	7.3	10.92	- 49.67
Pcs	3.5	0.291	91.70
Pima	–	0.575	–
Uis	3.5	0.381	89.11

7 Conclusion

The confidentiality of data is fundamental for cloud users. The cloud services should provide security of data storage and processing at any moment. In this paper, we propose a data confidentiality logistic regression algorithm for cloud service with homomorphic encryption based on a residue number system. We provide an experimental evaluation of its performance with several configurations and datasets of different domains in medicine (diabetes, cancer, drugs, etc.) and genomics. The training, testing, and prediction process are performed with ciphertexts. We show that LR-RNS has similar quality results in accuracy and AUC compared with the state of the art of homomorphic encryption LR algorithms but with a considerable time decrease of the training. However, further study is required to assess its actual efficiency and effectiveness in real systems. This will be the subject of future work on a real cloud environment.

Acknowledgment. This work is partially supported by the Russian Foundation for Basic Research (RFBR), project No. 18-07-01224.

References

1. Google. https://cloud.google.com/products/ai. Accessed 13 Mar 2020
2. Microsoft. https://azure.microsoft.com/en-us/services/machine-learning. Accessed 13 Mar 2020
3. Amazon. https://aws.amazon.com/machine-learning. Accessed 13 Mar 2020
4. CSO. https://www.csoonline.com/article/3441477/enabling-public-but-secure-deep-learning.html. Accessed 13 Mar 2020
5. PALISADE. https://palisade-crypto.org/community. Accessed 13 Mar 2020
6. Halevi, S., Shoup, V.: Algorithms in HElib. In: Garay, J.A., Gennaro, R. (eds.) CRYPTO 2014, Part I. LNCS, vol. 8616, pp. 554–571. Springer, Heidelberg (2014). https://doi.org/10.1007/978-3-662-44371-2_31
7. HEANN. https://github.com/snucrypto/HEAAN. Accessed 13 Mar 2020
8. SEAL. https://github.com/Microsoft/SEAL. Accessed 13 Mar 2020

9. Chervyakov, N., Babenko, M., Tchernykh, A., Kucherov, N., Miranda-López, V., Cortés-Mendoza, J.M.: AR-RRNS: configurable reliable distributed data storage systems for Internet of Things to ensure security. Futur. Gener. Comput. Syst. **92**, 1080–1092 (2019). https://doi.org/10.1016/j.future.2017.09.061
10. Aono, Y., Hayashi, T., Trieu Phong, L., Wang, L.: Scalable and secure logistic regression via homomorphic encryption. In: Proceedings of the Sixth ACM on Conference on Data and Application Security and Privacy - CODASPY 2016, pp. 142–144. ACM Press, New York (2016). https://doi.org/10.1145/2857705.2857731
11. Bonte, C., Vercauteren, F.: Privacy-preserving logistic regression training. BMC Med. Genomics **11**, 86 (2018). https://doi.org/10.1186/s12920-018-0398-y
12. Kim, A., Song, Y., Kim, M., Lee, K., Cheon, J.H.: Logistic regression model training based on the approximate homomorphic encryption. BMC Med. Genomics **11**, 83 (2018)
13. Cheon, J.H., Han, K., Kim, A., Kim, M., Song, Y.: A full RNS variant of approximate homomorphic encryption. In: Cid, C., Jacobson, Jr. M. (eds.) Selected Areas in Cryptography – SAC 2018. LNCS, vol. 11349, pp. 347 – 368. Springer, Cham (2019). https://doi.org/10.1007/978-3-030-10970-7_16
14. Cheon, J.H., Kim, D., Kim, Y., Song, Y.: Ensemble method for privacy-preserving logistic regression based on homomorphic encryption. IEEE Access **6**, 46938–46948 (2018)
15. Yoo, J.S., Hwang, J.H., Song, B.K., Yoon, J.W.: A bitwise logistic regression using binary approximation and real number division in homomorphic encryption scheme. In: Heng, S.-H., Lopez, J. (eds.) ISPEC 2019. LNCS, vol. 11879, pp. 20–40. Springer, Cham (2019). https://doi.org/10.1007/978-3-030-34339-2_2
16. Tchernykh, A., et al.: Towards mitigating uncertainty of data security breaches and collusion in cloud computing. In: 2017 28th International Workshop on Database and Expert Systems Applications (DEXA), pp. 137–141. IEEE (2017). https://doi.org/10.1109/DEXA.2017.44
17. Tchernykh, A., et al.: Performance evaluation of secret sharing schemes with data recovery in secured and reliable heterogeneous multi-cloud storage. Cluster Comput. **22**(4), 1173–1185 (2019). https://doi.org/10.1007/s10586-018-02896-9
18. Babenko, M., et al.: Unfairness correction in P2P grids based on residue number system of a special form. In: 2017 28th International Workshop on Database and Expert Systems Applications (DEXA), pp. 147–151. IEEE (2017)
19. Babenko, M., et al.: Positional characteristics for efficient number comparison over the homomorphic encryption. Program. Comput. Softw. **45**(8), 532–543 (2019). https://doi.org/10.1134/S0361768819080115
20. Tchernykh, A., et al.: AC-RRNS: anti-collusion secured data sharing scheme for cloud storage. Int. J. Approx. Reason. **102**, 60–73 (2018). https://doi.org/10.1016/j.ijar.2018.07.010
21. Smith, J.W., Everhart, J.E., Dickson, W.C., Knowler, W.C., Johannes, R.S.: Using the ADAP learning algorithm to forecast the onset of diabetes mellitus. In: Proceedings of the Annual Symposium on Computer Application in Medical Care, p. 261 (1988)
22. Tchernykh, A., et al.: Scalable data storage design for non-stationary IoT environment with adaptive security and reliability. IEEE Internet Things J. **7**, 1 (2020). https://doi.org/10.1109/JIOT.2020.2981276

Replication of "Tail" Computations in a Desktop Grid Project

Evgeny Ivashko[1,2] and Natalia Nikitina[1(✉)]

[1] Institute of Applied Mathematical Research, Karelian Research Centre of RAS, Petrozavodsk, Russia
{ivashko,nikitina}@krc.karelia.ru
[2] Petrozavodsk State University, Petrozavodsk, Russia

Abstract. The paper addresses the problem of accelerating the "tail" stage of a computational experiment in a Desktop Grid. We provide the mathematical model of a "tail" stage, describe the setting of simulation experiments and provide their results. Task replication in "tail" phase proves to be efficient in decreasing the duration of "tail" by orders of magnitude.

Keywords: Desktop Grid · Volunteer computing · BOINC · Replication

1 Introduction

Desktop Grid is a high-throughput computing paradigm which is based on use of idle time of non-dedicated geographically distributed general purpose computing nodes (usually, personal computers) connected over the Internet or by a local access network.

Desktop Grids are an important part of a high-performance computing domain; this concept is used in volunteer computing projects to perform large-scale scientific computations. The most powerful volunteer computing project Folding@home gathers resources exceeding 2 ExaFLOPS[1].

BOINC is an Open Source software platform for Desktop Grid deployment; it is a de-facto standard for volunteer computing. BOINC has a client-server architecture; the server consists of a number of parallel services sharing a database (see [1] for detailed architecture). A client software connects to the server to request tasks, then performs computations and returns the results. The server checks correctness and validates the results, then aggregates them to obtain the solution of the initial problem.

One of the important issues and significant difficulties of a Desktop Grid project implementation is unreliability of clients. A computing node can leave the project after receiving tasks without notification of the server. This leads to losing the tasks. To cope with unreliability, deadline and replication mechanisms

[1] https://stats.foldingathome.org/os.

V. Voevodin and S. Sobolev (Eds.): RuSCDays 2020, CCIS 1331, pp. 611–621, 2020.
https://doi.org/10.1007/978-3-030-64616-5_52

are used. Deadline is the expiration date of a task. If the node has not returned the result before this date, the task comes as lost and is requeued (and to be resent to another node) by the server.

Replication is the form of redundant computing; each task is replicated with a certain factor to be sent to several nodes. The results of computations by several nodes are compared to guarantee the correctness. In addition to correctness, replication is also used to speed up the results of certain tasks receiving having in mind unreliability of computing nodes.

Unreliability is a significant problem at the final stage of a computational experiment, when the number of tasks is less than the number of computing nodes. Then, because of multiple possible deadline violations and requeueing the tasks, the obtaining of the results could be significantly unnecessarily extended. Use of replication at the final stage of an experiment allows to significantly accelerate the experiment. But it is in question what a replication strategy should be used. In this paper we propose a mathematical model of the final stage of an experiment, describe four possible strategies and provide the results of numerical experiments demonstrating replication efficiency.

The structure of the paper is the following. Section 2 presents motivation and related works. In Sect. 3, we describe the mathematical model of the final stage of an experiment. Section 4 gives numerical experiments overview and results. Finally, in Sect. 5, we give the concluding remarks and discuss the results.

2 Motivation and Related Work

The standard computing process of BOINC does not involve dealing with unreliable computing nodes (the ones that never completed the computation of the received tasks and left the Desktop Grid). Instead, the concept of a deadline for computations is used, in case of violation of which the task is considered lost and re-queued for computations. This approach works well in most cases—when one needs to perform an infinite (or very large) number of tasks, mutually independent, of equal value from the point of view of a computational experiment.

However, there are also smaller-scale problems (solved, for example, within the framework of "umbrella" projects or in the Enterprise Desktop Grid), consisting of a sequence of dependent or independent computational experiments, where the results of each individual experiment can be evaluated only when all its tasks are completed. For such problems, the presence of unreliable computational nodes significantly affects the duration of a computational experiment. As the researchers note, the length of the "tail" of the experiment can be several times longer than the final computation period of an individual task (see, for example, [2,3]). This is due to the fact that the "tail" of the computations accumulates unreliable nodes and the nodes that have left the Desktop Grid. Therefore, in practice, it is important to develop methods for reducing this stage of computations.

Replication is the basic method of reducing the "tail" of computations. Being a form of redundant computing, this mechanism serves several purposes. Firstly, replication can increase the likelihood of obtaining the correct result in time, even if some nodes become unavailable without completing the computations; secondly, this mechanism allows to increase the efficiency of the system in terms of throughput of correct results. However, redundant computing significantly reduces the available processing power of the Desktop Grid. Therefore, the replication coefficient (the number of copies of a task simultaneously processed on computing nodes) should be as low as possible. The question of finding optimal replication parameters is studied, for example, in [4]: on the basis of a fairly simple mathematical model, the author determines the most suitable parameter values using the work history of the NetMax@home project.

In one of the pioneering works on Desktop Grids [2] (the paper uses the term SNOW—a shared network of workstations), the effect of replication on the speed of task execution is studied. The task is considered completed if it was not interrupted by the owner of the workstation for a certain period of time. If the task is interrupted, it starts firstly on the same or another workstation. Two types of applications are considered: strongly connected (interrupting any task requires a restart of all application tasks) and weakly connected (interrupting a task requires restarting only this task and does not affect the rest). From a mathematical point of view, the joint distribution of N independent identically distributed random variables is analyzed. In this setting, the authors prove that to speed up the application, it is better to make one replica for k tasks than k replicas of a single task. It is also shown that an additional workstation is better be used to replicate a task that has the least number of replicas. In addition, the paper presents an analysis of the trade-off between increased concurrency and replication. It has been shown that in a number of cases, for strongly related tasks, replication is more preferable than parallelization.

A large number of research articles have been devoted to solving the problem of reducing the "tail" of computations. In [3], the following methods for solving this problem are noted:

- redundant computations (replication): several copies of the same task are transferred to different nodes with the expectation that at least one of them will solve it in time;
- scheduling based on reliability and availability assessment: when assigning tasks, preference is given to reliable computing nodes with high availability;
- re-submission: problem tasks are detected and transferred to other nodes for computation.

In [5], the author considers two, previously known, solutions to the problem of reducing the "tail" of computations proposed for the GridBot:

- aggressive replication: replication of tasks in excess of the necessary quorum in case of inactive or too slow computing nodes. However, such an approach, as the author points out, leads to significant losses in computing power and processor time due to redundancy of computations;
- node selection: the use of slow or weakly active computing nodes in the "tail" of computations is inefficient from the point of view of computation time,

therefore, when processing the latest tasks, one needs to select nodes with good characteristics, ideally powerful and continuously available ones.

Along with the two indicated methods, the author of the work offers a new strategy (called "dynamic slack") related to the dynamic setting of the deadline for a task. When transferring the task to the node, the server sets the computation period depending on the completeness of the entire computational experiment and the capacity of the node. At the final stage, a shorter computation period is set, which allows previously inactive nodes to be detected. In this case, it is necessary to maintain a balance, because too short deadline will not allow a reliable node to complete the task in time and will lead to the loss of both computing power and time.

To evaluate the effectiveness of the methods considered in the work, the author conducted a series of experiments. To perform the experiments, the SimGrid simulator was used, in which data on the availability of 100 computing nodes of the SETI@home project were downloaded; three sets of experiments were carried out using various combined strategies for calculating the "tail":

- default strategy without any speedup methods;
- dynamic deadline and node selection;
- dynamic deadline, node selection and aggressive replication.

The results of the experiments showed that the use of a dynamic deadline and selection of a node can significantly reduce the time for calculating the "tail" of a computational experiment, while aggressive replication, on the contrary, leads to a loss of productivity and a delay in computations.

The work [6] addresses the problem of "tail" computation for distributed data processing systems of the MapReduce/Hadoop type. The main difference of such systems from the Desktop Grid in the considered formulation is the ability to remotely interrupt the computation of the task on the node (while, as in the case of Desktop Grid, there is no information about the current state of computations). Also, the work does not limit the number of available nodes, but all replicas start at the same time. It is noted that the following three questions are of interest:

1. What should be the fraction of incomplete tasks for the start of replication?
2. Which number of replicas for each incomplete task to launch?
3. When replicating, whether to cancel the initial incomplete task or not?

The paper considers a strategy with one replication start point in two versions:

- saving the original copy of the incomplete task upon replication;
- aborting the original copy of the incomplete task upon replication.

The main attention is paid to the analysis of the relationship between the completion time of a set of tasks and the expected amount of computations.

Let r be the number of additional replicas of a task. The completion time is determined as follows:

$$E[T] = E[\max_{i \in \{1,2,\ldots,n\}} T_i],$$

where $T_i = \min_{0 \leq j \leq r} (\tau + X_{ij})$ is the minimal time of completion of either one of the replicas or the initial, ith, task (if it was not aborted), $\tau = 0$ for the initial task (if it was not aborted) or the replication moment. X_{ij} is a random variable from the distribution function of task runtime on the node.

Expected amount of computations is determined as follows:

$$C = \frac{1}{n} \sum_{i=1}^{n} \sum_{j=0}^{r} (T_i - t_{ij}),$$

where n is the number of incomplete tasks.

The author notes that the relationship between the completion time of a set of tasks and the expected amount of computation (cost) depends on two key characteristics of the distribution function of task service time on a node: 1) whether the distribution tail is heavy, light or exponential, and 2) whether the distribution belongs to the type "new faster than old" or "new slower than old".

The authors of work [7], in addition to prioritizing resources and excessive replication, also propose two strategies for excluding nodes from the computation process: based on a simple performance threshold (for example, excluding all nodes with a clock rate less than a given one) and based on a forecast of task completion time (exclusion of nodes that would not complete the task in time if it was assigned to them). In this way, the risk of violating the deadline for computations and delaying computations is reduced. The authors experimentally test the proposed strategies.

From the point of view of resource prioritization, the use of only static information about the clock rate has led to improved application performance, despite the large number of factors affecting the operation of the computing node. This is partly due to the fact that the clock rate correlates with several performance indicators, such as the frequency and time of tasks completion. At the same time, the use of more dynamic information, such as statistics on the availability of a computing node, does not bring the expected benefits due to its weak predictability (complex periodicity, changes in the behavior of the node, etc.). The performance prioritization benefits can be significantly limited when the queue reaches nodes with a relatively low clock speed. Therefore, authors use resource exclusion so that these slow nodes do not delay application execution.

The use of a fixed threshold for clock rate to exclude weak computing nodes has proven itself well in systems with a large dispersion of clock rates. However, in systems with a relatively small dispersion, the exclusion of resources often affected the computational nodes which could be useful in solving the problem. As an alternative to the resource exclusion strategy based on the clock rate, a resource exclusion heuristic was proposed based on the forecasting of the working period, which proved to be better in experiments.

In their work, the authors also explore replication strategies: proactive, reactive, and hybrid ones. Proactive replication aims to replicate all tasks; reactive one—to replicate those tasks whose computation time exceeded the predicted one; finally, the hybrid strategy aims to replicate those tasks for which there is a high probability of exceeding the predicted completion time. The authors use a rather simple prediction of completion time as a ratio of the complexity of the task to the performance of the computing node (the complexity of the task is considered known, and the availability of the node is not taken into account).

The result of the work was the development of a scheduler for the Desktop Grid XtremWeb combining clock prioritization, adaptive exclusion of resources to predict task completion time and reactive replication. The experiments conducted by the authors showed a significant superiority of the developed scheduler over the standard First-Come-First-Served scheduler used in the Desktop Grid.

In this paper we use the same approach as described in [8,9]. We present the more general mathematical model, introduce four heuristic strategies and provide a numerical analysis.

Based on the analysis of research works, we can conclude that the use of intuitive heuristics allows one to get a significant advantage in time compared to the computations without replication. However, a stable time benefit with low redundancy of computations is possible only if the characteristics of computing nodes are taken into account. The main characteristics of the nodes are availability and reliability, defined in this case as follows:

- *reliability* is the ability of a computational node to finish the task before the given time moment (deadline);
- *availability* is the proportion of time periods when the node computes the tasks and the ones when it does not.

The indicated parameters can be expressed by numerical characteristics (see, for example, [10,11]) or, more adequately, with the corresponding statistics—in the form of empirical distribution functions of the task completion time.

Depending on the interrelation of parameters, the following heuristic strategies are possible:

- **Strategy 1** *for available and reliable nodes*: the nodes will compute the tasks in due time, and quickly, therefore, it is necessary to replicate the tasks for which it is possible to reduce the computation time (one by one, starting from expectedly the longest to compute).
- **Strategy 2** *for available but unreliable nodes*: the longer the task has been computing, the more is the probability that the node exited the Desktop Grid, so it is necessary to replicate the longest running tasks.
- **Strategy 3** *for unavailable but reliable nodes*: the longer the task has been computing, the more is the probability that the task will finish soon, therefore, it is necessary to replicate the "earliest" tasks.
- **Strategy 4** *for unavailable and unreliable nodes*: a node may fail to finish the task with high probability, so it is necessary to replicate the tasks with the highest probability to miss a deadline.

Next, we consider the mathematical models of the listed strategies.

3 Mathematical Models of the Experiment Completion

Let us consider Desktop Grid at the moment t_0, when a system consists of n nodes, there is no queue, $n - 1$ nodes are computing, and one of the nodes (released or new) requests a task from the server.

We will assume that the computation time of any task is random, independently and equally distributed. Equal distribution means that the distribution function of the task completion time on each node is the same for all tasks (all tasks have the same computational complexity). Independence means that the time for obtaining the result does not depend on the time at which the computations were performed, and the same task is not assigned to the same node.

Suppose that for each node, we know the distribution function of time to complete the computations

$$F_i(x) = F_{i,X}(x) = P(X \leq x),\ i = 1, \ldots, n.$$

Then at the moment of time t_0, the expected time to complete the experiment (i.e. finish all the tasks present in the system) is defined as

$$T_{exp} = \max_{i=1,\ldots,n-1} ET_i(t_0).$$

Formally, $ET_i(t_0)$ is written as follows:

$$ET_i(t_0) = \int_{t_0}^{d_i} x dF_i(x|t_0) + \overline{F}_i(d_i|t_0) G(d_i, t_0). \qquad (1)$$

Here,

- $F_i(d_i|t_0)$ is the distribution function of time to complete the computations upon condition that they have not finished at the moment t_0.
- $\overline{F}_i(d_i|t_0) = 1 - F_i(d_i|t_0)$ is, at the moment t_0, the probability of the node i to miss a deadline.
- d_i is the due date of computations on the node i (the earlier started the current task, the less)
- $G(d_i, t_0)$ is, calculated at the moment t_0, expected at the moment d_i time to recompute the task again after missing a deadline.

Note that T_{exp}, $T_i(t_0)$, t_0 and d_i are completely different time instances. In its turn, the expected time for recomputing the task again after missing a deadline $G(d_i, t_0)$ consists of the waiting time for the release of the node and the average computation time in case of either completion of the computations in time or the new expected time of recomputing the task again after missing a deadline:

$$G(d_i, t_0) = W(d_i, t_0) + \int_0^d x dF_a(x) + \overline{F}_a(d) \left(d + G\left(t_0 + d_i + W(d_i, t_0) + d\right)\right). \qquad (2)$$

Here, the expected time of releasing the node is

$$W(d_i, t_0) = \max\left(d_i, \min_{j=1,\dots,n; j\neq i} \int_{t_0}^{d_j} x dF_j(x)\right), \quad (3)$$

the index of the completing node is

$$a = \arg \min_{j=1,\dots,n; j\neq i} \int_{t_0}^{d_j} x dF_j(x) \quad (4)$$

It should be noted that this formula does not take into account the possibility of a new node appearing in the Desktop Grid, as well as the fact that the released node may be busy computing another task for which the deadline has been missed. Furthermore, the calculation of the completion time does not take into account overhead costs, such as the time between the end of the computation and the beginning of the computation of a new task.

Employment in the formulas (1), (2), (3), and (4) of several expected values of various random variables, as well as the above assumptions, lead to the fact that the calculated value of the task completion time on the node i (according to the formula (1)) will significantly differ from the real one. In addition, calculations by the above formulas are quite resource intensive. Therefore, from a practical point of view, it makes sense to move to the averaged values in the formulas (2) and (3). Then

$$G(d_i) = (d_i - t_0) + W' + \int_{0}^{d} x dF(x) + \overline{F}(d)G', \quad (5)$$

where

$$G' = d + W' + \overline{F}(d)G' = \frac{d + W' + \int_{0}^{d} x dF(x)}{F(d)} \quad (6)$$

and $W' = \frac{1}{2} \int_{0}^{d} x dF(x)$.

4 Experimental Results

The experiments were performed on a Desktop Grid simulator program. The simulator is aimed at research of the characteristics of scheduling and replication algorithms in Desktop Grids. It has the form of a simulation model of the process of computational experiment in a Desktop Grid with heterogeneous unreliable computational nodes and heterogeneous by complexity tasks.

The key features of the simulator program are: simulation of the computational process on the BOINC platform; implementation of various probability characteristics of the process; implementation of various heuristics of replication.

The main part of the simulator program is a computational module that simulates tasks distribution over Desktop Grid nodes, collection and processing of results. The varied characteristics of the simulated Desktop Grid are being identified before simulation. The program is implemented in C++.

Empirical distribution functions of task execution time on the nodes and of the task complexities were calculated using statistics of RakeSearch project [12] during its first computational experiment. For the completed tasks, the database field *elapsed_time* was used to construct an individual distribution function for every individual node.

In order to model unreliability of the nodes, we distinguished the following result outcomes that are set by the client and appeared at least once during the RakeSearch experiment:

- the task was completed successfully;
- the task could not be sent to the client;
- the task started on a node but failed to complete;
- the task was lost (deadline was missed);
- the task was completed, but the result was considered invalid.

The simulation experiments were conducted on a Desktop Grid model with 301 nodes and 300 tasks. The deadline was set to one week, the quorum to one. We considered unreliable nodes with the probability 0.1% of missing a deadline, 0.1% of a computational error or a validation error.

In an experiment, up to 2/3 tasks were replicated. Thus, with 300 initial tasks, up to 500 replicas were generated. All events of missing deadline/computation error/validation error occurred in the 1000 experiments with a proper frequency to evaluate the effect of replication.

In Table 1, we provide the experimental results. The table presents tail length with and without replication, the average lost CPU time on a single node and the total number of lost replicas (i.e. not needed ones), under different replication strategies. The CPU time is considered lost if a node spent it on executing a task already completed by someone else. According to the BOINC mechanisms, we suggested that the server does not notify the nodes about such tasks immediately. Although in BOINC such "never needed" results are rewarded as usual ones to encourage participants, we consider this CPU time as wasted.

Overall, replication in "tail" phase proves to be efficient at a cost of wasted CPU time for computing extra replicas. With the maximal number of replicas $R_{max} = 2$ (i.e. each task is additionally replicated at most once), "tail" length decreases by a factor of 50–100. With further increase of R_{max}, "tail" length and the average lost time do not show significant increase because the extra replicas are actually never generated.

The strategies (1) and (2) perform better in terms of wasted replicas. They experimentally proved to select for the replication such tasks that were highly possible to fail.

Table 1. Results of simulations averaged by 1 000 experiments. Considered replication strategies: (1) – replication of a task that would finish the latest; (2) – replication of a task with the largest expected computational time; (3) – replication of a task that started the first; (4) – replication of a task that started the latest; (5) – replication of a random task.

Replication strategy	Default tail length	Optimized tail length	Average lost CPU time	Replicas lost in tail
(1)	607 905	9 454 (2%)	3 468	181
(2)	607 974	9 552 (2%)	3 476	181
(3)	608 047	11 972 (2%)	3 527	197
(4)	607 992	11 756 (2%)	3 533	197
(5)	607 984	11 019 (2%)	3 506	194

More experiments will follow to determine the best replication strategy for a set of Desktop Grid nodes with given probabilistic characteristics and their subsets basing on analytical investigation of the model and simulation experiments.

5 Conclusion

The problem of "tail" completion is urgent in Desktop Grids due to the unreliability of their nodes. With multiple possible deadline violations and requeueing the tasks, the results can be significantly delayed, negatively impacting the research progress. Replication at the "tail" stage of an experiment allows to speed up its completion by orders of magnitude.

In this paper we address the problem of accelerating the "tail" stage of a computational experiment. We formalize the mathematical model of a final stage of a computational experiment in a Desktop Grid, describe the setting of simulation experiments and provide their results.

Task replication in "tail" phase proves to be efficient in decreasing the duration of "tail" by orders of magnitude. The choice of the particular replication strategy for a certain Desktop Grid requires further investigation, while all considered heuristics demonstrate high efficiency in simulation experiments.

Acknowledgements. This work was supported by the Russian Foundation of Basic Research, project 18-07-00628.

References

1. Anderson, D.P.: BOINC: a platform for volunteer computing. J. Grid Comput. **18**, 99–122 (2020)
2. Ghare, G.D., Leutenegger, S.T.: Improving speedup and response times by replicating parallel programs on a SNOW. In: Feitelson, D.G., Rudolph, L., Schwiegelshohn, U. (eds.) JSSPP 2004. LNCS, vol. 3277, pp. 264–287. Springer, Heidelberg (2005). https://doi.org/10.1007/11407522_15

3. Kovács, J., Marosi, A.C., Visegrádi, Á., Farkas, Z., Kacsuk, P., Lovas, R.: Boosting gLite with cloud augmented volunteer computing. Future Gener. Comput. Syst. **43**, 12–23 (2015)
4. Kurochkin, I.: Determination of replication parameters in the project of the voluntary distributed computing NetMax@ home. Sci. Bus. Soc. **1**(2), 10–12 (2016)
5. van Amstel, D.: Scheduling for volunteer computing on BOINC server infrastructures. http://helcaraxan.eu/content/pdf/M2_internship_report_VAN_AMSTEL.pdf (2011)
6. Joshi, G.: Efficient redundancy techniques to reduce delay in Cloud systems. Ph.D. thesis, Massachusetts Institute of Technology (2016)
7. Kondo, D., Chien, A.A., Casanova, H.: Scheduling task parallel applications for rapid turnaround on enterprise desktop grids. J. Grid Comput. **5**(4), 379–405 (2007)
8. Kolokoltsev, Y., Ivashko, E., Gershenson, C.: Improving "tail" computations in a BOINC-based desktop grid. Open Eng. **7**(1), 371–378 (2017)
9. Ivashko, E.: Mathematical model of a "tail" computation in a desktop grid. In: Proceedings of the XIII International Scientific Conference on Optoelectronic Equipment and Devices in Systems of Pattern Recognition, Image and Symbol Information Processing, pp. 54–59 (2017)
10. Essafi, A., Trystram, D., Zaidi, Z.: An efficient algorithm for scheduling jobs in volunteer computing platforms. In: 2014 IEEE International Parallel & Distributed Processing Symposium Workshops, pp. 68–76. IEEE (2014)
11. Miyakoshi, Y., Watanabe, K., Fukushi, M., Nogami, Y.: A job scheduling method based on expected probability of completion of voting in volunteer computing. In: 2014 Second International Symposium on Computing and Networking, pp. 399–405. IEEE (2014)
12. Manzyuk, M., Nikitina, N., Vatutin, E.: Start-up and the Results of the Volunteer Computing Project RakeSearch. In: Voevodin, V., Sobolev, S. (eds.) RuSCDays 2019. CCIS, vol. 1129, pp. 725–734. Springer, Cham (2019). https://doi.org/10.1007/978-3-030-36592-9_59

Risky Search with Increasing Complexity by a Desktop Grid

Ilya Chernov(✉) and Evgeny Ivashko

Institute of Applied Mathematical Research, Karelian Research Center of the Russian Academy of Sciences, Petrozavodsk, Russia
{chernov,ivashko}@krc.karelia.ru

Abstract. A common problem solved by high-performance computing is a search problem, when the unique object needs to be found among other objects. With a huge number of objects to examine and computationally hard examination of each, the search problem requires a lot of computing resources. However, the problem becomes even harder if an examination might give the wrong results with some probability. Such problem appears in unreliable high-throughput computing environments like Desktop Grids. In this paper, we present a mathematical model of such search problems, derive the optimal strategy of task assignment that minimizes the expected cost of examinations and thus reduces consumption of computing resources. We show that in a rather general case the optimal strategy is the "no-replication" one, i.e., all objects should be examined once, then for the second time if no target has been obtained, etc. We reveal the cases when this strategy is not optimal. Also, the expected costs of finding the target are obtained for a practical case of object-dependent cost.

Keywords: Desktop Grid · Volunteer computing · BOINC · Urn problem · Needle in a haystack

1 Introduction

Many scientific problems can be reduced to search problems, i.e., examining objects from a finite (though extremely large) set. Most objects are disposed being useless, rare (or unique) valuable targets are to be found. Such problems are often called "needle-in-a-haystack" problems; they consist of multiple relatively simple tasks (*Bag of Tasks*) and can be efficiently solved using Desktop Grids. Here are several examples:

- recovering a password given its hash. For any possible password, its hash is calculated and compared with the known value. The password is obtained if the hashes coincide. There are incredibly many possible passwords with the unique correct password to be found. Such problem is solved, e.g., in a distributed Desktop Grid computing project *distributed.net* (subproject RC5), aimed at attacking the symmetric block cipher RC5[1].

[1] http://www.distributed.net/RC5/.

V. Voevodin and S. Sobolev (Eds.): RuSCDays 2020, CCIS 1331, pp. 622–633, 2020.
https://doi.org/10.1007/978-3-030-64616-5_53

- Finding integer roots of a system of equations (Diophantine equations). Here there is either the unique set of integers (roots) that satisfy all equations of the system, or there may be several solutions; the search set grows exponentially with respect to the number of variables. Such problem was solved in the subproject Euler (6,2,5) of a volunteer computing project Yoyo@home[2]. Also, in 2019 the three cubes problem was closed for all positive integers up to 100 (not equal to 4 or 5 modulo 9) by decomposing the remaining number 42 using the Charity Engine BOINC-based distributed computing project[3].
- One more example is looking for a counter-example for a conjecture. Although often reduced to a Diophantine equation, sometimes it is more convenient to test cases directly. Such problem is being solved in the BOINC-based Collatz Conjecture[4] project. The Collatz conjecture is the (not yet proved nor disproved) statement that the recurrent sequence

$$x_{n+1} = 3x_n + 1 \text{ for } x_n = 2k + 1, \tag{1}$$

$$x_{n+1} = x_n/2 \text{ for } x_n = 2k \tag{2}$$

  converges to the loop 4, 2, 1 for any positive integer x_0.

Several volunteer computing scientific search projects have been launched in Russia[5], including RakeSearch (searches for orthogonal latin squares and investigates the structure of their space), Gerasim@home (solves various search problems of discrete math), Amicable Numbers (looks for pairs of large amicable numbers), to mention a few.

Desktop Grid is a form of distributed high-throughput computing system that uses idle time of non-dedicated geographically distributed computing nodes connected over a low-speed network. By "low-speed" we mean relatively low connection compared to special interconnect like Infiniband, Angara or other supercomputer interconnect. Under the popular middleware BOINC[6], only idle resources are used. Another popular middleware is HTCondor[7], which is also able to make use of idle computers. Desktop Grids need very low capital expenses, can accumulate large computing power, and are very flexible. However, their structure is highly heterogeneous and volatile, the nodes may be unreliable (both in the sense of risk of getting a wrong answer or not getting an answer at all), performance significantly depends on the scheduling algorithm and policy. In case of a search problem unreliability leads to a risk of missing the unique object, so to recheck all the space of objects.

One of the main tools for improving reliability in a Desktop Grid is replication. Replication is solving each task two or more times independently, comparing the results. It is sacrificing performance for reliability. Replication may be quite

---

[2] http://www.rechenkraft.net/yoyo/.
[3] https://www.charityengine.com.
[4] https://boinc.thesonntags.com/collatz/.
[5] https://boinc.ru/category/proekty/.
[6] https://boinc.berkeley.edu/.
[7] https://research.cs.wisc.edu/htcondor/.

effective in counteracting malefactors, but also can be used to reduce the average waiting time. However, using replication for reliability is sometimes questionable because it significantly reduces real performance compared to the peak of the desktop grid.

Given a bag of tasks, each of which is an unreliable examination of a single object in order to find the unique target, one needs to choose a replication policy. Each task, for example, can be examined a given number of times by different independent computing nodes, in parallel or sequentially, either voting for the result or examining up to a fixed number of the same results (the quorum approach). The quorum or size of the voting group may change depending on the complexity of a single examination. Alternatively, replication may be not used; instead, all tasks are re-examined as soon as all have been examined once. Rather counter-intuitively, this is often the optimal strategy, even if the risk of error is high. By induction, it remains optimal if the target is not unique.

In this paper, we consider a general needle-in-a-haystack problem, as follows. We formulate a mathematical model which describes unreliability and could be used to compare different strategies. We found the optimal strategy and the expected number of checks to find the target. We show that in a rather general case the optimal strategy is the "no-replication" one, i.e., all objects should be examined once, then for the second time if no target is obtained, etc. We reveal the cases when this strategy is not optimal. Also we consider the case of ball-dependent costs which is useful in solving practical tasks.

The rest of the paper is organized as follows. The problem is stated mathematically and solved in Sect. 2: we reveal the general optimal strategy of choosing the next examined object, then we consider practically interesting cases of object dependent complexity of a single examination (Sect. 2.1) and examination-dependent cost, equal for all objects (Sect. 2.2). Finally, we turn to the case of finite number of examinations, where the ultimate examination is error-free (Sect. 2.3). This is typical for Desktop Grid (based on BOINC or HTCondor, for example), where trusted and/or reliable (though usually expensive) resources may be used for examining objects. In Sect. 3, we discuss applications of the results to the theory of distributed computing including volunteer computing, using Desktop Grids. The relation of our work with other researchers is discussed in Sect. 4. Finally, in Sect. 5 we describe several other sub-problems based on the initial one and give the concluding remarks.

2 The Problem

Consider the countable number of boxes numbered by $i = 0, 1, \dots$ The boxes contain n balls in total, the balls are numbered $(1, \dots, n)$. A box i contains k_i balls. One of the balls is the *target*. One *move* is choosing a box i, drawing the ball with the lowest number and examining it. This takes some time and/or resources, so the cost of examination $C_{m,s}$ of a ball depends on its number m and the box number s and is increasing with respect to m (however, the cost need not to be strictly increasing with respect to s: it may be decreasing, if later checks are easier, or be non-monotone). Assume that for the ball 1 in the box 0 this cost is the unit: $C_{1,0} = 1$.

If a ball is not the target, this is established for sure; however, the target can be missed with probability $q > 0$. The ball, unless identified as the target, is put into the next box $i + 1$. So, each box is a queue of balls: an arriving ball has the next number, the removed ball has the least number.

It is clear why the ball with the lowest number is drawn: all balls within the box are equally likely to be the target, while the cost is increasing with respect to the ball's number, so the lowest number means the cheapest ball.

Let us denote the set of all k_i, $i \geq 0$, by $\{k\}$. Given this set, one can easily reconstruct the numbers of balls contained in boxes: the box 1 contains balls from $n - k_1 + 1$ to n (here we assume that 'from a to b' means 'no balls' if $a > b$), the box 2 does from $n - k_1 - k_2 + 1$ to $n - k_1$, etc. Thus, if a box s is chosen, the number m_s of the ball drawn (the smallest number among all balls in the box) is well-defined provided that the distribution $\{k\}$ is given.

Drawing a ball from a box s (a move) $\{k\}^s$ changes this distribution: $k_s \to k_s - 1$, $k_{s+1} \to k_{s+1} + 1$, other k_i remain unchanged. Note that moves commute provided that both are possible: $\{k\}^{s,t} = \{k\}^{t,s}$ if $k_s > 0$ and $k_t > 0$.

Let the probability of correct identification of the target be $p = 1 - q$. Note that the probability that the target reaches the box j (without being found) is q^j.

To find the target at the lowest possible cost (on the average), one should define a strategy which box to check first, which box to choose next if the target has not been found, and so on. More formally, a *strategy* is a rule of choosing the box j given the distribution $\{k\}$ of the balls in the boxes. We need the strategy that minimizes the expected cost of moves up to finding the target.

Any strategy generates, for any distribution, an infinite *sequence of moves*, which is interrupted when the target is found. Note that after each move the probabilities P_j that the target is in the box j change.

At any time, the state of the system is defined by the distribution of the balls in the boxes $\{k\}$. This means that the system has the Markov property: the state of the system depends only on the distribution of the balls and does not depend on the sequence of moves that bring the system to this state. The initial distribution is $k_0 = n$, $k_i = 0$ for $i > 0$, i.e., all balls are in the box 0.

Let us express the number of the ball to be drawn from a box s:

$$m_s = 1 + \sum_{j=1}^{s-1} k_j. \tag{3}$$

Now, we derive and analytically prove the optimal strategy and give an explicit formula for the expected number of checks needed to find out the target.

Let us see how this probability $\hat{P}_j$ changes after a move. If a ball is moved from a box j to the next box $j + 1$, the numerator remains the same while the denominator is added the quantity $q^{j+1} - q^j = -pq^j$ and, therefore, strictly decreases.

Obviously, the optimal strategy checks all non-empty boxes, sooner or later: otherwise it may never find the target.

Let us denote the expected cost of moves needed to find a target, provided that the initial distribution of balls is $\{k\}$ and the optimal strategy is used, by $E(\{k\})$. It is clear that after discarding an empty box 0, we return to exactly the same problem.

For the sake of brevity, denote

$$\Sigma = \sum_{i=0}^{\infty} k_i q^i. \tag{4}$$

Let $E^s(\{k\})$ denote the expected cost of moves needed to find the target, provided that the initial distribution of balls is $\{k\}$, the first move is s, and the optimal strategy is used after the first move.

Let us call the strategy of drawing balls from the first non-empty box until it is empty, the 'strategy F'. Now we are ready to reveal the conditions when the strategy F is the optimal strategy.

Theorem 1. *The optimal strategy is as follows: Given a distribution $\{k\}$, choose the box t, such that*

$$\frac{C_{m_t,t}}{q^t} < \frac{C_{m_s,s}}{q^s} \tag{5}$$

for all $s \neq t$.

If two boxes are equivalent, i.e.,

$$\frac{C_{m_t,t}}{q^t} = \frac{C_{m_s,s}}{q^s}, \tag{6}$$

then any can be chosen.

2.1 Ball-Dependent Cost

Let us consider the case when the cost depends on the ball, but not on the box: $C_{m,s} = C_m$.

The following facts immediately follow from the proved result:

- After the first move (examining the ball 1 from the box 0), we need to decide whether to examine this ball again or pass to the ball 2; the condition of passing to the next ball is $C_2 q < C_1$, which is $C_2 < q^{-1}$.
- The strategy F is optimal for the constant cost ($C_m \equiv 1$) and remains optimal for slowly growing C_m with respect to m. By 'slowly' or 'quickly' we mean (hereafter) 'compared to the progression q^{1-m}'. In the this case of 'slow' C_m we examine the ball 1 once, then we do the ball 2 once, and so on till the ball n; after that the box 0 is empty, all balls are in the box 2. The boxes can be re-numbered and the search is resumed from the same position.

- Let $C_m = q^{1-m}$. Then after n moves (provided that the target has not been found before that) a ball m is the only one in the box $n - m$. The strategy is as follows: examine the ball 1, then the balls 1 and 2 (in any order), then the balls 1, 2, and 3 in any order, etc.

Let us accept the convention that in case of ambiguity (when two or more boxes can be chosen following the optimal strategy), the ball with the largest number is drawn; e.g., if the ball 1 can be examined from the box 1 or the ball 2 in the box 0, preference is given to the ball 2. The optimal strategy for quickly growing C_m looks as follows (h_m is the box that is containing the ball m):

1. Examine the ball 1 until its 'effective cost' $C_1 q^{-h_1}$ exceeds C_2.
2. Examine the ball 2 once.
3. Examine the ball 1 again once.
4. Repeat steps 1, 2 until either $C_1 q^{h_1} > C_3$ or $C_2 q^{h_2} > C_3$.
5. Examine the ball 3 once.
6. Continue examining the first three balls in turn until the ball 4 joins the cycle.
7. etc.

Note that in case of exponentially growing C_m the distance between the balls examined in turn (measured in empty boxes between the balls) is almost constant, $\pm$ one box. For slow C_m the distance is zero, i.e., all balls are in the same box, $\pm$ one box. For a C_m that grows more quickly than any progression, the distance between the balls increases.

Note that the strategy strongly depends on the relation between the cost C_m and error risk q; more precisely, on the behavior of $C_m q^m$. For reliable (relative to the complexity) examination of the balls, when this quantity is constant or decreasing, the strategy F is optimal. Also it is optimal if the complexity C_m grows slowly relative to q^{1-m}.

Corollary 1. *For polynomial and exponential algorithms the strategy F is optimal.*

Indeed, polynomial algorithms take time proportional to some power of $\log m$, which is definitely slower compared to the exponent q^{1-m}. For exponential algorithms the complexity is proportional to $\exp(T \log^D m)$ for some T and $D > 0$, while $q^{-m} = \exp\Big(\log(1/q)\exp(\log m)\Big)$, which is growing more quickly.

Now let us evaluate the expected cost $E\big(\{k\}\big)$ of finding the target for the strategy F.

Theorem 2. *The expected cost*

$$E(n, 0, 0, \dots) = \sum_{m=1}^{n} \frac{n+1-m}{n} C_m + \frac{q}{p} \sum_{m=1}^{n} C_m \tag{7}$$

provided that the optimal strategy is the strategy F and it is followed.

2.2 Box-Dependent Cost

Now let is consider another partial case of box-dependent complexity:

$$C_{m,s} = Q_s. \tag{8}$$

This means, if Q_s is a decreasing sequence, that further examination of a ball is cheaper compared to the first one. The condition of optimality of a strategy is to choose the box t with

$$\frac{Q_t}{q^t} < \frac{Q_s}{q^s} \tag{9}$$

for all $s \neq t$. In particular, the strategy F is optimal for any non-decreasing Q_s and also for slowly decreasing Q_s in the sense that $Q_s q^{-s}$ is non-decreasing. However, for quickly decreasing complexity Q_s the optimal strategy becomes, paradoxically, useless: it is to examine the ball 1 infinitely many times (unless it is finally recognized to be the target). Indeed, after the initial examination of the ball it is placed to the box 1, where complexity of examination is so low that it is examined again and put to the box 2, etc.

Now let us evaluate the expected cost E_i of finding the target under the optimal strategy F when all the balls are initially in box i.

Theorem 3. *Let the cost Q_t of examination of any ball depends on the box number t and the series $\sum Q_t q^t$ converges. Let the strategy F be followed. Then the expected cost*

$$E_i = q^{-i}\left(p\frac{n+1}{2} + qn\right)\sum_{j=i}^{\infty} Q_j q^j. \tag{10}$$

Note that both in the case $Q_i = 1$ and in the case $C_m = 1$ we get the solution

$$E = \frac{n+1}{2} + n\frac{q}{p}, \tag{11}$$

where the first term stands for error-safe average search steps, while the second expresses the additional average number of checks due to possible mistakes. This can be rewritten as follows:

$$E = n\frac{1}{p} - \left(n - \frac{n+1}{2}\right), \tag{12}$$

and interpreted as the average value of tries (i.e., boxes) distributed geometrically with the success change p, when each try means examination of n balls; in the box where the target is finally found, it is, on the average, in the middle, so about only a half of the box is examined.

2.3 Finite Number of Boxes

Let us consider the case when we have a finite number of boxes. This means, that the ultimate examination of a ball is error-free, so that after this examination no further ones are necessary. Assume that $C_{m,s}$ depends either only on m, or only on s, i.e., the cost is either ball-dependent, or box-dependent (or just constant).

Let us consider only two boxes, i.e., the second (box 1) is the ultimate one. In this case, the condition of optimality for the strategy F is

$$q \leq \frac{\sum_{j=1}^{n}\sum_{i=1}^{j-1} C_{i,2}}{n\sum_{i=1}^{n} C_{i,1} - \sum_{j=1}^{n}\sum_{i=1}^{j-1}\left(C_{i,1} - C_{i,2}\right) - \sum_{j=1}^{n} C_{j,1}}. \quad (13)$$

For ball-dependent cost $C_{m,i} = C_i$ this reduces to

$$q \leq \frac{n}{n-1} - \frac{\sum_{j=1}^{n} jC_i}{(n-1)\sum_{i=1}^{n} C_i}. \quad (14)$$

and if $C_i \equiv \text{const}$,

$$q \leq \frac{n}{n-1} - \frac{(1+n)}{2(n-1)} = \frac{1}{2}. \quad (15)$$

For box-dependent cost, $C_{i,m} = Q_m$ we have:

$$q \leq \frac{Q_2}{Q_1 + Q_2}. \quad (16)$$

So, this probability is determined by the cost ratio Q_1/Q_2. In particular, if $Q_1/Q_2 \to 0$, which means very expensive second examination, the strategy F is always optimal. This the case, when we have a reliable trusted device of limited computing power and use it if no target has been found in a Desktop Grid search.

Another limit case is $Q_2 = 0$, i.e., the second check is free. Then, obviously, every ball is examined twice always.

For the case of equal cost, $Q_1 = Q_2$, we again have $q < 0.5$. Note that the strategy does not depend on the number of balls already examined: either the strategy F is used, or the alternative one, depending on the risk level.

Now let us get the average cost of finding the target for the optimal strategies, for the box-dependent cost $C_{i,m} = Q_m$. If the optimality condition holds, we have

$$E_n = \frac{1+n}{2}(pQ_1 + qQ_2) + nqQ_1. \quad (17)$$

The other strategy (twice-check) gives:

$$E_n = (Q_1 + Q_2)\frac{n-1}{2} + Q_1 + qQ_2. \quad (18)$$

Their difference is as follows:

$$\Delta = Q_2\frac{n-1}{2}\left(1 - q\left(1 + \frac{Q_1}{Q_2}\right)\right). \quad (19)$$

The loss due to the wrong strategy choice is thus linear with respect to quite a large factor $n - 1$ and can be significant.

3 Application to Desktop Grids

As it has been already stated, Desktop Grid is a form of distributed high-throughput computing system. Reliability, including counter-sabotage, is an important problem within Desktop grid computing, in particular, for volunteer computing projects. There are some techniques, e.g., replication, blacklisting, voting, quorum, reputation, that have been studied in literature (see the review [6]). In particular, much effort has been paid to optimize replication of the desktop grid computing using special aspects of specific problems [1,5,8,13–15]. Kondo [10] discusses error that occur in Desktop Grid computing. We [2] also studied the quorum replication for search-type problems with binary answers that may be of different value.

The results of this article show that, at least in many cases, replication (as a form of reliable computing) is not needed. Instead, all available power should be used for primary examination of all objects, then for secondary, and so on. This may seem counter-intuitive, especially in the case of low reliability and variable cost of an examination.

The risk-free "ultimate box", however, may change the situation. For a Desktop Grid, such ultimate box is a trusted reliable computing resource that can solve a task risk-free and which is usually expensive. Its high cost is due to its scarcity. Then, in the notation used herein, Q_2 is high compared to Q_1, so that $\varepsilon = Q_1/Q_2$ is small. The threshold error risk level is then

$$q^* = \frac{Q_2}{Q_2 + Q_1} = \frac{1}{1+\varepsilon} \approx 1 - \varepsilon = 1 - \frac{Q_1}{Q_2}. \tag{20}$$

In this case, the strategy F is optimal: all tasks must be solved on the cheap available resources, and only if the target is not found, it should be searched on a dedicated resource. However, for comparable cost ($q^* \approx 0.5$), using of highly unreliable nodes seems questionable.

Also the optimal strategy may change if complexity of an examination grows quickly (compared to the progression with factor q^{-1}) with respect to the object or decrease quickly (in the similar sense) with respect to the examination number. The paradoxical case of infinite check of an object is theoretically possible if complexity of the next check decreases to zero quickly enough; this absurd case should be avoided, especially if decisions are taken by a scheduling server algorithm. However, this is impossible provided that the complexity is bounded from below by any positive number.

Also, we derive the expected cost (or number of examinations when appropriate) needed to find the target.

Basing on the proposed mathematical models, a software module for BOINC software platform have been developed. This module allows to estimate the expected number of checks needed to solve the search problem with a BOINC-based project. Having an access to the BOINC database, the software module calculates the probability of error using the history of computing nodes and the number of tasks. This information is used to calculate the expected number of checks (tasks to be solved) to find the target.

4 Related Work

Our recent work is devoted to estimation of a Bag of Tasks runtime estimation in Desktop Grids [7].

The presented search problem belongs to the well-known class of urn problems [9]. In this class of problems real objects are modelled by balls of different colours placed in several urns (or boxes). For example, the coupon collector's problem is somehow resembled: there we estimate the expected number of draws until each ball is drawn at least once; however, the balls are placed back into the same box. In the 'Balls-into-bins' problem balls are randomly placed into boxes and the highest load is of interest. This problem also has applications in computer science.

Craswell [3] considers the problem of looking for the target in a fixed number of locations with known probabilities of finding it for each box. For the case when no location needs to be checked twice, so that error risk is zero, the problem is analytically solved. The proof technique is the same as we use in Theorem 1 and is based on comparing permutations in the search plan. For a more realistic case when a search in the right place fails with some probability, an example is analysed.

Pelc [12] considers three problems of a search with an error probability: on a real-line segment, where the hidden number needs to be found with a given accuracy, discrete bounded, and discrete unbounded set. The search is a game between the Questioner and the Responder. Questions are in the form of inequalities. Answers can be, possibly, wrong.

Marks and Zaman [11] study the problem of looking for a person in a social network. The graph structure of the network is modelled as an urn scheme with urns as vertices and marbles in them as neighbours. Although different from the problem we are considering, the strategy turns out to be similar: one urn, the most promising, needs to be checked completely before passing to the next.

The authors of [4] consider the box "treasure-hunt" problem with adversaries who act to complicate the search by multiple agents.

5 Conclusion

We have considered a rather general search problem with uncertainty, where objects are examined in a chosen order, multiple times, until the single target is found. Complexity of the examination may depend both on the object and on the examination number. We describe the strategy which is optimal in many cases, and reveal the conditions when it is not optimal. It turns out, that in most practical conditions all objects should be examined once, and, if the target has been missed, examined again, etc. This may be quite counter-intuitive, especially if risk of missing the target is high. Applied to Desktop Grid, this result means that replication gives little to efficiency of search: having free resources, it is better to use them for studying new objects, not for double-check.

It would be interesting to generalize the problem for multiple targets (with either known or unknown number of targets in advance); the strategy remains the same, though, the cost of finding all of them is not easy to express. Also, we have such an expression for the single target for some partial cases, given that no object has been examined beforehand. It is important to consider a general case, when each object has been examined zero or more times.

Finally, we assumed that useless objects are always recognized correctly, only the target may be missed with some known probability. However, the case when the error of the other kind has its own positive probability, is also of interest.

Acknowledgements. Supported by Russian Foundation for Basic Research, project 18-07-00628.

References

1. Ben-Yehuda O., Schuster, A., Sharov, A., Silberstein, M., Iosup, A.: ExPERT: Pareto-Efficient Task Replication on Grids and a Cloud. Parallel and Distributed Processing Symposium (IPDPS), pp. 167–178 (2012)
2. Chernov, I., Nikitina, N.: Virtual screening in a desktop grid: replication and the optimal quorum. In: Malyshkin, V. (ed.) PaCT 2015. LNCS, vol. 9251, pp. 258–267. Springer, Cham (2015). https://doi.org/10.1007/978-3-319-21909-7_25
3. Craswell, K.: How to find a needle in a haystack. Two-Year College Math. J. **4**(3), 18–22 (1973). http://www.jstor.org/stable/302651
4. Dobrev, S., Královic, R., Pardubská, D.: Treasure hunt with barely communicating agents. In: Aspnes, J., Bessani, A., Felber, P., Leitão, J. (eds.) 21st International Conference on Principles of Distributed Systems (OPODIS 2017), Leibniz International Proceedings in Informatics (LIPIcs), vol. 95, pp. 14:1–14:16. Schloss Dagstuhl-Leibniz-Zentrum fuer Informatik, Dagstuhl, Germany (2018). https://doi.org/10.4230/LIPIcs.OPODIS.2017.14. http://drops.dagstuhl.de/opus/volltexte/2018/8634
5. Ghare, G.D., Leutenegger, S.T.: Improving speedup and response times by replicating parallel programs on a SNOW. In: Feitelson, D.G., Rudolph, L., Schwiegelshohn, U. (eds.) JSSPP 2004. LNCS, vol. 3277, pp. 264–287. Springer, Heidelberg (2005). https://doi.org/10.1007/11407522_15
6. Ivashko, E., Chernov, I.A., Nikitina, N.: A survey of desktop grid scheduling. IEEE Trans. Parallel Distrib. Syst. **29**(12), 2882–2895 (2018). https://doi.org/10.1109/TPDS.2018.2850004
7. Ivashko, E., Litovchenko, V.: Batch of tasks completion time estimation in a desktop grid. In: Voevodin, V., Sobolev, S. (eds.) RuSCDays 2018. CCIS, vol. 965, pp. 500–510. Springer, Cham (2019). https://doi.org/10.1007/978-3-030-05807-4_42
8. Jimènez-Peris, R., Patiño Martìnez, M., Alonso, G., Kemme, B.: Are quorums an alternative for data replication? ACM Trans. Database Syst. (TODS) **28**(3), 257–294 (2003)
9. Johnson, N., Kotz, S.: Urn Models and Their Application: An Approach to Modern Discrete Probability Theory. Wiley series in probability and mathematical statistics. Wiley, New-York (1977)
10. Kondo, D., et al.: Characterizing result errors in internet desktop grids. In: Kermarrec, A.-M., Bougé, L., Priol, T. (eds.) Euro-Par 2007. LNCS, vol. 4641, pp. 361–371. Springer, Heidelberg (2007). https://doi.org/10.1007/978-3-540-74466-5_40

11. Marks, C., Zaman, T.: A multi-urn model for network search. ArXiv e-prints (2016)
12. Pelc, A.: Searching with known error probability. Theoret. Comput. Sci. **63**, 185–202 (1989)
13. Rumyantsev, A., Chakravarthy, S., Morozov, E., Remnev, S.: Cost and effect of replication and quorum in desktop grid computing. In: Dudin, A., Nazarov, A., Moiseev, A. (eds.) ITMM/WRQ 2018. CCIS, vol. 912, pp. 143–156. Springer, Cham (2018). https://doi.org/10.1007/978-3-319-97595-5_12
14. Sangho Y. Kondo, D., Bongjae, K.: Using replication and checkpointing for reliable task management in computational grids. In: International Conference on High Performance Computing and Simulation, pp. 125–131 (2010)
15. Storm, C., Theel, O.: A general approach to analyzing quorum-based heterogeneous dynamic data replication schemes. In: Garg, V., Wattenhofer, R., Kothapalli, K. (eds.) ICDCN 2009. LNCS, vol. 5408, pp. 349–361. Springer, Heidelberg (2008). https://doi.org/10.1007/978-3-540-92295-7_42

Running Many-Task Applications Across Multiple Resources with Everest Platform

Oleg Sukhoroslov(✉), Vladimir Voloshinov, and Sergey Smirnov

Institute for Information Transmission Problems of the Russian Academy of Sciences, Moscow, Russia
{sukhoroslov,vv_voloshinov}@iitp.ru, sasmir@gmail.com

Abstract. Distributed computing systems are widely used for the execution of loosely coupled many-task applications, such as parameter sweeps, workflows, distributed optimization. These applications consist of a potentially large number of computational tasks that can be executed more or less independently. Since the application users often have an access to multiple computing resources, it is important to provide a convenient and efficient environment for execution of applications across the user-defined heterogeneous resource pools. The paper discusses the related challenges and presents an approach for solving them based on Everest, a web-based distributed computing platform. The presented solution supports reliable and efficient execution of many-task applications, while taking into account resource performance, adapting to queuing delays and providing a mechanism for communication between tasks.

Keywords: Distributed computing · Many-task applications · Parameter sweep · Resource pool · Scheduling · Messaging

1 Introduction

Many-task applications [13] are loosely-coupled parallel applications consisting of potentially large number of computational tasks that can be executed more or less independently. Multiple classes of such applications are widely used in science and technology. Bag-of-tasks applications [6], such as parameter sweeps, Monte Carlo simulations, image rendering, have no dependencies between the tasks. Workflows [24], which are used for automation of complex computational and data processing pipelines, consist of multiple tasks with control or data dependencies between them. There is also a special class of many-task applications that require cooperation between the running tasks. For example, the distributed implementation of the branch-and-bound method [15] requires the exchange of incumbent values between the tasks concurrently processing different subtrees.

While many-task applications are naturally suited for execution on distributed computing resources, there exists a number of challenges related to the

V. Voevodin and S. Sobolev (Eds.): RuSCDays 2020, CCIS 1331, pp. 634–646, 2020.
https://doi.org/10.1007/978-3-030-64616-5_54

efficient execution of such applications such as management of a large number of tasks, accounting for dependencies between tasks, implementing coordination and data exchange, use of multiple independent computing resources, task scheduling, accounting for local resource policies and dealing with failures. Some of these challenges were addressed by existing solutions such as meta-schedulers [9,25], user-level frameworks [5,12], workflow management systems [24], grid portals [10,26] and science gateways [3,11]. However, there is still a lack of convenient tools and environments that do not require a considerable effort to master and use them, thereby allowing users to focus on problems being solved.

In particular, this paper considers the following use case. A user has accounts on multiple computing resources, such as an institution cluster and supercomputing centers. The user wants to run some many-task application by leveraging all available resources in a most efficient manner to obtain the results as quickly as possible. The desired solution should support reliable execution of long-running computations spanning multiple resources, while taking into account resource characteristics and policies. In addition, it should not require complex deployment and configuration, while allowing the users to adapt and run their applications with a minimal effort via a convenient user interface. However, there is a lack of solutions that meet all these requirements, which hinders their adoption by users with less technical background in distributed computing.

In this paper, we present an approach for addressing the aforementioned issues and requirements using Everest [1,20], a web-based distributed computing platform. Everest implements the Platform as a Service (PaaS) model by providing its functionality via remote user and programming interfaces. The platform allows the users to attach their computing resources, publish applications and run them on arbitrary combinations of resources. These features enable Everest to serve multiple distinct groups of users while satisfying the above requirements.

In particular, the paper describes the recent platform improvements and features related to the use of multiple resources, scheduling and execution of many-task applications. The major contributions are the resource pool mechanism (Sect. 2), task scheduling improvements (Sect. 3) and a generic mechanism for communication between tasks (Sect. 5). An experimental evaluation of presented solutions on several application cases is also included (Sect. 6).

2 Many-Task Applications and Resource Pools

Everest provides several tools for execution of many-tasks applications. A general-purpose service for execution of bag-of-tasks applications [27] enables users to describe parametrized computations using a declarative format. Execution of workflows is supported by external orchestration using a scripting language or via a general-purpose service integrated into the platform [22]. Finally, the low level programming interface enables platform clients to dynamically manage a set of tasks within a job by sending commands and receiving notifications [23].

Instead of using a dedicated computing infrastructure, Everest performs the execution of application tasks on external resources attached by users. The platform implements integration with standalone machines and clusters through a developed agent [16]. The agent runs on a resource and acts as a mediator between it and Everest enabling the platform to submit and manage tasks on the resource. The platform also supports integration with grid infrastructures [16], desktop grids [21] and clouds [28].

Everest users can flexibly bind their resources to applications. In particular, a user can specify multiple resources, possibly of different type, for running an application, which is especially important for many-tasks applications. In this case, the platform performs a dynamic scheduling of application tasks across the specified resources However, when submitting such jobs, the user had to manually form a list of used resources. In addition, there was no way to pass the user preferences for the use of individual resources. This complicated the use of Everest for running applications on multiple resources.

To solve the aforementioned problems, a concept of resource pool has been introduced and implemented in Everest. A resource pool is a virtual resource comprised of several regular resources. The user can configure the set of resources included in the pool, as well as the priorities of individual resources. The priorities may correspond to user preferences, relative performance or cost of the corresponding resources. These priorities are used by the platform during the task scheduling as described in Sect. 3. The user can also temporary forbid running tasks on a resource by setting its priority to zero. The changes of pool configuration made during the application execution are automatically picked up by the scheduler allowing the user to manually control the execution.

The resource pool can be used in the same way as regular resources when starting jobs and configuring applications. The pool owner can also configure a list of users and groups that are allowed to use the pool. This enabled Everest users to flexibly configure and conveniently use personal and collective computing infrastructures consisting of several resources for their computations.

3 Task Scheduling

Everest implements a flexible multi-level approach for scheduling jobs, corresponding to application runs, and individual tasks with jobs. The global *job scheduler*, which is periodically invoked with information about current jobs and resource states, fairly distributes the available resource slots among the jobs. Each job encapsulates an *application-level task scheduler*, which is invoked to select the tasks for running on the resources offered by the job scheduler. Finally, a complex resource, such as the one representing a grid or a cloud infrastructure, implements an internal *resource-level task scheduler* that is used to distribute the tasks assigned to the resource among a pool of dynamically allocated machines.

The same approach has been used to implement the scheduling of tasks assigned to resource pools. If a pool is specified as a resource for running a job, then the platform distributes the job tasks among the pool resources using the newly developed resource-level scheduler associated with the pool. By default, this scheduler assigns the tasks to resources with idle slots in the order of user-defined priorities specified in the pool configuration.

However, the relative resource performance often depends on an application, and the provided priorities may not allow for efficient task distribution. Also, for non-dedicated resources shared among many users, such as supercomputers, the presence of idle slots does not guarantee that the scheduled task will start immediately. In practice, the tasks can be arbitrary delayed in the resource queues and blocked by the pending reservations, which complicates the task scheduling and may negatively impact the application execution time. These issues are addressed by the following task scheduling improvements.

The relative performance of resources for a given many-task application can be taken into account in runtime by collecting the execution times of completed tasks. Since these tasks are different, this is not a real benchmarking. However, many-task applications are often comprised of one or several groups of tasks with similar characteristics, which justifies the use of this approach. The mean of execution times of completed tasks on a given resource is used as an estimate of a task execution time during the task scheduling. The mean is computed on a subset of recent measurements (10 latest values currently) to better account for the applications that consist of several "waves" of tasks, which characteristics differ between the waves, and for the variation of resource performance in time.

Another possible approach is to run an application-specific benchmark on all resources prior to the application execution and use the results to compute the resource priorities. However this approach requires more time and effort from the user, while not solving completely the problem of task execution time estimation and not taking into account the variation of resource performance in time. A more promising approach, which can be used to improve the current implementation, is to use the information on task execution collected during the previous application runs, as was previously demonstrated for workflows in [22].

To account for the wait times in resource queues, a similar approach is used to estimate the task wait time on a given resource during the scheduling. The wait times of completed tasks are collected during the application execution. Note that this information is not specific for a given application and is collected and shared across all jobs using the particular resource. A mean of recent measurements (10 latest values currently) is used as an estimate of a task wait time on a given resource during the task scheduling.

The described estimates for task execution and wait times are used during the task scheduling to compute the estimated task finish time for each resource in the pool. A task is assigned to a resource that provides the earliest task finish time. In the beginning of application execution, when these estimates are not available, the user-specified resource priorities are used as described above.

The presented approach allows to take into account the heterogeneous and dynamic characteristics of resources within a pool by using the available information and adapting to changes. However, after being scheduled, a task may still stuck in the resource queue or run extremely slowly on a malfunctioning machine, thereby severely delaying the application finish time. The so called "long tail" or "stragglers" problem is often observed for many-task applications, when the finish time is determined by the "slow" resources computing the several remaining tasks. While the previously described approach can partially alleviate this problem, it cannot eliminate it completely due to the dynamic nature of the execution environment. Therefore an additional mechanism is needed to reschedule the stuck tasks if it will improve the application finish time.

To address this issue, a task migration mechanism has been implemented inside the pool task scheduler. It periodically checks all tasks in a pool and cancels a task if it is waiting on a resource and there is another idle resource with a better estimate of the task finish time. The cancelled tasks are then rescheduled using the previously described approach. The migration of already running tasks is not enabled by default since it can lead to a bouncing of long-running tasks between resources. A user can optionally enable it when running an application with uniformly sized tasks, to protect against the slowly running machines. Besides, the task execution and wait time statistics are updated for the original resource if the cancelled task experienced longer wait or execution times than was expected, to better account for a sudden resource degradation. Finally, in the case of resource failure, Everest performs the rescheduling of tasks assigned to the resource, which can be considered as a special migration case.

The adaptive task scheduling and migration have been previously studied and implemented in the context of grid computing [4,9]. However, while the previous work was mostly focused on detecting the performance degradation and on opportunistic migration, the presented approach takes into account the resource queuing delays and leverages the recent execution history. Another approach widely used in grids is "pilot jobs" [14], which dynamically deploys the execution agents on the resource nodes as regular jobs and then directly assigns tasks to the agents bypassing the resource queues. While this approach allows to avoid the queuing delays, it requires a direct connection from the resource nodes to the external scheduler, which is often restricted on shared HPC resources and complicates the deployment. Also, this approach has an inevitable trade-off between stopping an idle agent and wasting the occupied resources. Another remedy for varying queuing delays is to simultaneously submit the same task to multiple resources [7], however this would complicate the task dispatching and introduce an additional load on the resource queues.

4 Supporting Communication Between Tasks

As described in Sect. 2, Everest supports execution of different types of many-task applications. In many cases, such as parameter sweep experiments, these tasks are executed completely independently of each other. In the case of workflows, the dependencies between the tasks correspond only to the use of the results of one task when starting another task. Therefore, there is no communication between the running tasks in the aforementioned cases.

However, there exist important cases of many-task applications that require coordination and data exchange between the tasks in the process of their execution. These are not only traditional tightly coupled parallel applications, such as MPI programs, but also loosely coupled applications, which permit non-simultaneous execution of tasks and have moderate communication requirements. The former applications are suited for running inside a single HPC system and their support in Everest has been described in previous work [17]. Here we focus on the latter applications that allow execution across multiple resources.

A typical example of a loosely coupled many-task application is a parallel version of the branch and bound method for solving optimization problems, in which the tasks process different subtrees of the whole search tree. In this case, it is necessary to implement the exchange of the best found solutions (incubment values) between the tasks to speed up the search process. Note that the tasks need not to be started and executed simultaneously, they can start upon the resource availability and can also fail independently. However, upon the startup a task should be able to obtain the current best incumbent value. This calls for a reliable storage of such values independent of the tasks.

The required interaction between the tasks can be implemented by the application developer in ad-hoc manner. However, this involves a solution of a number of difficult technical problems. For example, when the tasks are executed on multiple resources, it may not be possible to establish a direct connection between the tasks. Also, as was exemplified before, it may be necessary to reliably store and forward the previously transmitted data to the newly started tasks. Solving such problems requires the implementation and deployment of a rather complex separate component for organizing the data exchange between the tasks.

To simplify the implementation of the considered class of applications on Everest, a general-purpose mechanism for communication between the tasks was implemented at the platform level. The implemented mechanism is based on a two-way message exchange between the tasks and the platform through the Everest agent running on the resource and managing local tasks. When the task starts, the agent passes through the environment variable the port number of the local socket, by connecting to which the task can send and receive messages. The agent forwards the messages received from the task including the task identifier to the platform via the WebSocket connection. Similarly, the platform can send the messages back to the running tasks through the corresponding agents.

Using the messaging support, the following lightweight model for communication between tasks based on shared variables is implemented. Tasks can write and read the values of arbitrary named variables by sending special messages *VAR_SET* and *VAR_GET*. The current values of the variables are stored on the platform side and represent an analogue of the shared memory available to all tasks within a job. The variables belonging to different jobs are stored separately, such that the tasks of different jobs cannot share the data. When the value of a variable is changed, the new value is sent to all tasks of the job. Thus, this mechanism combines the shared memory and publish-subscribe models.

In addition to the basic operations for reading and writing the values of the variables, the conditional write operations *VAR_SET_MI* and *VAR_SET_MD* are implemented. These operations atomically change the value of the variable only if the passed value is greater or less than the current one. These operations, guaranteeing a monotonous increase or decrease of the variable values in the presence of concurrent updates, are used to exchange the incumbent values between the tasks in a distributed implementation of the branch and bound method [15]. In the future, the set of supported operations can be expanded. For example, an atomic compare and swap operation can be implemented in a similar way.

5 Experimental Evaluation and Applications

In this section we present experiments performed to evaluate the described results and demonstrate the use of Everest for running many-task applications across a pool of multiple HPC resources.

The following resources were used in the experiments:

- **HSE:** Supercomputing complex of NRU Higher School of Economics,
- **HPC4:** HPC-4 cluster at NRC Kurchatov Institute,
- **Lomonosov:** Lomonosov-1 supercomputer at Lomonosov Moscow State University,
- **Govorun:** Govorun supercomputer at Joint Institute for Nuclear Research.

5.1 Parameter Sweep Applications

The computational workload was represented by two real-world parameter sweep applications consisting of a large number of independent tasks. The first application is from the life sciences domain and represents the virtual screening of 1000 ligand molecules against the same protein using the molecular docking program Autodock Vina. The second application is from the geophysics domain and consists of 670 tasks performing tabulation of a complicated multidimensional function for solving an inverse problem. Both applications were run via the generic Parameter Sweep service [27] implemented on Everest. The experiments for each application were run in succession with a minimal interval between the runs in order to have the similar conditions on the resources. The number of simultaneously running tasks on each resource was limited as follows: HSE - 88, HPC4 - 64, Lomonosov - 32, Govorun - 32.

Figure 1 (top) displays the number of running tasks across the used resources during the execution of the virtual screening application without using the new resource pool functionality and task scheduler. The platform managed to reach the specified limits on allocated resources given their availability. The periodic drops of utilization are due to the completion of task batches and the queuing delays. However, it can be seen that the application run time suffers from the mentioned "tail problem" - the last batch of tasks was processed very slowly on Govorun. Actually this resource was the fastest one in terms of the task

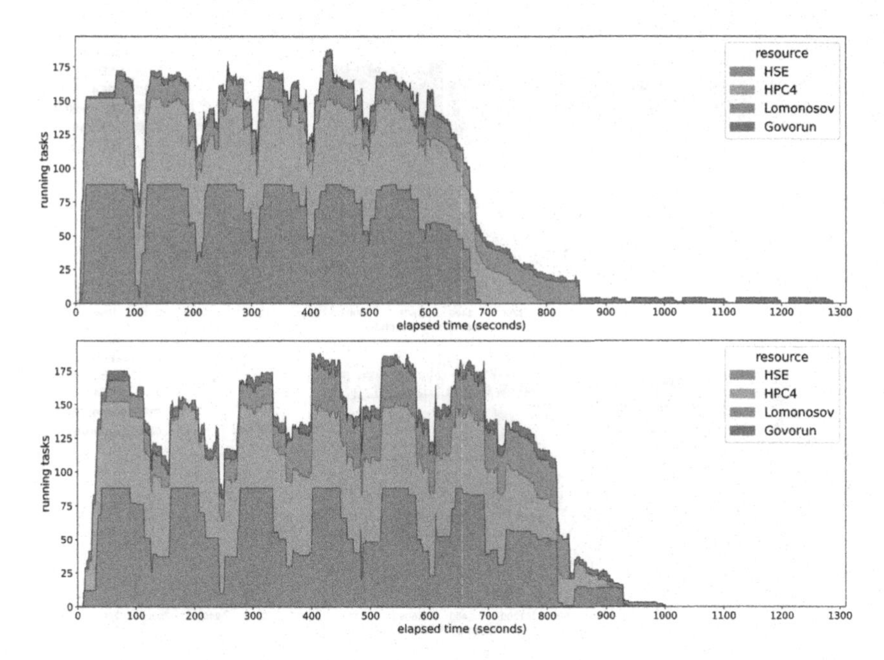

Fig. 1. Executions of the virtual screening application (top - no optimizations, bottom - with optimizations)

execution time, but the task wait times were very high in comparison to the other resources, which explains the observed slow progress.

Figure 1 (bottom) displays the execution of the same application using the new resource pool scheduler. The "tail" has been largely eliminated by both taking into account the collected task wait and execution times during the task scheduling and by migrating 14 tasks from Govorun to HSE (the second fastest resource) in the end of the computations. These optimizations reduced the application execution time from 1288 to 1001 s.

The execution of the geophysics application (see Fig. 2) has been similarly improved. In this case, the "tail" was caused by several factors - the long processing of a batch of tasks by the slowest resource (HPC4) and the high wait times on Govorun and Lomonosov. The improved task scheduling and migration of tasks from these resources to HSE helped to eliminate the problem and reduce the application execution time from 2987 to 2514 s.

The following subsections present other application use cases that are currently leveraging the described implementation to perform computations on the pools of HPC resources.

5.2 Coarse-Grained Parallelization of Branch-and-Bound

The first use case, where the integration of muliple resources via Everest can help to solve the hard optimization problems, concerns the so called coarse-grained

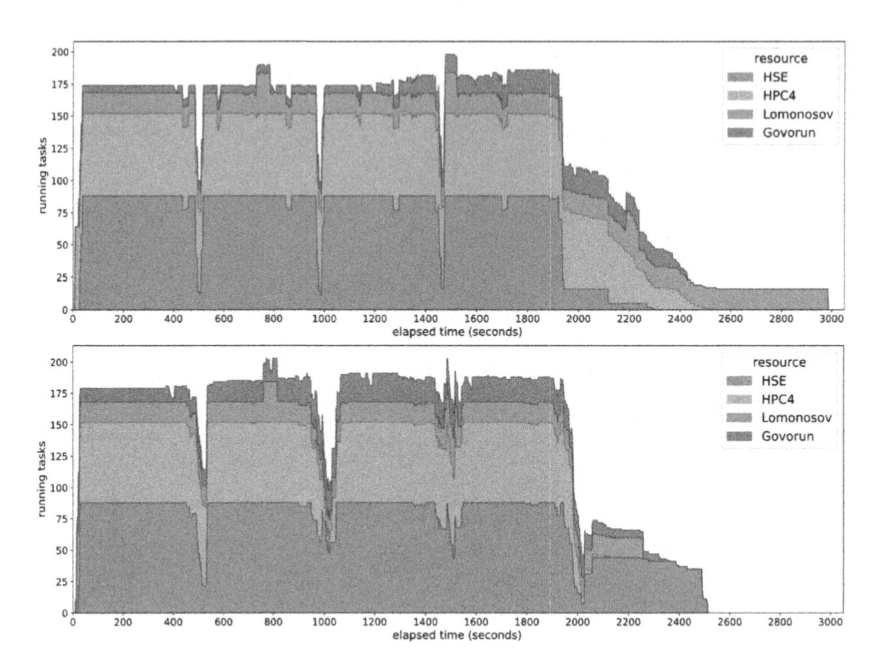

Fig. 2. Executions of the geophysics application (top - no optimizations, bottom - with optimizations)

parallelization of the branch-and-bound (BnB) algorithm [15,18,29]. Here the original problem is decomposed into sub-problems by some decomposition of its feasible domain. Then these sub-problems can be solved in parallel by a pool of existing general-purpose BnB-solvers running on multiple machines and exchanging the incumbent values they found via the communication mechanism described in Sect. 4. In addition to the decomposition, the concurrent approach can be used where the same sub-problems are being solved by multiple BnB-solvers with different algorithm options [18]. The developed Everest application *DDBNB*[1] (Domian Decomposition BnB) enables to combine the "decomposition" and "concurrent" modes of operation. The current implementation is based on SCIP solver [8].

Figure 3 displays the execution of tasks during a *DDBNB* run on two resources for solving the following instance of the Traveling Salesman Problem (TSP) from the TSPLIB[2] collection: *ch150* (150 cities), 64 subsets of the feasible domain and 7 sets of the solver options (448 tasks in total). The used limits on the number of simultaneously running tasks: HSE - 88 tasks, HPC4 - 48 tasks. On the top graph, each task corresponds to a horizontal line from the task submission time to the task completion time. The tasks are colored according to the used resource, the intense color corresponds to the actual task

[1] https://everest.distcomp.org/apps/ssmir/DDBNB.

[2] http://elib.zib.de/pub/mp-testdata/tsp/tsplib/tsp.

execution. It can be seen that the task execution times vary greatly due to the different complexity of the corresponding sub-problems.

5.3 Balanced Identification of Mathematical Models

The second use case relates to the method of balanced identification with regularization of mathematical models by experimental datasets. This method is also called the SvF-technology [19] and is based on the bi-level optimization with a set of independent mathematical programming problems to be solved at the lower level. All these problems (can be hundreds in practice) can be solved by the general-purpose solvers in parallel, and Everest provides the capabilities to do this on multiple resources. Currently the following solvers are supported: IPOPT [30], to find a local optimum in nonlinear optimization problems; SCIP if a global optimum is needed. The current implementation of SvF-technology is based on the Everest application *SSOP*[3] (Solve Set of Optimization Problems), which solves a batch of independent problems in parallel by using the available computing resources.

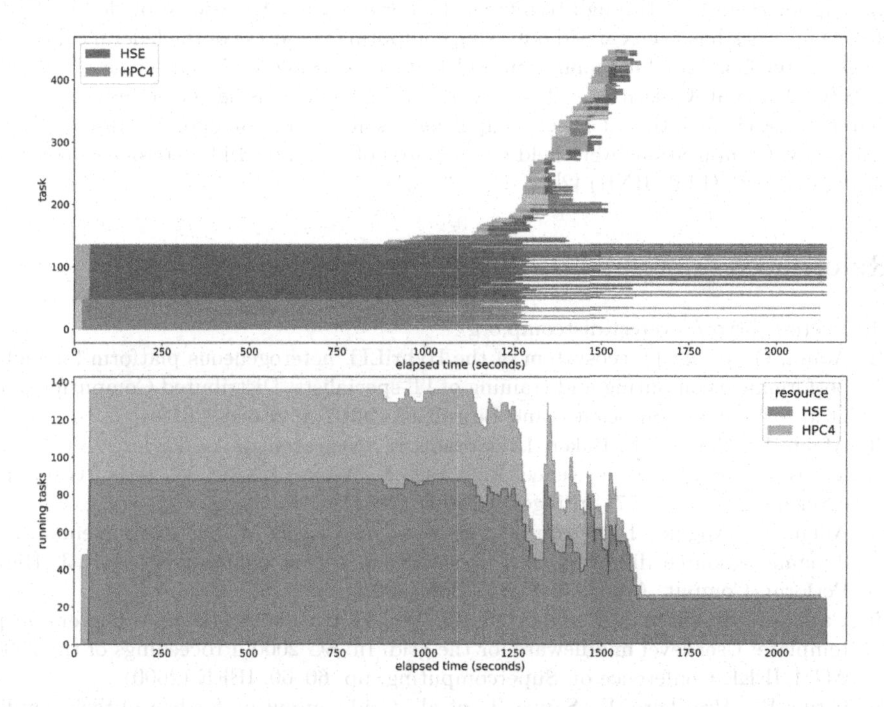

Fig. 3. Execution of *DDBNB* application for solving a TSP instance on HSE and HPC4 clusters (top - task executions in time, bottom - number of running tasks)

[3] https://optmod.distcomp.org/apps/vladimirv/solve-set-opt-probs.

6 Conclusion and Future Work

The paper presented a ready-to-use solution for performing computations across the user-defined resource pools consisting of separate HPC resources. The presented solution supports reliable and efficient execution of many-task applications, while taking into account resource performance, adapting to queuing delays and providing a mechanism for communication between tasks. In contrast to related work, the presented solution is built on a web-based platform and does not require complex deployment and configuration, while allowing the users to adapt and run their applications with a minimal effort via a convenient user interface. The public instance of the platform [1] is available online for all interested users. Future work will focus on improving and extending the presented functionality. We also plan to conduct and report extended large-scale experiments for the mentioned application use cases.

Acknowledgments. This work is supported by the Russian Science Foundation (Project 16-11-10352 - Sects. 4 and 5.2) and the Russian Foundation for Basic Research (Project 20-07-00701 - Sect. 5.3, Project 18-07-00956 - all other sections). This research was supported in part through resources of supercomputer facilities provided by NRU HSE. This work has been carried out using computing resources of the federal collective usage center Complex for Simulation and Data Processing for Mega-science Facilities at NRC "Kurchatov Institute". The research is carried out using the equipment of the shared research facilities of HPC computing resources at Lomonosov Moscow State University. Computations were held on the basis of the HybriLIT heterogeneous computing platform (LIT, JINR) [2].

References

1. Everest. http://everest.distcomp.org/
2. Adam, G., et al.: IT-ecosystem of the HybriLIT heterogeneous platform for high-performance computing and training of IT-specialists. Distributed Computing and Grid-technologies in Science and Education. **2267**, 638–644 (2018)
3. Afgan, E., Goecks, J., Baker, D., Coraor, N., Nekrutenko, A., Taylor, J.: Galaxy: a gateway to tools in e-science. In: Yang, X., Wang, L., Jie, W. (eds.) Guide to e-Science, pp. 145–177. Springer, London (2011)
4. Allen, G., Angulo, D., Foster, I., et al.: The Cactus Worm: experiments with dynamic resource discovery and allocation in a grid environment. Int. J. High Perform. Comput. Appl. **15**(4), 345–358 (2001)
5. Casanova, H., Berman, F., Obertelli, G., Wolski, R.: The AppLeS parameter sweep template: User-level middleware for the grid. In: SC 2000: Proceedings of the 2000 ACM/IEEE Conference on Supercomputing, pp. 60–60. IEEE (2000)
6. Cirne, W., Brasileiro, F., Sauve, J., et al.: Grid computing for bag of tasks applications. In: Proceedings of the 3rd IFIP Conference on E-Commerce, E-Business and EGovernment. Citeseer (2003)
7. Dean, J., Barroso, L.A.: The tail at scale. Commun. ACM **56**(2), 74–80 (2013)
8. Gamrath, G., Anderson, D., Bestuzheva, K., et al.: The SCIP Optimization Suite 7.0. Technical report, Optimization Online, March 2020

9. Huedo, E., Montero, R.S., Llorente, I.M.: A framework for adaptive execution in grids. Softw. Pract. Exp. **34**(7), 631–651 (2004)
10. Kacsuk, P.: P-GRADE portal family for grid infrastructures. Concurrency Comput. Practice Exp. **23**(3), 235–245 (2011)
11. McLennan, M., Kennell, R.: HUBzero: a platform for dissemination and collaboration in computational science and engineering. Comput. Sci. Eng. **12**(2), 48–53 (2010)
12. Moscicki, J.T.: Diane - distributed analysis environment for grid-enabled simulation and analysis of physics data. In: 2003 IEEE Nuclear Science Symposium. Conference Record (IEEE Cat. No. 03CH37515), vol. 3, pp. 1617–1620. IEEE (2003)
13. Raicu, I., Foster, I.T., Zhao, Y.: Many-task computing for grids and supercomputers. In: 2008 Workshop on Many-Task Computing on Grids and Supercomputers, pp. 1–11. IEEE (2008)
14. Sfiligoi, I., Bradley, D.C., Holzman, B., Mhashilkar, P., Padhi, S., Wurthwein, F.: The pilot way to grid resources using glideinWMS. In: 2009 WRI World Congress on Computer Science and Information Engineering, vol. 2, pp. 428–432. IEEE (2009)
15. Smirnov, S., Voloshinov, V.: On domain decomposition strategies to parallelize branch-and-bound method for global optimization in Everest distributed environment. Procedia Comput. Sci. **136**, 128–135 (2018)
16. Smirnov, S., Sukhoroslov, O., Volkov, S.: Integration and combined use of distributed computing resources with Everest. Procedia Comput. Sci. **101**, 359–368 (2016)
17. Smirnov, S., Sukhoroslov, O., Voloshinov, V.: Using resources of supercomputing centers with everest platform. In: Voevodin, V., Sobolev, S. (eds.) RuSCDays 2018. CCIS, vol. 965, pp. 687–698. Springer, Cham (2019). https://doi.org/10.1007/978-3-030-05807-4_59
18. Smirnov, S., Voloshinov, V.: Implementation of concurrent parallelization of branch-and-bound algorithm in Everest distributed environment. Procedia Comput. Sci. **119**, 83–89 (2017)
19. Sokolov, A., Voloshinov, V.: Balanced identification as an intersection of optimization and distributed computing. arXiv preprint arXiv:1907.13444 (2019)
20. Sukhoroslov, O., Volkov, S., Afanasiev, A.: A web-based platform for publication and distributed execution of computing applications. In: 14th International Symposium on Parallel and Distributed Computing (ISPDC), pp. 175–184, June 2015
21. Sukhoroslov, O.: Integration of Everest platform with BOINC-based desktop grids. In: Third International Conference BOINC:FAST 2017, pp. 102–107 (2017)
22. Sukhoroslov, O.: Supporting Efficient Execution of Workflows on Everest Platform. In: Voevodin, V., Sobolev, S. (eds.) RuSCDays 2019. CCIS, vol. 1129, pp. 713–724. Springer, Cham (2019). https://doi.org/10.1007/978-3-030-36592-9_58
23. Sukhoroslov, O.: Supporting efficient execution of many-task applications with Everest. In: Proceedings of GRID 2018, pp. 266–270 (2018)
24. Taylor, I.J., Deelman, E., Gannon, D.B., Shields, M.: Workflows for e-Science: Scientific Workflows for Grids. Springer Publishing Company, Incorporated (2014)
25. Thain, D., Tannenbaum, T., Livny, M.: Distributed computing in practice: the Condor experience. Concurrency Comput. Practice Exp. **17**(2–4), 323–356 (2005)
26. Thomas, M., Burruss, J., Cinquini, L., et al.: Grid portal architectures for scientific applications. In: Journal of Physics: Conference Series, vol. 16, p. 596. IOP Publishing (2005)

27. Volkov, S., Sukhoroslov, O.: A generic web service for running parameter sweep experiments in distributed computing environment. Procedia Comput. Sci. **66**, 477–486 (2015)
28. Volkov, S., Sukhoroslov, O.: Simplifying the use of clouds for scientific computing with Everest. Procedia Comput. Sci. **119**, 112–120 (2017)
29. Voloshinov, V., Smirnov, S., Sukhoroslov, O.: Implementation and use of coarse-grained parallel branch-and-bound in Everest distributed environment. Procedia Comput. Sci. **108**, 1532–1541 (2017)
30. Wächter, A., Biegler, L.: On the implementation of an interior-point filter line-search algorithm for large-scale nonlinear programming. Math. Program. **106**(1), 25–57 (2006)

Solving the Problem of Texture Images Classification Using Synchronous Distributed Deep Learning on Desktop Grid Systems

Ilya I. Kurochkin[1,2(✉)] and Ilya S. Kostylev[2]

[1] Institute for Information Transmission Problems, Russian Academy of Sciences, Moscow, Russia
qurochkin@gmail.com
[2] The National University of Science and Technology MISiS, Moscow, Russia
mr.ilyakos@gmail.com

Abstract. The problem of classifying a large set of texture images using deep neural networks is considered. To reduce the learning time of the neural network, it is proposed to use a desktop grid system. The deep neural network architecture is selected and its implementation is described. A method for organizing deep learning based on the data separation approach and synchronous parameter updates during distributed learning is presented. The features of deep learning on a desktop grid system are discussed and the results of a computational experiment are presented.

Keywords: Synchronous distributed deep learning · Texture-CNN · Desktop grid · BOINC

1 Introduction

Deep learning plays a huge role in machine learning tasks such as computer vision and pattern recognition. With a sharp jump in available computing power and an increase in data sets, as well as improved software for their training, it became possible to change the very model of using networks. From using manual setting of features and statistical classifiers, we have moved to models where features and classifiers are initialized themselves during training [1]. This approach requires much fewer parameters for manual configuration, since the model gets most of the necessary information from the data it is trained on [2]. The availability of a variety of data sets, specialized libraries, and frameworks for deep learning (Caffee, TensorFlow, Keras and Theano) has increased the popularity of using a variety of deep learning methods. The development of deep learning methods has made it possible to improve the solution of many problems in the field of computer vision in comparison with classical methods. In fact, in computer vision tasks such as image classification, object detection and recognition, and semantic segmentation, the vast majority of methods that show the best results are based on deep learning.

V. Voevodin and S. Sobolev (Eds.): RuSCDays 2020, CCIS 1331, pp. 647–657, 2020.
https://doi.org/10.1007/978-3-030-64616-5_55

It is proposed to solve the problem of classifying texture images first on the reference set, and then on the selected application problem with a large number of images. A special convolutional neural network (CNN) will be used to solve the problem. CNN learning is planned to be conducted on a distributed computing system. Options for the efficient use of computing resources are considered taking into account the features of desktop grid systems.

2 Classification of Images Textures

It would be wrong to design a universal deep neural network (DNN) for image classification. As a rule, such tasks have features that should be considered when choosing the DNN architecture. The problem of classifying texture images will be considered. The solution of such tasks is necessary for various fields from medicine and recognition of materials to the analysis of satellite images. To solve applied problems, quite large sets of images are often generated—from several thousand to hundreds of millions of images. When solving the problem of classifying textures in images for a neural network, image fragments are represented, since the source images can have a large size of 1920×1080 pixels (FullHD), 4096×3072 pixels (4 K) or more. But the input layer of neural networks will have a dimension on each side of about 200–500 pixels.

We describe the method of forming the size of a set of images for training a neural network. Let there be a set of similar images in the size of 1920×1080 pixels, which consists of 12 classes of 100 images each. Let DNN have an input layer of dimension 200 $\times$ 200 pixels, then each 1920×1080 pixels image will have $9*5 = 45$ non-overlapping fragments, and with a 50% overlap, one will get $18*9 = 162$ fragments. Since the orientation of the texture on the image is unknown, each fragment must be represented from a set of fragments rotated at different angles: when you rotate one fragment of the image at an angle from 0° to 345° with a step of 15°, we get 24 fragments. As a result, the dataset for neural network training formed from 1200 source images will consist of 4,665,600 fragments.

3 Texture-CNN

DNN tend to increase their depth and thus be able to detect any complex features of images (for example, the presence of a human face in the image). When classifying texture images, it is not so important to identify some specific complex features, but it is necessary to identify some repeating patterns that are less complex. The Texture CNN (T-CNN) architecture is adapted to this type of task. This neural network is based on the well-known architecture AlexNet, which, at one time, was a breakthrough in the field of computer vision. However, T-CNN has some differences that allow it to more accurately classify textures, while significantly reducing the resource intensity of the task.

The creators of T-CNN developed a new energy layer (see Fig. 1) and used many configurations with different numbers of convolutional layers [3]. Each feature map obtained from the last convolutional layer of the network is grouped by calculating the average value from the output of its activation function. This result, represented as a single number for each feature map (that is, a vector whose length is equal to the number of filters of the last convolutional layer), is similar to the energy response to the use of a set of filters, where the detected features of different complexity are calculated in contrast to the usual predefined filters.

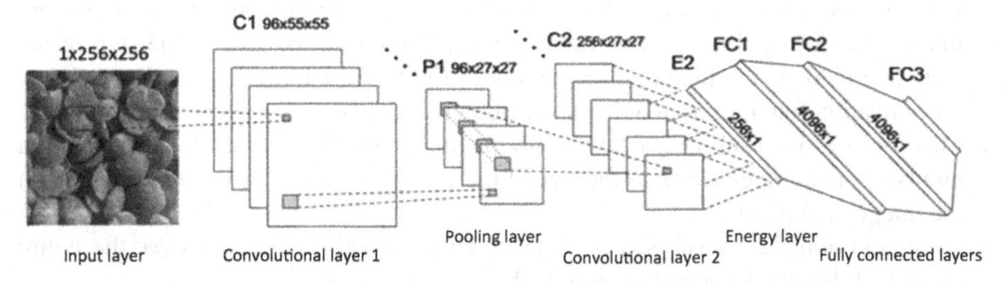

Fig. 1. T-CNN architecture.

Thus, we can note the main differences between T-CNN and AlexNet:

- variation in the number of convolution layers;
- presence of a new layer (energy layer);
- significant reduction of network parameters (due to the new layer).

The developers of this architecture have shown that a convolutional neural network with a relatively small number of layers (from 7 to 10) and 20 million configurable parameters is able to cope with the task of classifying textures on images as well as networks with a large number of parameters. The paper [3] compares the results of classification of networks of the T-CNN family and the AlexNet network.

AlexNet, with 2.5–3 times more configurable parameters (61 million parameters for AlexNet compared to 21–25 million for T-CNN, depending on the selected T-CNN option), shows comparable classification results on texture image datasets such as Kylberg [4], KTH-TIPS2-b, and ImageNet [5] image subdatasets (ImNet-T, ImNet-S1, and ImNet-S2) with textures, and in most tests of the T-CNN architecture, they have greater accuracy. The classification accuracy on the Kylberg dataset for AlexNet is 98.9% against the best result for T-CNN at 99.2%, on the KTH-TIPS2-b dataset, AlexNet's accuracy is 47.6%, and T-CNN's is 48.7%.

As noted earlier, there are several types of T-CNN that differ only in the number of convolutional layers. For our problem, we will use the T-CNN2 using two convolution layers, which will allow us to achieve a high classification result and, at the same time, optimize the cost of computing resources.

In general, it can be noted that this solution is a certain balance in terms of computing resources and the accuracy of classification of texture images.

The popular open source machine learning library TensorFlow [6] developed by Google was used to implement the T-CNN architecture. We also used an addon for TensorFlow-Keras, which allows one to quickly prototype the architecture of a convolutional neural network. This library contains numerous implementations of the "building blocks" of neural networks, such as layers of neural networks, optimizers, target and transfer functions, metrics, and many other tools to simplify working with images and text.

In this implementation of T-CNN, the following layers were used:

- Conv2d—convolutional layer, which parameters are the number of filters (feature maps), the size of the receptive field (convolution core), the size of the step of the receptive field on the image, initialization of the initial values of the layer weights, the activation function used at the layer output, and other parameters.
- MaxPooling2d—a subsample layer with maximization of the value, which also has a number of parameters, such as the size of the subsample window, the window step on the image, and others.
- AveragePooling2d—a sub-selection layer with the value averaged over the entire window. It has the same parameters as the previous layer.
- Dense—a fully connected layer characterized by the number of neurons, as well as the activation function.
- Flatten—an intermediate layer, which is a layer between the classifier itself and the convolutional part of the neural network, which converts a multidimensional vector into a one-dimensional one.
- BatchNormalization—this layer accelerates the learning of the neural network by normalizing the input data before each layer of the network (getting zero expectations and unit variance).

During training, the Adam learning algorithm [7] was used, with the criterion of early stopping of training when using a validation subset.

4 Datasets

The task of classifying textures in images is widespread. There are many datasets used for training and validating classifiers.

One of the most famous texture datasets is the Brodatz image set, which consists of 112 classes, instances of which are represented in gray scale and have a resolution of 512 × 512 pixels. All images are presented in excellent quality, due to special shooting conditions that exclude noise and negative effects of various nature.

An equally common set of textures is KTH-TIPS. This dataset consists of images taken from three different camera positions and a light source, as well as at 9 different distances from the material being captured. Thus, each class is represented by 81 instances, represented by color images, the vast majority of which have a resolution of 200 × 200 pixels.

Later, this dataset was updated with new texture classes and expanded with new lighting conditions. The expanded version is called KTH-TIPS2 and consists of 11 classes of 4 sets of images. Each set contains 108 images. A popular method for evaluating the classifier's performance on this dataset is to train the model using cross-validation: 4 iterations, in each of which training takes place on one of the four sets, and testing on the remaining three, with the calculation of the average accuracy index for all passes.

Also of note is the Amsterdam library of textures, which contains 250 different classes. Each class consists of a set of texture images obtained from three angles using eight different lighting sources, which, in turn, also have three positions. These survey factors allow us to determine the stability of the classifier to change parameters such as the color of the light source radiation and the reflection of light from the material.

The Kylberg Texture Dataset v.1.0 [4] was used to train and test our neural net-work model. The dataset consists of 28 classes with 160 images in each class. Images in gray scale have a size of 576 × 576 pixels. This set of textures is presented in two versions: images of textures without rotation (rotation angle 0°) and 12 positions of each image with a rotation of 30°. In this article, a dataset without image rotations was used. To perform a comparative analysis with the results of [3], the images of the Kylberg dataset were divided into 4 disjoint fragments with the size of 256 × 256 pixels. To reduce the computational complexity of the learning task, only 12 classes were taken. This results in 160*4 = 640 precedents in each class and 12*640 = 7680 precedents in total.

Testing of the T-CNN-2 DNN on the Kylberg-12 reference set showed a classification accuracy of 100%, which corresponds to the results in [3].

5 Distributed Methods of Deep Learning

With the development of deep learning and an increasing of data used, developers of neural network models faced the problem of training neural networks. Training such networks required huge data storage resources and also took a lot of time. In order to achieve a significant reduction in learning time, the learning process itself can be distributed among multiple computing nodes. Thus, we can assume that as the number of computing nodes increases, the network learning time should decrease.

There are two main approaches of distributed stochastic gradient descent (SGD) for training deep neural networks: synchronous AllReduce-SGD, which is based on rapid collective communication of system nodes [8–10], and asynchronous SGD via parameter servers [11, 12].

Both methods have their drawbacks when scaling. Synchronous SGD loses performance when there is slowdown at least one node, it does not fully use computing resources, and is not tolerant of failure of computing nodes. In turn, the asynchronous approach uses parameter servers, thereby creating a communication problem of bottleneck and unused network resources, thereby slowing down the convergence of the gradient.

Grid systems are characterized by failures of computing nodes and their different computing capacity. Based on these features the use of a pure synchronous approach for distributed deep learning on grid systems will be accompanied by downtime of computing nodes.

There are two approaches that can be used to distribute the learning of a DNN model: dividing the training dataset into different parts for different computational nodes (data parallelism) [13], or dividing the model itself between different nodes (model parallelism).

When using the model parallelism of a neural network, the model is distributed between different nodes. This approach is used if the model is too large to be fit on a single node.

When data is parallelized, the entire source training set is divided into n parts, which are sent to n nodes of the distributed system. Each of these nodes contains a copy of the neural network model and the parameters of this W_n model. After the calculation, the results are sent to the server and aggregated there.

In the next section, we will only discuss the specifics of implementing data parallelism. If the results of local models are aggregated only at the end of training, this method is called one-shot averaging. The learning process can be iterative. The aggregation process for local model results can occur after each iteration of the SGD method, however, in this case, the overhead of transferring results and input data between the computing nodes and the server can be very high compared to the computational complexity of a single iteration of the SGD method. For the reduction of the share of data-exchange overhead, it is proposed to aggregate results after performing several iterations of the SGD method.

Based on the differences in the properties of communication channels and the computing power of nodes, the speed of calculating the results will be different. The main problem with iterative distributed learning is the issue of synchronizing the transfer of results from computing nodes to the server. There are three different methods for updating results (gradients) on the server: synchronous, asynchronous, and delayed updates.

When updating synchronously, the server does not update the global network model until it receives gradients from all nodes in each iteration. This leads to the problem of losing the efficiency of the grid system—fast nodes are idle part of the time and the total time of distributed learning increases. One of the well-known implementations of synchronous update is bulk synchronous parallel (BSP) [14]. A characteristic feature of synchronous mode is that the server will always get the latest gradients of all nodes, which do not affect the model's convergence.

Asynchronous algorithms such as Hogwild [15] try to solve the problem of fast node downtime. As for asynchronous updates, fast nodes do not wait for slow ones. One node can send its local gradients to the server, while others calculate their gradients.

However, asynchronous updates have a problem with data being out-of-date; because fast nodes may use outdated parameters that compromise model convergence. In addition, the fastest node updates its local parameters only for its training subset, which causes the local model of the fastest node to deviate from the global model.

To overcome the disadvantages of asynchronous updates, attempts have been made to limit the outdated parameters. With this update, fast nodes will use outdated parameters, but this out-of-date (delay) may be limited [16]. This restriction mitigates the problem of non-relevance to some extent and increases the speed of model learning. Selecting an out-of-date limit is an important parameter, because if it is too large, it means complete asynchrony, and if it is too small, it corresponds to synchronous model updates.

6 Organization of Distributed Deep Learning on a Desktop Grid

To implement distributed learning of a deep neural network, a grid system of personal computers based on the BOINC platform was deployed [17]. The computing application was implemented as a Docker-image. This implementation allows one to use modern tools for training neural networks without installing them on computing nodes. Tasks for computing nodes are represented as Docker images based on the python image: 3-slim (lightweight Ubuntu 18.04 with Python 3.6 pre-installed with the TensorFlow, Keras, pillow, pandas, and other Python libraries installed).

In addition, each image contains:

- Module for creating a local training set for a local DNN model;
- Module for uploading images from an FTP server;
- Module for preprocessing the images from local training set;
- The current parameters of the global model;
- DNN learning module based on a local training sample.

Due to the fact that the training dataset is very large and consists of 4.5 million images, its location on the same server, where the server part of the BOINC project is deployed, is undesirable and can lead to denial of service for computational nodes. As a result, the dataset was located on an FTP server. The required images were requested by grid system nodes after receiving the task (Fig. 2)

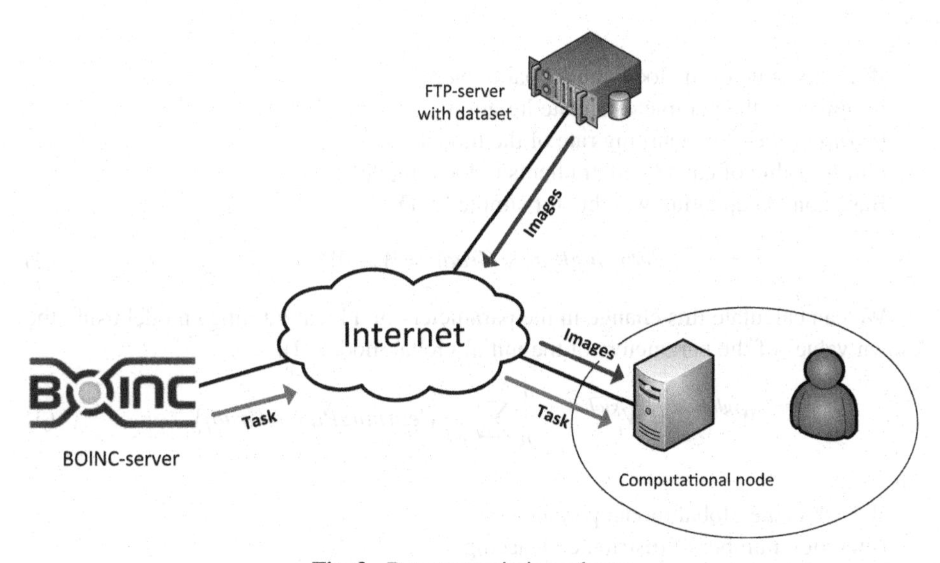

Fig. 2. Data transmission schema.

The method of distributed learning of a DNN with data parallelism was chosen. The distributed learning process is iterative with synchronous updating of neural net-work parameters. The neural network model is trained over several epochs during a single iteration of distributed learning on a computing node.

Since computing nodes may fail, and their number in the grid system may change significantly during the training of a single neural network, it doesn't make sense to divide the dataset into certain parts, because some parts of the dataset may not participate in training for several iterations. When forming a training set on a computing node, it is suggested to use a certain number of random images. The number is determined based on the computing capabilities of grid system nodes. There are 4500–6000 images in a local training set on the node for a desktop grid system. With this scheme for creating local training sets, all tasks in the i-th iteration are the same, since random selection of images is performed on nodes.

For the elimination of the disadvantages of synchronous updates and reduction node downtime, we introduce the concept of a minimum quorum (the minimum number of results) to complete the iteration. The number of generated tasks for the i-th iteration must be 10–25% greater than the minimum quorum value. In this case, a long calculation of multiple tasks will not delay the end of the iteration of distributed learning and aggregation of results. As a result of aggregation of local models results, the current parameters of the global model are calculated. The next iteration of tasks for computing nodes is generated based on the current parameters of the global model.

Results that are sent to the server after aggregation of results are ignored and discarded.

The global model parameters are updated according to the following rule.

$$W_{i+1} = W_i - learningRate * Grad_i \quad (1)$$

where:

W is the value of the local model parameters;

i-number of the parameter update iteration;

$learningRate$—the learning rate of the model;

$Grad$—value of calculated gradients in local models.

Based on (1) updating weights for simple SGD.

$$learningRate * Grad_i = W_i - W_{i+1} \quad (2)$$

We can calculate this change in the parameters of a locally trained model using the known values of the parameters of the initial global model (2).

$$W_{i+1}^{global} = W_i^{global} - \frac{1}{n}\sum\nolimits_{j=1}^{n} \left(learningRate * Grad_j\right) \quad (3)$$

where

W^{global} is the global model parameters;

i-iteration number of distributed learning;

n—the number of local models.

And using these changes from each received local model, aggregating them [18], we get updated parameters of the global model according to (3).

After getting new parameters for the global model, the current parameters are saved on the server, and the new parameters are used in creating tasks for the next iteration of distributed learning.

7 Results

When organizing a computational experiment, the following parameters were determined:

- Size of the local training set of images: 6000;
- Image size: 256 × 256 pixels;
- Number of classes: 3;
- Size of the test set on the server: 2700;
- The local DNN model is trained for 10 epochs in one iteration of distributed learning;
- Minimum quorum to complete the iteration: 10;
- Number of generated tasks at the beginning of the iteration: 15;
- Estimated time for calculating a task on a personal computer: 2–4 h;
- Number of iterations of distributed learning: 18.

The desktop grid system consisted of 10 nodes (8 active nodes). Each node could simultaneously calculate several tasks on the CPU. The minimum requirements for running one task were 2 CPU cores, 5.5 GB of memory, 4 GB on disks.

After conducting distributed training, the result was obtained in the form of a trained global model. The model was tested on a pre-loaded test set.

The graph of accuracy dynamics (Fig. 3) shows that significant improvement in classification accuracy occurs in the first iterations of training. The accuracy of classification on the training set increases, which means that the global DNN model is successfully trained. It should be noted that a small number of tasks were generated per iteration since no more than 10 computing nodes participated in the experiment. Therefore, less than 2% of the total number of images in the dataset was used in each iteration.

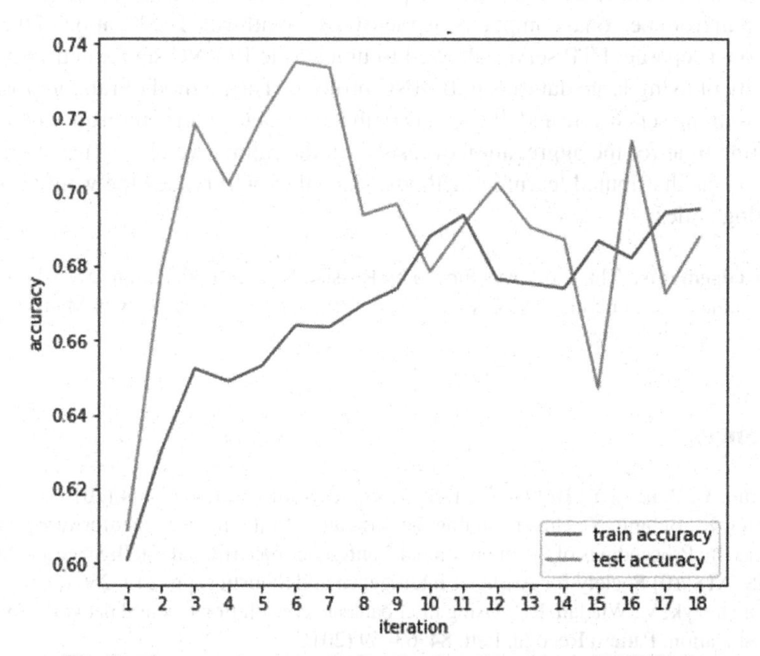

Fig. 3. The accuracy dynamic for distributed learning of T-CNN-2.

Table 1 show that the classification accuracy is different for each class. This feature of the results is a consequence of the complexity of the classification problem for this particular dataset [19].

Table 1. Classification quality metrics.

Name	Class C	Class H	Class N
Accuracy	0.874	0.756	0.664
Precision	0.81	0.65	0.50
Recall	0.82	0.58	0.54
F-measure	0.81	0.61	0.52

In general, the results show that it is possible to successfully train DNN on a grid system, but it is necessary to fine-tune the parameters of distributed training and use more than 80–90% of the dataset in each iteration for the accuracy improvement.

8 Conclusions

The experiment confirmed that there is a possibility of distributed training of DNN on a grid system from personal computers on the BOINC platform. The location of the image dataset on a separate FTP server allowed to unload the BOINC server and showed the possibility of using large datasets in BOINC projects. The method of random formation of local training sets was tested. It was shown that the usage of minimum quorum reduced the waiting time for the aggregation of results. In the future, we plan to pay attention to asynchronous distributed learning methods, since they will reduce the waiting time for computing nodes.

Acknowledgements. This work was funded by Russian Science Foundation (№16-11-10352).

The authors are grateful to the Crystal Dream team and its founder Maxim Manzyuk for the provided computing nodes.

References

1. LeCun, Y., Bengio, Y., Hinton, G.: Deep learning. Nature **521**, 436–444 (2015)
2. Glorot, X., Bengio, Y.: Understanding the difficulty of training deep feedforward neural networks. In: Proceedings of the International Conference on Artificial Intelligence and Statistics (AISTATS'10) Society for Artificial Intelligence and Statistics, pp. 249–256 (2010)
3. Andrearczyk, V., Whelan, P.F.: Using filter banks in convolutional neural networks for texture classification. Pattern Recogn. Lett. **84**, 63–69 (2016)
4. Kylberg, G.: Kylberg Texture Dataset v. 1.0. Centre for Image Analysis, Swedish University of Agricultural Sciences and Uppsala University (2011)
5. Krizhevsky, A., Sutskever, I., Hinton, G.E.: ImageNet classification with deep convolutional neural networks. Adv. Neural. Inf. Process. Syst. **25**, 1097–1105 (2012)

6. TensorFlow homepage. https://www.tensorflow.org, last accessed 2020/06/29
7. Kingma, D.P., Ba, J.: "Adam: A method for stochastic optimization." arXiv preprint arXiv: 1412.6980 (2014)
8. Das, D., et al.: "Distributed deep learning using synchronous stochastic gradient descent." arXiv preprint arXiv:1602.06709 (2016)
9. Chen, J., et al.: "Revisiting distributed synchronous SGD." arXiv preprint arXiv:1604.00981 (2016)
10. Jeffrey, D., et al.: Large scale distributed deep networks. In: Advances in Neural Information Processing Systems, vol. 25, pp. 1223–1231 (2012)
11. Li, M., et al.: Scaling distributed machine learning with the parameter server. In: 11th Symposium on Operating Systems Design and Implementation, pp. 583–598 (2014)
12. Li, M., Andersen, D.G., Smola, A.J., Yu, K.: Communication efficient distributed machine learning with the parameter server. In: Advances in Neural Information Processing Systems, pp. 19–27 (2014)
13. Ben-Nun, T., Hoefler, T.: Demystifying parallel and distributed deep learning: An in-depth concurrency analysis. ACM Comput. Surv. (CSUR) **52**(4), 1–43 (2019)
14. Valiant, L.G.: A bridging model for parallel computation. Commun. ACM **33**(8), 103–111 (1990)
15. Recht, B., et al.: "Hogwild: A lock-free approach to parallelizing stochastic gradient descent." Advances in neural information processing systems (2011)
16. Ho, Q., et al.: More effective distributed ml via a stale synchronous parallel parameter server. In: Advances in neural information processing systems, pp. 1223–1231 (2013)
17. Anderson, D.P.: BOINC: a platform for volunteer computing. J Grid Comput. **18**, 99–122 (2020). https://doi.org/10.1007/s10723-019-09497-9
18. Su, H., Chen, H.: "Experiments on parallel training of deep neural network using model averaging." arXiv preprint arXiv:1507.01239 (2015)
19. Kolosova, O.Y., Kurochkin, I.N., Kurochkin, I.I., Lozinsky, V.I.: Cryostructuring of polymeric systems. 48. Influence of organic chaotropes and kosmotropes on the cryotropic gel-formation of aqueous poly (vinyl alcohol) solutions. European Polymer Journal, vol. 102, pp. 169–177 (2018)

Author Index

Printed in the United States
By Bookmasters